PRINCIPLES OF METAL REFINING AND RECYCLING

Principles of Metal Refining and Recycling

Thorvald Abel Engh
Geoffrey K. Sigworth
Anne Kvithyld

OXFORD
UNIVERSITY PRESS

OXFORD

UNIVERSITY PRESS

Great Clarendon Street, Oxford, OX2 6DP,
United Kingdom

Oxford University Press is a department of the University of Oxford.
It furthers the University's objective of excellence in research, scholarship,
and education by publishing worldwide. Oxford is a registered trade mark of
Oxford University Press in the UK and in certain other countries

© Oxford University Press 2021

The moral rights of the authors have been asserted

First Edition published in 2021

Impression: 1

Published in the United States of America by Oxford University Press
198 Madison Avenue, New York, NY 10016, United States of America

British Library Cataloguing in Publication Data
Data available

Library of Congress Control Number: 2021939033

ISBN 978–0–19–881192–3

DOI: 10.1093/oso/9780198811923.001.0001

Printed and bound by
CPI Group (UK) Ltd, Croydon, CR0 4YY

Links to third party websites are provided by Oxford in good faith and
for information only. Oxford disclaims any responsibility for the materials
contained in any third party website referenced in this work.

Preface

An important field of study of industrial and commercial activity lies between extractive metallurgy and physical metallurgy. This book is an attempt to fill the gap. This field, which we call metal refining and recycling, has become especially important due to the use of lower grade ores, increased recycling, and higher quality requirements. Our aim is to understand the principles that guide today's operation and to develop new solutions to tomorrow's problems.

During the Industrial Revolution, a gap developed between research with a professor and assistant in the university and practice in industry. This became especially severe in Scandinavia for the iron and steel industry due to its strong growth. It was problematic for the universities to train a sufficient number of scientists and workers with a knowledge of extractive metallurgy, thermodynamics, physical metallurgy, economics, etc. After the Second World War an attempt was made to cover the shortage of trained workers and scientists by establishing institutes supported by both industry and government and expanding the education of PhD students trained in the needed disciplines. This book presents the scientific understanding of metal refining and recycling that resulted from this research and development activity.

To solve many industrial problems, it is necessary to draw on scientific principles from 'non-traditional' sources. For example, fluid mechanics and the behaviour of atoms at the solid–liquid interface are of paramount importance. These fields provide the link between physical metallurgy and process metallurgy. Measuring and controlling impurities in molten metals is necessary. For instance, Si for solar cells must be refined down to very low levels of impurities.

Changes and improvements during the past fifty years has made it possible to provide metals with low amounts of impurities. Why is it important to have a clean and properly alloyed metal? Answers to this question are presented in Chapter 1, which gives motivation for the rest of the book. The remaining chapters can be divided into three parts:

1. Fundamentals: Thermodynamics, physical and transport properties, mixing, mass transfer, and numerical models are described in detail in Chapters 2 and 3.

2. Problems and methods: The removal of dissolved impurity elements, particles and inclusions, and refining during solidification are covered in Chapters 4, 5, and 6, respectively.

3. Applications: Remelting and the addition of alloys, refining challenges and specific processes for each metal, with a focus on steelmaking, and recycling are considered, respectively, in Chapters 7, 8, 9, and 10.

Metals are essential for the technological society in which most of us live. To ensure future availability, it is important to conserve resources and take care of our environment. Previously, it was sufficient for a worker to manage one field. Today, a broader understanding is required. Knowledge in different scientific disciplines is often required to solve problems. Social, economic, environmental, and political (legal) factors must also be considered. To cover all of this properly is a problem. The authors have in sum more than 100 years of experience, but this was not sufficient to cover all important aspects of metal refining and recycling. Fortunately, we have been assisted and supported by a number of eminent colleagues.

We hope this book will serve two purposes. First, it should provide a detailed survey of the present state of the art. Each chapter has an extensive bibliography, so that a researcher may easily use the chapter as a starting point, or 'launching pad', for further activity. And second, we have paid special attention to the principles underlying the science, so the book may also be used by advanced students. Our primary intention has been to provide a comprehensive book. It was not mainly designed to serve as a textbook manual of instructions. It may be more appropriate to select portions suitable for the desired course of study. For example, parts of Chapters 2–6 should provide a course of study focusing on refining methods.

In practice, the difference between the two goals is somewhat artificial. An effective worker must always be ready to learn and discover new things both theoretical and practical. One should always be a student.

As you read this, we welcome you a colleague and co-worker in the field of refining and recycling of metals. We hope you will discover topics of interest and utility. We also hope this book will 'stand the test of time', be useful for our world, and add to society's knowledge of metal refining and recycling.

Extracting and recycling metals with minimal damage to the environment will become increasingly important. There will certainly be many interesting challenges to address in the future. One illustration of the challenge before us is the activity in the deep-sea mining of minerals. It would be an interesting challenge to utilize the principles in this book to process and refine metals obtained in this way.

Acknowledgements

The authors wish to acknowledge the encouragement and cooperation of a number of graduates, PhDs, and colleagues. With great pleasure and profound gratitude, we note that a number of our colleagues have generously prepared chapters, or parts of chapters, for this book:

- Christian Julius Simensen presents most of Chapter 1. Otto Lunder helped with the section on corrosion, and Oddvin Reiso with edge cracking.
- Gabriella Tranell and Erlend Bjørnstad produced the section in Chapter 4 on the oxidative ladle refining of silicon.
- Roderick Guthrie and Mihaiela Isac expanded Chapter 5, with a contribution by Sarina Bao on a ceramic foam model.
- Martin Syvertsen worked on Chapter 7, as well as the section in Chapter 4 on the pick-up of hydrogen from water vapour in the atmosphere.
- Eivind Johannes Øvrelid provided Chapter 6.
- Christina Meskers wrote Chapter 8.
- Olle Wijk wrote Chapter 9.
- Per Bakke provided the extensive treatment on magnesium recycling, which appears in Chapter 10.

Students at NTNU—Sigvart Eggen, Ivar Furu, and Magnus Skramstad—are to be thanked for presenting a useful and relevant student's perspective, for most of the chapters. Alejandro Abadías Llamas, Ivan Belo Fernandes, Harmen Ooterdoom, and Markus Reuter are thanked for their support in Chapter 8, and Arne Petter Ratvik for support assistance reviewing the same chapter. Ketil Motzfeldt, who initiated the thermodynamic presentation in Chapter 2, presented in the first book in 1992, is also gratefully remembered.

Arjan Ciftja with metallurgical insight and illustrative skill has produced and improved a number of figures and diagrams. Tone Heggenhougen has also cleverly assisted with redrawing some of the figures and given design input.

We would also like to thank our colleagues at NTNU and SINTEF in general who have shared their knowledge, given their time, and supported the work.

And last, but certainly not least, we extend our heartfelt gratitude to Ingrid Gamst Page for her dedication in the preparation of the manuscript. Truly, this book would not exist without her capable assistance and insight.

We also thank our spouses and families that have put up with our 'important' book, neglecting the short valuable time here on Earth together.

This book has been partly funded by the SFI Metal Production (Centre for Research-based Innovation, 237738). The authors gratefully acknowledge the financial support from the Research Council of Norway and the partners of the SFI Metal Production.

Contents

Notations and Units

ΔA	interfacial contact area in control volume	m^2
A	interfacial contact area, cross-sectional area or integration constant, or constant	m^2
$\dot{A}$	rate of production of gas bubble surface area	m^3/s
A_j	area in contact with phase j	m^2
A_M	Madelung constant	
A_M^S	Madelung constant for a surface layer	
A_p	surface area of particle or of injected powder	m^2
A_s	top surface area of bath	m^2
A_R	surface area of bubbles with radius R	m^2
A_v	vertical projection	m^2
A_{12}, A_{13}, A_{23}	contact areas between phases 1, 2, and 3	m^2
A_j^σ	interfacial area occupied by one mole of species j	m^2/mol
a	constant or radius of inclusion, length of crack, distance between atoms	m
a_1, a_2, a_3	constants	
a_c	reference particle radius	m
a	activity coefficient	
a_i	activity of component i	
a_s	specific surface area	m^{-1}
B	constant	
Bi, Bi^*	Biot number (eq. 7.19)	
b, b^*	constant, ratio of projected area to total surface area or thickness or slope of liquidus line	m K/wt%
b_j	constants	
C	constant or number of components	

c	concentration, total concentration	$kmol/m^3$, kg/m^3
	or volume fraction	or m^{-3}
$c*$	concentration of nuclei	$number/m^3$
$\bar{c}$	mean relative impurity content or refining ratio	
$c(a)$	number of inclusions of diameter $2a$ per unit volume	m^{-3}
c_b	bulk concentration (of inclusions)	$number/m^3$
	also bulk concentration of solute	$kmol/m^3$
c_c	volume fraction cake	
c_E	eutectic composition	$kmol/m^3$
c_{in}	inlet concentration, or initial concentration	$kmol/m^3$
c_j, c_i	concentration of component j or i	$kmol/m^3$
c_l	concentration of solute in liquid (in eq. 6.12)	$kmol/m^3$
c_e	concentration in liquid, or of solid in eq. 6.12	$kmol/m^3$
c_n	concentration of nuclei	$number/m^3$
c_o	outlet concentration	$number/m^3$
c_p	specific heat capacity of a substance	$J/(mol\ K)$
c_r	concentration in reactor	$kmol/m^3$
c_s	concentration in solid or surface concentration	$kmol/m^3$
c_δ	concentration of inclusions at distance δ from wall	$number/m^{-3}$
$[\%C]$	concentration of carbon	$100\ kg\ C/kg\ melt$
$[\%C]_e$	hypothetical concentration	$100\ kg\ C/kg\ melt$
	of carbon in melt in equilibrium with gas phase	
D	diffusion coefficient	m^2/s
D_j	diffusion coefficient of component j	m^2/s
D_E	eddy diffusivity for mass transfer	m^2/s
D_e	diameter of spherical bubble having an equivalent volume	m
d	diameter	m
d_c	collision diameter	m
d_c	diameter of surface deformation	m
d_p	particle diameter	m
d_{pc}	critical particle diameter	m

E	filtration or removal efficiency	
E	Young's modulus (modulus of elasticity)	N/m^2
$\dot{E}$	stirring power	W, W/m^3
E_0	viscosity group in eq. 3.80 or Eötvös number (p. 203)	
e	electronic charge	1.6×10^{-19}C
	or protrusion height (roughness)	m
	or factor	
e_i^j	interaction coefficient for effect of component j on i	(wt%)$^{-1}$
F	Helmholtz free energy	J or J/mol
	or degrees of freedom	
	or force	N
F_d	drag force	N
F_g	gravity force	N
f	function	
	or fraction liquid	
	or frequency	l/s
f_1, f_3, f_5, f_7	functions used in the stream function (Appendix 3)	
f_i	activity coefficient of component i	
f_i^σ	interfacial activity coefficient of component i	
$f_N(a)$	number-size distribution of inclusions	m^{-4}
f_R	number-size distribution of bubbles	m^{-4}
f_s	fraction solid	
f_x	activity coefficient for component x in melt	
G	Gibbs energy	J/mol, J
G_v	Gibbs energy per unit volume	J/m^3
G^σ	Gibbs energy of surface	J/mol, J
ΔG	change of Gibbs energy (free energy)	J/mol
ΔG_{mix}	change of Gibbs energy for mixture	J/mol, J
ΔG_{mix}^{id}	change of Gibbs energy for ideal mixture	J/mol, J
ΔG_{mix}^{xs}	excess Gibbs energy of mixing	J/mol, J

ΔG^0	standard Gibbs energy, Gibbs energy of adsorption	J/mol
ΔG_n	standard Gibbs energy for nucleus	J/mol
ΔG_n^*	critical Gibbs energy of nucleation	J
ΔG_v	Gibbs energy of nuclei	J/m^3
ΔG_{lens}^*	critical Gibbs energy of lens	J
$\dot{G}$	flow of gas	kmol/s, Nm^3/s
$\dot{G}_x$	removal rate of component x	kmol/s, Nm^3/s
g	Gibbs energy per unit area	J/(mol/m^2)
	or acceleration of gravity	m/s^2
	or volume fraction	
	or function	
g^σ	Gibbs energy per unit surface area	J/m^2
$g(r)$	pair distribution function	
$[\%H]_v$	hydrogen percentage in melt in equilibrium with	
	water vapour	per cent
H	distance from gas inlet to bath surface	m
	or thickness of filter	m
	or enthalpy	J
	or dimensionless group	
H_s	enthalpy that enters addition during shell period	J
H_i or H_μ	activation energy for viscous flow	J/mol
H^σ	difference between molar enthalpy on the	J/mol
	surface and in the bulk	
$\Delta H_m, \Delta H_m^*$	heat of melting	J/kg
h, h^*	heat transfer coefficient	J/(m^{2}K s)
$\bar{h}$	mean heat transfer coefficient	J/(m^{2}K s)
ΔH	change of enthalpy	J/mol
$\overline{\Delta H^0}$	average values of enthalpy change	J/mol
ΔH_v	enthalpy change of evaporation	J/mol
i	component i or symbol for interface	
J	integral (defined eq. 1.11)	

j	component j	
$[\%j]$	mass per cent of component j	100 kg j/kg melt
$[\%j]^{\sigma}$	mass per cent of component j at surface	100 kg j /kg melt
K, K'	equilibrium constants, inverse equilibrium constants	
	or stress intensity factor	$N/m^{3/2}$
K_{1C}	critical stress intensity factor	$N/m^{3/2}$
K_p	equilibrium constant for phosphorus equilibrium	
K_P	equilibrium constant for powder	
K_j	equilibrium constant for component j	
	or coefficient for electrical resistivity	Ωm
k	mass transfer coefficient for melt	m/s
	or thermal conductivity	J/(m K s)
	or number of components	
	or Boltzmann's constant 1.3806×10^{-23}	J/K
	or turbulent kinetic energy per unit mass	m^2/s^2
	or partition ratio (distribution coefficient) during solidification	
k^o	equilibrium partition ratio or coefficient	
k_{eff}	effective partition ratio or coefficient	
k_t	coefficient for transfer of particles (eqs. 5.14 and 5.27)	m/s
k_j	mass transfer coefficient for component j	m/s
k_g	mass transfer coefficient for gas	m/s
k_H	Henry's constant	
	or mass transfer coefficient for hydrogen	m/s
k_{H_2}	mass transfer coefficient for H_2 in gas	m/s
k_{H_2O}	mass transfer coefficient for H_2O in gas	m/s
k_s	thermal conductivity of shell	J/(m Ks)
k_{slag}	mass transfer coefficient for slag	m/s
k_t	total mass transfer coefficient	m/s
	for dissolved component or for inclusions	m/s
k_1	mass transfer coefficient for short contact time	m/s

k_2	mass transfer coefficient for long contact times	m/s
k_{tp}	total mass transfer coefficient for injected powder	m/s
k_{ts}	total mass transfer coefficient for top surface	m/s
L	characteristic vessel dimension, length of stirrer blades, distance	m
	or Lorenz number (eq. 2.148)	$W\Omega/K^2$
l	Prandtl mixing length or length of molten zone	m
l_e	characteristic length of energy containing eddies	m
M	mass of bath	kg
	or symbol for alloying element	
	or Morton number (eq. 3.65)	
M_j	molar weight of component j	g/mol or kg/kmol
M_{Na}	molar weight of sodium	g/mol or kg/kmol
M_s	mass of slag	kg
M_p	mass of powder	kg
M_1, M_2	molar weight of component 1, 2	g/mol or kg/kmol
$\dot{M}$	flow rate of metal into reactor, or circulation flow in a reactor	kg/s or kmol/s
$\dot{M}_p$	flow rate of injection of powder	kg/s
m	mass of atom or molecule	kg
	or number of broken bonds, or missing nearest neighbors	
$\dot{m}$	mass flux	kg/ (m^2 s)
N	number of back-mix tanks or number of bubbles present in melt or number of atoms or particles	
$\dot{N}$	rate of transfer of a species in moles	kmol/s
	or rate of creation of bubbles in gas purging	s^{-1}
N_A	Avogadro's number	mol^{-1}
	or ionic fraction of component A	
N_B	ionic fraction of component B	
N_I	injection number	
N_R	number of bubbles of radius R	
N_V	dimensionless volume	

Nu	Nusselt number (eq. 7.20)	
N_1, N_2	number of atoms	
	or number of kmoles per second	kmol/s
$\dot{N}_R$	number of bubbles of radius R created per second	1/s
n	integer or number	
	or rotations per sec	1/s
	or number of kmoles or moles	kmol, mol
n_i	number of molecules of component i	
	or number of moles of component i	mol
n_i^σ	number of moles of component i at surface	mol
n_j	stoichiometric factor	
$\dot{n}$	molar flux	kmol/(m^2 s)
$\dot{n}_1, \dot{n}_2, \dot{n}_3$	molar flux from phases 1, 2, 3	kmol/(m^2 s)
$\dot{n}_a$	flux of inclusions of size a (in Appendix 2)	1/(m^{2}s)
$[\%O]_i$	oxygen in inclusions	per cent
$[\%O]$	dissolved oxygen in steel	per cent
P	number of phases	
p	pressure	N/m^2
	or probability	
p_{bj}	partial pressure in bulk gas of component j	N/ m^2
p_{H_2O}	partial pressure of H$_2$O in gas	N/ m^2
p_i	pressure at interface or pressure of component i	bar
p_{in}	inlet pressure	bar
p_j	partial pressure of component j	N/ m^2
p_{ji}	partial pressure of component j at interface	N/ m^2
p_j^0	partial pressure of j that leaves metal	bar
	or pressure over a pure melt of component j	N/ m^2
p_{atm}	pressure above bath	bar
p	pressure	bar, N/ m^2
p_{CO}	partial pressure of CO gas	bar
p_{inert}	partial pressure of inert gas	bar

p_o	outlet pressure	bar
p_0, p_i^0	vapour pressure of pure substance i	bar
p_{tot}	total pressure	bar
p_x	partial pressure of component x	bar
p_x^0	partial pressure of component x in bubbles when leaving melt	bar
$\Delta p, \Delta p_c, \Delta p_f$	total pressure drop, pressure drop in cake, and in filter	N/m^2
p_{inert}	partial pressure of inert gas	bar
p_1, p_2	pressure	N/m^2
Δp	pressure drop	N/m^2
P_A, P_A	penetration depths of region A and B	m
Pe	Peclet number (eq. 7.32)	
Pr	Prandtl number, defined as the ratio of momentum diffusivity to thermal diffusivity (eq. 7.74)	
Q	heat	J
$\dot{Q}$	circulation flow of melt	m^3/s
$\dot{Q}_1, \dot{Q}_2$	volume flow of gas	m^3/s, Nm^3/s
$\dot{Q}_g$	flow rate of gas	m^3/s, m^3/hr, Nm^3/s
$\dot{Q}O_2(j)$	volume of oxygen that reacts with element j per unit time	Nm^3/s
$\dot{Q}_{CO}, \dot{Q}_{Ar}$	volume flow of CO, argon respectively	Nm^3/s
$\dot{Q}_{O_2}$	total flow of oxygen	Nm^3/s
$\dot{q}$	flow of liquid	m^3/s
	or heat flux	$J/(m^2\ s)$
R	radius of collector or addition, or bubble	m
	or universal gas constant	8.314 J/mol K
	or number of restrictions	
Ra	Rayleigh number (p. 377)	
Re	Reynolds number is defined as the ratio of inertial forces to viscous forces within a fluid (page 191)	

Re_c	Reynolds number for fluid flow in a collector (filter element)	
r	radial distance or radius	m
	or ratio	
r_c, r_f	resistance in cake and filter, respectively	m^{-2}
r_0	distance between ions	m
r_0^S	distance between surface layers	m
r^*	critical radius of nuclei	m
r_j^i	second order interaction coefficient, for the effect of element i on the activity of element j	$(wt\%)^{-2}$
S	entropy	J/(mol K), J/K
S, S_0	hydrogen solubility in molten aluminum	$cm^3/100g$
$S_1(x), S_2(x)$	sum functions	
S_n	fatigue limit	N/m^2
S^σ	difference between the molar entropy of the interface and the two bulk phrases	J/(mol K), J
s^σ	surface entropy per unit area	$J/(mol\ K\ m^2)$, J/m^2
ΔS	change of entropy	J/(mol K), J/K
ΔS_m	entropy of fusion (melting)	J/(mol K)
ΔS_n	entropy of nuclei	$J/(m^3\ K)$
$\overline{\Delta S^0}$	average value of entropy	J/(mol K)
ΔS_{mix}	entropy of a mixture	J/(mol K), J/K
ΔS_{mix}^{id}	entropy of an ideal mixture	J/(mol K), J/K
Sh	Sherwood number (p. 377)	
s	length or position of solidification front	m
Sc	Schmidt number is the ratio of the kinematic viscosity to the diffusion coefficient (p.213)	
Sf	Stefan number (eq. 7.33)	
T	temperature	°C, K

T_b	temperature of bulk	°C, K
T_{in}	initial temperature	°C, K
T_l	temperature of liquid	°C, K
T_W	temperature of wall	°C, K
T_m, T_m^*	melting temperature	°C, K
T_0	room or ambient temperature	°C, K
T_P	temperature of plate plunged into molten aluminum	°C, K
t	time, residence time	s
t_d	dissolution time	s
t_c	circulation time or critical time	s
t_a	melting down time	s
t_0	shell time	s
t_{tot}	total time	s
$t_{1/2}$	time to reduce concentration by one half	s
U	internal energy	J/mol, J
	or total energy	J
U_e	elastic strain energy	J, J/m^3
U_s	surface energy	J/m^2
u	velocity of particle	m/s
u_b	velocity of bubbles relative to melt	m/s
u_{br}	velocity of bubbles relative to particles	m/s
u_m	mean flow velocity of liquid	m/s
u_0	shear velocity along a wall, also	m/s
	z-component of the eddy velocity near boundary layer	m/s
u_r	velocity of particle relative to melt	m/s
V	volume or volume per mole	m^3, m^3/mol
V^*	critical volume of nucleus	m^3
V_m	volume at melting point or volume per mole at melting point	m^3, m^3/mol
V_3	volume of lens	m^3
$\dot{V}, \dot{V}_S$	volume that bubbles sweep through per second	m^3/s

$\dot{V}_e$	effective volume purified per second	m^3/s
v	average velocity of melt or	m/s
	velocity of solidification front	m/s
v_b	velocity of bubble	m/s
	or velocity outside boundary layer	m/s
v_g	velocity of gas phase	m/s
v_m	mean velocity (in melt)	m/s
v_0	shear velocity along wall	m/s
	or eddy velocity in bulk melt normal to free surface	m/s
$\bar{v}_x$	average velocity along wall in x-direction	
v_x	eddy velocity in x-direction	m/s
v_z	eddy velocity in z-direction	m/s
v_θ	velocity along meridian	m/s
v_{δ_1}	velocity at distance δ_1 from wall	m/s
W	thermodynamic probability	
	or work	J
We	Weber number (eq. 3.83)	
Z	number of nearest neighbours	
	or number of electronic charges	
X	variable	
	or inclusion spacing	m
X_a	dimensionless group for addition	
x	component x in melt	
	or coordinate in axial direction	m
	or molar fraction	
	or variable	
x_i	mole fraction of component i	
x_n	mole fraction in plate n	
x_{in}	inlet mole fraction	
$[\%x]$	mass percentage of x in melt	100 kg x/kg melt

$[\%x]_e$	hypothetical percentage of x in melt	100 kg x/kg melt
$[\%x]_i$	concentration at interface	
$[\%x]_{in}$	initial mass percentage of x in melt	100 kg x/kg melt
$[\%x]_\infty$	final mass percentage of x in melt	100 kg x/kg melt
$(\%x)_i$	mass percentage of x in slag at interface	100 kg x/kg slag
$(\%x)_{in}$	initial mass percentage of x in slag	100 kg x/kg slag
$(\%x)_p^0$	exit mass percentage of x in slag	100 kg x/kg slag
y	spacial coordinate	m
y_{exit}	exit mole fraction	
y_j	mole fraction of component j in gas phase	
y_n	mole fraction in gas phase above plate n	
Z, Z_1, Z_2	efficiency of gas purging for monoatomic gas or number of neighbours, or number of charges	
Z^2	efficiency of gas purging for two-atomic gas	
z	spatial coordinate normal to interface	m
α	thermal diffusivity	m^2/s
	or constant	
α_{ij}	dimensionless parameter	
α_s	thermal diffusivity of shell	m^2/s
$\dot{\alpha}_e$	constant	s^{-1}
β	constant, volumetric shrinkage during solidification	
$\beta_1, \beta_2, \beta_3$	constants	
γ_i	activity coefficient of component i in slag on mass per cent basis or Raoultian activity coefficient	
γ_i^0	Raoultian activity coefficient at infinite dilution	
γ_{ip}	activity coefficient of component i in powder on mass per cent basis	
Γ_i	concentration at surface of component i	mol/m^2
δ	collision diameter	m
	or thickness of diffusion boundary layer	m

δ_b	thickness of turbulent wall boundary layer	m
δ_u, δ_T	viscous and thermal boundary layers in section 7.4	m
δ_l	thickness of restricted turbulence boundary layer	m
δ_2	thickness of viscous (laminar flow) boundary layer	m
ε	an energy parameter in eq. 2.63 (page 134)	
	also void fraction in a filter	
ϵ	void fraction, also an integration constant used in Appendix 4	
$\dot{\varepsilon}$	specific power input (or dissipative rate of distribution energy)	m^2/s^2
ε_o	dielectric constant, 8.854×10^{-12}	$C^2/(Nm^2)$
ε_i^j	interaction coefficient for effect of component j on i	
ζ	dimensionless spacial coordinate ($\zeta = z/b$) in section 7.4	
ζ	$= 3\pi\mu d$	N·s/m
η	collision efficiency	
η	dimensionless spacial coordinate ($\eta = y/b$) in section 7.4	
η_i, η_r	collisions efficiency due to interception and	
	relative velocity, respectively	
Θ	dimensionless temperature introduced in section 7.4	
θ	polar angle or angle; contact angle	
κ	constant	°C/s
	or viscosity ratio	
	or compressibility	m^2/N
λ	distance between collisions	m
	or smallest eddy size	m
	or relative resistance or ratio	
	or dimensionless parameter	
μ	viscosity	Ns/m^2
μ_i	chemical potential of component i	J/mol, J
	or viscosity of component i	Ns/m^2

μ_i^0	chemical potential of reference state	J/mol, J
μ_i^σ	chemical potential of component i at a surface	J/mol, J
μ_t	turbulent viscosity	Ns/m^2
ν	kinematic viscosity or momentum diffusivity	m^2/s
	or a characteristic frequency of vibration	s^{-1}
ξ	dimensionless spacial coordinate ($\xi = x/b$) in section 7.4	
Π	dimensionless penetration depth ($\Pi = P/b$) in section 7.4	
Π_A, Π_B	dimensionless penetration depth of regions A and B	
ρ, ρ^*	density of melt or shell	kg/m^3
	or distance	m
	or resistivity	Ωm
ρ_b, ρ_p	density of particles	kg/m^3
ρ_i	density of inclusion	kmol/m^3
ρ_μ	density of solid alloy addition	kmol/m^3
ρ_O	density of oxygen in inclusion	kmol O/m^3
ρ_0	resistivity of pure metal	Ωm
ρ_{slag}	density of slag	kg/m^3
σ	surface tension	N/m
	or electrical conductivity	1/(Ωm)
	or stress	N/ m^2
σ_F	fracture stress	N/ m^2
σ_l	surface tension of pure solvent	N/m
σ_m	tensile strength	N/ m^2
	or surface tension of liquid at melting point	N/m
σ_p	plastic energy required to extend crack	J/ m^2
σ_s	surface energy of solid	J/ m^2
σ_{sl}	interfacial tension between solid and liquid	N/m
σ_{sl}^p	interfacial tension between pure solid and liquid	N/m

$\sigma_{13}, \sigma_{12}, \sigma_{23}$	interfacial, tensions between phases 1 and 3, 1 and 2, and 2 and 3 respectively	N/m
τ	shear stress	N/m^2
	or residence time	s
	or time constant	s
	or dimensionless time	
τ_a	dimensionless shell time	
τ_b	residence time of bubbles in melt	s
τ_e	lifetime of the smallest eddy	s
τ_m	time constant for mixing	s
τ_0	time constant	s
	or shear stress at wall	N/m^2
	or dimensionless shell time	
τ_p	time constant for injection of powder	s
	or 'relaxation time' after a collision	s
τ_R	residence time for bubbles of size R	s
τ_{xy}	shear force	N/m^2
ϕ	mean potential energy	J/mol
	or friction factor equal to v_0^2/v_b^2	
$\phi(x), \phi^*(x)$	functions	
ϕ_A	dimensionless group for contact area of bubbles	
ϕ_p	dimensionless group for contact area of powder	
ϕ_s	dimensionless group for contact area of top slag	
$\psi/[\%H]$	hydrogen dimensionless group for bubble contact area	
ψ	stream function	m^3/s
	or ductility, defined as percentage reduction of area at fracture	
$\psi(x)$	function	
Ω_μ	collision integral for viscosity	

Useful SI (Système International) units, conversion factors, and physical constants

Fundamental units

Physical quantity	Unit	Abbreviation
Length	meter	m
Mass	kilogram	kg
Time	second	s
Electric current	ampere	A
Temperature	kelvin	K
Amount of substance	mole	mol

Other physical quantities

Physical quantity	Unit	Abbreviation	In fundamental units
Mass	gram	g	10^{-3} kg
Mass	tonne	–	10^{3} kg
Volume	litre	L	10^{-3} m^3
Force	newton	N	mkg/s^2
Energy	joule	J	m^2 kg/s^2
Power	watt	W	m^2 kg/s^3
Pressure	pascal	Pa	kg/(ms^2)
Electric charge	coulomb	C	As
Electric potential	volt	V	W/A
Electric resistance	ohm	Ω	V/A
Viscosity	–	Pas	kg/(ms)
Surface tension		–	kg/s^2

Multiples or fractions of SI units

Fraction	Prefix	Symbol
10^{-2}	centi	c
10^{-3}	milli	m
10^{-6}	micro	μ
10^{-10}	–	Å
10^{3}	kilo	k
10^{6}	mega	M
10^{9}	giga	G

Conversion factors

1 mol of gas	$= 22.41$ Nl
1 kmol of gas	$= 22.4$ Nm^3
1 atm	$= 1.013\,25 \times 10^5$ Pa $= 1.013\,25$ bar $= 760$ torr (mm Hg)
1 calorie	$= 4.184$ J
1 kcalorie	$= 4.184$ kJ
0 K	$= -273.15°$C
1 electron volt (eV)	$= 1.602\,18 \times 10^{-19}$ J

Notes: The volume of ideal gas contained in 1 m^3 at 0°C and 1 bar is 1 Nm^3.
The volume of ideal gas contained in 1 l at 0°C and 1 bar is 1 Nl.

Physical constants

Avogadro's number	$N_A = 6.022\,14 \times 10^{23}$ atoms/mol
Boltzmann's constant	$k = 1.380\,66 \times 10^{-23}$ J/K
dielectic constant (permittivity of free space)	$\epsilon_0 = 8.854 \times 10^{-12} C^2/(\text{Nm}^2)$
electron charge	$e = 1.602\,18 \times 10^{-19} C$
Faraday's constant	$F = 96485.3$ C/mol
gas constant	$R = 8.314\,51$ J/(K mol)
mass of electron	$m_e = 9.109\,39 \times 10^{-31}$ kg
gravitational acceleration at sea level and 45° latitude	$g \approx 9.806$ m/s^2
Planck's constant	$h = 6.626\,08 \times 10^{-34}$ J s
speed of light	$c = 2.997\,924\,6 \times 10^8$ m/s

1

The Effect of Dissolved Elements and Inclusions on the Properties of Metal Products

Christian J. Simensen

1.1 Introduction

Metals have been in service for mankind since the early days of civilization. Tools were made from meteorites as far back as 10 000 BC, and copper was in use by the Sumerians in around 4500 BC. The use of light metals is of a much more recent date. The mass production of aluminium started in 1886 with the invention of the Hall–Héroult process, while magnesium was mass-produced only after World War I. The range of products is vast. Metals are in use in high-technology environments in satellites and aircraft and in common products such as kitchenware, refrigerators, and vacuum cleaners. All these products have one thing in common: certain impurities are prohibited as they ruin the metal quality, while other elements are added in order to meet specific requirements. Some of the major effects of impurities on product quality are outlined in this chapter in order to emphasize the necessity of chemical control and refining of the solvent metal and additions.

The earliest reported alloy development which was indirectly caused by impurities is the development of bronzes in the Middle East around 3500 BC. The first copper produced contained large quantities of arsenic which increased the strength of the metal. However, strengthening was more readily obtained by adding 5–10 per cent tin. Therefore, products where arsenic was removed and tin was added were developed over the centuries. The result may be seen today in precious artefacts and weapons.

The oldest metal products contained large quantities of material foreign to and insoluble in the matrix—inclusions—typically in the range 1–20 per cent. Modern production of metals gives a much lower content of inclusions and unwanted dissolved elements. The new refining and production techniques have resulted in impurity levels ranging from 2 to 3 per cent to a few parts per million (ppm). Extreme requirements on purity have led to the development of advanced refining processes such as vacuum techniques and zone refining.

Principles of Metal Refining and Recycling. Thorvald Abel Engh, Geoffrey K. Sigworth, and Anne Kvithyld, Oxford University Press.
© Oxford University Press 2021.
DOI: 10.1093/oso/9780198811923.003.0001

Examples of impurities in various commercial metal products are summarized in Table 1.1.[1] It is evident that metals like magnesium, zinc, and copper are available with a low content of impurities, ≈ 0.1 per cent, while iron is difficult to make without some carbon, and aluminium contains some iron and silicon. Re-melted metal has an even higher content of impurities.

Table 1.1 *A survey of common impurities in some metals*

Raw material:	Mg	Al	Ti Electr.	Mn Electr.	Fe	Cu	Zn
Content (%) of solvent element	99.8	99.7	99.8	99.93	96	99.9	99.9
Impurity	Typical concentration (mass ppm)						
H	0.2–3	0.2–0.5	—	150	—	—	—
C	—	—	—	—	0.3–1.5%	—	—
Mg	—	10–30	—	—	—	—	—
Al	50–150	—	—	—	—	—	—
Si	20–60	300–700	20	—	0.6–1.2%	—	—
S	—	—	—	—	100–350	—	—
Ca	14	3–10	—	—	—	—	—
Ti	<1	30–50	—	—	—	—	—
V	—	100–200	—	—	—	—	—
Mn	50–120	10–30	—	—	0.4–2%	—	—
Fe	100–400	500–2000	—	—	—	6	500–2000
Ni	1–5	10–80	40	25	—	3	—
Cu	14	5–30	—	—	—	—	—
Cd	7	—	—	—	—	—	100–300
Other elements:	—	Ga	—	Co, As, Mo	—	Ag	—
Carbides	1.5–7	1–5	—	—	—	—	—
Oxides	1–30	0.3–3	370	—	*	50–300	*
Nitrides	10–60	0.1–2	—	—	—	—	—
Salts	*	—	—	—	—	—	—
Sulfates	—	—	—	140	—	—	—
Sulfides	—	—	—	180	—	—	—

* Magnesium contains some salt; iron and zinc contain oxides.

The literature, especially concerning steel, is so vast that no attempt has been made to summarize the various effects of dissolved elements and inclusions on the properties. Only a rudimentary outline and some examples are given. However, this chapter should provide the student with an idea of how properties can be modified by refining and micro-alloying. Because inclusions and properties in steel are described at such length in the literature, many examples presented here concern other metals like aluminium and copper. Nevertheless, at least some of the examples should also be interesting for people working with steel.

Impurities can be classified into four main groups. These include volatile elements, reactive elements, non-reactive elements, and inclusions. Volatile elements are elements such as hydrogen and sodium with a high vapour pressure. Reactive elements are those which can be removed from melts by adding oxygen, oxides, chlorides, water, salts, boron, or other chemicals. These impurities include calcium, lithium, and titanium. Examples of non-reactive elements are iron in aluminium which is very difficult to remove by ordinary processes. One way to refine aluminium is to use three-layer electrolysis where the impurities are absorbed into another liquid. This aluminium product has a very low content of non-reactive impurities, namely less than 10 ppm, as shown in Table 1.2.[1]

The inclusions (see Figs 5.2 and 5.3) can be in both the liquid state and the solid state. Impurities in metal products lead to all kinds of changes, from an increase in the modulus of elasticity to a decrease in fatigue strength and ductility, an increase in corrosion, and a reduction in electrical conductivity. Some of the major problems are dealt with in this chapter in order to underline the importance of molten melt treatment and chemical and physical control of metal quality.

Table 1.2 *A comparison between aluminium metals from electrolysis cells and three-layer electrolysis with respect to impurities*

	Typical concentration (mass ppm)	
Impurity	**Electrolysis cells**	**Three-layer electrolysis**
Mg	10–40	< 1
Si	300–700	1–5
Ti	30–50	< 1
V	100–200	< 1
Fe	500–2000	1–10
Ni	10–80	< 1
Cu	5–30	< 1
Zn	20–200	< 1
Ga	80–180	1

1.2   Porosity

When a metal is held at high temperatures, some elements will evaporate and oxidize in air. Typical examples are hydrogen and sodium in aluminium. However, if the surroundings (air, furnace walls, slag, dross, or electrolyte) have a high content of such elements, the metal may absorb these impurities instead. The result can be porosity in the metal, and a reduced ductility. This section concerns the formation of pores and the deterioration of metal quality. Generally, metals have a higher solubility of hydrogen in the liquid state than in the solid state. For example, the equilibrium content of hydrogen in magnesium at the melting point and 1 bar H_2 is 50 ppm in the liquid and 37 ppm in the solid state (Fig. 1.1).[2–6] The corresponding values for aluminium are 0.64 ppm in the liquid and as low as 0.03 ppm in the solid state (see Fig. 4.20). When a metal with a high content of hydrogen is cast, hydrogen pores will be formed during solidification. The number of pores depends on the casting process and the content of hydrogen. An example is given for a series of aluminium alloys in Fig. 1.2.[7] It is evident that the porosity increases drastically with only small increases in the hydrogen content. Turner and Bryant have shown that this porosity causes a reduction in short-transverse tensile properties of semicontinuously cast ingots of an Al–Cu–Mg–Si–Mn alloy (Fig. 1.3).[8] Another effect is the blistering of sheet metal during hot rolling or high-temperature annealing as revealed in Fig. 1.4. In this case hydrogen pores are nucleated on inhomogeneities such as grain boundaries and inclusions.[9] The pores grow by diffusion of hydrogen during high-temperature treatment. The result is blisters caused by hydrogen pores formed close to the metal surface.

In base metals, such as aluminium and magnesium, hydrogen alone creates the pores. In other metals pores may be caused by several dissolved elements. This is the case for copper alloys where both hydrogen and oxygen can be in solution. The content of the elements in molten copper at 1150 °C is shown in Fig. 1.5.[10] It is evident that there is a relation between

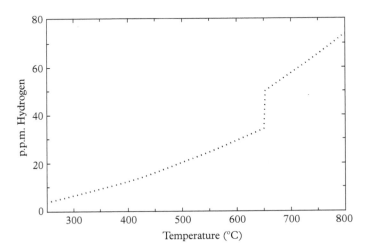

Fig. 1.1 *Solubility of hydrogen in magnesium at 1 bar H_2.*

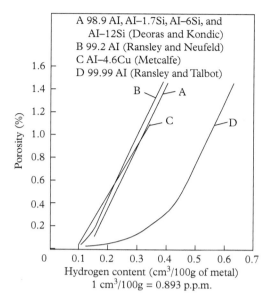

Fig. 1.2 *Relation between hydrogen content and porosity for sand-cast bars, 25 mm in diameter, of aluminium and aluminium alloys.*

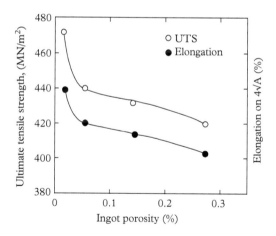

Fig. 1.3 *Effect of ingot porosity on short-transverse tensile properties of 7.5-cm-thick plate hot rolled from semicontinuously cast ingots of Al–4% Cu–0.6% Mg–0.6% Si–0.7% Mn with 28 × 112 cm cross-section.*[8]

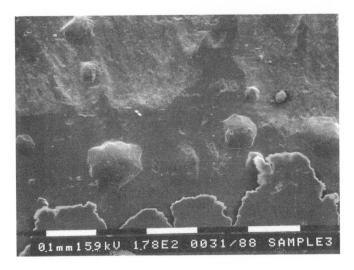

Fig. 1.4 *Blisters in an aluminium sheet (photo: T. Ustad, Hydro Aluminium).*

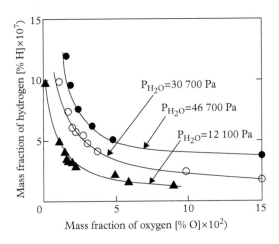

Fig. 1.5 *Dissolved oxygen and hydrogen in molten copper at 1150 °C.10*

the solubility of the two elements. As shown by Allen and Hewitt,[10] water vapour bubbles may form in molten copper alloys by the reaction

$$H_2O\ (gas) = 2\underline{H} + \underline{O} \tag{1.1}$$

The underlining indicates that the element is dissolved in the metal. Equation (1.1) is discussed in more detail in Chapter 3. Water vapour steam embrittlement can also occur in solid copper which contains oxygen. When the metal is heated above 400 °C in an atmosphere containing hydrogen, the metal can be damaged beyond recovery by grain boundary cracking.[11] The

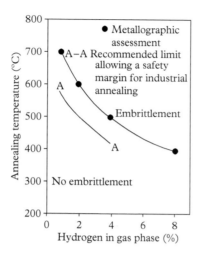

Fig. 1.6 *Limiting conditions of temperature and gas-phase composition for steam embrittlement of tough-pitch copper containing 0.03 mass% O annealed for 1.5 hours.*[12]

mechanism is that hydrogen diffuses into the metal and reacts with cuprous oxide particles, generating steam according to

$$Cu_2O + 2\underline{H} = 2Cu + H_2O \ (gas). \tag{1.2}$$

Fortunately, however, even though hydrogen is a small atom and has a high diffusion coefficient in solid metals, it is possible to anneal copper before too much harm has been done. Mattsson and Schiickher have investigated the reaction and have given the limiting conditions for safe 1.5 hours annealing of tough-pitch copper containing 0.03 per cent oxygen.[12] Their results are outlined in Fig. 1.6.

1.3   Hydrogen Embrittlement of Metals

As already mentioned, the solubility of hydrogen in solid metals increases with temperature. The solubility is high for metals like alpha-titanium which is 0.17 mass% (8 at%) at 325 °C and low in iron (0.32 mass ppm at 400 °C). Metals such as iron, alpha-titanium, uranium, and alpha-zirconium do have one common property: they all become extremely brittle when the hydrogen content surpasses a critical value. Rylski has measured the mechanical properties of pure titanium as a function of hydrogen at room temperature.[13] As shown in Fig. 1.7 he found that both the notch impact energy and the elongation decreased rapidly with increased hydrogen content. The drop in notched impact strength was substantial even for contents of 0.005 mass%. Microstructure investigations have revealed that insoluble hydride platelets are present when the hydrogen content is larger than 0.002 mass%.[14,15] As long as this phase is present in a finely dispersed form, the mechanical properties can be improved. However, the

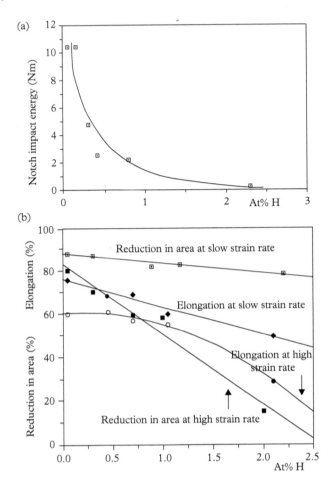

Fig. 1.7 *The variation of the mechanical properties of pure titanium as a function of the hydrogen content and the testing speed at room temperature.*[13] *(a) Notch-impact energy; (b) elongation and reduction in area.*

phase grows quickly even at room temperature and forms massive particles. The result is a deterioration in impact and tensile properties as shown in Fig. 1.7. Therefore, the brittleness in pure titanium is strongly related to the number and size of the hard hydride platelets.

Commercial titanium has a much lower solubility of hydrogen than high purity material. The detrimental properties of hydrogen are then evident for far lower contents of hydrogen. It has been found that beta-titanium is present at grain boundaries in such materials.[15] Since the solubility of hydrogen is substantially higher in beta-titanium than alpha-titanium, most of the hydrogen is present in the beta phase, and hydrides are not present at low hydrogen levels (0.005 mass% H).

For instance, Craighead, Lenning, and Jafee tested the mechanical properties of Ti–8% Mn–H alloys.[16,17] They measured a substantial drop in ductility at about 1 atomic% H

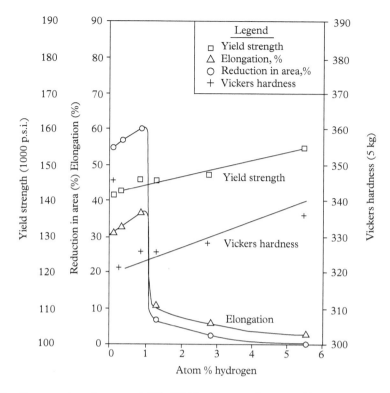

Fig. 1.8 *Tensile properties of commercial Ti–8% Mn alloy as a function of hydrogen content.*[16]

(= 0.02 mass% H) as shown in Fig. 1.8. It has been found that the ductility is highly dependent upon the strain rate for this class of materials, and that the ductility is reduced for low strain rates. Furthermore, Koffila and Burte have shown that the effect is dependent upon temperature and disappears at about 90 °C for a Ti–140 A alloy (which contains traces of iron, chromium, molybdenum, and carbon).[18] The onset of deformation within the metal causes hydrogen to diffuse from an interstitial solution within the beta phase to certain preferred sites within the specimen. This microsegregation under influence of applied stresses can then initiate microcracks. The propagation of these cracks may then cause brittle failure. This segregation of hydrogen, being a diffusion-controlled process, requires time in order to be damaging. Thus, when the strain rate is low, a considerable degree of microsegregation of hydrogen may occur with the result that the material fractures in a brittle manner. The effect has been found to increase with increasing hydrogen content.

The embrittlement of hydrogen in steels shows similar features. However, in this case the ductility decreases progressively as the amount of hydrogen increases from infinitesimal values to a certain upper limit. Above this limit the ductility is unaffected by the hydrogen content and remains low. In steels, embrittlement is most likely to be caused by the segregation of hydrogen towards points in a triaxial stress state in the metal.[19] Furthermore, these regions are created by a network of internal voids acting as internal stress raisers. Hydrogen is normally uniformly distributed throughout the metal lattice and is not damaging because its concentration is so low. But when hydrogen is segregated in highly stressed regions by stress-induced diffusion, the cohesive strength of the metal is reduced locally by the hydrogen atoms. The result is crack propagation at voids and an aggravated embrittlement.

1.4   Electrical Conductivity

Both aluminium and copper are used in electrical devices and wires due to their high electrical conductivities. However, the impurities have a strong effect. For small concentrations of elements in solid solution in aluminium,[20] it has been found that the resistivity changes according to

$$\rho = \rho_0 + \sum_j K_j [\%j] \tag{1.3}$$

where ρ_0 is the resistivity of the pure metal, K_j is the change of electrical resistivity in the presence of 1 per cent of element j and $[\%j]$ is the concentration of element j. Values for the parameter K_j are given for a series of elements in Table 1.3. This demonstrates that elements like iron, titanium, vanadium, and chromium, all with several electrons in their valence band,

Table 1.3 *The effect of impurity elements on the electrical resistivity of aluminium*

Element	Resistivity increase ($\Omega m \times 10^{-8}$) by adding 0.1% of an impurity element	
	In solid solution	As secondary phase
B	0.08	—
Cr	0.40	0.02
Cu	0.032	0.003
Fe	0.26	0.006
Mg	0.05	0.02
Mn	0.29	0.03
Ni	0.08	0.006
Si	0.10	0.009
Ti	0.29	0.01
V	0.40	0.03
Zn	0.01	0.002
Zr	0.17	0.004

have the largest effect. Willey found that the effect of an element was a decade smaller if the element was present as intermetallic particles instead of in solid solution.[21] This is why boron is added to molten aluminium so that titanium and other elements can be removed as borides or are present only in very small quantities as particles in the solid.

The change in resistivity in copper by adding impurity elements is shown in Fig. 1.9.[22] The resistivity increases linearly for most elements in copper in agreement with Eq. (1.3). Exceptions are iron, aluminium, and tin, which may exhibit some deviation from linearity. The effect of iron, phosphorous, and silicon on electrical resistivity is surprisingly large. Concentrations of only 0.1 per cent cause an increase in resistivity or a reduction in conductivity of more than 20 per cent. Therefore, copper in telephone wires is commonly alloyed with cadmium, which improves the mechanical properties without reducing the conductivity more than 5–10 per cent.

Copper melts are often deoxidized with phosphorus in order to avoid water vapour in the products which are destined for welding or thermal treatment in hydrogen atmospheres.

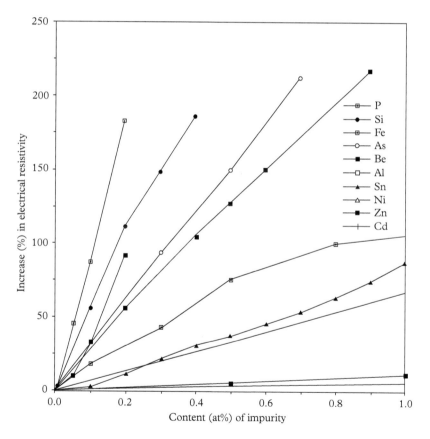

Fig. 1.9 *The effect of impurities on the electrical resistivity of copper.*[22]

However, such materials will have an excess content on the order 0.04 % P. The result is also a reduction of 25 per cent in conductivity which makes the material completely unsuitable for electrical purposes. It is necessary to employ high-purity electrolytic copper in such applications.

1.5   Magnetic Hysteresis and Particles in Steel

Sheets of low-alloy steels are commonly used as material in electro-engines. One of the problems with these materials is that there is an energy loss, that is magnetic hysteresis, when the material is in a changing magnetic field. It is found that the material consists of magnetic domains. When a magnetic field is applied, two independent processes are operating as suggested by Becker and Döring.[23] There is an increase in the volume of domains which are favourably oriented with respect to the field at the expense of unfavourable ones. The other mechanism is that the direction of magnetization is rotated towards the direction of the field. Investigations of various steels have revealed that the hysteresis increases with the content of nitrides and carbides. An example is shown in Fig. 1.10. Herø and Simensen aged steels with coarse grains containing 0.01 % C.[24] The aging temperature was varied between 100 and 300 °C. The power loss in the aged materials was measured using a magnetic field of 13 000 gauss. It was found that the energy loss had a maximum after 1000 hours at 150 °C and 30 hours at 300 °C. Transmission electron microscopy investigations revealed that thin discs of cementite were formed on dislocations at high temperature, while ϵ-carbides were nucleated at low temperatures ($\leq$200 °C). The latter were transformed to cementite during

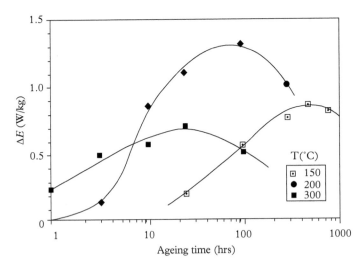

Fig. 1.10 *Power loss, ΔE, in low-carbon steel as a function of preheat treatment between 150 and 300 °C. Magnetic flux density: 13000 gauss.*

ageing. The maximum energy loss was related to thin discs, 1–1.5 μm in diameter. Leslie and Stevens have obtained similar results for Fe–0.45% Mn–0.017% C,[25] and Nacken and Rahmann for steels containing nitrides.[26] The former showed that large carbides or nitrides obstruct the movement of the domain walls and secondary domains are formed if the discs are larger than 0.6 μm. The detrimental effects could be reduced by decreasing the carbon content to very low values of less than 0.003 % C.

1.6 The Effect of Impurities on Hot Ductility of Steels

Ljungstrom has investigated the effect of impurities on the hot ductility of steels in the temperature range 1050–1250 °C.[27] He found that small amounts of lead, bismuth, tellurium, and tin in austenitic 17Cr–13Ni–3Mo steels reduced the hot ductility substantially as shown in Fig. 1.11. Comparison between the different elements showed that a lead equivalent could be defined empirically as[28]

$$Pb_{eq} = \%Pb + 4\%Bi + 0.01\%Sn + 0.02\%Sb + 0.007\%As. \tag{1.4}$$

Thus, bismuth and lead have far the greatest effect. Even a content of 1 ppm Bi or 21 ppm Pb gives a severe reduction of the hot ductility at 1150 °C. Microstructure investigations revealed that fractures were due to grain boundary embrittlement. Lead atoms segregated to the grain boundaries, and a detrimental effect was especially noticeable on those physical properties which depend on grain boundary conditions. The effects were accelerated by co-segregation of nickel, molybdenum, and chromium. A clean steel should therefore have a low content of

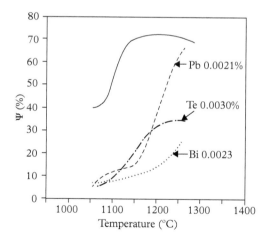

Fig. 1.11 *Decrease in hot ductility at different temperatures of austenitic 17Cr–13Ni–3Mo steel caused by trace elements Pb, Te, and Bi.[28] (The ductility Ψ is defined as the percentage reduction of diameter at fracture.* Ψ = 100 (d_0 − d)/d_0.)

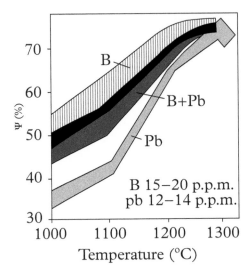

Fig. 1.12 *Trace amounts of boron can counteract the adverse effects of lead on the ductility of 18Cr–15Ni–3Mo steel.[29] The cleaner steel had inferior ductility. (The ductility Ψ is defined as the percentage reduction of diameter at fracture. Ψ = 100 (d₀ – d)/d₀.)*

the elements given by formula (1.4). The necessary cleanliness for stainless steels with a good hot ductility seems to be around 10 ppm lead, but this depends upon the type of steel.

The detrimental effect of lead can, however, be counteracted by certain other elements. These are boron, zirconium, hafnium, cerium, and magnesium, which are therefore deliberately added as trace elements. This is demonstrated in Fig. 1.12 showing how small additions of boron improve the hot ductility in a stainless ferritic-austenitic steel with 12–14 ppm Pb.[29] The effect is possibly caused by a segregation of boron to the grain boundaries which thereby prevents segregation of other elements. Additions of trace elements up to about 0.1 per cent are generally called micro-alloying. Micro-alloying has become an important part of material development during the past few decades.

1.7   The Effect of Intermetallic Phases on Macroproperties

The level of impurities in most metals is low as shown in Table 1.1. The exceptions are iron and silicon in aluminium, iron in zinc, and carbon, manganese and silicon in iron. Another raw material which is rather impure is commercial silicon. It contains commonly 0.1–1% Al, 0.1–1% Fe, 0.1–0.5% Ca, ≈ 0.1% Ti, and ≈ 0.1% C. Alloys which comprise one or several of these raw materials form intermetallic phases or carbides during solidification. An example is particles of Al₃Fe and α-AlFeSi and other phases in aluminium. The effect of intermetallic particles is rather similar to inclusions treated in Section 1.8. However, there is one main important difference. The intermetallic phases are chemically more active than most of the inclusions as they can go partly or totally into solid solution during annealing of the metal at high temperature. One important example is the effect of large particles on fatigue of high-alloy aluminium.

1.7.1   Fatigue in Al–Cu–Mg–Mn Alloys

High-strength Al–Cu–Mg–Mn alloys (AA2024) have the best combination of toughness, fatigue, and strength properties in natural aged conditions. These materials are commonly used in dynamically loaded parts in aircraft. However, development of new commercial long-service aircraft, designed on the basis of the damage tolerance principle, has required further improvement in properties such as fatigue.

A series of Al–Cu–Mg–Mn alloys was studied by Lechiner and Kovalyov[30] in order to describe the effect of iron impurities. The content of iron was varied from 0.05 to 0.24 per cent, while the contents of silicon, manganese, and magnesium were kept at 0.2, 0.4, and 1.3 per cent, respectively. Since the insoluble Al–Fe–Si phases in the material also contain copper, the copper concentration was varied between 3.8 and 4.6 per cent. The analyses of the materials were carried out on 30-mm-thick plates which had been commercially rolled and natural aged. Microstructure investigation revealed that the insoluble particles formed during casting were in the size range 5–20 μm. All the materials consisted of around 75 per cent 5-μm particles, 20 per cent of 5- to 10-μm particles, and 5 per cent of particles larger than

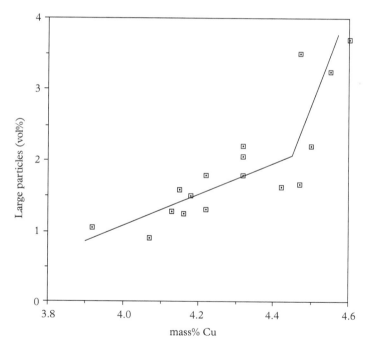

Fig. 1.13 *Volume fraction of 'excessive particles' as a function of the copper content in Al–Cu–Mg–Mn alloys.*[30]

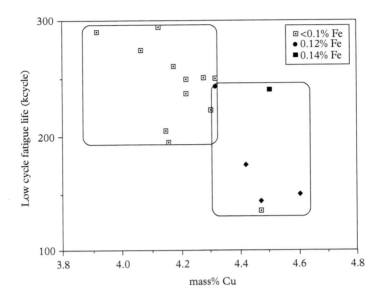

Fig. 1.14 *Low cycle fatigue life versus the composition of Al–Cu–Mg–Mn plates. The materials with the lowest iron and copper content have the greatest fatigue life in these materials (f = 3 Hz and axial tension stress is 155 MPa).*[30]

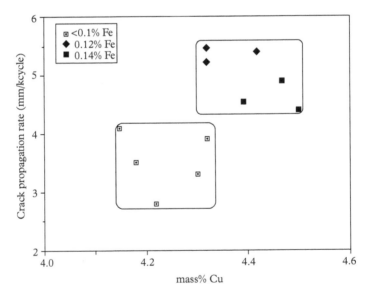

Fig. 1.15 *Fatigue crack propagation rate versus the composition of Al–Cu–Mg–Mn plates (stress: 80 MPa, f = 5 Hz).*[30]

10 μm. These large excessive particles were identified as Al–Fe–Mn–Cu–Si particles, while the small particles were Al–Cu–Mg phases without Fe. The content of the former phase increased rapidly with the copper content as shown in Fig. 1.13.

Low-cycle fatigue resistance was tested on notched specimens in axial tension with a frequency of 3 Hz and a maximum stress of 155 MPa. The results are summarized in Fig. 1.14. The low cycle life decreased for increasing volumes of large intermetallic particles and therefore also with increasing iron and copper content. The fatigue life was reduced from about 250 to 150 kc when the iron content increased from 0.07 to 0.15 per cent. Cracking resistance values were estimated for specimens with dimensions 10 mm × 500 mm × 1500 mm. Specimens were cut in the longitudinal direction from the central layers of the plates. Cracks 50 mm in diameter were made. Fatigue crack propagation was measured using a stress amplitude of 80 MPa and a frequency of 5 Hz. The crack propagation rate increased with increasing content of large Al–Cu–Fe–Si particles, that is, with the contents of copper and iron as shown in Fig. 1.15. These results clearly demonstrate the necessity of keeping the number of large intermetallic particles under control in advanced aluminium materials. Such investigations have led to the development of a new generation of alloys low in iron and also silicon with enhanced fracture toughness.

1.8   Inclusions and Mechanical Properties

There is a vast difference in the content of inclusions in commercial materials of various metals. Zinc and aluminium are typically in the range 0.1–10 ppm, magnesium 10–200 ppm, and steel 100–500 ppm inclusions (Table 1.1). The effect on mechanical properties are correspondingly much more pronounced in steels than in the other materials.

However, even the very low number of inclusions in aluminium has an effect. For instance, they cause pinholes in thin foils with thickness less than 50 μm, and wires to fracture during the late stages of drawing. Examples of harmful inclusions are shown in Figs. 1.16 and 1.17.[31] An Al–2% Mg alloy was coldrolled, and large cracks were detected on the surface (Fig. 1.16). Microstructure investigation of the surface and of cross-sections through the defects revealed the presence of MgO and Al_2MgO_4 particles containing traces of chlorides (Fig. 1.17). These particles together with the common α-AlFeSi particles were situated in the cracks 0.1–0.5 mm below the rolled sheet. A simple explanation for the defect is that microvoids are formed at the oxide inclusions situated close to the surface. When the strain becomes too large for the material between the defect and the metal surface, cracks grow along the intermetallic particles, causing small thin sheets of metal to peel off.

The cleanliness of steel with respect to non-metallic inclusions has always been of great concern to steelmakers. Very early the observation was made that inclusions had detrimental effects on properties and in particular ductility, fatigue, toughness, and machinability. The inclusions are mainly oxides and sulfides. Most of the oxides are small particles less than 0.2 μm in diameter, but it is the large particles with diameters above 20 μm which give the harmful effects. Since the inclusions originate from the raw materials or are formed during melt operations, the primary objectives for steelmakers have been to develop processes for removal of inclusions from melts, and to introduce methods for controlling their cast product.

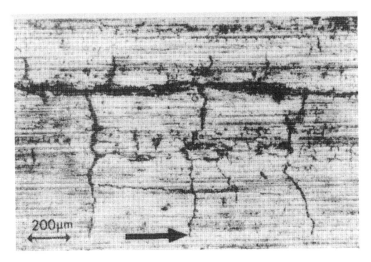

Fig. 1.16 *Cracks along the rolling direction (arrow) and normal to the rolling direction in a Al–2% Mg sheet.*[31]

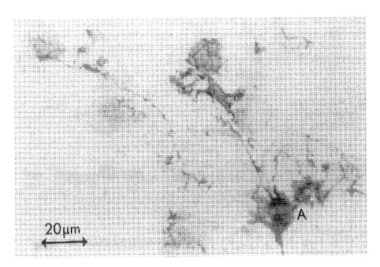

Fig. 1.17 *Large oxide films and pores (A) at the cracks in the cold rolled sheet of Al–2% Mg alloy.*[31]

This development has advanced significantly, and very impressive low levels of impurity content have been attained. One might ask why inclusions cannot be removed completely. In Chapters 2 and 5 it will become apparent that this ideal is very difficult to attain in practice.

The shape of inclusions in wrought products is largely controlled by their behaviour during steel processing, that is, whether they are harder or softer than the steel matrix. The behaviour of different types of inclusions are schematically illustrated in Fig. 1.18.[32] Examples of hard inclusions are globular vitreous silicates (type (a)), crystalline agglomerates of high alumina

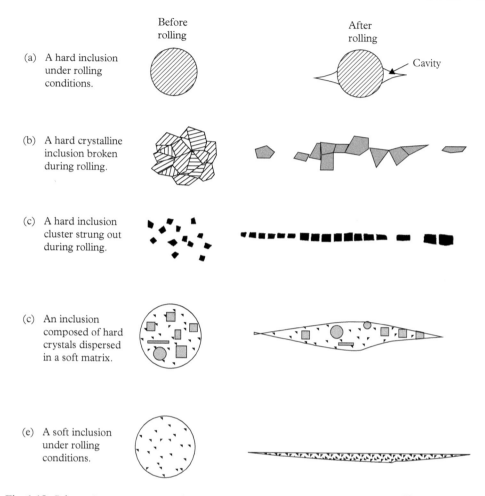

Fig. 1.18 *Schematic representation of inclusion shapes before and after deformation.*[32]

calcium aluminates (type (b)), and 'loose' clusters of alumina (type (c)). In each case the matrix has to deform around the inclusions during rolling, creating cavities and deformation damage. The inclusions which have the most deleterious effect on toughness and ductility are those which deform with the matrix (types (d) and (e) in Fig. 1.18). It is especially the properties in the transverse direction which become rather poor. Mixtures of MnS and oxides belong to type (d) while pure MnS inclusions give type (e). When an attempt is made to modify the inclusions in the refining of steel, the aim is to replace the soft inclusions with hard inclusions. In applications such as wire drawing, however, the soft inclusions (type (e)) are preferred.

It has been noticed that inclusions can also have a positive effect. The first development in this area was free-cutting steels where, for instance, sulfur was added to improve their machinability. This section, however, is focused on the more severe and damaging properties of common inclusions. The survey is partly based upon the work by Lagneborg.[33]

1.8.1   Ductile Fracture

Fracture in bulk deformation processes such as forging, extrusion, and rolling usually occurs as ductile fracture, rarely as brittle fracture. However, depending on temperature and strain rate, the details of the ductile fracture mechanism vary. At temperatures below about half of the melting point of a given material (below the hot working region), a typical dimpled rupture type of mechanism occurs.[34]

It is well established that ductile fracture in steels is due to inclusions. Fracture is initiated by the formation of voids at the inclusions. Voids appear because particles do not deform, which forces the ductile matrix around the particle to deform more than normal. This in turn produces more strain hardening, thus creating a higher stress in the matrix near the particles. When the stress becomes sufficiently great, the interface may separate (decohesion) or the particle may crack. During continued straining these voids grow and develop cavities elongated in the straining directions, and the ligaments of matrix material between the voids become thinner. Finally, the voids coalesce by localized deformation of the bridges between the voids, and the material fractures. Hence, we can distinguish three stages of the fracture process: void formation, void growth, and void coalescence. Experimentally it has been found that the fracture strain decreases strongly with increasing volume fraction of inclusions or content of carbon, oxygen, and sulfur as shown in Figs. 1.19[35] and 1.20.[36] The shape of the inclusions also has a strong effect. The elongation of inclusions during hot working, which takes place for many phases, has proved to be of particular significance. Thus, inclusions elongated in the tensile directions are far less harmful for ductility than discs and rods which are oriented perpendicular to the tensile direction. It should be noted that the content of inclusions in Figs 1.19 and 1.20 is very high compared to the mean level of inclusions encountered in commercial steels. However, in steel products there may be areas containing

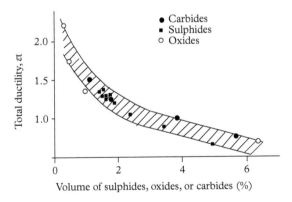

Fig. 1.19 *Effect of second-phase particles on the total ductility of steels.*[35]

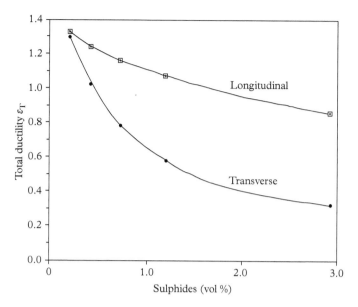

Fig. 1.20 *Effect of the volume fraction of sulphides on the total ductility of steels for longitudinal and transverse testing.*[36]

an excessive amount of inclusions. If these inclusions are located in a region of high stress, failure can take place.

1.8.2 Toughness

Toughness is the ability to absorb energy before fracture occurs. With respect to toughness, inclusions in metals behave like small cracks. The reason is that the bonding between the surrounding metal matrix and the inclusions is commonly weak, and that there exist small cracks at the interface even after casting. Thus, theories concerning the dependence of toughness on cracks can readily be applied to inclusions.

When a material is brittle, Griffith's energy criterion can be used to describe catastrophic propagation of pre-existing cracks in the material.[36] According to this criterion, a crack will propagate in a brittle material when the decrease in the elastic strain energy is at least equal to the energy required to create the new crack surface. Using this criterion, it is possible to determine the tensile stress necessary for crack propagation. For a thin sheet containing an edge crack of length a (or equivalently an internal crack or an inclusion of length $2a$) which is subjected to an axial stress σ as illustrated in Fig. 1.21, the decrease in the elastic strain energy per unit thickness due to the crack is[37]

$$U_e = -\frac{\pi a^2 \sigma^2}{E} \tag{1.5}$$

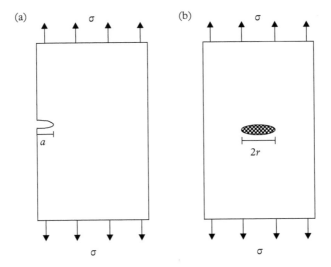

Fig. 1.21 *A thin sheet with (a) an edge crack and (b) an inclusion subjected to an axial stress (a = r).*[37]

where E is Young's modulus (modulus of elasticity). The increase in the surface energy due to the crack is $U_s = 4a\sigma_s$, where σ_s is the surface energy per unit area. The surface energy per unit area is discussed in Chapter 2. It should not be confused with stress, σ, given as force per unit area. Young's modulus, E, gives the ratio between the stress and the elongation per unit length in a tensile test up to the elastic limit of the material. E has the same units as the stress σ. Values of E are included in Appendix 5 where the key properties of metals are listed.

The total change in energy is

$$U = U_e + U_s. \tag{1.6}$$

According to the criterion an incremental increase in the crack length produces no change in the energy of the system. Then $dU/da = 0$. Here the case when the applied stress σ is constant is considered. From the above expressions one obtains

$$\sigma = \left(\frac{2E\sigma_s}{\pi a}\right)^{\frac{1}{2}}. \tag{1.7}$$

Orowan has suggested that localized plasticity at the crack tip could be taken into account with an additional term σ_p, which is a measure of the plastic: work required to extend the crack.[38] Thus

$$\sigma = \left(\frac{2E\left(\sigma_s + \sigma_p\right)}{\pi a}\right)^{\frac{1}{2}}. \tag{1.8}$$

This equation can be generalized using linear elastic fracture mechanics as shown by Knott.[39] From linear elastic theory the stresses in the region of the crack tip are given by

$$\sigma_{ij} = \frac{K}{\left(2\pi\alpha_{ij}(\theta)r\right)^{\frac{1}{2}}} \tag{1.9}$$

where r and θ are the cylindrical coordinates of a point with respect to the crack tip, i and j indicate the directions, $\alpha_{ij}(\theta)$ is in general a dimensionless parameter dependent upon the geometry of the specimen, and K is called the stress intensity factor. By comparison with Eq. (1.9) it is seen that K is proportional to the square root of $E(\sigma_s + \sigma_p)$. It therefore depends on the physical properties of the metal, on Young's modulus, the surface energy, and the plastic work.

The fracture toughness, K_{IC}, is the critical stress intensity factor in the opening mode, and this may be measured. The dimensions of the specimen are in this case selected so that there is plane strain at the crack tip. Then for the precracked specimen the fracture stress, σ_F, is given by

$$\sigma_F = \frac{K_{IC}}{\left(\pi\alpha r\right)^{\frac{1}{2}}} \tag{1.10}$$

where the factor α depends on the geometry of the specimen.

According to Lagneborg,[33] inclusions may promote fracture by two fundamentally different mechanisms. The first mechanism is that an inclusion of diameter $2r$ acts as a crack with diameter $2r$ from which a fracture develops. It has been verified experimentally that this equivalence between an inclusion and a crack is usually a good approximation. The second mechanism is that K_{IC} depends not only on the metal matrix but also on the inclusions.

Kiessling and Nordberg have shown that cracks are initiated at inclusions and that spherical or cylindrical inclusions may be regarded as sharp cracks.[40] A prerequisite for this is that small border cracks or cracks in the inclusion are formed at the early stages of loading the material. When such a sharp crack has been formed, the type of inclusion and its properties should have little influence. In fact, it has been shown experimentally for rolled steels that small cracks are frequently observed at hard inclusions, such as silicates, as well as at plastic inclusions like manganese sulfides. According to the second mechanism K_{IC} also depends on the inclusions if cracks are not present at the interface between inclusion and matrix. It is pointed out that the tendency for borderline cracks to form depends on both cleavage strength and the strength of the inclusion–matrix interface.

Knowing the fracture toughness, K_{IC}, the critical inclusion size, r, can be determined by using Eq. (1.10). The critical defect size for a design stress which has for simplicity been made equal to the yield stress has been calculated by Lagneborg.[33] He found that the critical inclusion size decreased with the yield stress through both the direct influence of this stress and the decreasing fracture toughness, K_{IC}. The critical size varied from 0.18 to 1 mm for different steels. When the temperature was decreased, the yield strength increased, and the fracture toughness decreased and resulted in a decrease in the critical inclusion size.

Lagneborg found that of all the measured inclusion parameters, it was only the size of the largest ones which counted and which controlled the brittle fracture.

Inclusions also influence growth propagation which leads to ductile fracture. Assume that a load is applied to a steel containing a sharp precrack. As the load increases and the crack opening displacement, COD, increases as illustrated in Fig. 1.22, the inclusions near the crack tip experience increasing strains. Then voids are formed at low strains around the inclusions as the bonds between the matrix and the inclusions are weak. Void growth is limited until the voids are subjected to high plastic strains and stresses. Initiation of fracture may then take place at the points where the blunting tip first coalesces with the growing void nearest to the crack tip.

The initiation process of ductile fracture at the tip of a precrack has been modelled by several investigators. One of the main problems has been to describe conditions near the crack tip. However, this difficulty can be avoided by considering the local stress field around the crack tip. Thus the J integral for stress energy is a surface integral that encloses the crack tip from one side of the crack to the other.[41]

The critical value of J required to initiate crack extrusion[42] is

$$J = 0.46 \tau X \tag{1.11}$$

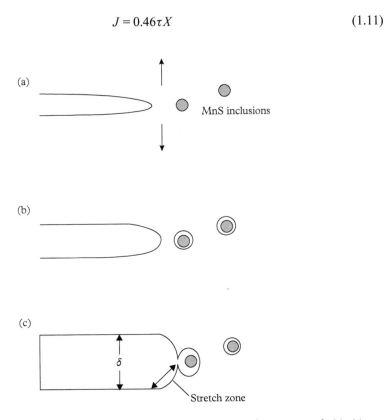

Fig. 1.22 *Schematic illustration of the internal necking process at MnS inclusions in steel: (a)–(c) represent increasing strains which cause crack blunting, void growth, and eventual coalescence of crack tip and void. δ = COD at initiation of the coalescence.*[33]

where τ is the shear yield stress and X is the inclusion spacing. The crack opening displacement, COD, may be calculated from the J integral. Figure 1.23 shows results of calculations of the dependence of the COD as a function of the spacing and radius of the particles.[43] Experimental results for steels tabulated in Table 1.4 are inserted in the diagram. There is fairly satisfactory agreement for steels with a low strength and a high work hardening capacity.[33] It is also evident that the theory is less applicable for high-strength steels. It is, for instance, typical that the critical COD decreases with decreasing tempering temperature in steel En25. Knott has shown that these low COD values appear to be related to shear cracks emerging from the tip of the precrack before a void has fully coalesced with the crack tip.[44] He has furthermore demonstrated that these shear cracks in quenched and tempered steels seem to run from carbide to carbide by decohering the carbide–matrix interface. From these observations it is evident that the spatial distribution, as well as the inclusion spacing and the size distribution of inclusions, has a significant effect upon the ductile fracture. However, these effects are not satisfactorily accounted for in the present theories.

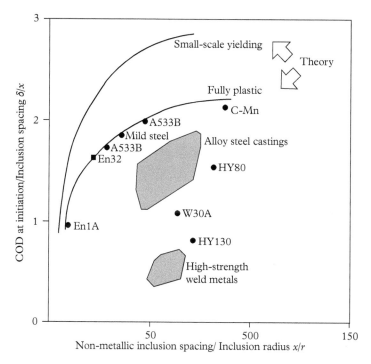

Fig. 1.23 *Crack tip ductility for a series of steels33 listed in Table 1.4. The results are compared with the theoretical curves determined by Rice and Johnson.*[43] *There is a reasonable agreement for the transverse directions of steels with high working capacity: En1A, En32, A533B, mild steel and C–Mn steel, but other steels fail to demonstrate the predicted ductility.*

Table 1.4 *Composition of steels used in toughness measurements illustrated in Fig. 1.23*

Name	Concentration (mass%)					
	C	**Mn**	**Ni**	**Cr**	**Others**	
En 1A	0.12	0.90	—	—	S:	0.24
En32	0.17	0.74	0.02	0.01	S:	0.03
A33B	0.24	1.42	0.70	—	Mo:	0.48
C-Mn	0.10	1.38	—	—	S:	0.009
Alloy castings	0.23	0.9	1.8–2	0.9	Mo:	0.4
HY80	0.17	0.34	2.36	1.23	Mo:	0.34
Ducol W30A	0.16	1.3	0.2	0.6	Mo:	0.23
Hy130	0.10	0.83	4.93	0.50	Mo:	0.53
Weld metals	0.1	1–1.6	2.3–3	0.5–1	Mo:	0.5
En25	0.34	0.72	2.72	0.62	Mo:	0.55

1.8.3   Fatigue

Failure in fatigue of materials is very often due to inclusions. Inclusions reduce the fatigue strength substantially. This is why inclusion studies have always been necessary to rate the quality of steels subjected to fatigue in service, such as ball-bearing, flapper valve, wheel, and shaft steels. It has been established that the fatigue strength can be predicted from the tensile strength and the type, content, and size of inclusions.

For materials without inclusions it has been found empirically that the fatigue limit, S_n, is related to the tensile strength, σ_m:[43]

$$S_n \approx 0.5\sigma_m. \tag{1.12}$$

This equation is obeyed for low-strength steels, while high-strength steels have a tendency to give lower values.

The first, more quantitative, treatment of the relationship between inclusions and fatigue properties has been given by Peterson.[45,46] He considers the inclusions as internal holes and calculates the stress concentrations by elastic theory. He states that if a volume around the inclusion larger than a critical value was stressed beyond the ideal fatigue strength as a result of the stress concentration, the fatigue limit of the whole material is reached. The inclusion will in this way influence the fatigue strength through its size, shape, and orientation relative to the applied stress. On this basis it may be shown that a steel containing a spherical inclusion of radius r would have a fatigue limit[47]

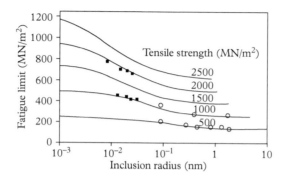

Fig. 1.24 *Comparison between the experimental and predicted fatigue limit in steel.*[47]

$$S_n = \frac{0.5\sigma_m}{1 + (1 + A/r)^{-1}} \qquad (1.13)$$

where $0.5\sigma_m$ is the ideal fatigue limit and A represents a measure of a zone around the inclusion where the material is stressed beyond the ideal fatigue strength. Peterson finds, as expected, that A decreases with increasing strength of the steel. He proposes the relation

$$A = 0.025(2070/\sigma_m)^{1.8} \qquad (1.14)$$

where A is in mm and σ_m is in N/ mm^2.

Thus, provided that it is the size of the inclusions that will cause failure, Eqs (1.13) and (1.14) are able to predict the fatigue strength. Nordberg has compared the prediction by these equations with actual measurements of fatigue strength and the size of the defect that initiated failure.[47] The defect represents a series of various types of inclusions as well as pores. As shown in Fig. 1.24 the agreement between Nordberg's measurements and the theory is very good, especially since Eqs (1.13) and (1.14) are only a first approximation. Despite the success of the theory outlined, there is clear evidence that the internal properties of inclusions have an effect as well. Hence Brooksbank and Andrews have shown that inclusions with thermal expansion smaller than that of the steel matrix, typically most oxide inclusions, affect the fatigue strength detrimentally and that inclusions with larger thermal expansion, like many sulfides, are not detrimental.[48] These authors argue that the effect is due to the fact that the former inclusions set up tensile stresses. This hypothesis has been supported by the measurements of Hauser and Wells, who have found that initiation of cracks in transverse fatigue testing of high-strength steels always starts at inclusions other than manganese sulfides, despite the fact that these are longer than the initiating inclusions.[49]

1.8.4 Machinability

Machining of metal pieces is a common standard industrial process. An example of machining of a metal workpiece is shown in Fig. 1.25. The purpose of such an operation is to remove

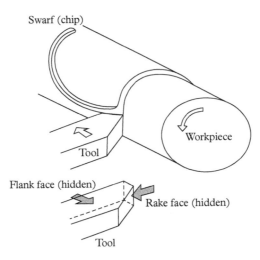

Fig. 1.25 *Machining of a cylindrical metal piece. The tool is drawn twice to illustrate where the flank face and the rake face are on the tool. Chip is a short swarf.*

the outer part and produce a metal piece that has a smooth surface of specified geometry and size. It is important to avoid any abrasion of the tool. When the process is running with low maintenance costs and with a low percentage of workpiece failures, the material is said to have a good machinability.

It is well established that non-metallic inclusions have a strong influence upon the machinability of steels. The effect is closely related to the inherent properties of the inclusions. As a result of this, free-cutting or free-machining grades of steel have been developed where advantage has been taken of the effect of special additions. One classical example is resulfurized steels with sulfur additions of up to about 0.35 per cent sulfur, containing mainly manganese sulfides. The non-metallic inclusions affect machinability in a complex manner. Kiessling has summarized the conditions that should be met in order to improve the machinability as follows:[50]

1. The inclusions should act as stress raisers in the shear plane of the swarf, so initiating crack formation and embrittling chips. The chip–tool contact length then decreases, which is beneficial. Inclusions should not, however, be such strong stress raisers that the work piece cracks.

2. The inclusions should participate in the flow of metal in the flow zone, increasing the shear of the metal. They should not cut through the plastic flow of metal and thus damage the tool surface.

3. The inclusions should form a diffusion barrier on the rake face (see Fig. 1.25) of the tool at the temperature of the tool–chip interface. This temperature depends on several variables, especially the cutting speed.

4. The inclusions should give a smooth piece surface and not act as abrasives on the flank face of the tool.

An example is MnS particles in steels. These inclusions act as stress raisers in the shear plane of the swarf and thereby improve the machinability. Large inclusions have a better effect than the smaller ones. Ollilainen has improved the machinability of steels by adding calcium and forming oxides with a low melting point and a low hardness.[51]

1.9   Corrosion

When materials are in water or air, they will corrode. A typical example is a car that has been used in winterly condition on salted roads. They will get a surface layer of oxide or hydroxide. The important issue is to get a very thin layer which protect the product from further oxidation. The yearly production of important metals (in million tonnes) are steel (2 538) aluminium (63.1), copper (19.1), Zn (13.2), lead (5.2), nickel (2.07), and magnesium (1.02). We have selected examples of corrosion of aluminium alloys and magnesium alloys which are two of the important, worldwide materials used. These materials combine light weight, low cost, and high strength.

1.9.1   Corrosion of Al and Mg Alloys—Electrochemical Aspects

Any corrosion process of a metal in aqueous solution must involve *oxidation* (anodic reaction) of the metal and *reduction* (cathodic reaction) of a species in solution. The anodic and cathodic reactions occur at the same rate and simultaneously on the metal surface, provided there is no electrical contact to another conducting material. Corrosion of metals that form an insulating (passive) oxide film on the surface, such as stainless steel, aluminium, magnesium, and titanium, is generally of a local nature. Corrosion can occur when the protective oxide film breaks down locally, e.g. due to presence of aggressive species such as chloride ions.

The basic anodic reaction during corrosion of aluminium is metal dissolution:

$$Al \rightarrow Al^{3+} + 3 \ e^- \tag{1.15}$$

The electrons released are consumed by the reduction of oxygen

$$O_2 + 2 \ H_2O + 4 \ e^- \rightarrow 4 \ OH^- \tag{1.16}$$

or by the reduction of hydrogen ions (protons) to form hydrogen gas in acidic solutions:

$$2 \ H^+ + 2 \ e^- = H_2 \ (gas) \tag{1.17}$$

The aluminium ions from the anode reaction (1.15) may precipitate as a white corrosion product (Al hydroxide) on the metal surface:

$$Al^{3+} + 3\,H_2O \rightarrow Al\,(OH)_3 + 3\,H^+ \qquad\qquad (1.18)$$

In contrast to carbon steel, corrosion of aluminium is primarily determined by the properties and distribution of intermetallic particles (typically up to a few micrometers in size) in the metal microstructure. Initiation of corrosion attacks invariably occurs at weak spots adjacent to these particles, caused by flaws in the oxide at the interface between the particles and surrounding aluminium. Since the aluminium oxide has very low electronic conductance, the reduction reactions (1.16) and (1.17) predominantly occur on the intermetallic particles, which are electrochemically nobler than the aluminium matrix.

1.9.2   Effect of Intermetallic Particles

It is the interaction between local cathodes and anodes on the surface that leads to nearly all forms of localized corrosion of aluminium alloys, including pitting corrosion, intergranular attack, exfoliation corrosion, and filiform corrosion of coated surfaces. The separation of anodic and cathodic sites causes acidification of the solution in anodic areas due to hydrolysis reaction (1.18), while a high pH environment is formed at the surface of the cathodic intermetallic particles, as illustrated in Fig. 1.26.

Intermetallic particles in aluminium alloys are classified into *precipitates* (formed by nucleation and growth from a supersaturated solid solution), *dispersoids* (formed at high

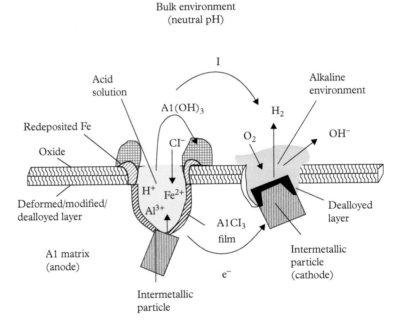

Fig. 1.26 *Conceptual sketch of localized corrosion on aluminium.*[53] *(Printed with permission from Professor K. Nisancioglu.)*

temperatures and are present to control grain size and recrystallization behaviour), and *constituent particles* (formed during alloy solidification and unable to re-dissolve during subsequent heat treatment).[52]

Precipitates range in size from nanometers to fractions of a micrometer and are essential in providing strength to heat treatable Al alloys. The alloying elements leading to precipitation include Cu, Mg, Si, and Zn. When concentrated on grain boundaries, they may affect intergranular corrosion and stress corrosion cracking susceptibility. Dispersoids are small particles comprising alloying elements that are highly insoluble in aluminium. Particle sizes typically range from 0.05 to 0.5 μm. Cr, Ti, Zr, and Mn are common dispersoid formers. The dispersoids are generally quite passive in environments where Al alloys are regularly used. Constituent particles typically contain iron, which is a common impurity element in aluminium alloys. The low solubility in aluminium causes Fe to combine with aluminium and alloying elements such as Mn, Si, and Cu, forming various Fe-rich constituent particles. Characteristic particle dimensions range from 0.1 to 10 μm. Rolling and extrusion tends to break up and align constituent particles into bands within the alloy. The constituent particles are nobler than aluminium and increase the corrosion rate as they effectively catalyse the cathodic reactions (1.16) and (1.17).

The structure and electrochemical properties of particles and their roles in localized corrosion in Al alloys have been extensively studied.[54–56] Tabulated data for some intermetallic compounds synthesized in bulk form in Table 1.5 indicate that their corrosion potentials may vary over several hundreds of mV, reflecting the electrochemical heterogeneity

Table 1.5 *Galvanic series of some intermetallic phases in 0.1 M NaCl solution*[54,56]

Phase	Potential (V_{SCE})	Electrochemical behaviour vs pure aluminium
Cu (99.9%)	−0.23	Cathodic
Si (99.9%)	−0.44	Cathodic
$Al_3Cu_2Mg_8Si_6$ (Q)	−0.51	Cathodic
Al_3Fe	−0.54	Cathodic
Al_7Cu_2Fe	−0.55	Cathodic
Al_2Cu	−0.63	Cathodic
Al_6Mn	−0.78	Cathodic
$Al_{12}Mn_3Si$	−0.81	Cathodic
Al (99.999%)	−0.82	—
$MgZn_2$	−1.03	anodic
Mg_2Si	−1.54	anodic

Note: Potentials are measured with respect to a saturated calomel reference electrode (SCE).

in many commercial Al alloys. Intermetallic phases containing Cu and/or Fe are nobler than aluminium. The relatively few phases that are less noble than aluminium typically contain Mg or Zn.

It should be noted that the data in Table 1.5 do not provide information about the *kinetics* of the electrochemical reactions. As the corrosion potential of an aluminium alloy is determined mainly by the aluminium-rich solid solution, the nobler intermetallic particles present in the Al alloy become *polarized* in the negative direction. The cathodic current density on the intermetallic particles at the potential of the alloy does not always reflect the position in the galvanic series in Table 1.5.[54] Moreover, the cathodic efficiency may change with time due to modification of their surface composition during the corrosion process.

The presence of noble metal impurities (Fe, Cu, Ni) has a particularly detrimental effect on the corrosion resistance of magnesium alloys in a chloride environment. Treatment of the magnesium melt with Mn is used to precipitate Fe in the form of various Al–Mn–Fe intermetallic phases that settle to the bottom of the melt. The presence of Al–Mn–Fe particles remaining in the cast product may affect the corrosion rate significantly, depending on the average Fe/Mn ratio of the intermetallic particles.[57,58] To ensure high corrosion resistance of cast magnesium alloys, 'tolerance limits' for Fe, Cu, and Ni at the ppm level have been specified (ASTM B94-88).

1.9.3   Elements in Solid Solution

Elements present in solid solution in commercial aluminium alloys typically include one or more of the element Si, Mg, Mn, Cu, and Zn. While Mn, Si, and Cu in solid solution shifts the corrosion potential in a positive (noble) direction, Mg and Zn shifts the potential in a negative direction,[59] as seen from Fig. 1.27. Local variations in the solute concentrations in the alloy microstructure may therefore induce micro-galvanic effects leading to localized corrosion. For example, the Si-depleted precipitate free zone (PFZ) along the grain boundaries of age-hardened AlMgSi alloys may constitute an anodic region causing increased susceptibility to intergranular corrosion.[56,60]

Copper is added to many aluminium alloys in order to improve the mechanical properties. Although Cu in solid solution makes the aluminium more noble from an electrochemical point of view, Cu may be detrimental for the corrosion properties even when present at low concentrations (< 0.2%). The potential of a corroding aluminium sample is normally too low for the Cu in solid solution to oxidize. As a result, Cu becomes enriched at the surface by a dealloying mechanism, forming a thin layer of metallic Cu which effectively increases the cathodic activity on the surface, and hence the corrosion rate.

1.9.3.1 Trace Elements

The amount of trace elements in commercial aluminium alloys is expected to increase in the future due to the recycling of post-consumed scrap and the use of less pure bauxite sources in primary aluminium production. Some of these elements may affect the corrosion properties significantly, even when present at ppm level. Especially Group 13 and 14 elements like Pb and Sn are of concern as the combination of a low melting point and low solid solubility in aluminium lead to significant enrichment at the grain boundaries and surface by

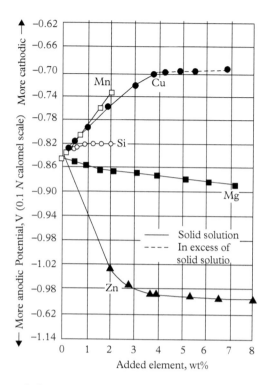

Fig. 1.27 *Effect of principal alloying elements on the corrosion potential of aluminium. Potentials are for solution-treated and quenched high-purity binary alloys in 53 g/l NaCl + 3 g/l H_2O_2 solution. (Reprinted from Hollingsworth and Hunsicker[59] with permission from ASM International.)*

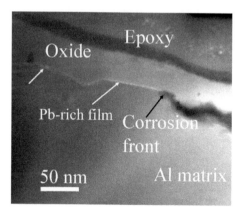

Fig. 1.28 *Dark-field scanning transmission electron microscope (STEM) image of a corrosion attack propagating along a thermally segregated Pb film under the thermal oxide of a binary Al–20 ppm Pb alloy.[61]* (*© IOP Publishing. Reproduced with permission. All rights reserved.*)

heat treatment, causing anodic activation.[61–63] The activation is attributed to the formation of metallic Pb (and/or Sn) nanofilms at the metal–oxide interface, giving a large area of aluminium surface wetted by a very small amount of segregated trace element. The film effectively weakens the passivity of the surface in the presence of chloride, causing a superficial attack of the aluminium just beneath the oxide film (Fig. 1.28).

Surface activation associated with the presence of low-melting-point trace elements can be utilized in a positive way in certain applications, such as in the development of sacrificial anodes, and in aluminium electrolytic capacitors. Their effect in architectural, packaging, and automotive applications are often undesirable because of increased corrosion susceptibility.

1.9.4   Pitting Corrosion

Parts made of steel and aluminium, copper, titanium, and magnesium alloys are common in products such as cars, buildings, and oil rigs. These devices are designed to have a long service life and therefore are highly resistant to corrosion. There are a variety of ways in which products of metal alloys can corrode. This section concerns only corrosion related to impurities. In most cases it turns out that corrosion is associated with impurity phases formed either in the melt or during casting. However, since the mechanisms are not known in detail in most cases, only a few examples are presented here.

Magnesium is a metal which corrodes rather easily. The presence of more than 0.016 % Fe or 0.004 % Ni aggravates the corrosion rate of magnesium considerably, as shown in Fig. 1.29.[64] These tests were carried out in sodium chloride solutions. Microstructure investigations revealed the presence of intermetallic phases. Simensen and Oberlander have shown that these are α-Fe and Fe_3Si in commercial Mg–Fe alloys,[65] and that these phases are formed in the melt for iron contents as low as 0.01% Fe. The solubility of nickel in solid magnesium is also extremely low, and Mg_2Ni is formed. Hanawalt et al. have shown that this corrosion can be reduced substantially by adding manganese (Fig. 1.30), which causes the formation of other intermetallic phases high in manganese.[66] In both examples the intermetallic phases are the origin of pitting corrosion. Pitting corrosion caused by inclusions in steel is discussed in some detail in the following example.

It has been known for a long time that non-metallic inclusions may have a harmful effect on the corrosion resistance of steel. The correlation between inclusions and corrosion is particularly obvious in localized corrosion, such as pitting. As one might expect, this effect strongly depends on the chemistry of the inclusions. It was found that various oxide inclusions do not appear to have any effect. Microstructure investigations by means of scanning electron microscopy have revealed that sulfide inclusions are the most likely initiation sites for pitting in stainless steels. These sulfides are MnS with varying amounts of dissolved iron and chromium. On the basis of such observations, Eklund has proposed the following mechanism: sulfides are in general good electronic conductors, and accordingly the manganese sulfides in stainless steel can be polarized to the potential of the passive steel surface.[67,68] At this potential MnS, as shown both experimentally and theoretically by Eklund, is not thermodynamically stable and will tend to dissolve in a surrounding salt solution. By the dissolution of sulfides, ions of manganese as well as iron and chromium form locally.

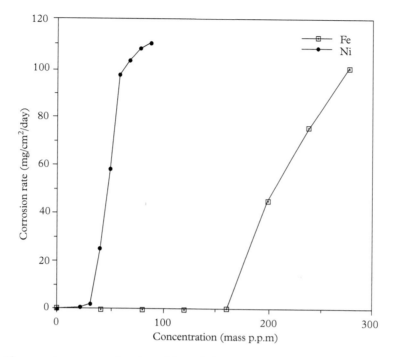

Fig. 1.29 *The corrosion rate as a function of the nickel content in a Mg–0.6% Zr alloy and iron in magnesium when tested in a 3 % NaCl solution.*[64]

They form, in turn, hydrogen ions and hydroxy and chloride complexes. One of the important reactions which takes place is

$$MnS + 4H_2O \Rightarrow Mn^{2+} + SO_4^{2-} + 8H^+ + 8e^-. \qquad (1.19)$$

From the reaction in Eq. (1.19) it is evident that the solution in the micropit becomes more and more acidic. When a certain degree of acidity is obtained, the uncovered metal surface around the sulfide will no longer be able to passivate, and metal ions from the steel will enter the solution. At this moment the electrochemical potential drops dramatically. Then the dissolution of sulfide will be terminated as the potential will now be within the region of thermodynamic stability of the sulfide. If the formation of hydrogen ions from the hydrolysis of the increased metal ion concentration is high, the dissolution of the metal matrix will continue as the solution remains quite acidic and a self-sustaining pit has developed. Otherwise, the solution will be neutralized by the formation of H_2S gas and the uncovered metal will passivate by the formation of a hydroxide layer.

An interesting alloy development has been made based on this model for pitting corrosion. Steels low in manganese have been produced, containing chromium sulfides instead of manganese sulfides.[69,70] Compared with manganese sulfides, this phase is much less soluble in acids. The result is a steel which is far more resistant to pitting corrosion.

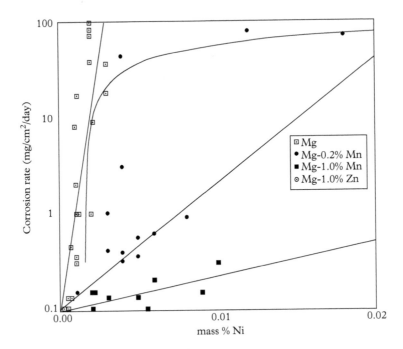

Fig. 1.30 *The corrosion rate of pure Mg, Mg–Mn, and Mg–Zn alloys as a function of the nickel content.*[66] *Corrosion is drastically reduced by alloying with Mn or Zn.*

Table 1.6 *Elements with a low melting point*

Element	Melting point (°C)	Element	Melting point (°C)
K	63.4	F	−219.5
Na	97.7	Cl	−101.5
Li	182.5	P	44.2
Pb	327.5	S	115.2
Zn	419.5	Mg	650.0

1.10   The Effect of Molten Particles in Aluminium Alloys

When aluminium alloys are cast, the latest phases to solidify are those with a low melting point. Such phases are found at dendrite boundaries, grain boundaries, and triple junctions in the solid metal. Table 1.6 show impurity elements with a low melting point.

When these impurities are present in aluminium alloys, they may form inclusions that can be harmful even if they are present only in a small amount. Na and K mixtures are extreme as they will be liquid well below 100 °C (Fig. 1.31). This was one of the reasons why we

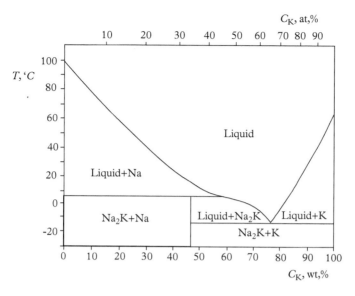

Fig. 1.31 *Na–K phase diagram.*

analysed aluminium materials by the use of Tof-SIMS (time of flight–secondary ion mass spectrometry) and NanoSIMS. The instrument detection limit is 1 ppm to 1 ppb for most elements, a few atom layers in thickness, and 0.3 μm lateral resolution.

1.10.1   Impurities in Al–Si Castings

A large variety of Al–Si alloys is used by the aluminium industry. The microstructure is a mixture of small Al grains and eutectic Al–Si grains. The metal is grain refined using Al–Ti–B additions. A eutectic structure with small dispersoids was obtained by adding small amounts of Na or Sr. This work is an investigation of the microstructure where the phases were determined optically and using microprobe analysis (EPMA). The impurities have been studied using SIMS. Furthermore, it was simple to change from detection of positive ions to negative ions using Tof-SIMS. (The particles shifted only a few micrometers in positions which we simply corrected.) The main purpose with this investigation was to determine where the impurities were located in or at intermetallic phases and to show their appearance.

The first material investigated was an Al–7.2 wt% Si–0.49 wt% Mg–34 wtppm Na–9 wtppm Sr–0.11 wt% Fe.[71] Figure 1.32 is an optical micrograph showing cut-through dendrites and eutectic consisting mainly of small α-Si. These particles will be much larger if sodium or strontium is removed.[71] The intermetallic particles formed are α-Si, gray β-AlFeSi (b), gray π-AlSiMgFe (π), and black Mg_2Si. In some places the intermetallic particles are larger. Such an area was analysed as shown in Figs. 1.33 and 1.34. Some of the observations were that Ca is not present in α-Si, but in the π-phase and in matrix. Na is detected as a thin layer on α-Si and in a solution of α-Si particles and π-$Al_8Si_5Mg_3Fe$, and Fe is detected in π-$Al_8Si_5Mg_3Fe$.

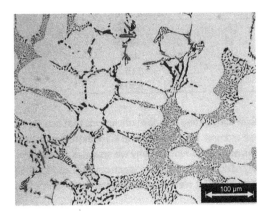

Fig. 1.32 *Optical micrograph of Al–7wt%Si–0.49wt%Mg–0.11 wt%Fe–34 wtppm Na. The microstructure consists of Al dendrites and mainly fine-dispersed eutectic. Some areas show larger α-Si and intermetallic particles.*

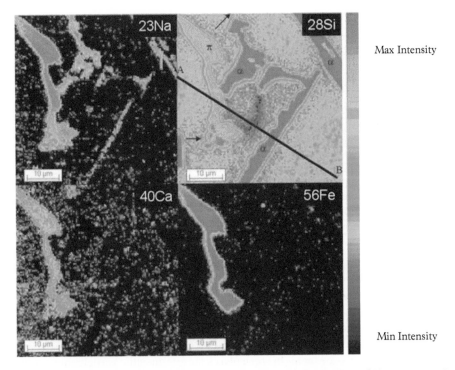

Fig. 1.33 *NanoSIMS of Na, Si, Ca, and Fe. Particles of α-Si, π-Al$_8$Si$_5$Mg$_3$Fe, and Na are present in addition to the Al matrix. The π-phase has Ca and Na impurities. A thin layer of Na is on the surface of α-Si particles.*

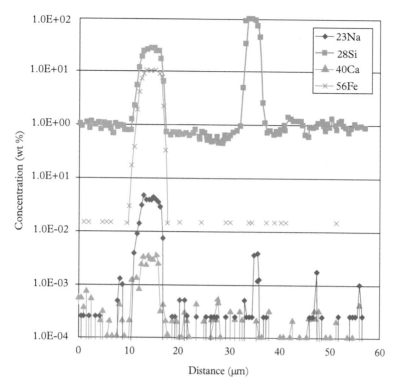

Fig. 1.34 *Line scan between point A and B in Fig. 1.33. A peak in Na is found after c.35 μm on the surface of α-Si particles. Π-phase is present at a distance of 10–17 μm, containing c.0.03% Na–0.003%Ca.*

A line scan was made between points A and B (Fig. 1.34). It shows that the surface of a large α-Si particle is covered by a thin layer of Na (plus some K atoms?). Such a layer is found on several α-Si-particles in this material. So if the material is heated above 100 °C, it is expected that Na will melt, and a crack will form if the material is stressed. The result is a decrease in hot ductility.

The next material investigated was an Al–7 wt% Si with 1 ppm Na. All particles seen in Fig. 1.35 are larger in this alloy than in the alloy with 34 wtppm Na seen in Fig. 1.32. A SIMS analysis of the material with less than 1 ppm Na showed that the Sr (9wtppm) was located in the α-Si-particles, Ca was unevenly distributed in the matrix and not in α-Si-particles, and Na was found in minute particles in the matrix (Fig. 1.36).

Al–10 wt% Si with either 160 wtppm Na or 520 wtppm Sr was analysed next. Optical microscopy and microprobe analysis revealed that the eutectic was mainly aluminium and small dispersoids of α-Si. Some aluminium grains contained larger α-Si and large intermetallic particles. The matrix had the composition Al–1.7 at% Si and a small amount of Na. The following types of particles were detected using a microprobe:

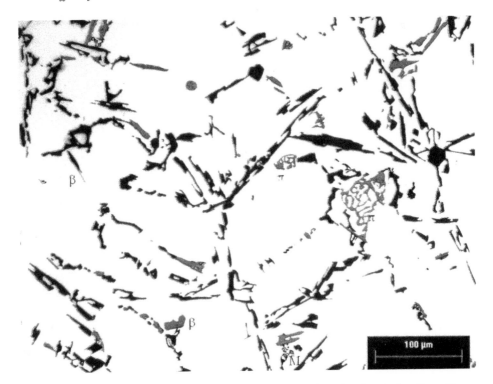

Fig. 1.35 *Al–7 wt% Si with 1 ppm Na. All particles are larger in this alloy than in the alloy with 34 wtppm Na.*

- δ-Al_3Si_2Fe rods (major type containing Fe)

  Al–31.2 at% Si–14.8 at% Fe–0.16 at% Mn–(0.1–0.07) at% Ni
  Other elements: 0.02–0.09 at% Mg–0.03 at% Na

- β-AlFeSi rods

  Al–17.9 at% Si–13.5 at% Fe–0.26 at% Ni–0.09 at% Mn
  Others: 0.03 at% Mg–0.02 at% P–0.01 at% Na

- α-Al(Fe,Mn,Cu,Ni)Si small agglomerates

  Al–9.2 at% Si–12.6 at% Fe–0.25 at% Ni–0.07 at% Mn–0.09 at% Cu
  Trace element: 0.02 at% Na

- α-Si: Si–1-4 at% Al(?)–0.25 at% Na–9 wtppm Sr

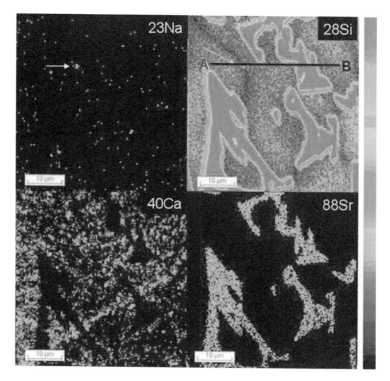

Fig. 1.36 *NanoSIMS analysis of Al–7 wt% Si with less than 1 wtppm Na. The arrow points at a particle rich in Na and Ca.*

NanoSIMS analysis revealed that most of volumes were eutectic with relatively small α-Si particles containing Na and K impurities (Figs. 1.37, 1.38). However, other phases were also mixed in the eutectic. Fe was found in rods together with Al, some Si and Mg, and traces of K, Na, and Ca. However, the sizes of Na, K, and Mg were larger than that of Fe. Therefore, this phase (probably impure β-AlFeSi) is surrounded by a phase containing Mg, Na, and K: a phase containing about 50% Si and no Fe also contains some Al, Na, and K. Finally, there are a few small particles ($\approx$1 μm) rich in Na or Ca.

NanoSIMS analysis revealed that most of the volumes were eutectic with relatively small α-Si particles containing Na and K impurities (Fig. 1.39). However, other phases were also mixed into the eutectic:

(a) Fe was found in rods together with Al, some Si and Mg, and traces of K, Na and Ca. However, the size of Na, K, and Mg was larger than that of Fe. Therefore, this phase (probably impure β-AlFeSi) is surrounded by a phase containing Mg, Na, and K.

(b) A phase containing about 50% Si and no Fe. It contain some Al, Na, and K. Finally, there are a few small partiles ($\approx$1 μm) rich in Na or Ca.

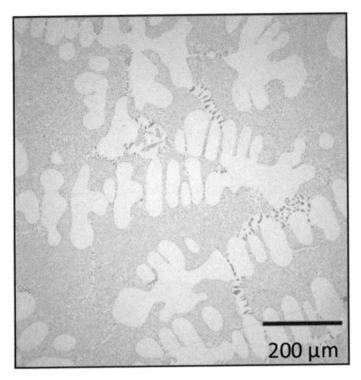

Fig. 1.37 *Al–10% Si–Na alloy. Light Al dendrites and darker eutectic.*

The second area analysed was a mixture of large particles and the Al matrix. It contained salt particles, impure α-Si particles, and δ-AlSiFe particles, as shown in Figs. 1.37, 1.38, and 1.40. Some of the particles were labelled as follows: salt, (Na,Mg)Cl; α, α-Si + impurities of Na and K; X, $Si_{24}(Na,K)_4$; and δ, δ-AlFeSi containing traces of Mg, Ca, Na, and K. The number of salt particles in this material was low, and their size was on the order of 10 µm. The next analysis was of Al–10 wt% Si–520 ppm Sr (see Fig. 1.41).

The next analyses of this alloy are shown in Figs. 1.42 and 1.43. Figure 1.42 is a SIMS analysis of a eutectic with a large number of small particles plus a salt particle, while Fig. 1.43 comprises a large, 40-µm-long rod plus a large number of small particles, a mixture of Al_2Si_2Sr and α-Si particles. Al_2Si_2Sr is probably formed first followed by nucleation of α-Si. The salt is the final phase formed since it is a phase with a low melting temperature.

It is evident from Fig. 1.43 (Mg, Si, Fe, Sr, Na, K, Ca) that this part of the material contains large intermetallic particles mixed with impurities that will give a reduction in mechanical and corrosive properties. The basis phase of the 40-µm-long rod is δ-AlFeSi (Fig. 1.41). However, nanoSIMS analysis of such a particle reveals that there are traces of Mg, Na, K, Ca. Sr in the thin sheet (a few-atoms-thick layer). In the initial stage we removed atoms from different phases. These phases were sputtered at different velocities, depending on the

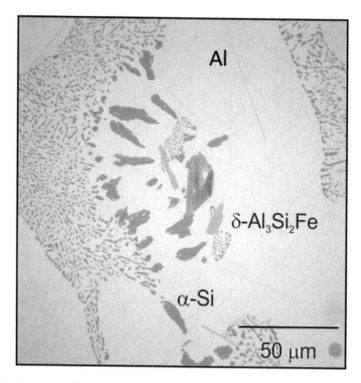

Fig. 1.38 *Al–10% Si–Na alloy. Al dendrites, fine dispersed eutectic, light grey δ-phase, dark α-Si in dendrites.*

hardness and bonding of the phases. Aluminium is soft and is sputtered at a high velocity, δ-AlFeSi, Al_2Si_2Sr, and α-Si at a much lower speed. Furthermore, the small Ca, Na, and K phases also have a different sputtering rate. Thus, the surface of the sample is rather 'hilly'. This explains why we get large variations in the intensity of all analysed elements, because the surface is inhomogeneous and some of the phases present are extremely thin. This also explains why we got a different result in the local Sr intensity in two successive analyses of the element's small particles (compare the two parts of Fig. 1.43).

The original surface was planar after polishing. Then we removed the top surface and the surface became somewhat 'hilly' because the Al-matrix was softer than the other phases in the alloy. When our analysis of the sample resulted in Fig. 1.43, a top layer was removed. The long rod (length *c*.50 µm) contained a lot of iron, which indicates that the basic particle is δ-AlFeSi. The presence of a small amount of Na, K, Ca, and Mg indicates that the particle is surrounded by a layer of these elements. Attached to the long particle are some α-Si particles (A) and some small Al_2Si_2Sr particles (B). There are also some small Mg_2Si (C) and small Al_2Si_2Sr particles in the area.

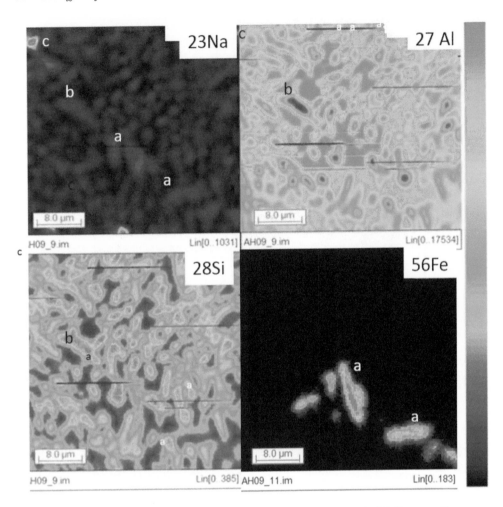

Fig. 1.39–1 *Eutectic of Al–10 wt% Si–Na with the following particles: (a) δ-AlFeSi + some Na, K + trace of Mg, Ca, (b) α-Si + Na, K, Al, and (c) Na-rich particle + Al, Si.*

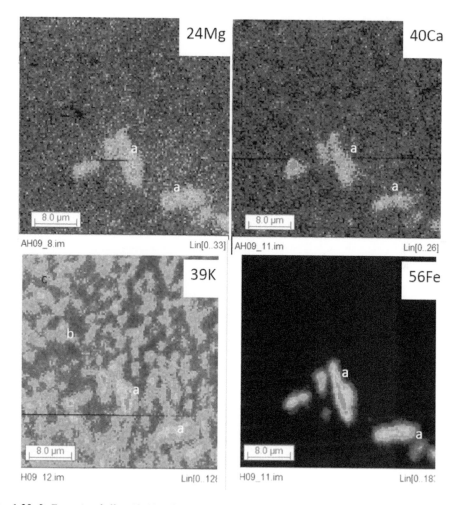

Fig. 1.39–2 *Eutectic of alloy Al–10 wt% Si–Na with the following particles: (a) δ-AlFeSi + Na, Mg, K, Ca, (b) α-Si +Na, K, and (c) Na-rich particle + Al, Si.*

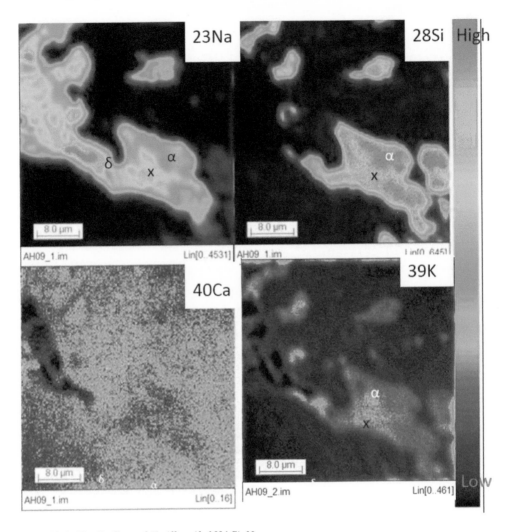

Fig. 1.40–1 *Na, Si, Ca, and K. Alloy Al–10% Si–Na.*

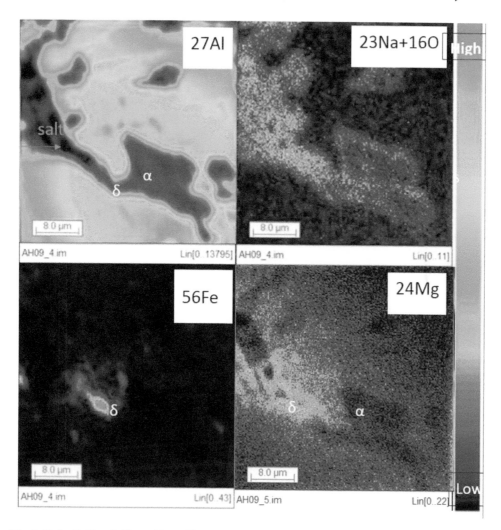

Fig. 1.40–2 *Al, Na +O, Fe, and Mg. Alloy Al–10% Si–Na.*

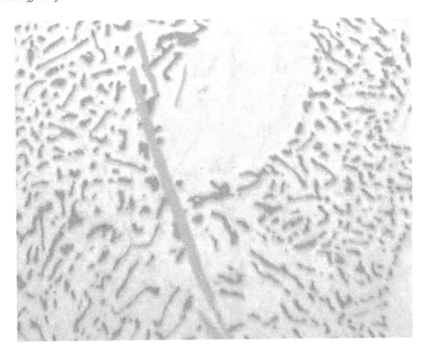

Fig. 1.41 *Optical micrograph of Al–10 wt% Si–0.052 wt% Sr showing small grey α-Si particles in the eutectic, one 20-μm-long δ-AlFeSi rod plus a few shorter, light grey ones. Some small α-Si particles are grown on δ-AlFeSi rods. Aluminium dendrites are also present.*

1.10.1.1   General Remarks about the SIMS Method Used

NanoSIMS and Tof-SIMS have the advantage that 25 elements and molecules can be analysed simultaneously or in sequence in a material. The methods have the advantage that the detection limit is extremely low, that is, less than 1 wtppm and 0.03–0.1 μm lateral resolution. Both positive and negative ions can be detected. Due to the difference in charge, the detection of negative ions must be slightly moved compared with the positive ions. We had to make a line scan in addition to the area detection (compare Figs. 1.33 and 1.34) in order to determine which element was the main element and which were trace elements like Na, K, and Ca. This was how we could trace the effect of minority elements in Al–Si alloys.

It was determined that Na and K can be found on the surface of particles in Al–Si castings. These phases will melt at a low temperature and can cause fracture at relatively low temperature.

Na and K will react with Cl and F and make salt, which can also react with Mg, Sr, and other elements like P and make extremely complex phases. Figure 1.44 shows Tof-SIMS analysis of salt particles in a modified aluminium alloy. Such phases may easily corrode and thereby reduce the quality of the material.

1.10.2   Edge Cracking in Hot-Rolled Materials of Al–Mg Alloys

A standard practice in the aluminium industry is to cast Al–Mg alloys as large, thick sheets and then to hot-roll the material into thin sheets. If the quality of the material is not good enough, the sheets can get edge cracks during the hot rolling. Such material is a failure and is dismissed.

Thomson and Burman did a systematic study on both industrial and laboratory sheets of Al–Mg and Al–Mg–Mn alloys that were hot-rolled from about 530 °C with a reduction rate of 2, 5, or 8 per cent in several passes.[72] The industrial materials were alloys ADC 5052, ADC 5082, and ADC 5454 with the composition outlined in Table 1.7. Three ingots were used in more detail (Table 1.8). The laboratory materials were an aluminium–4.5 wt% manganese alloy with a composition given in Table 1.9.

All ingots had a surface segregation layer with a high magnesium layer. Eight millimetres was removed by machining in order to get a surface free of this surface layer. Then the blocks received the following heat treatment: heat to 260 °C in 5 h; soak at 260–300 °C for 5 h; heat

Table 1.7 *Nominal chemical composition of industrial aluminium–magnesium alloys*

Alloy	wt% Si	wt% Fe	wt% Mn	wt% Mg	wt% Cr
ADC 5052	0.45	0.45	0.1	2.2–2.8	0.15–0.35
ADC 5082	0.40	0.40	0.5–1.0	4.0–4.9	0.05–0.25
ADC 5454	0.40	0.40	0.5–1.0	2.4–3.0	0.05–0.20

Reprinted from Thomson and Burman[72] with permission from Elsevier.

Table 1.8 *Analysis of alloys used in industrial trials*

Block number	wt% Si	wt% Fe	wt% Mg	wt% Mn	wtppm Na
17	0.14	0.19	4.20	0.10	12–19
18	0.17	0.23	4.45	0.09	14–26
19	0.12	0.19	4.20	0.10	<1.0

Reprinted from Thomson and Burman[72] with permission from Elsevier.

Table 1.9 *Analysis of alloys used in laboratory hot-rolling*

Alloy	wt% Si	wt% Fe	wt% Mg	wt% Mn	wtppm Na
D00148	0.18	0.19	4.45	0.06	<1.0
D00149	0.17	0.18	4.55	0.05	4
D00150	0.16	0.18	4.43	0.04	15

Reprinted from Thomson and Burman[72] with permission from Elsevier.

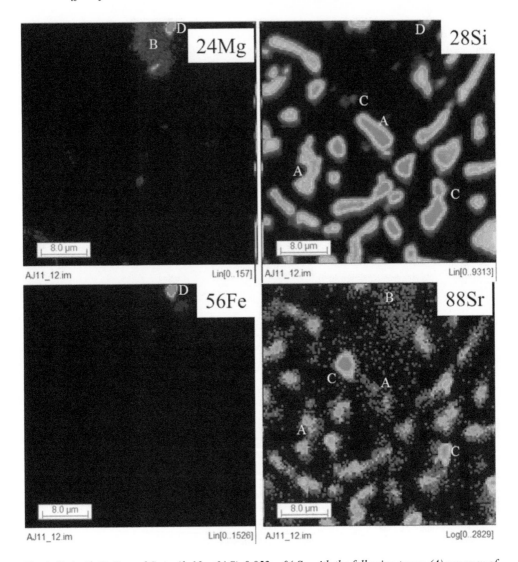

Fig. 1.42–1 *Al, Si, Fe, and Sr in Al–10 wt% Si–0.052 wt% Sr, with the following types: (A) a survey of the dominant α-Si containing traces of Sr and C; (B) salt particle (c.15 μm × 7 μm) of Na, K, and traces of Sr; (C) Al_2Si_2Sr with 0.04 at% Na–1.2 at% P–0.07 at% Ca–0.02 at% Ni (microprobe analysis); and (D) probably δ-AlFeSi with Mg impurities (2 μm in diameter).*

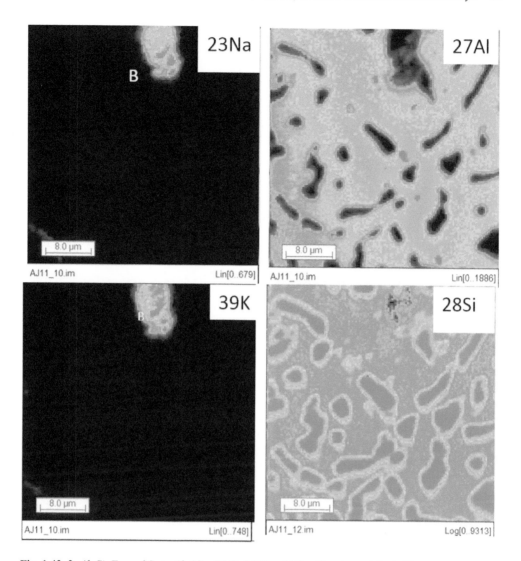

Fig. 1.42–2 *Al, Si, Fe, and Sr in Al–10 wt% Si–0.052 wt% Sr: (B) salt particle (c.15 μm × 7 μm) of Na, K, and traces of Sr.*

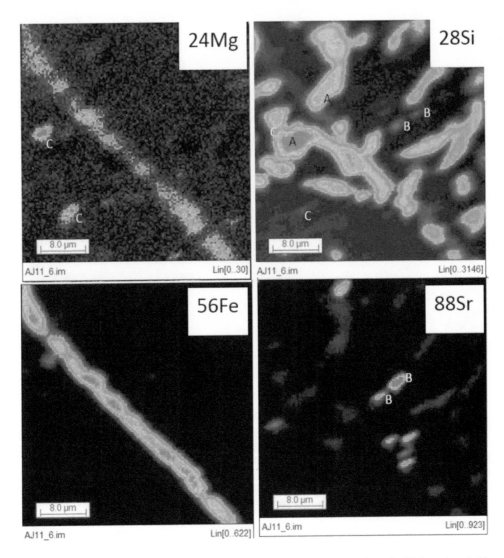

Fig. 1.43–1 *Al–10 wt% Si–Sr in an area with large particles: 30- to 40-μm-long δ-AlSiFe rod probably with impurities of Mg, Na, K, Ca on the surface. Other phases are (A) α-Si containing Sr impurities, (B) Al$_2$Si$_2$Sr, and (C) Mg–K–Na–Ca–Si–Al–Sr phase.*

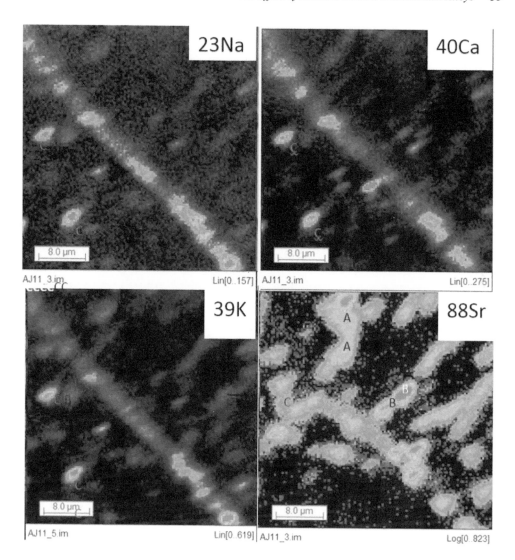

Fig. 1.43–2 *Area of an Al–10 wt% Si–Sr alloy.*

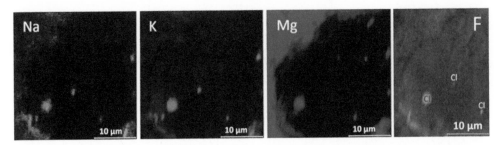

Fig. 1.44 *Tof-SIMS analysis of salt particles in modified alloy AA5005. The largest salt particle in the picture is a (Na,K,Mg)Cl, c.2 μm in diameter. There are also NaF and (Na,K)Cl particles. The SIMS pictures of F and Cl are moved slightly as negative ions hit the screen during the analysis, as compared with the positive ions of Na, K, and Mg.*

Fig. 1.45 *Small numerous cracks initiated and propagated within the massive primary segregation band (photograph reduced c.20%). Segregation cracking in the unscalped edges of hot-rolled Al–Mg alloy after 89% reduction (RD, rolling direction). (Reprinted from Thomson and Burman[72] with permission from Elsevier.)*

to 300–510 °C in 8 h; soak at 510–30 °C for 4 h. The temperature of the blocks was measured with a contact pyrometer. After the machining, the industrial ingots were 3.0 m long, 0.87 m wide, and 0.39 m thick.

The rolling was performed on a two-high mill with 159-mm rolls at pass reductions of 2, 5, and 8 per cent. At 8 per cent reduction it was found that only one or two passes could be made before reheating the specimens, whereas at 2 per cent reduction three passes were possible.

Examination of edge-cracked samples from a large number of hot-rolled Al–Mg alloy ingots showed that there were three major classes of edge cracks, as shown in Figs. 1.45, 1.46, and 1.47, respectively.

Microstructure investigations revealed that cracks were related to the presence of agglomerates of intermetallic particles in the segregation zone. They were of two types, Mg_2Si clusters and Al–Fe or Al–Fe–Si particles. This fits with the observation of C. J. Simensen and

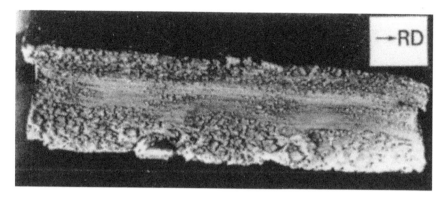

Fig. 1.46 *Large edge cracks of varying severity initiated in the segregation band but propagated into the bulk of a c.5-mm-thick ingot. Major edge cracks in the hot-rolled Al–Mg alloy ADC 5052 after 87% reduction (RD, rolling direction). (Reprinted from Thomson and Burman[72] with permission from Elsevier.)*

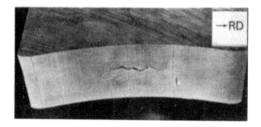

Fig. 1.47 *Massive cracks, although first observed on the edge of the ingot, were shown to have initiated internally. Massive crack formed at the scalped edge of the hot-rolled aluminium alloy ADC 5454 after 15% reduction (RD, rolling direction) (photograph reduced to 25% of full size). (Reprinted from Thomson and Burman[72] with permission from Elsevier.)*

co-authors that intermetallic particles can be covered by a thin layer of Na or Na–K. These layers will melt at a low temperature around 100 °C. Then cracks will form during hot-rolling.

Because the segregation band had been removed from the hot-rolling alloys studied in the laboratory, no segregation cracking was expected nor was it observed. Edge cracks of various sizes occurred in a number of rolled blocks. Of the cracks formed at small reductions, some were transverse, but all assumed a longitudinal path after the initial passes. The majority of these cracks were small. It was found that cracks formed at low total reductions were partially closed or completely healed at higher total reductions. This phenomenon was also observed during industrial rolling. As with the blocks hot-rolled under industrial conditions, the edge cracks produced in the laboratory specimens occurred at random along the edges.

The lengths of the edge cracks in the material rolled at a nominal temperature of 500 °C ranged up to a maximum of 10 mm, but the majority of cracks were less than 5 mm long. The numbers of cracks in each block in the arbitrary size ranges 0–5 and 5–10 mm were counted and the frequency of cracking (calculated as the number of new cracks in two blocks after each

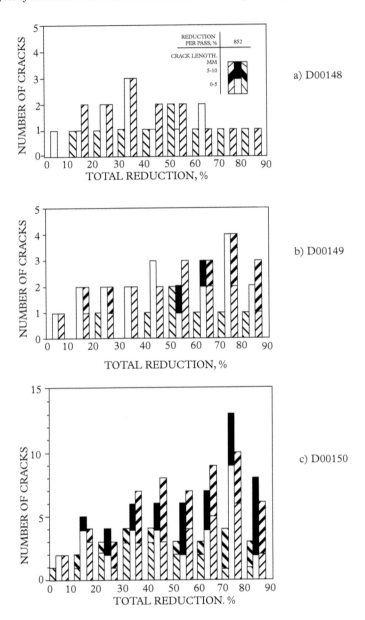

Fig. 1.48 *The frequency of cracking in the laboratory hot-rolling of an Al–Mg alloy at a nominal rolling temperature of 500°C. (The bars show the number of cracks observed after increments of approx. 10% in total deformation with pass reduction of 2, 5, and 8%.) (Reprinted from Thomson and Burman[72] with permission from Elsevier.)*

10 per cent increment in the total reduction of each alloy). The results of these measurements are shown in Fig. 1.48. The effect of increased sodium concentration can be seen generally to increase both the size and the number of edge cracks, but the absence of edge cracks in some specimens of the high sodium alloy indicates that sodium was not only requirement for their occurrence. There was some evidence that the frequency and severity of cracking depended on the reduction per pass.

The total hydrogen concentration was determined in the three laboratory alloys (Table 1.9). High values of hydrogen were numbers above 0.4 ml H_2 per 100 g aluminium, and low values were less than 0.1 ml H_2 per 100 g aluminium. The largest and most numerous cracks occurred in samples with high sodium and hydrogen levels, but there were some unexplained exceptions in which no cracking was observed even when the concentrations in both hydrogen and sodium were high.

Thomson and Burman came to the conclusion that the laboratory trials indicated that sodium and hydrogen interacted in the mechanism of edge cracking although they were unable to determine the exact nature.[72]

1.10.3   Extrusion of AlMgSi Alloys and the Melting of Secondary Phase Particles

Reiso made several series of extrusions of Al–Mg–Si alloys at different temperatures. His work was published in a series of papers and was the base for his Dr. Techn. Thesis.[73] The process of producing extruded products is outlined in Fig. 1.49. He added AlTi5B1 material to all the investigated alloys in order to obtain materials with small equiaxed grains. A sketch of the temperature and time profile is outlined in Fig. 1.50. He concentrated on the effect of silicon and magnesium in his first paper and later on the effect of the cooling rate after the homogenization.[74] He used mainly three alloys in his extrusion experiments, as shown in Table 1.10. Then he varied the cooling rate between 500 and 300 °C, as summarized in Table 1.11.

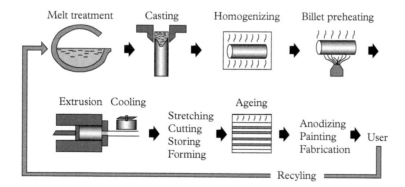

Fig. 1.49 *Sketch of the process chain for the production of aluminium extrusions.*[75]

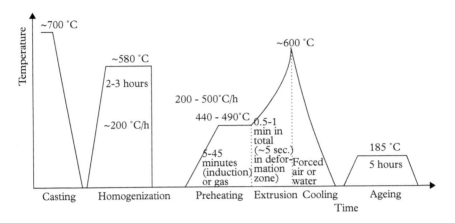

Fig. 1.50 *Sketch of the temperature/time profile for the production of aluminium extrusions. Typical temperatures and times used for AlMgSi alloys are indicated.*[75]

Table 1.10 *Chemical composition of material used in paper 1*

Alloy	wt% Mg	wt% Si	wt% Fe
I	0.45	0.44	0.20
II	0.66	0.45	0.20
III	0.45	0.79	0.19

Reprinted from Reiso[74] with permission from the Aluminum Association.

The results of Reiso showed that the maximum extrusion speed was reduced with increasing content of both magnesium and silicon. He found that the tearing initiating during extrusion was of two types. Type 1 resulted in cracks on the extruded materials at maximum extrusion speed (Fig. 1.51a). Type 2 occurred in materials with large Mg_2Si particles where phases were melted during extrusion and voids were formed, as shown in Fig. 1.51b. Type 2 took place in materials that experienced a low cooling rate and had a high content of magnesium and/or silicon. Type 1 took place in materials which experienced a high cooling rate (see Fig. 1.52). It was found that the tearing initiation of type 2 took place in materials with large Mg_2Si particles in the extruded materials (see Table 1.12).

The extrusion dies used in his next tests are shown in Fig. 1.53, where he mainly tested two alloys as shown in Table 1.13.[76] The first alloy was close to the quasi $Al–Mg_2Si$ and the second had a surplus of silicon. The temperature on the surface of the extruded material was also measured in the second test for alloy 2. Test 1 was at a 30MN press with a gas-fired preheating furnace. Heating time for the billets was approximately 35 minutes. The container temperature was ca 370 °C, and the reduction ratio was 52.

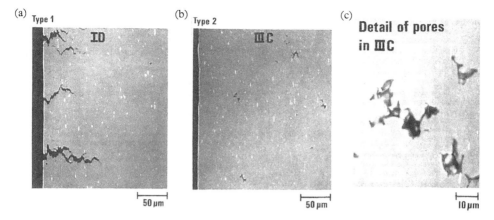

Fig. 1.51 *(a) Tearing initiation in material ID. Type 1 with cracks in the surface of the extruded material. (b) Tearing initiation in material III C. Type 2 with internal black cracks. (c) Details of the internal pores in the material. Type 2 tearing initiation. (All reprinted from Reiso[74] with permission from the Aluminum Association.)*

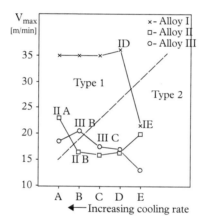

Fig. 1.52 *Tearing initiation in the ribs of the extrusion related to the extrusion speed and cooling speed. Note that type 1 is related to high cooling rate and large extrusion speed. (Reprinted from Reiso[74] with permission from the Aluminum Association.)*

Test no 2 was done at 20MN press with an induction preheating furnace. Heating time of the billets was approximately 5 minutes. The container temperature was $c.450\,°C$, and reduction ratio was 40. The term 'extrudability' is defined as the maximum speed (V_{max}) before tearing occurred in the ribs of the section.

The billet temperatures were measured at the entrance of the press in the front, middle, and back of each billet. The reported billet temperatures are the front temperature. (The temperature difference was typically $25–35\,°C$ in test 2.)

Table 1.11 *Cooling rate of different alloys between 500 and 300°C*

Alloy	Cooling rate (°C/h)				
	A	B	C	D	E
I	Water quenched	350	210	115	20
II	Water quenched	540	205	120	20
III	Water quenched	540	205	120	20

Reprinted from Reiso[74] with permission from the Aluminum Association.

Table 1.12 *Number of Mg_2Si in an area of 0.51 mm^2 in the extrusions*

Alloy	Cooling rate after homogenization				
	A	B	C	D	E
I	8	15	28	30	217
II	9	26	151	339	329
III	6	28	45	89	249

Reprinted from Reiso [74] with permission from the Aluminum Association.

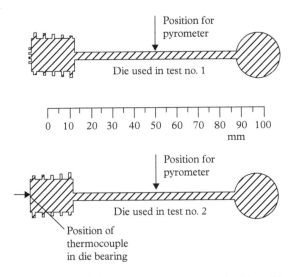

Fig. 1.53 *Extrusion dies used to establish maximum extrusion speed, with positions for temperature measurements indicated. (Reprinted from Reiso[76] with permission from the Aluminum Association.)*

Table 1.13 *Spectrographic analysis of alloys*

Test	Alloy	wt% Fe	wt% Mg	wt% Si	wt% Mn
1	1	0.17	0.60	0.47	—
2	2	0.20	0.56	0.90	0.50

Reprinted from Reiso[76] with permission from the Aluminum Association.

1.10.3.1 Results of Test 2

The results from test 2 are plotted in Fig. 1.54. The plot of V_{max} versus billet temperature (Fig. 1.54a) confirms that that the anticipated upward shift in the extrusion limit diagram does occur, and that the overheating and cooling of the billets gives increased extrusion speeds relative to billets heated directly to the same temperature. For this alloy the shift in speed appeared at a billet temperature of about 530 °C and the gain was somewhat lower than was found for alloy 1. (Overheating was done by solutionizing the billets at 550 °C and then cooling the billets down to a wanted temperature prior to extrusion.)

Figure 1.54b shows the extrusion surface temperatures at the event of tearing as recorded both by the pyrometer situated 0.7 m behind the die and by the thermocouple located on the bearing surface in the die. The pyrometer records an increasing surface temperature with increasing billet temperature, as well as higher temperatures for the overheated billets. (The results were similar in the test of alloy 1.) The thermocouple in the die, however, seems to record two different but relative constant surface temperatures at tearing, independent of the billet temperatures. These temperatures were 584 and 590 °C.

Figure 1.54c shows an increase in mechanical properties for those with increasing lower temperatures. At the highest billet temperatures a rather big scatter was recorded. The overheated billets have equal or better mechanical property than those billets heated directly to the same temperature.

Optical and SEM analyses of the microstructure of the extrusions of the two alloys showed that β-AlFeSi and Mg_2Si are present in alloy 1 and α-Al(Fe,Mn)Si and Mg_2Si in alloy 2 (Reiso found particles with varying amounts of Mg and Si, but that they are most likely Mg_2Si that has corroded during polishing). Most of the extruded samples investigated had cracks on the surface (type 1). Type 2 with tearing initiation in the presence of large Mg_2Si particles was detected in materials directly heated to a low temperature less than 500 °C and a mixture of the two types in material heated to 540 °C.

The results from the tests with regard to extrudability may be summarized in the form of a modified extrusion limit diagram as shown in Fig. 1.55. The diagram is applicable to all alloys (not only AA 6xxx alloys) and production practices which include incipient melting due to precipitated phases that are soluble in the Al matrix.

The fully drawn line in Fig. 1.55 represents billets heated directly to temperature. On the right-hand side of the diagram an upward shift in extrusion speed appears for those billets at temperatures high enough to dissolve the low melting point phases (Mg_2Si in 6xxx alloys). This sharp increase in extrusion speed results from type 2 to type 1 tearing initiation, as

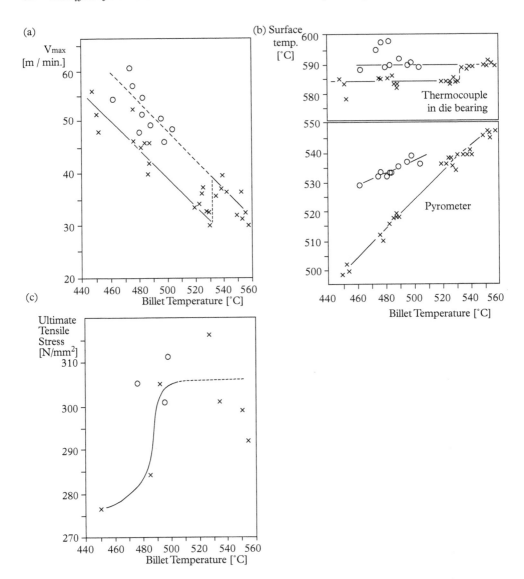

Fig. 1.54 *Test no. 2: (a) Maximum extrusion speed for different billet temperatures. (b) Surface temperature at tearing (pyrometer and thermocouple in die bearing). (c) Mechanical properties versus billet temperature. x = billets heated directly to temperature; o = billets cooled down from 550°C. (Reprinted from Reiso[76] with permission from the Aluminum Association.)*

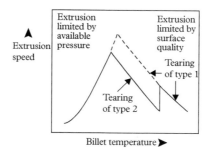

Fig. 1.55 *Modified extrusion limit diagram with indications of type tearing initiation for directly heated billets (fully drawn line) and overheated and cooled billets (dashed line). (Reprinted from Reiso[76] with permission from the Aluminum Association.)*

Table 1.14 *Chemical compositions of the alloys' wt%[77]*

Alloy	Mg	Si	Mg$_2$Si	ΔSi[a]
1	0.57	0.33	0.90	—
2	0.60	0.70	0.95	0.35

[a] ΔSi is excess Si over the necessary amount to form Mg$_2$Si.
Reprinted by permission from Springer Nature: copyright 1993.

indicated in Fig. 1.55. The reason for type 1's higher speed is that the material can now withstand a higher critical surface temperature before tearing.

The dashed line in Fig. 1.55 represents billets which have first been heated above the solvus temperature long enough to dissolve the Mg$_2$Si particles (in 6xxx alloys) and then allowed to cool down to respective temperatures before extrusion.

1.10.3.2 Melting of Secondary Phase Particles in AlMgSi Alloys

Reiso, Ryum, and Strid investigated two Al–Mg–Si alloys (Table 1.14) at temperatures above 500 °C.[77] (The materials were made from high-purity raw materials of Al, Mg, and Si.) They homogenized the alloys for 24 hours at 540 °C. Then the materials were quenched to room temperature, heated for 1 hour at 540 °C, and slowly cooled at a rate of 1 °C/h down to 100 °C. These materials had large particles of Mg$_2$Si in alloy 1 and Mg$_2$Si and α-Si in alloy 2 (Fig. 1.56). The next step was to rapidly heat the alloys in a salt bath to temperatures between 550 and 600 °C and to anneal for a shorter period. Only Mg$_2$Si particles were detected in the quasi-binary alloy 1. These particles were found in the eutectic of the materials upquenched to 593 °C (Fig. 1.57).

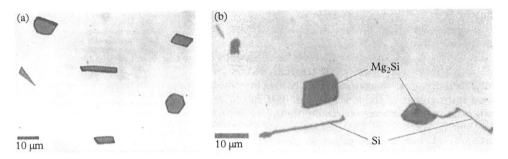

Fig. 1.56 *(a) Microstructure after slow cooling of alloy 1. Mg_2Si-particles. (b) Microstructure after slow cooling of alloy 2. Mg_2Si and Si particles.*[77] *(Reprinted by permission from Springer Nature: copyright 1993.)*

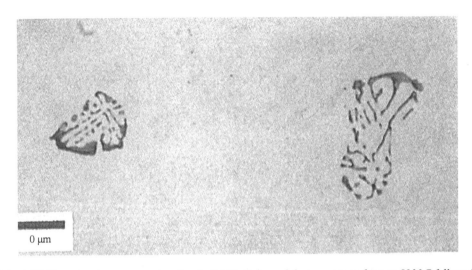

Fig. 1.57 *Eutectic microstructure in alloy 1 of AlMgSi formed during upquenching to 593 °C followed by quenching in air.*[77] *(Reprinted by permission from Springer Nature: copyright 1993.)*

The microstructure was more complex in alloy 2 after upquenching and cooling in air. Figure 1.58a shows a fine ternary Al–Mg_2Si–Si eutectic after 0 seconds (no holding time) at 560 °C. Figure 1.58b show Al–Si after upquenching to 577 °C, and Fig. 1.58c shows an Al–Mg_2Si eutectic after upquenching to 593 °C.

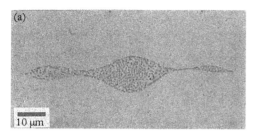

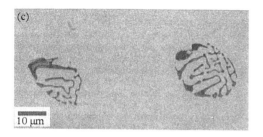

Fig. 1.58 *Microstructures in alloy 2 after upquenching and subsequently cooling in air.*[77] *(a) Ternary Al–Mg₂Si–Si eutectic after upquenching to 560 °C. (b) Binary Al–Si eutectic after upquenching to 577 °C. (c) Al–Mg₂Si eutectic after upquenching to 593 °C. (Reprinted by permission from Springer Nature: copyright 1993.)*

The next series show the development of the microstructure in the Al–Mg$_2$Si eutectic when alloy 2 is upquenched to 600 °C and annealed for different holding times. Figure 1.59a is after 0 seconds, showing a binary Al–Mg$_2$Si eutectic. Figure 1.59b shows some ternary Al–Mg$_2$Si-Si eutectic after 20 seconds, while Fig. 1.59c shows more ternary eutectic, and finally only ternary eutectic are present after 600 seconds (Fig. 1.59d).

Reiso et al. showed that three eutectic processes took place in alloy 2:

1) α-Al + Mg$_2$Si $\rightarrow$ liquid at 593 °C or higher temperatures
2) α-Al + Si $\rightarrow$ liquid at 577 °C or higher temperatures
3) α-Al + Mg$_2$Si +Si $\rightarrow$ liquid at 559 °C or higher temperatures

These reactions were responsible for the local melting in alloy 2 during the extrusion and caused the incipient tearing type 2. It is to be expected that similar effects may occur in other aluminium alloys containing phases with a low melting point. Examples are aluminium with a high content of Na and/or K. Then the local melting can be initiated at a very low temperature below even 100 °C.

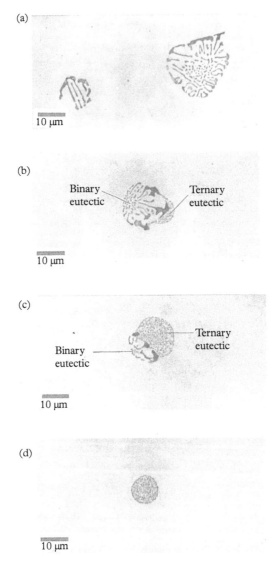

Fig. 1.59 *Changes in microstructure in alloy 2 after upquenching to 600°C and annealing at 600°C,[77] for (a) 0 seconds (binary Al–Mg2Si eutectic), (b) 20 seconds (mixture of binary and ternary eutectic), (c) 60 seconds (mixture of binary and more ternary eutectic), and (d) 600 seconds (pure ternary eutectic). (Reprinted by permission from Springer Nature: copyright 1993.)*

1.11 Concluding Remarks

This chapter has concentrated on materials commonly used in industrial countries, namely steel and alloys of Al, Mg, and Cu. It is expected that there will be analogous effects in other metals such as alloys of Ag, Au, Ni, Mn, Cr, Co, Ti, and Be.

We have used dynamic SIMS, NanoSIMS, and Tof-SIMS to analyse trace elements in aluminium. However, the technique has also been used frequently for other materials such as steel,[78] silicon, copper,[79] and nickel.[80] Other important topics where SIMs have been applied are minerals, meteorites, semiconductors, and cancer tumors.

All elements in the periodic table and also molecule fragments can be analysed using SIMS. We have mainly been interested in bulk properties. Therefore, surface layers of $c.1$ µm have initially been removed. The sputtering rate is dependent upon hardness and the chemical bonding of the phases. Thus, the aluminium matrix is commonly sputtered faster than hard particles of TiB_2 and covalent bonded atoms of Si and Fe in α-AlFeMnSi. It can be difficult to distinguish between molecule fragments with total atomic weight similar to that of element isotopes. In such cases it may be necessary to analyse the materials using several techniques like SIMS, microprobe, and optical X-ray diffraction.

Our strategy has been to initially analyse the material using emission spectrometer and to examine optically using either SEM or microprobe. Then a survey is made in SIMS where the the intensity of most elements is detected. Then the main elements and important trace elements are selected. The important industrial properties are measured and compared with microstructure observations. SIMS became important when trace elements were a key factor.

REFERENCES

1. Simensen CJ. Sources of impurities in aluminium melts and their control. In: Nilmani M, eds, *Aluminium melt refining and alloying.* University of Melbourne, Australia; 1989: C l-C 20.
2. Watanabe T, Huang YC, Komatsu R. Solubility of hydrogen in magnesium. *J. Jpn. Inst. Light Metals*, 1976; 26(2): 76–81.
3. Shapavalov VI, Serdyuk NP, Semik OP. Magnesium-hydrogen and aluminium-hydrogen phase diagrams. *Dop. Akad. Nauk Ukr. RSR. Ser. A Fiz.-Mat. Tekh. Nauki*, 1981; 6: 99–101 (in Russian).
4. Popovic ZD, Piercy GR. Measurement of the solubility of hydrogen in solid magnesium. *Metall. Trans. A*, 1976; 6: 1915–17.
5. Koeneman J, Metcalfe AG. The solubility of hydrogen in magnesium. *Trans. ASM*, 1959; 51: 1072–82.
6. Chernega DF, Gotvjanskij UA, Prisjasznjok TN. Diffusion and solubility of hydrogen in liquid magnesium. *Vestn. Kiev. Politekh. In-ta. Kiev*, 1979; 16: 56–9.
7. Talbot DEJ. Effects of hydrogen in aluminium, magnesium, copper and their alloys. *Int. Met. Rev.*, 1975; 20(1): 166–84.
8. Turner AN, Bryant AJ. The effect of ingot quality on the short- transverse mechanical properties of high-strength aluminium alloy thick plate. *J. Inst. Metals*, 1967; 95: 353–9.
9. Kostron H. Aluminium and gas. *Z. Mkde.*, 1952; 43: 269–84 and 373–87.

10. Allen NP, Hewitt T. The equilibrium of reaction between steam and molten copper. *J. Inst. Metals*, 1933; 51: 257–75.

11. Harper S, Calicut VA, Townsend DW, Eborall R. The embrittlement of tough-pitch copper during annealing or preheating. *J. Inst. Metals*, 1962; 90: 422–9.

12. Mattsson E, Shuckher F. An investigation of hydrogen embrittlement in copper. *J. Inst. Metals*, 1959; 87: 241–7.

13. Rylski O.Z. P.M. 203. Dept, of Mines and Technical Surveys, Mines Branch, Ottawa, Canada.

14. Liue TS, Steinberg MA. The mode of hydride precipitation in alpha titanium and alpha titanium alloys. *Trans. Amer. Soc. Metals*, 1958; 50: 455–77.

15. Cotterill P. The hydrogen embrittlement of metals. *Progr. Mater. Sci.*, 1961; 9: 201–304.

16. Craighead CM, Lenning GA, Jafee RI. Hydrogen embrittlement of beta-stabilized titanium alloys. *J. Metals N.Y.*, 1956; 8(8): 923–8.

17. Jafee RI, Lenning GA, Craighead CM. Effect of testing variables on the hydrogen embrittlement of titanium and a Ti-8pct Mn alloy. *J. Metals N.Y.*, 1956; 8(8): 907–13.

18. Koffila RJ, Burte HM. Hydrogen contamination in titanium and titanium alloys. Part I: Hydrogen embrittlement on alpha-beta titanium alloys. USAEC Report No. WADC-TR-54-616 (unclassified).

19. Troiano AR. The role of hydrogen and other interstitials in the mechanical behaviour of metals, Campbell memorial lecture. *Trans. ASM*, 1960; 52: 54–80.

20. Mondolfo LF. *Aluminium alloys, structure and properties*. London: Butterworths; 1976.

21. Willey LA as quoted by Dean WA. In: Horn KR, ed., *Aluminium* 1, p. 174. Metals Park, OH: ASM; 1967.

22. Higgins RA. *Engineering metallurgy* 1. *Applied physical metallurgy*. London: Hodder and Stoughton; 1993.

23. Becker R, Doring W. *Ferromagnetismus*. Berlin: Springer; 1939.

24. Simensen CJ, Herø H. Magnetic ageing of sheet metals of Fe-Si-C alloys. SI report 730112 (in Norwegian), 1974.

25. Leslie WC, Stevens DW. The magnetic ageing of low-carbon steels and silicon irons. *Trans ASM*, 1964; 57: 264-83. Also Leslie WC. The quench-ageing of low carbon iron and iron-manganese alloys. *Acta Met.*, 1961; 9(11): 1004–22.

26. Nacken M, Rahmann J. Untersuchungen zur Abschreckalterung stickstoffhaltiger Stahle durch Messung der Dampfung, Koerzitivfeldstarke und Rückbildung. *Archiv für das Eisenhüttenwesen*, 1962; 33: 131–40.

27. Ljungstrom LG. The influence of trace elements on the hot ductility of austenitic 17 chromium–13% nickel–molybdenum–steel. *Scand. J. Metall.*, 1977; 6(4): 176–84.

28. Effects of impurities on hot ductility of austenitic stainless steels. Stockholm: STU information; 1978: No. 105.

29. Kiessling R. Clean steel—a debatable concept. In: Nordberg H, Sandberg R. eds, *Swedish Symposium on non-metallic inclusions in steel*, pp. 7-18. Hagfors, Sweden: Uddeholms AB; 1981.

30. Leschiner LN, Kovalyov VG. Effect of Fe and Si on properties and structure of Al-Cu-Mg alloys. In Kovacs I, ed., *Effect of iron and silicon in aluminium and its alloys : Proceedings of the International Workshop, held in Balatonfured, Hungary*. Zurich, Switzerland: Trans. Tech. Publ.; 1989.

31. Simensen CJ. Investigations of cracks in rolled materials. SI report 720717-7 (in Norwegian), 1974.

32. Hilty DC. Inclusions in steel. In: Taylor CR, ed. *Electric furnace steelmaking*, pp. 237–51. Chelsea, MI: AIME Iron and Steel Soc.; 1985.

33. Lagneborg R. The influence of non-metallic inclusions on properties in steel—a review. In: Nordberg H, Sandstrom R, eds. *Swedish symposium on non-metallic inclusions in steel*, pp. 285–352. Hagfors, Sweden: Uddeholms AB; 1981.

34. Dieter GE. Bulk workability testing. In: *Metals Handbook*, Vol. 8 *Mechanical Testing*, 9th edn., pp. 571–3. Metals Park, Ohio: ASM; 1985.

35. Pickering FB. High-strength low-alloy steels—a decade of progress. In: *Microalloying 75, Int. Symp.*, p. 1. New York: Union Carbide Corp.; 1975.

36. Pickering FB. Towards improved ductility and toughness. *Climax Molybdenum Comp. Conf., Kyoto; 1971*, p. 9.

37. Dodd B, Bai Y. *Ductile fracture and ductility.* London: Academic Press; 1987.

38. Orowan E. *Fatigue and fracture of metals.* Cambridge, MA: MIT Press; 1950.

39. Knott JF. *Fundamentals of fracture mechanics.* London: Butterworths; 1973.

40. Kiessling R, Nordberg H. Influence of inclusions on mechanical properties in steel. Critical inclusion size. *Rep. Royal Swedish Acad. Eng. Sci.*, 1971; 169(1): 159–69.

41. Andersson H. Analysis of a model for void growth and coalescence ahead of a moving crack tip. *J. Mech. Phys. Solids*, 1977; 25(3): 217–33.

42. Rice JR. A path independent integral and the approximate analysis of strain concentration by notches and cracks. *J. Appl. Mech.*, 1968; 35: 379–86.

43. Rice JR, Johnson MF. The role of large crack tip geometry changes in plane strain fracture. In: Kanninen MF et al., eds, *Inelastic behaviour of solids*, p. 641. New York: McGraw-Hill; 1970.

44. Knott JF. Micromechanisms of fibrous crack extension in engineering alloys. *Metal Science*, 1980; 14: 327–36.

45. Peterson RE. *Proc. symp. fatigue of aircraft structure, 1959*; WADC TR 59-507, p. 273.

46. Peterson RE. In Sines G. ed. *Metal fatigue.* New York: McGraw-Hill; 1959.

47. Nordberg H. The effect of notches and non-metallic inclusions on fatigue properties of high strength steel. Internal Report, IM-1450, Swedish Institute for Metal Research. 1980.

48. Brooksbank D, Andrews KW. Stress fields around inclusions and their relation to mechanical properties; production and application of clean steels. In: *Proceeding of production and application of clean steels, Balatonfured, Hungary*, pp. 186–98. London: Iron and Steel Institute; 1970; Publ. 134.

49. Hauser JJ, Wells MGH. The effect of inclusions on fatigue properties in high-strength steels. In *Mechanical working and steel processing*, p. 186. New York: AIME; 1976.

50. Kiessling R. *Non-metallic inclusions in steel.* London: Metals Society; 1978.

51. Ollilainen V. The effect of Ca treatment on the machinability of steel. In: Nordberg H, Sandstrom R, eds, *Swedish symposium on non-metallic inclusions in steel*, pp. 429–49. Hagfors, Sweden: Uddeholms AB; 1981.

52. Sukiman NL, Zhou X, Birbilis N, et al. Durability and corrosion of aluminium and its alloys: overview, property space, techniques and developments. In: Ahmad Z, ed., *Aluminium alloys: new trends in fabrication and applications*, pp. 47–97. London: IntechOpen; 2012; http://dx.doi.org/10.5772/53752

53. Nisancioglu K, Nordlien JH, Afseth A, Scamans GM. Significance of thermomechanical processing in determining corrosion behavior and surface quality of aluminum alloys. *Mater. Sci. Forum*, 2000; 331–7: 111–26.

54. Birbilis N, Buchheit RG. Electrochemical characteristics of intermetallic phases in aluminum alloys. *J. Electrochem. Soc.*, 2005; 152: B140–51.

55. Zhu Y, Sun K, Frankel GS. Intermetallic phases in aluminum alloys and their roles in localized corrosion. *J. Electrochem. Soc.*, 2018; 165: C807–20.

56. Kairy SK, Rometsch PA, Davies CHJ, Birbilis N. On the intergranular corrosion and hardness evolution of 6xxx series Al alloys as a function of Si:Mg ratio, Cu content, and aging condition. *Corrosion*, 2017; 73: 1280–95.

57. Lunder O, Nisancioglu K, Hansen RS. Corrosion of die cast magnesium-aluminium alloys. SAE Technical paper series, International Congress and Exposition, Detroit, Michigan, March 1-5, 1993.

58. Lunder O, Videm M, Nisancioglu K. Corrosion resistant magnesium alloys. SAE Technical paper series, International Congress and Exposition, Detroit, Michigan, Feb 27–March 2, 1995.

59. Hollingsworth EH, Hunsicker HY. Corrosion of aluminum and aluminum alloys. In *ASM Handbook*, vol. 13 *Corrosion*. ASM International: 1992.

60. Larsen MH, Walmsley JC, Lunder O, Nisancioglu K. Effect of excess Si and small Cu content on intergranular corrosion of 6000-series aluminum alloys. *J. Electrochem. Soc.*, 2010; 157: C61–8.

61. Anawati, Graver B, Nordmark H, et al. Multilayer corrosion of aluminum activated by lead. *J. Electrochem. Soc.* 2010; 157(10): C313–20.

62. Tan J, Nisancioglu K. Effect of small amounts of alloyed tin on the electrochemical behavior of aluminium in sodium chloride solution. *Corrosion Science*, 2013; 76: 219–30.

63. Kurt K, Diplas S, Walmsley JC Nisancioglu K. Effect of trace elements lead and tin on anodic activation of AA8006 aluminum sheet. *J. Electrochem. Soc.*, 2013; 160: C542–52.

64. Emley EF. *Principles of magnesium technology.* Oxford: Pergamon Press; 1966.

65. Simensen CJ, Oberlander BC. A survey of inclusions in magnesium. *Z. Prakt. Metallographies*, 1980; 17: 125–36.

66. Hanawalt JD, Nelson CE, Peloubet JA. Corrosion studies of magnesium and its alloys. *Trans. AIME*, 1942; 147: 273-99.

67. Eklund G. Relation between pitting and non-metallic inclusions. *Jernkont. Ann.*, 1971; 155: 637–42.

68. Eklund G. Initiation of pitting at sulfide inclusions in stainless steel. *J. Electro-Chem. Soc.* 1974; 121: 467–73.

69. Szummer A, Lublinska K, Janik-Czachov M. Low manganese corrosion resistant steels. *ISIJ*, 1991; 31(2): 239–40.

70. Konig U, Bier M, Schultze JW. The passivation of iron and iron/chromium alloys in acid and neutral solution. *ISIJ*, 1991; 31(2): 142–8.

71. Simensen CJ, Hillion F, Nielsen Ø, Nogita K, Dahle AK. Analysis of trace elements in Na-modified Al-10 wt%Si using EPMA and NanoSIMS. In: Jones H (ed), *Proceedings of the 5th decennial international conference on solidification processing, 2007*, pp 244–8.

72. Thomson PF, Burman NM. Edge cracking in Hot-rolled Al-Mg alloys. *Mater. Sci. Eng.*, 1980; 45(1): 95–107.

73. Reiso O. The effect of microstructure on extrudability of some aluminium alloy. Dr. Techn. Thesis. Trondheim: Norwegian Institute of Technology; 1992.

74. Reiso O. The effect of composition and homogenization treatment on extrudability of AlMgSi alloys. *Proc. 3rd international extrusion seminar*, vol. 1, pp. 31–40. Atlanta GA: Aluminium Association. 1984.

75. Reiso O. Extrusion of AlMgSi alloys. In: Nie JF, Morton AJ, Muddle BC, eds, *Proceedings of the 9th international conference on aluminium alloys*. Brisbane: Institute of Materials Engineering Australasia Ltd; 2004.

76. Reiso O. The effect of billet preheating practice on extrudability of AlMgSi alloys. *Proc. 4th international aluminium extrusion technology seminar*, vol. 2, pp. 287–95. Chicago, Il: Aluminium Association, 1988.

77. Reiso O, Ryum N, Strid J. Melting of secondary-phase particles in Al-Mg-Si alloys. *Metall. Mater. Trans. A Phys.*, 1993; 24A: 2629–41.

78. Tanaka T, Hayashi S. Analysis of the distribution of light elements in steels by the time-of-flight secondary ion mass spectrometry (Tof-SIMS). Nippon Steel & Sumitomo Metal Technical Report 118; 2018.
79. Carvalho ML. Corrosion of copper alloys in natural seawater. PhD thesis. Paris: Université Pierre et Marie Curie; 014.
80. Gerlach RL, Davis LE. Semiquantitative analysis of alloys with SIMS. *J. Vac. Sci. Technol.*, 1977; 14: 339.

RECOMMENDED FURTHER READING

Simensen CJ, Berg G. A survey of inclusions in aluminum. *Aluminium*, 1980; 56: 335–340.
Simensen CJ, Nielsen Ø, Hillion F, Voje J. NanoSIMS analysis of trace element segregation during the Al-Si Eutectic Reaction. *Metall. Mater. Trans. A Phys. Metall. Mater. Sci.*, 2007; 38A: 1448–1451.
Simensen CJ, Kubowicz S, Holme B, Graff JS. Analysis of an aluminium alloy containing trace elements. In: Ratvik AP (ed.), *Light Metals*, pp. 265–72. New York: Springer; 2017.

2

Thermodynamics and Transport Properties

2.1   Thermodynamics

2.1.1   Introduction

Thermodynamics and kinetics are the tools we use to describe how impurities enter the melt and how they can be removed. Refining metallurgy is a never-ending battle of trying to remove impurities and inclusions from the molten metal. These impurities come from various sources. Partly, they stem from the raw materials (ores) used in extractive metallurgy, and partly they enter the melt during the alloying and refining treatment. The melt picks up impurities from the containing walls (refractories), from the atmosphere, and even from the phase used to extract the impurities (slag). The slag and the walls must stand up to the high temperatures and chemical attack from the melt. The walls are therefore composed of refractory materials such as Al_2O_3 or MgO in the case of steelmaking. These refractory materials must be relatively inexpensive; they are therefore never 'pure'. They also contain binding materials such as SiO_2, B_2O_3, and carbon-based binders.

Finally, oxygen, hydrogen, and nitrogen enter the melt due to contact with the atmosphere. This problem of 'pick-up' is illustrated in Fig. 2.1.

When an alloy is produced, one starts with a more or less pure molten metal and then adds the various alloying elements in controlled amounts. Thus for instance, Si, Mn, and Zn are added to molten aluminium to create the specified aluminium alloy composition. Similarly, iron is the starting point for iron-based alloys and copper for copper-based alloys. The alloy additions may contain a number of undesired impurities and inclusions. Production of iron, aluminium, copper, and so on is the subject of extractive metallurgy. In Chapter 7 production of the metals and alloy addition is discussed only in so far as light is thrown on the origin and amounts of dissolved impurities and inclusions.

In high-temperature metallurgy (pyrometallurgy) thermodynamics tells us what processes or reactions are possible. Furthermore, thermodynamics give us the 'driving force' for the description of the rates of reactions (kinetics). From thermodynamics we may map out both the positive and negative aspects of a refining process. Usually, we assume that the

Principles of Metal Refining and Recycling. Thorvald Abel Engh, Geoffrey K. Sigworth, and Anne Kvithyld, Oxford University Press.
© Oxford University Press 2021.
DOI: 10.1093/oso/9780198811923.003.0002

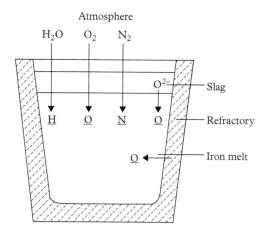

Fig. 2.1 *Pick-up of impurities in steel from the atmosphere and refractory.*

pressure and temperature are constant and apply thermodynamics to study equilibria at the interface between two or more phases. The interfaces may be slag–metal, refractory–metal, inclusion–metal, or atmosphere–metal. In pyrometallurgy, due to the high temperatures and correspondingly high reaction rates, we often assume that the chemical reaction rates are not rate-limiting (see Chapter 4). Mass and heat transfer to and from the interfaces determine the kinetics. It is then very important that the thermodynamics (the equilibria) are given correctly for the interfaces.

For practical experience in the application of thermodynamics and kinetics it would be a good idea for the student to study carefully the problems and examples given throughout the book.

Even at high temperatures it is possible that the equilibria are not attained at the interfaces. This may be the case in problems concerning contact between more than two interfaces or when contact times are very short. Examples of this are removal of inclusions to walls, bubbles, or filters; wetting problems; and rapid solidification. Towards the end of this chapter, surface and interfacial tension problems are discussed. Wetting is partly treated in Chapter 5.

In the following, thermodynamics is presented in a form tailor-made for the description of molten low-alloy metals. It is assumed that the student has some previous knowledge of thermodynamics.

2.1.2   Enthalpy, Entropy, and Gibbs Energy

As is well known, Thomson and Berthelot believed that the tendency of a reaction to take place is determined by the heat of reaction. For a reaction to occur, the heat given off from the system had to be positive. In our present terminology we would say that the *enthalpy change* of the reaction, ΔH, would have to be negative. While this is true for a number of spontaneous and well-known reactions, it is also easy to find examples of spontaneously occurring processes where the system absorbs heat, that is, ΔH of the reaction is positive. Thus, there must be another and more subtle quantity which plays a part in determining the tendency for a reaction to take place, or, more generally, the direction of a spontaneous process. This quantity is the *entropy*, or rather the *entropy change (ΔS)* of the process.

The simplest definition of the entropy content of a body is the reversibly absorbed heat (or enthalpy) divided by the absolute temperature. When a body or a system absorbs heat at a constant temperature, such as, for example, when a chunk of ice is melting, the entropy change is simply

$$\Delta S = \frac{\Delta H}{T}. \tag{2.1}$$

However, there is also another aspect of entropy: it is related to probability, or rather to disorder in the system. The relation to the *thermodynamic probability W* was first proposed by Boltzmann:

$$S = k \ln W. \tag{2.2}$$

Here $k = 1.3806 \times 10^{-23}$ J K^{-1}, which is identical to the gas constant divided by the number of particles in a mole. The thermodynamic probability, W, may be defined as the total number of possible and distinguishable microstates of a system, subject to the condition of constant total energy, volume, and amount of substance. In a perfect crystal at absolute zero, the atoms can be arranged in one way only, that is, $W = 1$, and $S = k \ln 1 = 0$. (This, in fact, is the essential content of the third law of thermodynamics.) As the crystal is heated, the atoms start to vibrate, and according to quantum theory they are distributed over a certain set of energy levels. With increasing temperature, which means increased energy, the atoms may be distributed over a wider range of energy levels, which means increasing disorder, which in turn means increasing entropy. With precise knowledge of these energy levels, both the entropy and the enthalpy of pure substances may, in principle, be calculated from statistical thermodynamics. In practice, this is generally not possible for solids and liquids, whereas the data for gases are usually acquired in this way. Thus, statistical thermodynamics has contributed significantly to our store of thermodynamic data. It does not, however, have any significant impact on our use of this data and hence, for our purposes, we will not pursue the subject of statistical thermodynamics. Nevertheless, it is important to have at least a qualitative understanding of the connection between the disorder in a system and its entropy.

The entropy resulting from heating is called *thermal entropy*. But increased disorder may also result when particles (atoms or molecules of different kinds) exchange positions in space. This gives rise to *configurational entropy*. We will give a simple example of the statistical calculation of the latter. Suppose we have N_1 atoms of substance A and N_2 atoms of substance B, and that these substances are in liquid or gaseous form so they may mix readily. We assume furthermore that there is no specific interaction between atoms A and B (or more properly, that the bond energy of the bond A–B is equal to the average of the bonds A–A and B–B). In this case the most probable state is that of random mixing of A and B atoms. The total number of atoms is $N_1 + N_2$ and the total number of different ways these atoms may be arranged in a quasi-lattice is $(N_1 + N_2)!$ Permutation of identical atoms does not, however, lead to a new and distinguishable microstate; hence, we have to reduce the total number by $N_1!N_2!$ From this the entropy change is given by

$$\Delta S = k \ln W = k \ln \frac{(N_1 + N_2)!}{N_1 \, ! \, N_2!} \tag{2.3}$$

When N_1 and N_2 are large numbers we may use Stirling's approximation: $\ln (N!) = N \ln N - N$. After some rearrangement, this leads to

$$\Delta S = -(N_1 + N_2) k \left[\frac{N_1}{N_1 + N_2} \ln \frac{N_1}{N_1 + N_2} + \frac{N_2}{N_1 + N_2} \ln \frac{N_2}{N_1 + N_2} \right].$$

The fractions are nothing but the mole fractions. Dividing through by the number of atoms in a mole, we get the familiar equation for the entropy change when n_1 moles of substance A and n_2 moles of substance B form an ideal mixture:

$$\Delta S_{mix}^{id} = -(n_1 + n_2) R (x_1 \ln x_1 + x_2 \ln x_2). \tag{2.4}$$

Now let us suppose that the mixture consists essentially of substance A, with only a little of substance B dissolved. It is then realized that when the added amount of substance B tends towards zero, then the entropy change per mole of B added tends towards infinity because of the logarithmic relationship. Due to this peculiar property of the entropy, a pure metal is highly improbable, and in fact does not exist. Nature tends to create disorder, to mix up the elements, and this behaviour generates an abundance of difficulties for the worker in refining metallurgy.

Let us now return to the general question of 'the driving force' of a process or a reaction. From what has already been said about entropy and disorder we may assume that a spontaneous process will always occur in the direction of increasing entropy (which is one way of stating the second law of thermodynamics). That is, the total change in entropy must be positive. This total is the sum of the change in the system under consideration and the change in the environment. Denoting the environment by e, we have

$$\Delta S_{tot} = \Delta S + \Delta S_e \geq 0, \tag{2.5}$$

where the inequality sign denotes a spontaneous process while the equality stands for equilibrium.

We usually prefer to take into account only the changes taking place within our system, leaving the rest of the universe out of consideration; hence we would like to eliminate ΔS_e. Let us consider a system in which a process, for example a chemical reaction, takes place at constant temperature and pressure. In the system, then, we have the enthalpy change ΔH and the entropy change ΔS. The environment essentially plays the part of a heat reservoir; thus, by the first law of thermodynamics, $\Delta H_e = -\Delta H$. Furthermore, from

Eq. (2.1), the entropy change in the environment is $\Delta S_e = \Delta H_e/T = -\Delta H/T$. Inserted into Eq. (2.5) this gives

$$\Delta S - \frac{\Delta H}{T} \geq 0.$$

By a slight rearrangement

$$\Delta H - T\Delta S \leq 0$$

We now introduce another function, the Gibbs energy (formerly known as the Gibbs free energy), $G = H - TS$, named in honour of J. Willard Gibbs (1839–1903). The criterion for spontaneous change, respectively equilibrium, in a system is then[1]

$$\Delta G = \Delta H - T\Delta S \leq 0. \tag{2.6}$$

Again the inequality refers to a spontaneous change and the equality refers to equilibrium (a reversible change). The change in Gibbs energy thus has replaced the *heat of reaction* as the correct measure of the driving force (strictly speaking, this is true for reactions taking place at constant temperature and external pressure, which is usually the case).

The change in Gibbs energy may most often be considered the net outcome of two opposing tendencies, as appears from the equation for ΔG. We may illustrate this with the freezing of water as a very simple example. In order for freezing to occur heat must be given off and ΔH is negative. But freezing leads to increased order on a molecular scale, which means that ΔS is also negative. Thus, the two terms ΔH and $T\Delta S$ work in opposite directions. Which of the two terms will 'win' is then determined by the temperature. In thermodynamic calculations we use the absolute temperature, degrees Kelvin, whose scale starts at absolute zero (the lowest entropy state). The freezing point of water (0 degrees Centigrade) is equal to 273 degrees Kelvin. So, at $T < 273$ K, ΔH wins, and water freezes. At $T > 273$ K, $T\Delta S$ wins, and ice melts.

2.1.3   The Effect of High Temperature on Molten Metals

Usually in a molten metal alloy we assume that the various elements are present only as atoms. Since the dissolution of one substance in another can occur only if there is a decrease in the Gibbs energy, it follows that, generally speaking, gases and refractory compounds do not dissolve in molten metals as readily as do other metals. To understand this, the dissolution of a solid compound can be visualized as occurring in three steps as follows.

1. The compound is broken down into its various elements.
2. The elements are brought into liquid form, which means that they must be melted if they are solid, or condensed if they are gaseous.
3. The liquid elements are dissolved in the liquid metal.

Steps 1 and 2 will always have positive ΔG values; therefore, gases and chemical compounds are generally less soluble than metals.

For solid compounds, the positive Gibbs energy of melting and of atomization decreases with rising temperature; at the same time the term $T\Delta S_{mix}$ tends to decrease the Gibbs energy of step 3. Thus the solubility of a solid in a liquid metal rises with temperature. The same is true for the solubility of the diatomic gas H_2. Two examples of this effect on gas solubility are given in Figs 2.2 and 2.3. This behaviour is very important. When the temperature in a metal is increased, various elements and gases from the refractory, atmosphere, slag, and so on are

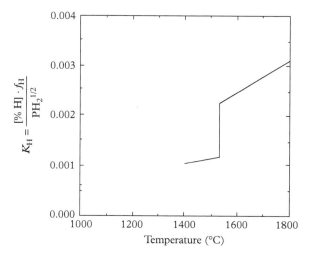

Fig. 2.2 *The solubility of hydrogen in iron.*

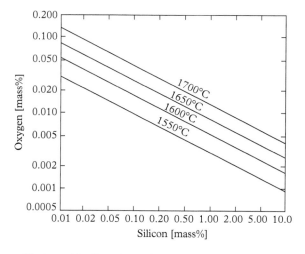

Fig. 2.3 *Oxygen in equilibrium with silicon in steel at various temperatures.*

more easily dissolved in the melt. Conversely, during solidification solubility decreases, the processes are reversed, and inclusions (particles) and gas (H_2 pores) are precipitated. This effect is one of the major problems in refining metallurgy.

It should be mentioned that if the process of mixing produces heat (ΔH_{mix} negative), this may compensate for steps 1 and 2 and give a high solubility. In this case the solubility may in fact decrease with increasing temperature.[2]

2.1.4   Chemical Potentials and Activities

Let U denote the internal energy of a system, Q the heat content, and W the work supplied to the system. The first law of thermodynamics in differential form may then be written

$$dU = \partial Q + \partial W,$$

where the symbols indicate that ∂Q and ∂W are not exact differentials, but merely small amounts of heat and work. If we consider an equilibrium process in a closed system, with volume work as the only form of work, these two quantities may be replaced by

$$dU = TdS - pdV. \tag{2.7}$$

Now let us extend this fundamental equation of thermodynamics to include changes in the chemical composition of the system. We note that various forms of work and energy may be written as the product of an intensive quantity (such as electrical potential, pressure, and mechanical force) and an extensive quantity (such as electrical charge, volume, and distance). By analogy we extend this to the chemical energy of a substance. The extensive quantity in this case is the amount of substance, usually given in terms of the number of moles, n_1. The intensive quantity is simply named the *chemical potential*, with the symbol μ_i, for substance i. Thus Eq. (2.7) is extended to

$$dU = TdS - pdV + \sum_i \mu_i dn_i, \tag{2.8}$$

where the values of dn_i may be either positive or negative. In order to gain insight into the nature of this chemical potential, we again write the definition of Gibbs energy:

$$G = U + pV - TS.$$

Taking the total differential, the following is obtained:

$$dG = dU + pdV + Vdp - TdS - SdT.$$

Introducing dU from Eq. (2.8) gives

$$dG = V\,dp - S\,dT + \sum_i \mu_i dn_i. \tag{2.9}$$

For a process occurring at constant pressure and temperature, dp and dT are zero, and the equation reduces to $dG = \sum_i \mu_i dn_i$. From this we have

$$\mu_i = \left(\frac{\partial G}{\partial n_i}\right)_{p,T}. \tag{2.10}$$

In other words, at constant temperature and pressure, the chemical potential of a substance is the *partial molar Gibbs energy* of that substance. It could be written $\overline{G}$, as for other partial molar quantities, but because of its significance in chemical thermodynamics, the special symbol μ_i is used. (For a pure substance, the term 'partial' is superfluous, and the chemical potential is simply the Gibbs energy per mole.)

As is well known, the internal energy, U, of a substance cannot be stated as an absolute quantity; it can only be stated relative to some chosen reference state. The same is then true for the Gibbs energy, and hence for the chemical potential. The chemical potential of a substance in its reference state is usually designated μ_i^0. Thus, when substance i is transformed from its reference state to a mixture, the change in its chemical potential is $(\mu_i - \mu_i^0)$.

Next we introduce yet another concept, the *activity* a_i defined by the equation

$$a_i = \exp\left(\frac{\mu_i - \mu_i^0}{RT}\right)$$

or more conveniently

$$RT \ln a_i = \mu_i - \mu_i^0. \tag{2.11}$$

Note that in the reference state where $\mu_i = \mu_i^0$ the activity, a_i, becomes equal to 1. It is clear from the defining equation that the physical significance of the activity depends on the choice of reference state. This will be discussed in the following sections.

2.1.5   The Pure Substance as Reference State, and Raoult's Law

Before proceeding, it is necessary to make a semantic note: 'mixture' and 'solution' are synonyms in the field of chemical and metallurgical thermodynamics. We often talk of mixtures when the components are completely or nearly completely miscible, and of solutions when some components (the solutes) are present only in low concentrations in the main component (the solvent). In principle, however, there is no difference and we will use one or other of the two terms without discrimination.

We will now consider a liquid mixture of the two substances A and B. As reference states we choose the pure liquid substances A and B at the temperature of the mixture. Now it is common knowledge from phase diagrams that one or both of the components in pure form may be solid at a temperature when their mixture is liquid. Thus the pure, liquid component may not be the thermodynamically stable form at the relevant temperature; nevertheless, this is the reference state we have chosen.

For mixtures in which both components are present in comparable amounts, the mole fractions x_1 and x_2 are the most reasonable expressions for the composition. We will assume, as in Section 2.1.2, that the two substances mix in all proportions without any specific interaction. This means that we assume the enthalpy of mixing $\Delta H_{mix} = 0$. The entropy of mixing in this case is the same as that given in Eq. (2.4) from statistical arguments. It follows that

$$\Delta G_{mix} = \Delta H_{mix} - T\Delta S_{mix} = RT\,(n_1\ln x_1 + n_2\ln x_2). \tag{2.12}$$

This sort of mixture, in which the enthalpy of mixing is zero and the entropy corresponds to completely random mixing, is called an *ideal mixture*. Focusing on, for example, component B, we get from Eq. (2.12) by differentiation

$$\frac{\partial \Delta G_{mix}}{\partial n_2} = \mu_2 - \mu_2^0 = RT\ln x_2. \tag{2.13}$$

By comparison with Eq. (2.11) it is seen that the activity of a component in an ideal solution is equal to its mole fraction. When $x_2 \to 1$, then $a_2 = 1$. The pure component is the standard state. This is the virtue of the concept of activity in general: it becomes equal to the chosen measure of concentration in the limiting case.

Most mixtures do not show ideal behaviour, and the activity of a component in a mixture generally must be found by experiment. When it is found that the activity departs from the mole fraction, this is expressed by an *activity coefficient* γ_i such that

$$a_i = \gamma_i x_i. \tag{2.14}$$

In a system with several phases at equilibrium, the chemical potential of any component of the system must be the same in all the phases. It follows that the activity of the component must also be the same, provided that the activity in the different phases is referred to the same reference state.

In the above presentation, the ideal solution and the activity were introduced on the basis of a statistical picture of mixing. Textbooks on thermodynamics usually follow a slightly different course by first considering the isothermal expansion of an ideal gas, and next the vapour pressure of substances. Thus, if the vapour pressure of the pure substance (the reference state) is p_i^0 and the vapour pressure of that substance in a mixture is p_i, then its activity in the mixture is $a_i = p_i/p_i^0$, provided the vapour behaves as an ideal gas. The identity with the activity as defined above comes about because the chemical potential of the substance must be the same in the mixture and in the vapour at equilibrium, and because ideal gases form ideal mixtures as defined above.

The French chemist F.-M. Raoult (1830–1901) found for a number of liquid mixtures that the vapour pressure of a component is proportional to the mole fraction of that component in the mixture, that is, $p_i = x_1 p_i^0$. This is known as *Raoult's law*. It is strictly valid for an ideal mixture as defined above. In modern usage it is not only restricted to vapour pressures; whenever the activity equals the mole fraction we say that Raoult's law is obeyed.

As indicated above, most mixtures do not show ideal behaviour. Thus for activity greater than the mole fraction we talk about positive deviation from Raoult's law; when it is smaller than the mole fraction we call it negative deviation. It is important to note, however, that regardless of deviations over most of the composition range, Raoult's law always holds as a limiting law for the main component of a mixture (the solvent) when its mole fraction tends towards unity. This is stated here without proof and shown graphically in Fig. 2.4. We will return to it in Section 2.1.7.

2.1.6 The Dilute Solution and Henry's Law

The English physician and chemist W. Henry (1774–1836) studied the solubility of gases in liquids, and found that the pressure of the gas over the solution was proportional to its concentration in the solution. This is called *Henry's law*. With concentration expressed in terms of mole fraction we may write it

$$P_i = k_H x_1, \qquad (2.15)$$

where k_H is an empirical constant. We will say that Henry's law is valid for a solute whenever the activity of that solute is proportional to its concentration. This is true for any solute at very low concentration (that is, at 'infinite dilution') in any solvent.

The mole fraction, however, may not be the best or most practical choice of expression for concentration in dilute solutions. For example, in salts and other substances in aqueous solution, concentration is usually expressed in terms of moles per litre. Correspondingly, the reference state for a solute is chosen as its solution in water at 1 mol/l. For solutions in liquid metals, concentration is more conveniently expressed as mass per cent (mass%) solute, with standard state 1 mass% in the relevant solvent metal. Then Henry's law becomes

$$a_i = c_i. \qquad (2.16)$$

The proportionality constant in Henry's law has disappeared, and it says simply that the activity of a component at low concentration is equal to its concentration. In some solutions, however, the activity may deviate from the concentration even at concentrations lower than that of unit activity. This is definitely the case, for example, in aqueous solutions of electrolytes, where marked deviations from Henry's law appear already at concentrations below 0.1 mol/l. In such cases, at the chosen reference state of unit concentration the unit activity only represents a hypothetical state.[3] This near-trivial problem is not discussed here because for solutions in liquid metals the reference state of 1 mass% usually represents such a low concentration that Henry's law is reasonably well obeyed. At larger concentrations, deviations may occur.

When deviations occur, the activity may be expressed in terms of the concentration multiplied by an activity coefficient:

$$a_i = f_i [\%i] . \tag{2.17}$$

In this equation, the symbol [%i] stands for the concentration expressed as mass% of component i. Note that different symbols are used for the activity coefficient, depending upon whether the activity is expressed 'according to Raoult' or 'according to Henry'. The use of γ and f as introduced here is opposite to convention in chemistry where f is used for the activity coefficient with reference to Raoult's law, and γ with reference to Henry's law. The use of γ and f in the present book, however, has a long standing in the field of metallic solutions. For the activity itself, however, the same symbol a is commonly used regardless of the reference. For this reason, it is important to emphasize the reference state that is used, whenever doubt may occur.

The relation between Raoult's and Henry's laws is shown graphically in Fig. 2.4. It is seen that, on a mole fraction scale, Raoult's law says that the activity as a function of concentration has a slope of unity. Henry's law, on the other hand, says that, starting from zero or near zero, the activity curve has a constant slope, but it says nothing about the magnitude of this slope. It is also seen that a mixture which shows positive deviation according to Raoult shows

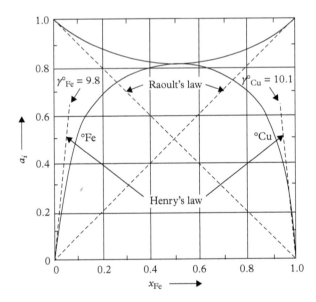

Fig. 2.4 *Activities in liquid mixtures of copper and iron at 1550°C. The Henry's law tangent is indicated for copper dissolved in iron and iron dissolved in copper.*[3]

negative deviation according to Henry. Correspondingly, negative deviation from Raoult's law is accompanied by positive deviation from Henry's law.

Equation (2.17) is used when mass percent describes the composition of an alloy. When mole fraction is used, it is convenient to apply Henry's law by defining an activity coefficient at infinite dilution:

$$\gamma_i^o = \frac{a_i}{x_i} \quad (as \ x_i \to 0). \tag{2.18}$$

In other words, the activity coefficient of the dissolved element i becomes a constant value (γ^o) in dilute solutions. There is a free energy change associated with the dissolution of this element, equal to

$$\Delta G_i^o = RT ln \gamma_i^o. \tag{2.19}$$

When using weight percent as the composition coordinate, the standard free energy of solution becomes[4]

$$\Delta G_i^o = RT \ln \left(\gamma_i^o \cdot [M_1 / 100 M_i] \right), \tag{2.20}$$

where M_1 is the molecular weight of the nearly pure solvent, and M_i is the molecular weight of the element i.

Information on the thermodynamics of metal alloys is often useful. For example, in Chapter 6 the removal of alkali and alkaline earth metals from aluminium is considered. Section 4.12 touches briefly on the chemistry of hydrogen removal by purging with inert gases and an analysis is given of the refining of aluminium by vacuum treatment. There are also many other uses for information on the chemistry of metal alloys.

Unfortunately, the necessary thermodynamic data are usually difficult to obtain. The experimental data are frequently 'buried' in obscure publications. Another problem is that information may be presented in various formats, some of which are not very useful for working calculations. In particular, different composition coordinates are employed. Extracting the information needed is often difficult and time consuming. Hence, we include in the following data for the solution of impurity and alloying elements in important commercial metals. For most calculations it is more convenient to use weight percent as the composition coordinate, and so thermodynamic data are presented in this format.

Consider, for example, aluminium-based alloys. In this case aluminium is the major constituent. This means the chemical reactivity of aluminium in our alloys will be 90–95 per cent that of pure aluminium. This is very close to 1.0, and assuming a unit activity for aluminium in thermodynamic calculations causes only a negligible error. For this reason we are concerned only with the chemical properties of elements, which are present in aluminium alloys in relatively much smaller quantities. Table 2.1 shows the free energy of solution of elements in liquid aluminium.

In most tables of thermodynamic data, the pure component is the reference state. To allow us to use these data with our reference state of 1 mass%, the Gibbs energy change for

Table 2.1 *Standard Gibbs energies ΔG_i^0 for solution of elements to 1 mass%
solution in liquid aluminium*[5,6]

Element, i		ΔG_i^0 (joules/mole)
Ag	(s)	$18620 - 73.676\ T$
Au	(s)	$(-57175 - 87.782\ T)$
B	(s)	$61700 - 67.622\ T$
Ba	(l)	$-106200 - 27.856\ T$
Be	(s)	$39\ 346 - 40.210\ T$
Bi	(l)	$22210 - 48.928\ T$
C	(s)	$(112681 - 66.816\ T)$
Ca	(l)	$-99579 - 20.852\ T$
Cd	(l)	$29710 - 52.316\ T$
Ce	(s)	$-170450 - 8.127\ T$
Co	(s)	$-138655 + 0.687\ T$
Cr	(s)	$-64600 - 29.216\ T$
Cu	(s)	$-4623 - 68.578\ T$
Fe	(s)	$-83717 - 34.710\ T$
Ga	(l)	$3\ 481 - 48.354\ T$
Ge	(s)	$25392 - 81.982\ T$
$\frac{1}{2}H_2$	(g)	$51534 + 25.270\ T$
In	(l)	$28450 - 55.350\ T$
La	(s)	$-198464 + 36.742\ T$
Li	(l)	$-26990 - 9.714\ T$
Mg	(l)	$-8800 - 37.417\ T$
Mn	(s)	$-63940 - 35.147\ T$
Mo	(s)	$-82460 - 18.791\ T$
Na	(l)	$34430 - 21.064\ T$
Nd	(s)	$-177010 + 2.145\ T$
Ni	(s)	$-122290 - 46.590\ T$
Pb	(l)	$41710 - 53.714\ T$
Sb	(l)	$54810 - 90.352\ T$

Table 2.1 *Continued*

Element, i		ΔG_i^0 (joules/mole)
Sc	(s)	$-92900 - 34.151\ T$
Si	(s)	$40080 - 85.969\ T$
Sn	(l)	$24455 - 59.995\ T$
Sr	(s)	$-100000 - 30.547\ T$
Ti	(s)	$-99850 - 13.053\ T$
V	(s)	$-60950 - 33.501\ T$
Y	(s)	$(-153700 + 7.799\ T)$
Zn	(l)	$10620 - 49.857\ T$
Zr	(s)	$-136740 + 18.481\ T$

Notes. s denotes solid, l denotes liquid, and g denotes gas.

transfer between the two reference states, Eq. (2.20), was employed to give the values given in Tables 2.1–2.4 for the solvent metals aluminium,[5,6] iron,[7–9] copper,[4] and cobalt.[10] That is, the tables give the change in Gibbs energy, ΔG_i^0, that occurs when 1 mole of the pure substance i, initially in its reference state at 1 bar pressure and at temperature T, is dissolved in the solvent metal to give a solution with [mass%] = 1 at the same temperature and pressure. It may be added that this ΔG_i^0 is essentially the change in partial molar Gibbs energy of component i as a result of the dissolution process. In these and in the tables that follow, values in parentheses are estimates, or are considered to be approximate values.

Table 2.2 *Standard Gibbs energies for solution of elements to a 1 mass% solution in liquid iron*[7–9]

Element, i		ΔG_i^0 (joules/mole)
Ag	(l)	$82\ 430 - 43.76\ T$
Al	(l)	$-63\ 180 - 27.91\ T$
B	(s)	$-65\ 270 - 21.55\ T$
C	(graphite)	$22\ 590 - 42.26\ T$
Ca	(g)	$-39\ 460 + 49.37\ T$
Ce	(l)	$-16\ 700 - 46.44\ T$
Co	(l)	$1000 - 38.74\ T$
Cr	(l)	$-37.70\ T$

(Continued)

Table 2.2 *Continued*

Element, i		ΔG_i^0 (joules/mole)
Cr	(s)	19 250 – 46.86 T
Cu	(l)	33 470 – 39.37 T
½H$_2$	(g)	36 480 + 30.46 T
La	(l)	125 900 – 94.60 T
Mg	(l)	96 780 – 31.37 T
Mg	(g)	– 20 660 + 54.99 T
Mn	(l)	4080 – 38.16 T
Mo	(l)	– 42.80 T
Mo	(s)	27 610 – 52.38 T
½N$_2$	(g)	3600 + 23.89 T
Nb	(l)	– 42.68 T
Nb	(s)	23 010 – 52.30 T
Ni	(l)	– 20 920 – 31.05 T
½O$_2$	(g)	– 117 110 – 3.39 T
½P$_2$	(g)	– 157 700 + 5.40 T
Pb	(l)	212 550 – 102.32 T
½S$_2$	(g)	– 125 100 + 18.50 T
Si	(l)	– 131 500 – 17.24 T
Sn	(l)	15 980 – 44.56 T
Ta	(l)	(– 50 200 – 48.06 T)
Ti	(l)	– 46 020 – 37.03 T
Ti	(s)	– 31 130 – 44.98 T
U	(l)	– 56 070 – 50.21 T
V	(l)	– 42 260 – 35.98 T
V	(s)	– 20 710 – 45.61 T
W	(l)	– 42.68 T
W	(s)	46 960 – 61.23 T
Zr	(l)	– 51 050 – 42.38 T
Zr	(s)	– 34 730 – 50.0 T

Table 2.3 *Standard Gibbs energies for solution of elements to 1 mass% solution in liquid copper*[4,8]

Element, i		ΔG_i^0 (joule/mole)
Ag	(l)	16 320 – 44.02 T
Al	(l)	– 36 110 – 57.91 T
As	(g)	– 93 510 – 39.50 T
Au	(l)	– 19 370 – 50.58 T
Bi	(l)	24 940 – 63.18 T
C	(graphite)	35 770 + 50.21 T
Ca	(l)	– 92 890 – 34.31 T
Cd	(g)	– 107 530 + 53.14 T
Cd	(l)	– 7 780 – 42.68 T
Co	(s)	33 470 – 37.66 T
Cr	(s)	46 020 – 36.48 T
Fe	(s)	54 270 – 47.45 T
Fe	(I)	38 910 – 38.79 T
Ga	(l)	– 45 190 – 36.32 T
Ge	(l)	– 66 940 – 32.89 T
½H$_2$	(g)	43 510 + 31.38 T
In	(l)	– 39 960 – 23.35 T
Mg	(g)	– 168 200 + 63.18 T
Mg	(l)	– 36 280 – 31.51 T
Mn	(l)	(– 8 160 –36.94 T)
Mn	(s)	(6 490 – 46.61 T)
Ni	(l)	9 790 – 37.66 T
Ni	(s)	27 410 – 48.12 T
½O$_2$	(g)	– 85 350 + 18.54 T
Pb	(l)	36 070 – 58.62 T
Pd	(s)	3 350 – 42.26 T
Pt	(s)	(– 42 680 – 43.81 T)
½S$_2$	(g)	– 119 660 + 25.23 T

(Continued)

Table 2.3 *Continued*

Element, i		ΔG_i^0 (joule/mole)
Sb	(l)	− 52 300 − 43.51 T
Se	(g)	− 76 150 − 40.08 T
Si	(l)	− 62 760 − 31.38 T
Si	(s)	− 12 130 − 61.42 T
Sn	(l)	− 37 240 − 43.52 T
Te	(g)	− 41 840 − 44.06 T
Tl	(l)	28 160 − 49.12 T
V	(s)	117 570 − 75.73 T
Zn	(l)	− 23 600 − 38.37 T

Table 2.4 *Standard Gibbs energies for solution of elements to 1 mass% solution in liquid cobalt*[8, 10]

Element, i		ΔG_i^0 (joule/mole)
Al	(l)	− 91 210 − 27.15 T
C	(graphite)	20 880 − 20.42 T
Fe	(l)	− 12 050 − 27.49 T
½H$_2$	(g)	35 560 + 31.51 T
Mn	(l)	− 37.70 T
½N$_2$	(g)	41 840 + 21.92 T
Ni	(l)	1260 − 44.14 T
½O$_2$	(g)	− 61 630 − 9.71 T
½S$_2$	(g)	− 135 140 + 30.25 T
Si	(l)	− 146 440 − 7.03 T
Ti	(s)	− 94 060 − 44.43 T
V	(s)	− 25 690 − 44.61 T

To show the conversion from mole fraction to mass per cent, we consider a binary mixture of A and B, with mole fractions x_1 and x_2. We then have

$$[\%B] = \frac{100x_2 M_2}{x_1 M_1 + x_2 M_2},$$ (2.21)

where M_1 and M_2 are the atomic weights of the two components. At infinite dilution the equation reduces to

$$[\%B] = 100x_2 \frac{M_2}{M_1}.$$ (2.22)

The ratio between the Henrian activity a_2^H and the Raoultian activity a_2^R at infinite dilution is then given by

$$\frac{a_2^H}{a_2^R} = \frac{[\%B]}{\gamma_2^0 x_2} = \frac{100 M_2}{\gamma_2^0 M_1}.$$ (2.23)

In the latter equation, γ_2^0 is the Raoultian activity coefficient for component B at infinite dilution, which is the same as the slope of the Henry's law tangent at $x_2 = 0$.

As stated in the introduction to this section, Henry's law in its original form says that the concentration of a dissolved gas is proportional to its pressure over the solution, cf. Eq. (2.15). This is not always true, however. When a diatomic gas such as hydrogen dissolves in a liquid metal, it simultaneously dissociates:

$$\frac{1}{2} H_2(g) = H \text{ (in metal)}.$$

The equilibrium constant has the form

$$K = \frac{a_H}{\sqrt{p_{H_2}}}.$$

At low concentrations

$$[\%H] = K \sqrt{p_{H_2}}.$$ (2.24)

Equation (2.24) is commonly referred to as Sievert's law, generally valid for the dissolution of diatomic gases in liquid metals.

In the cases where Henry's law does not hold for the solute, Raoult's law does not hold for the solvent. This is the case whenever the solute dissociates in the solvent. Hydrogen in a liquid metal is one example; another is the dissolution of a salt in water. Both 'laws' depend on the condition that the dissolved substance is present in the same molecular form in the solution and in the standard state.

For mass transfer between metal and a second phase, resistance to transfer is often on the metal side. Then it will be seen in Chapter 4 that the notation is simpler if the inverse equilibrium constant is employed. In this case, the symbol 'A' is also used for the inverse equilibrium constant.

2.1.7   Gibbs–Duhem's Law

From Eq. (2.9), for a system at constant pressure and temperature, we have

$$dG = \sum_i \mu_i dn_i. \tag{2.25}$$

This equation is valid for an infinitesimal amount of a system containing a number of components. Now we may conceptually increase the amount of the system while keeping its composition constant, which means that all of the μ_i values remain constant. This sort of 'chemical integration' gives

$$G = \sum_i \mu_i n_i. \tag{2.26}$$

Differentiation of Eq. (2.26) gives

$$dG = \sum_i \mu_i dn_i + \sum_i n_i d\mu_i. \tag{2.27}$$

By comparison with Eq. (2.25) it is seen that we must have

$$\sum_i n_i d\mu_i = 0. \tag{2.28}$$

This is the basic form of the Gibbs–Duhem equation, which may take various forms depending upon the application. The equation gives a relation between the chemical potentials or activities of the various components. We will apply it to one mole of a binary mixture. The combination of Eqs. (2.28) and (2.11) gives

$$x_1 d\ln a_1 + x_2 d\ln a_2 = 0. \tag{2.29}$$

Equation (2.29) may be used to evaluate the activity of a component in a solution when that of the other is known over a range of compositions. The integration meets with some difficulty, however, because in the limit where $a_1 \to 0, \ln a_1 \to -\infty$. To avoid this difficulty, we note that $x_1 + x_2 = 1$, $dx_1 + dx_2 = 0$, and hence also

$$x_1 d\ln x_1 + x_2 d\ln x_2 = 0. \tag{2.30}$$

When we subtract Eq. (2.30) from Eq. (2.29) we get

$$x_1 d \ln \gamma_1 + x_2 d \ln \gamma_2 = 0. \tag{2.31}$$

Thus the Gibbs–Duhem equation may also be applied to the activity coefficients, and these do not go to zero at any point. We will not, however, pursue the integration of Gibbs–Duhem's equation; this is discussed in textbooks on thermodynamics such as that of Gaskell.[3]

We will instead use Eq. (2.31) to demonstrate the connection between Raoult's and Henry's laws. For this purpose, we will regard component 2 as the minor constituent, and we will assume that Henry's law is valid for it. Composition is expressed in terms of mole fractions; thus, Henry's law is written

$$a_2 = k_H x_2,$$

which gives

$$d \ln a_2 = d \ln x_2 = \frac{dx_2}{x_2}$$

It is noted also that $dx_2 = - dx_1$. With these substitutions, Eq. (2.29) yields

$$x_1 d \ln a_1 - dx_1 = 0$$

or

$$d \ln a_1 = d \ln x_1$$

Integrated, it becomes

$$\ln a_1 = \ln x_1 + \ln \text{const.}$$

or

$$a_1 = \text{const.} x_1. \tag{2.32}$$

But by definition, $a_1 = x_1$ for $x_1 = 1$; hence, we must have const. = 1, and thus $a_1 = x_1$. In other words, *in the concentration range where Henry's law is valid for the solute, Raoult's law is valid for the solvent*, which is what we set out to prove.

For metals there is always a range at high dilution (low concentration) where Henry's law is valid for the solute, and as a consequence, Raoult's law must always be valid for the solvent when its mole fraction approaches unity. This was stated without proof at the end of section 2.1.5, and it is demonstrated graphically in Fig. 2.4. If it is found experimentally for some systems that Raoult's law is not obeyed as $x_1 \rightarrow 1$, this means that the calculation is based on an erroneous assumption regarding the formula of the dissolved species. This would be the case if one assumes, for example, that hydrogen dissolves in a molten metal as H_2, or that sodium chloride dissolves in water as $NaCl$ molecules. However, in the example, if H is selected as the species present in the metal, Eq. (2.32) applies and Raoult's law is valid. Conversely, agreement with Raoult's law for the solvent may be used to determine the molecular identity of the solute. This, in essence, is the basis for the classical method of determining molecular weights by freezing-point depression.

2.1.8   Gibbs Energies of Solution

To illustrate the use of the Gibbs energy of solution tables, the equilibrium constants for the dissolution of H_2 and Na (gas) in molten aluminium are worked out. These equilibrium constants are employed in Chapter 4.

For the reaction from Table 2.1,

$$\frac{1}{2}H_2 = \underline{H}$$

$$K_H = \frac{P_{H_2}^{\frac{1}{2}}}{a_H} = exp\left(\frac{\Delta G_H^0}{RT}\right) = exp\left\{\left(\frac{\Delta H^0}{T} - \Delta S^0\right)\bigg/ R\right\}$$

At 760 °C from Table 2.1,

$$\frac{\Delta G_H^0}{RT} = \left(\frac{51534}{760+273} + 25.27\right)\bigg/ 8.314$$

Then

$$a_H = P_{H_2}^{\frac{1}{2}}\bigg/ K_H \quad \text{and} \quad [\%H] = \frac{P_{H_2}^{\frac{1}{2}}}{K_H f_H},$$

where

$$K_H = 0.84 \times 10^4.$$

The solubility in aluminium is given in Fig. 4.20. Hydrogen solubility curves usually have the same general shape for the various metals. Figure 2.2 shows the hydrogen solubility in iron. For low-alloy steels $f_H \approx 1$. This will be discussed later.

Next, for

$$Na\ (l) = \underline{Na}$$

$$K_{Na} = \frac{1}{a_{Na}} = exp\left(\frac{\Delta G_{Na}^0}{RT}\right).$$

At 1100 K

$$\frac{\Delta G^0_{Na}}{RT} = \left(\frac{34430}{1100} - 21.064\right) \bigg/ 8.3145 = 1.19$$

and $K_{Na} = 3.29$. Thus, in equilibrium with liquid sodium, the sodium activity in aluminium (and its solubility), given in per cent, is $0.30 = 1/3.29$.

$$[\%Na] = a_{Na}/f_{Na}.$$

If the content of alloy additions is low, $f_{Na} \sim 1$.

The thermodynamic data given in the tables are valid only in a limited temperature range. This range is found in the references. Usually, the tables must be used in conjunction with tables of standard Gibbs energy data.[11–13]

Appendix 1 contains a selection of the Gibbs energy of reaction. As an example of the use of Appendix 1 and the tables, let us look at the dissolution of SiO_2 in molten iron.

From Appendix 1:

for:	$\Delta G°$ (J/mol)
SiO_2 (s) = Si(s) + O_2	$906000 - 176\ T$
Si(s) = Si(l)	$50540 - 30.0\ T$

and from Table 2.3:

Si(l) = $\underline{Si}$	$-131\,500 - 17.24\ T$
O_2(g) = $2\underline{O}$	$-234\,300 - 6.78\ T$

Addition of the above gives

SiO_2(s) = $\underline{Si}$ + $2\underline{O}$	$590\,740 - 230.02\ T$

Then

$$\frac{a_{Si}a_O^2}{a_{SiO_2}} = exp\left\{\left(-\frac{590740}{T} + 230.02\right)\bigg/8.3145\right\}$$

$$= exp\left(-\frac{71049}{T} + 27.66\right)$$

This equation is presented in Fig. 2.3, where corresponding values of [% Si] and [% O] are plotted for various steelmaking temperatures. The difference between activities and concentrations is disregarded in Fig. 2.3. The ratios between activity and concentration can be calculated from the interaction coefficients given later. It is found that the differences are small for Si and O in iron.

2.1.9   Interaction Coefficients

Henry's rule applies only in a limited low concentration range. As the concentration of component i increases, the activity a_i deviates from the line given by Henry's rule. In the following it will be seen how corrections can be introduced to describe the deviations. When Henry's rule was discussed, it was assumed that there was no interaction between the solute atoms i. Also it was assumed that no other solute atoms j were present in the melt.

However, the i atoms will in fact interact with each other and with the j atoms. This will change the enthalpy of solution for the i atoms and also the structure around the i atoms, and thus effect the entropy of the i atoms. Then we write for the chemical potential

$$\mu_i \approx \mu_i \text{ (no interaction)}$$
$$+ \sum_{j=2}^{k} \frac{\partial \mu_i}{\partial x_j} x_j \qquad (2.33)$$

Here μ_i (no interaction) is the chemical potential if there had been no interaction between the solute components, and $\partial \mu_i / \partial n_j$ is the effect of adding component j on the activity of i. The term $\partial \mu_i / RT \partial x_j = \varepsilon_i^j$ is called the first-order interaction coefficient. From Eqs. (2.9) and (2.10) we see that

$$\frac{\partial \mu_i}{\partial n_j} = \frac{\partial G}{\partial n_i \partial n_j} = \frac{\partial \mu_j}{\partial n_i}$$

Thus, it is realized that

$$\varepsilon_i^j = \varepsilon_i^j. \qquad (2.34)$$

From Eqs. (2.11) and (2.14) it is seen that Eq. (2.33) becomes

$$\ln \gamma_i = \ln \gamma_i^0 + \sum_{j=2}^{k} \varepsilon_i^j x_j, \qquad (2.35)$$

where γ_i^0 is the activity coefficient (at infinite dilution) if there had been no interaction between the solute components. It therefore corresponds to Henry's law (cf. Fig. 2.4).

In practical applications, as mentioned previously, it is usually preferable to employ concentration units of mass per cent and the corresponding activity coefficient

$$f_i = \frac{a_i}{[\%i]}. \qquad (2.36)$$

The interaction coefficients e_i^j in this case are defined by the equation

$$log_{10}f_i = \sum_{j=2}^{n} e_i^j[\%j].$$ (2.37)

Note that in Eqs. (2.35) and (2.37) component number 1 is the solvent.

Often we wish to convert from the ε_i^j to the e_i^j coefficients. To accomplish this it is pointed out that the concentration units are related by

$$[\%i] = \frac{100x_iM_i}{x_iM_i + \sum_{j=2}^{k}x_jM_j}.$$ (2.38)

Furthermore, the definitions of the Henrian activity a_i^H and Raoultian activity, a_i^R, are not identical. To find the conversion it is recalled that at infinite dilution $a_i^H = [\%i]$. Then from Eqs. (2.38) and (2.14) the conversion is obtained by comparing two calculations of the mole fraction x_i at infinite dilution:

$$x_i = \frac{M_1a_i^H}{100M_i} = \frac{a_i^R}{\gamma_i^0}$$

Thus, the relation between the Henrian and Raoultian activities is given by

$$\frac{M_1a_i^H\gamma_i^0}{100M_ia_i^R} = 1.$$ (2.39)

From the definitions of the activity coefficients we get

$$1 = \frac{M_1a_i^H\gamma_i^0}{100M_ia_i^R} = \frac{M_1f_i[\%i]\gamma_i^0}{100M_i\gamma_ix_i}.$$ (2.40)

Finally, insertion of Eq. (2.38) into the right-hand side of Eq. (2.40) gives

$$1 = \frac{f_i\gamma_i^0}{\gamma_i\left(1 + \sum_{j=2}^{k}x_j(M_j/M_1 - 1)\right)}.$$ (2.41)

If we now take the logarithm of the two sides of the equation, apply Eqs. (2.37) and 2.35) for $\log f_i$ and $\ln\gamma_i/\gamma_i^0$, respectively, and insert x_j instead of $[\%j]$ using Eq. (2.38), the first-order x_j terms in the Taylor expansion should vanish. This gives[2]

$$e_i^j = 230\frac{M_j}{M_1}e_i^j + \frac{M_1 - M_j}{M_1}.$$ (2.42)

The term $(M_1 - M_j)/M_1$ has often been mistakenly omitted in the literature. This equation also applies when $i = j$. In other words, conversion of the self-interaction coefficients is given by

$$\varepsilon_i^i = 230\frac{M_i}{M_1}e_i^i + \frac{M_1 - M_i}{M_1}. \tag{2.43}$$

Orders of magnitude for the interaction coefficients should be kept in mind. The ε_i^j coefficients usually vary from -10 to $+10$. The corresponding e_i^j coefficients are 100 times smaller. Ohtani and others suggest that the ε_i^j change in a systematic manner with the atomic number.[14] For instance, for iron as the solvent it has been possible to construct graphs of ε_i^j as a function of atomic number and position in the periodic table of elements.[15] Values of ε_i^j not available experimentally can then be obtained by interpolation or extrapolation of the graphs. In this book, only a selection of experimentally determined values of ε_i^j are presented for aluminium, iron, copper, cobalt, and silicon as solvent metals. We note that for other solvent metals the data are extremely limited. In Tables 2.5 to 2.10 the known interaction coefficients are given for binary aluminium,[5,6,16] iron,[2,7,9] copper,[4] and cobalt[8,10] -based alloys.

To illustrate the use of the interaction coefficients for ternary systems we calculate the activity coefficient for hydrogen, f_H, for one of the aluminium alloys given in Table 4.2. With 6% Si and 3.5% Cu, we obtain from Table 2.6, $e_H^{Si} = 0.019$ and $e_H^{Cu} = 0.033$. Then log $f_H = 0.019 \times 6 + 0.033 \times 3.5 = 0.23$; and $f_H = 1.696$. (One commonly finds the use of an alloy correction factor, C, which is equal to the reciprocal of f_H. For this example $C = 0.59$. That is, the solubility of hydrogen in this alloy is 59 per cent of the value in pure aluminium.)

It is important to recognize the limits of interaction coefficients. Equations (2.35) and (2.37) arise from a Taylor series expansion of the excess free energy of solution, starting from a pure metal solvent. These equations are therefore only valid for 'dilute', or nearly pure, solutions. They are not accurate when applied to highly alloyed materials. One approach to this limitation is to include the second-order terms in the Taylor series expansion. Second-order coefficients that give corrections proportional to the second power of the solute concentrations have been obtained in the literature.[2] These second-order terms are included in this book only in Chapter 9 for stainless steels, where the alloy content is high.

There are also other ways to extend the range of application of interaction coefficients. These methods have been reviewed by Pelton.[17] Instead of using Eqs. (2.35) and (2.37), the interaction coefficients can be used in other, more robust, solution models. Pelton also provides ways for estimating ternary interaction parameters from associated binary system coefficients.

In recent years there has been an increased demand for high purity silicon for photovoltaic solar cells. Traditionally, solar grade silicon has been manufactured from scrap and leftover materials used to make semi-conductors. The availability of this material is limited, however, and there is an interest in the refining of metallurgical grade silicon. The latter material is much more abundant and less costly. This situation has led metallurgists to examine possible new ways of refining silicon.[18,19]

Of course, any new process will be constrained by the chemical properties of silicon and its impurities. Hence, it will be useful to provide a concise summary of thermodynamic

Table 2.5 *Self-interaction coefficients in liquid Al-i binary alloys*[5,6]

Element, i	ε_i^i	e_i^i	$T(K)$
Ag	0	0.0033	1 000
Au	(9.5)	(0.0092)	1 000
Ba	13	0.014	1 000
Ca	15	0.044	1 000
Cd	−5.0	−0.0019	1 373
Ce	28	0.026	1 000
Co	17	0.035	1 000
Cr	10	0.024	1 000
Cu	2.2	0.0065	1 000
Fe	13	0.029	1 000
Ga	−0.3	0.0023	1 000
Ge	3.1	0.0076	1 000
In	−4.9	−0.0015	1 000
La	32	0.029	1 000
Li	0	−0.012	1 000
Mg	2.0	0.009	1 000
Mn	18	0.039	1 000
Mo	20	0.027	1 000
Ni	(0)	(0.002)	1 000
Sc	9.8	0.027	1 000
Si	12	0.049	1 000
Sn	−5.5	−0.0018	1 000
Sr	20	0.029	1 000
Ti	16.8	0.042	1 000
V	5	0.013	1 000
Y	(23)	(0.032)	1 000
Zn	−1.2	0.0005	1 000
Zr	8	0.013	1 000

Table 2.6 *Interaction coefficients in ternary aluminium alloys*[5,16] *(see Eqs. (2.35) and (2.37))*

Part A	Metallic alloys	Temperature (K)
$\varepsilon_{Zn}^{Si} = 2.2$	$e_{Zn}^{Si} = 0.0094$	963–1053
$\varepsilon_{Si}^{Zn} = 2.2$	$e_{Si}^{Zn} = 0.0065$	
$\varepsilon_{Li}^{Sn} = -16.0$	$e_{Li}^{Sn} = -0.012$	949
$\varepsilon_{Sn}^{Li} = -16.0$	$e_{Sn}^{Li} = -0.28$	
$\varepsilon_{Na}^{Si} = (-12)$	$e_{Na}^{Si} = (-0.05)$	973
$\varepsilon_{Si}^{Na} = (-12)$	$e_{Si}^{Na} = (-0.062)$	
$\varepsilon_{Mg}^{Si} = (-9.0)$	$e_{Mg}^{Si} = (-0.037)$	973–1073
$\varepsilon_{Si}^{Mg} = (-9.0)$	$e_{Si}^{Mg} = (-0.044)$	
$\varepsilon_{Sn}^{Pb} = (-1.5)$	$e_{Sn}^{Pb} = (0.003)$	973–1073
$\varepsilon_{Pb}^{Sn} = (-1.5)$	$e_{Pb}^{Sn} = (0.0019)$	

Part B Al–H–*j* alloys

Element, *j*	ε_H^j	e_H^j	Temperature (K)
Ce	(−29)	(−0.08)	1073
Cr	−1	0	973–1073
Cu	17	0.033	973–1023
Fe	31	0.066	973–1023
Li	−15	−0.25	973–1023
Mg	−14	−0.066	973–1023
Mn	(28)	(0.06)	973–1073
Ni	(20)	(0.04)	973–1073
Si	4.7	0.019	973–1023
Sn	0.6	0.004	973–1023
Th	−20	−0.006	1073–1173
Ti	(−43)	(−0.1)	1073–1173
Zn	10.7	0.016	973–1023

Table 2.7 Interaction coefficients e_i^j in ternary iron-based alloys[7,2,9] (see Eq. (2.59))

	Ag	Al	B	C	Ca	Ce	Co	Cr	Cu	H	La	Mn	Mo
Ag	−0.04	−0.08	—	0.22	—	—	—	−0.0097	—	—	—	—	—
Al	−0.017	0.045	—	0.091	−0.047	—	—	—	—	0.24	—	—	—
B	—	—	0.04	0.22	—	—	—	—	—	0.58	—	−0.0009	—
C	0.028	0.043	0.24	0.24	−0.097	−0.003	0.008	−0.024	0.016	0.67	0.0066	−0.008	−0.014
Ca	—	−0.072	—	−0.34	−0.002	—	—	—	—	—	—	—	—
Ce	—	—	—	−0.077	—	0.004	—	—	—	−0.6	—	0.13	—
Co	—	—	—	0.021	—	—	0.0051	—	—	−0.14	—	−0.0042	—
Cr	−0.0024	—	—	−0.012	—	—	−0.02	−0.0003	0.02	−0.33	—	0.004	0.002
Cu	—	—	—	0.066	—	—	—	0.018	−0.023	−0.19	—	—	—
H	—	0.013	0.06	0.06	—	—	0.002	−0.0024	0.0013	—	−0.027	−0.002	0.0029
La	—	—	—	0.03	—	—	—	—	—	−4.3	−0.008	0.28	—
Mn	—	—	−0.024	−0.054	—	0.054	−0.0036	0.0039	—	−0.34	0.11	0	0.0046
Mo	—	—	—	−0.14	—	—	—	−0.0003	—	−0.13	—	0.0048	0.012

(*Continued*)

Table 2.7 *Continued*

	Ag	Al	B	C	Ca	Ce	Co	Cr	Cu	H	La	Mn	Mo
N	—	0.01	0.094	0.13	—	—	0.011	−0.046	0.009	—	—	−0.02	−0.011
Nb	—	—	—	−0.49	—	—	—	—	—	−0.7	—	0.0093	—
Ni	—	—	—	0.032	−0.067	—	—	−0.0003	—	−0.36	—	−0.008	—
O	−0.011	−3.90	−0.31	−0.45	—	—	0.008	−0.055	−0.013	0.73	−5	−0.021	0.005
P	—	0.037	0.015	0.13	—	—	0.004	−0.018	−0.035	0.33	—	−0.032	0.001
Pb	—	0.021	—	0.066	—	—	—	0.02	−0.028	—	—	−0.023	0
S	—	0.041	0.13	0.11	(−110)	−9.1	0.0026	−0.011	−0.008	0.41	−18.3	−0.026	0.003
Si	—	0.058	0.20	0.18	−0.067	—	—	−0.0003	0.014	0.64	—	−0.015	2.4
Sn	—	—	—	0.18	—	—	—	0.015	−0.024	0.16	—	—	—
Ta	—	—	—	−3.5	—	—	—	—	—	−0.47	—	0.0016	—
Ti	—	—	—	—	—	—	—	0.055	—	−1.1	—	−0.043	—
V	—	—	—	−0.14	—	—	—	0.012	—	−0.59	—	0.0056	—
W	—	—	—	−0.15	—	—	—	—	—	0.088	—	0.014	—
Zr	—	—	—	—	—	—	—	—	—	−1.2	—	—	—

Table 2.7 *Continued*

	N	Nb	Ni	O	P	S	Si	Sn	Ta	Ti	V	W	Zr
Ag	—	—	—	-0.1	—	—	—	—	—	—	—	—	—
Al	0.015	—	—	-6.6	0.033	0.035	0.056	—	—	—	—	—	—
B	0.074	—	—	-0.21	0.008	0.048	0.078	—	—	—	—	—	—
C	0.11	-0.06	0.012	-0.32	0.051	0.046	0.08	0.022	-0.23	—	-0.03	-0.0056	—
Ca	—	—	-0.044	—	—	(-140)	-0.1	—	—	—	—	—	—
Ce	—	—	—	—	—	(-40)	—	—	—	—	—	—	—
Co	0.037	—	—	0.018	0.004	0.0011	—	—	—	0.059	0.012	—	—
Cr	-0.18	—	0.0002	-0.016	-0.033	-0.17	-0.0043	0.009	—	—	—	—	—
Cu	0.026	—	—	-0.065	-0.076	-0.021	0.027	-0.011	—	—	—	—	—
H	—	-0.0033	-0.0019	0.05	0.015	0.017	0.027	0.0057	0.0017	-0.019	-0.0074	0.0048	-0.0088
La	—	—	—	(-43)	—	(-79)	—	—	—	—	—	—	—
Mn	-0.091	0.0073	-0.0072	-0.083	-0.06	-0.048	-0.033	—	0.0035	-0.05	0.0057	0.0071	—
Mo	-0.091	—	—	0.0083	-0.006	-0.0005	8.05	—	—	—	—	—	—

(Continued)

Table 2.7 *Continued*

	N	Nb	Ni	O	P	S	Si	Sn	Ta	Ti	V	W	Zr
N	—	−0.07	0.007	−0.12	0.059	0.007	0.047	0.007	−0.049	−0.6	−0.12	−0.002	−0.63
Nb	−0.47	0	—	−0.72	−0.045	−0.047	−0.01	—	—	—	—	—	—
Ni	0.015	—	0.0009	0.01	0.0018	−0.0037	0.0057	—	—	—	—	—	—
O	−0.14	−0.12	0.006	−0.2	0.07	−0.133	−0.066	−0.011	−0.10	−1.12	−0.14	0.0085	(−3.0)
P	0.13	−0.012	0.003	0.13	0.054	0.034	0.10	0.013	—	−0.04	−0.24	−0.23	—
Pb	—	—	−0.019	—	0.048	−0.32	0.048	0.057	—	—	—	−0	—
S	0.01	−6.0	0	−0.27	0.035	−0.046	0.075	−0.0044	−0.019	−0.18	−0.19	0.011	−0.21
Si	0.09	0	0.005	−0.12	0.09	0.066	0.11	0.017	0.04	1.23	0.025	—	—
Sn	0.27	—	—	−0.11	0.036	−0.028	0.057	0.0016	—	—	—	—	—
Ta	—	—	−0.69	−1.2	—	−0.13	0.23	—	0.11	—	—	—	—
Ti	−2.08	—	—	−3.4	−0.06	−0.27	2.1	—	—	0.42	—	—	—
V	−0.46	—	—	−0.46	−0.042	−0.033	0.42	—	—	—	0.031	—	—
W	−0.07	—	—	0.052	−0.16	0.043	—	—	—	—	—	—	—
Zr	−4.1	—	—	(−23)	—	−0.16	—	—	—	—	—	—	—

Table 2.8 *Self-interaction coefficients in binary liquid copper alloys*[4]

Element, i	e_i^i
Ag	−0.005
Al	0.14
Au	0.008
Bi	−9/T + 0.0052
Ca	(0.14)
Fe	−0.029
Ga	0.028
Ge	0.051
H	(0)
Mg	(0.1)
Mn	0.03
O	−414/7 + 0.122
Pb	−0.0006
S	−179/7 -0.004
Sb	(0.04)
Sn	0.025
Ti	−0.008
Zn	0.017

information useful for silicon refining. For more detailed information, it is possible to consult several detailed reviews.[20–23] Since the interest is in relatively pure alloys, we are concerned only with dilute alloys of silicon (i.e., with alloys containing more than about 98–9% Si).

Table 2.11 gives the free energy of solution and self-interaction coefficients for a number of elements dissolved in liquid silicon. In contrast to tables presented for other metals, this table provides the free energies of dissolution when using mole fraction as the composition coordinate. It also lists the values of activity coefficients at 1700 K. (This is equal to 1427 °C, a temperature just above the melting point of silicon.) This is because mole fraction and related coordinates are commonly used for silicon. One coordinate is atom percent. This needs no explanation. It is simply the mole fraction times 100. Another coordinate is atoms per cubic centimeter. This requires a bit of calculation. From the density of liquid silicon at its melting point, and from Avogadro's number, we find that one cm^3 has 5.294×10^{22} atoms. Hence, atoms/$cm^3 = X_i \times 5.294 \times 10^{22}$.

Table 2.9 *Interaction coefficients in ternary alloys of copper*[4]

Element, j	e_H^j	e_O^j	e_S^j
Ag	0.0006	0	—
Al	0.0058	—	—
Au	0.0003	0.015	0.012
Co	0.015	−0.32	−0.023
Cr	0.0092	—	—
Fe	−0.015	−20 000/T + 10.8	−125/T + 0.042
Mn	−0.006	—	—
Ni	−0.026	−169/T + 0.079	−159/T + 0.069
P	0.088	−6230/T + 3.43	—
Pb	0.031	−0.007	—
Pt	−0.0084	0.057	0.019
S	0.073	−0.164	—
Sb	0.031	—	—
Si	0.042	(−62)	0.062
Sn	0.016	−0.009	—
Te	−0.012	—	—
Zn	0.029	—	—

Tables 2.12 and 2.13 show the known first-order interaction coefficients in silicon-based solutions.

When metals solidify there is a segregation of impurity elements. Impurities usually collect in the liquid phase. Several processes use solidification processing to produce silicon of higher purity. The Czochralski (CZ) and float zone (FZ) processes are the most common. These processes produce crystals of high purity silicon, which are grown from less pure molten silicon. Fractional crystallization and other processes have also been employed.

The distribution of impurity elements between solid and liquid silicon during solidification is characterized by a coefficient,

$$k_i^o = \frac{X_i^S}{X_i^L},$$

(2.44)

where X_i^S is the mole fraction of element i in the solid phase and X_i^L is the mole fraction of element i in the liquid phase.

Table 2.10 *Interaction coefficients in ternary cobalt-based alloys for H, N, O, and S at 1600°C[8, 10]*

Element, j	e_H^j	e_N^j	e_O^j	e_S^j
Al	0.014	0.04	—	—
Au	—	—	0.0026	—
B	0.09	—	—	—
Cr	0	−0.043	−0.07	—
Cu	−0.0031	−0.009	−0.009	−0.016
Fe	0.0014	−0.01	−0.019	−0.0034
Mn	−0.0035	—	−0.20	—
Mo	0.0041	−0.008	−0.0011	—
Nb	0	−0.042	—	—
Ni	−0.002	0.024	—	−0.007
S	—	—	−0.133	—
Si	0.021	0.12	—	—
Sn	0.006	—	—	—
Ta	0.0054	−0.029	—	—
Ti	—	−0.45	—	—
V	—	−0.1	−0.28	—
W	0.0064	−0.007	—	—

The true equilibrium value of the distribution coefficient is often hard to measure. Some elements have a very small solubility in solid silicon, and the small quantities are difficult to quantify accurately. Some elements segregate strongly to grain boundaries and other defects, which lowers the measured segregation in crystalline materials. And finally, the determination of k^o from phase diagrams is complicated by the retrograde solubility (for which a thermodynamic explanation has been provided by McKelvey[44]) of most impurities in solid silicon (Figure 2.5).

Table 2.14 shows values for the thermodynamic distribution coefficients, together with reported effective values in the CZ refining process. It may be worth noting that there is a strong correlation between the maximum solubility of most elements (with the exception of carbon, nitrogen, and oxygen) in solid silicon and its distribution coefficient.[45] That is,

$$X_{max} \cong 0.1k^o. \tag{2.45}$$

Table 2.11 *Thermodynamics of binary silicon-based alloys*

Solution Reaction	γ^o (1700 K)	ΔG^o for X (J/mole)	ε_i^i	ΔG^o for wt. % (J/mole)	e_i^i	Data source (ref.)
Ag (l) → Ag	2.83	$-83\ 700 + 57.884T$	(−2)	$-83\ 700 + 8.409\ T$	(0.004)	21, 24
Al (l) → Al	0.32	$-9\ 790 - 3.740T$	2.3	$-9\ 790 - 41.694\ T$	0.010	21[b]
As (l) → As	0.33	(−15 500)	—	$(-15,500 - 46.445\ T)$	—	20
Au (l) → Au	0.206	−22 300	3.2	$-22\ 300 - 54.481\ T$	0.040	21[a]
B (s) → B	3.9	$2\ 400 + 9.894\ T$	—	$2\ 400 - 20.456\ T$	—	20, 23, 27
Bi (l) → Bi	28.7	$61\ 650 - 8.353\ T$	−6.7	$61\ 650 - 63.327\ T$	−0.001	21[b]
C (s) → C	99.4	$63\ 785 + 0.715\ T$	—	$63\ 785 - 30.510\ T$	—	22, 28–30
Ca (l) → Ca	1.7×10^{-4}	$-118\ 900 - 2.078\ T$	9.9	$-118\ 900 - 43.322\ T$	0.045	23, 41
Ce (l) → Ce	(2.3×10^{-6})	(−183 600)	(4.9)	$(-183\ 600 - 51.65\ T)$	(0.039)	22
Co (l) → Co	4.6×10^{-3}	$-121\ 342 + 26.688\ T$	1.24	$-121\ 342 - 17.761\ T$	0.010	22
Co (s) → Co	4.9×10^{-3}	$-104\ 188 + 16.985\ T$	1.24	$-104\ 188 - 27.464T$	0.010	22
Cr (l) → Cr	6.1×10^{-3}	−72 000	−3.2	$-72\ 000 - 43.408\ T$	−0.010	22
Cr (s) → Cr	8.2×10^{-3}	$-51\ 500 - 9.623\ T$	−3.2	$-51\ 500 - 53.031\ T$	−0.010	22
Cu (l) → Cu	0.30	$-15\ 670 - 0.682\ T$	2.6	$-15\ 670 - 45.758\ T$	0.017	23
Fe (l) → Fe	0.024	$-107\ 530 + 32.175\ T$	−13.9	$-107\ 530 - 11.827\ T$	−0.056	22
Fe (s) → Fe	0.025	$-92\ 340 + 23.765\ T$	−13.9	$-92\ 340 - 20.237\ T$	−0.056	22
Ga (l) → Ga	1.61	$14\ 844 - 4.75\ T$	−0.96	$14\ 844 - 50.597\ T$	0.002	20, 21[b]
½H₂ (v) → H	—	$102\ 080 + 9.764\ T$	—	$102\ 080 - 0.86\ T$	—	22

Table 2.11 *Continued*

Solution Reaction	γ° (1700 K)	ΔG° for X (J/mole)	e_i^i	ΔG° for wt. % (J/mole)	e_i^i	Data source (ref.)
In (l) → In	5.15	46 900 − 13.97 T	−3.3	46 900 − 63.964 T	−0.001	20, 21[b]
Mg (l) → Mg	0.272	−73 850 + 32.612 T	2.6	−73 850 − 4.473 T	0.011	21, 32, 33[b]
Mg (v) → Mg	2.3	−202 350 + 125.99 T	2.6	−202 350 + 88.908 T	0.011	21, 32, 33[b]
Mn (l) → Mn	0.022	−78 000 + 13.968 T	8.1	−78 000 − 29.898 T	0.039	22, 28
Mo (l) → Mo	(0.064)	(−38 900)	—	(−38 900 − 48.501 T)	—	22
Mo (s) → Mo	(0.14)	(−11 070 − 9.623 T)	—	(−11 070 − 58.124 T)	—	22
½N₂ (v) → N	—	−17 570 + 60.766 T	—	−17 570 + 28.23 T	—	34, 35
Nb (s) → Nb	(1.0 × 10⁻⁴)	(−130 200)	—	(−130 200 −48.234 T)	—	23
Ni (s) → Ni	(1.6 × 10⁻³)	(−194 300 + 61.0 T)	—	(−194 300 + 16.585 T)	—	23
½O₂ (v) → O	—	−290 700 + 75.287 T	—	−290 700 + 41.678 T	—	22, 36, 37
½P₂ (v) → P	—	−99 500 + 68.561 T	3.0	−99 500 + 29.46 T	0.013	23, 38, 39
Pb (l) → Pb	44.2	66 630 − 7.700 T	−7.2	66 630 − 62.602 T	−0.0036	23, 41
Pd (s) → Pd	0.03	−73 420 + 14.00 T	3.6	−73 420 − 35.363 T	0.028	21[b]
Sb (l) → Sb	5.3	15 464 + 4.805 T	−3.2	15 464 − 45.677 T	0.0006	20, 23
Sc (l) → Sc	(6.7 × 10⁻⁴)	(−103 200)	—	(−103 200 − 42.199 T)	—	22[a]
Sc (s) → Sc	(7.3 × 10⁻⁴)	(−85 750 − 9.623 T)	—	(−85 750 − 51.822 T)	—	22[a]
Sn (l) → Sn	5.0	31 160 − 4.910 T	−3.4	31 160 − 55.181 T	−0.001	21[b]
Ta (s) → Ta	(7.9 × 10⁻⁴)	(−156 800 + 32.87 T)	—	(−156 800 − 20.906 T)	—	23
Ti (l) → Ti	4.04 × 10⁻⁴	−110 450	4.0	−110 450 − 42.72 T	0.020	40[a]
Ti (s) → Ti	4.36 × 10⁻⁴	−93 000 − 9.623 T	4.0	−93 000 − 52.343 T	0.020	40[a]

(Continued)

Table 2.11 *Continued*

Solution Reaction	γ^o (1700 K)	ΔG^o for X (J/mole)	ε_i^i	ΔG^o for wt. % (J/mole)	e_i^i	Data source (ref.)
Tl (l) → T̲l̲	925	69 500 + 15.9 T	–13	(69 500 – 38.889 T)	–0.029	20[b]
V (l) → V̲	(1.2 × 10⁻³)	(–95 400)	—	(–95 400 – 43.237 T)	—	22[a]
V (s) → V̲	(1.7 × 10⁻³)	(–72 555 – 10.431 T)	—	(–72 555 – 53.668 T)	—	22[a]
W (s) → W̲	(5.3 × 10⁻³)	(–133 760 + 35.10 T)	(11)	(–133 760 – 18.808 T)	(0.072)	23
Y (l) → Y̲	(3.2 × 10⁻⁵)	(–146 100)	—	(–146 100 – 47.868 T)	—	22[a]
Y (s) → Y̲	(3.5 × 10⁻⁵)	(–128 750 – 9.623 T)	—	(–128 750 – 57.491 T)	—	22[a]
Zn (l) → Z̲n̲	1.44	29 200 – 14.13 T	—	29 200 – 59.442 T	—	21
Zr (l) → Z̲r̲	(3.6 × 10⁻⁶)	(–177 000)	—	(–177 000 – 48.08 T)	—	22[a]
Zr (s) → Z̲r̲	(4.9 × 10⁻⁶)	(–156 550 – 9.623 T)	—	(–156 550 – 57.70 T)	—	22[a]

Notes: The thermodynamic temperature, T, is in degrees Kelvin. The liquid phase is metastable for the elements As, Co, Cr, Fe, Mg, Mo, Sc, Ti, V, Y and Zr at a temperature of 1700K. For example, arsenic is under pressure (>40 atm) to be liquid at 1700K; and iron is solid at this temperature. Values in parentheses are estimated or approximate values.

[a] A regular solution has been assumed. For calculations at temperatures near the melting point of silicon, any resulting error will be small.
[b] Heat of solution and interaction coefficients were calculated from Eqs. (9) of ref. [21].

Table 2.12 *Interaction coefficients in liquid silicon*

Mole ratio	Weight percent	Temperature (K)	Source
$\varepsilon_{Pb}^{Al} = 2.34$	$e_{Pb}^{Al} = 0.011$	1723	41
$\varepsilon_{Pb}^{Ca} = -9.0$	$e_{Pb}^{Ca} = -0.026$	1723	41
$\varepsilon_{Pb}^{Mg} = 10.2$	$e_{Pb}^{Mg} = 0.051$	1723	41
$\varepsilon_{Ca}^{Al} = 6.46$	$e_{Ca}^{Al} = 0.030$	1723	41
$\varepsilon_{Ti}^{Ca} = 25.3$	$e_{Ti}^{Ca} = 0.078$	1723	41
$\varepsilon_{Fe}^{Ca} = 2.25$	$e_{Fe}^{Ca} = 0.008$	1723	41
$\varepsilon_{Fe}^{Mn} = (99)$	$e_{Fe}^{Mn} = (0.22)$	1700	22
$\varepsilon_{Mn}^{Fe} = (99)$	$e_{Mn}^{Fe} = (0.22)$	1700	22
$\varepsilon_{P}^{Fe} = 7.43$	$e_{P}^{Fe} = 0.018$	1723	42
$\varepsilon_{P}^{Mn} = 12.0$	$e_{P}^{Mn} = 0.029$	1723	42

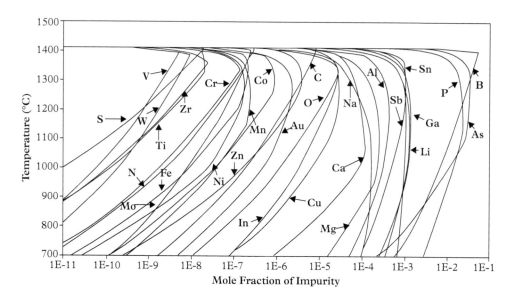

Fig. 2.5 *Solubility of impurities in solid silicon.*[43] *[Reprinted by permission from Springer Nature © 2009.]*

Table 2.13 *Interaction coefficients in liquid Si–C–i,[43] Si–O–I,[22] and Si–H–i[22] alloys (at temperatures near the melting point of silicon)*

Element, i	ε^i_C	e^i_C	ε^i_O	e^i_O	ε^i_H	e^i_H
Al	−0.72	−0.004	−118	−0.56	—	—
As	−1.58	0.000	(−115)	(−0.19)	—	—
B	−4.02	−0.052	−2.04	−0.03	—	—
C	—	—	0	−0.006	—	—
Ca	5.46	0.018	<0	<0.0013	—	—
Co	—	—	—	—	4.5	0.012
Cr	10.72	0.027	—	—	—	—
Cu	5.24	0.012	—	—	—	—
Fe	3.50	0.010	−6	−0.011	3.5	0.010
H	—	—	—	—	—	—
Mn	−1.58	−0.001	—	—	1.4	0.005
N	0.55	0.000	—	—	—	—
Ni	2.28	0.007	—	—	—	—
O	18.5	0.138	—	—	—	—
P	−5.67	−0.022	−45	−0.177	—	—
Sb	—	—	−23.3	−0.020	—	—
Zn	−3.39	−0.004	—	—	—	—
Zr	−7.35	−0.007	—	—	—	—

In practice, the actual distribution coefficient may be different from the true, thermodynamic coefficient. The effective distribution coefficient has been shown[46] to be

$$k_{eff} = k^o \div \left[k^o + (1 - k^o) \, exp\left(-\frac{v\delta}{D}\right) \right],$$

(2.46)

where v is the growth rate or solidification rate of silicon, δ is the diffusion boundary layer thickness, and D is the diffusivity of the impurity in liquid silicon.

A number of processing variables may influence the effective distribution coefficient. This is particularly so for reactive elements, such as C, O, and P. Carbon and oxygen may react to form carbon monoxide gas, especially under reduced pressures. Hence, the distribution between liquid and solid may be different from that predicted by Eqs. (2.44) and (2.46).

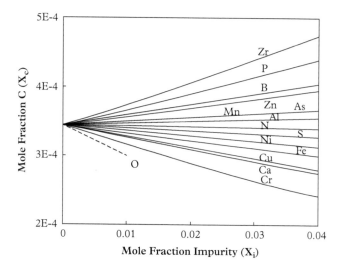

Fig. 2.6 *The effect of dissolved elements on carbon solubility in liquid silicon.*[43] *[Reprinted by permission from Springer Nature © 2009.]*

One must also consider the possible changes caused by chemical interactions. These have been determined in most detail for the carbon solubility in liquid silicon (Figure 2.6). Elements that increase carbon solubility would tend to lower the distribution coefficients shown in Table 2.14. Unfortunately, the effects of most elemental interactions in solid silicon are unknown. The quantities dissolved are usually very small, however; so these interactions are probably of little practical importance in most cases.

It is worth making an observation on the elements boron and phosphorus. These have a strong effect on the properties of silicon, and are rather difficult to remove by solidification refining. However, the phase diagrams for the Si–B and Si–P systems suggest that the removal in the CZ and FZ processes could be much improved. The equilibrium distribution coefficient for boron is about 0.25, whereas the effective distribution coefficient observed is 0.8. For phosphorus, the equilibrium distribution coefficient may be as low as 0.03—one tenth of the value observed in the CZ process. More study on the kinetics and thermodynamics of refining of these elements is strongly recommended. Carbon and oxygen are other elements that could use additional study.

2.1.10   Equilibria between Particles (Inclusions) and Melts; Precipitation Deoxidation

Precipitation deoxidation of molten steel is described here partly due to its importance and partly to illustrate the use of thermodynamic data to calculate equilibria between two immiscible phases. Oxygen dissolved in steel increases with temperature at equilibrium with atmosphere, slag, or refractories. When the temperature is decreased during teeming and

Table 2.14 *Distribution coefficients for impurity elements in silicon*

Element	k^0	Source	K_{eff} (CZ)	source
Al	0.0020	47	0.002–0.03	46
Ag	5.0×10^{-5}	28	—	—
As	0.3	47	0.30	46
Au	2.5×10^{-5}	28, 47	2.5×10^{-5}	46
B	0.80	47	0.73–0.80	46
B	0.20-0.25	23, 27	—	—
Bi	7×10^{-4}	28, 47	—	—
C	0.07	28, 36, 29	0.050–0.058	46
Co	8×10^{-6}	47	$8–10 \times 10^{-6}$	46
Cr	1.1×10^{-5}	28	1.1×10^{-5}	46
Cu	4×10^{-4}	47	$4–8 \times 10^{-4}$	46
Fe	6.4×10^{-6}	28, 48	$6.4–8 \times 10^{-6}$	46
Ga	0.0080	28, 47	0.008	46
Ge	0.33	47	0.6	46
In	4×10^{-4}	28, 47	4×10^{-4}	46
Li	0.01	47	—	—
Mn	1.3×10^{-5}	28	—	—
N	7.5×10^{-4}	49	8×10^{-4}	46
Ni	1.5×10^{-4}	47	—	—
O	0.955	36, 37	0.5–1.3	46
P	0.35	47, 49	0.35	46
P	0.03–0.11	23, 38	—	—
Sb	0.023	28, 47, 49	0.002–0.023	—
Sn	0.016	28, 47	—	—
Ta	10^{-7}	47	—	—
Ti	2×10^{-6}	28, 48	2×10^{-6}	46
Zn	1×10^{-6}	28, 47	—	—

solidification, this process is reversed; the solubility of oxygen decreases. Then oxygen may react with an alloying element, M, in the steel,

$$x\underline{O} + y\underline{M} = M_yO_x, \qquad (2.47)$$

where the reaction product M_yO_x is a solid, liquid, or gaseous phase. For instance, M may be Si, Mn, C, or the solvent Fe. Oxygen often reacts simultaneously with the various components and complex inclusions may form. In the case of carbon, CO bubbles are created. If the inclusions or CO bubbles are trapped in the solidified metal, they may be harmful to the mechanical properties. This was discussed in Chapter 1.

How do we try to reduce the amount and size of such inclusions? Before casting, a deoxidation element is introduced with a much greater affinity for oxygen than the other alloying elements. Then ideally the content of dissolved oxygen is lowered to such low levels that it does not react with the melt components during solidification. However, there is a catch to this procedure; the deoxidation product must not remain in the melt! For instance, if Al is used to precipitate oxygen as follows,

$$2\underline{Al} + 3\underline{O} = Al_2O_3,$$

the hard and brittle Al_2O_3 particles must be removed. In Chapter 5 this removal of inclusions is treated in detail.

From standard Gibbs energy data[2,7] and the Gibbs energies of solution given in Table 2.2 one can calculate the diagrams presented in Fig. 2.7 that relate oxygen activity and activity of the various deoxidation elements. An example of the calculations will be given. The activity of calcium is so low that it is not included in the diagram. Zirconium and calcium are the most efficient deoxidation elements. The reason why Al is preferred in practice is partly the lower price of Al and partly problems of separating the ZrO_2 inclusions from the melt. The use of calcium will be discussed in section 2.1.11.

Carbon and silicon seem to be poor deoxidizers according to the diagram. However, it should be remembered that Fig. 2.7 is calculated with the activity equal to 1 for the deoxidation product and $p_{co} = 1$ bar for carbon. If p_{co} is much less than 1 bar, the line for carbon moves downwards (see Fig. 2.8) and carbon may become a useful deoxidant. How can we reduce the partial pressure of CO? One way is to dilute the CO gas with an inert gas such as argon or nitrogen. Another solution is to evacuate the gas above the melt so that the total pressure is less than 1 bar. In practice there is a limit to how low we can reduce oxygen levels by this approach. The explanation lies in the realm of kinetics treated in Chapter 7.

As for carbon, it is observed from the equilibrium

$$\underline{Si} + 2\underline{O} = SiO_2$$

that silicon can become an effective deoxidant if the activity of SiO_2 is reduced. As will be seen from Fig. 2.16, a_{SiO_2} is far less than 1 in a $CaO-Al_2O_3-SiO_2$ slag high in lime (basic slag).

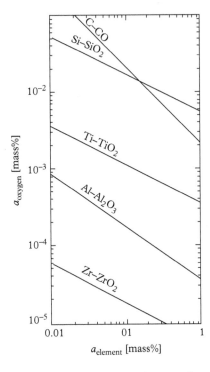

Fig. 2.7 *The activity of oxygen in steel at 1600°C in equilibrium with various deoxidation elements.*

It will be shown how the Al line in Fig. 2.7 is obtained. From Appendix 1 and Table 2.2:

	ΔG^0 (J/mol)
$\frac{1}{3}Al_2O_3 = \frac{2}{3}Al(l) + \frac{1}{2}O_2$	562 433 – 109 T
$\frac{1}{2}O_2 = \underline{O}$	–117 110 – 3.39 T
$\frac{2}{3}Al(l) = \frac{2}{3}\underline{Al}$	54 953 – 29.17 T
$\frac{1}{3}Al_2O_3 = \frac{2}{3}\underline{Al} + \underline{O}$	500 276 – 141.56 T

or at $T = 1873$ K and $a_{Al_2O_3} = 1$

$$\frac{a_{\underline{Al}}^{\frac{2}{3}}a_{\underline{O}}}{a_{Al_2O_3}^{\frac{1}{3}}} = K = 2.77 \cdot 10^{-7}.$$ (2.48)

Note that the scales in Fig. 2.7 are logarithmic and that the slope is $-2/3$.

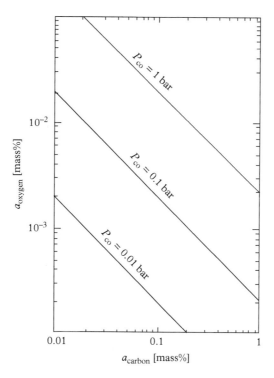

Fig. 2.8 *Activity of oxygen in steel in equilibrium with carbon at 1600 °C for various pressures of CO.*

From the figure it might be imagined that dissolved oxygen could be brought down to whatever levels we desire by the addition of a strong deoxidizer. However, this is not the case. A deoxidizer, M, has a strong affinity for oxygen and this results in an interaction coefficient e_O^M which is large and negative. This gives a minimum for [% O] at low values of M. We will show this for $M =$ Al. The logarithm of the equilibrium equation gives

$$\frac{2}{3} \log a_{Al} + \log a_O = \log K(T)$$

or, introducing activity coefficients,

$$\frac{2}{3} (\log f_{Al} + \log [\%Al]) + \log f_O + \log [\%O] = \log K(T). \tag{2.49}$$

The minimum is found from

$$\frac{\partial [\%O]}{\partial [\%Al]} = O.$$

From the definition of the activity coefficients (see Eq. (2.37))

$$e_O^{Al} = \frac{\partial \log f_O}{\partial [\%Al]}$$

and

$$e_{Al}^{Al} = \frac{\partial \log f_{Al}}{\partial [\%Al]}.$$

The minimum is given by differentiating Eq. (2.49) with respect to [% Al]. Note that $\log[\%\,Al] = \ln[\%\,Al]/\ln 10 = 0.434 \ln[\%\,Al]$. Then

$$\tfrac{2}{3}\left(e_{Al}^{Al} + \tfrac{0.434}{[\%Al]}\right) + e_O^{Al} = O$$

$$[\%Al] = -\frac{0.434}{\tfrac{3}{2}e_O^{Al} + e_{Al}^{Al}}.$$

This determines the minimum for [% O]. From Table 2.7 of interaction coefficients $e_O^{Al} = -3.90$ and $e_{Al}^{Al} = 0.045$ so that [% Al] = 0.075. Furthermore, since $e_O^O = -0.20$ and $e_{Al}^O = -6.6$

$$\log f_O = e_O^O[\%O] + e_O^{Al}[\%Al] \cong e_O^{Al}[\%Al] = -3.90 \times 0.075,$$

which gives

$$\log f_{Al} = e_{Al}^{Al}[\%Al] + e_{Al}^O[\%O] \approx e_{Al}^{Al}[\%Al] = 0.045 \times 0.075,$$

so that $f_{Al} \approx 1$.

Finally, we find from the equilibrium that the lowest oxygen value attainable by Al deoxidation with Al_2O_3 as the reaction product is [% O] =0.0004 or 4 ppm. However, if Al_2O_3 is mixed with other compounds so that the activity is less than 1, lower oxygen concentrations can be attained. Figure 2.9 gives [% O] as a function of [% Al] assuming $a_{Al_2O_3} = 1$. The oxygen values must not be confused with total oxygen which also includes oxygen in the form of Al_2O_3 inclusions. In steelmaking it is usually difficult to attain total oxygen below 10–20 ppm. Removal of inclusions of Al_2O_3 will be discussed in Chapter 5.

Due to pick-up of oxygen from the surroundings of the melt, there will be a steady loss of deoxidizer. Thus, the Al is often added to the melt in excess of the amount corresponding to 0.075 per cent.

In equilibrium with a slag inclusion where the Al_2O_3 activity is less than 1, Eq. (2.48) may be used to obtain a_O. For instance, $a_{Al_2O_3} = 0.085$. At a temperature of 1823 K, the equilibrium constant in Eq. (2.48) is reduced to $K = 1.85 \times 10^{-5}$. Then $a_O = 0.0002 \times 1.85/3.7 \times 0.085^{1/3} = 0.00004\%$. Thus, by dilution of the Al_2O_3 reaction product we can attain very low oxygen levels. This last result is used in section 2.1.11 on Ca addition.

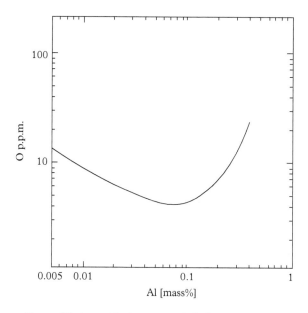

Fig. 2.9 *Oxygen in steel in equilibrium with aluminium. Only first-order interaction coefficients are taken into account.*

2.1.11 Modification of Inclusions; Ca Additions to Steel

In precipitation deoxidation, even if removal of the inclusions to the top slag and walls proceeds at a high rate, there will always be present a certain number of particles. The various removal, pick-up, and nucleation processes that may occur are illustrated in Fig. 2.10. Harmful effects of the inclusions were reviewed in Chapter 1. In rolling and other deformation processes it is found that the particles should not be too hard. The inclusions will then destroy the matrix surrounding them, and voids are created and grow during the deformation process. On the other hand, if the inclusions are too soft, as may be the case with MnS, for instance, they elongate and give poor mechanical properties in the transverse direction.

Furthermore, it has been found that solid oxide or sulfide inclusions tend to clog teeming or casting nozzles.[50,51] Figure 2.11 gives the temperature–composition diagram for the system CaO–Al_2O_3. Typically, there is a molten region at about equal amounts of CaO and Al_2O_3. Silica also increases the extent of the molten region. Clogging is reduced when the inclusions contain roughly equal amounts by mass of CaO, Al_2O_3, and SiO_2.

How do we change the composition of Al_2O_3 inclusions to inclusions that also contain CaO? One approach is to add calcium to the melt, despite the fact that calcium is an expensive metal and also has a high vapour pressure at steelmaking temperatures. Calcium is dissolved into the melt by injection of Ca and $CaSi^{52,53}$ or by wire feeding. This calcium together with aluminium may precipitate liquid calcium aluminate inclusions. At the same time precipitation of solid CaS should be prevented.

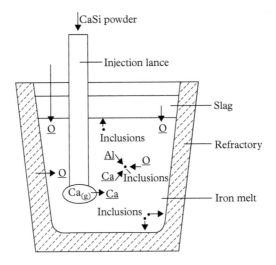

Fig. 2.10 *The various removal, pick-up, and nucleation processes that occur when Ca is added to steel.*

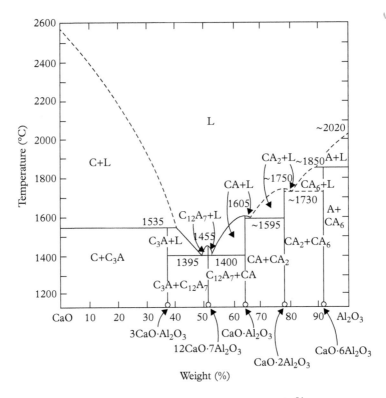

Fig. 2.11 *Temperature–composition diagram for the system CaO–Al₂O₃.*[54]

The objective outlined above presents a formidable problem from both a practical and a theoretical point of view. In the following we will look at the various equilibria to outline the thermodynamics involved in creating liquid calcium aluminate inclusions with no CaS precipitates.

The phase rule is a useful tool in the treatment of the problem. Due to its general importance it is discussed in section 2.1.12.

2.1.12   The Phase Rule Applied to the Problem of Calcium Addition

When dealing with a system comprising several phases and a number of components, application of the phase rule may give useful information about the behaviour of the system. We will briefly review the background of this rule.

Assume that we have a system consisting of a number C of different components, distributed over a number P of phases. At equilibrium, the chemical potential of each component must be the same in all of the phases. For each component this gives $P - 1$ relations regarding the chemical potentials, and thus a total of $C(P - 1)$ equations for the whole system.

In each phase there are $C - 1$ composition variables, and thus a total of $P(C - 1)$ composition variables, plus the two variables defining temperature and pressure. Thus, we have

Number of variables	$PC - P + 2$
minus number of equations	$-PC + C$
Number of independent variables:	$= C - P + 2$

This is Gibbs' phase rule. The equation looks simple, but its application quite often requires some thought. A main question is this: how do you count the components? To cut short the discussion, we may state that in metallurgy it is basically a good idea to take the number of elements in the system as the number of components. In so doing, however, one has to look out for possible fixed relations between the amounts of different elements, which eventually represent restrictions on the number of free variables. This is notably the case in systems in which each element is present in one oxidation state only, since there must be a balance between positive charges (cations) and negative charges (anions) in such a system. Thus the more complete phase rule may be written $F = C - P + 2 - R$, where R stands for the number of restrictions. It may have the values 0 or 1, and rarely 2.

Turning to the system presented in section 2.1.11, we disregard silicon, and thus we have $C = 5$ (Fe, O, S, Al, Ca). There is a metal phase, an aluminate phase, and a calcium sulfide phase; thus, $P = 3$. There is no restriction, $R = 0$, since several of the elements occur both in their elemental forms and in compounds. Thus we have the number of free variables:

$$F = 5 - 3 + 2 = 4.$$

If pressure and temperature are given, this leaves two degrees of freedom. To use these we can adjust Al and O. Finally, hopefully, S can be kept at such low levels that CaS does not form. This last point means, in terms of the phase rule, that P is reduced by one. Then there is an additional degree of freedom and we can adjust, for instance, S.

Next we will turn our attention to the aluminate inclusions by considering separately the system $CaO–Al_2O_3$.[55] Using the 'rule of elements' this is a three-component system, but there is also a restriction (fixed oxidation numbers) so that $F = 3 – P+2 – 1= 4 – P$. (We would of course get the same result if we considered the system as composed of the two components CaO and Al_2O_3.)

With pressure and temperature given, one finds $F = 0$ for $P = 2$ in this system. That is, equilibrium at a fixed temperature between a solid and a liquid, or two solid phases, requires a definite composition in each of the phases, as is readily apparent from the phase diagram in Fig. 2.11. At temperatures around $1500\,°C$ and about equal mass fractions of alumina and calcium oxide, the system comprises only a liquid phase in which the composition may be varied. On the CaO-rich side the melt eventually precipitates $3CaO \cdot Al_2O_3$, while on the Al_2O_3-rich side $CaO \cdot Al_2O_3$ is precipitated. In the following we will discuss whether it is practical to deoxidize with a combination of aluminium and calcium, to remain within the region where the oxide inclusions are entirely liquid at $1500\,°C$, and simultaneously to prevent precipitation of CaS.

To give inclusions that contain CaO, calcium must be able to compete with aluminium as a deoxidizer. At the same concentration (activity) levels, Ca is in fact a better deoxidizer than Al. The problem in practice may become one of reaction rates and mechanisms involved, but still thermodynamics yields important information. What will happen when solid calcium is added to molten steel at $1500\,°C$? First of all, it melts (m.p. $838\,°C$). Next we may find its boiling point from the equation in Appendix 1:

$$Ca\,(l) = Ca\,(g); \quad \Delta G^0 = 156800 – 89.5T\,(J/mol).$$

That is, the boiling point is $156\,800/89.5 = 1752\,K = 1479\,°C$. More accurate data give 1755 K, or $1482\,°C$. (The discrepancy is due to the use of a simple linear function for ΔG^0 as given in Appendix 1.) In fact, calcium is added to molten steel at even higher temperatures, and it is then realized that severe losses by volatilization of calcium is to be expected. How should this problem be reduced and the yields of calcium increased? In practice two approaches are in use. If Ca is injected deep down in the melt, pressure is higher and boiling is prevented. Furthermore, one may employ an alloy of calcium where the activity relative to pure calcium is less than 1. Such an alloy is CaSi. Use of this alloy also gives the modified inclusions containing SiO_2.

Let us assume that a vapour pressure of calcium higher than 1 bar cannot be sustained. What then is the activity of calcium in the melt? For the reaction according to Table 2.2:

$$Ca\,(gas) = \underline{Ca}; \Delta G^0 = -39460 + 49.37T \quad J/mol.$$

Then at 1482 °C

$$\frac{a_{Ca}}{p_{Ca}} = 0.039$$

so that the activity of calcium is 0.039 per cent at 1 bar Ca pressure. This value is roughly the same as the activity of the aluminium in the melt. Then mass transfer rates by diffusion to the new modified inclusions created during teeming and solidification should be roughly the same for calcium and aluminium. Thus conditions are present for producing new $CaO \cdot Al_2O_3$ inclusions.

Mass transfer problems are studied further in Chapters 5 and 7. However, there is one point remaining. What are the conditions for avoiding precipitation of CaS? The relevant equilibrium is

$$CaO + \underline{S} = CaS + \underline{O}.$$

From the thermodynamic data (Appendix 1 and Table 2.2):

	ΔG^0 (J/mol)
$CaO\,(s) = Ca\,(l) + \frac{1}{2}O_2$	$640\,000 - 108.6\,T$
$\frac{1}{2}O_2 = \underline{O}$	$-117\,110 - 3.39\,T$
$Ca\,(l) + \frac{1}{2}S_2\,(g) = CaS\,(s)$	$-548\,000 + 104\,T$
$\underline{S} = \frac{1}{2}S_2\,(g)$	$125\,100 - 18.50\,T$
$CaO\,(s) + \underline{S} = CaS\,(s) + \underline{O}$	$99\,990 - 26.49\,T$

Then at 1550°C (1823 K)

$$\frac{a_S}{a_O} = 30.3\frac{a_{CaS}}{a_{CaO}}.$$

For equal amounts by mass of CaO and Al_2O_3 at 1550 °C it is found from Elliott etal.[54] that $a_{CaO} = 0.4$ and $a_{Al_2O_3} = 0.085$. Previously it was shown for Al deoxidized steel in equilibrium with this slag that $a_O = 0.00004\%$. Then with $a_{CaS} = 1$

$$a_s = 0.00004 \times 30.3/0.4 = 0.0034.$$

Thus a_s should be less than 0.0030 per cent or 30 ppm in order to prevent precipitation of CaS. Due to the very low oxygen activity, the steel must be desulfurized to low levels prior to the calcium treatment in order to prevent formation of CaS.

2.1.13   The Regular Solution Model and Molten Salts

In the preceding sections emphasis has been placed on activities in dilute metallic solutions, where concentration is expressed in terms of mass per cent and Henry's law forms the basis for the definition of activity. On the other hand, when the solution can no longer be regarded as dilute, as, for example, in mixtures of molten salts, composition is better expressed in terms of mole fractions with activities related to Raoult's law.

In Section 2.1.5 an ideal solution was defined as one which obeys Raoult's law over the whole composition range. In practice, however, this is a rare exception; most mixtures show deviations from ideality. In order to discuss these deviations, we introduce the concept of *excess* quantities. An excess quantity is the difference between the real quantity and the corresponding ideal value. We will look at the excess Gibbs energy of mixing, ΔG_{mix}^{xs}.

For the Gibbs energy of mixing for a binary solution we have in general

$$\Delta G_{mix} = RT\,(n_1 \ln a_i + n_2 \ln a_2). \tag{2.50}$$

The corresponding expression for an ideal solution is

$$\Delta G_{mix}^{id} = RT\,(n_1 \ln x_1 + n_2 \ln x_2). \tag{2.12}$$

Taking the difference, and noting that $\ln a_1 - \ln x_1 = \ln(a_1/x_1) = \ln \gamma_1$, we have

$$\Delta G_{mix}^{xs} = RT\,(n_1 \ln \gamma_1 + n_2 \ln \gamma_2).$$

From this follows

$$\frac{\partial \Delta G^{xs}}{\partial n_1} = RT \ln \gamma_1; \quad \frac{\partial \Delta G^{xs}}{\partial n_2} = RT \ln \gamma_2. \tag{2.51}$$

Equations (2.51) are quite general. Now we will introduce a class of solutions in which the enthalpy of mixing deviates from the ideal value (which is zero), while the entropy of mixing still has very nearly the ideal value (as given by Eq. (2.4)). This class, introduced by J. H. Hildebrand in 1929, is called a *regular solution*. Because of the ideal entropy of mixing, we have for regular solutions

$$\Delta G_{mix}^{xs} = \Delta H_{mix}. \tag{2.52}$$

Next we will go one step further, and assume that the enthalpy of mixing is symmetrical with respect to the two components. We also know that we must have $\Delta H_{mix} = 0$ for $x_1 = 0$ and for $x_2 = 0$. Furthermore because of the symmetry it must have a maximum for $x_1 = x_2 = 0.5$. A simple function which meets these criteria is

$$\Delta G_{mix}^{xs} = \Delta H_{mix} = (n_1 + n_2)\,\alpha x_1 x_2 = \alpha\,\frac{n_1 n_2}{n_1 + n_2},$$

where α is an empirical constant. By differentiation as in Eq. (2.51), this gives the expressions for the activity coefficients:

$$\ln \gamma_1 = \frac{\alpha}{RT}x_2^2; \quad \ln \gamma_2 = \frac{\alpha}{RT}x_1^2. \tag{2.53}$$

These equations are representative of a model which we will call the *symmetrical regular solution*.

For molten salt mixtures it is often assumed that the different types of cations A^+ and B^+ are randomly distributed among the anions and the anions X^-, Y^- are randomly distributed among the cations (the Temkin model). For a binary salt mixture $AX–BX$ with a common ion X, Eq. (2.53) then becomes

$$\ln \gamma_{AX} = \frac{\alpha}{RT}N_B^2 \tag{2.54}$$

$$\ln \gamma_{BX} = \frac{\alpha}{RT}N_A^2, \tag{2.55}$$

where the N_A and N_B are the ionic fractions

$$N_A = \frac{x_A}{x_A + x_B}$$

$$N_B = \frac{x_B}{x_A + x_B}. \tag{2.56}$$

The cryolite melt employed in aluminium electrolysis cell may, disregarding the presence of the AlF_6^{3-} ion, be regarded as composed of NaF and NaAlF$_4$ in the molar ratio 2:1, with Na$^+$ as the common ion. Then[56]

$$RT\ln \gamma_{NaF} = \alpha(1 - x_{NaF})^2, \tag{2.57}$$

where

$$\frac{\alpha}{RT} = -\frac{51000}{8.3145 \cdot 1285} = -4.77.$$

The validity of this equation is tested in Fig. 2.12. The value 51 000 J is found from the slope of the experimental points.

To pursue this example further, Na contamination in the Hall–Héroult cell is studied. The content of sodium is determined in molten aluminium in equilibrium with cryolite of the stoichiometric composition Na$_3$AlF$_6$ at the melting point 1285 K (sodium causes edge cracks in rolled products; for some 5xxx alloys sodium levels below 1 ppm are required):

$$x_{NaF} = 0.67 \quad \text{and} \quad x_{NaAlF_4} = 0.33$$

$$\ln \gamma_{NaF} = -4.77 \cdot 0.33^2 \quad \text{and} \quad \gamma_{NaF} = 0.595.$$

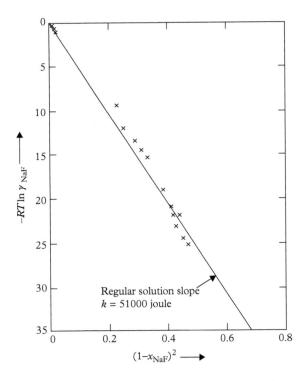

Fig. 2.12 *Example of regular solution behaviour. The system is Na_3AlF_6 at the liquidus temperature,*[56] *1285 K.*

From the free energy tables, ΔG^0 for the exchange reaction between Na and Al may be obtained:

	ΔG^0 (J/mol)
$2NaF(l) = 2Na(g) + F_2(g)$	$1248000 - 296.0\,T$
$Na(g) = Na(l)$	$-101300 + 87.9\,T$
$Na(l) = \underline{Na}$	$34430 - 21.38\,T$
$F_2 + \frac{1}{3}\,Al(l) + Na(g) = \frac{1}{3}\,Na_3AlF_6(l)$	$-1126667 + 207.8\,T$
$2NaF(l) + \frac{1}{3}\,Al = \underline{Na} + \frac{1}{3}\,Na_3AlF_6(l)$	$54463 - 21.68\,T$

Then at 1285 K:

$$\ln\left(\frac{f_{Na}}{a_{Al}^{\frac{1}{3}}a_{NaF}^2}\right) = -\frac{-(54463 - 21.68 \cdot 1285)}{8.3145 \cdot 1285} = -2.49$$

or

$$f_{Na}[\%Na] = 0.083 \cdot 1 \cdot (0.67 \cdot 0.595)^2 = 0.0132 \text{ or } 132 \text{ p.p.m.}$$

In practice the electrolyte is higher in AlF_3 than corresponds to Na_3AlF_6. This lowers the sodium content to about 100 ppm.

Removal of sodium from aluminium is discussed in detail in Chapter 4.

2.1.14 Slags

In the ores that serve as a basis for metal production, the metal or metal compound that we wish to extract is intimately mixed with various oxides and sulfides. Silicates, aluminates, and phosphates are examples of such components. In the pyrometallurgical processes they remain as a molten oxide phase insoluble in the primary metal.

Sometimes the slag also contains significant amounts of the metal we are producing, for instance due to oxidation of the metal. This is the case when phosphorus or carbon is removed from steel under oxidizing conditions.

Slags can be employed to extract unwanted elements from the metal, for instance P from steel as just mentioned. Another impurity transferred to slag from steel is sulfur. However, sulfur removal is most efficient under reducing conditions, that is, when the activity of oxygen in the steel is low. Then an exchange reaction takes place between sulfur in the steel and oxygen in the slag. (This is discussed later.)

The present knowledge of the structures of slags is based primarily on X-ray and neutron diffraction studies of slags and glasses.[57,58] The structures of silicate, phosphate, and borate glasses are conceived as random networks in which the small atoms with a high positive oxidation number, where B(+III), Si(+IV), and P(+V), the 'network formers', are tetrahedrally surrounded by oxygen atoms, O(−II), which form bridges between the positive atoms. The most typical example of such structures is afforded by silica and silicates. In pure silica, the SiO_4 tetrahedra share all four corners with neighbouring tetrahedra, forming a three-dimensional network. With the addition of a basic oxide such as Na_2O or CaO, oxygen bridges are broken, and the structure is changed to one in which the SiO_4 tetrahedra form sheets, and then chains (or rings) as illustrated in Fig. 2.13. With the addition of sufficient basic oxide the structure finally contains isolated SiO_4^{4-} ions, such as in calcium orthosilicate, Ca_2SiO_4. Along with the breaking of oxygen bridges, the viscosity of the molten slag or oxide mixture decreases.

Various attempts have been made to interpret the thermodynamic properties of slag melts in terms of the ionic models.[59,60] These models were applied to the equilibrium relations for slag–metal exchange reactions in terms of the concentrations of ionic species. A problem is that the activity coefficients of the various ionic species can only be determined experimentally. Therefore, there is a limit to how far the theory can be put to use. Ultimately, for a given slag–melt system experiments must be performed, for instance to determine the partition coefficient for an impurity. This will be discussed shortly. The theoretical studies are helpful in a general way; they throw light on the concept of basic and acid slags. Also they suggest the variables to use in describing the results of the experimental studies.

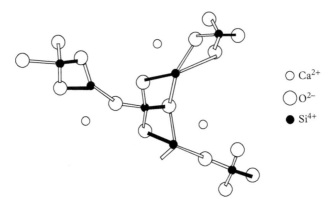

Fig. 2.13 *Silicate chains in a slag partly broken by the addition of CaO.*

An acid slag, containing a large amount of network formers, cannot be used to extract impurities such as oxygen or sulfur from a melt. One reason is thermodynamic: such a slag has a low capacity for these impurities. Furthermore, acid slags have high viscosity so that the mass transfer rates are low.

The typically basic oxides are essentially ionic compounds with relatively large cations of low charge. Examples, in order of increasing 'basicity', are MgO, FeO, MnO, CaO, PbO, BaO, Na_2O, and K_2O. Magnesium and iron oxides are not very effective in breaking oxygen bridges, while the oxides of sodium and potassium have a strong ability to break the network. The action of the basic oxides depends upon their ability to contribute oxygen ions, which in turn react with the acid oxide. The reaction to complete the breaking of the network may be written as follows for the various acid oxides:

$$SiO_2 + 2O^{2-} = SiO_4^{4-}$$
$$P_2O_5 + 3O^{2-} = 2PO_4^{3-}$$
$$B_2O_3 + 3O^{2-} = 2BO_3^{3-}.$$

Further addition of basic oxide beyond this point results in free oxygen ions. It is realized that the content of oxygen ions plays an important role in determining the slag properties. The ionic model described above allows us to calculate the content of O^{2-} ions in the slag. Thus, if a slag is made up of w moles of SiO_2; x moles of P_2O_5, B_2O_5, or Al_2O_3; y moles of MgO, FeO, MnO, CaO, PbO, or BaO (we write M for these cations); and z moles of Na_2O or K_2O, we have the following equation for the number of moles of O^{2-}:

$$wSiO_2 + x(P_2O_5, B_2O_3, Al_2O_3) + yMO + z(Na_2O, K_2O)$$
$$= wSiO_4^{4-} + 2x(PO_4^{3-}, BO_3^{3-}, AlO_3^{3-}) + (y+z-2w-3x)O^{2-} + yM^{2+} + 2z(Na^+, K^+). \quad (2.58)$$

This equation fulfils the requirement that the number of each atom species is conserved and that electrical neutrality is retained. It is seen that the basic components increase O^{2-} while the acid components (including Al_2O_3) consume O^{2-} ions.

2.1.15 The Equilibrium between Sulfur in Steel and in a Basic Slag

Removal of sulfur from steel to a basic slag is a very important part of steelmaking. Therefore the equilibrium between steel and slag is discussed here. Another reason is to illustrate the use and limitations of the ionic theories outlined earlier.

In the slag, sulfur is assumed to be present as S^{2-} similar to the O^{2-} ion. The sulfur, in moving from metal to slag, must acquire two electrons, since the components in the metal are all in atomic form. Then the only manner in which S transfer can take place is by an exchange reaction with a negatively charged ion from the slag. The only candidate seems to be O^{2-}:

$$\underline{S} + O^{2-} = S^{2-} + \underline{O}.$$

Sulfur is removed from the melt and in return we are getting oxygen in the metal! To drive the equilibrium to the right it is seen that O^{2-} should be high (basic slag) and $\underline{O}$ should be low (reducing conditions).

For the exchange equilibrium we may write

$$\frac{a_O a_{S^{2-}}}{a_S a_{O^{2-}}} = K_S(T),$$

where the equilibrium constant $K_S(T)$ depends not only on the temperature but also on the composition of the slag. The simplest manner of eliminating $a_{O^{2-}}$ is to assume that there is equilibrium between oxygen in the melt and in the slag:

$$\underline{O} + Fe = O^{2-} + Fe^{2+}.$$

If the two reactions are added, one obtains

$$\underline{S} + Fe = S^{2-} + Fe^{++}.$$

This equation probably does not reflect the mechanism of sulfur removal, but can be used to determine the equilibrium. Then

$$\frac{a_{S^{2-}} a_{Fe^{2+}}}{a_S a_{Fe}} = K'_s(T).$$

From a fundamental point of view one would wish to obtain the 'sulfur capacity' of the slag given as the ratio

$$\frac{x_{S^{2-}}}{[\%S]} = \frac{K'_S(T) f_s}{\gamma_{S^{2-}} \gamma_{Fe^{2+}} x_{Fe^{2+}}},$$

where $x_{S^{2-}}$ refers to the anion mole fraction and $x_{Fe^{2+}}$ refers to the cation mole fraction. However, the activity coefficients $\gamma_{S^{2-}}$ and $\gamma_{Fe^{2+}}$ are unknown and the factor $K'_S(T)/\left(\gamma_{S^{2-}} \gamma_{Fe^{2+}}\right)$

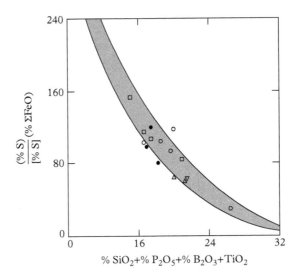

Fig. 2.14 *The sulfur partition coefficient (%S)/[%S] between slag and steel as a function of acid components in slag at temperatures from 1550 to 1650°C.*[8]

must be determined experimentally. Furthermore, from a practical point of view, it is the ratio (% S)/[% S] that is of most interest. Figure 2.14 shows some industrial results[8] for the sulfur capacity (partition coefficient) K_s = (% S)/[% S], where (%S) is the sulfur content in the slag in mass per cent. $x_{Fe^{2+}}$ has been replaced by (% FeO*) where the asterisk indicates that in the chemical analysis of the slag small amounts of Fe^{3+} have also been included. The figure shows that (%S)/[% S](% FeO*) depends primarily on the content of network formers summed up as mass percentages. As this sum increases (as the slag becomes more acid) the sulfur capacity goes down. Also the sulfur capacity is reduced as the conditions become more oxidizing, that is, as (% FeO) and [% O] become higher.

One would expect that summation of mole percentages corresponding to the ions $SiO_4^{4-}, TiO_4^{4-}, PO_4^{3-}$, and BO_3^{3-}, respectively, weighted by the factors $2w$ and $3x$ from Eq. (2.58) would give a better fit. However, a closer look shows that the improvement would be marginal especially since the content of SiO_2 dominates. Therefore, preference has been given to the industrially more practical units of mass per cent.

2.1.16   The Equilibrium between Phosphorus in Steel and in a Basic Slag

For many years it has been a problem to obtain a satisfactory description of the phosphorus equilibrium between steel and slag.[61] Here, the application of ionic theories to the problem will be outlined. Then a more direct approach will be given.

As previously mentioned, phosphorus is mainly present in a basic slag as PO_4^{3-}. To produce a phosphate ion, PO_4^{3-}, with its negative charge, P̲ in the melt must unite with O^{2-} ions.

If phosphorus, oxygen, and negative charge are balanced, the following equation for the phosphorus exchange is obtained,

$$\underline{P} + \frac{5}{2}\underline{O} + \frac{3}{2}O^{2-} = PO_4^{3-},$$

and the equilibrium constant for the reactions becomes

$$K_P = \frac{a_{PO_4^{3-}}}{a_P a_O^{\frac{5}{2}} a_{O^{2-}}^{\frac{3}{2}}}$$

$$\frac{x_{PO_4^{3-}}}{[\%P]} = \frac{f_P K_P a_O^{\frac{5}{2}} a_{O^{2-}}^{\frac{3}{2}}}{\gamma_{PO_4^{3-}}}. \tag{2.59}$$

It is seen that the distribution of phosphorus between slag and melt (partition coefficient) is proportional to $a_O^{\frac{5}{2}}$ (oxidizing conditions) and with $a_{O^{2-}}^{\frac{3}{2}}$ (basic slag, high lime content). Probably partly due to the complex form of the relation and the difficulties in determining $a_{O^{2-}}^{\frac{3}{2}}$ and $\gamma_{PO_4^{3-}}$, Eq. (2.59) has not proved satisfactory in industrial applications. Instead a more direct empirical approach may be useful. Selin has taken slags of the composition employed industrially, equilibrated them with steel melts, and measured the equilibrium phosphorus contents in the steel.[62] The reference slag was CaO–FeO–MgO + 0.4% P held in an MgO crucible. The properties of more complex slags containing additions of Al_2O_3, TiO_2, and VO_4 were determined relative to the corresponding properties of the reference system.

One may wonder about the number of variables involved in the description of the reference system. To get an idea of this, we invoke the Gibbs phase rule again,

$$F = C + 2 - P,$$

where C is the minimum number of components required to create the system and P is the number of phases. The laboratory system consists of four phases: slag, metal, atmosphere, and MgO crucible so that $P = 4$. The minimum number of components is $C = 7$: Al, Ca, Fe, Mg, Si, P, and O. This gives five degrees of freedom: $F = 7 + 2 - 4 = 5$. At fixed temperature and pressure (1 bar) and fixed phosphorus content in the slag there only remains two degrees of freedom. For these independent variables CaO and FeO* were chosen. Then on a diagram with CaO and FeO* on the ordinates one can describe properties of the system, such as the phosphorus content of the melt. Figure 2.15 gives lines for constant phosphorus distribution:

$$K_p = \frac{(\%P)}{[\%P]}.$$

The results are in agreement with the predictions from Eq. (2.59); the value of K_p increases with increasing FeO* in the slag (oxidizing conditions) and with increasing CaO (basic slag).

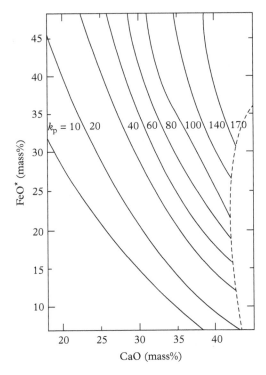

Fig. 2.15 *Lines for constant partition coefficient (%P)/[%P] between slag and steel as a function of lime content and FeO* content in MgO-saturated slag (see text) at 1600°C.*[62]

2.1.17 Activities of Slag Components

It was seen in the previous section that it is difficult to determine the phosphorus capacity of steelmaking slags. To give a more balanced picture it is worth pointing out that there exists a wealth of experimental data concerning activities in slag systems.[54,62,63] This is especially the case for 'blast furnace' type slags containing SiO_2, Al_2O_3, and CaO. Figure 2.16 gives SiO_2 activities for this system while Fig. 2.17 provides the CaO and Al_2O_3 activities.

2.2 Physical and Transport Properties of Molten Metals and Gases

In describing and modelling the refining of metals, the interaction between thermodynamics and physics and transport properties is very important. In all areas it is important to proceed beyond the use of qualitative and approximate data for properties and parameters. The field of refining of metals is wide and expanding. It is not easy for the worker or student to guess which metal or alloys will come into the 'limelight'. Therefore, from an understanding and teaching point of view it is necessary to focus on what the various metals and processes

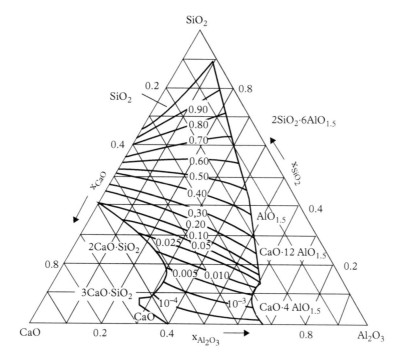

Fig. 2.16 *SiO$_2$ activities in a CaO–Al$_2$O$_3$–SiO$_2$ slag at 1550°C.*[54] *Concentrations are given in mole fractions. Activity is referred to the pure solid oxide.*

have in common. Ideally, a balance must be struck between general knowledge and studies in depth (research, PhD). General knowledge (physics and transport phenomena) should keep up with research (experimental, thermodynamics, computer modelling). A goal with this section is that knowledge of transport phenomena does not become (or remain) a bottleneck in the development of new refining processes or materials. Perhaps this too brief outline of transport phenomena of molten metals can motivate workers to enter this field.

To describe the various processes that take place in refining reactors it is necessary to possess physical data concerning:

(1) viscosity,
(2) diffusion coefficients,
(3) surface tension (interfacial tension),
(4) thermal conductivity, and
(5) electrical conductivity.

The order given above reflects the importance of the various physical quantities. Viscosity is important in the study of flow and mixing, while the diffusion coefficients enter into the

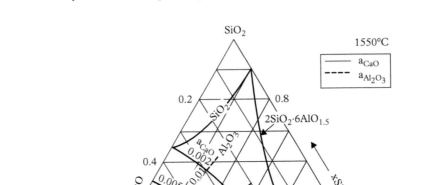

Fig. 2.17 *CaO and Al₂O₃ activities in a CaO–Al₂O₃–SiO₂ slag at 1550°C.[54] Concentrations are given in mole fractions. Activity is referred to the pure solid oxide.*

relations for the mass transfer coefficients. Surface tension is very important in the study of wetting and contacting problems. Interfacial tension and thermal conductivity must be known in order to describe nucleation and solidification phenomena. Other key subjects in connection with the use of such data are grain refining, recrystallization, grain growth, and fracture of metals and alloys. In the processing field one may mention gas purging, removal of inclusions, soldering and welding, and composites produced, for instance, by impregnation with a liquid metal binder of a porous high-melting framework.

The thermal conductivity of liquid metals is roughly proportional to the electrical conductivity.[64] Reliable data on electrical conductivity of various liquid metals are often available.

Data concerning the properties of gases are of interest in the study of various melt/gas operations such as gas purging and vacuum treatment. There are a number of analogies between the behaviour of molten metals and gases. Gases are discussed here also to throw light on some analogous problems concerning diffusion in melts.

Some references to salts and slags are given to assist the reader to find specific data.

2.2.1   Viscosity of Gases

Consider a fluid (liquid or gas) flowing past a stationary wall. At a certain distance from the wall, the fluid has a velocity v_x where x represents the direction of flow. Closer to the wall, the velocity will be less. That is, there is a gradient in velocity, dv_x/dz, where z represents the direction perpendicular to the flow. This gradient results in a shear force within the liquid

$$\tau_{zx} = -\mu \frac{dv_x}{dz}$$

This is Newton's law for viscous flow in its simplest form, and the proportionality constant is called the viscosity of the fluid. The equation may alternatively be interpreted in terms of momentum transfer. A fluid moving in the x-direction imparts some of its momentum to an adjacent 'layer' of liquid, causing it also to move in the x-direction. Hence, x-momentum is transmitted through the fluid in the z-direction. Consequently, τ_{zx} may be interpreted as the viscous flux of x-momentum in the z-direction. This interpretation ties in better with the molecular (atomic) nature of the momentum transport process. Momentum flux is in the direction of the negative velocity gradient. A velocity gradient can thus be thought of as a 'driving force' for momentum transport.

In a gas at low density, momentum is transferred by collisions between the molecules or atoms in the gas. Since each molecule has an average kinetic energy of $\frac{1}{2}mv^2 = \frac{3kT}{2}$, the velocity is proportional to $(kT/m)^{1/2}$. Now the number of collisions is proportional to the number of molecules per unit volume, N, and to the velocity of the molecules. The distance travelled between two collisions, λ, is inversely proportional to N and inversely proportional to the cross-sectional area for collision of two molecules, πd_c^2 (see Fig. 2.18). It is assumed that all molecules have velocities representative of the region in which they last collided. The velocity difference between two consecutive collisions will then be proportional to $\lambda(dv_x/dz)$ and the momentum flux τ_{zx} will be proportional to

$$-Nm\left(\frac{kT}{m}\right)^{\frac{1}{2}} \lambda \frac{dv_x}{dz} \propto -\frac{(kTm)^{\frac{1}{2}}}{d_c^2} \frac{dv_x}{dz}.$$

According to kinetic gas theory[65] the proportionality constant is $5/\left(16\pi^{\frac{1}{2}}\right)$ so that

$$\tau_{zx} = \frac{5(mkT)^{\frac{1}{2}}}{16\pi^{\frac{1}{2}}d_c^2} \frac{dv_x}{dz}. \tag{2.60}$$

This gives a viscosity of

$$\mu = \frac{5(mkT)^{\frac{1}{2}}}{16\pi^{\frac{1}{2}}d_c^2}, \tag{2.61}$$

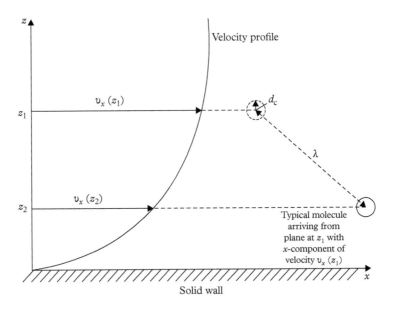

Fig. 2.18 *Molecular transport of x-momentum from z_1 to z_2.*

or, since Boltzmann's constant $k = R/N_A$ and the atomic mass $m = M/N_A$, where N_A is Avogadro's number, $N_A = 6.022 \times 10^{23}$ atoms/mol, and M is the mass of one mole,

$$\mu = \frac{5(MRT)^{\frac{1}{2}}}{16\pi^{\frac{1}{2}} N_A d_c^2}, \tag{2.62}$$

which represents the viscosity of a gas composed of hard spheres at low density. Note that measurements are needed to determine the collision diameter d_c. Equation (2.62) indicates that μ is proportional to $T^{\frac{1}{2}}$. However, data for various gases show that μ varies slightly more rapidly than $T^{\frac{1}{2}}$. To predict the temperature dependence of μ accurately, one has to replace the rigid sphere model with a more realistic molecular force field. Such a rigorous kinetic theory of gases was developed by Chapman and Enskog.[65] This theory gives expressions for the transport coefficients in terms of the interaction between a pair of molecules in the gas. This potential energy ϕ is related to the force of interaction F by the relation $F = -d\phi/dr$ in which r is the distance between the molecules. If one knew exactly how the forces depend on the distance between molecules, then one could substitute these relations into the Chapman–Enskog formulas and calculate the transport coefficients.

The Lennard–Jones potential

$$\phi(r) = 4\varepsilon \left\{ \left(\frac{\delta}{r}\right)^{12} - \left(\frac{\delta}{r}\right)^6 \right\} \tag{2.63}$$

Table 2.15 *Intermolecular force parameters.*[66]

Elements	δ $(10^{-10}$ m$)$	$\frac{\varepsilon N_A}{R}K$
H_2	2.915	38.0
Ar	3.418	124
N_2	3.681	91.5
Air	3.617	97.0
O_2	3.433	113
CO	3.590	110
CO_2	3.996	190
SO_2	4.290	252
Cl_2	4.115	357

has proved to be very useful in giving such a description of mono- and polyatomic gases. Note that it displays the characteristic features of molecular interactions: strong repulsion at small separations (roughly proportional to r^{-12}) and weak attraction at large separations (proportional to r^{-6}). δ is a characteristic diameter of the molecule called the collision diameter. Values of δ and ε are known for many substances; a partial list is given in Table 2.15. When values of δ and ε are not known for monoatomic gases, they may be estimated from the empirical relations:

$$\frac{\varepsilon}{k} \approx 1.92 T_m \ and \ \delta \approx 1.2 \times 10^{-8} V_m^{\frac{1}{3}}, \tag{2.64}$$

where V_m is the molar volume of the liquid at the melting point. The viscosity becomes

$$\mu = \frac{0.844 \times 10^{-24}}{\Omega_\mu \delta^2}(MT)^{\frac{1}{2}}, \tag{2.65}$$

where Ω_μ is a slowly varying function of temperature. $RT/(\varepsilon N_A)$ is given in Table 2.16. If the gas were made up of rigid spheres of constant diameter δ, then Ω_μ would be unity. Ω_μ may therefore be interpreted as giving the deviation from rigid sphere behaviour. Note that δ replaces d_c used in Eq. (2.62). If $\Omega_\mu = 1$, then $d_c = \delta$ and Eq. (2.65) becomes identical to Eq. (2.62).

Equation (2.64) for δ inserted in Eq. (2.65) gives

$$\mu = \frac{5.9 \times 10^{-9}(MT)^{\frac{1}{2}}}{V_m^{\frac{2}{3}}\Omega_\mu}. \tag{2.66}$$

To illustrate the use of Eq. (2.65), the viscosity of CO_2 is calculated at 300 and 1000 K. From Table 2.15 $\varepsilon N_A/R = 190$, $\delta = 3.996 \times 10^{-10} m$. The molecular weight of CO_2 is 44.01 g/mol = 44.01×10^{-3} kg/mol. At 300 and 1000 K, $RT/\varepsilon N_A = T/190 = 1.58$ and 5.26, respectively. This gives from Table 2.16 $\Omega_\mu = 1.286$ and 0.9196, respectively. Then at 300 K

$$\mu = \frac{0.844 \times 10^{-24}(44.01 \times 10^{-3} \times 300)^{\frac{1}{2}}}{3.996^2 \times 10^{-20} \times 1.286} = 1.494 \cdot 10^{-5} Ns/m^2$$

or

$$\mu = 1.494 \times 10^{-2} mPas.$$

This may be compared with the experimental value of 1.495×10^{-5} Ns/m². At 1000K the viscosity becomes

$$\mu = \frac{0.844 \times 10^{-24} \times 44.01^{\frac{1}{2}}}{3.996^2 \times 10^{-20} \times 0.9196} = 3.814 \times 10^{-5} Ns/m^2$$

The change of Ω_μ from 1.2866 to 0.9196 takes into account the fact that the viscosity increases slightly more strongly than $T^{\frac{1}{2}}$.

The following correlation has been found to represent, with a reasonable degree of accuracy, the viscosity of multicomponent gas systems, with the exception of hydrogen-rich mixtures:[67]

$$\mu = \frac{\sum_{i=1}^{k} x_i M_i^{\frac{1}{2}} \mu_i}{\sum_{i=1}^{k} x_i M_i^{\frac{1}{2}}}, \tag{2.67}$$

where x_i and M_i are the mole fraction and the molecular mass fraction of i, respectively.

2.2.2   Introduction, the Pair Distribution Function

In order to describe the transport behaviour of liquids (liquid metals) it is necessary to study the structure of liquids. However, a liquid lacks a regular arrangement of its molecules. That is, it lacks the regular crystalline order of a solid. Probability plays a central role in describing the molecular structure of molten metals. The problems in a general approach have proved to be very great. Therefore, it has been very useful and necessary to concentrate efforts near the melting points. Near the melting points the liquid—to a considerable degree—still 'remembers' its history as a solid. The pair distribution function $g(r)$ is of central importance in the modern theory of liquids. Consider a liquid in equilibrium and any atom with its centre at $r = 0$. The number of atoms, dN, in the spherical shell between radial distances r and $r + dr$ from the origin atom is

Table 2.16 *Values of Ω_μ for viscosity based on the Lennard–Jones potential*[66]

$\frac{RT}{\varepsilon N_A}$	Ω_μ
0.3	2.785
0.4	2.492
0.5	2.257
0.6	2.065
0.7	1.908
0.8	1.780
0.9	1.675
1.0	1.587
1.2	1.452
1.4	1.353
1.6	1.279
1.8	1.221
2.0	1.175
4.0	0.970
6.0	0.8963
8.0	0.8538
10	0.8242
20	0.7432
40	0.6718
60	0.6335
80	0.6076
100	0.5882
200	0.5320
400	0.4811

$$dN = 4\pi r^2 dr \frac{N}{V} g(r), \qquad (2.68)$$

where N is the total number of atoms in the volume V. The pair distribution function $g(r)$ is proportional to the probability that an atom is found at distance r. If there is no correlation between atoms, as in the case of an ideal gas consisting of point atoms, $g(r) = 1$. Then the probability of finding another atom is the same for all distances r.

For real liquids the pair distribution function becomes complicated due to the forces of attraction and repulsion. When r is less than the collision diameter, $g(r)$ is essentially zero, since two molecules cannot overlap. At slightly larger separations there is a more than average chance of finding a molecule; then $g(r)$ is greater than 1. This enhancement arises mainly from the geometric necessity of there being many pairs at such separations since the liquid density is only slightly less than that of a solid. Minor maxima and minima at larger separations are the remnants in the liquid of the much stronger extrema found in the solid (for which the pair distribution function $g(r)$ along a crystal axis consists of a series of narrow peaks separated by regions in which it is zero). Thus a liquid lacks the long-range order of a solid but shows appreciable order at short separations. Figure 2.19 gives the pair distribution function of iron near the melting point as an example.[68] All pair distribution functions for liquid metals have the same general shape.

A knowledge of the pair distribution function is the minimum information needed for the discussion of the properties of a liquid since molecules interact predominantly in pairs through the potential energy $\phi(r)$ and since $g(r)$ describes the statistics of such pairs. Therefore, knowledge of $\phi(r)$ and $g(r)$ is necessary in order to link the microscopic energies and forces and the macroscopic properties. There are two methods of measuring the pair distribution function: by X-ray or neutron diffraction[69] and by computer simulation of the molecular structure and motions in a liquid. In the first, the liquid is exposed to a specific monochromatic radiation. The intensity of the scattered radiation is measured for various angles. Then a Fourier transform treatment gives $g(r)$.[70] The second method of obtaining the pair distribution function $g(r)$ supposes that $\phi(r)$ is known. A computer model of the liquid is set up in which

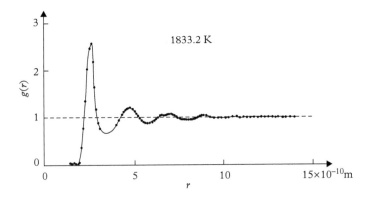

Fig. 2.19 *The pair distribution function, g(r), of iron near the melting point.*[68]

between 100 and 4000 molecules are contained within a cube. There are now two methods of proceeding: by Monte Carlo calculation[71] or by molecular dynamics. In the latter each molecule is initially assigned a random velocity and position, and Newton's equations of motion are solved to calculate the path of each molecule in the changing field of all the others. After a time the mean positions are those appropriate to the density and to the mean kinetic energy of the liquid under equilibrium conditions. Functions such as the pair distribution function can now be evaluated by taking suitable averages as the system evolves in time.

Note that $g(r)$ in Fig. 2.19 has a very strong peak. This peak is found at a distance, r, which gives room for two atoms, $r = 2$ atom diameters.

As already mentioned, the potential energy $\phi(r)$ and the pair distribution function $g(r)$ can be employed to derive various thermodynamic and macroscopic physical properties. The use of $g(r)$ and $\phi(r)$ to describe properties of a liquid is illustrated by deriving relations for the internal energy U (in J/mol) and for the enthalpy of vaporization. For a monoatomic liquid the internal energy per mole is given by

$$U = \frac{3}{2}RT + \phi, \tag{2.69}$$

where $\frac{3}{2}RT$ is the kinetic energy and $<\phi>$ is the mean potential energy per mole obtained by integrating from $r = 0$ to $r = \infty$:

$$\phi = \frac{N_A}{2}\int_0^\infty \phi(r)dN_A = \frac{N_A^2}{2}\int_0^\infty \phi(r)g(r)4\pi r^2 dr/V_A, \tag{2.70}$$

where V_A is the molar volume, and N_A is the number of molecules per mole (Avogadro's number). The factor ½ prevents us from counting each molecule twice in the summation of the potential energies.

The molar enthalpy of vaporization ΔH_v is obtained from Eq. (2.69) and

$$\Delta H_v = U_g - U + RT, \tag{2.71}$$

where RT is the work performed against the external pressure during vaporization. Since $U_g = \frac{3}{2}RT$ (for an ideal monoatomic gas) we get from Eqs. (2.69) and (2.70)

$$\Delta H_v = RT - \phi = RT - \frac{N_A^2}{V_A}2\pi\int_0^\infty \phi(r)g(r)r^2 dr. \tag{2.72}$$

2.2.3   The Viscosity of Liquids

Born and Green using kinetic theory have derived an equation for the viscosity in terms of the pair distribution function $g(r)$ and the pair interatomic potential $\phi(r)$,[72]

$$\mu = \frac{2\pi}{15}\left(\frac{M}{RT}\right)^{\frac{1}{2}}\left(\frac{N_A}{V_A}\right)^2 \int_0^\infty g(r)\frac{d\phi}{dr}r^4 dr, \tag{2.73}$$

where V_A is the volume per mole. It is not a simple matter to derive Eq. (2.73). However, it is seen that, reasonably enough, the interatomic force $d\phi/dr$ is multiplied by a probability density $g(r)\cdot 4\pi r^2 dr/V$ and by an area per volume of an atom $r^2 N_A/V_A$.

There has been some discussion as to the validity of this equation. It gives results that coincide roughly with experimental data for metals, but it is difficult to determine whether the discrepancies are due to errors in the experimental measurements or errors in $g(r)$ or $\phi(r)$ employed in the calculations. The theory assumes that the atoms interact only in pairs so that the use of $g(r)$ and $\phi(r)$ suffice in order to describe the system. Furthermore, Eq. (2.73) uses a hard-sphere model. It does not take into account the soft, long-range part of the potential. A number of references concerning molten metals in general are given.[73–78]

Chapman[79] and others, after starting with Eq. (2.73), have attempted to calculate viscosity of molten metals by introducing universal functions such as

$$\phi(r) = \varepsilon\phi^*(r/\delta, T^*), g(r) = g^*(r/\delta, T^*), T^* = \frac{kT}{\varepsilon},$$

This has only been partially successful. For most metals a general description of the viscosity of pure metals is possible. However, this is accomplished at the expense of not getting accurate values for a given metal. For accurate values one may have to resort to experiments and computer-aided analysis and prototyping.

A problem with solutions such as that proposed by Born and Green, Green-Kubo,[80,81] or the Ashcroft model[82,83] is that the calculations depend on two complicated and approximate relations for force or potential and distribution of atoms, $g(r)$. The solutions only apply to a limited selection of elements and range of temperatures. The goal of deriving a general theory for the viscosity of metals seems to be difficult to reach. Perhaps it would be fruitful to base the solution only on forces or only on the density.[84]

The thermodynamic route[85] must be mentioned. It relates the diffusion coefficient, D, to the excess entropy S_E of a liquid. S_E is equal to the total entropy minus the ideal mixing contribution.

For the general description of the viscosity of pure metals it has been possible to replace the universal function approach with only dimensional considerations. According to Andrade,[86] the atoms in the liquid state at the melting point vibrate about equilibrium positions, just as if they were still in the solid state (i.e. Einstein oscillators). Viscosity is produced by transfer of momentum of atomic vibrations from one layer to a neighbouring one.

Andrade[86] proposed the following equation for the melting point viscosity of simple (or monoatomic) liquids,

$$\mu_m = \frac{4}{3}\frac{v\cdot m}{a}\Big|_{T=T_m}, \tag{2.74}$$

where v is a characteristic frequency of vibration and a is an average distance between the atoms. The proportionality factor is a rough estimate. To obtain v Andrade used a relation proposed by Lindemann. Lindemann[87] considered the melting process from the standpoint of the solid. He suggested that a solid shakes itself to pieces and thus changes itself into a liquid once the amplitude of vibration reaches a certain fraction of the average interatomic distance a.

A recent estimate[88] of this distance is

$$\delta = 0.08a.$$

It is assumed that the solid consists of atoms of mass m, vibrating harmonically and isotropically, independent of each other, with the same frequency about the equilibrium positions (i.e. Einstein model). This results in a frequency $v^{88,8}$, given by

$$v = C_L \left(\frac{T_m}{MV_m^{2/3}} \right)^{\frac{1}{2}}, \tag{2.75}$$

where $C_L = 12.5 N_A^{1/3} R^{\frac{1}{2}} / \left(\sqrt{2\pi} \right)$.

Finally, inserting this in Eq. (2.74), and setting $\left(\frac{V_m}{N_A} \right)^{\frac{1}{3}} = a$, we obtain

$$\mu_m = \frac{1.34 \times 10^{-7} (MT_m)^{1/2}}{V_m^{2/3}}.$$

The factor 4/3 in Eq. (2.74) and 1.34 in the above equation must not be taken too seriously. An average value has been found by dividing experimental values by $\frac{(MT_m)^2}{V_m^{2/3}}$ for the various liquid metals where data is available. This gives[74]

$$\mu_m = 1.8 \times 10^{-7} \frac{(MT_m)^{\frac{1}{2}}}{V_m^{2/3}}, \tag{2.76}$$

The similarity with Eq. (2.66) for gases should be noted. This equation can be applied as a very rough approximation to most elements.[89]

Equation (2.76) is compared with experimental results in Fig. 2.20. It is seen that the correspondence is acceptable except for Al, Bi, Ge, Ti, Zn, Sb, and Zr. Of these, Ge and Sb are metalloids.

The model sketched above probably only describes the final stage before formation of a liquid. The melting of solids is initiated by formation and migration of point defects or dislocations.[76,77] For many metals it is found that the vacancy concentration at the free surface is as high as 10 per cent at the equilibrium melting point. Then the surface is not able to support the shearing stress. It is tempting to cite Max Born:[78] 'The difference between a solid and a liquid is that the solid has resistance against shearing stress while the liquid has not.'

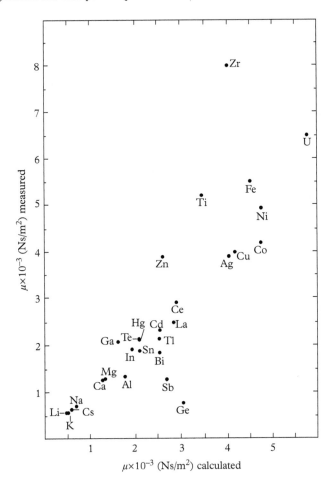

Fig. 2.20 *Comparison of viscosity calculated from Eq. (2.76) with experimental values*[91] *at the melting points.*

To illustrate the use of Eq. (2.76), the viscosity of iron at the melting point is calculated. $M = 55.85 \times 10^{-3}$ kg/mol, $T_m = 1808$ K, and the density of iron at the melting point is 7030 kg/m^3. Then

$$V_m = \frac{55.85 \times 10^{-3}}{7030} = 7.94 \times 10^{-6} m^3/mol.$$

This gives

$$\mu_m = \frac{1.8 \times 10^{-7}\left(55.85 x 10^{-3} \times 1808\right)^{\frac{1}{2}}}{\left(10^{-6} \times 7.94\right)^{\frac{2}{3}}} = 4.54 \times 10^{-3} Ns/m^2.$$

Similarly for Ga

$$\mu_m = \frac{1.8 \times 10^{-7}(69.7 \times 10^{-3} \times 303)^{\frac{1}{2}}}{(10^{-6} \times 11.4)^{\frac{2}{3}}} = 1.63 \times 10^{-3} Ns/m^2.$$

These values probably lie within the errors in the experimental values.[68] For dilute systems where the solvent dominates one may as a rough approximation take viscosity values valid for the pure solvent.

It is not possible in this book to dwell on the subject of viscosity of salt and slag systems. For salts the viscosity is surprisingly low;[68,92] not much higher than in water or molten metals. The viscosity of slags[54,93,94] increases strongly with the content of network formers such as SiO_2 and is in the range 1–100 Pas. As an illustration, isoviscosity lines for the system CaO–Al_2O_3–SiO_2 at 1500 °C are given in Fig. 2.21.

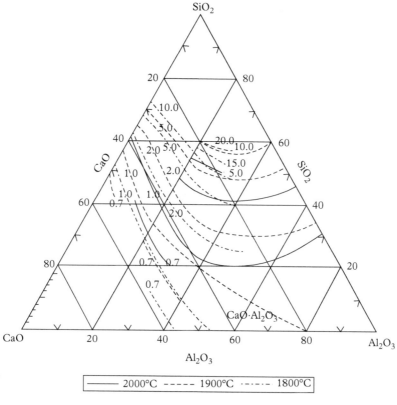

Fig. 2.21 *Isoviscosity lines for the system CaO–Al₂O₃–SiO₂ given in poise (1 poise = 0.1 Ns/m²)*[54]

2.2.3.1   Estimation of Viscosities at Higher Temperatures

To predict the viscosity of molten metal for various temperatures, George Kaptay has proposed[95]

$$\mu = \frac{A \cdot (MT)^{1/2}}{V^{2/3}} \exp\left(B\frac{T_m}{T}\right), \tag{2.77}$$

where

$$A = (1.80 \pm 0.39)\, 10^{-8} \left(J/Kmol^{1/3}\right)^{\frac{1}{2}}$$

$$B = 2.34 \pm 0.20$$

Note that at the melting point

$$\mu = \frac{\mu_m}{10} \exp(B) \approx \frac{\mu_m}{10} \cdot 10 = \mu_m,$$

where μ_m is given by Eq. (2.76).

Corresponding values of log inverse absolute temperatures $1/T$ and the logarithm of viscosity for any pure metal, i, with melting point $T_m(i)$, may be plotted in a x–y diagram. The points will tend to lie on a straight line.[74] Then we get for pure metal, I,

$$\mu_i = A_i \exp\left(\frac{\beta_i}{T}\right),$$

or introducing the 'activation energy' for viscous flow, H_i:

$$\beta_i = \frac{H_i}{R}$$

$$\mu_i = A_i \exp\left(\frac{H_i}{RT}\right). \tag{2.78}$$

Corresponding values of log $T_m(i)$ and log (H_i) are plotted in Fig. 2.22. Also here the points tend to lie close to a straight line. Regression analysis[74] gives

$$\frac{H_i}{T_m(i)} = 2.32 T_m^{0.29}(i)$$

$$H = 2.32 T_m^{1.29} \tag{2.79}$$

$$\frac{\mu}{\mu_m} = \exp\left(\frac{H_i}{R}\left(\frac{1}{T} - \frac{1}{T_m}\right)\right) \tag{2.80}$$

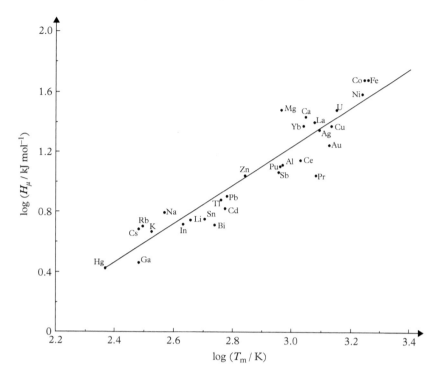

Fig. 2.22 *Plot of* log H_μ *vs* log T_m *for 30 liquid metallic elements.*[74] [*Reproduced with permission of Oxford Publishing Limited through PLSclear.*]

Example: Application of Eq. (2.80)
   At 700 °C, the viscosity of Al is 1.29 mPas. What is the viscosity at 850 °C?
   The melting point is 660 °C.
Solution:
   We get

$$\frac{\mu\,(850\ °C)}{\mu\,(700\ °C)} = \exp\left(\frac{H_i}{R}\left(\frac{1}{1123} - \frac{1}{973}\right)\right)$$

$$H_i = 2.32 \times 933^{1.29} = 15727\ J/mol$$

$$R = 8.314\ J/mol$$

$$\frac{\mu\,(850\ °C)}{\mu\,(700\ °C)} = \exp(-0.2597) = 0.771$$

$$\mu(850) = 1.29 \times 0.771 = 0.995\ \text{m Pas}$$

The measured value was 1.04 m Pas.[95]

2.2.3.2   Estimation of Viscosities of Liquid Alloys

Hirai has also studied the viscosity of binary alloys.[96] He introduces

$$V_m = \frac{M}{\rho_m}.$$

In Eq. (2.76). This gives

$$\mu_m = 1.7 \times 10^{-7} \rho^{2/3} M^{-\frac{1}{6}} T_m^{1/2}.$$

Note that 1.7 replaces 1.8 in Eq. (2.76). The value of 1.7 gives a better fit to the experiments. T_m was taken from the liquidus line at the alloy composition. The ρ_m and M of the alloys were calculated by a simple weighted average of the two metals. To simplify the calculations, ρ_m was treated as the density, ρ, at room temperature. It is found that a given binary alloy has behaved as if it was a pure metal with density ρ and melting point T_m. Seven alloys at three compositions each have been studied.

For the activation energy one gets

$$H = 2.65 T_m^{1.27} \tag{2.81}$$

This is close to the value $H = 2.32 T_m^{1.29}$ obtained for pure metals.

2.2.4   Surface Tension of Pure Molten Metals

The surface tension, σ, may be defined in terms of the work $\sigma\, dA$ required to create a surface area dA. Alternatively, σ may be thought of as the force per unit length needed to stretch the surface.

From statistical mechanics Fowler has derived the following equation for the surface tension:[97]

$$\sigma = \frac{\pi}{8} \left(\frac{N_A}{V}\right)^2 \int_0^\infty g(r) \frac{\partial \phi}{\partial r} r^4 dr. \tag{2.82}$$

In the development of the equation it was assumed that density changes discontinuously at a mathematical surface that separates the liquid and gas phases. This approximation has also been successfully employed in the study of the surface tension of alloys. This point will be discussed later.

The similarity of Eq. (2.82) to the equation for viscosity is striking; the integrals in the two expressions are identical. This may be explained in part by pointing out that in viscous flow work is required to overcome the interatomic forces and move the atoms relative to each other. Similarly, it will be seen that work is needed to displace atoms from the bulk melt to the surface. Surface tension, however, is a purely thermodynamic quantity and should not depend on atomic mass m or on M. In fact, Eq. (2.82) does not contain M. Comparison of Eqs. (2.73) and (2.82) gives

$$\frac{\sigma}{\mu} = \frac{15}{16}\left(\frac{RT}{M}\right)^{\frac{1}{2}}, \tag{2.83}$$

At the melting point it is found from Eqs. (2.75) and (2.83) that

$$\sigma_m = 5.9 \times 10^{-8}\frac{RT_m}{V^{\frac{2}{3}}}, \tag{2.84}$$

Comparison with experiments (see Fig. 2.23) and with Eq. (2.84) indicates that agreement is better with a coefficient of 5 instead of 5.9. The discrepancy may be due partly to the discontinuity assumption in the derivation of Eqs. (2.73) and (2.82). However, the effect of impurities on the surface tension of metals should also be emphasized. Surface active elements all reduce the surface tension.

Equations (2.83) and (2.84) may be compared by using the experimental result that the molar enthalpies of evaporation of metallic elements at their boiling points are approximately proportional to their melting points:[74]

$$\Delta H_V \approx 29RT_m \tag{2.85}$$

Thus

$$\sigma_m \approx 5 \times 10^{-8}\frac{RT_m}{V_m^{2/3}} = \frac{5}{29} \times 10^{-8}\frac{\Delta H_V}{V_m^{2/3}} = 1.7 \times 10^{-9}\frac{\Delta H_V}{V_m^{2/3}}.$$

As an example, the surface tension of Fe is calculated at the melting-point. From Eq. (2.84) with coefficient 5:

$$\sigma_m = \frac{5 \times 10^{-8} \cdot 8.3144 \cdot (1536 + 273)}{\left(7.94 \times 10^{-6}\right)^{\frac{2}{3}}} = 1.89 N/m,$$

This may be compared with the published results of $\sigma_m = 1.88$ N/m.[98]

For various salts it is found[68] that σ_m is proportional to $T_m/V_m^{\frac{2}{3}}$. Thus, for the molten alkali halides

$$\sigma_m = \frac{1.3 \times 10^{-7}T_m}{V_m^{\frac{2}{3}}}.$$

Similarly for other molten halides

$$\sigma_m = \frac{2 \times 10^{-7}T_m}{V_m^{\frac{2}{3}}},$$

while for molten salts of sulfates, carbonates, nitrates, and so on, the proportionality factor in front of $T_m/V_m^{\frac{2}{3}}$ is 2.7×10^{-7}.

For non-ideal binary systems the surface tension may vary in a complex manner.[90] The problem of the interfacial tension between the solid and melt of a pure metal is treated. Furthermore, it is indicated how one can estimate the surface tension of an ionic solid.

2.2.5   Thermodynamics of Interfaces

The plane of separation of two phases is known as a surface or interface. From a mechanical point of view it may be regarded as having infinitesimal thickness. With such a surface we associate an interfacial tension, σ. σ is equal to the Gibbs energy required to stretch a unit area of the interface at constant pressure p and constant temperature with no exchange of atoms or molecules with the surroundings:

$$\sigma = \left(\frac{\partial G}{\partial A}\right)_{T,p,n_i}.$$

Strictly speaking a system should be regarded as composed of two bulk phases and a region of space between them whose properties are different from those of the bulk phases. At the extremities the properties in the interphase layer should merge with the properties in the bulk phases. The interphase is composed of a multilayer of atoms. However, such a treatment is beyond the scope of this book. Here it is only assumed that the interphasial layer is only one atom or molecule thick, or that there is a discontinuity of properties at the interface. This is the approximation made by Fowler in the derivation of Eq. (2.82). The Gibbs energy, G^σ, for the interphasial layer is given by

$$G^\sigma = G - G_1 - G_2,$$

where G is the Gibbs energy for the total system, and G_1 and G_2 refer to the two bulk phases. If one of the phases (index 2) is a vacuum or an inert gas we speak of surface tension instead of interfacial tension. Then

$$G^\sigma = G - G_1.$$

Here G takes into account conditions on the surface while G_1 assumes constant properties of the bulk all the way to the surface.

One mole of atoms in a monolayer approximately covers a surface

$$A \approx N_A \left(\frac{V_m}{N_A}\right)^{\frac{2}{3}} = N_A^{\frac{1}{3}} V_m^{\frac{2}{3}}, \tag{2.86}$$

where V_m is the volume of one mole of the substrate and $(V_m/N_A)^{\frac{2}{3}}$ is roughly the area covered by an atom. For a closest packed configuration[2]

$$A = 1.09 N_A^{\frac{1}{3}} V_m^{\frac{2}{3}}. \tag{2.87}$$

The surface tension for a one-component system is

$$\sigma = \frac{G^\sigma}{A}. \tag{2.88}$$

Now

$$G^\sigma = H^\sigma - TS^\sigma. \tag{2.89}$$

H^σ is the difference between the molar enthalpy on the surface and in the bulk. To estimate H^σ for a metal, one compares the number and type of bonds in the bulk phase and on the interface. The surface tension is then obtained by counting the number of broken bonds, m, when an atom or molecule is brought to the surface.

This procedure ignores the matter of surface distortion. The surface atoms, being subject only to a one-sided attraction, lie closer to their underlying neighbours than they do in the bulk. Consequently, surface atoms should be bound to the atoms immediately beneath the surface by greater binding energies. The energies of the bonds at the surface are approximately 13 per cent greater.[2]

Thus, if Z is the number of neighbours to the atom in the bulk,

$$\sigma = \frac{\Delta H_V 1.09 \times 10^{-8}}{V_m^{\frac{2}{3}}} \left(\frac{Z - (Z-m) \, 1.13}{Z} \right), \tag{2.90}$$

ΔH_V is the molar enthalpy of evaporation if $m/Z = 0.25$.[2] Then

$$\sigma = \frac{\Delta H_V 1.7 \times 10^{-9}}{V_m^{\frac{2}{3}}}, \tag{2.91}$$

Figure 2.23 compares Eq. (2.91) using ΔH_V from Kubaschewski and Alcock[99] with experimental values of σ at the melting-point.[75,91]

Lu and Jiang[100] take the cohesion energy into account. They determine m from the distance between the surface atoms and the corresponding nearest neighbour atoms. This does not change the result given in Eq. (2.91).

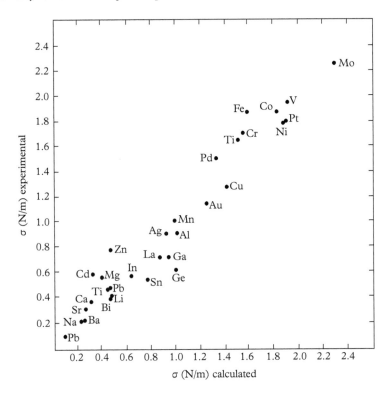

Fig. 2.23 *Comparison of calculated surface tension with experimental values of pure metals at the melting point.*

2.2.6   Surface Energy of Compounds

In order to study wetting it is important to know the surface energies. The surface energy of compounds may be obtained by a procedure similar to that applied to metals in the previous section. The surface energy H^σ is the difference between the energy in the surface layer of atoms or molecules and in the bulk. The bonding energy between atoms in compounds, based on the nature of the interactions of the electrons in the outer shells, or their electronic structure, can be divided into the general categories of (a) ionic, (b) covalent, and (c) metallic. In addition there are weak forces of attraction between atoms which are referred to as van der Waals forces. Actual bonds are mixtures of several types. Usually one of the types is dominant.

Ionic bonds are easier to understand than other types of bonding. Typically, they form between strongly electropositive elements (most metals) and electronegative ones (non-metals, F, O, S). In sodium one electron is held in the outer shell of the atom and is held by an effective potential of a single charge. It loses this outer electron comparatively easily to become positively charged Na^+. To remove an electron from each atom requires 5×10^5 J/mol.[101] For fluorine the situation is quite different. Each of the seven outer electrons is attracted to a kernel with a charge + 7 so that the ionization energy of F is very high,

17×10^5 J/mol. On the other hand, there is room for an eighth electron in the outer shell of F. Adding an electron yields 3.5×10^5 J/mol. To convert a mole of Na atoms and a mole of F atoms to Na^+ and F^- requires an energy $(5 - 3.5) \times 10^5$ J/mol. The resulting ions are capable of forming ionic bonds whose energy yield comes from the electrostatic forces of attraction. From this energy of attraction must be subtracted 1.5×10^5 J/mol necessary to form the ions.

The energy of an ionic bond is calculated from simple electrostatic forces. Consider an ion with charge Z_1e attracting an ion with a charge $-Z_2e$. The force of attraction is $Z_1Z_2e^2/r^2$, where r is the distance between the ions. The attractive energy as the ions approach each other from infinity is

$$U = \frac{e^2 Z_1 Z_2}{4\pi \varepsilon_0} \int_\infty^{r_0} \frac{dr}{r^2} = -\frac{e^2 Z_1 Z_2}{4\pi \varepsilon_0 r_0}.$$

ε_0 is the dielectric constant, 8.854×10^{-12} C^2/Nm^2. As the ions approach each other their electron clouds overlap, building up a very strong repulsive potential which may be taken to be proportional to exp $(-r/\rho)$ and is called the central field repulsive potential.[102] ρ is called the screening length. At the equilibrium distance, r_0, the forces of repulsion balance the forces of attraction. At this point the energy is at a minimum as seen in Fig. 2.24, which gives a schematic plot of energy, U, versus interionic distance r.

In calculating the energy of an ionic crystal, it is necessary to sum up all the attractive and repulsive forces between all the ions. Various elegant methods exist.[103] The original approach according to Madelung[104] considered, for a NaCl structure, an infinite number of charges with periodic repetition after a length $2r_0$ in all three normal directions. The potential at a charge in a neutral row of alternating charges e and $-e$ is

$$U = \frac{2e^2}{4\pi \varepsilon_0 r_0} \left(\frac{1}{1} - \frac{1}{2} + \frac{1}{3} - \frac{1}{4} + \right) = -\frac{2e^2}{4\pi \varepsilon_0 r_0} \ln 2.$$

For a crystal of NaCl we get

$$U = -\frac{e^2}{4\pi \varepsilon_0 r_0} \times 3.495$$

If all the N_A positive and negative ions are taken into account, the electrostatic energy per mole becomes

$$H = \frac{N_A U}{2}. \tag{2.92}$$

The electrostatic factor ½ arises from the fact that in the double summations each pair of ions is counted twice. Then

$$H_{\text{electrostatic}} = -N_A \frac{e^2 1.74756}{r_0 4\pi \varepsilon_0},$$

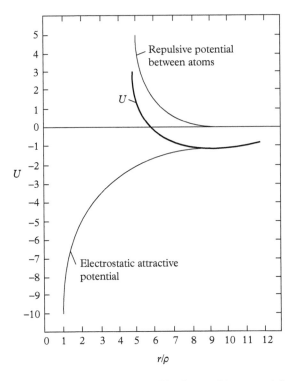

Fig. 2.24 *The energy, U, between atoms is determined by the repulsive potential and electrostatic attractive potential.*

where the factor $A_M = 1.74756$ is called the Madelung constant for NaCl. Values of the Madelung constant for various ionic structures are given in the literature.[103]

From the electrostatic energy must be subtracted the Coulomb energy of repulsion from the overlapping of electron clouds around the ions. This energy, which usually amounts to only about 10 per cent of the electrostatic energy, is obtained from experimental data. The crystal energy is

$$H = -\frac{A_M N_A e^2}{4\pi\varepsilon_0 r_0} + B\exp(-r_0/\rho). \tag{2.93}$$

The values of B (the repulsive coefficient) and ρ which take screening into account must be determined. ρ can be obtained from compressibility data. The values obtained for various crystals cluster around 0.33 Å $= 0.33 \times 10^{-10}$ m. For NaCl $\rho = 0.321$ Å.[102] B is proportional to the number of nearest neighbours, Z. One relation is fixed from the condition that r_0 is the equilibrium distance, so that

$$\frac{dH}{dr} = 0 \ \text{at} \ r = r_0. \tag{2.94}$$

This gives

$$\frac{A_M N_A e^2}{4\pi \varepsilon_0 r_0{}^2} - \frac{B}{\rho} \exp\left(-r_0/\rho\right) = 0. \tag{2.95}$$

Equations (2.93) and (2.95) give

$$H = -\frac{A_M N_A e^2}{4\pi \varepsilon_0 r_0} \left(1 - \frac{\rho}{r_0}\right). \tag{2.96}$$

This equation agrees with experimental data within 2 per cent for essentially ionic crystals. For instance, for NaCl the crystal energy is 0.75×10^6 J/mol. With $r_0 = 2.82 \times 10^{-10}$ m,[102] Eq. (2.96) gives

$$H = \frac{1.748 \times 6.022 \times 10^{23} \times 1.6022^2 \times 10^{-38}}{4\pi \times 8.854 \times 10^{-12} \times 2.82 \times 10^{-10}} \left(1 - \frac{0.321}{2.82}\right)$$
$$= 0.763 \times 10^6 J/mol.$$

In the calculation, 1.6022×10^{-19} is the electron charge in coulombs.

To calculate H^σ one may note that H^σ is the difference between the molar enthalpy on the surface and in the bulk. On the surface the electrostatic energy is less than in the bulk since there are no charges above the surface layer. Therefore, a Madelung constant A_M^S is introduced for the surface layer, where $A_M^S < A_M$. For instance for NaCl for the face-center $A_M^S = 1.68155$:

$$\frac{2\left(A_M - A_M^S\right)}{2A_M} = \frac{0.132}{3.495} = 0.0378.$$

On the other hand, the surface layer is drawn closer to the sublayer since the atoms in the surface layer are not attracted from above. Thus, the distance between the two layers, r_0^S, becomes less than r_0. This increases the electrostatic energy stored in the surface. To calculate r_0^S, Eq. (2.94) is applied to the two surface layers,

$$\frac{\left(A_M^S + A_M\right) N_A e^2}{4\pi \varepsilon_0 \left(r_0^S\right)^2} - \frac{B}{\rho}\left(\frac{Z-m}{Z} + 1\right) \exp\left(-\frac{r_0^S}{\rho}\right) = 0, \tag{2.97}$$

where Z is the number of nearest neighbours and m is the number of missing nearest neighbours for the top layer. For NaCl, $Z = 6$. If the surface lies along a crystal plane, $m = 1$.

Equations (2.95) and (2.97) give

$$\left(1+\frac{A_M^S}{A_M}\right)=\left(1+\frac{Z-m}{Z}\right)\exp\left(\frac{r_0-r_0^S}{\rho}\right)\left(\frac{r_0^S}{r_0}\right)^2.\tag{2.98}$$

For NaCl one obtains $r_0^S = 2.80$ Å from Eq. (2.98).

For the surface it is assumed that the screening term is proportional to the number of remaining bonds:

$$H^\sigma = H^S - H = -\frac{A_M^S N_A e^2}{4\pi\varepsilon_0 r_0^S}\left(1-\frac{(Z-m)\rho}{Zr_0^S}\right)+\frac{A_M N_A e^2}{4\pi\varepsilon_0 r_0}\left(1-\frac{\rho}{r_0}\right)$$

or

$$\frac{H^\sigma}{H}=1-\frac{A_M^S r_0}{A_M r_0^S}\left(\frac{1-\frac{(Z-m)\rho}{Zr_0^S}}{1-\frac{\rho}{r_0}}\right).\tag{2.99}$$

For NaCl

$$\frac{H^\sigma}{H}=1-\frac{3.363\times2.82}{3.495\times2.80}\left(\frac{1-\frac{5}{6}\times\frac{0.321}{2.80}}{1-\frac{0.321}{2.82}}\right)=0.011$$

or

$$H^\sigma = 0.011\times0.765\times10^6 = 8415\ J/mol.$$

One mole occupies an area

$$A\approx N_A^{\frac{1}{3}}V_m^{\frac{2}{3}}=\left(6.022\times10^{23}\right)^{\frac{1}{3}}\left(\frac{58.45}{2.17\times10^6}\right)^{\frac{2}{3}}=76000m^2/mol$$

Then the surface tension becomes

$$\sigma=\frac{H^\sigma}{A}=0.11\ N/m.\tag{2.100}$$

This value may be compared with $\sigma = 0.11 J/m^2$ obtained by fracture of NaCl.[105] The correspondence between the computed value and the experimental value is probably fortuitous.[106-8] It should be recalled that the calculation is only approximate. Several corrections have been omitted deliberately to obtain equations in simple forms. These corrections include the van der Waals attraction and the repulsive forces between non-nearest neighbours.

Bahadur gives a value of 0.114 N/m for NaCl–air.[109] Bo Li suggests a best estimate for the surface energy of NaCl(001) of $(140\text{–}240) \times 10^{-3}$ N/m.[110]

In this book we do not pursue the theory further. The main point is to illustrate that the surface energy is given by the difference between the energy associated with the surface and that with the bulk. For ionic crystals it is given by the difference between the electrostatic energy associated with the surface and that with the bulk. From this difference we subtract the repulsion energy difference between bulk and surface. It is recognized that the numerical problems are considerable: the differences between the various energies are small. Furthermore, the electrostatic terms and the repulsion energy terms are on the same order of magnitude. The net result is that the surface energy is only around 1 per cent of the crystal energy. From these facts it is recognized that the surface energy (surface tension) is sensitive to minor changes in structure and crystal radii caused by the presence of small amounts of impurities in the surface layer.

2.2.7   Interfacial Tension of Liquids with Several Components

Only for one-component systems is the interfacial tension σ equal to the Gibbs energy per unit area $g^\sigma = G^\sigma / A$. For systems with several components, σ is in general different from g^σ.[111] Then

$$\sigma = \left(\frac{\partial G^\sigma}{\partial A}\right)_{n_i} = \left(\frac{\partial (A g^\sigma)}{\partial A}\right)_{n_i} = g^\sigma + A\left(\frac{\partial g^\sigma}{\partial A}\right)_{n_i}. \tag{2.101}$$

σ is equal to g^σ only if $(\partial g^\sigma / \partial A)_{n_i} = 0$, that is, only if g^σ does not change when the surface area is changed. In general this is the case only for a one-component system. If the concentration of the various components at the surface is given by

$$\Gamma_i = n_i^\sigma / A,$$

then

$$\left(\frac{\partial g^\sigma}{\partial A}\right)_{n_i} = \sum_{i=1}^{k} \frac{\partial g^\sigma}{\partial \Gamma_i}\left(\frac{\partial \Gamma_i}{\partial A}\right)_{n_i},$$

where $\partial g^\sigma / \partial \Gamma_i = \mu_i^\sigma$ is the chemical potential of component i at the interface. Under equilibrium conditions $\mu_i^\sigma = \mu_i$, that is, μ_i^σ is equal to the bulk chemical potential:

$$\left(\frac{\partial \Gamma_i}{\partial A}\right)_{n_i} = \left(\frac{\partial (n_i^\sigma / A)}{\partial A}\right)_{n_i} = n_i^\sigma \frac{\partial 1/A}{\partial A} = \frac{-n_i^\sigma}{A^2} = -\frac{\Gamma_i}{A}.$$

Then Eq. (2.101) becomes

$$\sigma = g^{\sigma} - \sum_{i=1}^{k} \mu_i^{\sigma} \Gamma_i. \tag{2.102}$$

This shows that in general the interfacial tension σ cannot be identified with the interfacial free energy per unit area g^{σ}. σ depends also on the chemical potential at the surface μ_i^{σ} and on the surface coverage $\Gamma_i = n_i^{\sigma}/A$. It will be seen that the ratio between the n_i^{σ} at the surface in general is different from the molar ratios in the bulk. Electronegative elements in metals such as F, O, and S tend to accumulate at the surface. These elements are therefore called surface active. We have seen that the interfacial tension is given by the difference between the potential for the surface and the potential for the bulk. The electronegative elements give a deeper surface potential. This reduces the surface tension.

The interfacial (surface) tension of the pure metal is often known and one would like to determine the effect of various alloy additions and impurity levels. Then differentiation of Eq. (2.102) gives

$$d\sigma = dg^{\sigma} - \sum_{i=1}^{k} \mu_i^{\sigma} \, d\Gamma_i - \sum_{i=1}^{k} d\mu_i^{\sigma} \, \Gamma_i. \tag{2.103}$$

g^{σ} depends on T, P, and the Γ_i. It has a total differential

$$dg^{\sigma} = \left(\frac{\partial g^{\sigma}}{\partial T} \right)_P dT + \left(\frac{\partial g^{\sigma}}{\partial p} \right)_T dp + \sum_{i=1}^{k} \mu_i^{\sigma} \, d\Gamma_i.$$

If the two equations are added, the following is obtained,

$$d\sigma = \left(\frac{\partial g^{\sigma}}{\partial T} \right)_P dT + \left(\frac{\partial g^{\sigma}}{\partial p} \right)_T dp - \sum_{i=1}^{k} \Gamma_i d\mu_i^{\sigma}.$$

where $(\partial g^{\sigma}/\partial T)_p = -s^{\sigma}$. Usually the effect of pressure can be neglected. Then

$$d\sigma = -s^{\sigma} \, dT - \sum_{i=1}^{k} \Gamma_i d\mu_i^{\sigma}.$$

At equilibrium, where $\mu_i{}^\sigma = \mu_i$ and at constant temperature[112]

$$d\sigma = -\sum_{i=1}^{k} \Gamma_i d\mu_i. \tag{2.104}$$

If there is only one component so that k = 1, $d\mu_i = RTd\,(\ln a_1) = 0$ since $a_i = 1$ and $\sigma = g = G/A$. This equation has been used previously in the discussion of one-component systems.

To obtain the interfacial tension σ at equilibrium when there are several components in addition to the solvent metal, the area occupied by one mole of species i in the monolayer, A_i^σ, is introduced. Then

$$\Gamma_i = \frac{n_i^\sigma}{\sum\limits_{j=1}^{k} n_j^\sigma A_j^\sigma},$$

or in terms of mass per cent at the surface

$$\Gamma_i = \frac{[\%i]^\sigma / M_i}{\sum\limits_{i=1}^{k} [\%j]^\sigma A_i^\sigma / M_j}.$$

An *interfacial activity coefficient* f_i^σ is defined as

$$f_i^\sigma = \frac{a_i}{[\%i]^\sigma}, \tag{2.105}$$

where the activity in the bulk a_i is given in mass per cent so that f_i^σ is dimensionless. Note that f_i^σ compares activity in the bulk liquid with the concentration in the interface. It will be seen that f_i^σ is a useful tool in the derivation that follows. Finally, f_i^σ can be eliminated.

If the liquid is in contact with a vacuum or an inert gas, the mass percentages on the surface sum up to 100 per cent:

$$\sum_{i=1}^{k} [\%i]^\sigma = 100. \tag{2.106}$$

For a liquid in contact with a solid or another liquid the percentage on the right-hand side is an unknown quantity, A, less than 100. In terms of activities and surface activity coefficients Eq. (2.106) becomes

$$\sum_{i=1}^{k} \frac{a_i}{f_i^\sigma} = 100. \tag{2.107}$$

Similarly for the bulk phase

$$\sum_{i=1}^{k} \frac{a_i}{f_i} = 100.$$

According to the Gibbs–Duhem equation (Eq. (2.26))

$$\sum_{i=1}^{k} x_i \frac{da_i}{a_i} = 0.$$

Now for the mole fraction

$$x_i = \frac{[\%i]/M_i}{\sum_{i=1}^{k} [\%j]/M_j} \tag{2.108}$$

Inserted in the Gibbs–Duhem equation this gives

$$\sum_{i=1}^{k} \frac{[\%i]\, da_i}{M_i a_i} = 0$$

or

$$\sum_{i=1}^{k} \frac{da_i}{M_i f_i} = 0 \tag{2.109}$$

To obtain an equation for the surface tension, the expressions for Γ_i and $d\mu_i = RT\,d\ln a_i$ are introduced in Eq. (2.104). Furthermore, concentrations are replaced by activities. This gives

$$\frac{d\sigma}{RT} = \frac{-\sum_{i=1}^{k} \frac{a_i}{M_i f_i^\sigma} d\,(\ln a_i)}{\sum_{i=1}^{k} \frac{a_i A_i^\sigma}{M_i f_i^\sigma}},$$

or, since $a_i\, d\,(\ln a_i) = da_i$

$$\frac{d\sigma}{RT} = \frac{-\sum_{i=1}^{k} da_i/\left(M_i f_i^\sigma\right)}{\sum_{i=1}^{k} a_i A_i^\sigma /\left(M_i f_i^\sigma\right)}. \tag{2.110}$$

We note that if $f_i^\sigma = f_i$ then according to the Gibbs–Duhem Eq. (2.109), $d\sigma = 0$. This means that if the activity coefficients were the same in the bulk as on the surface, surface tension would not change with composition. In general $f_i^\sigma \neq f_i$. In this case Eq. (2.110) may be written

$$\sum_{i=1}^{k} \frac{a_i A_i^\sigma}{M_i f_i^\sigma} + \sum_{i=1}^{k} \frac{da_i}{d\sigma} \frac{RT}{M_i f_i^\sigma} = 0. \tag{2.111}$$

We wish to study the effect of the solute additions. Therefore, the solvent component, $i = 1$, is eliminated from Eq. (2.111). To accomplish this, Eq. (2.109) is multiplied by $f_1 RT / \left(f_1^\sigma d\sigma \right)$ to give

$$\sum_{i=1}^{k} \frac{da_i}{d\sigma} \frac{RT f_1}{f_1^\sigma M_i f_1} = 0. \tag{}$$

Furthermore, Eq. (2.107) is multiplied by A_1^σ / M_1 to give

$$\sum_{i=1}^{k} \frac{A_1^\sigma a_i}{M_i f_i^\sigma} = 100 \frac{A_1^\sigma}{M_1}. \tag{}$$

If these two equations are subtracted from Eq. (2.111), one obtains

$$\frac{100 A_1^\sigma}{M_1} + \sum_{i=2}^{k} \left\{ \left(A_i^\sigma - \frac{A_1^\sigma M_i}{M_1} \right) a_i + RT \left(1 - \frac{f_1 f_i^\sigma}{f_1^\sigma f_i} \right) \frac{da_i}{d\sigma} \right\} \frac{1}{M_i f_1^\sigma} = 0. \tag{2.112}$$

In the following only a binary system is studied. For a dilute binary system f_2 and f_i may as an approximation be regarded as constant with respect to a_2 according to Henry's and Raoult's laws. However, the concentration of component 2 may be high on the surface. Therefore, f_2^σ and f_i^σ depend on a_2. This relationship must be determined before Eq. (2.112) can be integrated. A Langmuir-type model is employed for this purpose. According to the model adsorption is limited to a monolayer on the surface and gives an 'exchange reaction':

$$\text{(component 2 in bulk)} + \text{(component 1 on surface)} = \text{(component 2 on surface)} + \text{(component 1 in bulk)}. \tag{2.113}$$

The equilibrium constant for the exchange reaction is expressed by ΔG^0, the Gibbs energy gained by moving one mole of component 2 from the bulk to the surface, the *Gibbs energy of adsorption*,

$$\frac{a_2 [\%1]^\sigma}{[\%2]^\sigma a_1} = \exp\left(\frac{\Delta G^0}{RT} \right), \tag{2.114}$$

where a_i and a_2 are the solvent and solute activities in solution, and the activities in the adsorbed layer are approximated by the respective mass percentages. Note that Eq. (2.113) and (2.114) differ from the Langmuir model in that the size of species 1 and 2 is only taken into account in ΔG^0. ΔG^0 must be determined experimentally.

Equation (2.114), combined with the definition of the surface activity coefficients, Eq. (2.105), gives, when $[\%1]^{\sigma}$ and a_2 are eliminated using Eq. (2.106) and $f_i = a_i / [\%1]$ and $f_2 = a_2 / [\%2]$,

$$f_2^{\sigma} = \frac{a_2}{100} \left(1 - \frac{f_1}{f_2} \exp\left(\frac{\Delta G^0}{RT} \right) \right) + f_1 \exp\left(\frac{\Delta G^0}{RT} \right). \tag{2.115}$$

If Eq. (2.114) is written in activities only, we find that

$$\frac{f_2^{\sigma}}{f_1^{\sigma}} = \exp\left(\frac{\Delta G^0}{RT} \right). \tag{2.116}$$

Equation (2.112) for a two-component system ($k = 2$) can then be integrated to give

$$\sigma = \sigma_1 - \frac{M_1}{A_1^{\sigma} M_2} \frac{RT}{b_2} \left(1 - \frac{f_1}{f_2} \exp\left(\frac{\Delta G^0}{RT} \right) \right)$$
$$\times \ln\left(b_2 \frac{f_2}{f_1} \exp\left(-\frac{\Delta G^0}{RT} \right) a_2 + 1 \right), \tag{2.117}$$

where

$$b_2 = \frac{A_2^{\sigma} M_1}{A_1^{\sigma} M_2} - \frac{f_1}{f_2} \exp\left(\frac{\Delta G^0}{RT} \right) \quad a_2 = \frac{[\%2]}{100}. \tag{2.118}$$

σ is now given in terms of $[\%2]$ and the surface tension of the pure solvent σ_1. This may be regarded as a further development of an equation given by Belton.[113] In the following we set both f_2 and f_1 equal to 1.

So far we have considered a melt surface in contact with a vacuum. In principle the same argument can be applied for a melt surface in contact with a solid or with a slag.[114] Then

$$\sigma_{sl} = \sigma_{sl}^p - \frac{M_1}{A_1^{\sigma} M_2} \frac{RT}{b_2} \left(1 - \frac{f_1}{f_2} \exp\left(\frac{\Delta G^0}{RT} \right) \right)$$
$$\times \ln\left(b_2 \frac{f_2}{f_1} \exp\left(-\frac{\Delta G^0}{RT} \right) \frac{[\%2]}{A} + 1 \right), \tag{2.119}$$

where σ_{sl}^p is the interfacial tension between the pure melt and the solid and σ_{sl} is the interfacial tension when the melt contains $[\%2]$ mass per cent component 2. b_2 is defined by analogy with Eq. (2.118). ΔG^0 must of course be determined experimentally for each solid–melt system.

A is the percentage of the interface occupied by the liquid. For pure metals at the melting point it is possible to determine A as indicated in Section 2.2.8.

For metallic additions the metallic radius is used to calculate A_2^σ. For surface-active elements such as oxygen or sulfur in iron, the ionic radius is used to calculate A_2^σ. For oxygen in iron the ionic radius, r_i is 1.40 Å.[115] Then an estimate for the area occupied by one oxygen atom is

$$\frac{A_2^\sigma}{N_A} = \pi r_1^2 = \pi \times 1.4^2 = 6.16 \text{Å}^2. \tag{2.120}$$

The metallic radius of iron is 1.24 Å. The area occupied by a neutral iron atom is (see Table 2.14)

$$\frac{A_2^\sigma}{N_A} = \pi r^2 = \times 1.24^2 = 4.83 \text{Å}^2.$$

For various solvent metals with impurities of oxygen, sulfur, selenium, and tellurium Eq. (2.117) has been compared with experimental data,[113,116] and the values of ΔG^0 that give the 'best fit' for Eq. (2.117) have been determined. The ΔG^0 values are shown in Fig. 2.25. Figures 2.26 and 2.27 compare Eq. (2.117) and experimental data for the two important systems Fe–O and Cu–S, respectively. For the iron–oxygen system at 1550 °C

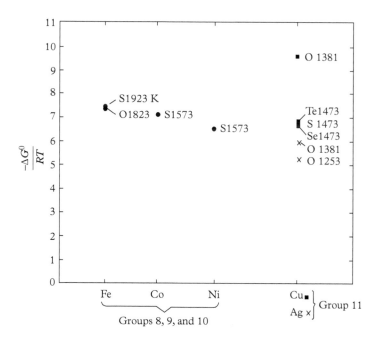

Fig. 2.25 *Gibbs energy of adsorption from bulk to surface, ΔG^0, of O, S, Se, and Te in solvent metals Fe, Cu, and Ag. R is the gas constant. The absolute temperatures T are indicated.*

$$\sigma_1 = 1.8 \, \text{N/m},$$

$$b_2 = \frac{6.16 \times 55.85}{4.83 \times 16} - \exp\left(\frac{-111156}{8.3145 \times 1823}\right) = 4.44.$$

$$A_1^\sigma = 4.83 \times 10^{-20} \times 6.022 \times 10^{23}$$

Equation (2.119) gives

$$\sigma = 1.8 - 0.409 \left(\ln\left(68 \, [\%O] + 1\right)\right). \tag{2.121}$$

Similarly for sulfur in copper an ionic radius for sulfur[117] of 1.84 Å and a metallic radius of copper of 1.28 Å[64,118] gives, using $-\Delta G^0 = 84\ 100$ J/mol (see Fig. 2.25),

$$\sigma = 1.244 - 0.191 \left(\ln\left(39.3 \, [\%S] + 1\right)\right). \tag{2.122}$$

Figures 2.26 and 2.27 are typical examples that show good correspondence between equation and experiments when employing the 'fitted' values of ΔG^0 as the one adjustable parameter.

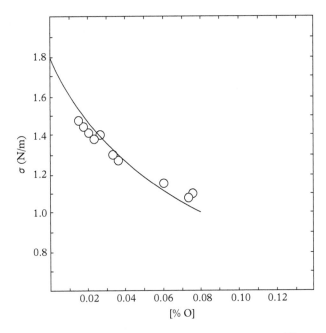

Fig. 2.26 *The depression of the surface tension of iron by oxygen at 1550°C[113] compared with Eq. (2.121).*

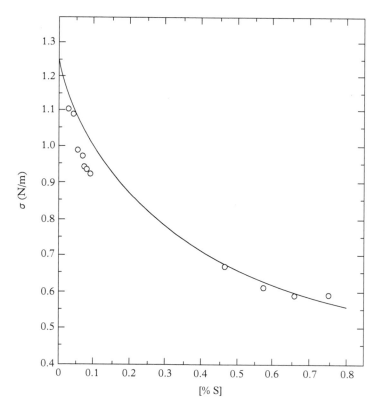

Fig. 2.27 *The depression of the surface tension of copper by sulphur at 1200°C*[119] *compared with Eq. (2.122).*

2.2.8 Solid–Liquid Free Energy of Close-Packed Metals

The solid–liquid Gibbs energy is important to determine the micro-structure and mechanical properties of metals. Turnbull[120,121] has found experimentally for many pure metals that the interfacial tension, σ_{sl}, at the melting point is given by

$$\sigma_{sl} = \frac{0.45\,\Delta H_m}{N_A^{\frac{1}{3}} V_m^{\frac{2}{3}}}, \qquad (2.123)$$

where ΔH_m is the molar enthalpy of melting and $N_A^{\frac{1}{3}} V_m^{\frac{2}{3}}$ is roughly equal to the area of one mole of atoms in a monolayer. This result was obtained from surface tensions calculated from the observed supercooling of small drops of liquid metals.

It will be instructive to derive Eq. (2.123). We introduce the monolayer model and for the interface assume that this layer is a mixture (configuration) of atoms in the liquid and solid state, with mole fractions x_l and x_s of the layer belonging to the liquid and solid phases, respectively.

We have for an ideal mixture of solid and liquid atoms,

$$nA\sigma_{sl} - nS^\sigma T_m = -RT_m [n_s \ln x_s + n_l \ln x_l], \qquad (2.124)$$

where $n = n_s + n_l$ and n is the number of atoms on the surface of the nucleus. n is taken to be a constant determined by a constant nucleus size. n_s is the number of atoms in the solid state in the interface; n_l is the number of atoms in the liquid state: $n = n_s + n_l$.

The Gibbs energy, ΔG, of the interface relative to the bulk liquid is given by

$$n\Delta G = RT_m [n_s \ln x_s + n_l \ln x_l] - n_s \Delta H_m \qquad (2.125)$$

At equilibrium, $n = $ constant and $\Delta G = $ constant. Then

$$\frac{\partial (n\Delta G)}{\partial n_s} = 0$$

$$n_l = n - n_s \text{ and } \frac{\partial n_l}{\partial n_s} = -1 \qquad (2.126)$$

$$0 = RT_m [\ln x_s - \ln x_l] - \Delta H_m$$

$$\ln x_s - \ln x_l = \frac{\Delta H_m}{RT_m} = \frac{\Delta S_m}{R}.$$

Also

$$\frac{n_s}{n} = x_s; \frac{n_l}{n} = x_l \text{ and } x_s + x_l = 1. \qquad (2.127)$$

Equations (2.126) and (2.127) give

$$x_l = \frac{\exp(-\Delta S_m/R)}{1 + \exp(-\Delta S_m/R)} \qquad (2.128)$$

$$x_s = \frac{1}{1 + \exp(-\Delta S_m/R)} \qquad (2.129)$$

$$b = \exp(-\Delta S_m/R).$$

Then Eq. (2.124) becomes

$$A\sigma_{sl} = T_m R \left(\ln(1+b) - \frac{b}{1+b} \ln b \right) + S^\sigma T_m, \qquad (2.130)$$

where b is a constant.

According to Richard's rule[99] that $\frac{\Delta H_m}{T_m} \approx 9.2\ J/mol$

$$\frac{\Delta S_m}{R} = \frac{\Delta H_m}{RT_m} \approx \frac{9.2}{8.3145} = 1.1,$$

so that

$$b = 0.33$$

and

$$A\sigma_{sl} = T_m R \cdot 0.56 + S^o\, T_m$$

If the Richard's rule approximation is employed once more,

$$\sigma_{sl} = 0.51 \Delta H_m / A + \frac{S^o\, T_m}{A} \tag{2.131}$$

Using Eq. (2.86) for A gives

$$\sigma_{sl} = \frac{0.46 \Delta H_m}{N_A^{\frac{1}{3}} V_m^{\frac{2}{3}}} + \frac{S^o\, T_m}{A} \tag{2.132}$$

This equation may be compared with experimental data[122] where the proportionality factor is 0.436, 0.490, 0.439, 0.445, 0.436, 0.444, and 0.450 for Au, Co, Cu, Fe, Ga, Ni, and Pd, respectively. $S^o = 0$ gives good correspondence with experiments.

For Si, Ge, and Bi which are non-metals in the solid state and acquire metallic properties on melting, the above theory cannot be applied. The number of nearest neighbours is about four in the solid state and increases to about eight on melting. A rough estimate of the proportionality constant may be carried out as follows. For the two phases the 'average' number of bonds broken between bulk phase and the interface is $(8 - 4)/2 = 2$. The enthalpy contribution of the surface energy is then $\Delta H_m 2/8 = 0.25 \Delta H_m$ while the entropy contribution is given by the right-hand side of Eq. (2.130). Richard's rule cannot be applied to the metalloids. For instance for Si, $\Delta H_m / RT_m = 3.62$ so $b = \exp(-3.62) = 0.0267$ and the right-hand side of Eq. (2.130) becomes $0.12 RT_m$. Then for Si

$$\sigma_{sl} = \left(0.25 + \frac{0.12}{3.62} \right) \frac{\Delta H_m}{A}$$

And

$$\sigma_{sl} = \frac{0.26 \Delta H_m}{N_A^{\frac{1}{3}} V_m^{\frac{2}{3}}}$$

For Ge and Bi a similar calculation gives 0.26 and 0.34, respectively, for the proportionality constant. For these elements it is found experimentally that the proportionality constant in Eq. (2.132) is 0.32–0.35.[122,123]

A more accurate calculation that takes into account both the heat of melting and melting temperature[124] may be obtained from Eq. (2.130) by inserting $b = \exp\left(-\frac{\Delta H_m}{RT_m}\right)$.

Then

$$A\sigma_{sl} = T_m R \ln\left(1 + \exp\left(-\frac{\Delta H_m}{RT}\right) + \frac{\exp\left(-\frac{\Delta H_m}{RT_m}\right)\frac{\Delta H_m}{RT_m}}{1 + \exp\left(-\frac{\Delta H_m}{RT_m}\right)}\right),$$

2.2.9   Diffusion in Molten Metals

It is recalled that the diffusion coefficient D enters into Fick's law of diffusion,

$$\dot{n}_2 = -D\frac{\partial c_2}{\partial z}, \tag{2.133}$$

where $\dot{n}_2$ is the flux of solute particles and c_2 is their concentration. Equation (2.133) applies when there is no convection. The remarkable characteristic of diffusion in molten metals, and liquids in general, is the fact that the values of the diffusion coefficient D are almost all on the same order of magnitude, 10^{-8}–10^{-9} m²/s, even though their solid state properties differ widely.[68,112,117]

The Stokes–Einstein equation[102] is a very useful starting point in the description of diffusion in molten metals and other liquids. This is the case despite the fact that the equation is derived for particles moving randomly in a continuum (called Brownian motion). The assumption is that the particles are much larger than the particles in the surrounding continuous liquid. However, the diffusing atoms in a molten metal are about the same size as the solvent atoms; they may even be smaller as is the case with carbon in iron.

According to Einstein the mean squared displacement (z^2) in a random walk[102] is

$$z^2 = \frac{2kT}{\varsigma}t, \tag{2.134}$$

where t is the time elapsed from position $z = 0$. Equation (2.134) is valid only after a large number of collisions has taken place between the Brownian particle and the surrounding particles (continuum). (z^2) is the square of the distance that the particle has drifted in the direction z. The total distance travelled is of course much greater. ς is the factor given by the ratio between the force F acting on the Brownian particle and the velocity of the particle:

$$\varsigma = \frac{F}{u}. \tag{2.135}$$

For a large (Brownian) particle moving in a continuum we have the relation first obtained by Stokes,[66] $F = 3\pi \mu d$, so that

$$\varsigma = 3\pi \mu d, \qquad (2.136)$$

where μ is the viscosity of the fluid and d is the particle diameter. Stokes' equation is based on the boundary condition (no-slip condition) that there is no relative motion between the particle and the fluid (continuum) immediately in contact with it. This condition cannot be applied to the case of collisions between a particle and neighbouring particles of greater or equal mass. In this case we assume—somewhat arbitrarily—that the tangential forces are negligible when the masses are equal. Then, if there are no tangential forces, the factor 2 replaces 3 in Eq. (2.136).[125]

The diffusion coefficient D may be obtained by first using Eq. (2.133) to derive an equation for the concentration of particles, c_z, then calculating z^2, and finally comparing the result with Eqs. (2.134) and (2.136).

The accumulation of c_2 per unit time is equal to the difference between the flux in and out of the control volume element. This equation gives the diffusion equation

$$\frac{\partial c_2}{\partial t} = D \frac{\partial^2 c_2}{\partial z^2} \qquad (2.137)$$

(see also Chapters 4, 7, and 8). If n_2 moles of particles are added at position $z = 0$ at time $t = 0$, it is found[102] that the solution of Eq. (2.137) is

$$c_2 = \frac{n_2}{2} \exp\left(-\frac{z^2}{4Dt}\right)(\pi Dt)^{-\frac{1}{2}}.$$

Then

$$z^2 = \frac{\int_{-\infty}^{\infty} x^2 c_2(x,t)\, dx}{n_2} = 2Dt. \qquad (2.138)$$

Comparison of Eq. (2.138) with Eqs. (2.134) and (2.136) gives for Brownian motion

$$D = \frac{kT}{3\pi \mu d}. \qquad (2.139)$$

Equation (2.139) can be employed for particles or inclusions where the diffusing particle is much larger than metal atoms.

When the diffusing particle radius is approximately equal to that of the medium, it is suggested that we can use the solution with no tangential force

$$D = \frac{kT}{2\pi \mu d}. \qquad (2.140)$$

It is not clear whether the diffusing particle is in the form of an atom or ion in metallic liquids. The Goldschmidt–Pauling radius, etc., can be used for $d/2$.[2] We choose the hard-sphere model with packing fraction 0.46. Then

$$\left(\frac{V}{N_A}\right)^{1/3} = 1.05d. \tag{2.141}$$

The hard-sphere radius is close to the metallic radius. Suggested radii are given in Table 2.17. For the non-metals it is proposed that the Goldschmidt atomic radii be employed:[91] $r = 0.46$, 0.77, 0.71, and 0.60 Å for H, C, N, and O, respectively, $r = 1.17$, 1.09, and 1.4 Å for Si, P, and S.

If Eq. (2.76) for the viscosity $\mu_m = 1.8 \times 10^{-7}(MT)^{1/2}/V_m^{2/3}$ is inserted, we get

$$D = 1.08 \times 10^{-9}\left(\frac{T_m}{M}\right)^{1/2} V_m^{1/3} \tag{2.142}$$

at the melting point.

Iida and Guthrie,[74] minimizing the mean square deviation between model and experiments, get

$$D = 1.20 \times 10^{-9}\left(\frac{T_m}{M}\right)^{1/2} V_m^{1/3}. \tag{2.143}$$

From self-diffusion experiments it is found that in plotting log D against $1/T$ generally straight lines are obtained,

$$D \approx D_0 \exp\left(-\frac{H_D}{RT}\right), \tag{2.144}$$

where D_0 and H_D are called Arrhenius parameters. Figure 2.28 shows a plot of log H_D versus T_m for various liquid metallic elements. This line is described by

$$\log\frac{H_D}{T_m} = 0.10\log T_m + 1.13 \tag{2.145}$$

or

$$H_D = 13.5 T_m^{1.10}. \tag{2.146}$$

At the melting point $D_m = D_0 \exp\left(-1.62 T_m^{0.10}\right)$.

Table 2.17 *Valence, relative atomic mass, metallic radius, electronegativity, and reduction potential in aqueous solution (at 25 °C) of the elements*[115]

	1	2	3	4	5	6	7	8	9	10	11	12	13	14	15	16	17
	Li	Be											B	C	N	O	F
Valence	1	2	—	—	—	—	—	—	—	—	—	—	3	4	3	2	1
atomic mass	6.94	9.01											10.82	12.01	14.01	16.00	19.00
radius, Å	1.52	1.12											—	—	—	—	—
electronegativity	0.98	1.57											2.04	2.55	3.04	3.44	3.98
potential	−3.0	−1.97															2.87
	Na	Mg											Al	Si	P	S	Cl
Valence	1	2	—	—	—	—	—	—	—	—	—	—	3	4	3	2	1
atomic mass	22.99	24.31											26.98	28.09	30.97	32.06	35.45
radius, Å	1.86	1.60											1.43	—	(1.10)		
electronegativity	0.93	1.31											1.61	1.90	2.19	2.58	3.16
potential	−2.71	−2.36											−1.66			−0.48	1.36
	K	Ca	Sc	Ti	V	Cr	Mn	Fe	Co	Ni	Cu	Zn	Ga	Ge	As	Se	Br
Valence	1	2	3	4	5	3, 6	4, 6	2, 3	2, 3	2	1, 2	2	3	4	3	2	1
atomic mass	39.10	40.08	44.96	47.90	50.94	52.00	54.94	55.85	58.93	58.70	63.55	65.38	69.72	72.59	74.92	78.96	79.90
radius, Å	2.27	1.97	1.61	1.45	1.31	1.25	1.37	1.24	1.25	1.25	1.28	1.33	1.22	1.23	(1.22)	(1.16)	—
electronegativity	0.82	1.0	1.36	1.54	1.63	1.66	1.55	1.83	1.88	1.91	1.90	1.65	1.81	2.01	2.18	2.55	2.96
potential	−2.92	−2.97						−0.41	−0.28	−0.23	0.52	−0.76	−0.56			−0.92	1.07

(Continued)

Table 2.17 *Continued*

	1	2	3	4	5	6	7	8	9	10	11	12	13	14	15	16	17
	Rb	Sr	Y	Zr	Nb	Mo	Tc	Ru	Rh	Pd	Ag	Cd	In	Sn	Sb	Te	I
Valence	1	2	3	4	5	6	7	4	3	2	2	2	3	4	3	2	1
atomic mass	85.47	87.62	88.91	91.22	92.91	95.44	(97)	101.07	102.91	106.4	107.87	112.41	114.82	118.69	121.75	127.60	126.90
radius, Å	2.48	2.15	1.78	1.59	1.43	1.36	1.35	1.33	1.34	1.38	1.44	1.49	1.36	1.51	1.45	1.43	—
electronegativity	0.82	0.95	1.22	1.33	1.6	2.16	1.9	2.2	2.28	2.20	1.93	1.69	1.78	1.96	2.05	2.1	2.66
potential	−2.92	−2.89								0.92	0.80	−0.40	−0.34			−0.95	0.54
	Cs	Ba	La	Hf	Ta	W	Re	Os	Ir	Pt	Au	Hg	Tl	Pb	Bi	—	—
Valence	1	2	3	4	5	6	7	4	2	2	2	2	3	2, 4	3		
atomic mass	132.91	137.33	138.91	178.49	180.95	183.85	186.2	190.2	192.2	195.1	196.97	200.59	204.37	207.2	208.98		
radius, Å	2.65	2.17	1.87	1.56	1.43	1.37	1.37	1.34	1.36	1.39	1.44	1.50	1.70	1.75	1.55		
electronegativity	0.79	0.89	1.10	1.3	1.5	2.36	1.9	2.2	2.20	2.28	2.54	2.00	2.04	2.33	2.02		
potential	−2.92	−2.90	−2.52							1.2	1.4	0.85		−0.13			

The reduction potential is for the lowest valence in the table. For instance for iron $Fe^{2+} + 2e^- = Fe(s)$.

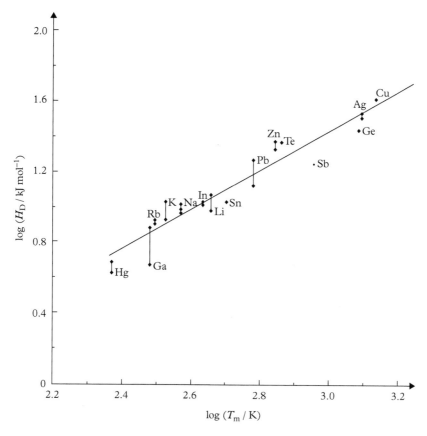

Fig. 2.28 *Log H_D vs log T_m. Points linked by vertical line represent data from two or three different experimental data for a given metal.*[74] *[Reproduced with permission of Oxford Publishing Limited through PLSclear.]*

Then Eq. (2.144) becomes

$$D = D_m \exp\left(1.62 T_m^{0.10}\right) \exp\left(\frac{-13.5 T_m^{1.10}}{RT}\right). \tag{2.147}$$

Experimental values given in the literature vary widely, even by a factor of 10. In calculations of mass transfer D usually enters as the square root. Thus, the factor of 10 can lead to an error of 3 in mass transfer calculations. Probably this is the most serious problem we face in the development of quantitative models of refining operations.

2.2.10   Thermal and Electrical Conductivity

At low temperatures the thermal conductivity of metals may drop with increasing temperatures. The reason is that for impure metals or disordered alloys the contribution from phonons (lattice waves) can be comparable to the electronic contribution. The contribution of the lattice waves to the thermal conductivity drops with temperature. In normal pure metals even at room temperature the electrons carry almost all the heat current. The contributions from electrons increase with temperature; therefore at higher temperatures, for both pure and impure metals, the electronic contribution overwhelms the phonon contribution and then the thermal conductivity finally increases with temperature.

Although there are several compilations of transport properties of elements, no comprehensive survey has been made for metals and alloys. Only a few studies for predicting metal thermal conductivities have been carried out.[126,127] At high temperatures the Wiedemann–Franz law which gives a relation between the thermal and electrical conductivities of metals applies. On combining this law and measurements of electrical conductivities determined experimentally, the prediction of thermal conductivities of solid or liquid metals and alloys is possible, since reliable data on electrical conductivities are often available. Various data on electrical conductivities (or on the reciprocal quantity, electrical resistivity) are found in *Smithells' Metals Reference Book*[91] and in a textbook by Meaden.[128]

The Wiedemann–Franz law states that for metals at temperatures which are not too low the ratio of the thermal conductivity to the electrical conductivity is directly proportional to the absolute temperature. The constant of proportionality is independent of the particular metal. The basis for the law is that the electrons transport charge and energy by the same mechanism. The ratio between the thermal conductivity k and electrical conductivity σ is[64]

$$\frac{k}{\sigma} = \frac{\pi^2}{3}\left(\frac{k}{e}\right)^2 T = LT. \tag{2.148}$$

L, called the Lorenz number, is

$$L = \frac{\pi^2}{3} \times \left(\frac{1.38x10^{-23}}{1.6x10^{-19}}\right)^2 = 2.45 \times 10^{-8}\ watt-ohm/K^2.$$

Table 2.18 compares the thermal conductivity and electrical resistivity at the melting-point of various metals. It is seen that, apart from the metalloids, the ratios between thermal and electrical conductivities are not far from 2.45×10^{-8} WΩ/K^2.

The thermal conductivity of steel alloys becomes close to that for iron above 850 °C. Above 1000 °C the Lorenz ratio for steel alloys is around 2.55×10^{-8} WΩ/K^2.

Table 2.18 *Melting-point, T_m, electrical resistivity, thermal conductivity, and Lorenz number at the melting-point for various molten elements*[91]

Metal	Melting point (K)	$1/\sigma \ (10^{-8}\Omega m)$	k (W/mK)	$L \ (10^{-8}W\Omega/K^2)$
Ag	1234	17.25	175	2.45
Al	933	24.25	94	2.44
Au	1337	31.25	104.4	2.44
Bi	544	129	17.1	4.05
Cd	594	33.7	42	2.38
Cs	301.5	37.0	19.7	2.42
Cu	1358	20.0	165.6	2.44
Ga	302.8	26	25.5	2.19
Hg	234.1	90.5	6.78	2.62
In	429.6	32.3	(42)	3.16
K	336.7	13.65	53.0	2.15
La	1193	138	(21.0)	2.43
Li	454	24.0	46.4	2.45
Mg	923	27.4	78	2.31
Na	370.8	9.64	89.7	2.33
Pb	600.5	94.85	15.4	2.43
Rb	311.9	22.83	33.4	2.44
Sb	903.5	113.5	21.8	2.74
Sn	504.9	47.2	30	2.80
Te	723	550	2.5	1.90
Tl	576.5	731	(24.6)	3.12
Zn	692.4	37.4	49.5	2.67

REFERENCES

1. Denbigh K. *The principles of chemical equilibrium*, 4th edn. Cambridge, UK: Cambridge University Press; 1981.
2. Lupis CHP. *Chemical thermodynamics of materials*. New York: North-Holland; 1983.

3. Gaskell DR. *Introduction to metallurgical thermodynamics*, (2nd edn). New York: Hemisphere Publ. Co./McGraw-Hill; 1981.

4. Sigworth GK, Elliott JF. The thermodynamics of dilute liquid copper alloys. *Can. Metall. Q.*, 1974; 13: 455–61.

5. Sigworth GK, Engh TA. Refining of liquid aluminium—a review of important chemical factors. *Scand. J. Metall.*, 1982; 11(3): 143–9.

6. Sigworth GK. Refining of Secondary Aluminum: Important Chemical Factors. *JOM (Jounal of Metals)*, 2021; 73(9): 2594–602. https://doi.org/10.1007/s11837-021-04774-z.

7. Sigworth GK, Elliott JF. The thermodynamics of liquid dilute iron alloys. *Met. Sci.* 1974; 8: 298–310.

8. Turkdogan ET. *Physical chemistry of high temperature technology*. New York: Academic Press; 1980.

9. *Steelmaking Data Sourcebook*. New York: Gordon and Breach; pp. 278–93.

10. Sigworth GK, Elliott JF. The thermodynamics of dilute liquid cobalt alloys. *Can. Metall. Q.*, 1976; 15: 123–7.

11. JANAF Thermochemical Tables, 3rd edn. Chase MW et al. eds. American Chem. Soc. and the American Institute of Physics for the National Bureau of Standards. Michigan, USA. *J. Phys. Chem. Ref. Data* vol 14, 1985; Supplement No. 1.

12. Barin I. *Thermochemical data of pure substances*, parts I and II. Weinheim, Germany: VCH Verlagsgesellschaft; 1989.

13. Hultgren H, Desai PD, Hawkins DT, Gleiser M, Kelley KK, Wagman DD. *Selected values of the thermodynamic properties of the elements*. Metals Park, OH: Am. Soc. Met; 1973.

14. Ohtani M, Gokcen NA. Thermodynamic interaction parameters of elements in liquid iron. *Trans. Metall. Soc. AIME*, 1960; 218: 533–40.

15. Bodsworth C, Bell HB. *Physical chemistry of iron and steel manufacture*. London: Longman; 1972.

16. Sigworth GK. Gas fluxing of molten aluminum: part 1, hydrogen removal. *Light Metals*, 1999; 641–8.

17. Pelton AD. The polynomial representation of thermodynamic properties in dilute solutions. *Metal. Mater. Trans. B*, 1997; 28: 869–76.

18. Ciftja A, Engh TA, Tangstad M. Refining and recycling of silicon: a review. Trondheim, Norway: Report of Norwegian University of Science and Technology; 2008.

19. Safarian J, Tranell G, Tangstad M. Processes for upgrading metallurgical grade silicon to solar grade silicon. *Energy Procedia*, 2012; 20: 88–97.

20. Mostafa A, Medraj M. Binary phase diagrams and thermodynamic properties of silicon and essential doping elements (Al, As, B, Bi, Ga, In, N, P, Sb and Tl). *Materials*, 2017; 10(6): 676.

21. Safarian J, Kolbeinsen L, Tangstad M. Thermodynamic activities in silicon binary melts. *J. Mater. Sci.*, 2012; 47: 5561–80.

22. Bakke P, Klevan O-S, Thermodynamics of liquid silicon based alloys. In: *Proc. Process Technology Conference*, Vol. 14, *Ladle processing: metallurgy and operations, Orlando, Florida, USA, 12-15 Nov., 1995*. Warrendale, PA: Iron and Steel Society/AIME; 1995.

23. Sigworth GK. Thermodynamics of dilute liquid silicon alloys. *Canad. Metal. Q.*, 2020; 59(3): 251–61.

24. Robinson VS, Tarby SK. The thermodynamic properties of liquid Ag-Si alloys. *Metal. Trans.*, 1971; 2(5): 1347–52.

25. Tanahashi M, Fujisawa T, Yamauchi C. Thermodynamics of the Si-B-N system. *Proc. Conference: Value-Addition Metallurgy, San Antonio, Texas, USA, 16-19 Feb., 1998*. Warrendale, PA: Minerals, Metals and Materials Society of AIME, 1998; pp. 103–9.

26. Tanahashi M, Fujisawa T, Yamauchi C. Activity of boron in molten silicon. *J. Min. Mater. Inst. Japan* (Shogen-to-Sozai), 1998; 114: 807–12.
27. Zaitsev AJ, Kodentsov AA. Thermodynamic properties and phase equilibria in the Si-B System. *J. Phase Equilibria*, 2001; 22(2): 126–135.
28. Yoshikawa T, Morita K, Kawanishi S, Tanaka T. Thermodynamics of impurity elements in solid silicon. *J. Alloys Compd.*, 2010; 490: 31–41.
29. Dalaker H, Tangstad M. Time and temperature dependence of the solubility of carbon in liquid silicon equilibrated with silicon carbide and its dependence on boron levels. *Mater. Trans.JIM*, 2009; 50(5): 1152–6.
30. Nozaki T, Yatzurugi Y, Akiyama N. Concentration and behavior of carbon in semiconductor silicon. *J. Electrochem. Soc.*, 1970; 117(12): 1566–8.
31. Witusiewicz V, Arpshofen I, Sommer F. Thermo-dynamics of liquid Cu-Si and Cu-Zr alloys. *Z. Metallkunde.*, 1997; 88: 866–72.
32. Eldridge JM, Miller E, Komarek KL. Thermodynamic properties of liquid magnesium-silicon alloys: discussion of the Mg-Group IVB systems. *Trans. Met. Soc. AIME*, 1967; 239(6): 775–81.
33. Rao YK, Belton GR. Thermodynamic properties of Mg-Si system. In: Gocken NA (ed.), *Metallurgy—a Tribute to Carl Wagner*. Warrendale, PA: TMS-AIME; 1981: pp. 75–96.
34. Dalaker H, Tangstad M. Temperature dependence of the solubility of nitrogen in liquid silicon equilibrated with silicon nitride. *Mater. Trans., JIM*, 2009; 50(11): 2541–4.
35. Stull DR, Prophet H, eds. *JANAF Thermochemical Tables*, 2nd edn. Washington, DC: US. Department of Commerce, National Bureau of Standards; 1971.
36. O'Mara WC. Oxygen, carbon and nitrogen in silicon. In: O'Mara WC, Herring RB, Hunt LP, eds. *Handbook of Semiconductor Silicon Technology*, pp. 451-549. Park Ridge, NJ: Noyes; 1990.
37. Carlberg T. Calculated solubilities of oxygen in liquid and solid silicon. *J. Electrochem. Soc.* 1986; 133(9): 1940–2.
38. Liang S-M, Schmid-Fetzer R. Modeling of thermodynamic properties and phase equilibria of the Si-P system. *J. Phase Equilibria Diffus.*, 2014; 35(1): 24–35.
39. Zaitsev AI, Litvina AD, Shelkova NE. Thermodynamic properties of Si–P melts. *High Temp.*, 2001;39(2): 227–32.
40. Miki T.Morita K.Sano N. Thermodynamic properties of titanium and iron in molten silicon. *Metal. Mater. Trans. B*, 1997; 28(5): 861–7.
41. Miki T, Morita K, Sano N. Thermodynamic properties of Si-Al, -Ca, Mg binary and Si-Ca-Al, -Ti-Fe ternary alloys. *Mater. Trans., JIM*, 1997; 40: 1108–16.
42. Wagner C. *Thermodynamics of Alloys*, p. 51. Reading, MA: Addison-Wesley; 1962.
43. Tang K, Øvrelid EJ, Tranell G, Tangstad M. Thermochemical and kinetic databases for the solar cell silicon materials. In Nakajima K, Usami N, eds. *Crystal Growth of Si for Solar Cells*, Advances in Materials Research series, pp. 219–50. Berlin and Heidelberg: Springer; 2009.
44. McKelvey AL. Retrograde solubility in semiconductors. *Metall. Mater. Trans. A*, 1996; 27(9): 2704–7.
45. Fischler S. Correlation between maximum solid solubility and distribution coefficient for impurities in Ge and Si. *J. Appl. Phys.*, 1962; 33: 1615.
46. Pizzini S. *Physical chemistry of semiconductor materials and processes*, pp. 290–5. West Sussex, UK: Wiley; 2015.
47. Trumbore FA. Solid solubilities of impurity elements in germanium and silicon. *Bell Syst. Tech. J.*, 1960; 39(1): 205–33.
48. Morita K, Miki T. Thermodynamics of solar-grade silicon refining. *Intermetsallics*, 2003; 11(11–12): 1111–17.

49. Coletti G, McDonald G, Yang D. Role of impurities in solar silicon. In: Pizzini S, ed., *Advanced silicon materials for photovoltaic applications*, pp. 79–125. Chichester: Wiley; 2011.

50. Wilson FG, Heesom MJ, Nicholson A, Hills AWD. Effect of fluid flow characteristics on nozzle blockage on aluminium-killed steels. *Iron-mak. Steelmak.*, 1987; 14: 296–309.

51. Saxena SK, Sandberg H, Waldenstrom T, Persson A, Steensen S. Mechanism of clogging of tundish nozzles during continuous casting of aluminium killed steel. *Scand. J. Metall.*, 1978; 7: 126–33.

52. Hilty DC, Farrell JW. Steel flow through nozzles: influence of deoxidizers. *Iron Steelmaker*, 1975; 2(4): 17–22.

53. Vainola R, Karvonen P, Helle LW. Establishment of calcium treatment practices at Ovako— historical development and evaluation of alternative methods. *Proceedings 4th International Conference on Injection Metallurgy, MEFOS, Luleoa, Sweden, 1986*; 23: pp. 1–19.

54. Elliott JF, Gleiser M, Ramakrishna V. *Thermochemistry for steelmaking*. Reading, MA: Addison-Wesley; 1963.

55. Sharma RA, Richardson FD. Activities in lime-alumina melts. *J. Iron Steel Inst. London*, 1961; 198: 386–90.

56. Dewing EW. Thermodynamics of the system NaF-AlF3. Part 3: Activities in liquid mixtures. *Metall. Trans. B*, 1972; 3(2): 495–501.

57. Warren BE. X-ray determination of the structure of liquids and glass. *J. Appl. Phys.* 1937; 8: 645.

58. Wong J, Angell CA. *Glass structure by spectroscopy*. New York: Marcel Dekker; 1976.

59. Flood H, Forland T. The acidic and basic properties of oxides. *Acta Chem. Scand.* 1947; 1(6): 592–604.

60. Richardson FD. Thermodynamic aspects of slags and glasses. In *Vitreous State*. The Glass Delegacy of the University of Sheffield, Sheffield; 1955.

61. Shirota Y, Katohgi K, Klein K, Engell H-J, Janke D. Phospate capacity of $FeO–Fe_2O_3–CaO–P_2O_5$ and $FeO–Fe_2O_3–CaO–CaF_2–P_2O_5$ slags by levitation melting. *Trans ISIJ.*, 1985; 25: 1132–40.

62. Selin R. *The role of phosphorus, vanadium and slag forming oxides in direct reduction based steelmaking*. PhD thesis, Royal Inst. of Technology, Stockholm, Sweden; 1987.

63. Muan A, Osborn EF. *Phase equilibria among oxides in steelmaking*. Reading, MA: Addison-Wesley; 1965.

64. Kittel C. *Solid state physics*, 6th. edn. New York: Wiley; 1976.

65. Chapman S, Cowling TG. *The mathematical theory of nonuniform gases*, 3rd edn. Cambridge, UK: Cambridge University Press; 1970.

66. Geiger GH, Poirier, DR. *Transport phenomena in metallurgy*. Reading, MA: Addison-Wesley; 1973.

67. Herning F, Zipperer L. Beitrag zur Berechnung der Zahigkeit technischer Gasgemische aus den Zahigkeitswerten der Einzelbestandteile. *Gas- und Wasserfach.* 1936; 79: 49–54, 69–73.

68. Iida T, Guthrie RIL. *The physical properties of liquid metals*. Oxford: Clarendon Press; 1988.

69. Eisenstein H, Gingrich NS. The diffraction of x-rays by argon in the liquid, vapor and critical regions. *Phys. Rev.*, 1942; 62: 261–9.

70. Ocken H, Wagner CN J. Temperature dependence of the structure of liquid indium. *Phys. Rev.*, 1966; 149: 122–30.

71. Longuet-Higgins HC, Pople JA. Transport properties of a dense fluid of hard spheres. *J. Chem. Phys.*, 1956; 25: 884–9.

72. Born M, Green HS. A general kinetic theory of liquids III. Dynamical properties. *Proc. Roy. Soc., A*, 1947; 190: 455–74.

73. Iida T, Guthrie R. Performance of a modified Schytil model for the surface tension of liquid metallic elements at their melting point temperatures. *Metall. Trans. B*, 2010; 41B(2): 437–47.

74. Iida T, Guthrie R. *The thermophysical properties of metallic liquids*, Vol. 1, pp. 236–8. Oxford New York: University Press; 2015.

75. Mills KC, Su YC. Review of surface tension data f or metallic elements and alloys. *Int. Materials Rev.*, 2006; 51(6): 329–51.

76. Mei OS, Lu K. Melting of metals role of concentration and migration of vacancies at surfaces. *Philos. Mag. Lett.*, 2008; 88(3): 203–11.

77. Samanta A, Tuckerman ME, Yu T-Q, Weinan E. Microscopic mechanisms of equilibrium melting of a solid. *Science*. 2014; 346(6210): 729–732.

78. Born M. Thermodynamics of crystals and melting. *J. Chem. Phys.*, 1939; 7: 591–603.

79. Chapman TW. The viscosity of liquid metals. *AICHE*, 1966; 12: 395–400.

80. Khusnutdinoff RM, Mokshin AV, Beltyukov AL, Olyanina NV. Viscosity and structure configuration properties of equilibrium and supercooled liquid cobalt. *Phys. Chem. Liquids*, 2018; 56(5): 561–70.

81. Hansen JP, McDonald IR. *Theory of simple liquids*. New York: Academic Press; 2006.

82. Samsonnikov AV, Muratov AS, Roik AS, Kazimirov VP. Molecular dynamics simulation of the structure of aluminum in the liquid and supercooled states. *Russ. Metall.*, 2013; 2013(5): 367–74.

83. Heine V, Cohen M, Weir D. *The pseudopotential concept; the fitting of pseudopotential to experimental data and their subsequent application; pseudopotential theory of cohesion and structure*. New York: McGraw-Hill; 1970.

84. Jakse N, Pasturel A. Liquid aluminum: atomic diffusion and viscosity from ab initio molecular dynamics. *Sci. Rep.* 2013; 3(3135).

85. Rosenfeld YA. A quasi-universal scaling law for atomic transport in simple fluids. *J. Phys. Condens. Matter*, 1999; 11(28): 5415–27.

86. Andrade EN. A theory of the viscosity of liquids—Part I. *Philos. Mag.*, 1934; 17: 497-511.

87. Lindemann FA. The calculation of molecular vibration frequencies. *Phys. Z.*, 1910; 11: 609–12.

88. Lawson CC. Physics of the Lindemann melting rule. *Philos. Mag.*, 2009; 89(22–4): 1757–70.

89. Iida T, Guthrie IL. *The thermophysical properties of metallic liquids*, Vol 1, pp. 14, 236–8. Oxford: Oxford University Press; 2015. (See ref. 73.)

90. Yeum KS, Speiser R, Poirier DR. Estimation of the surface tensions of binary liquid alloys. *Met. Trans. B*, 1989; 20(5): 693–703.

91. Smithells CJ, Brandes EA. *Smithells metals reference book*, 6th edn. London: Butterworths; 1983.

92. Janz GJ. *Moltensaltshandbook*. New York: Academic Press; 1967.

93. Turkdogan ET, Bills PM. A critical review of viscosity of CaO–MgO–Al$_2$O$_3$–SiO$_2$ melts. *Am. Ceram. Soc. Bull.* 1960; 39: 682.

94. Kozakevitch P. (1961). Viscosity of lime-alumina-silica melts between 1600 and 2100 °C. In: St Pierre GR (ed.), *Physical chemistry of process metallurgy*, p. 97. New York: Wiley Interscience; 1961.

95. Kaptay G. A unified equation for the viscosity of liquid metals. *Z.Metallkd.* 2005; 96(1): 24–31.

96. Hirai M. Estimation of viscosities of liquid alloys. *ISIJ Int.*, 1992; 33(2): 251–8.

97. Fowler RH. A tentative statistical theory of Macleods equation for surface tension and the parachor. *Proc. Roy. Soc., A*, 1937; 159: 229–46.

98. Nogi K, Ogino K, McLean A, Miller WA. The temperature coefficient of the surface tension of pure liquid metals. *Met. Trans. B*, 1986; 17: 163–70.

99. Kubaschewski O, Alcock CB. *Metallurgical thermochemistry*, 6th edn. Oxford: Pergamon Press; 1983.

100. Lu HM, Jiang Q. Surface tension and its temperature coefficient for liquid metals. *J. Phys. Chem. B*, 2005; 109(32): 15463–8.

101. Hultgren RR. Interatomic forces and chemical bonding energies. In: Pask JA (ed.), *An atomistic approach to the nature and properties of materials*, Chapter 3. New York: Wiley; 1967.
102. Berry RS, Rice SA, Ross J. *Physical chemistry*. New York: Wiley; 1980.
103. Sherman J. Crystal energies of ionic compounds and thermochemical applications. *Chem. Rev.*, 1932; 11: 93–170.
104. Madelung E. Das elektrische Feld in Systemen von regelmassig angeordneten Punktladungen. *Phys. Z*, 1918; 19: 524.
105. Adamson AW. *Physical chemistry of surfaces*. New York: Wiley; 1982.
106. Mulheran PA. Surface free-energy calculations and the equilibrium shape of NaCl crystals. *Modelling Simul. Material Sci. Eng.* 1994; 2(6): 1123–9.
107. Huang B. *Work of fracture of sodium chloride single crystals*. Iowa state university: retrospective thesis and dissertations; 1986.
108. Baker AD, Baker MD, Hanusa CRH. Corner ion, edge-center ion, and face-center ion Madelung expressions for sodium chloride. *J. Math. Chem.*, 2011; 49(6): 1192–8.
109. Bahadur R, Russell LM, Alavi S. Surface tensions in NaCl-water-air systems from MD simulations. *J. Phys. Chem. B*, 2007; 111(41): 11989–96.
110. Li B. Density-functional theory and quantum chemistry studies on 'dry' and 'wet' NaCl(001). PhD Technische Universität Berlin; 2009.
111. Defay R, Prigogine I, Bellemans A. *Surface tension and adsorption*, trans. Everett DH. London: Longmans, Green; 1966.
112. Elliott JF. Physical chemistry of liquid metal solutions. In: Tien JK, Elliott JF, eds. *Metallurgical Treatises*. Warrendale, Pa: Metallurgical Soc. of AIME; 1981.
113. Belton GR. Langmuir adsoprtion, the Gibbs adsorption isotherm and interfacial kinetics in liquid metal systems. *Met. Trans. B*, 1976; 7(1): 35–42.
114. Ogino K, Hara S, Miwa T, Kimoto S. The effect of oxygen content in molten iron on the interfacial tension between molten iron and slag. *Trans. ISIJ*, 1984; 24: 522–31.
115. Aylward GH, Findlay TJV. *SI chemical data*. Chichester, UK: Wiley, 1974.
116. Sahao P, Debroy T, McNallan MJ. Surface tension of binary metal-surface active solute systems under conditions relevant to welding metallurgy. *Met. Trans. B*, 1988; 19: 483–90.
117. Lide DR. *CRC handbook of chemistry and physics*, 71st edn. Ann Arbor, MI: CRC Press; 1990–1.
118. Darken LS, Gurry RW, Bever MB. *Physical chemistry of metals*. New York: McGraw-Hill; 1953.
119. Frisvold F. *Filtration of aluminium: theory, mechanisms, and experiments*. Dr. Ing. Thesis, Trondheim, Norway: The Norwegian Institute of Technology; 1990.
120. Turnbull D. Formation of crystal nuclei in liquid metals. *J. Appl. Phys.*, 1950; 21: 1022–8.
121. Turnbull D. Correlation of liquid-solid interfacial energies calculated from supercooling of small droplets. *J. Chem. Phys.*, 1950; 18: 769.
122. Nishi Y, Mikagi K, Igarashi A. Solid-liquid interfacial energy of non-metallic bonded elements. *J. Mater. Sci.* 1988; 23(2): 400–4.
123. Nishi Y, Igarashi A, Kubo Y, Ninomiya N, Mikagi K. Solid–liquid interfacial energy of Pd-16.5 at.% Si glasses. *Mater. Sci. Engng.* 1988; 97: 199–201.
124. Jones. H. The solid–liquid interfacial energy of metals: calculations versus measurements. *Mater. Lett.* 2002; 53: 364–6.
125. Lamb H. *Hydrodynamics*, 6th edn. New York: Dover; 1932.
126. Viswanath, DS, Mathur, BC. Thermal conductivity of liquid metals and alloys. *Met. Trans. B*, 1972; 3: 1769–72.
127. Ho CY, Powell RW, Liley PE. Thermal conductivity of elements: a comprehensive review. *J. Phys. Chem. Ref. Data: Suppl.3.* Washington, DC: American Chemical Society for the National Bureau of Standards; 1974.

128. Meaden GT. *Electrical resistance of metals*. New York: Plenum Press; 1965.

129. Barin I, Knacke O. *Thermodynamic properties of inorganic substances*. Berlin: Springer Verlag; 1973.

130. Barin I, Knacke O, Kubaschewski O. *Thermochemical properties of inorganic substances, supplement*. Berlin: Springer Verlag; 1977.

131. Baker EM. The strontium oxide-carbon dioxide system in the pressure range 0.3–400 atmospheres. *J. Chem. Soc.* 1963; Part I: 339–46.

132. Chang YA, Ahmad N. *Thermodynamic data on metal carbonates and related oxides*. Warrendale, PA: Metallurgical Society of AIME; 1982.

133. Schurmann E, Schmidt R. Vapor pressure equations and caloric variables of pure liquid and solid (β) calcium. *Arch. Eisenhüttenwesen*, 1975; 46: 773–5.

134. Kubaschewski O. The thermodynamic properties of double oxides. *High Temp–High Press.*, 1972; 4: 1–12.

135. Hills AWD. Equilibrium decomposition pressure of calcium carbonate between 700 and 900 °C. *Inst. Min. Metall Trans.* 1967; Sect. C76: 241–5.

136. Turkdogan ET, Rice BB, Vinters JV. Sulfide and sulfate solid solubility in lime, magnesia, and calcined dolomite: Part I. CaS and $CaSO_4$ solubility in CaO. *Metall. Trans. B*, 1974; 5(7): 1527-35.

137. Gschneider KA, Kippenhan N. *Thermochemistry of the rare earth carbides, nitrides and sulfides for steelmaking*. IS-RIC-5, 1971; Rare Earth Inf. Cent., Iowa State Univ., Ames.

138. O'Dell KD, Hensley EB. The decomposition pressure, congruent melting point, and electrical resistivity of cerium nitride. *J. Phys. Chem. Solids*, 1972; 33(2): 443–9.

139. Baker FB, Holley CE. Enthalpy of formation of cerium sesquioxide. *J. Chem. Eng. Data*, 1968; 13: 405–8.

140. Fruehan RJ. The free energy of formation of Ce_2O_2S and the non-stoichiometry of cerium oxides. *Metall. Trans.*, 1979; B10: 143–8.

141. Rosenqvist T. A thermodynamic study of the iron, cobalt and nickel sulfides. *J. Iron Steel Inst.*, 1954; 176: 37–57.

142. Schwerdtfeger K. The nitriding of chromium in N_2–H_2 gas mixtures at elevated temperatures. *Trans Metall. Soc. AIME*, 1967; 239: 1432–8.

143. Mazandarany FN, Pehlke RD. Standard free energy of formation of Cr_2O_3. *J. Electrochem. Soc.*, 1974; 121: 711–14.

144. Toker NY. Equilibrium phase relations and thermodynamics for the systems chromium-oxygen and iron-chromium-oxygen in the temperature range from 1500 to 1825 °C. PhD thesis, Pennsylvania State University.1978.

145. Hager JP, Elliott JF. The free energy of formation of CrS, Mo_2S_3 and WS_2. *Trans. Metall. Soc. AIME*, 1967; 239: 513–28.

146. Jacob KT, Rao DB, Nelson HG. Stability of chromium (III) sulfate in atmospheres containing oxygen and sulfur. *Metall. Trans. A*, 1979; 10: 327–31.

147. Kellogg HH. Thermochemical properties of the system Cu-S at elevated temperatures. *Can. Metall. Q.*, 1969; 8: 3–23.

148. King EG, Mah AD, Pankratz LB. *Thermodynamic properties of copper and its inorganic compounds*. INCRA series on the metallurgy of copper, Monograph II.Washington, DC, US Bureau of Mines; 1973.

149. Pemsler JP, Wagner C. Thermodynamic investigations on chalcopyrite. *Metall. Trans. B*, 1975; 6: 311–20.

150. Kellogg HH. A critical review of sulfation equilibria. *Trans. Metall. Soc. AIME*, 1964; 230: 1622–34.

151. Chipman J. Thermodynamics and phase diagram of the Fe-C system. *Metall. Trans. B*, 1972; 3: 55–64.

152. Hillert M, Jarl M. A thermodynamic analysis of the iron-nitrogen system. *Metall. Trans. A*, 1975; 6: 553–9.

153. Hillert M, Staffanson L-I. An analysis of the phase equilibria in the Fe-FeS system. *Metall. Trans. B*, 1975; 6: 37–41.

154. Jacob KT, Alcock CB. The oxygen potential of the systems $Fe-FeCr_2O_4 + Cr_2O_3$ and $Fe + FeV_2O_4 + V_2O_3$ in the temperature range 750-1600 °C. *Metall. Trans. B*, 1975; 6: 215–21.

155. Skeaff JM, Espelund AW. An E.M.F. method for the determination of sulfate-oxide equilibria results for the Mg, Mn, Fe, Ni, Cu and Zn systems. *Can. Metall. Q*, 1973; 12: 445–58.

156. Alcock CB, Zador S. Thermodynamic study of the manganese/ manganese-oxide system by the use of solid oxide electrolytes. *Electrochem. Acta*, 1967; 12: 673–7.

157. Schwerdtfeger K, Muan A. Equilibria in the system Fe-Mn-O involving (Fe, Mn) O and (Fe, Mn)$_3$O$_4$ solid solutions. *Trans. Metall. Soc. AIME*, 1967; 239: 1114–19.

158. Ingraham TR. Thermodynamics of the Mn-S-O system between 1000 K and 1250 K. *Can. Metall. Q.*, 1966; 5: 109–22.

159. Turkdogan ET, Olsson RG, Vinters JV. Sulfate and sulfide reactions in the Mn-S-O system. *Metall. Trans. B*, 1977; 8: 59–65.

160. Kellogg HH. Thermodynamic properties of the oxides of copper and nickel. *J. Chem. Eng. Data*, 1969; 14: 41–4.

161. Pehlke RD, Elliott JF. High-temperature thermodynamics of the silicon, nitrogen, silicon-nitride system. *Trans. Metall. Soc. AIME*, 1959; 215: 781–94.

162. Rosenqvist T, Tungesvik K. Thermodynamics of silicon sulfides and zinc sulfide. *Trans. Faraday Soc.*, 1971; 67: 2945–51.

163. Baker, EM. The strontium oxide carbon dioxide system in the pressure range 0.3-400 atmospheres. *J. Chem. Soc.*, 1963; Part 1: 339–46.

164. Ackermann RJ, Rauh EG. Thermodynamic properties of ZrO (g) and HfO (g); a critical examination of isomolecular oxygen-exchange reactions. *J. Chem. Phys.*, 1974; 60: 2266–71.

165. Gupta DK, Seigle LL. Free energies of formation of WC and W2C, and the thermodynamic properties of carbon in solid tungsten. *Metall. Trans. A*, 1975; 6: 1939–44.

RECOMMENDED FURTHER READING:

• Rosenqvist T. *Principles of extractive metallurgy*, 2nd edn. New York: McGraw- Hill; 1988.
• Flynn H, Morris AE, Carter D. An iterative gas-phase removal version of SOLGASMIX. *Proc. the 25th CIM conference of metallurgists, TMS-CIM*, Toronto, Canada; 1986.
• Tipp J-E. *The development of a thermochemical database and its application in metallurgical process modeling*. Teknisk doktor Thesis. Stockholm, Sweden. Institutionen for Teoretisk Metallurgy, KTH: 1984.
• Bale CW, Eriksson G. Metallurgical thermochemical databases—a review. *Can. Metall. Q.*, 1990; 29: 105–32.
• Sundman B. *Application of Computer Techniques on the Treatment of the Thermodynamics of Alloys. Doctoral thesis, Royal Institute of Technology, Stockholm Sweden, March 9, 1981.*

- Iguchi Y. KorosPJ, St. Pierre GR. The thermochemistry of ferrous melts. in. *Proc of The Elliott Symposium on Chemical Process Metallurgy*. Warrendale, PA: ISS-AMIE; 1990.
- Lupis CHP. On the use of polynomials for the thermodynamics of dilute metallic solutions. *Acta Met.* 1968; 16(11), 1365–75.
- Lupis CHP, Elliott JF. The relationship between the interaction coefficients ε and e. *Trans. Met. Soc. AIME*, 1965; 233: 257–8.
- Chhabra RP. Application of Hildebrands fluidity model to molten metallic alloys. *Z. Metallkd.*, 1989; 80(9): 658–62.
- Sigworth GK, Engh TA. Chemical and kinetic factors related to hydrogen removal from aluminium. *Met. Trans. B*, 1982; 13: 447–60.

3

Mixing, Mass Transfer, and Numerical Models

3.1 Introduction

Mixing and stirring are terms widely used when discussing melt treatment. These terms are useful in that they have an immediate intuitive meaning. However, a worker in refining metallurgy very early finds it necessary to give them more substance. In Section 3.2 an attempt is made to explain and clarify the meaning of these terms.

Mass transfer coefficients, k, at various interfaces will be introduced. The mass transfer coefficient relates a driving force to a mass transfer flux. The driving force is typically either a difference across an interface or a gradient across a phase in the chemical potential or concentration of a compound. In the examples, values of k are often presented as purely empirical quantities. It should be noted that we very often have assumed that at gas–melt and solid–melt interfaces the mass transfer coefficient in the melt is rate limiting. Here, one seeks to relate the mass transfer coefficients to quantities such as the diffusion coefficients, D, and the velocity along the interface. The effect of turbulence on mass transfer is discussed. Finally, resistances to mass transfer in the melt and gas phases are compared.

In Chapters 4 and 5 flow and mixing are described in terms of idealized extreme situations: plug flow (no mixing) and complete mixing in so-called back-mixed tanks or reactors. In real life we attempt to develop reactors that approach the idealized case.

However, in the planning stage or in operation the question arises about how close we are to the idealized reactors, and what steps can be taken for improvement. It becomes necessary to perform a mathematical analysis of the turbulent flow and to carry out extensive numerical calculations to compare with physical models and molten metal experiments. This chapter won't delve too deeply into the difficult fields of fluid flow and numerical analysis. Some basic concepts in turbulence theory are mentioned and the potential benefits of numerical calculations are sketched. Some references to the specialized literature are given.

3.1.1 Mass Transfer Coefficient

In many situations, if not all, where a certain volume is subjected to an external force or influence, it will take some time for a reaction inside the volume to occur. This is true when

Principles of Metal Refining and Recycling. Thorvald Abel Engh, Geoffrey K. Sigworth, and Anne Kvithyld, Oxford University Press.
© Oxford University Press 2021.
DOI: 10.1093/oso/9780198811923.003.0003

we are dealing with heat transfer, chemical reactions, mass transfer, etc. In all these examples some gradient will act as a driving force for the reaction to happen.

In this chapter we will focus on mass transfer, that is, the movement of some elements or compounds from one place (or phase) to another. In most cases we can consider a system in a steady state, when no time dependency is involved. For such a steady state the mass transfer is governed by Fick's first law:

$$\dot{n} = -D\frac{dc}{dx},\qquad(3.1)$$

where $\dot{n}$ (with dimensions such as moles per time and area) is the flux of something over an interface, and D is the diffusion coefficient for that substance (for many elements these are tabulated values with units of m^2/s), and dc/dx is the concentration gradient over a diffusion boundary layer. This concentration gradient is actually the driving force for the transport.

However, since the thickness of the boundary layer is unknown (or at least extremely difficult to measure), the value of the gradient is also unknown. To overcome this, we have introduced the mass transfer coefficient, k, which relates the concentration difference between the interface and the bulk phase to the transport flux:

$$\dot{n} = k\,(c_i - c_b),\qquad(3.2)$$

where the mass transfer coefficient, k, is in units of m/s, the concentration in the bulk phase is c_b, and c_i is a hypothetical concentration at the interface in equilibrium with the relevant concentration on the other side of the interface. For instance, if we are looking at removal of hydrogen from aluminium by gas purging, and the flux is the transport of dissolved hydrogen in the liquid to the gas bubbles, the interface concentration is given by the reaction and equilibrium:

$$\underline{H} = \frac{1}{2}H_2(g)\qquad(3.3)$$

$$K(T) = \frac{\sqrt{p_{H_2}}}{f_H c_i}.\qquad(3.4)$$

Then the flux of hydrogen to the gas bubbles is

$$\dot{n} = k\left(\frac{\sqrt{p_{H_2}}}{f_H K(T)} - c_b\right)\qquad(3.5)$$

It will be shown later how the mass transfer coefficient relates to the diffusion coefficient. This relation is dependent on the different physical situations; melt flow past a solid wall or a free surface will give different k versus D dependencies.

3.2   Mixing and Circulation Flow; Flow Models

To illustrate problems concerning mixing, the addition of alloys is examined. When alloys in lump, granule, or particle form have melted and partly dissolved into a surrounding melt, they still, to a degree, retain their identity as concentrated volumes of solute surrounded by melt. They still have not mixed properly into the melt, so that the concentration c at an arbitrary point is not equal to the average concentration of the melt, c_m. Such a state is, arguably, only a problem if sufficient mixing has not occurred by the time that solidification takes place.

Most alloy additions to steel are more reactive to oxygen (and even nitrogen) in the air than the melt. Also, they may react with the slag or refractory. This causes a loss of the alloy addition and even delivers an extra penalty in the form of unwanted reaction products such as oxide inclusions. Thus ideally, flow conditions and the method of addition should prevent concentration build-up near interfaces, the pick-up of impurities, and increased losses. Now this sounds like a different task, especially if one considers that convection seems to be necessary in order to attain good mixing. Furthermore, there are only four main ways of stirring: pumping, bubbling, electromagnetic, and mechanical (see Chapter 5). At high temperatures mechanical stirring is a problem due to the attack on the impeller material.

Figure 3.1 illustrates how difficulties with unwanted reactions may be solved. A ceramic ring keeps the slag away from an 'eye' swept open by rising bubbles from a porous plug. The alloy additions are fed at a moderate rate into the melt through the eye. The bubbling action gives good circulation of the melt around the alloy additions and dispenses the alloy throughout the melt. Attack is concentrated on the simple refractory ring which can be replaced when necessary. The potentially oxidizing slag is unable to react with the alloy, and the high circulation rates of the melt and moderate addition rates of the alloy ensure that no excessively high alloy concentrations occur at any time.

The vacuum circulation RH (Rheinstahl-Heraeus) reactor shown in Fig. 3.2 operates on a similar principle with regard to alloy additions. The increased yield is due to the controlled addition of alloys in the vacuum chamber environment. To be fair it should be mentioned that the penalty is that the RH is susceptible to attack on the refractories, especially in the legs and bottom surface. This is due either to the effect of the alloy additions or to erosion from the shear forces created by the flowing metal and rising bubbles.

When the mechanism or process of interest takes place in a limited region of the melt, or in a special reactor as described in the two examples given above, it is often helpful to describe mixing in terms of a circulation flow, $\dot{Q}$,[1,2] with units of m³/s. We show that sometimes $\dot{Q}$ can be determined from rather simple reasoning or experiments. For the RH reactor, $\dot{Q}$ may be taken as the flow that circulates through the vacuum unit. Additions of Si have been used to measure this $\dot{Q}$ in an RH reactor.[3] If c_{up} and c_{down} are the Si concentrations (in kg/m³) in the up and down leg, respectively, then

$$\left(c_{down} - c_{up}\right) \dot{Q} = \dot{M}_{Si}, \tag{3.6}$$

where $\dot{M}_{Si}$ is the feed rate in kg/s of Si added in the chamber in the RH. When c_{down} and c_{up} are measured, they increase linearly with time and with the same constant slope. This allows $\dot{Q}$ to be determined from Eq. (3.6). Various cold model and plant experiments[2,3] indicate that

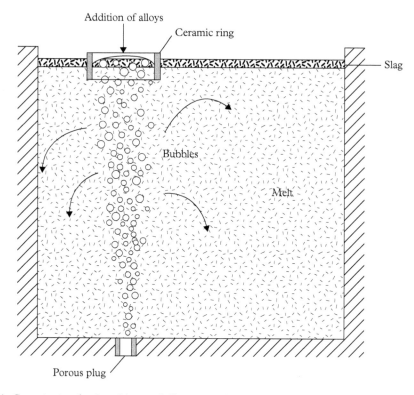

Fig. 3.1 *Ceramic ring for the addition of alloys to steel.*

$$\dot{Q} \propto \left\{ \dot{Q}_g d^4 \ln (p_{in}/p_0) \right\}^{\frac{1}{3}}. \tag{3.7}$$

Here is the flow rate of gas in Nm^3/s, d is the diameter of the up-leg (snorkel), and p_{in} and p_0 are the (static) pressures where the stirring gas enters and exits, respectively.

Equation (3.7) can be derived from elementary energy considerations. The rate at which mechanical energy is irreversibly converted to thermal energy in the up and down legs is

$$\dot{E} = e\rho \dot{Q} u_m^2, \tag{3.8}$$

where the factor e depends on geometry and, slightly, on the Reynolds number. e is on the order of magnitude 1. By comparison, for a rounded 90° valve 0.4 <e< 0.9.[4] u_m is the average velocity over a cross-section of a leg:

$$\dot{Q} = \frac{\pi d^2}{4} u_m. \tag{3.9}$$

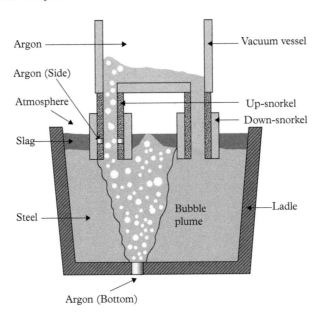

Fig. 3.2 *An RH (Rheinstahl-Heraeus) degasser.*

The stirring power $\dot{E}$ is supplied by the expansion energy of the bubbles:

$$\dot{E} = \dot{Q}_g p_{in} \ln(p_{in}/p_0). \tag{3.10}$$

Here $\dot{Q}_g$ is the gas supply at the inlet point at pressure p_{in}. Equation (3.10) is valid for isothermal (cold) conditions. In hot conditions, the thermal expansion term must also be taken into account. Equations (3.8)–(3.10) give

$$\dot{Q} = \left(\frac{\pi^2 d^4}{16e}\right)^{\frac{1}{3}} \left(\frac{\dot{Q}_g p_{in} \ln(p_{in}/p_0)}{\rho}\right)^{\frac{1}{3}}. \tag{3.11}$$

Comparing Eq. (3.11) with the plant experiments reported by Kawabara et al.[3] gives $e = 0.55$. However, they do not seem to have taken the thermal expansion term of the gas into account. If such a correction is performed, e would be somewhat larger.

In the RH chamber only a small fraction of the metal is vacuum treated at a time. Therefore, it is important at a high rate, $\dot{Q}$, to circulate metal from the ladle in and out of the vacuum chamber. It may be instructive to point out the various mechanisms and challenges encountered in carbon removal in the RH process. Dissolved carbon is removed by reaction with dissolved oxygen according to

$$\underline{C} + \underline{O} = CO(g). \tag{3.12}$$

With the equilibrium constant given by

$$K = \frac{p_{CO}}{a_O \cdot a_C}. \qquad \qquad (3.13)$$

To get low in carbon, the partial pressure p_{CO} should be as small as possible. This is achieved by introducing pure argon gas (or another inert gas) and ensuring there is a large bubble contact area between the bubbles and melt. In addition, there will be some removal at the top surface of the melt as long as the evaporated CO gas is removed from the melt surface inside the vacuum chamber.

Even though the equilibrium (Eq. (3.13)) is independent of the total gas pressure in the bubbles, the removal of the dissolved carbon and oxygen will increase by reducing the total pressure inside the bubbles. The reason for this is that inside the bubbles, the gas pressure p_{tot} is equal to the sum of the partial pressure of CO and argon gas. This total pressure is also equal to the ferrostatic pressure ($\rho g h$) plus the vacuum pressure above the melt p_V:

$$p_{tot} = p_{CO} + p_{Ar}$$
$$= \rho g h + p_V.$$

Now, imagine a bubble fixed at a position in the melt. If the pressure is reduced, and $\rho g h$ is kept constant, the total pressure inside the bubbles is reduced. Since CO and argon both feel the pressure reduction, p_{CO} will be reduced and to keep the equilibrium (3.13) fulfilled; also the $\underline{C}$ and $\underline{O}$ must be reduced.

In all models it has been assumed that there is equilibrium at the gas–metal interface; i.e. Eqn. (3.13) holds. Rates are normally determined by mass transfer in the metal at the interface. However, we will see that in some cases rates are controlled in the gas phase.

The operation of the RH process meets various challenges:

- Short circuiting' of carbon in the ladle between up and down snorkel legs; that is, complete mixing in the main chamber is required.
- The circulation flow, $\dot{Q}$, is not sufficient to provide a good mixing of carbon in the ladle. The gas-lift pump in the up leg does not function properly; if an excess of argon is employed, circulation rates and gas–metal contact areas reach saturation.
- The evacuation capacity must be sufficient to handle argon flow from the up leg (argon added at the bottom of the vacuum chamber) and CO produced, and to give a low pressure above the melt.

At low carbon levels CO bubbles cannot form in the bulk melt due to the high surface energy of the initial bubbles. Argon must be present for reaction (3.12) to proceed.

In many cases it is difficult to obtain $\dot{Q}$ from elementary considerations. It is, for instance, a complex matter to determine $\dot{Q}$ for a bubble-stirred metal bath.[5,6] However, $\dot{Q}$ may be measured in tracer experiments.[1,7]

The inverse value of $\dot{Q}$ may be regarded as a resistance that must be added to the mass transfer resistance. As we will see in Section 4.3 by analogy with electrical circuits, the mass transfer coefficient is the inverse of the mass transfer resistance. Two or more resistances can be combined into a total resistance which again allows us to find the total mass transfer coefficient. Let us illustrate this by constructing a simple example where carbon is removed by reaction with oxygen in an RH operating as part of a future continuous steelmaking process. We assume complete mixing in the RH chamber, so that the concentration in this unit is c_r and complete mixing in the ladle with concentration c (kmol/m^3). $\dot{Q}$ is the volume flow of melt through the up and down snorkels. The vacuum–melt interface area is A in the chamber with the total mass transfer coefficient k_t for carbon across the interface. Then the number of moles of carbon transferred per second from ladle to chamber is

$$\dot{N} = \dot{Q}(c - c_r), \tag{3.14}$$

and the moles per second removed from the melt in the chamber to the vacuum system is

$$\dot{N} = k_t A c_r. \tag{3.15}$$

If Eq. (3.14) is divided by $\dot{Q}$ and (3.15) by $k_t A$, and the resulting two equations are summed, one obtains

$$\dot{N} \left(\frac{1}{\dot{Q}} + \frac{1}{k_t A} \right) = c. \tag{3.16}$$

It is seen that $\frac{1}{\dot{Q}}$ must be added to the mass transfer resistance $1/(k_t A)$ to give the total resistance. Thus, if impurities are extracted in only part of our system, the transportation resistance $\frac{1}{\dot{Q}}$ must also be taken into account. This is one reason why vacuum units in some cases may be less efficient than gas purge units. The impurity is removed only from a surface layer. If the circulation, $\dot{Q}$, between the bulk melt and the surface is poor in the vacuum system, the resistance $\frac{1}{\dot{Q}}$ is large. A mass balance for carbon in our continuous steelmaking process gives

$$\dot{q} c_{in} = \dot{q} c + \dot{N}$$

or

$$\dot{q} c_{in} = \dot{q} c + \frac{c}{\frac{1}{k_t A} + \frac{1}{\dot{Q}}}, \tag{3.17}$$

or

$$\frac{c}{c_{in}} = \frac{1}{1 + \left(\frac{\dot{q}}{k_t A} + \frac{\dot{q}}{\dot{Q}}\right)^{-1}},$$

where $\dot{q}$ is the volumetric flow of metal in and out of the whole reactor. This means that the 'dimensionless' resistance

$$\frac{\frac{1}{k_t A} + \frac{1}{\dot{Q}}}{\frac{1}{\dot{q}}} \ll 1 \tag{3.18}$$

in order to give good carbon removal. This result may be compared with the case of the simple completely mixed reactor ($\dot{Q} \rightarrow \infty$) in Chapter 4. There it is only required that

$$\frac{\dot{q}}{k_t A} \ll 1. \tag{3.19}$$

In the equations in Chapter 4, $\dot{Q}$ can be taken into account by replacing $k_t A / \dot{q}$ with

$$\left(\frac{\dot{q}}{k_t A} + \frac{\dot{q}}{\dot{Q}}\right)^{-1}.$$

Similar relations are obtained for batch reactors, for instance for a conventional RH/ladle system. However, the mathematics and models become more complex.

Both in continuous reactors and batch reactors, there will be gradients in concentration, zones which are near-stagnant where the exchange flow with the other zones is small, and there will be short-circuiting effects. Short-circuiting here means that part of the metal flow circumvents the reaction zone and is not treated in the reactor. Both the stagnant zones and short-circuiting can be determined from tracer experiments.[1,7,8] Once this is known together with the total mass transfer coefficients k_t and contact areas A, the conversion in an extractive reactor[6,9] or the fraction removed in a refining unit can be calculated. A fundamental problem is that the reactor must physically exist before tracer measurements can be carried out. Alternatively, tracer experiments can be carried out on a smaller-scale reactor or in a water model. The water model may be built in full scale. Laboratory measurement results must be scaled up for the melt tracer experiments, or transformed from water to molten metal in the full-scale experiments. Scaling requires a good understanding of the mechanisms and phenomena that occur and may lead to extensive calculations. Furthermore, it is often only practical to physically model the flow aspects of a process. Phenomena such as a change of phase (solidification), heat, and mass transfer can usually be included only in a mathematical model. As a result, it is tempting to develop 'complete' models that include flow. Such models should apply to both small and large scales and to water and molten metal. This allows them to be checked and adjusted by comparison with model, laboratory, and industrial experimental results.

3.3   Mass Transfer to Walls

In Chapter 5, where we study the transfer of inclusions to a wall, the viscous (laminar flow) boundary layer δ_2 is neglected. The justification was that the inclusions penetrated this layer simply due to their size. This is, of course, not the case for dissolved elements. In fact, for dissolved elements mass transfer is governed by both diffusion in $x = \delta_2$ and transport by eddies in the restricted turbulence layer. Equation (A2.6 in Appendix 2) gives the diffusion coefficient as

$$D_E = \frac{1.4 \times 10^{-3} u_0^3 x^3}{v^2}.$$

The total transport is taken to be the sum of the laminar and turbulent contributions and the transport rate is given by

$$-\dot{n} = \left(D + \frac{1.4 \times 10^{-3} u_0^3 x^3}{v^2} \right) \frac{dc}{dx}$$

$$-\frac{\dot{n}}{D} = \left(1 + a^3 x^3 \right) \frac{dc}{dx},$$

where

$$a^3 = \frac{1.4 \times 10^{-3} u_0^3}{Dv^2}. \tag{3.20}$$

By changing variable $y = ax$, we get

$$-\frac{\dot{n}}{Da} = \left(1 + y^3 \right) \frac{dc}{dy}.$$

Rearranging and integration through the boundary layers, we get

$$\int_{y=0}^{\infty} \frac{dy}{1+y^3} = -\frac{Da}{\dot{n}} (c_\infty - c_0), \tag{3.21}$$

where c_∞ is the concentration of the solute far from the wall and c_0 the concentration at the wall. We have that

$$\int \frac{dy}{1+y^3} = \frac{1}{6}\ln\left[\frac{(1+y)^2}{y^2-y+1}\right] + \frac{1}{\sqrt{3}}\tan^{-1}\frac{2y-1}{\sqrt{3}} + Const.$$

When inserting the integration limits, the ln term goes to zero for both $y=0$ and $y=\infty$ while

$$\tan^{-1}(\infty) = \frac{\pi}{2}$$

and

$$\tan^{-1}\left(-\frac{1}{\sqrt{3}}\right) = -\frac{\pi}{6}.$$

Thus, Eq. (3.21) becomes

$$-\frac{Da}{\dot{n}}(c_\infty - c_0) = \frac{2\pi}{3\sqrt{3}},$$

or introducing the mass transfer coefficient

$$k = -\frac{\dot{n}}{c_\infty - c_0} = \frac{3\sqrt{3}Da}{2\pi} = 0.0925\left(\frac{D}{\nu}\right)^{2/3}u_0. \tag{3.22}$$

Here the shear velocity $u_0 = \sqrt{\tau_0/\rho}$ may be obtained directly from flow calculations, determined from the specific stirring power $\dot{\varepsilon}$ (Eq. (A2.21)), or we make use of equations such as (A2.16) and (A2.17) in Appendix 2. Then for a smooth pipe:

$$k = 0.0185\left(\frac{D}{\nu}\right)^{\frac{2}{3}}u_m Re^{-\frac{1}{8}}, \tag{3.23}$$

where u_m is the mean velocity in the pipe.

The Reynolds number, Re, is one of several dimensionless numbers used in this book. $Re = du/\nu$ is the ratio between the inertial forces and viscous forces within a fluid, and indicates whether the flow is laminar or turbulent, or a mixture of these two flow patterns. In the Reynolds number, d is a characteristic length of some geometry (e.g. pipe diameter, particle diameter), u is a velocity, and ν is the viscosity. If the viscous forces dominate (i.e. low Re) the flow is laminar, and if the inertial forces dominate (i.e. high Re) the flow is likely to be turbulent.

Equation (3.23) applies not only to dissolved elements but also to inclusions in the laminar boundary layer. To give an example of the calculation of the diameter, where $d_{pc} = \delta_2$, we return to Appendix A2.3 on nozzle blockage and use the data given there. From Section 2.2.9 we have $D = k_B T / 3\pi \rho v d_p$, so that Eq. (3.21) gives on setting $d_p = \delta_2$ and $u_0 = \phi^{\frac{1}{2}} u_m$

$$d_{pc}^{\frac{4}{3}} = 8.9 \left(\frac{k_B T}{3\pi \rho v} \right)^{\frac{1}{3}} \frac{v^{\frac{2}{3}}}{\phi^{\frac{1}{2}} u_m}$$

or for steel at $1600\,^{\circ}\mathrm{C}$

$$d_{pc}^{\frac{4}{3}} = 8.9 \left(\frac{1.38 \times 10^{-23} \times 1873}{3\pi \times 7030 \times 0.8 \times 10^{-6}} \right)^{\frac{1}{3}} \frac{(0.8 \times 10^{-6})^{\frac{2}{3}}}{0.0036^{\frac{1}{2}} \times 2}$$

This gives $d_{pc} = 0.6 \times 10^{-6}$ m.

3.4   Mass Transfer in Liquids to a Clean Free Surface

Mass transfer between a melt and a gas phase plays a very important role in extractive and refining metallurgy. For instance, vacuum treatment is based on removal of unwanted elements through a free surface. Metal in holding furnaces picks up hydrogen from moisture in the atmosphere above the melt.

In gas purging, impurity elements are transferred to the bubbles through the melt–gas interface.

In chemical engineering systems the presence of impurities at the interface can greatly influence mass transfer.[10,11] The same problems are encountered with metal melts.[12–15] In addition, metal oxide films may play a role.

In this section we attempt to take a close look at liquid-side mass transfer to a clean interface. Conditions are very different from those at a solid surface (cf. Section A2.2 in Appendix 2). As opposed to a solid surface, there remain (if the surface is clean) very considerable average movements along the surface. In this section we introduce eddies in the models. Basically, an eddy is the swirling of a fluid and show up as reversed currents in a turbulent flow. These movements persist right into the surface and may be observed clearly by sprinkling a powder on to the surface. In the z-direction (normal to the horizontal surface) the turbulent eddies are restrained by surface tension and gravity effects, so that at $z = 0$, we may as a first approximation set the eddy velocity $u_z = 0$. To obtain a relation for u_z we proceed by the Levich method.[10] There is a zone of thickness δ_1 near the surface within which turbulence in the z-direction is damped. At the lower limit (that is, far enough into the melt), the eddy velocity is

$$u_z = u_0, \tag{3.24}$$

where u_0 is the z-component of the eddy velocity in the bulk melt near the boundary layer. Since the velocity along the surface, u_x, does not depend on z, the continuity equation (cf. Section A2.2 in Appendix 2) gives that

$$u_z = \frac{z u_0}{\delta_1}. \tag{3.25}$$

Thus, the eddy velocity decreases linearly with z instead of with z^2 as was the case for a solid surface. The application of Eq. (3.25) depends on calculating the value of δ_1; that is, the thickness of the restricted turbulence layer and the z-component of the eddy velocity u_0 at $z = \delta_1$.

As an eddy approaches from the bulk, the surface becomes deformed. The pressure fluctuations p' are proportional to the square of the velocity fluctuations so that

$$p' = C\rho u_0^2, \tag{3.26}$$

where C is about 1.[10] For isotropic turbulence at high Reynolds numbers,[10] $C = 0.7$. In general, turbulence is not isotropic close to the surface and C will depend on geometry and the method of stirring. The pressure fluctuation is restrained by surface tension and gravity effects. Thus, p' may be compared to the excess pressure in a soap bubble, for example. One obtains[10]

$$p' = \frac{2}{R}\left(\sigma + \frac{d_e^2 \rho g}{16}\right), \tag{3.27}$$

where R is the radius of curvature of the deformation, σ is the surface tension, and d_e is the diameter of the surface deformation at $1/e$ of the maximum height. Experiments show that roughly (for $C = 1$)

$$\delta_1 = \frac{R}{2}. \tag{3.28}$$

Equations (3.25)–(3.28) give

$$u_z = \frac{z\rho u_0^3}{\sigma + d_e^2 \rho g/16}. \tag{3.29}$$

Then from the Prandtl mixing length model with eddy length $l = 0.4z$ (see Section 3.11.2), if it is assumed that eddy viscosity and eddy diffusivity are equal, the diffusion coefficient becomes

$$D_E = u_z l = \frac{0.4 z^2 \rho u_0^3}{\sigma + d_e^2 \rho g/16}. \tag{3.30}$$

The thickness of the laminar flow boundary layer δ_2 is found by setting D_E equal to D and solving Eq. (3.30) with respect to $z = \delta_2$:

$$\delta_2 = \left(\frac{D\left(\sigma + d_e^2 \rho g/16\right)}{0.4\rho u_0^3} \right)^{\frac{1}{2}}. \tag{3.31}$$

As in the case of transport to walls, the total transport is given by the sum of pure diffusion and transport due to eddies

$$\dot{n} = -(D + D_E)\frac{dc}{dz}. \tag{3.32}$$

With the use of Eqn. (3.30) and (3.31) this can be rewritten to

$$\dot{n} = -D\left(1 + \frac{z^2}{\delta_2^2}\right)\frac{dc}{dz}. \tag{3.33}$$

Integration from bulk phase ($z = -\infty$ and $c = c_B$) to surface ($z = 0$ and $c = c_i$)

$$\frac{\dot{n}}{D}\int_{-\infty}^{0} \frac{dz/\delta_2}{\left(1 + \frac{z^2}{\delta_2^2}\right)} = -\int_{c_B}^{c_i} \frac{dc}{\delta_2} \tag{3.34}$$

gives

$$\frac{\dot{n}}{D}\left[\tan^{-1}(0) - \tan^{-1}(-\infty)\right] = \frac{c_B - c_i}{\delta_2} \tag{3.35}$$

and

$$\dot{n} = \frac{2D}{\pi \delta_2}(c_B - c_i). \tag{3.36}$$

Then again, comparing the above to the definition of the mass transfer coefficient in Eq. (3.2), and noting that we have transport in the opposite direction, we get

$$k = \frac{2D}{\pi \delta_2}. \tag{3.37}$$

3.4.1   Mass Transfer to a Moving, Clean, Free Surface

Figure 3.3 illustrates a typical situation where a crucible is placed in a vertical tube furnace with heating from the side. Due to thermal gradients in the melt, the melt will rise along the

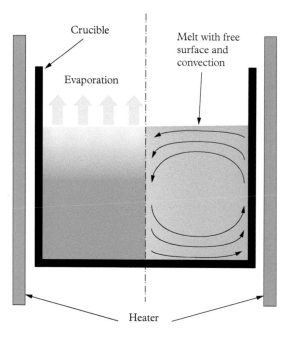

Fig. 3.3 *Schematic of a melt positioned inside a cylindrical tube furnace with vertical heating elements. The right-hand side shows a thermal gradients flow pattern, while the left-hand side shows solute concentration after some time of evaporation.*

crucible wall and sink along the centre line. This means that a fluid element approaching the melt surface on the edge will be transported along the melt surface to the centre where it will sink again. During the time the fluid element is in contact with the surface, it will experience a driving force and some solute will evaporate if the concentration of solute in the melt is higher than the equilibrium partial pressure in the atmosphere above the melt.

The liquid is free to move along the surface in the x-direction. Then the conservation equation for the diffusing component gives

$$\frac{\partial c}{\partial t} = D\frac{\partial^2 c}{\partial z^2}.$$

(3.38)

The left-hand side of Eq. (3.38) is the rate of accumulation of component c in a unit volume of liquid moving along the surface. The right-hand side gives the net diffusion into the volume element in the z-direction. Diffusion in the other directions is neglected since concentration changes strongly only in the z-direction.

Figure 3.3 also gives the boundary conditions for our solution of Eq. (3.38). The unit volume of fluid approaches the surface from the bulk melt at time $t = 0$. Thus for

$$t = 0, \quad c = c_b.$$

(3.39)

At the surface where

$$z = 0, \quad c = c_s. \tag{3.40}$$

Furthermore, far from the surface

$$z \to \infty, \quad c = c_b. \tag{3.41}$$

A solution of Eq. (3.38) that fulfils the boundary conditions (Eqs. (3.39)–(3.41)) is[16,17]

$$\frac{c - c_s}{c_b - c_s} = \text{erf}\left(\frac{z}{2(Dt)^{\frac{1}{2}}}\right). \tag{3.42}$$

The so-called error-function is defined as

$$\text{erf}(x) = 2\pi^{-\frac{1}{2}} \int_0^x e^{-w^2} dw.$$

Equation (3.42) represents a solution for a semi-infinite body and is shown graphically in Fig. 3.4. It can be seen from the figure that for small x, the curve is close to linear. A series expansion of the error function is given by[18]

$$\text{erf}(x) = \frac{2}{\sqrt{\pi}}\left(x - \frac{x^3}{1! \cdot 3} + \frac{x^5}{2! \cdot 5} - \frac{x^7}{3! \cdot 7} + \dots\right), \tag{3.43}$$

which means that for small x, that is small z and/or t, Eq. (3.42) can be simplified to

$$\frac{c - c_s}{c_b - c_s} \approx \frac{z}{\sqrt{\pi Dt}}, \tag{3.44}$$

which also is plotted in Fig. 3.4. Comparing Eq. (3.44) to Eq. (3.42) shows that for $x \lesssim 0.2$, the error by using Eq. (3.44) is less than 1%.

Another example is mass transfer to a bubble, rising in a stagnant melt, as illustrated in Fig. 3.5. The concentration in a bulk melt solute is c_b and the concentration of the same solute at the bubble surface is c_s. The bubble, which has a diameter d, is rising through the melt with a velocity u. This gives a contact time between a melt element and the bubble of

$$t = \frac{d}{u}. \tag{3.45}$$

Due to the difference between the solute concentration in the bulk melt and at the bubble surface, $c_b - c_s$ (called the driving force), the solute will be transported to the bubble surface. This is indicated with the inward pointing arrows, and the speed of this transport represented

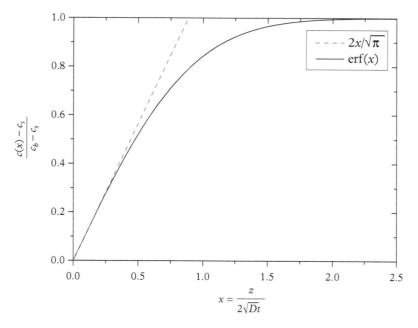

Fig. 3.4 *Graphical representation of diffusion in a semi-infinite system. The dashed line represents the first term in the series expansion of the error function.*

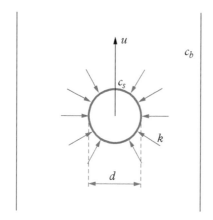

Fig. 3.5 *Transport process to a rising bubble in a melt. The rise velocity is u and bubble diameter d. The concentration of solute in bulk liquid and at the surface are represented by c_b and c_s, respectively, and the mass transfer coefficient is represented by k.*

by the mass transfer coefficient k. One example of this situation is the removal of dissolved hydrogen from molten aluminium by gas purging. This situation will be treated in Chapter 4.

The total mass transfer rate from the unit volume at time $\dot{n}(t)$ is obtained from Fick's law and Eq. (3.44):

$$\dot{n}(t, z = 0) = -D\frac{dc}{dz} = -\frac{D(c_b - c_s)}{(\pi Dt)^{\frac{1}{2}}}. \tag{3.46}$$

Then by integrating over time one obtains the number of moles of the solute per unit area which have crossed $z = 0$ between $t = 0$ and t:

$$N = \int_{t=0}^{t} |\dot{n}| \, dt = \int_{t=0}^{t} \frac{D(c_b - c_s)}{(\pi Dt)^{\frac{1}{2}}} dt$$
$$= (c_b - c_s)\sqrt{\frac{4Dt}{\pi}}. \tag{3.47}$$

The average mass transfer coefficient is obtained by dividing N by the elapsed contact time, t, and by the driving force for diffusion, $(c_b - c_s)$. This gives

$$k = \frac{N}{t(c_b - c_s)} \tag{3.48}$$

Thus,

$$k = \left(\frac{4D}{\pi t}\right)^{\frac{1}{2}}. \tag{3.49}$$

This equation is commonly used to describe mass transfer at a free surface.[4,10,11] As mentioned in Eqn. (3.45), one obtains the contact time, t, by dividing bubble diameter d by the relative velocity between bubble and melt, u. Then Eq. (3.49) becomes

$$k = \left(\frac{4Du}{\pi d}\right)^{\frac{1}{2}}. \tag{3.50}$$

3.4.2 Model Compared to Measurements of Mass Transfer

Figure 3.6 shows the results of measurements made in a system similar to the setup shown in Fig. 3.3.[19] The mass transfer coefficient, k, was measured by absorption/desorption of CO_2 in water in a small stirred laboratory unit. The residence time, t, in Eq. (3.49) was determined from near-surface velocity measurements using a laser-doppler unit. It is seen that the correspondence between measurements and the model in Fig. 3.6 is good.

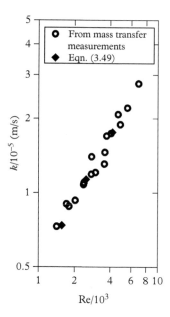

Fig. 3.6 *Comparison of experimental surface mass transfer coefficient, k, for an impeller-stirred melt, with k calculated from Eq. (3.49) using a time t obtained from measurements.*[19] *The Reynolds number is defined as Re = du/v.*

Keep in mind that if the melt is covered by even a thin slag layer, it may be appropriate to assume that the surface is solid and to apply Eqn. (3.22), where u_0 is now the shear velocity, $u_0 = (\tau_0/\rho)^{\frac{1}{2}}$. τ_0 is the shear force between the slag and melt.

3.5 Mass Transfer in Liquids to Bubbles, Droplets, and Particles

In Section 3.4.1 it was mentioned that

$$k = \left(\frac{4Du}{\pi d}\right)^{\frac{1}{2}} \tag{3.51}$$

was very often used to describe mass transfer to bubbles. When it was derived, we did not take viscosity into account. This might indicate that it is valid only for high bubble Reynolds numbers, $Re = ud/v$; that is, for large bubbles. We will see, however, that Eq. (3.51) is useful as a reference even for small bubbles and small Re.

Boundary layer theory[11] introduces a correction for the effect of viscosity:

$$k = \left(\frac{4Du}{\pi d}\right)^{\frac{1}{2}} \left(1 - \frac{2.89}{Re^{\frac{1}{2}}}\right)^{\frac{1}{2}}, \tag{3.52}$$

valid for $Re \geq 70$. At $Re = 70$ the correction factor to Eq. (3.52) becomes

$$\left(1 - 2.89/70^{\frac{1}{2}}\right)^{\frac{1}{2}} = 0.81.$$

For both bubbles and liquid drops with low viscosity, μ_p, in a fluid with viscosity μ, and $\kappa = \mu_p/\mu < 2$, Eq. (3.51) may be expanded to give[11]

$$k = \left(\frac{4Du}{\pi d}\right)^{\frac{1}{2}} \left(1 - \frac{2.89 + 2.15\kappa^{0.64}}{Re^{\frac{1}{2}}}\right)^{\frac{1}{2}}.$$ (3.53)

An empirical formula[11] applicable for $Re > 1$ is

$$k = \left(\frac{4Du}{\pi d}\right)^{\frac{1}{2}} \left\{1 - \frac{\frac{2+3\kappa}{3(1+\kappa)}}{\left(1 + \left[\frac{\left(2+3\kappa Re^{\frac{1}{2}}\right)}{(1+\kappa)(8.670+6.45\kappa^{0.64})}\right]^n\right)^{\frac{1}{n}}}\right\}^{\frac{1}{2}},$$ (3.54)

where $n = 3\kappa + \frac{4}{3}$. The correction factor to Eq. (3.51) for bubbles ($\kappa = 0$) at $Re = 1$ is 0.63.

For $Re < 1$ we have creeping, also called Stokes' flow. In this case, if the interfacial contamination is negligible, for bubbles or low-viscosity droplets with $\kappa \leq 2$,[11]

$$k = \left(\frac{4Du}{\pi d}\right)^{\frac{1}{2}} \left(\frac{1}{1+\kappa}\right)^{\frac{1}{2}} 0.58.$$ (3.55)

Bubbles smaller than 1 mm, bubbles in contaminated liquids, and high-viscosity droplets tend to give the same mass transfer relations as solid particles. Then the tangential velocity is zero at the interface and increases linearly with distance from the interface. The situation is very different from what we have on a free surface. Quite involved calculations are required to deal with this case. A convenient solution valid for $Re < 1$ is given by the relation between the Sherwood number ($Sh = kd/D$) and the Peclet number ($Pe^* = ud/D = Re \cdot Sc$).[11]

For bubbles in metals $Pe^* \gg 1$ so that

$$Sh = 1 + \left(1 + Pe^*\right)^{\frac{1}{3}} \approx Pe^{*\frac{1}{3}},$$ (3.56)

where the Sherwood number shows the ratio between mass transfer by bulk transport and diffusion rate, and the Peclet number is the ratio of mass transfer via bulk convection and diffusion.

In a stagnant bath where the relative velocity $u = 0$, Eq. (3.58) reduces to

$$Sh = 2 \quad \text{or} \quad k = \frac{2D}{d}. \tag{3.57}$$

This may be deduced directly from the diffusion equation for a sphere:

$$\frac{\partial \left(r^2 \frac{\partial c}{\partial r}\right)}{\partial r} = 0. \tag{3.58}$$

The boundary conditions are $c = c_s$ at $r = \frac{d}{2}$ and $c = c_b$ when $r \to \infty$. The solution is

$$c = c_b - (c_b - c_s)\frac{d}{2r}.$$

and the molar flux is

$$\dot{n} = -D\frac{dc}{dr} = -(c_b - c_s)\frac{2D}{d}.$$

since $r = \frac{d}{2}$.

By the definition

$$\dot{n} = -k\,(c_b - c_s).$$

one obtains the constant terms in Eq. (3.58).

For $0.25 < Sc < 100$ and $1 < Re \le 400$ numerical solutions available for mass transfer to rigid spheres can be correlated by the expression[11]

$$Sh = 1 + Re^{0.41}\left[Sc + \frac{1}{Re}\right]^{\frac{1}{3}}. \tag{3.59}$$

For high $Sc = v/D$, $Sc > 200$, and $100 < Re < 2000$, we have

$$Sh = 1 + 0.724Re^{0.48}Sc^{\frac{1}{3}}. \tag{3.60}$$

In extractive and refining metallurgy, we usually want to obtain a high value for the product of the mass transfer coefficient, k, and the contact area, A. As shown in Section 4.5, A for bubbles is inversely proportional to the diameter d and proportional to the residence time τ_b. Thus

$$A \propto \frac{1}{d}\tau_b. \tag{3.61}$$

Then the product kA is proportional to $d^{-\frac{1}{2}}u^{-\frac{1}{2}}$ for large bubbles (Eq. (3.51)) and to $d^{-\frac{5}{3}}u^{-\frac{2}{3}}$ for small bubbles (Eq. (3.56)). For small diameters, velocities u increase with diameter. In Fig. 4.23, this is illustrated for bubbles. Thus, kA increases with decreasing size of the dispersed phase. However, later we will discover that there are practical limits to how small the diameters of the bubbles or droplets may become.

3.6   Velocities of Bubbles, Droplets, or Particles and the Corresponding Mass Transfer Coefficients

It is apparent from Section 3.5 and Chapter 4 that we strive to achieve large contact areas in refining processes. This means that bubbles, droplets, or particles should be as small as possible. Sometimes in gas-purging processes a fraction of the gas is in the form of large bubbles. The contact area associated with this ineffective fraction is relatively small, and it is not very important to describe this mass transfer accurately. For these reasons, in the following attention is mainly given to obtaining velocities relative to the melt, of a dispersed phase with relatively small diameters. The friction factor f is introduced in Section 5.4 for solid particles. For bubbles and droplets

$$\phi = \frac{4}{\mathrm{Re}} \left(\frac{2 + 3\kappa}{1 + \kappa} \right) \tag{3.62}$$

is valid in the creeping flow region where $\mathrm{Re} = ud/v < 4$. The factor that contains κ takes the drag reduction into account due to circulation of the fluid in the droplet. For bubbles where $\kappa = 0$, $\phi = 8/\mathrm{Re}$. Equation (3.62) also applies to solids for $\mathrm{Re} < 1$. For solid particles the viscosity ratio $\kappa = \mu_{\mathrm{p}}/\mu \to \infty$. For $4 < \mathrm{Re} < 100$ the following correlation is suggested[11] for bubbles and droplets:

$$\phi = \frac{1.525 \left(783\kappa^2 + 2142\kappa + 1080 \right) \mathrm{Re}^{-0.74}}{(60 + 29\kappa)(4 + 3\kappa)}. \tag{3.63}$$

For solids ($\kappa \to \infty$) this equation would reduce to

$$\phi = 13.725 \mathrm{Re}^{-0.74}. \tag{3.64}$$

Above $\mathrm{Re} = 100$, bubbles in water tend to oscillate and this strongly affects the drag. Oscillation depends on the interplay between gravitational, viscous, surface tension and inertial forces. Two dimensionless groups have proved to be useful to describe the behaviour of bubbles and droplets. These are the Morton and Eötvös numbers defined, respectively, by

$$M = \Delta\rho\rho^2 g \frac{v^4}{\sigma^3} \tag{3.65}$$

and

$$Eo = \frac{\Delta \rho g d^2}{\sigma}.$$

For bubbles in Al

$$M = \frac{2350^3 \times 9.81 \cdot \left(0.55 \times 10^{-6}\right)^4}{0.9^3} = 1.6 \times 10^{-14}.$$

For bubbles of intermediate size (for instance, $d = 10$ mm)

$$Eo = \frac{2350 \times 9.81 \times 10^{-4}}{0.9} = 2.6.$$

Now if $M < 10^{-3}$ and $Eo < 40$, then for contaminated liquids[11]

$$\mathrm{Re}\, M^{0.149} + 0.857 = 3.42 H^{0.441} \quad \text{if } H > 59.6, \tag{3.66}$$

where

$$H = \frac{4}{3} Eo M^{-0.149} (\mu / \mu_w)^{-0.14}. \tag{3.67}$$

In practice we usually find that bubbles are so large that $H > 59.3$. In Eq. (3.67) μ_w is the viscosity of $H_2O = 0.9 \times 10^{-3}$ Ns/m^2. In the example

$$H = \frac{4}{3} \times 2.6 \times \left(1.6 \times 10^{-14}\right)^{-0.149} \left(\frac{0.9}{1.3}\right)^{0.14} = 374.$$

Then from Eqn. (3.66) we obtain

$$\mathrm{Re} = 5204$$
$$u = \frac{5204 \times 1.3 \times 10^{-3}}{2350 \times 10^{-2}} = 0.29 \, \tfrac{m}{s}.$$

Equations (3.60) and (3.61) are employed to obtain the bubble velocity on the right part of the graph in Fig. 4.23.

3.6.1   Removal of Mg from Molten Aluminium in a Continuous Gas-Purging Reactor

In theory, it should be possible to remove magnesium from molten aluminium by gas purging, either with a reactive gas (such as Cl_2) or with an inert gas such as argon since magnesium has a 'relative' high vapour pressure (compared to aluminium). In our example, we use a gas with more Cl_2 available than necessary for complete chemical removal of magnesium. This will simplify the driving force for the removal process and thus give the fastest kinetics.

We calculate the mass transfer coefficient for the bubbles. As an example Mg transfer to argon and Cl_2 bubbles with diameter $d = 10^{-2}$ m in Al is treated.

From literature,[20] we find that $D_{Mg} = 7 \times 10^{-9}$ m²/s. Equation (3.52) gives

$$k = \left(\frac{4 \times 7 \times 10^{-9} \times 0.29}{\pi \times 0.01} \right)^{\frac{1}{2}} \left(1 - \frac{2.89}{5204} \right)^{\frac{1}{2}} = 5.0 \times 10^{-4} \frac{m}{s}.$$

This mass transfer coefficient can then be used to calculate the removal efficiency for magnesium from molten aluminium.

Consider a gas purged back-mix tank with a melt height h, which uses $\dot{Q}$ Nl/min of gas and has a melt flow of $\dot{M}$ tonnes per hour melt flow. The total contact area of gas bubbles in the tank is simply the bubble area produced per time multiplied by the average residence time, τ, of the bubbles in the tank,

$$A = \dot{A}\tau = \dot{A}\frac{h}{u_b} = \frac{6\dot{Q}}{d}\frac{(T+273)}{273}\frac{h}{u_b}, \tag{3.68}$$

where the ratio $(T + 273)/273$ takes into account the expansion of the gas due to temperature difference between standard conditions (1 bar pressure and 273 K) and the melt temperature.

Conservation of the mass of magnesium requires that what comes into the reactor with the melt must be equal to what leaves the reactor with the melt plus the mass removed by the reactive gas,

$$\frac{\dot{M}c_{in}}{100} = \frac{\dot{M}c_{out}}{100} + \rho k A \frac{c_{out}}{100}, \tag{3.69}$$

where it is also assumed that the concentration of magnesium is uniform inside the back-mix tank and equal to the exit concentration. Introducing the removal efficiency and doing some rearranging gives

$$E = \frac{c_{in} - c_{out}}{c_{in}} = 1 - \frac{1}{1 + \rho k A / \dot{M}}. \tag{3.70}$$

A plot of the removal efficiency in such a back-mix tank is shown in Fig. 3.7, which demonstrates that by either increasing the mass transfer coefficient or contact area or decreasing the melt flow, the removal efficiency will increase.

Now for our back-mix tank we use the numbers given in Table 3.1 and Eqs. (3.68)–(3.70) to get

$$\frac{\rho k A}{\dot{M}} = 0.43,$$

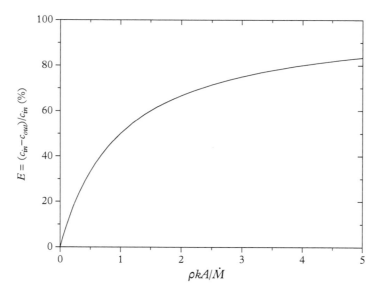

Fig. 3.7 *Model for removal of magnesium in a back-mix tank.*

Table 3.1 *Data for the back-mix tank used in the example*

h	=	1.0 m
ρ	=	2350 kg/m³
$\dot{M}$	= 60 t/h	= 16.7 kg/s
$\dot{Q}$	= 50 Nl/min	= 8.33×10^{-4} Nm³/s
K	=	5.0×10^{-4} m/s
D	= 10 mm	= 0.01 m
T	=	700 °C
u_b	=	0.29 m/s

and finally the removal efficiency of magnesium in such a reactor is

$$E = 30\%.$$

This requires that there is enough Cl_2 in the purge gas so that all of the Mg transported to the bubbles reacts chemically to $MgCl_2$.

3.7   Bubbles, or Droplets Dispersed in Molten Metal

3.7.1   Introduction

In Sections 3.6 and Chapter 4, the importance of attaining large contact areas *A* is emphasized. Large contact areas are necessary in order to speed up the kinetics and to achieve a high productivity and high yields. In Chapter 4 we see, taking bubbles as an example, that the contact area is inversely proportional to the dimensions of the dispersed phase and proportional to the residence times. The residence time depends on the size of the particle, droplet, or bubble, the point of entry into the melt, and convection and turbulence in the bath.

For solid particles we might expect that the particle size is known from a screening analysis taken before injection into the melt. However, this is not necessarily so. In particular, the smaller size fractions may not be able to penetrate the bubble–melt interface, as illustrated in Fig. 4.17. This is discussed in the following where mainly the large size fractions enter and these give smaller surface areas relative to their mass.

For bubbles and droplets the size must be determined on the basis of our knowledge of gas–liquid and liquid–liquid fluid dynamics and of the conditions in our metallurgical reactor. As we increase the amounts of gas blown into the reactor and increase the mixing and stirring of the slag and melt phases, the modelling task becomes progressively more difficult. The explanation is that research in this field has, reasonably enough, relied on the gradual accumulation of knowledge based first on single bubbles using low gas flow rates, then a train of bubbles, and finally the transition between bubbling and jetting. These are phenomena where the melt is the continuous phase and bubbles are the dispersed phase. Some process developments in the industry, however, seem to have outpaced research. For instance, decarburization reactors for stainless steel (AOD) and for low-alloy steel (LD and BOF) use such large amounts of oxygen and argon that it might be possible to describe them as units where gas is the continuous phase and metal and slag droplets are the dispersed phases. There seems to be very little research published in this area and this chapter will not deal with such systems. Fortunately, however, even without this knowledge concerning contact areas, useful process models can be developed as demonstrated in Chapter 9.

3.7.2   Penetration of Solid Particles into a Melt

The problem of penetration through a gas–melt interface by particles has attracted a number of workers.[21–4] Here their work will not be described in detail; only a few major points will be outlined. Finally, some remarks are made based on the practical industrial situation.

Initially, theoretical and experimental work studied the penetration of a single particle through a planar gas–liquid interface. It was found that the particle velocity must exceed a certain critical velocity in order to penetrate the interface; see also Fig, 4.17. This velocity is given in dimensionless form as a Weber number:

$$We = \frac{\rho u^2 d_p}{2\sigma}. \tag{3.71}$$

Another dimensionless group is the particle–melt density ratio:

$$\rho^* = \frac{\rho_p}{\rho},$$

(3.72)

where ρ^* is the ratio between the density of the particle and that of the melt. We·ρ^* gives the ratio between the kinetic energy of the particle and the surface tension energy. Furthermore, there is a group that characterizes the 'wetting' between a particle and the melt:

$$-\cos\theta = \frac{\sigma_{ls} - \sigma_{gs}}{\sigma_{gl}}.$$

(3.73)

Here σ_{ls} is the solid–liquid and σ_{gs} the solid–gas interfacial tension, and σ_{gl} is the surface tension. θ is called the contact angle. It is the angle between the liquid–gas and liquid–solid interfaces. If there is complete wetting between the particle and melt, $\theta = 0$. The models give the critical We as a function of ρ^* and $\cos\theta$. If the particles wet the melt ($\theta = 0$) and $\rho^* \geq 1$, the critical We number is zero. That is, the particles are immersed in the melt whatever the velocity. From the definition of the critical We number it is found that, for non-wetting particles, if the particle size d_p is reduced, we must compensate by increasing the particle velocity, u. The critical We number is found to be greater than 100 for $\rho^* < 0.5$. The various equations show that the critical We number decreases with improved wetting; that is, with decreased θ and with increased density of the particles relative to the melt (that is, increased ρ^*).

Numerical work[25] on the penetration of single particles has taken the viscosity of the melt into account by using a modified Morton number:

$$M^* = \frac{\rho^3 g \nu^4}{\sigma^3}.$$

(3.74)

For penetration into a slag, the critical We number increases by several orders of magnitude.

Studies on the injection of particles using a carrier gas show that for a given set of operating conditions, some particles penetrate and some do not. The fraction that penetrates increases with We, ρ, $180° - \theta$, and $1/M$, and depends on the mass flow ratio between particles and carrier gas. Furthermore, the fraction depends on velocities and flow patterns in the melt. For instance, during injection, particles may accumulate or 'pile up' near the lance tip and finally choke up the lance.[26] Perhaps the ratio between the mass flow of particles and mass flow of carrier gas

$$r = \frac{\dot{M}_p}{\dot{M}_g}$$

(3.75)

is the most important factor in determining the fraction that penetrates.[21] Model studies in the laboratory[25] indicate that if $r > 1$ then 80–90 per cent of the particles penetrate (for We > 200).

In powder injection at high solid to gas mass ratios,[21] a transition from cavity to jet penetration occurs when the number of particles in gas exceeds a critical value. This case has been handled numerically using smoothed particle hydrodynamics.

It should be pointed out that injection in industry is usually carried out for $1 < r < 100$ and We > 1000. This would indicate that recovery of particles is close to 100 per cent. However, even losses less than 1 per cent can be economically important. Furthermore, particles or dust in the exit gases often represent serious environmental problems and will have to be collected via filtration systems and treated separately.

3.7.3   Size of Bubbles and Droplets in Melts

In high-temperature melts, such as steel and copper, gases are fed into the melt only through lances, tuyeres, or porous plugs. Due to the aggressive environment impellers cannot usually be employed. For these high-temperature processes convection and stirring is caused only by the momentum and buoyancy effects of the gas. The bubble size is then determined mainly by the gas flow rate and the geometry of the tuyeres or nozzles.

In the following some remarks are made concerning our state of knowledge regarding bubble size in high-temperature melts and in melts at lower temperatures.

In low-temperature melts, such as aluminium and magnesium, small bubbles can be produced using a variety of impeller systems and nozzles. In these, shear forces probably play the dominating role in fixing the bubble size. To some extent such mechanisms have been treated in the chemical engineering literature.[27] There are, however, problems in the transfer of the results to molten metals due to the high surface tensions encountered for metals and the thermal expansion of the gas.

Bubble sizes from nozzles at low gas flow rates have been the object of extensive study in the literature.[5] In industry very high gas flow rates are commonly used, but such conditions are not easy to study in the laboratory for several reasons. One difficulty is the large tuyere size and gas volumes involved. Another problem is the unsteady and statistical nature of the bubbling or jetting that occurs at high gas flow rates. Finally, bubble size is different in various parts of the melt. For bubbling we may speak of at least three zones:[28]

(1) the primary (or large bubble) zone;

(2) the dispersed bubble zone; and

(3) the surface or disengagement zone where slopping or splashing occurs.

The primary or large bubble zone is of special importance in processes where the melt viscosity is high. Examples of such processes are slag reduction and slag fuming units for the recovery of zinc and tin.[29] Often there is a low liquid seal depth and shallow tuyere submergence. As a consequence, the large primary bubbles formed at the tuyere tip may not have sufficient rise time to break into secondary bubbles.

For highly viscous liquids such as slags it is found that viscosity increases bubble size. In addition to viscosity, inertial and buoyancy effects must be taken into account at high gas flow rates. Bubbles are elongated in the axial direction of the nozzle axis due to the stagnation pressure of the gas jet inside the bubble. It has been possible to calculate bubble volumes based on a number of assumptions,[22] such as:

(1) constant flow of gas;

(2) no convective flow in the liquid;

(3) no effect of the preceding bubble; and

(4) the shape of the bubble at an orifice at high injection rates becoming that of a prolate ellipsoid (where the eccentricity changes during growth of the bubble).

For low-viscosity liquids such as water and molten metals at high injection rates it is found theoretically and experimentally for a nozzle that points upwards[30]

$$N_V = 0.5 N_1^{-\frac{1}{4}}.$$ (3.76)

N_v is a dimensionless volume:

$$N_V = \frac{V g^{\frac{3}{5}}}{\dot{Q}_g^{\frac{6}{5}}}$$ (3.77)

and N_1 is the so-called injection number[23] that takes the momentum of the gas into account,

$$N_1 = \frac{\dot{Q}_g^{\frac{4}{3}} \rho_g / \rho}{g^{\frac{2}{3}} r_0^2},$$ (3.78)

where r_0 is the nozzle radius. Equation (3.76) is valid for the range $0.017 \le N_1 \le 1$. The group N_V is recognized as that given by the well-known Davidson and Schüler equation[24] derived for gas flow rates so low that the bubble can be assumed to be spherical. Then

$$N_V = 1.378.$$ (3.79)

Davidson and Schüler considered the spherical bubble to detach when it is in tangential contact with the nozzle tip. The same criterion was employed here for the elliptical bubble used for high flow rates in the derivation of Eq. (3.76). The meaning of N_1 is perhaps best understood by taking the ratio between N_1 and N_V:

$$\frac{N_1}{N_V \pi} = \frac{\rho_g \dot{Q}_g^2}{\rho \pi r_0^2 g V} = \rho_g \left(\frac{\dot{Q}_g}{\pi r_0^2} \right)^2 \frac{\pi r_0^2}{\rho g V}.$$

This is equal to the ratio between the momentum of the gas and the buoyancy force of the bubble.

For viscous liquids an additional dimensionless group E_0 must be taken into account:[22]

$$E_0 = 177v \left(\frac{\rho g}{\rho \pi r_0^2 g \dot{Q}_g} \right)^{\frac{1}{3}}. \tag{3.80}$$

E_0 should not be confused with the Eötvös number.

Figure 3.8 gives N_v as a function of N_1 and $E_0 = 0$, 1.5, and 3.0. Agreement is acceptable with the model for high velocities and high viscosities. The model tends to give slightly higher values for the volume, probably due to use of the Davidson–Schüler solution as a limiting case for low viscosities.

3.7.4   Small Bubbles from Impeller

As discussed in Section 3.5, mass transfer increases with decreasing bubble size. It is tempting to employ a mechanical device to produce small bubbles. However, the problem of the break-up and coalescence of bubbles or a dispersed liquid phase is difficult from both the experimental and theoretical points of view. Here only an indication can be given of the possible mechanisms and the resulting relations for bubble or droplet size. Pressure or shear forces can be the cause of break-up. For instance, drops can break up in the turbulent flow along a wall due to shear forces.[11] Similarly, shear forces should break up and disperse the gas forced through a rotating impeller. Intuitively, a bubble or a droplet should break up if the dynamic forces are greater than the surface tension forces holding it together:

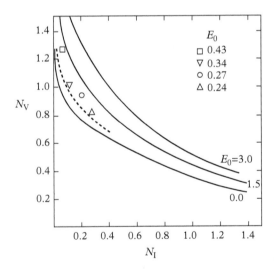

Fig. 3.8 *Dimensionless bubble volume N_v as a function of the injection number N_1 and a viscosity group E_0.[22] $E_0 = 0$ gives the inviscid case.*

$$\rho u^2 \pi \frac{d^2}{4} \gtrsim \sigma \pi d$$

$$d \gtrsim \frac{4\sigma}{\rho u^2}, \tag{3.81}$$

where u is an eddy velocity. Equation (3.81) is equivalent to stating that the bubbles break up at a Weber number:

$$We = \frac{\rho u^2 d}{\sigma} = 4$$

In a bubble plume, u may be rather large, on the same order of magnitude as the axial velocity.[30] For instance, model experiments in water[30] give $u = 0.2$ m/s. Equation (3.81) yields

$$d = \frac{4 \cdot 0.07}{1000 \cdot 0.04} = 7 \; mm$$

This is in reasonable accordance with visual observations. In steel, for the same eddy velocity, we would obtain

$$d = \frac{4 \cdot 1.8}{7030 \cdot 0.04} = 26 \; mm$$

Bubble sizes for impellers may be estimated by running the impeller in a full-scale water model at the same rotational speed as employed in the melt. If it is assumed that the gas in the industrial case undergoes thermal expansion while it is still in the impeller, the gas flow in the model compared to the melt should be increased in accordance with the gas law. Power consumption is roughly proportional to $\rho n^3 L^5$, where n is the number of rotations per second and L is the length of the impeller blades.[27] The energy dissipation $\dot{\varepsilon}$ in units of power consumption per unit mass of fluid should then be the same in the model and in the melt.

It is assumed that a major part of the power is consumed in producing fine bubbles. Then from the theory of homogeneous isotropic turbulence using the bubble diameter d as a characteristic length, one obtains

$$\left(\frac{\tau}{\rho}\right)^{\frac{3}{2}} \propto \dot{\varepsilon} d. \tag{3.82}$$

(Equation (3.82) may be compared with Eq. (A2.21) for flow in a pipe.) We can then define the Weber number as[27]

$$We = \frac{\tau d}{\sigma}, \tag{3.83}$$

where τ is the shear force acting on a bubble of diameter d, $\tau \approx \rho u^2$. As previously mentioned, the bubbles should break up in the impeller at the same critical Weber number in the model and in the melt. From Eqs. (3.82) and (3.83) one obtains on eliminating τ

$$\left(\frac{We\,\sigma}{d\rho}\right)^{\frac{3}{2}} \propto \dot{\varepsilon}d$$

or

$$\frac{\sigma}{\rho} \propto \frac{\dot{\varepsilon}^{\frac{2}{3}}d^{\frac{5}{3}}}{We}.$$

If $\frac{\dot{\varepsilon}^{\frac{2}{3}}}{We}$ is the same in H_2O and melt, the ratio between bubble size in the melt and in water is

$$\frac{d}{d_{H_2O}} = \left(\frac{\sigma\rho_{H_2O}}{\sigma_{H_2O}\rho}\right)^{\frac{3}{5}}. \tag{3.84}$$

For instance, for bubbles in aluminium compared with water model experiments:

$$\frac{d_{Al}}{d_{H_2O}} = \left(\frac{0.9}{0.072}\cdot\frac{1000}{2350}\right)^{\frac{3}{5}} = 2.7. \tag{3.85}$$

Thus, bubbles are larger and contact areas smaller in the molten metal than in a water model.

3.8   Gas-Side Mass Transfer Resistance

3.8.1   Introduction; Monoatomic Gases

It is usually assumed that resistance to mass transfer at a melt–gas interface is dominated by the melt boundary layer. Often this is justified, especially if the solubility in the melt is low; that is, the vapour pressure is high. Many gaseous impurities, fortunately, have low solubilities in molten metals. However, in some cases it is necessary to study the gas side. Vacuum systems are often employed when gas-side resistance is important. A chemical reaction between gas and melt may profoundly change the kinetics. Furthermore, often the gases are two-atomic, such as H_2, O_2, and N_2. Then resistance in the gas may become significant at low concentrations in the melt.

To begin with we look at monoatomic vapours and turn to Section 4.12 concerning vacuum treatment. The ratio, λ, between resistance in the gas phase and in the melt phase is found from Eq. (4.91),

$$\lambda = \frac{kRT}{k_gK_j\gamma_j}, \tag{3.86}$$

where the product of the inverse equilibrium constant K_j and the Raoultian activity coefficient γ_j is given by

$$K_j\gamma_j = \frac{p_j}{c_j}.$$

(3.87)

It is seen that a high solubility of component j, that is, a low value of $K_j\gamma_j$ tends to give a high λ. To determine the value of λ we must also obtain a relation for the ratio k_g/k. For gases the Schmidt number Sc is very close to 1. In this case the velocity and concentration profiles have the same shape and thickness. We may then apply the Reynolds analogy given by Eq. (A2.14) not only for the turbulent wall layer, but also for the restricted turbulence and the viscous boundary layers. For instance, from Eq. (A2.19) for pipe flow one obtains

$$\frac{k_g}{u_g} = \phi,$$

(3.88)

where u_g is the mean gas velocity in the pipe and ϕ is the friction factor described in Chapter 5 here applied to the gas flow. From Eq. (A2.17) we have

$$\phi = 0.04\mathrm{Re}_g^{-\frac{1}{4}},$$

(3.89)

where $\mathrm{Re}_g = 2u_gR/\nu$ now refers to the gas phase and the velocity, u_g, is the difference between the gas and melt surface velocities, and R the radius of the pipe.

The Schmidt number for gases is not exactly 1 but can vary between 0.5 and 2.5. Then, according to the Chilton–Colburn correlation,[4] Eq. (3.88) is replaced by

$$k_g Sc_g^{\frac{2}{3}}/u_g = \phi$$

(3.90)

Equation (3.89) inserted into Eq. (3.90) then gives

$$k_g = 0.04 Sc_g^{-\frac{2}{3}} u_g \mathrm{Re}_g^{-\frac{1}{4}}$$

(3.91)

or

$$\frac{k_g 2R}{D_g} = 0.04\mathrm{Re}_g^{\frac{3}{4}} Sc_g^{\frac{1}{3}}.$$

(3.92)

For engineering calculations with $3000 < Re_g < 40000$ the following equation[31] is often applied for wetted-wall columns:

$$\frac{k_g 2R}{D_g} = 0.033\mathrm{Re}_g^{0.77} Sc_g^{0.33}.$$

(3.93)

For the driving force the logarithmic mean at the two ends of the wetted-wall column is used.

For binary gas mixtures at low pressures, D_g is inversely proportional to the total pressure. D_g increases with increasing temperature and is almost independent of composition for a given gas pair.[4] As an example, for Ar–O_2 at 2000K and 0.1 atm pressure, $D_g = 5.17 \cdot \times 10^{-3}$ m^2/s. Then, if $Sc_g = 0.5$, $Re_g = 10^4$, and $2R = 0.1$, Eq. (3.93) gives $k_g = 1.6$ m/s.

From Eqs. (3.91) and (3.23) we can now determine the relative resistance λ:

$$\lambda = \frac{0.015}{0.04} \left(\frac{Sc_g}{Sc} \right)^{\frac{2}{3}} \frac{u_m RT}{u_g K_j \gamma_j} \left(\frac{Re_g}{Re} \right)^{\frac{1}{4}} Re^{\frac{1}{8}}. \tag{3.94}$$

Equation (3.94) is derived for pipe flow but should also be useful in the turbulent regime for other geometries, except that the factor $(Re_g/Re)^{\frac{1}{4}}$ may have a slightly different exponent. To obtain the order of magnitude of the various terms we note that $Sc_g \sim 1$, $u_g \sim 10^{-5}$ m^2/s, $v \sim 10^{-6}$ m^2/s, $Sc \sim 100$, and $RT \sim 10^7$ J/kmol. $Re^{\frac{1}{8}}$ is not much greater than 1. Then if the gas and melt velocities are roughly equal, the gas/liquid resistance ratio becomes

$$\lambda \sim \frac{10^5}{K_j \gamma_j}. \tag{3.95}$$

Note that $K_j \gamma_j$ is a ratio between pressure given in units of Pa and concentration in kmol/m³ (cf. Section 4.12).

Only if the vapour pressure is so high that $K_j \gamma_j$ is greater than 10^5 can resistance in the gas phase be neglected at one atmosphere pressure. We will see that this explains why it is a simple matter to remove Na from molten aluminium, while inert gas purging of Li and Ca proceeds much more slowly.[32,33] From Table 2.1 and Appendix 1 one obtains at 1000K $p_{Na}/[\%Na]$ = 1.06 (with 1 bar and 1% as reference states). This gives $K_j \gamma_j$ = $1.06 \cdot 10^5 \cdot 23 \cdot 100/2350 = 1.04 \cdot 10^5$. The factor 10^5 converts pressure from bar to Pa. The denominator 2350 kg/m³ is the density of molten Al and 23 kg/kmol is the atomic weight of Na. Thus, the factor $2350/(23 \cdot \times 100)$ converts concentration from mass% to kmol/m³.

For Li and Ca it is found that $K_j \gamma_j$ is far less than 10^5 so that resistance in the gas film dominates when we purge with inert gas. However, this resistance can be eliminated by using a gas such as Cl_2 that reacts with Li or Ca.

In Section 4.12 evacuation of Cu from molten iron is treated. Cu has a low vapour pressure and $K_{Cu} \gamma_{Cu} = 27.4$. For this low value, resistance in the gas dominates unless the pressure is very low. However, if a very high vacuum can be attained so that there are few or no collisions in the gas phase, resistance, k_g, in the gas is completely eliminated. To attain this, the mean free path of the atoms in the gas must be on the same order of magnitude as the dimensions of the vacuum chamber. This requires that the total pressure be reduced to around 10^{-6} bar or 10^{-1} Pa.[34] In this case if the atoms collide with cold walls, they are removed by condensation.

At this low pressure and pumping speed, removal rates are low. In Chapter 4 we look at the case where the chamber pressure is higher. A higher pressure allows for a much greater pumping speed and increased throughput of refined metal.

3.8.2 Gas-Side and Interfacial Resistance for Diatomic Gases

For monoatomic gases we found for low solubilities that the melt boundary layer determined the mass transfer resistance. A similar result is obtained for diatomic gases except that now the calculations become slightly more involved. Initially, we disregard any interfacial resistance and write by analogy with Eqs. (4.90) and (4.92) for the melt boundary layer

$$\dot{n}_j = k\left(c_j - c_{ji}\right) \tag{3.96}$$

and for the boundary layer in the gas

$$\dot{n}_j = \frac{2k_g}{RT}\left(p_{ji} - p_{bj}\right), \tag{3.97}$$

where now the equilibrium at the interface is given by

$$c_{ji}^2 K_j^2 \gamma_j^2 = p_{ji} \tag{3.98}$$

Equations (3.96) and (3.97) are altered to the form

$$c_{ji}^2 = \left(c_j - \frac{\dot{n}_j}{k}\right)^2$$

and

$$p_{ji} = p_{bj} + \frac{\dot{n}_j RT}{2k_g}.$$

Note that when we square Eq. (3.96) we introduce a false solution in addition to the correct solution.

Then the interfacial concentration and pressure c_{ji} and p_{ji} are eliminated from the two equations by employing Eq. (3.98). One obtains from the resulting second-degree equation

$$\frac{\dot{n}_j}{kc_j} = 1 + \frac{\lambda}{2c_j} \pm \sqrt{\left(\frac{\lambda}{2c_j}\right)^2 + \frac{p_{bj}}{K^2 \gamma_j^2 c_j^2} + \frac{\lambda}{c_j}}, \tag{3.99}$$

where

$$\frac{\lambda}{c_j} = \frac{kRT}{2k_g K_j^2 \gamma_j^2 c_j}. \tag{3.100}$$

If

$$\frac{\lambda}{c_j} = \frac{kRT}{2k_g K_j^2 \gamma_j^2 c_j} \ll 1, \tag{3.101}$$

Eq. (3.99) reduces to

$$\dot{n}_j = k\left(c_j - c_{je}\right), \tag{3.102}$$

where c_{je} is the hypothetical concentration in the melt in equilibrium with the bulk concentration in the gas, $c_{je} = p_{bj}^{\frac{1}{2}}/K_j \gamma_j$. The + sign in Eq. (3.99) has no physical meaning and is disregarded. Equation (3.102) may be compared with the left-hand side of Eq. (4.61). There hydrogen was taken as an example, $c_H = \rho[\% \text{ H}]/(100 \, m_H)$. If inequality (3.101) is fulfilled the total mass transfer coefficient k_t is equal to the mass transfer coefficient for the melt boundary layer, k. By comparison with Eqs. (3.86) and (3.95) we see that (3.101) is fulfilled if

$$\frac{10^5}{2K_j^2 \gamma_j^2 c_j} \ll 1.$$

It is realized that the inequality is not satisfied at sufficiently low values of c_j. At least in theory if not in practice, for sufficiently low values of c_j, the gas-side resistance (see Eq. (3.106)) dominates for any two-atomic gas. The reason is that the partial pressure, p_{ji}, becomes very low because it is proportional to the second power of c_{ji}. Thus, p_{ji} becomes a bottleneck for the mass transfer. It is possible to reduce gas-side resistance by the use of a vacuum.

For hydrogen in pure aluminium at 1000 K one obtains from Table 2.1 $p_{H_2}^{\frac{1}{2}}/[\%H] = 9.43 \times 10^3$ (with 1 bar and 1 mass% as the reference states). Then

$$K_H^2 \gamma_H^2 c_H = \frac{\left(9.43 \times 10^3\right)^2 \times 10^5 \times 100}{2350} [\%H] = 3.8 \times 10^{11} \, [\%H]$$

(with 1 N/m^2 and 1 kmol/m^3 as the reference states). Resistance to transfer is greater on the metal side than on the gas side for practical values of [%H]. For nitrogen in iron, Table 2.2 gives, at 1800 K

$$\frac{p_{N_2}^{\frac{1}{2}}}{[\%N]} = 22.5 \, \text{bar}^{\frac{1}{2}}.$$

Then

$$K_N^2 \gamma_N^2 c_N = \frac{22.5^2 \times 10^5 \times 100 \times 14}{7030} [\%N] = 1.01 \times 10^7 \, [\%N].$$

Thus, also, for nitrogen in steel, resistance in the melt boundary layer should dominate for $[\%N] > 0.01 = 100$ ppm.

However, for nitrogen in steel a very strong effect of surface-active elements such as oxygen and sulfur is described in the literature,[13,14] where it is often assumed that this chemical reaction at the interface is partially blocked by surface-active elements

$$N_2 = 2\underline{N}. \tag{3.103}$$

This slows down the mass transfer at higher oxygen and sulfur levels. Then Eq. (3.96) is replaced by two equations,

$$\dot{n}_N = k\left(c_N - c_{N_i}\right), \tag{3.104}$$

where c_{N_i} is the concentration in the melt at the interface. The chemical reaction is presented as 'second order':

$$\dot{n}_N = k_c\left(c_{N_i}^2 - c_{N_e}^2\right), \tag{3.105}$$

where k_c is the associated chemical rate constant. k_c must be determined empirically; its value drops strongly with increasing activity of oxygen and sulfur.[12,15] c_{N_e} is in equilibrium with p_{ji} so that $K_j^2 c_{N_e}^2 = p_{ji}$. Furthermore, there may be an interfacial resistance due to oxide films on surfaces exposed to oxidation (but not on pure argon bubbles). This complex behaviour of nitrogen is not pursued further here.

Until now we have discussed systems where the solubility of the diatomic gases is low. However, sometimes the solubility is high. For instance, for O_2 dissolved in pure iron,

$$\frac{1}{2}O_2 = \underline{O} \text{ at } 1800\text{K},$$

we find $\dfrac{p_{O_2}^{\frac{1}{2}}}{[\%O]} = 2.82 \times 10^{-4}$. This means that oxygen has a very high solubility in pure iron:

$$K_O^2\gamma_O^2 c_O = \frac{\left(2.82 \times 10^{-4}\right)^2 \times 10^5 \times 100 \times 16}{7030} [\%O] = 1.8 \times 10^{-3} [\%O].$$

Thus, there is no resistance in the molten iron. Another example is the high solubility of hydrogen in the platinum group metals. In these cases λ/c_j in Eq. (3.100) may be much greater than 1. Then it may be shown that $c_{ji} \approx c_j$ and

$$\dot{n}_j \approx \frac{2k_g}{RT}\left(c_j^2 K_j^2 \gamma_j^2 - p_{bj}\right), \tag{3.106}$$

which is the same as Eq. (3.97) with p_{ji} in equilibrium with the bulk concentration in the melt, c_j. Thus, for sufficiently high solubilities the resistance to mass transfer is determined by the boundary layer in the gas. Mass transfer of oxygen to molten steel is limited on the gas side or by resistance in a FeO layer. However, conditions are often, of course, far more complex than indicated here due to the presence of an FeO layer, the effect of carbon in the steel, and the simultaneous transfer of species such as CO and CO_2 in the gas phase.

3.9 Removal of Impurities by Reactive Gases and Compounds

In the previous section we saw that elements with low vapour pressures such as Li and Ca were removed very slowly from the surface of molten aluminium. The reason for this was the low vapour pressures of these elements; that is, low $K_j\gamma_j$; so that the gas film mass transfer resistance $RT/(k_g K_j \gamma_j)$ became very high. If there is a 'reactive' gas species present that joins with the elements to be removed, resistance in the gas phase can be eliminated. Table 3.2 calculated from the thermodynamic data gives minimum concentrations, in the presence of Cl_2 or F_2, of some solute elements in aluminium. Below these levels the reactive gas will prefer the parent metal. The table indicates that Cl_2 can suppress Li to 0.01 ppm and Ca even lower. Similarly, F_2 can depress Li down to 0.05 ppm.

At the interface the reactive gas reacts with the various elements, for instance Li:

$$\underline{Li} + \frac{1}{2}Cl_2 = LiCl\,(l).$$

Ideally, the flow of Cl_2 to the interface should just balance the transfer through the melt boundary layer of the various elements that we want to expel, Na, Li, and Ca:

$$\frac{k_{Na}\,[\%Na]}{100 \times 23} + \frac{k_{Li}\,[\%Li]}{100 \times 7} + \frac{2k_{Ca}\,[\%Ca]}{100 \times 40} = \frac{2k_g p_{Cl_2}}{RT}. \tag{3.107}$$

Thus, the pressure of Cl_2 in the gas should be proportional to the content of impurities. In practice an excess of chlorine is employed. This may result in the loss of alloy additions, such

Table 3.2 *Approximate minimum impurity levels attainable in aluminium by reactive gases*[35] *(T = 1000K)*

Element	Content remaining in ppm	
	Chlorine	Fluorine
Be	65	85
Ca	3×10^{-5}	5.7×10^{-4}
Li	0.01	0.05
Mg	1.4	4.9
Na	9×10^{-5}	0.19

as Mg. Also aluminium may be removed as $AlCl_3(g)$. This gas represents a problem from an environmental point of view since above the melt $AlCl_3$ reacts with oxygen or water vapour to Al_2O_3 (dust), Cl_2, or HCl.

Instead of bubbling an inert plus a reactive gas through a melt, one may inject or stir into the melt a reactive compound in the solid or liquid form. For instance, AlF_3 is used to remove Li from Al.[36] Then

$$AlF_3\,(s) + 3\underline{Li} = Al + 3LiF\,(s) \qquad (3.108)$$

3.10 Pick-up of Hydrogen from Water Vapour

3.10.1 Model

Previously we have studied the pick-up of monoatomic and diatomic gases from the atmosphere. Very often the mass transfer of a species was found to be limited by resistance in the melt boundary layer. Then the partial pressure of the species at the interface was equal to the pressure in the bulk gas. However, if a chemical reaction with the metal takes place, it may be necessary to determine the partial pressure of the transferred species at the interface. Only then does one know the driving force for diffusion through the melt boundary layer so that the mass transfer calculation can be performed. A very important example of this problem is hydrogen pick-up by molten metals from water vapour. We take pick-up by aluminium as an example. The chemical reaction at the surface is

$$H_2O + \frac{2}{3}Al\,(l) = \frac{1}{3}Al_2O_3\,(s) + H_2 \qquad (3.109)$$

From Appendix 1, the standard Gibbs energy for the reaction is found to be

$$\Delta G^0 = -314930 + 53.14T \qquad (3.110)$$

Thus, the equilibrium constant becomes

$$K = \frac{p_{H_2}}{p_{H_2O}} = \exp\left(-\frac{\Delta G^0}{RT}\right) \qquad (3.111)$$

Since H_2O and Al are separated by a boundary layer in the gas, we cannot obtain p_{H_2} at the interface simply from thermodynamic or stoichiometric considerations. It is useful to look at the various steps involved, indicated in Fig. 3.9:

1. Water vapour diffuses through the gas boundary layer to the metal surface.
2. The water molecules are adsorbed at the surface.
3. The adsorbed molecules react with Al according to Eq. (3.109).

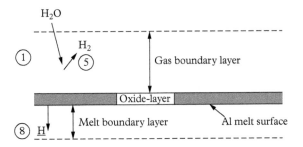

Fig. 3.9 *Some of the steps involved in the transfer of hydrogen from water vapour in the atmosphere to an aluminium melt.*

4. Hydrogen molecules are desorbed from the surface to the atmosphere.
5. Hydrogen molecules diffuse back out of the gas boundary layer.
6. Hydrogen molecules dissociate and form atomic hydrogen on the surface.
7. Hydrogen diffuses through the oxide layer.
8. Hydrogen atoms diffuse through the metal boundary layer.

The usual assumptions are made that the chemical reactions, dissociation, absorption, and desorption are rapid. This leaves steps 1, 5, and 8. In Section 3.8.2 we saw that step 8 is slow. However, steps 1 and 5 play an important role in determining p_{H_2} at the surface. Step 1 is described by

$$\dot{n}_{H_2O} = \frac{k_{H_2O}}{RT} \left(p_{H_2O} - p_{H_2O_i} \right).$$ (3.112)

Similarly, for step 5 we have for the flux of hydrogen molecules upwards from the interface

$$\dot{n}_{H_2} = \frac{k_{H_2}}{RT} \left(p_{H_{2i}} - p_{H_2} \right)$$ (3.113)

Here, p_{H_2} is the partial pressure of hydrogen in the atmosphere above the melt. Unless there is another source of H_2, we can usually set $p_{H_2} = 0$. $p_{H_2O_i}$ and $p_{H_{2i}}$ are the partial pressures at the interface. Their ratio is determined by the equilibrium given by Eq. (3.109). Finally step 8 gives

$$\dot{n}_H = k_H \left(c_{Hi} - c_H \right),$$ (3.114)

where the hydrogen concentration at the interface is given by the equilibrium constant for the reaction:

$$\frac{1}{2} H_2 = \underline{H}$$

$$K_H = \frac{f_H c_H}{\sqrt{p_{H_2}}}.$$
(3.115)

The f_H is the activity coefficient defined by Eq. (2.37). The mole balance for hydrogen for the reactions that take place at the surface is (noting that H_2O and H_2 both contain two hydrogen atoms)

$$2\dot{n}_{H_2O} = 2\dot{n}_{H_2} + \dot{n}_H$$
(3.116)

By subtraction of Eq. (3.113) from Eq. (3.112) one obtains from Eq. (3.116)

$$\frac{RT\dot{n}_H}{2} = k_{H_2O}p_{H_2O} - k_{H_2O}p_{H_2Oi} - k_{H_2}p_{H_2i}$$
(3.117)

p_{H_2Oi} is now eliminated using the equilibrium constant for Eq. (3.109):

$$\frac{p_{H_2i}}{p_{H_2Oi}} = K.$$
(3.118)

This gives

$$\frac{RT}{2}\dot{n}_H = k_{H_2O}p_{H_2O} - \left(\frac{k_{H_2O}}{K} + k_{H_2} \right) p_{H_2i}.$$
(3.119)

Then c_{H_i} and p_{H_2Oi} are eliminated from Eqs. (3.114) and (3.119) by using Eq. (3.115) in a similar way to the procedure employed for Eqs. (3.96) to (3.98). The solution of the second-degree equation leads to

$$\frac{\dot{n}_H}{k_H c_H} = -\left(1 + \frac{\lambda}{2c_H} \right) + \left\{ \left(\frac{\lambda}{2c_H} \right)^2 + \frac{k_{H_2O}p_{H_2O}K_H^2}{(k_{H_2O}K + k_{H_2})f_H^2 c_H^2} + \frac{\lambda}{c_H} \right\}^{\frac{1}{2}}$$
(3.120)

with

$$\frac{\lambda}{c_H} = \frac{k_H RTK_H^2}{2\left(\frac{k_{H_2O}}{K} + k_{H_2} \right) f_H^2 c_H}$$
(3.121)

Equation (3.120) may be compared with Eq. (3.99). The signs are different since the flow is now into the melt. The equilibrium constant K in Eq. (3.111) is very large so that the term $\frac{k_{H_2O}}{K}$ may be neglected. Again, as shown in Section 3.8.2, λ/c_H is very small as well. Then Eq. (3.120) reduces to

$$\dot{n}_H = k_H \left\{ \frac{\left(\frac{k_{H_2O}}{k_{H_2}} p_{H_2O} \right)^{\frac{1}{2}} K_H}{f_H} - c_H \right\}.$$
(3.122)

The result is the same as if we had assumed a partial pressure of hydrogen at the interface given by

$$p_{H_2} = \frac{k_{H_2O}}{k_{H_2}} p_{H_2O}. \tag{3.123}$$

A simple physical explanation is that water vapour is reduced to hydrogen at the surface. The factor k_{H_2O}/k_{H_2} takes into account diffusion of H_2 back out through the gas film. p_{H_2O} is then obtained either from direct measurements or from the relative humidity and temperature. The water vapour saturation pressure $p_{H_2O}^{sat}(T)$ is tabulated, and p_{H_2O} is equal to this pressure multiplied by the relative humidity.

3.10.2   Measurements of Absorption of Hydrogen to an Aluminium Melt

It has been known for a long time that the gas which is in contact with liquid aluminium can have a large impact on hydrogen concentration in the melt. This section describes measurements of water vapour in furnace atmosphere and hydrogen concentrations in the melt performed in 2017 and discusses mechanisms for the transfer of hydrogen to the melt.[37] In the aluminium industry, liquid aluminium is commonly kept in reverberatory furnaces for treatment and alloying before casting. The furnaces are heated by gas (or oil) which burns over the metal surface as illustrated in Fig. 3.10.

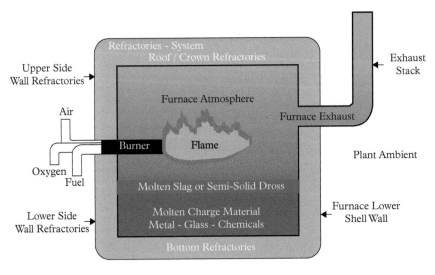

Fig. 3.10 *Schematic of a reverberatory furnace. The heat is mainly transported to the liquid by radiation from the flame and indirectly via the walls and ceiling. [Reprinted from Furu[38] with permission from Oak Ridge National Laboratory.]*

The heat from the flame is transported to the metal partly by gas convection and partly by radiation (including via furnace ceiling and walls). It has been shown that radiation accounts for at least 70 per cent of the heat transported to aluminium during melting.[38] The fuel in a gas burner is often methane or propane. For methane, the combustion reaction is given by

$$CH_4 + 2O_2 = CO_2 + 2H_2O, \tag{3.124}$$

with about 50 MJ of energy released per kg of methane. As seen from Eq. (3.124) two moles of water is produced for every mole of methane combusted. The water may in turn react with aluminium according to

$$H_2O + \frac{2}{3}Al = \frac{1}{3}Al_2O_3 + H_2, \tag{3.125}$$

Then there is hydrogen gas available at the aluminium surface which can dissolve into the metal. The question becomes: what is the hydrogen concentration in the melt in equilibrium with the amount of humidity in the atmosphere in a gas-fired furnace? To answer that question, the humidity in the off-gas from a furnace has been measured and is compared to hydrogen measurements in the aluminium melt.

Air is commonly used as the oxidizing gas for the combustion, so the off-gas will be diluted with nitrogen. The combustion reaction can then be written as

$$CH_4 + 2O_2 + 8N_2 = CO_2 + 2H_2O + 8N_2, \tag{3.126}$$

which means that there will be 'only' 18 per cent water in the atmosphere. Keep in mind that this is absolute humidity, and far above 2.1 per cent, which is the concentration of water vapour in air on a summer day at 30 °C and 50 per cent relative humidity.[39]

Due to leakages, industrial furnaces are never completely sealed. There are openings in tapping holes and not least around the main door where the charging of solid material and melt handling takes place. This will lead to entrainment of air into the chamber and reduce the water concentration in the furnace atmosphere.

Hydro Aluminium's reference centre in Sunndal has a melting furnace with a 20-tonne capacity, equipped with a launder and a metal pump. At the reference centre it is possible to adjust melt flow and melt temperature freely without considering the ingot yield. Figure 3.11 is the schematic of the loop system with measurement positions.

3.10.2.1   *Off-Gas and Hydrogen Solubility Measurements*

By inserting a probe into the furnace exhaust channel, gas samples can be extracted from the atmosphere by gas analyser equipment. One such system is the ProtIR 204M from Protea. This system is capable of measuring several gas species in addition to H_2O. However, only the water vapour is interesting here. This water vapour concentration was measured over two days. Part of the results are shown in Fig. 3.12.

With an AlSCAN instrument produced by BOMEM, the concentration of dissolved hydrogen in molten aluminium can be measured. Figure 3.13 shows a schematic of the

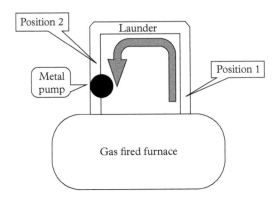

Fig. 3.11 *Schematic of the loop system at Hydro Aluminium's reference centre in Sunndal, Norway. The grey arrow indicates the metal flow direction in the launder.*

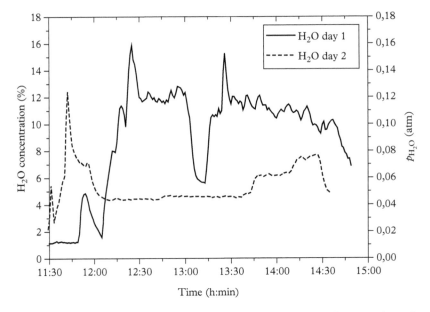

Fig. 3.12 *Measured water vapour concentration in the off-gas channel as a function of time from an air-fuel burner. Left-hand vertical axis is the measured water vapour concentration and the right-hand axis the corresponding partial pressure of H_2O. [Reprinted from Syvertsen et al.[37] by permission from Springer Nature: copyright 2019.]*

apparatus. The left-hand side illustrates how water vapour reacts chemically with aluminium to aluminium oxide and hydrogen gas at the melt surface. Some of the hydrogen gas will then go into solution in the melt until an equilibrium is obtained between the hydrogen gas in the atmosphere and the dissolved hydrogen in the melt. The right-hand side of the figure illustrates the principle of the AlSCAN instrument; pure nitrogen gas circulates in a loop from

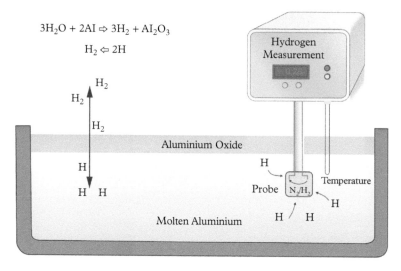

$$3H_2O + 2Al \Rightarrow 3H_2 + Al_2O_3$$
$$H_2 \Leftarrow 2H$$

Fig. 3.13 *Principle of AlSCAN hydrogen measurement. Based on the AISCAN Manual.*[40]

the instrument, down into a probe immersed in the melt, and up again into the instrument. Hydrogen from the melt will diffuse into the probe and be picked up by the circulating nitrogen until equilibrium between the partial pressure in the gas in the loop and dissolved hydrogen in the melt is obtained. The partial pressure of hydrogen in the gas mixture is then calculated by measuring the thermal conductivity of the gas mixture. This calculated partial pressure is then used together with measured melt temperature to calculate the concentration of dissolved hydrogen in the melt.

For historical reasons, the AlSCAN apparatus reports the hydrogen concentration in the melt in units of ml per 100 g molten aluminium. By using the ideal gas law, where 1 kmol of gas equals 22.4 m^3 at standard conditions (0 °C and 1 atm pressure), the relation between ml/100g Al and ppm (parts per million by weight) becomes:

$$1.00 \frac{\text{ml } H_2}{100 \text{ g Al}} = \frac{10^{-6}m^3 H_2 \times 2000 \frac{g}{\text{kmol } H_2}}{22.4 \frac{m^3}{\text{kmol}} \times 100 \text{ g Al}}$$

$$= 0.892 \times 10^{-6} \frac{\text{g } H_2}{\text{g Al}} = 0.892 \text{ ppm} \qquad (3.127)$$

The AlSCAN calculates the hydrogen concentration in the melt according to[40]

$$S = S_0 \times \sqrt{p_{H_2}} \times \text{CF}(A) \times \text{CF}(T), \qquad (3.128)$$

where $S_0 = 0.92$ ml/100g and is the solubility of hydrogen in pure aluminium in equilibrium with 1 atm partial pressure of hydrogen at $700\,°C$, $CF(T)$ the temperature (given in $°C$) correction factor given by

$$CF(T) = \exp\left(6.531\frac{T-700}{T+273}\right),\tag{3.129}$$

and $CF(A)$ an alloy correction factor defined by

$$CF(A) = 10^{0.0170\times\%Mg-0.0269\times\%Cu-0.0119\times\%Si}.\tag{3.130}$$

For the alloy used, which was a Hydro 300333 alloy containing 0.23% Mg, 0.09% Si, and 0.86% Cu, the alloy correction factor becomes

$$CF = 0.95.$$

It can be shown that Eq. (3.128), without the alloy correction factor, is a rewritten version of the relation published by Ransley and Neufeld in 1947.[41] The alloy correction factor is related to the hydrogen activity coefficient (as defined in Section 2.1).

Two AlSCAN units were used to measure the hydrogen concentration in the melt during the same time periods as the water vapour pressure was measured. These two units were placed in positions 1 and 2, where the first position was about 1 m from the furnace exit while the second point was about 6 m further down the launder, as indicated in Fig. 3.11. The measurements for the two days are shown in Figs. 3.14 and 3.15, respectively.

It can be seen from the AlSCAN measurements that even though the temperature drops about $10\,°C$ when the metal flows from position 1 to position 2, the hydrogen concentration is unchanged (at least within the measurement errors as given by the AlSCAN producer[40]). The fact that there is very little difference between the hydrogen levels in the melt at the two positions shows that the transport process of hydrogen in and out of the melt is slow.

Figure 3.16 shows both the water vapour and hydrogen measurements. This graph clearly shows that higher water vapour in the atmosphere gives higher hydrogen concentration in the melt.

Now, if the melt temperature is the same inside the furnace (where the pick-up of hydrogen from the water vapour occurs) and at the first AlSCAN position, then the partial pressure of hydrogen in equilibrium with the hydrogen content in the melt in the AlSCAN probe must be the same as the 'effective' partial pressure of hydrogen in equilibrium with the melt inside the furnace. Thus, by using Eqs. (3.128)–(3.130), the equilibrium partial pressure inside the probe can be calculated. Figures 3.17 and 3.18 show the measured partial pressure of water vapour in the off-gas together with the partial pressure of hydrogen in the AlSCAN probe at position 1 over day 1 and day 2, respectively. It is clear from the two graphs that only some of the water vapour goes into hydrogen in the melt (and hydrogen gas in the probe).

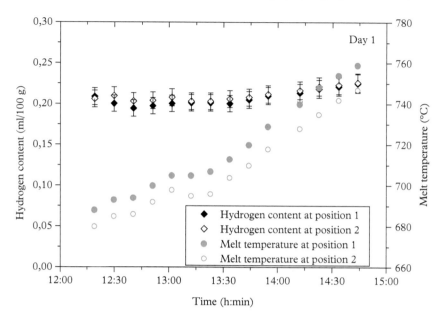

Fig. 3.14 *AlSCAN measurements of melt temperature and hydrogen content during the first day.*

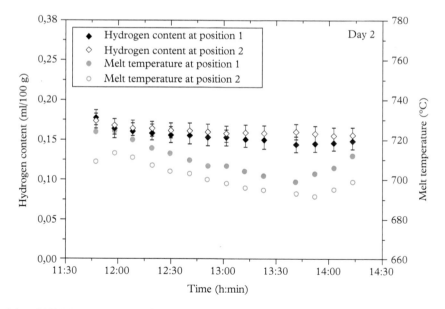

Fig. 3.15 *AlSCAN measurements of melt temperature and hydrogen content during the second day.*

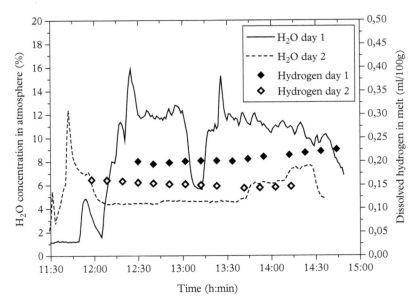

Fig. 3.16 *Combination of the water vapour measurements and the AlSCAN measurements at position 1.*

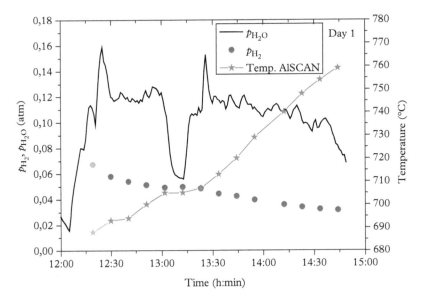

Fig. 3.17 *Calculated partial pressure of hydrogen in the AlSCAN instrument (green dots) and melt temperature (red stars) at position 1 together with measured water vapour for the first day.*

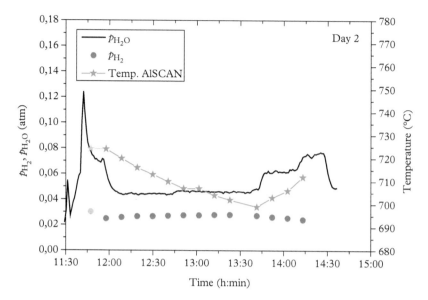

Fig. 3.18 *Calculated partial pressure of hydrogen in the AlSCAN instrument (green dots) and melt temperature (red stars) at position 1 together with measured water vapour for second day.*

In Section 3.10.1, it was found that the partial pressure of hydrogen is equal to the partial pressure of water vapour multiplied by the (inverse) ratio between their respective mass transfer coefficients:

$$p_{H2} = \frac{k_{H_2O}}{k_{H_2}} p_{H_2O}. \qquad (3.131)$$

In Section 3.3 it was found that for the transport of solute species in a gas phase onto a solid surface (the gas does not 'see' any difference between a solid or liquid surface), the mass transfer coefficient relates to the diffusion coefficient as

$$k \propto D^{\frac{2}{3}}. \qquad (3.132)$$

Then

$$\frac{k_{H_2O}}{k_{H_2}} = \left(\frac{D_{H_2O}}{D_{H_2}} \right)^{\frac{2}{3}} \qquad (3.133)$$

With the use of tabulated values for the diffusion coefficients, and the assumption that the ratio between them is not dependent on the temperature (according to *CRC Handbook of Chemistry and Physics*, the ratio is fairly constant for 100, 200, 300, and 400 °C; we assume it is constant also for 700–800 °C),[39]

$$D_{H_2O} \Big/ D_{H_2} = 0.36.$$

Then

$$p_{H_2} = 0.506 \cdot p_{H_2O}. \tag{3.134}$$

Then the measured water vapour measurements can be used directly in Eq. (4.68) for calculating the equilibrium hydrogen content in an aluminium melt. Figures 3.19 and 3.20 show these calculated hydrogen concentrations in the melt (blue squares), the measured water vapour concentration in the atmosphere in the furnace (black curve), and two calculated hydrogen concentrations in the melt (red and blue curves) for the two days, respectively. For the red curve, it is assumed that all water vapour dissociated at the melt surface acts as the source for dissolved hydrogen in the melt $(p_{H_2} = p_{H_2O})$. The blue curve also takes into account the transport of hydrogen gas away from the surface as assumed in the model. It can be concluded that the model fits the measurements well, at least for the water vapour concentrations measured and presented here.

Again, it is shown that the process for obtaining equilibrium between atmospheric water vapour and dissolved hydrogen is slow. For Fig. 3.20 where the temperature is relatively

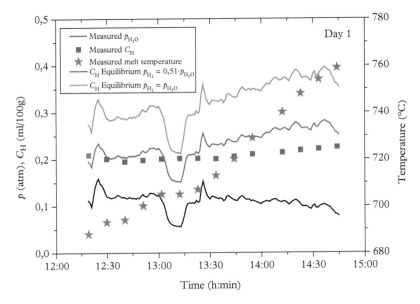

Fig. 3.19 *Different models for water vapour as the source for hydrogen in a melt during day 1. The blue squares and green stars are the measured hydrogen level and temperature in the melt, respectively. The black curve is the measured water vapour concentration in the atmosphere and the blue and red curves are the calculated hydrogen concentration in the melt (from water vapour measurement) with different models for $H_2O \Rightarrow \underline{H}$.*

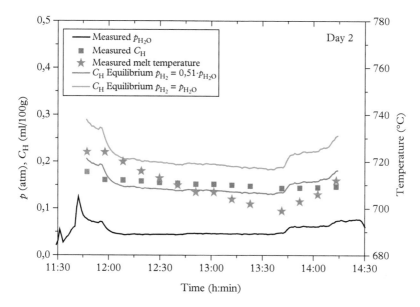

Fig. 3.20 *Different models for water vapour as the source for hydrogen in a melt during day 2. The blue squares and green stars are the measured hydrogen level and temperature in the melt, respectively. The black curve is the measured water vapour concentration in the atmosphere and the blue and red curves are the calculated hydrogen concentration in the melt (from water vapour measurement) with different models for* $H_2O \Rightarrow \underline{H}$.

constant, the model presented above fits the measured data well. However, in Fig. 3.19, where the temperature is rising relatively fast, the difference between the measured hydrogen concentration in the melt (blue squares) and the model (blue line) increases as the temperature increases.

From a geometric point of view it may be reasonable that only 50 per cent of the hydrogen acts as the source for dissolved hydrogen. When a hydrogen molecule is formed, and moves freely above the melt surface, it is 50 per cent likely that it moves away from the surface and 50 per cent likely that it moves toward the surface.

3.11   Fluid Dynamics

3.11.1   Introduction

In this book we frequently refer to conditions in the 'bulk liquid'. It is often assumed that we have complete mixing so that concentrations and temperatures are the same throughout the bulk melt. In Chapter 5 and in Section 3.4 the words 'eddy velocity', 'eddy viscosity', 'eddy diffusivity', and 'eddy length' are mentioned repeatedly. No attempt will be made here to justify the use of these concepts in any rigorous manner. Rather, the aim is to give the reader some background of, and understanding for, the underlying ideas and theories.

To start out it must be emphasized that in the majority of cases our problems are turbulent; that is, flow varies in an irregular and random manner with time and position. The mathematical complexity of turbulence precludes any exact analysis at this time. We have no other recourse than the idea of Osborne Reynolds to work with the time-averaged equations of motion. Velocities, pressures, and temperatures have a mean and a fluctuating component:

$$u = \bar{u} + u\prime$$
$$p = \bar{p} + p\prime. \tag{3.135}$$
$$T = \bar{T} + T\prime$$

The time average of the fluctuating component u', p', or T' is zero. The following rules of averaging apply for any two turbulent quantities f and g:[42]

$$\bar{f}' = 0, \ \bar{\bar{f}} = \bar{f}, \ \bar{\bar{f}}\bar{g} = \bar{f}\bar{g}$$
$$\overline{f'\bar{g}} = 0, \ \overline{f+g} = \bar{f} + \bar{g}.$$
$$\frac{\partial \bar{f}}{\partial x} = \frac{\partial \bar{f}}{\partial x}, \ \overline{fg} = \bar{f}\bar{g} + \overline{f'g'}$$

For incompressible flow the continuity equation becomes

$$\frac{\partial \bar{u}_x}{\partial x} + \frac{\partial \bar{u}_x}{\partial y} + \frac{\partial \bar{u}_z}{\partial z} = 0$$
$$\frac{\partial u_x'}{\partial x} + \frac{\partial u_x'}{\partial y} + \frac{\partial u_z'}{\partial z} = 0 \tag{3.136}$$

(In Chapter 4 we discard the dash for the fluctuating velocity in order to simplify the notation.) With these averaging rules the Navier–Stokes equation, for instance for the x-component of the velocity, becomes

$$\rho \frac{D\bar{u}_x}{Dt} + \rho \left(\frac{\partial \overline{u_x'^2}}{\partial x} + \frac{\partial \overline{u_x'u_y'}}{\partial y} + \frac{\partial \overline{u_x'u_z'}}{\partial z} \right) = -\frac{\partial \bar{p}}{\partial x} + \mu \nabla^2 \bar{u}_x, \tag{3.137}$$

where $\rho \overline{u_x'^2}, \rho \overline{u_x'u_y'}$ and $\rho \overline{u_x'u_z'}$ are called apparent, or Reynolds, stresses. These new terms are never negligible in any turbulent shear flow and are the source of our analytical difficulties, because their analytical forms are not known *a priori*. In essence, the time-averaging procedure has introduced new variables such as $\overline{u_x'u_y'}$ and $\overline{u_x'u_z'}$, which can be defined only through (unavailable) knowledge of the detailed turbulent structure. The variables are related not only to fluid physical properties but also to local flow conditions: velocity, geometry, surface roughness, and upstream history.

Equations (3.136) and (3.137) (in three dimensions) are the basic relations for continuity and the mean momentum. These relations contain many unknown terms and do not give much information about the evolution of the turbulent structure. Other relations might be useful and Prandtl suggested formulating a mean mechanical energy equation (see discussion in White[42]). He obtained a relation for the rate of change of what may be called the turbulent kinetic energy per unit mass:

$$k = \frac{1}{2}\left(u_x^2 + u_y^2 + u_z^2\right).$$ (3.138)

The terms in the equation are rather complex.[43] In words:

    The change in k per unit time

    = Turbulent and viscous diffusion of turbulent kinetic energy

    + Production (redistribution) $-\dot{\varepsilon}$. (3.139)

Here $\dot{\varepsilon}$ is the dissipation rate of turbulence energy, and has units of m^2/s^3.

Equations (3.138) and (3.139) contain unknown turbulence correlation terms. These terms must be modelled with semiempirical assumptions which will now be discussed.

3.11.2   Turbulence Modelling Assumptions

Perhaps the first move towards a model of turbulence can be attributed to Boussinesq.[44] More than a hundred years ago, he suggested that the effective shear stress, arising from the cross-correlation of fluctuating velocities, could be replaced by the product of the mean velocity gradient and a quantity termed the turbulent viscosity. $\tau = \rho \overline{u_z u_x}$ is replaced by

$$\tau = -\mu_t \frac{\partial \bar{u}_x}{\partial z}.$$ (3.140)

Unlike μ, the molecular viscosity, μ_t, is not a property of the fluid. It is determined by the structure of the turbulence and varies from point to point in the flow.

Pre-eminent among the models which employ algebraic relations for μ_t is Prandtl's proposal,[10,45]

$$\mu_t = \rho l^2 \left| \frac{\partial \bar{u}_x}{\partial z} \right|,$$ (3.141)

where l is the Prandtl eddy length. For near two-dimensional boundary layer flows the proposal has served very well. It is simple, is surprisingly accurate, and has a wide range of applicability. For example, the Prandtl eddy length concept is employed in Appendix 2 concerning inclusion removal to walls. Prandtl took into account that near a wall the velocity gradient $d\bar{u}_x/dz$ is large for the fluid in turbulent flow. To explain the model proposed by Prandtl we exploit the analogy with molecular transport by going back to Fig. 2.18 and replacing the molecule with a pulse (lump) of fluid. A pulse of fluid moving a negative distance l in the z-direction towards a wall will be entering a slower moving part of the stream where it will increase locally the velocity in the x-direction. Thus a negative u'_x should correlate with a positive u'_x. The value of u'_x depends on the distance of travel, l, before the lumps lose their identity and are mixed into the fluid at the new position. u'_x is proportional to l and to the mean velocity gradient: $u'_x = l d\bar{u}_x/dz, u'_z = -l d\bar{u}_x/dz$ and $\tau = -\rho l^2 \, | \, du_x/dz \, | \, d\bar{u}_x/dz$. l is effectively the mean free path of a pulse of liquid, and is thus a measure of the scale of the turbulence eddies. In Appendix 2 it is also assumed that the transfer of mass and momentum takes place by the same mechanisms. For the turbulent diffusion coefficient, $D_E = u'_z l$. This relation was also used in Section 3.4 concerning a free surface.

One problem with the proposal by Prandtl was to determine the 'length scale', l. It can be specified fairly realistically for boundary layers along solid walls, $l = 0.4 \, z$, where z is the distance from the wall. However, it varies in a complicated way in irregular channel sections, and in mixing and recirculating flows. For such flows it seems to be necessary to describe turbulence in terms of quantities such as k and $\dot{\varepsilon}$. Then Eq. (3.139) must be included for k. In addition an equation for $\dot{\varepsilon}$ can be derived.[43] This leads to the so-called k–$\dot{\varepsilon}$ model for turbulence. The k–$\dot{\varepsilon}$ model is typical of the multiequation models widely used in computer calculations of turbulent flow problems.[7,45,46] This also applies to turbulent flow in multiphase systems. Section 3.11.13 offers some comments concerning systems with several phases.

3.11.3 Multiphase Flows

In process metallurgy multiphase flow arises when, for example, a liquid bath is stirred by injection of an inert gas. Injection of solid particles into a melt by a carrier gas is another example. The clogging of nozzles, transfer of inclusions to walls and top-slag, and filtration represent multiphase situations. In fact, in the field of refining metallurgy, all problems associated with liquid or solid inclusions or particles and their removal are examples of multiphase flow. In this book, with some exceptions, it is assumed that the particles or bubbles are so far apart that each behaves as a single particle or bubble surrounded by a liquid of infinite extent. Usually, the behavior in this case is well known and allows us to describe the problem in a satisfactory manner. However, there are a number of cases when a more accurate analysis of two-phase flow would be desirable. For instance, in the description of gas-purging processes we would like to determine the residence time of the bubbles accurately. When we study the removal of inclusions by settling, we wish to characterize the behaviour of the slurry at the bottom of the settling unit. The inclusions in this part of the settling or holding furnace are so close together that they certainly interact. In filtration liquid inclusions may tend to coalesce to give a liquid film that escapes through the filter. This last example is perhaps particularly difficult in that the thermodynamics of surfaces is also involved.

For inclusions we are usually justified in assuming that the dispersed phase does not affect the flow of the continuous phase. Obviously, this is not the case for bubble stirring. Then the interaction between the bubble plume and the melt generates the circulating flow. The momentum acting on the liquid is equal but opposite in sign to the drag experienced by the bubbles. To model the properties of the bubble plume it has been postulated that the fluctuating liquid velocities which a bubble encounters are isotropic and possess a Gaussian probability distribution.[47] A bubble is assumed to interact with an eddy for a time equal to the eddy lifetime or bubble transit time, whichever is smaller.[47] These effects cause the bubbles to

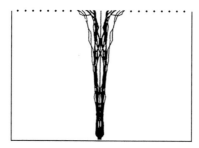

Fig. 3.21 *Predicted bubble trajectories for the situation depicted in Fig. 3.22. Bubble diameter is 2.2×10^{-2} m and gas flow rate is 30 Nl/min.*

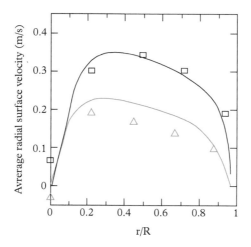

Fig. 3.22 *A comparison between experimental surface velocities and predicted velocities in steel.[47] The results for a gas flow rate of 30 Nl/min are given by triangles (experiments) and the red line (theory). The squares and black line show the experimental and theoretical results, respectively, for 75 Nl/min.*

spread out and generate the plume as shown in Fig. 3.21. Figure 3.22 gives the velocities obtained from the model[47] and they agree quite well with the measurements.[30,48] Such results indicate that 'turbulence models' can quite successfully represent the gross time-averaged aspects of many turbulent flow situations. This applies even to the multiphase case.

3.12   Numerical Solution

In this book we see that our problems are described in terms of conservation equations. Examples of such conservation equations are mass balances for impurities such as oxygen and hydrogen in metals. If the process is steady state, these mass balances can lead to simple algebraic equations. If time enters into the problem, the mass balances contain time-dependent accumulation terms, and we enter the difficult field of differential equations. To allow us to handle such problems, we often assume that there is complete mixing in the bulk melt. Then the melt is described by one set of time-dependent variables for concentrations and temperature and an exact analytical solution is often possible. In many cases this simplistic approach to our problems is valuable especially from the point of view of understanding the main features of a process. Once the initial obstacles of a pedagogic nature have been surmounted, one may wish to obtain a more accurate description. For instance, perhaps there are zones in the melt that are near-stagnant so that the assumption of complete mixing is not valid. In other zones circulation of the melt may change the residence time of bubbles, inclusions, and so on. Then it becomes imperative to calculate the flow patterns and velocities throughout.

Unhappily for the refining metallurgist, and perhaps especially for the engineer studying a stirred reactor, the numerical flow problems often belong to the most difficult class; the flow is two-way, recirculating.[43] Another name for this is that the problem is elliptic. Such problems require considerable computer storage and computer time.

A numerical solution of a flow problem consists of numbers that give the dependent variables such as velocities, concentrations, and temperatures in a set of points (grid points) placed throughout the melt. A numerical method includes the task of providing a set of algebraic equations for these unknown variables and of prescribing an algorithm for solving the equations. In the control volume formulation the melt is divided into a number of non-overlapping control volumes such that there is one control volume surrounding each grid point. Each of these control volumes behaves as a small (completely mixed) reactor. The behaviour of this small reactor is as described elsewhere in this book. Quantities such as mass, concentrations, momentum, and energy are exactly conserved for each control volume and of course over the whole calculation domain. This characteristic exists for any number of grid points, even when there is only one grid point corresponding to the assumption, often employed in this book, of complete mixing of the whole melt.

The algebraic equation that relate the various grid points are called discretization equations in the literature. Patankar[49] has formulated a number of rules for such equations in a readable and lucid manner.

The discretization equations are algebraic equations and are solved by the standard methods for linear algebraic equations. However, often one encounters non-linear situations, even in heat conduction. For instance, the thermal conductivity k may depend on temperature, T, or a source term (chemical reaction term) in a heat balance equation may be a non-linear function of T. Then the coefficients in a discretization equation for T will themselves depend on T. Such situations are handled by iteration. This process, called an algorithm, involves the following steps.

1. Start with a guess or estimate of T at all grid points.
2. From these guessed values of T calculate tentative values of the coefficients in the discretization equation.
3. Solve the nominally linear set of algebraic equations to get new values of T.

4. Return to step 2 using the values from step 3 until further repetitions (called iterations) cease to produce any significant changes in T.

It is a simpler task to solve heat diffusion problems than to calculate the velocity field. The real difficulty here lies in the unknown pressure distribution. The pressure gradient forms a part of the source term for a momentum equation. Once the pressure is known, there is no particular problem in solving the momentum equations. When the correct pressure field is substituted into the momentum equations, the resulting velocity field satisfies the continuity equation. There is no obvious equation for obtaining pressure. However, the pressure field is, indirectly, specified via the continuity equation. To obtain the correct pressure the main feature in the algorithm is a pressure correction equation based on the continuity equation. Again, the procedure is iterative and goes on until the pressure no longer changes, that is, until a converged solution is obtained.

It should be mentioned that a staggered grid is employed; the grid points are different for velocity and pressure. The important advantage of this is that the pressure difference between two adjacent grid points now becomes the driving force for the velocity component located between these two points.

This very brief outline of numerical flow calculations concludes this chapter on mixing, mass transfer, and numerical models. Without a doubt the reader will feel that this final section on fluid flow and numerical calculations was all too meagre and sketchy. Nevertheless, the author hopes the message has got across that numerical fluid flow calculations is a very useful field. Hopefully, the reader will proceed on his own to study the literature, such as the books by Patankar[49] and Guthrie.[7]

··

REFERENCES

1. Ueda T, Furuyama S, Nakagawa R. Model experiment and its analysis on circulating continuous degassing. *Trans. ISIJ*, 1981; 21(10): 716–25.
2. Seshadri V, DeSouza Costa S. Cold model studies of R.H. degassing process. *Trans. ISIJ*, 1986; 26: 133–8.
3. Kawabara T, Umezawa K, Mori K, Watanabe H. Investigation of decarburization behaviour in RH-reactor and its operation improvement . *Trans. ISIJ*, 1988; 28(4): 305–14.
4. Bird RB, Steward WE, Lightfoot EN. *Transport Phenomena*. New York: Wiley; 1960.
5. Brimacombe JK, Nakanishi K, Anagbo PE, Richards GG. Process dynamics: gas-liquid. In: Koros PJ, St. Pierre GR, eds, *Proc Elliott symposium on chemical process metallurgy:* Warrendale, Pa.; 1990; pp. 343–412.
6. Ogawa K, Onoue T. Mixing and mass transfer in ladle refining process. *ISIJ International*, 1989; 29(2): 148–53.
7. Guthrie RIL. *Engineering in process metallurgy*, rev. edn. Oxford: Clarendon Press; 1992.
8. Szekely S, Themelis NJ. *Rate phenomena in process metallurgy*. New York: Wiley-Interscience; 1971.

9. Engh TA, Grip C-E, Hanson L, Worner HK. A one-dimensional model of the WORCRA steel refining launder. *Jernkont. Ann.* 1971; 155(9): 553–64.
10. Davies JT. *Turbulence phenomena,* New York: Academic Press; 1972.
11. Clift R, Grace JR, Weber ME. *Bubbles, drops and particles*. New York: Academic Press; 1978.
12. Ito K, Amano K, Sakao H. Kinetic study on nitrogen absorption and desorption of molten iron. *Trans. ISIJ*, 1988; 28(1): 41–48.
13. Bruckner R. Interfacial convection and mass transfer. In: Dahl W, Lange WK, Papamentillos D, eds, *Kinetics of metallurgical processes in steelmaking*, pp. 459–91. Diisseldorf: Verlag Stahleisen; 1975.
14. Mori K, Sano M. Effect of surface agents on convection and mass transfer. In: Dahl W, Lange WK, Papamentillos D, eds, *Kinetics of metallurgical processes in steelmaking*, pp. 492–513. Düsseldorf: Verlag Stahleisen; 1975.
15. Belton GR. Langmuir adsorption, the Gibbs adsorption isotherm, and interfacial kinetics in liquid metal systems. *Met. Trans. B*, 1976; 7(1): 35–42.
16. Carlsaw HS, Jaeger JC. *Conduction of heat in solids*. Oxford: Clarendon Press; 1959.
17. Luikow AV. *Analytical heat diffusion theory*. New York: Academic Press; 1968.
18. Kreyzig E. *Advanced engineering mathematics*, 6th edn. New York: Wiley; 1988.
19. Dong L. *Flow generated by an impeller and mass transfer across gas-liquid interfaces in stirred vessels*. Dr.ing.Thesis. Trondheim, Norway: Norwegian Institute of Technology; 1991.
20. Roy AK, Chhabra RP. Prediction of solute diffusion coefficients in liquid metals. *Metall. Trans. A*, 1988; 19(2): 273–9.
21. Ozawa Y, Mori K. Critical conditions for penetration of solid particles in liquid metal. *Trans. ISIJ*, 1983; 23(9): 769–74.
22. Engh TA, Nilmani M. Bubbling at high flow rates in inviscid and viscous liquids (slags). *Metall. Trans. B*, 1988; 19(1): 83–94.
23. Wraith AE. Two stage bubble growth at a submerged plate orifice. *Chem. Eng.-Sci.*, 1971; 26(10): 1659–71.
24. Davidson JF, Schüler BOG. Bubble formation at an orifice in an inviscid liquid. *Trans. Instn. Chem. Engrs.*,1960; 38: 335–42.
25. Lee BW. Single and multiple particle penetration across a gas-liquid interface relevant to injection metallurgy. M.Eng.Sc. Thesis, Department of Chem. Engineering Science, University of Melbourne, Australia: 1990.
26. Engh TA, Johnston A. Injection of non-wetting particles into melts. In: Bickert C, ed., *Proc. of the international symposium on reduction and casting of aluminium*, pp. 11–20. Montreal: Pergamon Press; 1988.
27. Nagata S. *Mixing, principles and applications*. New York: Wiley; 1975.
28. Nilmani M, Engh TA, Gray NB, Floyd SM. Plant and laboratory investigations on injection and related phenomena in some smelting and refining processes. *Scand. J. Metallurgy*, 1991; 19: 278–87.
29. Waladan M, Firkin GR, Bultitude-Paull J, Lightfoot BW, Floyd JM. Top submerged lancing for recovery of zinc from ISF slag. In: Mackey TS, ed., *Lead-zinc '90*, pp. 607–28. Warrendale, PA: TMS; 1990.
30. Johansen ST, Robertson DGC, Woje K, Engh TA. Fluid dynamics in bubble stirred ladles: Part I. Experiments. *Metall. Trans. B*, 1988; 19(5): 745–54.
31. Sherwood TK, Pigford RL, Wilke CR. *Mass transfer*. New York: McGraw-Hill; 1975.
32. Patak F. *Untersuchungen zur natrium- und lithium- entfernung aus huttenaluminium*. Doktor-Ingenieurs dissertation. Aachen: Rheinisch-Westfalischen Technische Hochschule; 1983.

33. Pedersen, T. (1984). *Spylgassbehandling av aluminium for fjerning av hydrogen, løste alkalimetaller og inneslutninger*. Dr.ing. avhandling, Metallurgisk Institutt, Norges Tekniske, Høgskole, Trondheim.

34. Roth A. *Vacuum technology*. Amsterdam: North-Holland; 1976.

35. Sigworth GK, Engh TA. Refining of liquid aluminum—a review of important chemical factors. *Scand. J. Metallurgy*, 1982; 11(3): 143–9.

36. Gariepy B, Dube G, Simoneau C, Leblanc G. The TAC process: A proven technology. In: *Light Metals 84*, pp. 1267-79. Warrendale, PA: TMS-AIME; 1984.

37. Syvertsen M, Kvithyld A, Gundersen E, Johansen I, Engh TA. Furnace atmosphere and dissolved hydrogen in aluminium. In: *Light Metals 2019*, pp. 1051–6. Warrendale, PA: TMS-AIME; 2019.

38. Furu J. *An Experimental and Numerical Study of Heat Transfer in Aluminium Melting and Remelting Furnaces*. PhD-thesis NTNU; 2013:30. Also: Oak Ridge National Laboratory, 'Refractories for Industrial Processing: Energy Reduction Opportunities'. December 2004

39. Lide DR. *CRC Handbook of Chemistry and Physics*, 85th edn. Boca Raton, FL: CRC Press; 2004.

40. *AlSCAN Manual*, IMZ9165, Revision J September. ABB Group; 2015.

41. Ransley CE, Neufeld H. The solubility of hydrogen in liquid and solid aluminium. *J. Inst. Metals*, 1948; 74: 599–620.

42. White FM. *Viscous fluid flow*, 2nd edn. New York: McGraw-Hill, 1991.

43. Reynolds AJ. *Turbulent flows in engineering*. London: Wiley; 1974.

44. Hinze JO. *Turbulence*, 2nd edn. New York: McGraw-Hill; 1975.

45. Launder BE, Spalding DB. *Lectures in mathematical models of turbulence*. London: Academic Press; 1972.

46. Launder BE. Second-moment closure: present and future. *Int. J. Heat Fluid Flow.*, 1989; 10(4): 282–300.

47. Johansen ST, Boysan F. Fluid dynamics in bubble stirred ladles: Part 2, Mathematical modelling. *Metall. Trans. B*, 1988; 19(5): 755–64.

48. Hsiao T-C, Lehner T, Kjellberg B. Fluid flow in ladles- experimental results. *Scand. J. Metallurgy* 1980 9(3): 105–10.

49. Patankar SV. *Numerical heat transfer and fluid flow*. New York: CRC Press; 1980.

4

Removal of Dissolved Impurities from Molten Metals

4.1 Introduction

Impurities are present in the ores and raw materials used in the extractive metallurgical processes. Furthermore, impurities and inclusions are also introduced from refractories and atmosphere during the production of the metal and even in the refining processes. Thus oxygen, hydrogen, nitrogen, and carbon often enter the molten metal and affect the mechanical properties of the cast metal.

For many extractive processes it is not practical and economic to try to protect the melt from the environment. Instead, the impurities are removed in the final refining operations. Basically in the refining stages we reverse the transfer of impurities. For instance, hydrogen and nitrogen are eliminated by gas purging and vacuum treatment. Then the pressure of H_2 and N_2 is so low in the gas that H and N leave the melt and enter the bubble or vacuum phase. For the less noble metals, such as iron, oxygen cannot be removed as gas. Instead an element with a high affinity to oxygen, such as aluminium, is introduced into the melt. The resulting reaction product (Al_2O_3 inclusions) must then be transferred out of the melt. Similarly carbon dissolved in molten aluminium from the refractories may react with the melt as it cools down to give solid aluminium carbide inclusions. In many cases these inclusions must be removed before casting the metal. This removal of dissolved elements and inclusions is discussed in this chapter and in Chapter 5. In Chapter 1, we outlined the harmful effects of impurity elements and inclusions.

This chapter starts out by presenting the various methods that are available for removal. The concept of mixing or stirring is introduced. Then the important question of what determines the rate of removal is brought in. Are the components at an interface in the refining reactor in chemical equilibrium or is removal limited by slow mass transfer or slow chemical reaction rates? There may even be an inert layer that covers the interface, blocks mass transfer, and prevents the final attainment of equilibrium.

The principles behind the various reactors available for refining purposes are given. Some important applications concerning impurity removal in steel and aluminium are included as examples of the various methods and principles.

Principles of Metal Refining and Recycling. Thorvald Abel Engh, Geoffrey K. Sigworth, and Anne Kvithyld, Oxford University Press.
© Oxford University Press 2021.
DOI: 10.1093/oso/9780198811923.003.0004

The methods available for removal of dissolved elements may be listed as follows:

1. The impurity x is transferred to another phase where x has a high solubility. This second phase is not soluble in the molten metal. An example is the transfer of sulfur from molten iron to slag. A second example is the transfer of B from molten Si to FeSi solvent.

2. x reacts with a second phase that may be solid, molten, or gaseous. Figure 4.1 illustrates methods 1 and 2. For instance, sodium is removed from molten aluminium by burning with oxygen in the air at the bath surface; or by adding reactive salts to the melt surface.

3. x reacts with another element (added to the melt). The reaction product, whether solid, liquid, or gaseous, must be removed from the melt. This removal may be produced by the effect of gravity or buoyancy forces, inertial forces, or forces generated by stirring the melt. A well-known example is precipitation deoxidation of steel with aluminium. The oxidation product Al_2O_3 must be removed.

4. An electric potential is applied to the melt to remove x by electrolysis.

5. The molten metal is allowed to solidify partially or a solid is partially melted. x is enriched in the liquid phase. This phase or the last part to solidify is removed.

6. A vacuum is applied to remove x by vaporization.

All the methods outlined in points 1–6 above are applied industrially. Points 1 and 2 are the most commonly used. An example of 3 is precipitation deoxidation in steelmaking, which is discussed in Chapter 2. In the reviews in Chapter 8 reference is made to electrochemical methods of purification. Method 5 is discussed in Chapter 8.

An essential point is that usually two or more phases are to react. These phases can be metal, slag, sulfide, gas, and so on. A dissolved component (impurity) or inclusion is removed from the metal and transferred to the slag, second metal, or gas (atmosphere) phase.

An impurity element can never be completely eliminated from the metal. For instance, when sulfur in molten steel is removed by extracting it to the slag phase, some sulfur will always remain in the steel. The sulfur content may be as low as ten parts per million. Though

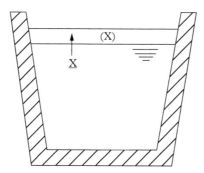

Fig. 4.1 *Transfer of dissolved element x from melt to top slag.*

low, this sulfur level may still play a role in determining the physical and mechanical properties of the final product as we saw in Chapter 1.

Now what determines the final levels of an impurity in a melt? Is it the equilibrium between the impurity content in the metal and in the phase that is extracting the undesirable component, or is it mass transfer? This is one of the main problems that we are addressing here. Whatever the case, some impurity will remain. If the equilibrium is controlling, the chemical activity of the impurity element x is the same in the two phases. To take the example of sulfur, even if one succeeds in obtaining a slag with a very low activity coefficient for sulfur (high sulfur capacity), some S must remain in the steel due to the equilibrium. If removal is determined by mass transfer, the removal rates are governed by the driving force. In the most favourable case it is proportional to concentration [%S]. We realize that as [%S] is reduced in a batch melt, removal becomes slower and slower and the time required for removal may be so long that the cost of attaining low impurity levels becomes prohibitive.

4.2 Mixing

Chemical reaction rates increase very strongly with temperature.[1] We remember the well-known Arrhenius relation that tells us that rates increase roughly exponentially with temperature. Therefore, in purification operations involving molten metals at high temperatures, the chemical reaction rates are usually not rate limiting. Hence the purification of molten metals will be controlled by mass transfer, and can be speeded up by increased stirring or 'mixing'. Mixing may be performed by the use of:

(1) mechanical stirrers;
(2) gas stirring (bubbling);
(3) electromagnetic stirring;
(4) pumps (mechanical or vacuum);
(5) pouring from one ladle to another (Perrin process); or
(6) oscillating ladles.

Mechanical stirrers (impellers) in steel melts are usually of a ceramic material. Often, they will not stand up to the combined chemical and erosion attack by the slag and metal phases. However, in an aluminium melt at temperatures below 800 °C we find that graphite impellers may operate satisfactorily.

Oscillating ladles are usually no longer employed. The mechanical problems involved in moving the whole ladle with its large mass of ceramic material and the wear on bearings, axles, and so on are too troublesome.

Gas stirring (gas injection) in a ladle using a porous plug or a lance is an effective and well-established operation, but has various limitations. For instance, we usually cannot employ this method to mix top-slag down into the melt. In converters, however, where the gas flows are very great, molten metal and slag are mixed.

Electromagnetic stirring often requires a high capital expenditure. Nevertheless this method may be competitive due to the fact that the stirrer is not in direct contact with the melt.

The Perrin process is effective, but it often does not give reproducible results. The melt is exposed to the atmosphere.

Mixing by pumps seems to be a very interesting field in high-temperature metallurgy, though it does not seem to have been properly explored as yet.

It should be mentioned that as a side effect of all methods unwanted 'stirring' is generated at the melt–refractory interfaces. This increases refractory wear.

In the following discussions of purification it will be assumed that mixing or stirring is sufficient. This means that the concentration of dissolved elements throughout the bulk metal bath is constant; that is, concentration does not depend on spatial coordinates. For a batch reactor, however, concentrations will change with time. Often to characterize 'mixing', the concept of mixing time τ_m is introduced:[2]

$$\tau_m \sim 100 \left(\frac{L^2}{\dot{\varepsilon}} \right)^{\frac{1}{3}}. \tag{4.1}$$

Equation (4.1) assumes that mixing takes place by 'eddy diffusion' and that the largest eddies play the dominant role in this process. Thus τ_m depends on a characteristic vessel dimension, L, for instance the diameter. Also, it is assumed that turbulence is homogeneous and isotropic. Then turbulence may be characterized by the energy dissipation $\dot{\varepsilon}$. Electromagnetic stirring of a steel ladle typically has $\dot{\varepsilon} \sim 0.05$ W/kg $= 0.05$ m^2/s.[3] For a ladle with $L = 1$ m this gives $\tau_m = 270$ s $= 4\frac{1}{2}$ min. If the treatment time for the melt is considerably greater than τ_m, we often employ the assumption that the melt is completely mixed. This means that we disregard resistance to mass transfer in the bulk melt (see also Section 3.2).

At the interface between two phases, such as metal and slag, movement is along the interface and not normal to it. This is illustrated in Fig. 4.2. Now the main mechanism for transport of a dissolved component or an inclusion is no longer convection or turbulent eddies. Instead transport to a large degree must depend on the slow mechanism of diffusion. Then the

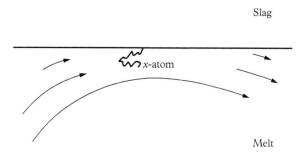

Fig. 4.2 *Velocities are low normal to the interface so that the transfer of x depends on the slow mechanism of diffusion.*

mass transfer coefficient for the melt, k, may be estimated by relations of the type found in penetration theory[4,5] (see Chapter 3):

$$k \sim \sqrt{D/t},$$

(4.2)

where D is the (binary) diffusion coefficient and t is the residence time of the melt at the interface.

It will be seen in the following that often the product of k and the interfacial contact area A determines the kinetics of the purification process. This is the case when kA is too small to allow us to reach equilibrium.

For transfer between two phases we must in general take into account the resistance to transfer on both sides of the interface. This is discussed in Section 4.3.

4.3   The Total Mass Transfer Coefficient, k_t

There is resistance to flow in the 'films' on both sides of the interface between the metal and the extractive phase (here taken as slag). Here it will be seen that the two resistances can be combined into a total resistance, k_t. By analogy with electrical circuit theory one might believe that the two resistances given as the inverse of the mass transfer coefficients $1/k$ and $1/k_{slag}$ could simply be added together. However, we will see that the procedure is not quite that simple. The reason is that the driving forces (voltages) in the rate equations for the two phases are not given in the same units.

For the metal phase, the molar flux of component x is in kmol/(m^2·s):

$$\dot{n} = \frac{k\rho}{100M_x} \left([\%x] - [\%x]_i \right).$$

(4.3)

Here M_x is the molar mass of component x in kg/kmol and ρ is the density of the metal. k is the mass transfer coefficient discussed in Chapter 3. The metal and slag concentrations at the interface, $[\%x]_i$ and $(\%x)_i$, are assumed to be in equilibrium (due to the high chemical reaction rates at the high temperature):

$$K = \frac{(\%x)_i \, \gamma_x}{[\%x]_i \, f_x}.$$

(4.4)

Equation (4.3) can be rewritten in terms of the driving force (potential, voltage); thus:

$$[\%x] - [\%x]_i = \frac{\dot{n} 100 M_x}{k\rho}.$$

(4.5)

Similarly for the slag phase one has

$$\dot{n} = \frac{k_{slag}\rho_{slag}}{100M_x}\left((\%x)_i - (\%x)\right).$$

From this one obtains

$$(\%x)_i = (\%x) + \frac{\dot{n}100M_x}{k_{slag}\rho_{slag}}. \tag{4.6}$$

As already pointed out the units for the driving force are mass per cent metal in Eq. (4.5) and mass per cent slag in Eq. (4.6). However, we notice that if we divide the two sides by the partition ratio

$$\frac{(\%x)_i}{[\%x]_i} = \frac{f_x K}{\gamma_x},$$

Eq. (4.6) becomes

$$[\%x]_i = \frac{(\%x)\,\gamma_x}{Kf_x} + \frac{\dot{n}100M_x\gamma_x}{k_{slag}\rho_{slag}Kf_x}. \tag{4.7}$$

Now the molar flux $\dot{n}$ can be obtained in terms of the sum of the 'resistances' in the two phases. We sum Eqs. (4.5) and (4.7) to give

$$[\%x] - \frac{(\%x)\,\gamma_x}{Kf_x} = \dot{n}100M_x\left(\frac{1}{\rho k} + \frac{\gamma_x}{\rho_{slag}k_{slag}Kf_x}\right). \tag{4.8}$$

We can finally define a total resistance $1/(\rho k_t)$, equal to the sum of the resistance for the metal phase $\frac{1}{(\rho k)}$ and for the slag phase $\gamma_x/\left(\rho_{slag}k_{slag}Kf_x\right)$:

$$\frac{1}{\rho k_t} = \frac{1}{\rho k} + \frac{\gamma_x}{\rho_{slag}k_{slag}Kf_x}. \tag{4.9}$$

We get

$$\dot{n} = \frac{\rho k_t}{100M_x}\left([\%x] - \frac{(\%x)\,\gamma_x}{Kf_x}\right), \tag{4.10}$$

where $(\%x)\,\gamma_x/(Kf_x)$ is the hypothetical concentration in the metal in equilibrium with the concentration in the slag phase. Figure 4.3 sketches the concentration profiles in the two phases. Equation (4.10) plays an important role in modelling purification processes. It is derived for metal and slag, but applies to any two-phase system. For instance, it is used in Section 4.4 for the melt–bubble interface.

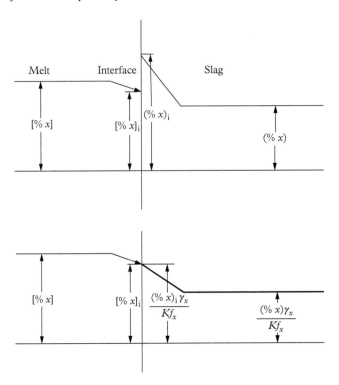

Fig. 4.3 *The concentration in the slag is converted to concentration in the melt by dividing with the partition coefficient Kf_x/γ_x.*

4.4   Equilibrium or Mass Transfer Control in Gas Purging; Reactive Gas

In Chapter 2, thermodynamic factors were considered. We now consider the rate at which a dissolved component x moves to the extracting phase. As an example, we study removal to ascending inert gas bubbles (gas purging). In general x is transferred by the following series of steps:

(1) Transport in the melt to the vicinity of a gas bubble by a combination of convection and diffusion;

(2) Diffusive transport through a thin layer of fluid, called a boundary layer, surrounding the bubble;

(3) Chemical adsorption on to the outside and subsequent desorption from the inside of the bubble surface; and

(4) Diffusion of x or x_2 as a gaseous species inside the bubble of purge gas.

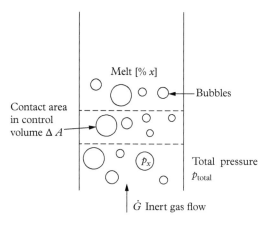

Melt [% x]

Bubbles

Contact area
in control
volume ΔA

p_x

Total pressure
p_{total}

$\dot{G}$ Inert gas flow

Fig. 4.4 *Gas-purging control volume.*

We now present a mathematical analysis of removal based on use of the total mass transfer coefficient. Four assumptions are made:

1. The metal bath is well stirred so that the concentration is essentially uniform.
2. The concentration is uniform inside the gas bubble.
3. The gas bubble surface area is not a function of height in the bath. This is tantamount to assuming that the total pressure in the gas bubble is constant.
4. The partial pressure of x in the bubbles is very small compared to the total pressure.

Assumption 1 should be met in any well-designed commercial process and is discussed in Chapter 3. Assumption 3 applies to systems where the bath is shallow. This is often the case, for instance, when aluminium melts are treated.

The third and fourth assumptions do not always apply for vacuum systems. For other systems the assumptions may allow us to discuss equilibrium and mass transfer in a simple manner. Once the main points are understood, it is not difficult using a computer to remove assumptions 3 and 4.

We consider the change in gas composition as a bubble ascends through the melt. Figure 4.4 considers a horizontal slice of a reactor. If we construct a mass balance equation for component x leaving this control volume of the melt and entering the gas phase, we find that

Moles of x transferred per second to bubbles in control volume = Increase of flow of x carried by bubbles through control volume

or

$$\frac{k_t\rho\left([\%x]-[\%x]_e\right)\Delta A}{100M_x}=\dot{G}\Delta\left(\frac{p_x}{p_{inert}}\right). \tag{4.11}$$

In accordance with assumption 4, p_{inert} = constant. We represent mass transfer with the melt-side total mass transfer coefficient k_t, having units of meters per second, the surface area of the bubbles ΔA, and the driving force for transfer $[\%x] - [\%x]_e$. Here $[\%x]_e$ is the hypothetical x concentration in the melt in equilibrium with the partial pressure p_x in the bubble. p_x is uniform throughout the bulk gas in the bubble

$$[\%x]_e \, f_x = \frac{p_x}{K} . \tag{4.12}$$

ΔA is the bubble surface area present in the volume element under consideration (see Fig. 4.4). According to the ideal gas law the number of moles of x is proportional to p_x. Thus $\dot{G}\Delta (p_x/p_{inert})$ is the incremental increase in the gas composition of x between bubbles entering and leaving the control volume element. $\dot{G}$ is the constant molar flow of inert gas.

We notice that the driving force for mass transfer to the bubble changes as it rises up through the melt since $[\%x]_e$ in Eq. (4.12) changes from zero to p_x^0/Kf_x traversing the melt. p_x^0 is the partial pressure of x when the bubble leaves the bath. We move the terms that depend on p_x over to the right-hand side of Eq. (4.11) and integrate from the bottom to top surface:

$$\frac{k_t \rho A p_{inert}}{100 M_x \dot{G}} = \int_0^{p_x^0} \frac{dp_x}{[\%x] - p_x/(f_x K)} .$$

This gives

$$\frac{k_t \rho A p_{inert}}{100 M_x \dot{G} f_x K} = \ln \left(\frac{[\%x]}{[\%x] - p_x^0/(f_x K)} \right) .$$

The left-hand side of the equation is a dimensionless group, ϕ_A, for the contact area A. We relate the exit pressure p_x^0 to the value at equilibrium, $f_x[\%x]K$, *the efficiency of gas purging*:

$$Z = \frac{p_x^0}{f_x [\%x] K} . \tag{4.13}$$

p_x^0 cannot exceed the value $f_x[\%x]K$ for equilibrium with the melt. Thus $Z \leq 1$. The antilogarithm of the previous equation gives

$$Z = 1 - \exp(-\phi_A) ,$$

where

$$\phi_A = \frac{k_t \rho A p_{inert}}{100 M_x \dot{G} f_x K} .$$

Figure 4.5 show the efficiency Z as a function of ϕ_A (dimensionless contact area).

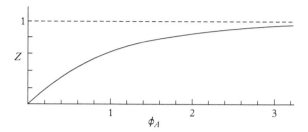

Fig. 4.5 *Efficiency Z versus dimensionless contact area ϕ_A.*

The removal rate is

$$\dot{G}_x \approx \frac{\dot{G}p_x^0}{p_{inert}} = \frac{\dot{G}f_x\,[\%x]\,K}{p_{inert}} \cdot Z,$$

where

$$\frac{\dot{G}f_x\,[\%x]\,K}{p_{inert}} = \textit{Maximum possible removal rate.}$$

To approach equilibrium ϕ_A must be greater than about 3. For $\phi_A = 3$, $Z = 0.95$. If $\phi_A = 2$, then $Z = 0.86$. ϕ_A increases with the ratio between contact area and gas flow rate, $\dot{G}$. Also, it is apparent that ϕ_A increases when the equilibrium constant K is small. When K is small, the bubbles at equilibrium will have a low partial pressure of x and the 'x-capacity' of the bubbles is low. Similarly, ϕ_A increases with p_{inert} which is approximately equal to the total pressure. When the total pressure is high, the gas volume available to be 'filled' with gaseous x is low.

When the dimensionless group for contact area ϕ_A is less than, say, 1/20, $\exp(-\phi_A)$ $\approx 1 - \phi_A$ and

$$Z \approx \phi_A.$$

Then this value for Z gives

$$\dot{G}_x = \frac{k_t \rho A\,[\%x]}{100 M_x}. \tag{4.14}$$

This is the case when the counter-driving force term $[\%x]_e$ in Eq. (4.11) can be neglected; the driving force is equal to the concentration $[\%x]$. In summary, for $\phi_A > 3$ equilibrium between bubble and melt controls removal by gas purging. For $\phi_A < 1/20$, mass transfer as given by Eq. (4.14) controls removal.

By way of illustration, we give a numerical example: sodium is purged from aluminium with an inert gas:

$$\dot{G} = 2Nm^3 \text{per hour} = \frac{2}{22.4 \times 3600} \text{ kmol/s}$$

$$\frac{1}{Kf_{Na}} = 0.29 \text{ at } 1100K \text{ (see Chapter 2)}$$

$$M_{Na} = 23 \text{ kg/kmol}$$

$$k_t = 5 \times 10^{-4} \text{ m/s (see Chapter 3)}$$

$$A = 0.3m^3$$

$$p_{inert} = 1 \text{ bar}$$

$$\rho = 2350 \text{ kg/m}^3.$$

This gives

$$\phi_A = \frac{5 \times 10^{(-4)} \times 2350 \times 0.3 \times 1 \times 0.29 \times 22.4 \times 3600)}{(100 \times 23 \times 2)} = 1.79,$$

which gives an efficiency for the purge gas of

$$Z = 1 - \exp(-1.79) = 0.83.$$

The contact area A of 0.3 m² is sufficient to give an efficiency of 83 per cent. We will see in the following that the bubble contact area A decreases if the bubble size increases.

Sodium is also removed at the top surface; sometimes we can even see sodium flames. This mechanism of removal is included later. Equation (4.14) is valid also when a gas is added to the purge gas that reacts with x. For instance, Cl_2 will react with $\underline{Na}$ to give $NaCl$:

$$\underline{Na} + \frac{1}{2}Cl_2 \rightarrow NaCl.$$

Then the counter-force term $[\%x]_e$ in Eq. (4.11) can be neglected since the partial pressure of Na in equilibrium with NaCl and Cl_2 (actually $AlCl_3$) is very small. Of course this is the case only if enough Cl_2 is added to the purge gas to react with the sodium in the bubbles.

If we inject Cl_2 into an aluminium alloy that contains Mg, we must expect that $AlCl_3$ first reacts with Mg since the content of Mg is much greater than that for Na:

$$2AlCl_3 + 3\underline{Mg} = 3MgCl_2 + 2Al.$$

Then $MgCl_2$ takes part in an exchange reaction with sodium:

$$MgCl_2 + 2\underline{Na} = 2NaCl + \underline{Mg}$$

since sodium has much greater affinity to chlorine than Mg (see Chapter 2 on thermo-dynamics). The rate of production of $MgCl_2$ depends on the bubble surface area, while removal of Na depends on the $MgCl_2$/melt contact area.

When stainless steel is decarburized down to very low levels by blowing O_2 and argon, the behaviour is analogous. Oxygen first reacts with iron and chromium and then, depending on the contact area between the oxides and carbon, carbon is removed.

4.5 Bubble Contact Area

To simplify matters we assume initially that we have spherical bubbles of radius R. Also we do not take into account that, according to the gas law, the bubble volume increases with decreasing pressure as the bubble rises up through the melt.

If the flow of gas is $\dot{Q}_g \, m^3$ per second, the number of bubbles created per second is

$$\dot{N} = \frac{\dot{Q}_g}{4\pi R^3 / 3}.$$

With a rise time (residence time) of τ_b this gives for the number of bubbles present in the melt

$$N = \dot{N}\tau_b.$$

The contact area now becomes

$$A = N4\pi R^2 = \frac{3\dot{Q}_g \tau_b}{R}. \tag{4.15}$$

This equation is important since it demonstrates that the contact area, A, is inversely proportional to the bubble size, R. Also A is proportional to the residence time τ_b. Often τ_b in the literature is obtained by assuming that bubble velocities are much greater than convective velocities and that the bubbles move independently of each other. Then

$$\tau_b = \frac{H}{v_b}, \tag{4.16}$$

where H is the vertical distance from the gas inlet to the bath surface and v_b is the velocity of a single bubble. Typically $H = 0.5$ m and $v_b = 0.5$ m/s so that $\tau_b = 1$ second.

If we assume that the gas instantly reaches the melt temperature in our previous example with the aluminium melt, then at 1100 K

$$\dot{Q}_g = \frac{(2 \times 1100)}{273 \times 3600} = 0.0022 m^3/s.$$

If bubbles with $R = 0.010$ m are produced, then

$$A = \frac{3 \times 0.0022 \times 1}{0.01} = 0.66 m^2.$$

In the example this contact area was sufficient to allow the sodium gas in the bubbles to approach equilibrium.

To increase the contact area one should, according to Eq. (4.15), either decrease the bubble size R or increase the residence time, τ_b, of the bubbles. Injection of gas through a lance is not a solution since the bubbles roughly have a diameter equal to the outer lance diameter.[5,6] A porous plug may give bubbles of size around 1 mm at low gas volume flow rates[7] in water. However, in molten metals, the material in the porous plug is often not wetted by the melt. Then the bubbles may become much larger.[8] Various impeller and jetting systems have been developed that stir the melt and produce small bubbles.[9-12] Figure 4.6 gives two examples of such systems.

In Eq. (4.15) it was assumed that all the bubbles were of the same size. In the real case there is a distribution of bubble sizes; a fraction f_R of the bubbles produced has size R. Then the number of bubbles of size R produced per second is

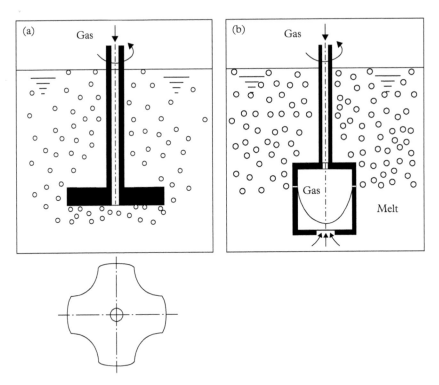

Fig. 4.6 (a) *Rotor for creating and dispersing bubbles in aluminium; (b) a rotor-pump system for creating and dispersing bubbles in the melt.*[10]

$$\dot{N}_R = \dot{N} f_R.\tag{4.17}$$

If these bubbles have a residence time τ_R, the number of bubbles of size R in the melt becomes

$$N_R = \dot{N} f_R \tau_R,$$

and the corresponding contact area is

$$A_R = \dot{N} f_R \tau_R 4\pi R^2.$$

The total contact area A is obtained by summation over all bubble sizes

$$A_R = \dot{N} \Sigma f_R \tau_R 4\pi R^2.$$

$\dot{N}$ is equal to the volume flow divided by the volume of the bubbles

$$\dot{N} = \frac{\dot{Q}_g}{\Sigma f_R 4\pi R^3 / 3}.$$

Inserted in the previous equation this gives for the contact area

$$A = \frac{3\dot{Q}_g \Sigma f_R \tau_R R^2}{\Sigma f_R R^3}.\tag{4.18}$$

f_R can be measured by employing electrical probes,[13] by using a pressure pulse technique,[14] or by photographic methods from water model studies. Equation (4.18) does not take into account the effect of coalescence and splitting of bubbles or the effect of pressure changes up through the melt.

Theory and experimental methods only allow us qualitatively to consider coalescence and splitting in practice and to scale from water model experiments to molten metals.

4.6 Continuous Back-Mix Reactors

Continuous reactors are conceptually simpler than batch reactors. Therefore, we discuss them first even though batch reactors are employed in most metal purification processes. Continuous reactors are perhaps mainly used in aluminium cast houses, for instance, for gas purging.

Usually it is a good approximation to assume that the melt is completely mixed as mentioned initially in this chapter. (This point is discussed further in Chapter 3.) To proceed we now refer to Fig. 4.7. Note that x is assumed to change discontinuously from $[\%x]_{in}$ in entering the reactor to $[\%x]$ constant throughout the melt. The molar balance for x is then

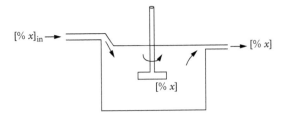

Fig. 4.7 *Reactor where melt is completely mixed.*

x (in kmols/s) following melt into reactor
$= x$ (in kmols/s) following melt out of reactor
$+ x$ removed from melt per unit time

or

$$\frac{\dot{M}[\%x]_{in}}{100M_x} = \frac{\dot{M}[\%x]}{100M_x} + \dot{G}_x. \tag{4.19}$$

Here $\dot{M}$ is the flow of metal into the reactor in kg/s.

Removal, $\dot{G}_x$, is often by several mechanisms. For instance when gas-purging aluminium, Na is removed partly as gas in the bubbles and partly by evaporation from the top surface and subsequent burning with oxygen in the atmosphere. Na has a great affinity to oxygen so that the reaction

$$2\underline{Na} + \frac{1}{2}O_2 \rightarrow Na_2O(s) \tag{4.20}$$

is displaced completely over to the right. Then $[\%x]_e \sim 0$ and the driving force for the transfer of Na through the top surface is taken to be equal to the concentration in the bath $[\%x]$.

According to the gas law, removal to the bubbles is equal to $\dot{G}p_x^0/p_{inert}$. This removal added to transfer at the top surface gives (see the definition of Z)

$$\dot{G}_x = \frac{\dot{G}Kf_x[\%x]Z}{p_{inert}} + \frac{k_{ts}A_s\rho[\%x]}{100M_x}, \tag{4.21}$$

where k_{ts} is the total mass transfer coefficient for the bath top surface, and A_s is the bath top surface area.

We notice that both terms in the equation for $\dot{G}_x$ are proportional to $[\%x]$. This allows us to calculate the reduction ratio $[\%x]/[\%x]_{in}$ for the reactor. Equations (4.21) and (4.19) give

$$\frac{[\%x]}{[\%x]_{in}} = \left(1 + \frac{\dot{G}f_xKZ100M_x}{p_{inert}\dot{M}} + \frac{k_{ts}A_s\rho}{\dot{M}}\right)^{-1}. \tag{4.22}$$

Due to the relatively low vapour pressure of Na, we will see that the term for removal with the bubbles is smaller than the last term for removal from the top surface. In this case the main effect of gas purging is to stir the top surface and there give a high value of the mass transfer coefficient, k_{ts}. It should be noted that with increasing throughput $\dot{M}$ the reduction ratio becomes less favourable (increases).

It may be useful to expand the numerical example used previously:

$$\dot{G} = 2Nm^3/hr = 2.48 \times 10^{-5} kmol/s$$
$$M_{Na} = 23kg/mol$$
$$\frac{1}{Kf_{Na}} = 0.29$$
$$p_{inert} = 1bar$$
$$Z = 0.83$$
$$\rho = 2350kg/m^3$$
$$A_s = 1m^3$$
$$\dot{M} = 0.5kg/s$$
$$k_{ts} = 1.5 \times 10^{-4} m/s \,(\text{see Chapter 3}).$$

This gives, on insertion in Eq. (4.22),

$$\frac{[\%Na]}{[\%Na]_{in}} = (1 + 0.33 + 0.71)^{-1} = 0.49.$$

The term 0.33 gives the removal by gas purging, while 0.71 is due to the top surface. We notice that inert gas purging does in fact play a role. However, in a practical system (see Chapter 3) the gas-purging efficiency Z will often be lower than indicated in this example.

Removal by gas purging can be enhanced if a reactive gas is added to the purge gas. This should give smaller bubbles and greater contact area and increase the mass transfer coefficient. In this case the driving force for transfer to bubbles is equal to the concentration $[\%x]$ (Eq. (4.14) applies).
$k_t = 5 \times 10^{(-4)} m/s$ and, say, $A = 3$ m^2 then gives

$$\frac{[\%Na]}{[\%Na]_{in}} = (1 + 7.1 + 0.71)^{-1} = 0.11.$$

The term for the bubbles is 7.1 as compared with 0.33 when only inert gas is employed. For impurities with a low vapour pressure, such as Li or Ca, a reactive gas must be added if the impurities are to be removed by bubbles.

In some cases, one wants to increase production (the throughput $\dot{M}$) and retain the low levels of sodium in the output. Equation (4.22) then tells us that $k_{ts}A_s\rho/\dot{M}$ should remain constant when $\dot{M}$ is increased. However, k_{ts} only increases slightly with an increase in the purge gas flow $\dot{G}$. Therefore the top surface contact area A_s must be increased roughly as $\dot{M}$.

Often it is not practical to rebuild the purification unit in order to increase A_s. Perhaps it is possible to install a similar unit in series with the first. Let us look at the case that $\dot{M}$ is to be doubled from $\dot{M} = 0.5$ kg/s to $\dot{M} = 1$ kg/s and that two identical units are to be installed. Then

$$\frac{[\%Na]_l}{[\%Na]_{in}} = \left(1 + \frac{0.33}{2} + \frac{0.71}{2}\right)^{-1} = 0.66,$$

where [%Na] is the output from the first unit. The output from the second unit is similarly

$$\frac{[\%Na]_2}{[\%Na]_1} = 0.66.$$

When multiplied, these two equations give

$$\frac{[\%Na]_2}{[\%Na]\,in} = 0.66 \times 0.66 = 0.44.$$

Thus if we double the throughput and install an extra unit, we not only retain the quality but in fact improve it: from $[\%Na]/[\%Na]_{in} = 0.49$ down to the ratio of 0.44.

It may be fruitful to speculate a little regarding the above result. Why is it better to treat the melt in two units in series instead of one? Removal is proportional to the concentration [%x]. When we mix the input $[\%x]_{in}$ into the back-mix tank, our concentration is reduced from $[\%x]_{in}$ to [%x]. Thus when we carry out the mixing operation, we lose the driving force for the removal of x. With two back-mix tanks in series, the concentration is reduced in two steps: from $[\%x]_{in}$ to $[\%x]_1$ and then to $[\%x]_2$ instead of directly to [%x]. Therefore, for the same total contact areas and purge gas flows, two back-unit tanks in series are better than one.

What happens in the idealized case when we have N instead of two identical back-mix tanks in series? In this case the flow of inert gas $\dot{G}$ is distributed to N tanks and the contact area is A_s/N in each tank. Then Eq. (4.22) is replaced by

$$\frac{[\%x]}{[\%x]_{in}} = \left(1 + \frac{a\dot{G}}{N\dot{M}} + \frac{bA_s}{N\dot{M}}\right)^{-N}, \tag{4.23}$$

where [%x] is the output from the Nth unit and

$$a = \frac{f_x Z 100 M_x K}{p_{inert}} \qquad b = k_{ts}\rho.$$

If we allow N to increase to infinity, a well-known mathematical result is that

$$\left(1 + \frac{a\dot{G}}{N\dot{M}} + \frac{bA_s}{N\dot{M}}\right)^{-N} \to \exp\left(-\frac{a\dot{G}}{\dot{M}} - \frac{bA_s}{\dot{M}}\right). \tag{4.24}$$

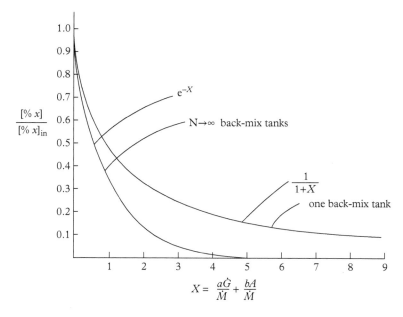

$$X = \frac{a\dot{G}}{M} + \frac{bA}{M}$$

Fig. 4.8 *Ratio between concentration [%x] and original concentration [%x]$_{in}$ as a function of X.*

In the numerical example, $[\%x]/[\%x]_{in} = \exp(-1.04) = 0.35$. Thus introduction of an infinite number of tanks instead of two has given a further improvement in purification from 0.44 down to 0.35. Now, this result does not seem to be very impressive. However, Fig. 4.8 shows that if reduction ratios $[\%x]/[\%x]_{in}$ less than ¼ are desired, the system with $N \rightarrow \infty$ gives a far better performance than a single back-mix tank.

The effect of a large number of back-mix tanks is to give the same residence time for all elements of fluid:

$$\tau = \frac{M}{\dot{M}}.$$

M is the total mass of the melt in all the N tanks. The behaviour is the same as for a reactor where the metal flows plug-like with residence time τ (or for a batch reactor operated for a time τ).

In practical high-temperature refining metallurgy, one hardly ever finds systems that behave like the plug-flow reactor or like N tanks in series. However, for systems where a low ratio $[\%x]/[\%x]_{in}$ is required, there is a strong economic incentive to attempt an approach to *the constant residence time continuous reactor*. Thus, in refining aluminium three units in series are in use.

4.7   Batch Reactors

As mentioned previously, batch reactors are still the most commonly used reactors in metal refining operations. For instance, in steelmaking all industrial units handle a batch of metal.

Our treatment of batch reactors is very similar to that given for continuous reactors. However, the mass balance given in Eq. (4.19), now becomes:

Reduction of x (in kmols/s) contained in the melt per unit time = x transferred (in kmols/s) from melt per unit time

or

$$-\frac{M}{100M_x}\frac{d[\%x]}{dt} = \dot{G}_x.$$

(4.25)

If we return to the example with gas purging

$$\dot{G}_x = (a\dot{G} + bA_s)\frac{[\%x]}{100M_x}.$$

(4.26)

Equation (4.25) then becomes

$$\frac{d[\%x]}{[\%x]} = -(a\dot{G} + bA_s)\frac{dt}{M}$$

or integrating from $[\%x] = [\%x]_{in}$ at time $t = 0$ to $[\%x]$ at time t

$$\int_{[\%x]_{in}}^{[\%x]}\frac{d[\%x]}{[\%x]} = -\int_0^t\frac{(a\dot{G} + bA_s)\,dt}{M}$$

For constant operating conditions so that $a\dot{G} + bA_s$ does not change with time, the above equation gives

$$\ln\frac{[\%x]}{[\%x]_{in}} = -(a\dot{G} + bA_s)\frac{t}{M}$$

or

$$\frac{[\%x]}{[\%x]_{in}} = \exp\left\{\frac{(-a\dot{G} - bA_s)t}{M}\right\}.$$

(4.27)

The reduction ratio drops exponentially with time.

This result may be compared with Eq. (4.24) valid for a plug-flow (constant residence time) continuous reactor. If we set $t = \tau = M/\dot{M}$ in Eq. (4.27), then Eqs. (4.24) and (4.23) are obtained. The reduction ratio is the same whether the fluid element is in a batch reactor or moving through a continuous plug-flow reactor.

We have seen that removal in a batch reactor is equal to the best attainable in a continuous system. Why then is there a strong trend in industry towards the development and use of continuous purification systems? The answer is probably that batch processes are messy and labour consuming due to the discontinuous operations of filling and tapping the refining ladle.

Also, batch systems are difficult to instrument and control due to the inherent time dependence involved.

4.8 Traditional Slag–Metal Refining in a Batch Reactor (Ladle)

One problem with refining metal by extraction to a second (slag) phase is that of mixing in the slag phase. Often the slag phase is relatively viscous so that it is difficult to obtain an even distribution of the impurity element in the slag. However, here we assume that also the slag phase is 'completely mixed'.

Equation (4.25) then becomes

$$-M\frac{d[\%x]}{dt} = k_t \rho A_s \left([\%x] - [\%x]_e\right),$$ (4.28)

where $[\%x]_e$ is the hypothetical concentration in the metal in equilibrium with the actual concentration in the slag $(\%x)$:

$$[\%x]_e Kf_x = \gamma_x (\%x).$$ (4.29)

To integrate Eq. (4.28) over time $[\%x]_e$ must be replaced by a function of $[\%x]$. $[\%x]_e$ may be obtained in terms of $[\%x]$ if we note that whatever leaves the melt enters the slag:

$$M\left([\%x]_{in} - [\%x]\right) = M_s (\%x),$$

where M_s is the amount of slag. Here we assume that the slag originally did not contain any component x. Thus, the above equation together with Eq. (4.29) gives

$$[\%x]_e = \frac{\gamma_x}{f_x K}\frac{M}{M_s}\left([\%x]_{in} - [\%x]\right).$$ (4.30)

The driving force in Eq. (4.28) becomes

$$[\%x] - [\%x]_e = [\%x]\left(1 + \frac{\gamma_x M}{Kf_x M_s}\right) - \frac{\gamma_x M}{Kf_x M_s}[\%x]_{in}.$$ (4.31)

The lowest value of $[\%x]$ attainable is when the driving force given by Eq. (4.31) becomes zero.

Then

$$[\%x] = [\%x]_\infty = \frac{M[\%x]_{in}}{M + Kf_x M_s/\gamma_x},$$ (4.32)

where $[\%x]_\infty$ is the value of $[\%x]$ when equilibrium between slag and melt is finally reached (at time $t \to \infty$). Using Eq. (4.32) we replace $[\%x]_{in}$ in Eq. (4.31) with $[\%x]_\infty$. This gives the driving force:

$$[\%x] - [\%x]_e = \left(1 + \frac{\gamma_x M}{Kf_x M_s}\right)([\%x] - [\%x]_\infty).$$
$$(4.33)$$

This equation, introduced into Eq. (4.28), gives

$$\int_{[\%x]_{in}}^{[\%x]} \frac{d[\%x]}{[\%x] - [\%x]_\infty} = -\int_0^t \frac{k_t \rho A_s}{M}\left(1 + \frac{\gamma_x M}{Kf_x M_s}\right) dt.$$
$$(4.34)$$

This becomes, on integration and assuming that k_t, γ_x, f_x, and so on do not change with time,

$$\frac{[\%x] - [\%x]_\infty}{[\%x]_{in} - [\%x]_\infty} = \exp\left\{-\frac{k_t \rho A_s t}{M}\left(1 + \frac{\gamma_x M}{Kf_x M_s}\right)\right\}.$$
$$(4.35)$$

In Fig. 4.9 we see that $[\%x]$ drops exponentially down to the value $[\%x]_\infty$, given by the equilibrium between x in the melt and in the slag. When

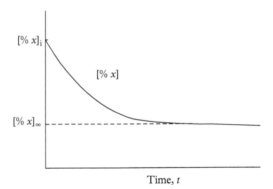

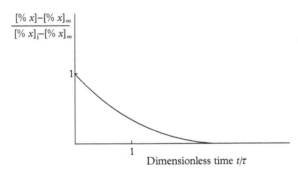

Fig. 4.9 *Concentration in the melt $[\%x]$ drops exponentially down to $[\%x]_\infty$.*

$$t = \tau = \cfrac{M}{k_t \rho A_s \left(1 + \frac{\gamma_x M}{K f_x M_s}\right)},\tag{4.36}$$

the difference between [%x] and the final value [%x]$_\infty$ is reduced to $1/e = 0.37$ of the difference between [%x]$_{in}$ and [%x]$_\infty$. As one would expect, τ is inversely proportional to the total mass transfer coefficient k_t. If complete mixing in the slag phase is not attained, the analysis above does not apply. [%x] will then remain higher than given by Eq. (4.35).

4.9 Removal of Ca and Al Impurities in MG-Si

4.9.1 An Industrial Example of Reactive Gas and Slag Refining

Metallurgical Grade silicon (MG-Si), i.e. commercial silicon grades containing more than 96 wt% Si, is used in a variety of applications such as an alloying element for aluminium, as raw material for production of silicones, and as a precursor for electronic/photovoltaic silicon grades. Industrial production of MG-Si takes place via carbothermic reduction of quartz in an electric furnace, followed by an oxidative ladle refining (OLR) process, solidification, and crushing, as illustrated in Fig. 4.10.

The carbothermally produced alloy is tapped into the ladle directly from the electric furnace. While electric furnace Si production is considered a 'slag-free' process, a small amount of slag, consisting of un-reduced impurities from the carbon, electrodes, and quartz raw materials, is generally tapped into the ladle together with the liquid Si. The composition of the tapped silicon is a function of the purity of raw materials, and as silicon forms comparatively stable oxides, only a limited number of impurity elements can be removed from silicon through an oxidative treatment. Common critical Si impurities, such as Fe and Ti, are not feasible for removal through such treatment and form different intermetallic phases with silicon as it solidifies, the types determined by the amounts and ratios of the impurities present.[17,18] The concentrations of impurities such as B and P are also not reduced by this treatment.[16] Hence, the content of these impurity elements is generally controlled through raw material selection.

The primary purpose of the OLR process, illustrated in Fig. 4.11, is to reduce the concentration of Ca and Al, elements whose oxides are more stable than Si, to meet alloy specifications. The compositions of two typical Si grades are given in Table 4.1.[19]

Throughout the tapping cycle of the electric furnace, the metal entering the ladle is purged by an air–oxygen mixture through a porous plug fitted at the bottom of the ladle, aiming to oxidize the impurities into a slag phase. Unlike many other oxidative slag refining processes, MG-Si refining *does not start with a slag*, but the slag is formed *in situ* by reaction between the gas, the silicon, and impurities in the molten alloy. While Section 4.4 describes the generalized transport of—and reaction between—an impurity element in the bulk metal phase and a reactive gas, the specifics of the silicon OLR process can be described in four steps:

1. Transport of the element from the bulk metal phase to the boundary layer at the bubble surface. This transport takes place via stirring/convection in the metal.

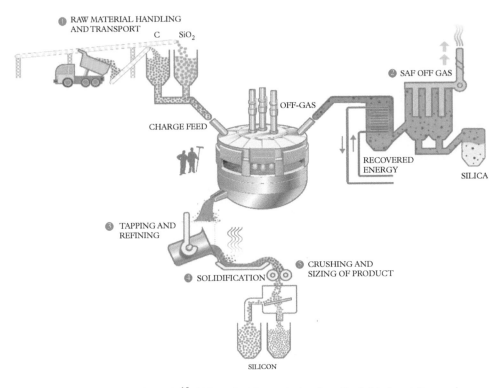

Fig. 4.10 *The production of MG-Si.*[15] *(Adapted with permission from Fagbokforlaget, Norway.)*

2. Diffusion of the element through the boundary layer.
3. Reaction of the element with oxygen at the surface of the bubble, or oxide on the interface with subsequent formation of a mixed oxide phase.
4. Ejection and transport of the oxide slag droplet from the bubble surface into the metal with subsequent coagulation/formation of a 'bulk' slag phase.

For step 3, since the most abundant element in the alloy melt is silicon, the first reaction to occur is the oxidation of silicon at the bubble interface (in the refining gas plume or on the entrained air bubbles in the tapping jet):

$$Si + O_2 = SiO_2. \tag{4.37}$$

The two most abundant impurities in the Si alloy (after Fe) are Ca and Al, which will react with the silica (SiO_2) formed according to

$$SiO_2(cr) + 2\underline{Ca}\,(\%\ \text{in Si}) \rightarrow 2CaO\,(s) + Si\,(l) \tag{4.38}$$

$$4\,\underline{Al}\,(\%\ \text{in Si}) + 3SiO_2(cr) = 2Al_2O_3 + 3Si\,(l). \tag{4.39}$$

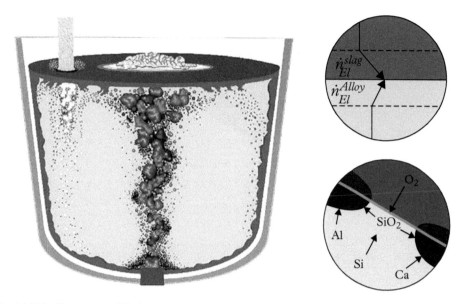

Fig. 4.11(a) *Illustration of the bubble and fluid flow in the ladle. (b) Element transport and reaction at the gas bubble–metal interface.*

Table 4.1 *Typical compositions of commercial metallurgical Si grades, based on data from Elkem Silicon Products*[19]

Grade	Application	Fe (wt%)	Ca (wt%)	Al (wt%)	Ti (ppmw)	P (ppmw)
Si97	Al alloys	2	0.02	0.3	700	50
Si99	Silicones and EG/PV-Si	0.3–0.5	0.015–0.1	0.2–0.3	200–300	10–20

The underlined elements ($\underline{Ca}$ and $\underline{Al}$) are dissolved in liquid silicon, and their composition and activity are given using weight percent as the composition coordinate. These reactions result in the formation of small, primarily SiO_2–CaO–Al_2O_3-based slag droplets at the bubble surface. (While Ca and Al are main slag forming impurities in the system, several other minor/trace impurities exist in the alloy that will simultaneously be oxidized.[16]) The respective interfacial tensions between the slag–gas bubble and the slag–alloy mean that the droplets are dragged from the gas bubble and dispersed in the alloy, where slag droplets coagulate through melt movement into bulk slag phases. The alloy–bulk slag equilibrium with respect to Ca and Al may also be approached through the exchange reaction

$$3\underline{Ca} + Al_2O_3 = 3CaO + 2\underline{Al}. \tag{4.40}$$

On a macro level, one first considers the chemical equilibrium between the Si alloy and the slag phase. This is accomplished using the thermodynamic theory and data from Chapter 2. We first consider the Gibbs free energy for the two refining reactions under consideration.

It will be instructive to show in detail how the free energy for reaction (4.38) is obtained. We write the corresponding sub-reactions and then add them up, together with the corresponding free energies. This process is shown in the following.

Reaction	Free energy (J)	Source of data
$2Ca(l) + O_2 \rightarrow 2CaO(s)$	$\Delta G^\circ = 2 \times (-640{,}000 + 108.6T)$	Appendix 1
$2\underline{Ca}(\% \text{ in Si}) \rightarrow 2\,Ca(l)$	$\Delta G^\circ = 2 \times (118{,}900 + 43.322T)$	Table 2.11
$SiO_2(cr) \rightarrow Si(s) + O_2$	$\Delta G^\circ = 906{,}000 - 176T$	Appendix 1
$Si(s) \rightarrow Si(l)$	$\Delta G^\circ = 50{,}540 - 30.0T$	Appendix 1
$SiO_2(cr) + 2\underline{Ca} \rightarrow 2CaO(s) + Si(l) \quad \cdot$	$\Delta G^\circ = -85{,}660 + 97.844T$	Sum of above

At the temperature of interest, 1550 °C or 1823 K, $\Delta G^\circ = 92\,710$ J/mol or 92.71 kJ/mol. In a similar fashion reaction (4.39) can be shown to have a free energy of -96.16 kJ/mol.

We may now find the equilibrium coefficients (note that the activity coefficients for aluminium and calcium has been assumed equal to 1, so their activities are equal to their concentration in weight percent; in view of the compositions of the alloys involved, this assumption will produce little error):

$$K_{4.38} = \exp\left(\frac{-\Delta G^\circ}{RT}\right) = 0.0026 = \frac{a_{Si(l)} \cdot a^2_{CaO(s)}}{\%Ca^2 \cdot a_{SiO2(s)}}. \tag{4.41a}$$

And since the silicon activity is very close to one, we find that

$$\%Ca = 19.6\frac{a_{CaO(s)}}{\sqrt{a_{SiO2(s)}}}. \tag{4.41b}$$

Similarly for the distribution of aluminium between slag and liquid silicon:

$$K_{4.39} = \exp\left(\frac{-\Delta G^\circ}{RT}\right) = 569 = \frac{a^3_{Si(l)} \cdot a^2_{Al2O3(s)}}{\%Al^4 \cdot a^3_{SiO2(s)}} \tag{4.42a}$$

and

$$\%Al = 0.205\frac{\sqrt{a_{Al2O3(s)}}}{\left(a_{SiO2(l)}\right)^{3/4}}. \tag{4.42b}$$

The activities in the slag phase are given in Figs. 2.16 and 2.17. Using these activities in Eqs. (4.41) and (4.42), the equilibrium concentrations of Ca and Al can be determined. The results are presented in Fig. 4.12, which shows the ternary SiO_2–CaO–Al_2O_3 phase diagram (1550 °C isotherm), where equilibrium Ca and Al contents in liquid silicon (in wt%) have been projected onto the diagram.

The diagram in Fig. 4.12 allows the calculation of the equilibrium slag composition required to produce an alloy with a certain composition. If we take the example of Si97 alloy in Table 4.1, with the compositional requirement of 0.02 wt% Ca and 0.3 wt% Al, the equilibrium slag composition will need to be approximately 45wt% SiO_2, 30wt% Al_2O_3, and 25wt% CaO. If the relative contents of Ca and Al in the tapped alloy does not allow for a final $CaO:Al_2O_3$ ratio consistent with that required to reach targeted alloy composition, slag formers such as limestone are added during the refining process to adjust the final slag composition.

It should be noted that the first slag to form in the contact between oxygen and alloy is usually high in CaO and Al_2O_3. This slag has a higher density than the alloy and tends to stay at the bottom of the ladle. As the process proceeds, the melt becomes purer (more refined), and the slag is richer in SiO_2. This slag has a lower density and usually (depending on alloy composition) floats on top of the alloy. The relative densities of slags in the SiO_2–CaO–Al_2O_3

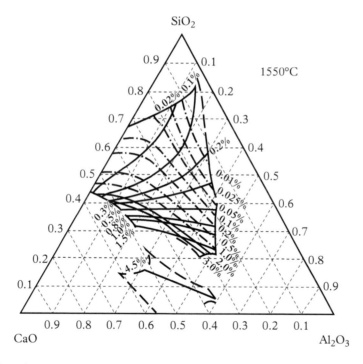

Fig. 4.12 *Phase diagram isotherm of the SiO_2–CaO–Al_2O_3 system at 1550 °C and 1 atm, with projected dissolved equilibrium concentrations, in wt%, of Al (unbroken) and Ca (dot-dash).*

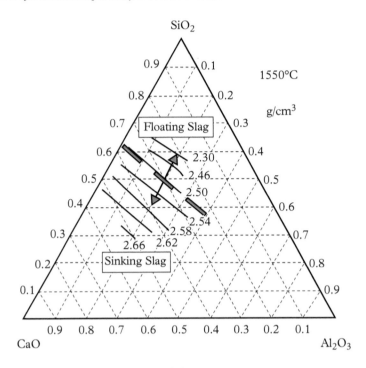

Fig. 4.13 *Densities of slags in the SiO₂–CaO–Al₂O₃ system.*

system are plotted in Fig. 4.13. (Comparing Fig. 4.13 with Fig. 2.16, one finds that floating slags will have a SiO_2 activity greater than 0.3 to 0.5.)

Studies have shown that the top and bottom slags are different in composition, and that the removal by oxidation of various impurity elements is different in the two slags. (See Fig. 10 of reference 16.)

In the above it was assumed that SiO_2 is formed by reaction between the liquid silicon and oxygen used to refine the ladle. We should note, however, that high temperatures favour the gas-forming reaction

$$Si\,(l) + 1/2O_2 = SiO\,(v). \qquad (4.43)$$

Since the oxidizing refining reactions cause the melt temperature to rise, cooling metal (typically fines of the alloy produced) are often added to control the metal temperature. This reduces the amount of fume produced by reaction (4.43).

We now consider the kinetics of the refining reactions in the OLR process. Unfortunately, an accurate characterization is extremely difficult, for the following reasons:

1. OLR is not strictly a batch process, as the amount of melt in the ladle changes continuously with time until it is full.

2. A reactive gas is used to refine the silicon alloy melt. However, the models developed for gas purging (in Section 4.4) do not apply to this process. Oxygen reacts with the liquid metal to produce droplets of slag. Most of the refining appears to result from reactions between the slag produced and the liquid metal, not as a result of a direct reaction with gaseous oxygen.

3. As noted above, the slag composition changes with time. That means the equilibrium concentration of impurity elements (Fig. 4.12) also changes with time.

4. The contact area between slag and metal changes with time.

Despite these difficulties, we note that the metal/slag ratio remains reasonably constant during the ladle refining process. Also, owing to the high temperature of the system, neither the convective transport of dissolved impurities in the alloy nor the chemical oxidation reaction at the bubble surface is likely to be rate determining. Instead, the diffusive transfer of species through the metal and slag boundary layers will determine the rate of refining. In other words, the kinetics of refining may be *approximated* by a batch process, which is controlled by diffusive mass transfer at the slag–metal interface.

An explanation of the model used is not given here. However, one may find details in recent publications.[16,20] Figure 4.14 shows the calculated concentrations of Ca and Al as a function of time. The concentrations shown have been normalized. That is, the concentration at any time has been divided by the original (starting) concentration.

The modelled behaviour can be seen to correspond very well with the industrially determined trends for Ca and Al refining, as measured by Kero et al.[16] These have been plotted in Fig. 4.15, together with the behaviour of boron, to demonstrate the difference in its behaviour compared to that of calcium and aluminium.

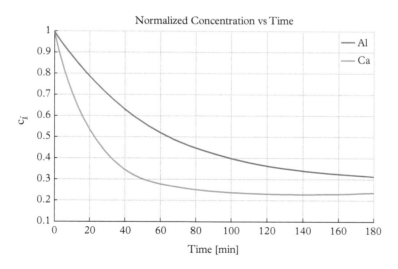

Fig. 4.14 *The calculated normalized concentrations of Ca and Al in the melt as a function of refining time.*

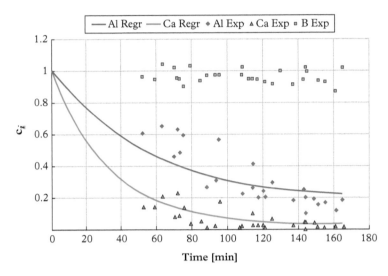

Fig. 4.15 *Refining kinetics of Ca and Al as measured in an industrial ladle. Values on the Y axis are normalized with respect to the initial concentration. Measured concentrations of B are also shown as a comparison. Based on data as presented in Kero et al.[16] (This work is licensed under a Creative Commons Attribution 4.0 generic license.)*

4.10   Injection

We have seen that mass transfer from the melt to the gas phase can be speeded up by bubbling gas into the melt. Transfer to the bubbles may be much greater than through the top surface of the melt. Similar advantages may be obtained for slag–melt processes by the injection of powder into the melt.[21–6] Usually in refining, this powder becomes a molten phase separate (immiscible) from the melt. This phase initially has its full thermodynamic driving force for removal.

There is an important difference in principle between melt–slag and melt–gas processes, however. Impurities in the gas phase are usually totally removed from the system, while for powder injection the impurities remain. Then they can revert to the melt. Here we assume that the impurities enter the slag phase.

Powder injection is usually pneumatic (often using inert gas such as argon). If high-pressure systems are employed (1–3 bar) it is possible to operate at very low mass flow ratios between gas and powder (down to 1–3 Nl per kg of powder). Usually the powder is conveyed down into the melt through a lance (see Fig. 4.16).

Injection of powder has been employed in commercial operations in industries ranging from steel and ferroalloys to aluminium. However, it seems that often the flow rates of powder and gas are not controlled and measured properly. Furthermore, in some cases the position of the lance tip is not well defined. Sometimes there are problems with clogging of the lance, vibrations of the lance, splashing, variations in yield, erratic refining efficiencies, and so on.

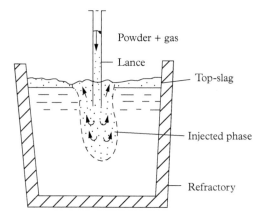

Fig. 4.16 *Injection with a lance.*

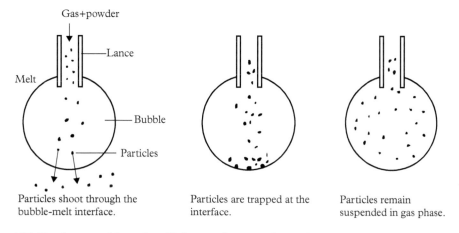

Particles shoot through the bubble-melt interface.

Particles are trapped at the interface.

Particles remain suspended in gas phase.

Fig. 4.17 *The three possible modes of behaviour for injected particles.*

The injected particles can penetrate through the bubble–melt interface, be trapped at the interface, or remain suspended in the gas phase. This is illustrated in Fig. 4.17. If particles remain in the gas phase, they are lost. In fact they will present a problem in the off-gas system from the reactor. Usually only particles smaller than 10 μm will remain in the gas. If particles are 'wetted'[27,28] by the melt, they will enter it. Non-wetting particles will penetrate the interface if their momentum on impact is sufficiently high. This is discussed in Chapter 3. For the case when particles remain in the interface and wet it, the melt–particle contact area may be taken to be equal to the melt–bubble contact area.[21–30]

Ideally impurities removed by powder injection should remain in the slag phase. Then the slag must have the capacity to retain the reaction products from the injection refining.

A problem may be that the reaction products are lighter than the metal and heavier than the slag so that they accumulate at the interface. In a batch reactor a high concentration of impurities may be initially trapped at the interface and increasingly revert to the metal when the impurity concentration declines in the metal. Solutions to the problem may be to create reaction products that are lighter than the slag so that they float up and/or that the kinetics of reversion are slow.

In the following injection and reversion will be studied in some detail. We calculate the removal of an impurity by the injected phase. The treatment is analogous to removal by bubbling. We assume that there are no concentration gradients in the melt or in the injected phase. Also there is initially no x in the powder. Then the term on the right-hand side of Eq. (4.11) is replaced by

$$\frac{\dot{M}_p}{100} \frac{\Delta(\%x)_p}{M_x},$$

where $\dot{M}_p$ is the rate of injection of powder in kg/s and $(\%x)_p$ is the mass fraction impurity removed to the injected phase. Thus, the mass balance for the control volume (compare Fig. 4.4) becomes

$$k_{tp}\rho\left([\%x] - [\%x]_e\right)\Delta A_p = \dot{M}_p\Delta(\%x)_p, \tag{4.44}$$

where

$$[\%x]_e f_x = \gamma_{xp}\frac{(\%x)_p}{K_p}. \tag{4.45}$$

As for bubbling, on performing the integration of Eq. (4.44) we obtain an efficiency of injection

$$Z_p = \frac{(\%x)_p^0\gamma_{xp}}{f_x K_p [\%x]}. \tag{4.46}$$

Here $(\%x)_p^0$ is the x-concentration in the powder when it leaves the metal phase:

$$Z_p = 1 - \exp\left(-\phi_p\right), \tag{4.47}$$

where

$$\phi_p = \frac{k_{tp}\rho A_p\gamma_{xp}}{\dot{M}_p f_x K_p}. \tag{4.48}$$

ϕ_p is a dimensionless contact area.

If we write for Eq. (4.48)

$$\phi_p = \frac{k_{tp}\rho A_p\,[\%x]}{\dot{M}_p f_x K_p\,[\%x]/\gamma_{xp}},$$

(4.49)

it is seen that

$$\phi_p = \frac{\text{Removal to powder}}{\text{Maximum capacity of powder for }x}.$$

Removal (in kmol/s) is

$$\dot{G}_x = \frac{\dot{M}_p(\%x)_p^0}{100M_x} = \frac{f_x K_p \dot{M}_p\,[\%x]\,Z_p}{100M_x\gamma_{xp}}.$$

(4.50)

For the batch reactor one obtains from a mass balance (compare Eq. (4.28))

$$-M\frac{d\,[\%x]}{dt} = \frac{f_x K_p \dot{M}_p\,[\%x]\,Z_p}{\gamma_{xp}} + k_{ts}A_s\rho\left([\%x]-[\%x]_e\right),$$

(4.51)

where $[\%x]_e$ depends on the slag concentration $(\%x)$ as given by Eq. (4.29). Now $[\%x]_e$ is eliminated by employing the mass balance for x from Eq. (4.30). It is pointed out that the amount of slag now is not constant but increases due to the injection $\dot{M}_p$ of powder (at a constant rate):

$$M_s(t) = M_o\,(t=0) + \dot{M}_p t.$$

(4.52)

Equation (4.30) introduced to Eq. (4.51) gives

$$-\frac{d\,[\%x]}{dt} = \left(\frac{1}{\tau} + \frac{\phi_s}{1+\frac{t}{\tau_p}}\right)[\%x] - \frac{\phi_s}{1+\frac{t}{\tau_p}}[\%x]_{in},$$

(4.53)

where

$$\frac{1}{\tau} = \frac{f_x K_p \dot{M}_p Z_p}{M\gamma_{xp}} + \frac{k_{ts}\rho A_s}{M}$$

(4.54)

$$\frac{1}{\tau_p} = \frac{\dot{M}_p}{M_o}$$

$$\phi_s = \frac{k_{ts}\rho A_s \gamma_x}{\dot{M}_o K f_x}.$$

(4.55)

The equation is similar to that for top-slag removal only, except for the injection term $f_x K_p \dot{M}_p Z_p / (M \gamma_{xp})$.

τ is a time constant for the combined removal to powder and top-slag. τ_p is a time that compares the injection rate $\dot{M}_p$ with the initial mass of slag. $(\phi_s \cdot \tau_p) = \phi$ is a dimensionless number similar to ϕ_p. However, $(\phi_s \tau_p)$ compares slag–metal mass transfer to removal by injection. In Eq. (4.53) $(-[\%x]_{in} + [\%x]) \frac{\phi}{t + \tau_p}$ takes into account the reduction of the driving force for removal (by reversion from the slag). That is, part of the impurity is transferred from slag to metal. The equilibrium constant K is assumed to be the same as for the original slag.

The solution of Eq. (4.53) is

$$Y = CY_0 + Y_0 \int_0^t \frac{1}{Y_0} \phi_s \frac{dv}{1 + \frac{v}{\tau_p}}, \tag{4.56}$$

where C is an integration constant and v is a substitution made to facilitate the integration (see Section 4.10.1: $z = \frac{t - v}{\tau}$ or $v = t - Z\tau$) and

$$Y = \frac{[\%x]}{[\%x]_{in}}.$$

Y_0 is the solution of the homogeneous equation:

$$-\frac{dY_0}{dt} = \left(\frac{1}{\tau} + \frac{\phi_s}{1 + \frac{t}{\tau_p}} \right) Y_0,$$

giving

$$Y_0 = \frac{\exp\left(-\frac{t}{\tau}\right)}{\left(1 + \frac{t}{\tau_p}\right)^{\tau_p}} \tag{4.57}$$

In the following we use: $\phi = \phi_s \cdot \tau_p$.

Then we get for the relative concentration of impurity:

$$Y = CY_0 + \phi Y_0 \int_0^t \exp(v/\tau) \left(1 + \frac{v}{\tau_p} \right)^{(\phi - 1)} dv.$$

The integration constant is determined by setting $t = 0$ and $Y = 1$. This gives

$$C = 1 \tag{4.58}$$

$$Y = \frac{\exp\left(-\frac{t}{\tau}\right)}{\left(1 + \frac{t}{\tau_p}\right)^{\phi}} + \frac{\phi_s \exp\left(-\frac{t}{\tau}\right)}{\left(1 + \frac{t}{\tau_p}\right)^{\phi}} \int_0^t \exp\left(\frac{v}{\tau}\right) \left(1 + \frac{v}{\tau_p} \right)^{\phi - 1} dv. \tag{4.59}$$

If $\phi_s = 0$, there is no transfer from slag to melt. The slag is inert. No reversion. Only injection plays a role.

Then

$$Y = \exp\left(-\frac{t}{\tau}\right) \tag{4.60}$$

for $\frac{t}{\tau} = 3$, $Y = 0.05$. If $\phi = 1$ and $\phi_s = \frac{1}{\tau_p}$, then

$$Y = \frac{\exp\left(-\frac{t}{\tau}\right)}{\left(1 + \frac{t}{\tau_p}\right)} + \frac{\exp\left(-\frac{t}{\tau}\right)}{\left(1 + \frac{t}{\tau_p}\right)} \cdot \frac{t}{\tau_p} \exp\left(\frac{t}{\tau}\right) - 1$$

$$= \frac{\exp\left(-\frac{t}{\tau}\right)}{t + \tau_p}\left(\tau_p - \tau\right) + \frac{\tau}{t + \tau_p}. \tag{4.61}$$

Note the following example:

$$\frac{t}{\tau} = 3, \frac{\tau_p}{\tau} = 1$$

$$Y = \frac{1}{\frac{t}{\tau} + \frac{\tau_p}{\tau}} = \frac{1}{3 + 1} = \frac{1}{4}$$

$\phi = 1$ may be regarded as a typical case of reversion. This may be compared with Eq. (4.58) when there is no reversion. Then the impurity concentration drops exponentially. The real case probably lies between these two extremes. Note that we have assumed there is complete mixing of impurity in the slag.

Reversion from the top-slag and removal by the injected powder is illustrated in Fig. 4.18.

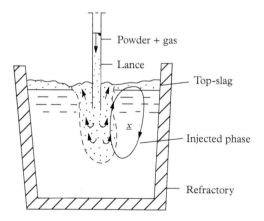

Fig. 4.18 *Conditions when the bath x content is lower than that corresponding to equilibrium with the top-slag (x reverts from top-slag).*

4.10.1 Details of Mathematical Treatment

Here we present a detailed description of the derivation of the preceding equations. It is necessary to solve this equation:

$$Y = \frac{\exp\left(-\frac{t}{\tau}\right)}{\left(1+\frac{t}{\tau_p}\right)^\phi} + \frac{\phi_s}{1+\frac{t}{\tau_p}} \cdot \int_0^t \exp\left(\frac{v-t}{\tau}\right)\left(\frac{v+\tau_p}{t+\tau_p}\right)^{\phi-1} dv. \tag{4.62}$$

It is convenient to introduce $z = \frac{t-v}{\tau}$ or $v = t - z\tau$. Then

$$Y = \frac{\exp\left(-\frac{t}{\tau}\right)}{\left(1+\frac{t}{\tau_p}\right)^\phi} + \frac{\phi\tau}{t+\tau_p} \cdot \int_0^{t/\tau} \exp(-z)\left(\frac{-\tau+\tau_p}{t+\tau_p}\right)^{\phi-1} dz \tag{4.63}$$

$$\int_0^{t/\tau} \exp(-z)\left(\frac{t-z\tau+\tau_p}{t+\tau_p}\right)^{\phi-1} dz$$

$$= \int_0^{t/\tau} \exp(-z)\left(1-\frac{z}{\frac{t}{\tau}+\frac{\tau_p}{\tau}}\right)^{\phi-1} dz$$

$$= \int_0^{t/\tau} \exp(-z)\left(1-\frac{z}{R}\right)^{\phi-1} dz.$$

The final value of the dimensionless concentration depends on t/τ, $R = \frac{t}{\tau} + \frac{\tau_p}{\tau}$; ϕ and $\left(1-\frac{x}{R}\right)^{\phi-1}$ may be expanded according to the binominal formula

$$\left(1-\frac{z}{R}\right)^{\phi-1} = 1 - \frac{z}{R}(\phi-1) + \frac{z^2}{R^2 2!}(\phi-1)(\phi-2)$$

$$- \frac{z^3}{R^3 3!}(\phi-1)(\phi-2)(\phi-3) + \ldots \ldots \tag{4.64}$$

$$\int_0^{t/\tau} \exp(-z)\,dz = 1 - \exp\left(-\frac{t}{\tau}\right)$$

$$\int_0^{t/\tau} \exp(-z)\cdot z\,dz = -\exp(-z)\cdot z \ \Big/ \ _0^{t/\tau} + \int_0^{t/\tau} \exp(-z)\,dz$$

$$= -\exp(-t/\tau)\frac{t}{\tau} + 1 - \exp\left(-\frac{t}{\tau}\right)$$

$$\int_0^{t/\tau} \exp(-z) z^2 dz = -\exp(-z) z^2 \Big/ \begin{matrix} t/\tau \\ 0 \end{matrix} + \int_0^{t/\tau} \exp(-z) \cdot z \, dz$$

$$= -\exp\left(-\frac{t}{\tau}\right) \frac{t^2}{\tau^2} + 2\left(-\exp\left(-\frac{t}{\tau}\right)\right) \frac{t}{\tau} + \left(1 - \exp\left(-\frac{t}{\tau}\right)\right)$$

Note the following example:

$$\phi = \frac{1}{2}, R = 4, \frac{t}{\tau} = 3, \frac{\tau_p}{\tau} = 1$$

$$\left(1 - \frac{z}{R}\right)^{\phi - 1} = 1 + \frac{z}{4} \cdot \frac{1}{2} + \frac{z^2}{16} \cdot \frac{3}{8} = 1 + \frac{z}{8} + \frac{3z^2}{128} + \ldots \ldots$$

$$\int_0^3 \exp(-z) \cdot \left(1 + \frac{z}{8} + \frac{3z^2}{128} + \ldots \ldots \right) dz = 1 - \exp(-3)$$

$$+ \frac{1}{8}\left(-\exp(-3) \cdot 3 + (1 - \exp(-3)) + \frac{3}{128}(-\exp(-3) \cdot 9 + z(-\exp(-3) \cdot 3 + 1 - \exp(-3)))\right)$$

$$= 1.077$$

$$\frac{[\%x]}{[\%x]_{in}} = Y = \frac{\exp(-3)}{4^{1/2}} + \frac{\frac{1}{2}}{4} \cdot 1.077 = 0.16$$

4.10.2 Concluding Comments

It has been attempted to describe combined injection refining and slag refining. With the assumption that reactions at the slag–metal interface are reversible this approach does not lead to impressive reductions of the concentration of impurities in the melt.

Useful conclusions regarding injection could be the following:

Ideally, the reaction product, x_I should be removed or become inert. When reducing the content of impurities down to low levels by injection, the presence of a standard refining slag may not be helpful.

We have assumed that there is complete mixing in the injected phase. This may be reasonable for gases or even for a powder that melts during injection, but the assumption is not applicable when the injected phase remains in the solid state. For instance, when CaC_2 is injected into steel, sulfur is removed only as a thin layer of CaS that covers the CaC_2 particles. In this case k_{tp} depends on diffusion in the solid state so that k_{tp} is small and $\phi_p \ll 1$. Then Eq. (4.47) must be employed.

4.11 Hydrogen Removal; Diatomic Gases

Hydrogen is probably the main source of the gas problem in the casting of aluminium. The gas present in metal prior to casting has a strong influence on how the casting feeds near the end of solidification. Gas content also influences the amount and location of shrinkage and porosity. The effect of gas on porosity is enlarged by solidification under vacuum as shown in Fig. 4.19. And since gas pores nucleate on inclusions suspended in the melt, filtration reduces the porosity observed in samples solidified under reduced pressures.

The reason for our gas problem lies in the chemical and physicochemical properties of aluminium discussed in Chapter 2. Aluminium is highly reactive and easily decomposes the oxide of hydrogen by the reaction:

$$H_2O(g) + \frac{2}{3}Al(1) \rightarrow \frac{1}{3}Al_2O_3 + 2\underline{H} \text{ (in Al)}. \tag{4.65}$$

This means that gas may enter the metal if steam or water vapour (in the air as humidity) is in contact with the melt. Secondly, the gas solubility in solid aluminium is much smaller than in the liquid form, so most of the hydrogen present in the liquid prior to casting is evolved as gas porosity during solidification.

The problem for most foundrymen is to remove the gas to acceptably low levels. As will be seen in the following treatment, several properties of aluminium determine the rate of degassing, and understanding these principles allows degassing to become much more effective.

Although what follows is addressed primarily to aluminium, the principles of gas removal apply to other metals as well. Hydrogen is removed by the reaction

$$\underline{H} = \frac{1}{2}H_2 \text{ (gas)} \tag{4.66}$$

with equilibrium constant

$$K_H = \frac{p_{H_2}^{\frac{1}{2}}}{f_H\,[\%H]} \tag{4.67}$$

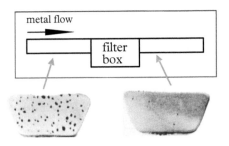

Fig. 4.19 *Example of H$_2$ pores in an aluminium alloy in a 'gassy' sample before filtration compared with after filtration. (The samples solidified under a reduced pressure, to enhance pore formation.)*

given by[31,32] (see also Table 2.1)

$$\ln K_H = \frac{5869}{T} + 3.282. \tag{4.68}$$

The activity coefficient, f_H, is introduced to account for the effect of other elements (such as Si, Cu, Mg, and Fe) on the solubility of hydrogen. General equations that can be used to calculate f_H are presented in Chapter 2. In Table 4.2 f_H is given for some common alloys. Silicon and copper both reduce hydrogen solubility in liquid aluminium, as do manganese and nickel. Chromium and iron have little effect, while magnesium increases the hydrogen solubility.

The hydrogen solubility in pure aluminium is shown as a function of temperature in Fig. 4.20. Note that only about 5 per cent of the gas present in the liquid remains in the solidified metal; the other 95 per cent is rejected into the liquid until the concentration increases to the point where gas bubbles may form. Also note that the scale used is semi-logarithmic. This means that the hydrogen content, at saturation, doubles for each 110°C (200°F) increase in the molten metal temperature. This dramatic increase is responsible for the increase in degassing problems at high melt temperatures. Low temperatures are an advantage not only in order to prevent hydrogen pick-up, but also with hydrogen removal in mind. With a decrease in temperature at a given [%H] the equilibrium value of p_{H_2} increases.

Important consequences can be drawn from Fig. 4.20 and Eq. (4.68). Our ability to degas to low hydrogen content can be no better than thermodynamic solubility will allow. In other words, we admit purge gas (nitrogen or argon) to the bottom of the melt, whereupon hydrogen starts to diffuse to ascending purge gas bubbles, and the hydrogen pressure in the bubbles (p_{H_2}) increases. To remove all the hydrogen possible (the case of 100 per cent efficiency) p_{H_2} in the bubble as it exits the melt must be in equilibrium with the metal. Then the volume of purge gas required to remove a certain volume of hydrogen can be easily calculated according

Table 4.2 *Hydrogen activity coefficients in aluminium and in common casting alloys*

f_H	Compostion of alloy
1	Pure aluminium
1.24	Al–7% Si–0.3% Mg
1.18	Al–7% Si–0.6% Mg
1.36	Al–4.5% Cu
0.47	Al–5.0%Mg
1.56	Al–6% Si–3.5% Cu
0.59	Al–4% Mg–2% Si

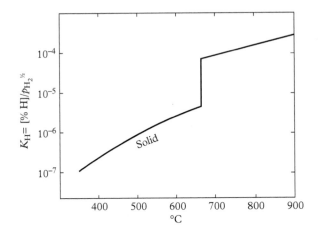

Fig. 4.20 *Solubility of hydrogen (at one bar pressure) in pure aluminium as a function of temperature.*

to a simple mass balance: hydrogen removed from the melt equals hydrogen escaping in the bubbles:

$$-\frac{Md[\%H]}{100dtM_H} = \frac{2\dot{G}p^0_{H_2}}{p_{inert}}.$$ (4.69)

Note that there are two hydrogen atoms for each H_2 molecule. For reasonable contents of [%H] the exit pressure of hydrogen is small ($p^0_{H_2} \ll p_{atmosphere}$). At atmospheric pressure then $p_{inert} \approx p_{atm}$. Equation (4.67) inserted into Eq. (4.69) then gives

$$-\frac{d[\%H]}{[\%H]^2} \approx \frac{2/100M_H\dot{G}K^2_Hf^2_H dt}{Mp_{atm}}.$$ (4.70)

If [%H] = [%H]$_{in}$ at time $t = 0$, integration results in

$$\frac{1}{[\%H]} = \frac{1}{[\%H]_{in}} + \frac{2 \times 100M_HK^2_Hf^2_H\dot{G}t}{Mp_{atm}}.$$ (4.71)

The rate of removal increases with the flow of the purge gas $\dot{G}$ and with K^2_H (decreasing temperature).

It is instructive to consider the effect of decreasing hydrogen content in the metal on the purge gas requirements for effective gas removal. As H drops, the pressure of hydrogen in the exiting purge gas bubbles decreases. This means that towards the end of degassing a lot of purge gas is required to remove a little bit of hydrogen. This ratio of the two gases, called the gas removal ratio, is

$$r = \frac{\text{volume of purge gas}}{\text{volume of } H_2 \text{ removed}} = \frac{1 - p_{H_2}}{p_{H_2}}. \tag{4.72}$$

In Fig. 4.21 the gas removal ratio is plotted versus hydrogen concentration. r is numerically equal to the volume of purge gas required to remove one volume of hydrogen. It becomes obvious from this figure that it is difficult to obtain low gas levels, since a large volume of nitrogen or argon is required to remove a small amount of hydrogen.

Actually, thermodynamic equilibrium was assumed in the calculation of the curve shown in Fig. 4.21, but in many cases 100 per cent efficiency is never obtained. We realize that when the bubbles of purge gas are large (the contact area between gas and melt is small), then kinetic factors determine the hydrogen removal rate, and we operate far below 100 per cent efficiency.

To study this case, we return to Section 4.4 concerning gas purging. Equation (4.11) for the kmole balance in the control volume becomes

$$\frac{k_t \rho \left([\%H] - [\%H]_c \right) \Delta A}{100 M_H} = 2\dot{G}\Delta \left(\frac{p_{H_2}}{p_{inert}} \right). \tag{4.73}$$

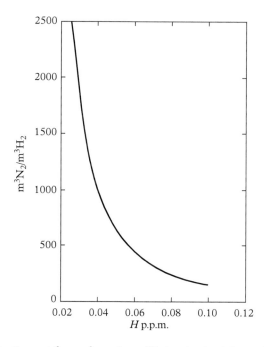

Fig. 4.21 *Gas removal ratio, r, at thermodynamic equilibrium in aluminium at 760°C. Initial concentration of H is 0.1 ppm.*

Again the number 2 on the right-hand side reflects the fact that hydrogen is removed as a diatomic gas. The counter-driving force term in Eq. (4.12) is replaced by

$$[\%H]_c f_H = \frac{p_{H_2}^{\frac{1}{2}}}{K_H}. \tag{4.74}$$

$[\%H]_c$ from Eq. (4.74) is seen to depend on the pressure p_{H_2} in the bubble. p_{H_2} changes from 0 to $p_{H_2}^0$ in traversing the melt.

We move the terms that depend on p_{H_2} over onto the right-hand side of Eq. (4.73) and integrate from the bottom to top surface:

$$\frac{k_t \rho \, [\%H] \, \Delta p_{inert}}{100 M_H 2 \dot{G}} = \int_0^{p_{H_2}^0} \frac{dp_{H_2}}{1 - \frac{p_{H_2}^{\frac{1}{2}}}{f_H [\%H] K_H}}. \tag{4.75}$$

This equation may be made dimensionless by relating the pressure p_{H_2} in the bubble to the value $f_H^2 [\%H]^2 K_H^2$ at equilibrium:

$$Z^2 = \frac{p_{H_2}}{f_H^2 [\%H]^2 K_H^2}. \tag{4.76}$$

$Z = 0$ when the inert gas bubble enters the melt. If the bubble attains saturation, $Z = 1$.

When Z is introduced instead of p_{H_2} in the integral, Eq. (4.75) becomes

$$\frac{\psi}{[\%H]} = \int_0^{Z^0} \frac{Z \, dZ}{(1 - Z)}. \tag{4.77}$$

The group on the left-hand side is called the dimensionless contact area:

$$\frac{\psi}{[\%H]} = \frac{k_t \rho A p_{inert}}{400 M_H \dot{G} f_H^2 [\%H] K_H^2}. \tag{4.78}$$

It is a simple matter to show that the integral is equal to $-Z^0 - \ln (1 - Z^0)$. Z^0 is obtained by setting $p_{H_2} = p_{H_2}^0$ in Eq. (4.76):

$$\left(Z^0 \right)^2 = \frac{p_{H_2}^0}{f_H^2 [\%H]^2 K_H^2}. \tag{4.79}$$

Equation (4.77) gives the degassing efficiency $(Z^0)^2$ as a function of $\psi/[\%H]$. Figure 4.22 shows this relationship graphically. Note from Eq. (4.78) that if $[\%H]$ is high, a large value of the contact area is needed for $\psi/[\%H]$ to have a value greater than 2. This reflects the fact that

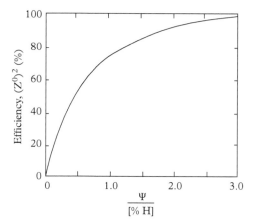

Fig. 4.22 *Degassing efficiency $(Z^0)^2$ versus the dimensionless bubble contact area $\psi/[\%H]$.*

a large contact area is required in order to saturate the bubbles in this case. The reason for this is that p_{H_2} at equilibrium is proportional to the second power of $[\%H]$.

For values of $\psi/[\%H]$ lower than about 0.3, we have the diffusion of hydrogen limiting the removal rate, and Eq. (4.14) applies. Then

$$-Z^0 - \ln\left(1 - Z^0\right) \approx \frac{(Z^0)^2}{2} = \frac{\left(p^o_{H_2}\right)}{2f^2_H[\%H]^2 K^2}.$$

This is equal to the right-hand side of Eq. (4.78):

$$\frac{k_t \rho [\%H] A p_{inert}}{400 M_H \dot{G} f^2_H [\%H] K^2_H}.$$

This gives

$$\frac{2\dot{G}p^o_{H_2}}{p_{inert}} = \frac{k_t \rho A [\%H]}{100}.$$

At values of $\psi/[\%H]$ greater than about 2, we reach the case of thermodynamic equilibrium ($Z^0 = 1$). Then $p^0_{H_2}$ is in equilibrium with the melt, so that Eq. (4.74) may be used for $p^0_{H_2}$. In between there is a transition regime.

What actually determines the performance of the purge gas is the size of its bubbles, which influences the total surface area, A, in Eq. (4.78). Not only do smaller bubbles give a larger surface area for a given volume of purge gas, but they rise more slowly (see Fig. 4.23). As bubble size decreases, $\psi/[\%H]$ increases sharply and we move closer to the case of 100 per cent efficiency.

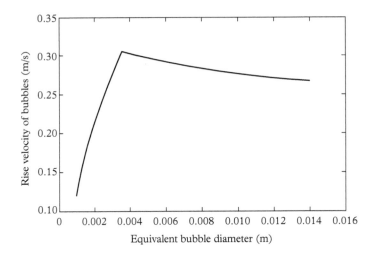

Fig. 4.23 *Steady-state free rise velocity of gas bubbles in aluminium at 760°C. The bubble size is plotted as equivalent spherical diameter, that is, the diameter of a sphere having the same volume as the bubble.*

To determine the dimensionless contact area $\psi/[\%H]$, we must determine the surface area of the bubbles in the melt as a function of bubble radius. If we assume that we have a single bubble size, and that they rise in the melt at the speed shown in Fig. 4.23 without interfering with one another and neglect convection, then as shown previously from Eqs. (4.15) and (4.16)

$$A = \frac{3\dot{Q}_g H}{v_b R}. \tag{4.80}$$

Taking v_b from Fig. 4.23, we can calculate $\psi/[\%H]$ as a function of bubble radius, which in turn allows us to determine the degree to which we approach chemical equilibrium. The results of the calculation are shown in Fig. 4.24 for $(Z^0)^2$. The importance of having small bubble sizes becomes very clear with this plot. Also note that we move to higher efficiencies as the hydrogen content drops—a fortunate event.

It should be noted that the calculations shown in Fig. 4.24 are approximate, since each bubble was assumed to rise at its steady-state velocity in stagnant aluminium, as shown in Fig. 4.23. Usually, however, the upward velocity is greater owing to convection. Thus when a flux tube or a porous plug is used, the bubbles rise in a plume above their point of entry into the melt. As a result, the melt develops a recirculating flow pattern, which makes the bubbles rise faster, thereby lowering the degassing efficiency.

In Fig. 4.24 it is somewhat confusing that the efficiency $(Z^0)^2$ increases with decreasing hydrogen content. This fact ties in with Fig. 4.22 where we again see that efficiency increases with decreasing [%H]. The explanation is in the definition of the efficiency, Eq. (4.79), that arises naturally in the mathematical solution, Eqs. (4.75) and (4.76). In the definition of efficiency, the H_2 pressure is compared with the equilibrium pressure $f_H^2 K_H^2 [\%H]^2$. For

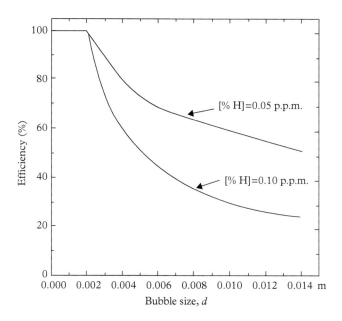

Fig. 4.24 *Calculated degassing efficiency $(Z^0)^2$ as a function of gas bubble size. The result is for two different gas contents: 0.05 and 0.1 ppm.*

instance, for the 1.4-cm diameter bubbles in Fig. 4.24 the efficiency is 25 per cent at hydrogen levels of 0.1 ppm and about twice as much at 0.05 ppm hydrogen. However, the ratio between the equilibrium pressures is 4. Thus, at 0.1 ppm dissolved hydrogen, the bubbles have a hydrogen pressure twice that of bubbles in a melt that contains 0.05 ppm. This rather complex behaviour is due to the fact that hydrogen is a diatomic gas and follows Sievert's law. At low hydrogen levels the bubbles do not need large contact areas to be saturated with H_2.

Undoubtedly some of the results of the calculations mentioned would come as no surprise to foundrymen, who instinctively know that lower temperatures and smaller bubbles give better results. Of primary importance at the higher hydrogen levels is the gas bubble size. If one uses a lance, large spherical cap bubbles (often several inches in diameter) are obtained, and the degassing process is inefficient, as seen from the results given in Fig. 4.24. Using a porous plug improves the efficiency by reducing the bubble size somewhat, but the bubbles rise in a plume above the plug, where they tend to coalesce to larger sizes even at moderate flow rates. Good results are obtained with a rotary impeller degasser, since such a unit may produce fine bubbles. The bubbles are also spread out more evenly along the bottom of the furnace. With this even dispersion efficiencies closer to 100 per cent are obtained. The results obtained with the three methods, a flux tube, a porous plug, and a rotary impeller,[32] are compared in Fig. 4.25. Three 200-kg batches of Al–7% Si–0.6% Mg alloy were degassed using the same flow rate of purge gas, and all conditions were maintained constant except for the method of gas introduction. The relevant data and physical constants are given in Table 4.3.

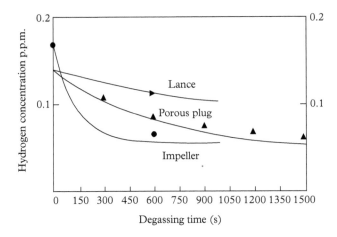

Fig. 4.25 *Results obtained with three different degassing techniques in a 200-kg (500-lb) furnace.*[32]

Table 4.3 *Data used in example calculations*

Metal temperature	$= 760\,°C = 1033\ K$
K_H	$= 0.78 \times 10^4$
f_H	$= 1.18$
k_t	$= 2.3 \times 10^{-4}$ m/s
ρ	$= 2350\ kg/m^3$
$\dot{G}$	$= 4.5 \times 10^{-6}$ kmol/s
Height of bath (H)	$= 0.5$ m

The porous plug works fairly well, but it takes nearly five times as long to produce the results obtained with a rotary impeller.

Many foundrymen use chlorine in their degassing cycle. However, due to environmental concerns, it is expected that the future use will be restricted. The benefits of chlorine are poorly understood. Thermodynamics tell us that $\underline{H}$ and Cl_2 (actually $AlCl_3$) cannot react to HCl at these temperatures.

If the degassing process is mass transfer controlled, that is, if bubbles are large, the use of chlorine may reduce the surface tension, bubble size, and rise velocity and give better results. If very small purge gas bubbles are present, however, chlorine has no effect on hydrogen removal. In this case, only the amount of gas is important (the mass transfer coefficient is not important because the efficiency is already nearly 100 per cent).

Many foundrymen are also aware of the effect oxides have on casting. One experiment in degassing is to take vacuum gas samples before and after a filter box. The result is similar to that shown in Fig. 4.19. The sample before the filter may be 'gassy'. Melters in the plant often see this and believe that the gas is 'filtered out' of the metal. However, pin samples taken at the same time and analysed for hydrogen by vacuum extraction will show that the filtered and non-filtered metal have the same gas content. Oxides therefore appear to help nucleate gas bubbles during solidification

This effect may perhaps be illustrated in a simple manner by carefully pouring a glass of carbonated beverage (beer is best) and then adding a small amount of table salt to the liquid. The supersaturated solution forms numerous bubbles on the salt crystals.

When porosity in a casting is found, this is probably due to both oxides and hydrogen. Thus, filtration of inclusions may help. If it is hot and humid it may be difficult to reduce the hydrogen concentration. This is due to the effect of 'pick-up' from the atmosphere through the top surface. The top surface transfer may be described by

$$\frac{k_{ts}A_s\rho\left([\%H]-[\%H]_v\right)}{100M_H},$$

where $[\%H] - [\%H]_v$ is the driving force and $[\%H]_v$ depends on p_{H_2O} in the atmosphere. H_2O is reduced to H_2 at the molten aluminium surface. A relation for $[\%H]_v$ is given in Chapter 3.

Often the atmosphere above the melt will be protected with a lid so that there should be little or no hydrogen pick-up from the air. In this case the gas composition is equal to that of the escaping bubbles so that $[\%H]_v = [\%H]Z^0$.

By blowing inert gas on the melt surface (combined with a lid), $[\%H]_v \approx 0$. $[\%H]_v$ can be determined experimentally for a given melt by letting it stand for a long time with no gas purging. Then $[\%H]_v = [\%H]$.

The mass balance equation for a batch reactor becomes:

> Reduction of content of hydrogen in batch
> = Hydrogen transferred to bubbles
> + Hydrogen transferred to atmosphere at bath surface

Thus,

$$-\frac{M}{100M_H}\frac{d[\%H]}{dt}$$
$$=\frac{2[\%H]^2 f_H^2 K_H^2 \dot{G} Z^2}{p_{inert}} + \frac{k_{ts}A_s\rho\left([\%H]-[\%H]_v\right)}{100M_H}. \tag{4.81}$$

Finally, it should be pointed out that the mass transfer coefficients k_t and k_{ts} are not known very accurately. There are several reasons for this. The mass transfer coefficient is proportional to the square root of the diffusivity of hydrogen, D_H, which is known only roughly. Also the mass transfer coefficients are proportional to the square root of the surface velocity which depends on geometry, the position of the impeller, and so on (see Chapter 3).

Figure 4.26 shows theoretical curves for hydrogen removal[31,32] with (a) the surface completely exposed to air with moisture; (b) a lid; and (c) blowing inert gas on the surface. We

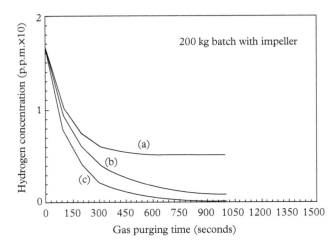

Fig. 4.26 *Calculated curves for hydrogen removal with (a) the surface completely exposed to air; (b) a lid; and (c) blowing inert gas on the surface (combined with a lid).*

see from Fig. 4.26 (b) that a lid greatly improves efficiency. Blowing inert gas on the surface seems to be interesting only if extremely low hydrogen levels are required.

4.12   Metal to Gas (Vacuum) Transfer

Dissolved elements with a high vapour pressure may be removed by vacuum or purge gas treatment. Previously we studied an example concerning Na removal by gas purging. Examples of vacuum refining are inductively stirred vacuum melts[33] and stream degassing.[34]

Vacuum treatment is carried out in order to reduce the concentration, c_j, of the impurity element (here given index j). Often the solvent metal also evaporates. For vacuum treatment to be interesting, treatment must result in lower levels of the impurity. Then

$$\frac{\dot{n}_j}{\dot{n}_1} > \frac{c_j}{c_1},$$ (4.82)

where $\dot{n}_j$ and $\dot{n}_1$ are kmoles of impurity and solvent metal distilled off per second and square meter, and c_j and c_1 are their concentrations (in kmol per m^3) in the melt.

Often the literature tabulates the vapour pressures p_j^0 and p_1^0 over the pure components j and 1. Then

$$p_j = a_j p_j^0$$
$$p_1 = a_1 p_1^0,$$ (4.83)

where a_j and a_1 are the activities

$$a_j = \gamma_j x_j$$
$$a_1 = \gamma_1 x_1.$$

Here γ_j and γ_1 are the Raoultian activity coefficients and x_j and x_1 are the mole fractions, respectively. Usually the metals contain only small amounts of impurities and alloying elements so that $\gamma_1 \approx 1$ and $x_1 \approx 1$ and $x_j = c_j M_1/\rho$. Then the partial pressure in the gas in equilibrium with the melt is

$$p_j = \gamma_j x_j p_j^0 = \gamma_j \frac{M_1 p_j^0 c_j}{\rho} \tag{4.84}$$

or

$$p_j = \gamma_j K_j c_j, \tag{4.85}$$

where $K_j = M_1 p_j^0/\rho$. Kinetic gas theory tells us that there is a flux,

$$\frac{p_j}{RT} \left(\frac{RT}{2\pi M_j} \right)^{\frac{1}{2}} \text{moles/s}$$

of component j in one direction across any plane in a gas. M_j is the molecular mass and p_j the partial pressure of component j in the gas. Under equilibrium conditions when p_j is the same throughout a horizontal plane, the same number of moles cross in the opposite direction. Then there is no net transport of the j component through the plane. We introduce an imagined horizontal plane; this does not change the flux downwards or upwards of component j. If the pressure in the gas is p_{ji} near the melt surface and the partial pressure of j in equilibrium with the melt is p_j, then the net flux from the surface is given by

$$\dot{n}_j = \frac{(p_j - p_{ji})}{RT} \left(\frac{RT}{2\pi M_j} \right)^{\frac{1}{2}}, \tag{4.86}$$

or, employing Eq. (4.85),

$$\dot{n}_j = (\gamma_j K_j c_{ji} - p_{ji}) \frac{1}{(2\pi M_j RT)^{\frac{1}{2}}}, \tag{4.87}$$

where p_{ji} is the actual partial pressure of j in the gas at the interface, and c_{ji} is the actual concentration in the liquid at the interface. Note that if p_{ji} and c_{ji} are in equilibrium so that $p_{ji}/\gamma_j c_{ji} = K$, then there is no net flux, $\dot{n}_j = 0$. Conversely, if there is a net flux of component j, the pressure in the gas, p_{ji}, and the concentration in the melt at the interface, c_{ji}, are not in equilibrium.

The flux $\dot{n}_j$ must pass through resistances in both metal and gas films on the two sides of the interface:

$$\dot{n}_j = k\left(c_j - c_{ji}\right) \tag{4.88}$$

$$\dot{n}_j = \frac{k_g\left(p_{ji} - p_{bj}\right)}{RT}. \tag{4.89}$$

p_{bj} is the pressure in the bulk gas away from the interface. k_g depends on the total pressure and increases with decreasing total pressure (see Chapter 7). As in our calculation of mass transfer from metal to slag we wish to obtain $\dot{n}_j$ in terms of the total driving force for transfer from the metal with concentration c_j to the gas with concentration p_{bj}/RT. Then we rewrite Eqs. (4.87)–(4.89) in terms of the driving forces:

$$\frac{\dot{n}_j}{k} = c_j - c_{ji}$$

$$\frac{\dot{n}_j\left(2\pi M_j RT\right)^{\frac{1}{2}}}{K_j \gamma_j} = c_{ji} - \frac{p_{ji}}{K_j \gamma_j}$$

$$\frac{\dot{n}_j RT}{k_g K_j \gamma_j} = \frac{p_{ji}}{K_j \gamma_j} - \frac{p_{bj}}{K_j \gamma_j}$$

Summation of the two sides of the three equations above gives

$$\dot{n}_j\left(\frac{1}{k} + \frac{\left(2\pi M_j RT\right)^{\frac{1}{2}}}{K_j \gamma_j} + \frac{RT}{k_g K_j \gamma_j}\right) = c_j - \frac{p_{bj}}{K_j \gamma_j}. \tag{4.90}$$

We see that a total mass transfer resistance, k_t, may be defined thus:

$$\frac{1}{k_t} = \frac{1}{k} + \frac{\left(2\pi M_j RT\right)^{\frac{1}{2}}}{K_j \gamma_j} + \frac{RT}{k_g K_j \gamma_j}. \tag{4.91}$$

$1/k_t$ is equal to the sum of the three resistances for the melt boundary layer, evaporation, and gas boundary layer. The resistance in the gas boundary layer can be reduced by lowering the total pressure, that is, by vacuum treatment. In Section 3.8 it is shown that k_g is inversely proportional to the total pressure. The ultimate total pressure in the vacuum chamber determines the final value of the gas film resistance, $1/k_g$. The relation for $1/k_t$ is similar to that developed for metal to slag mass transfer except that here an additional term appears for resistance to evaporation. If the vapour pressure of j is low (K_j is low), then this resistance is high and we have a low total mass transfer coefficient, k_t. The resistance in metal and gas in fact also depends on the physical properties of component j (on the diffusivity D_j). However, we have found that it is not helpful to include the index j for k_t and k_g.

Equations (4.90) and (4.91) give for the flux of component j

$$\dot{n}_j = k_t \left(c_j - \frac{p_{bj}}{K_j \gamma_j} \right). \tag{4.92}$$

$p_{bj}/\left(K_j \gamma_j\right)$ is the hypothetical concentration in the melt in equilibrium with the partial pressure p_{bj} in the bulk gas phase.

Vacuum refining has proved to be very useful for systems where the impurities have a high vapour pressure and the solvent metal has a low vapour pressure. Solutions employed for molten metals are often similar to those employed in the chemical industry. We explore the case that the solvent and impurity are in the same range of pressures. An important example of this situation is the need to remove Cu from steel scrap before the iron can be recycled.[35–7] Molten iron at 2000 K has a vapour pressure of 15 Pa. This may be compared with the vapour pressure of 0.5 mol% Cu in iron: $375 \times 8 \times 0.5 \times 10^{-2} = 15$ Pa.

It should be noted that the maximum allowable copper levels are low as seen in Table 4.4.

To estimate the loss of solvent metal, s, we employ the Hertz–Knudsen relation for pure metal.[39] An example illustrating problems and possibilities is presented: large unit and with high pumping speed. The amount of solvent lost is estimated.

Can Cu be extracted from a melt of iron by evacuation? Iron has a relatively high vapour pressure. Assume that there is complete mixing in the melt. The initial Cu concentration is $C_{Cu}^0 = 0.005$ mol fraction.

The Raoultian activity coefficient is $\gamma_{Cu} = 8$. The mass transfer coefficient in the melt boundary layer is $k = 10^{-4}$ m/s. The vapour pressure of pure Cu at 2000 K is $p_{Cu}^0 = 375$ Pa. The atomic weights of Cu and Fe are 63.5 and 55.85 kg/kmol, respectively. The density of the molten steel is 7030 kg/m³. The gas constant R is equal to 8.314×10^3 J/kmol K. The melt surface area is $A = 1$ m².[38]

Table 4.4 *Maximum allowable copper levels in steel*[36]

Application	%Cu
Deep drawing steel	0.06
Tin plate	0.06
Drawing quality	0.10
Steel forgings	0.35
Irradiated steels	0.10
Bar products	0.35

For the pure solvent we employ Eq. (4.87); set $c_{ji} \approx 1$ and get

$$\dot{n}_s = \frac{(\gamma_s K_s - p_{Si})}{(2\pi M_s RT)^{\frac{1}{2}}}. \tag{4.93}$$

This is the well-known Hertz–Knudsen equation.

For a solute, for instance Cu in Fe we have by analogy with Eq. (4.93), replacing $M_{solvent}$ with M_{solute}:

$$\dot{n}_{Cu} = \frac{\left(\gamma_{Cu} K_{Cu} C^l_{Cu} - p_{Cu\,i}\right)}{(2\pi M_{Cu} RT)^{1/2}}. \tag{4.94}$$

Note that p_{Si} and p_{Cui} are pressures in the gas near the melt surface. C^l_{Cu} is the concentration in the liquid at the surface. It is lower than the bulk concentration, C_{Cu}.

It is suggested to simplify matters that the vacuum pump chosen gives a constant pumping speed $\dot{V}_p = 72\,000\,\mathrm{m}^3/\mathrm{hr}$—in the pressure range of operation (Fig. 4.27). Then $\dot{V}_p = \mathrm{constant}$ for $p^*_{Cu} + p^*_{Fe} = p_{tot}$ from 15 to 200 Pa (0.2 to 2 mbar). p^*_{Cu} and p^*_{Fe} are pressures close to the pump inlet. The mass balance gives

$$RTA\dot{n} = \dot{V}_p p_{tot}$$

$$RTA\dot{n}_{Cu} = \dot{V}_p p^*_{Cu} \tag{4.95}$$

$$RTA\dot{n}_{Fe} = \dot{V}_p p^*_{Fe}. \tag{4.96}$$

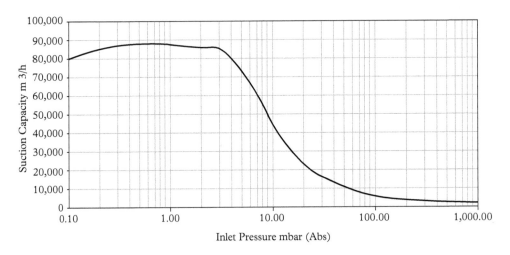

Fig. 4.27 *Pumping speed curve (m^3/hr) for a four-stage vacuum pump.*[41] *(Reprinted by permission from ELIVAC Swiss GmbH.)*

A is the 'effective cross-sectional area' for the flow of gas from the melt to the vacuum pump. $\dot{n}A$ gives the number of kmol per second transferred from melt to pump. To determine C'_{Cu} we must take resistance in the melt into account

$$\dot{n}_{Cu} = k_{Cu}\left(C_{Cu} - C'_{Cu}\right). \tag{4.97}$$

k_{Cu} is the mass transfer coefficient for the liquid side given in units of m/s.
   Equation (4.97), introduced into Eq. (4.94), gives

$$\dot{n}_{Cu} = \frac{\gamma_{Cu}K_{Cu}C_{Cu} - p_{Cu\ i}}{(2\pi M_{Cu}RT)^{1/2} + \frac{\gamma_{Cu}K_{Cu}}{k_{Cu}}}. \tag{4.98}$$

Note that C_{Cu} is the concentration in the bulk liquid and $p_{Cu\ i}$ is the concentration in the gas near the surface.
   For the unit described here it will be argued that pressure drop may be neglected from above the liquid to the inlet of the pump. This, of course, depends on the size, shape, etc. Nevertheless, a cylinder is used as a reference to indicate the magnitude of the pressure loss.
   For viscous flow we may employ the Hagen–Poiseuille equation,[40]

$$\frac{\Delta P}{L} = 32 \frac{\dot{V}_p\ \mu}{A\ D^2} \tag{4.99}$$

where D is the diameter of the cylinder, and the cross-section $A = \frac{\pi\ D^2}{4}$.
   For the viscosity we use

$$\mu = \frac{5}{16} \frac{(MRT)^{1/2}}{\pi^{1/2} \cdot N_A\ d_C^2}, \tag{4.100}$$

which represents the viscosity of a gas composed of hard spheres at low density. For the collision diameter, d_C, we have taken two atomic radii for Fe

$$\mu = \frac{5}{16 \cdot} \frac{(55.85 \times 8314 \times 2000)^{1/2}}{3.14^{1/2} \times 6.022 \times 10^{26} \cdot (2 \times 1.24)^2 \times 10^{-20}} = 0.145 \times 10^{-3} \frac{N_s}{m^2};$$

for Cu

$$\mu = 0.145 \times 10^{-3} \frac{N_s}{m^2}.$$

We set Eq. (4.99) as a case that can be handled with regard to practice, size, and cost:

$$\frac{\dot{V}_p}{A} = 20\ \frac{m}{s}.$$

$$A = \frac{\pi D^2}{4} = 1m^2$$

$$D = 1.13 \ m.$$

Then Eq. (4.97) gives

$$\frac{\Delta P}{L} = \frac{32 \times 20 \times 0.145 \times 10^{-3}}{4/\pi} = 0.073 \frac{Pa}{m},$$

where $L = 1m$, and $\Delta P = 0.073 Pa$.

This indicates that we can neglect the pressure drop in the space from above the melt to the inlet of the pump.

Thus, we find that the pressures at the pump inlet, P^*_{Cu} and P^*_{Fe}, are practically the same as pressures above the melt,

$$P_{Cu \ i} = P^*_{Cu}$$

$$P_{Fe \ i} = P^*_{Fe}, \tag{4.101}$$

and employing Eqs. (4.93) to (4.96) and (4.98) gives

$$\frac{RT\left(\gamma_{Cu}K_{Cu}C_{Cu} - P^*_{Cu}\right)}{(2\pi M_{Cu}RT)^{1/2} + \frac{\gamma_{Cu}K_{Cu}}{k_{Cu}}} = \frac{\dot{V}_P}{A}P^*_{Cu} \tag{4.102}$$

and

$$\frac{RT\left(\gamma_{Fe}K_{Fe} - P^*_{Fe}\right)}{(2\pi M_{Fe}RT)^{1/2}} = \frac{\dot{V}_P}{A}P^*_{Fe}. \tag{4.103}$$

We need to determine the resistance to mass transfer in the melt, $\frac{1}{k_{Cu}}$. We set $k_{Cu} = 10^{-4}$ m/s.[44] For the solvent we get

$$\frac{\left(15 - P^*_{Fe}\right)8314 \times 2000}{76368} = 20P^*_{Fe},$$

giving $P^*_{Fe} = 13.73 Pa$.

For Cu, $C_{Cu} = 0.5$ mol%, we get

$$\frac{\left(8 \times 375 \times 0.005 - P^*_{Cu}\right)8314 \times 2000}{(2\pi \ 63.55 \times 8314 \times 2000)^{1/2} + 8 \times 375 \times 10^4} = 20P^*_{Cu}$$

$$\frac{\left(15-P^*_{Cu}\right)831400}{81462+3\times10^7}=P^*_{Cu},$$

giving $P^*_{Cu}=0.4$ *Pa*.

To give a reasonable, high value for P^*_{Cu} the mass transfer coefficient k in the melt must be increased by a factor of 10.

We have a pump that retains its pumping speed of 72 000 m³/hr at a pressure of 15 Pa (see Fig. 4.27). Our case can probably be handled by a four-stage vacuum pump.[41] $\frac{1}{k_{Cu}}$ in Eq. (4.98) can be reduced by agitating the bath electromagnetically[42] or by argon gas purging through the bottom. However, gas purging adds an additional load to the pump. The value of k is discussed in Chapter 3, Eq. (3.51). There $k=10^{-4}$ m/s is found to be a rather high value.

The validity of the Hertz–Knudsen equation has been questioned.[39] A problem is that the gas is in various states: viscous, intermediate, and molecular.

4.12.1 Conclusions

– If the solvent has a vapour pressure of the same magnitude as the solute to be removed, the solute may be 'swamped' by solvent. Removal by vacuum is not possible unless mass transfer in the liquid is radically enhanced.

– For units where the size scale is 1 m, it is indicated that resistance in the space from the melt surface to the pump inlet is not a problem in viscous flow.

– Equation (4.98) should be useful for describing mass transfer of the solute at the melt surface. It may be regarded as a modified version of the Hertz–Knudsen equation.

4.13 Vacuum Refining of Aluminium

Some elements have a vapor pressure higher than aluminium, and they might be removed selectively by vacuum treatment, or by bubbling inert gases through the melt. This might be done intentionally to refine impure aluminium melts, or high vapor pressure may cause an unintentional loss of a desirable alloying element. We can examine this case by using the data in Table 2.1 together with information on the vapor pressure of the elements.[43] Consider the general reaction

$$nM \rightarrow M_n(v), \tag{4.104}$$

where M is the element dissolved in liquid aluminium. The equilibrium constant for this reaction is

$$K = P_{M_n}/(f_M \cdot \%_M)^n. \tag{4.105}$$

At the 'hypothetical' one weight percent standard state, the equilibrium product is numerically equal to the vapour pressure. This pressure has been plotted as a function of temperature

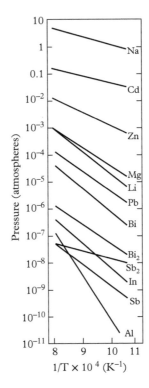

Fig. 4.28 *Calculated vapour pressures in a 'hypothetical' one weight percent solution.*

for selected elements in Fig. 4.28. The vapour pressure of aluminium is also shown for comparison, since it represents a limit on the vacuum that may be obtained to refine an alloy.

When a molten alloy is subjected to a vacuum, the most volatile elements tend to boil off first. As these are depleted, less volatile species are removed. This process continues until the pressure has decreased to the point where aluminium begins to be lost. Refining ends at this point, and may be roughly defined as when the pressure of M_n is equal to that of aluminium. This simple approximation was used to make calculations at a temperature of 1000 K (727 °C or 1340 °F). The resulting concentrations are given in Table 4.5.

In practice, dissolved impurity element levels would be higher than those calculated in Table 4.5, owing to kinetic factors. These have been discussed above and also in some detail by Richardson.[44]

Figure 4.28 helps to understand how some elements behave in molten aluminium. For example, if you try to add metallic sodium to modify aluminium–silicon alloys, you will probably find the Na recovery to be low, and that a good portion of the Na has gone into the plant's atmosphere—to create a nasty, irritating fume. Likewise, if zinc is alloyed at high metal temperatures, the Zn recovery is likely to be less than desired.

Table 4.5 *Approximate minimum concentrations to be realized by vacuum refining at 1000 K*

Dissolved Element	Composition, ppm
Na	3.7×10^{-7}
Cd	7×10^{-6}
Zn	3.3×10^{-4}
Mg	1.3×10^{-2}
Li	3.4×10^{-2}
Pb	0.11
Bi	0.41
Ca	39
In	110
Sb	310

4.14 Distillation

For many metals, on both a laboratory and industrial scale, purification by distillation is important. An example is separation of Zn, Pb, and Cd by distillation. Some indication is given here of how an absorption or distillation tower is designed. The goal in design is to describe the number of trays, their dimensions, the distance between each tray, and so on.[5,42,45–46] Here the discussion will be confined to the determination of the number of trays required to give the specified composition of the exit streams. The most important tool that we need is thermodynamic data that relates the gas composition of component $j(y_j)$ to the composition in the melt, X_j. Ideally the gas above a tray has been in intimate contact with the melt and is in equilibrium with it. Usually the assumption is made that the melt in a tray is completely mixed and that the gas above the tray is also completely mixed. The composition, x_n, in tray number n is different from tray $n + 1$ (the index j for component j has been dropped to simplify the notation). Similarly, the gas compositions above trays n and $n + 1$ are different. From Fig. 4.29 it is seen that y_n can be given in terms of x_{n+1} by setting up a mass (or molar) balance for the lower part of the tower:

$$\dot{G}_{in}y_{in} + \dot{M}_{n+1}x_{n+1} = \dot{G}_n y_n + \dot{M}_1 x_1 \tag{4.106}$$

or

$$y_n = \frac{\dot{M}_{n+1}}{\dot{G}_n}x_{n+1} + \frac{\dot{G}_{in}y_n - \dot{M}_1 x_1}{\dot{G}_n}. \tag{4.107}$$

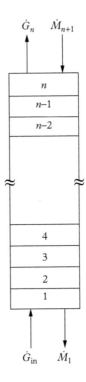

Fig. 4.29 *Schematic diagram of absorption tower.*

$\dot{G}_n$ is obtained from the overall material balance for the lower part of the tower

$$\dot{G}_{in} + \dot{M}_{n+1} = \dot{G}_n + \dot{M}_1. \qquad (4.108)$$

Equation (4.107) relates the gas composition above plate n to the composition in the melt in the plate above, $n + 1$. Equations (4.107) and (4.108) are used together with the equilibrium data and mass balance equations for the melt and gas at the various plates

$$\sum_{j=1}^{k} x_j = 1 \qquad (4.109)$$

$$\sum_{j=1}^{k} y_j = 1, \qquad (4.110)$$

where k is the number of components describing the distillation tower. Obviously, extensive numerical calculations employing a powerful computer are needed to handle Eqs. (4.107)–(4.110) for a tower with a large number of trays. In fact, no mention has been made that the heat balance for each tray must also be taken into account to calculate the temperatures. In turn, the temperature determines y_j in equilibrium with x_j.

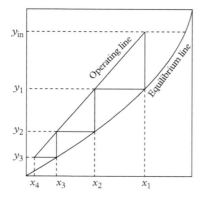

Fig. 4.30 *Graphical calculation of the required number of stages in gas absorption.*

If it is assumed that the temperature is constant and that the $\dot{G}_n$ and $\dot{M}_n$ are constant and equal to $\dot{G}$ and $\dot{M}$, the number of trays can be determined very simply from a diagram such as that shown in Fig. 4.30. In this case Eq. (4.107) reduces to

$$y_n = \frac{\dot{M}}{\dot{G}}x_n + y_{in} - \frac{\dot{M}}{\dot{G}}x_i. \qquad (4.111)$$

In Fig. 4.30 the gas composition is given on the y-axis and the composition in the melt on the x-axis.

Equation (4.111) is called the operating line. The slope of the operating line $\dot{M}/\dot{G}$ can often (within limits) be chosen. Once $\dot{M}/\dot{G}$ has been chosen, the operating line is determined by the process conditions and specifications. For instance in absorption of a component from a gas to the melt, the inlet composition of the gas, y_{in}, is known. Also the maximum allowed exit composition of the gas, y_{exit}, is specified. Furthermore, the inlet composition of this component in the melt, x_{in}, is known. As seen from Fig. 4.30, these data together with the slope determine the operating line. y_{in} determines the composition on the first plate, x_1. Now the melt composition on the second plate, x_2, is fixed by the gas composition below, y_1, according to the mass balance Eq. (4.111) (the operating line). y_1 is also assumed to be in equilibrium with x_1. Therefore, since (x_2, y_1) is on the operating line, we may proceed to draw a horizontal line from x_1 to x_2 as shown in the figure. We draw a line from x_2 on the equilibrium line to x_3 on the operating line and so forth. If $y_3 = y_{exit}$, only three trays are required. In practice, more than three trays would be needed. The reason is that the assumptions made here concerning attainment of equilibrium and complete mixing are only rough approximations.

4.15 Comparison of Different Methods for Refining Al Alloys

In practice, one must consider the chemical factors which control the possible elimination of undesirable impurities from an alloy. The important principles and necessary thermodynamic data were reviewed in detail in Chapter 2. These can be used together with standard free energies of formation (Appendix 1) to determine whether a proposed refining method is possible.

We conclude this chapter with a summary of a study of possible removal methods of common impurities in 8xxx alloys, which are used to manufacture aluminium foil. This information will illustrate how the principles of metal refining may be applied to an industrial problem.

First, it will be useful to consider the composition limits for an alloy commonly used as a feedstock for foil manufacture. The composition of 8111 alloy is shown in Table 4.6. Although not specified by the Aluminum Association (AA), because foil is often used in food preparation, the total heavy metals concentration is typically limited to a maximum total of 100 ppm.

The refining of aluminium alloys may be accomplished in four possible ways. They are:

1. Selective oxidation of reactive elements;
2. Vacuum refining of volatile elements;
3. Partial solidification, or fractional crystallization, using the segregation that occurs during freezing of an alloy; and
4. Addition of an element to form a reactive compound, followed by removal of the compound by sedimentation or filtration.

The results of a detailed study are summarized in Table 4.7, which illustrates the possibilities offered to us by thermodynamic laws and the chemical properties of the individual elements when dissolved in aluminium. Each element is discussed separately.

Table 4.6 *Composition limits for 8111 alloy*

	Si	Fe	Cu	Mn	Mg	Cr	Zn	Ti	Li	HM[a]	Others each	total
AA	0.50–0.90	0.6–1.0	0.10	0.20	0.05	0.05	0.10	0.08	—		0.05	0.15

[a] HM: Heavy metals—sum of Cd, Pb, Hg, and As—CONEG limit is 100 ppm

Table 4.7 *Evaluation of different refining methods*

Process \ Element	Fe	Mn	Cu	Zn	Mg	Li	Heavy metals
Selective oxidation							
Vacuum refining							
Fractional crystallization						???	
Compound formation		???	???	???	???	???	???

best			worst

4.15.1   Iron

Iron is much less reactive than aluminium, and it has a high melting point and low vapour pressure. It cannot be removed by either of the first two methods. However, during solidification the first solid formed has very little iron—about 2% of that present in the original liquid alloy. Hence, it is possible to purify Al–Fe alloys by fractional crystallization. A commercial process has been developed to remove Fe (and some other elements) from aluminium. However, anecdotal evidence suggests this process is rather expensive, on the order of $300–500 a tonne. The limit for iron is rather high in 8111 alloy. Also, the premium for low iron sow is usually much less ($50–100 a tonne). So, although favourable from a chemical point of view, there is no economic case for fractional crystallization. It would be less expensive to 'dilute' high iron materials by alloying with low iron sow.

Experimental studies have removed iron by formation of an iron phosphide. The process does not appear very promising, however. Only about half of the iron can be removed in this way.

4.15.2   Manganese

Manganese is similar to iron chemically; but it does not segregate during solidification. Hence, there is no way to remove it once it is dissolved in aluminium.

4.15.3   Copper

Copper is similar to iron chemically, but it does not segregate as strongly during solidification. Hence, a fractional crystallization process will be less efficient for copper. The above comments for iron also apply here, so one must conclude there is no inexpensive way to remove copper once it is dissolved in aluminium.

4.15.4   Zinc

Zinc is similar to iron chemically, but it segregates weakly during solidification. It has a relatively high vapour pressure, so a vacuum refining process could be used to remove zinc. Some zinc will also be lost to vaporization when a heat is held at high temperatures.

4.15.5   Magnesium and Lithium

These elements are more reactive than aluminium and are easily removed by oxidation. The usual method is to flux with chlorine, but Cl-containing fluxes or fluorides may also be used. There is off-the-shelf equipment that can be used to inject reactive fluxes into melting furnaces and which can remove these elements. The operating cost will be on the order of $20 a metric ton.

A flux that would be suitable for this treatment would be a eutectic mixture of KCl and NaCl, plus about 10% AlF_3 or sodium cryolyte. There are also a number of other commercial fluxes sold for this purpose.

If time permits, these elements can also be removed to low levels by allowing them to react with oxygen in the air, as they report to the oxide dross on the melt surface. The cost associated with this option would be longer furnace holding times and somewhat increased amounts of dross formed.

4.15.6   Heavy Metals

Some of the heavy metals may tend to 'boil off' with time, especially when the melt is held at high temperatures. This would be especially so for Hg, which has a very high vapour pressure at temperatures which melt aluminium. In the absence of thermodynamic data for their solution in aluminium, it would be useful to study the apparent loss of these elements with holding time in melting furnaces.

There is a fair amount of missing information, as indicated by the cells containing question marks in Table 4.7. More study, or some experimental work, is required to fill in these 'blanks'.

....

REFERENCES

1. Glasstone S, Laidler KJ, Eyring H. *The theory of rate processes*. New York: McGraw-Hill; 1941.
2. Nakanishi K, Fujii T, Szekely J. Possible relationship between energy dissipation and agitation in steel processing operations. *Ironmaking and Steelmaking* 1975; 3: 193–7.
3 Ghoti RA, Raman AAA, Ibrahim S, Buroutian SJ. Liquid–liquid mixing in stirred vessels: a review. *Chem. Eng. Commun.*, 2013; 200(5): 595–627.
4. Higbie R. The rate of absorption of a pure gas into a still liquid during short periods of exposure. *Trans AIChEJ*, 1935; 31(2): 365–89.
5. Danckwerts PV. Significance of liquid-film coefficients in gas absorption. *Ind. Eng. Chem.*, 1951; 43(6): 1460–7.
6. Clift R, Grace JR, Weber ME. *Bubbles, drops and particles*. New York: Academic Press; 1978.
7. Qin G, Weian L, Xingyan C, Keren C. 2nd Chin. Metal. Science Meeting, Shenyang, China. 1984; 5; 483.
8. Mori K, Ozawa Y, Sano M. Characterization of gas jet behaviour at a submerged orifice in liquid metal. *Trans ISIJ*, 1982; 22(5): 377–84.
9. Szekely AG. The removal of solid particles from molten aluminium in the spinning nozzle inert flotation process. *Trans AIME B*, 1976; 7(2): 259–70.
10. Myrbostad E, Pedersen T, Venas K, Johansen ST. The ASV inline system for refining of aluminium. In: *Light metals*, pp. 861–72. Warrendale, PA: Metallurgical Society of AIME; 1986,
11. Nagata, S. *Mixing, principles and applications*. New York: Wiley; 1975.
12. Bopp JT, Neff DV, Stankiewicz, EP. Degassing multicast filtration system—new technology for producing high quality metal. In: *Light metals*, pp. 729–36. Warrendale, PA: Metallurgical Society of AIME; 1987.
13. Kawakami M, Kitazawa Y, Nakamura T, Miyake T, Ito K. Dispersion of bubbles in molten iron and nitrogen transfer in the bubble dispersion zone at 1250 °C. *Trans ISIJ*, 1985; 25: 394–402.
14. Mori K, Sano M, Sato, T. Size of bubbles formed at single nozzle in molten iron. *Trans ISIJ*, 1979; 19(9): 553–8.

15. Kero I, Groadahl S, Tranell G. Airborne emissions from Si/FeSi production. *JOM*, 2017; 69: 365–80. Adapted from Schei A, Tuset JK, Tveit H. *Production of High Silicon Alloys*, 1st edn. Trondheim, Norway: TAPIR forlag; 1998.

16. Kero I, Næss MK, Andersen V, Tranell GM. Refining kinetics of selected elements in the industrial silicon process. *Metall. Mater. Trans. B*, 2015; 46: 1186–94.

17. H. Tveit, *Størkning av 75% ferrosilisium, forløp, struktur og styrke*. Trondheim, Norway: NTH; PhD Thesis (in Norwegian); 1986.

18. Førde Møll M. *Solidification of silicon: macro- and microstructure as functions of thermal history and composition*. Trondheim, Norway: Norwegian University of Science and Technology; PhD Thesis: 2014.

19. Elkem Silicon Products. http://www.elkem.com/silicon-materials

20. Olsen JE, Kero I, Engh TA, Tranell, GM. Model of silicon refining during tapping: removal of Ca, Al, and other selected element groups. *Metall. Mater. Trans. B*, 2017; 48: 870–7.

21. Engh TA, Sandberg H, Hultkvist A, Nordberg, LG. Si deoxidation of steel by injection of slags with low Si02 activity. *Scand. J. Metall.*, 1972; 1: 103–14.

22. Monden S, Tanaka J, Hisaoka I. Impurity eliminations from molten copper by alkaline flux injection. In: Sohn HY, ed., *Advances in sulfide smelting*, Vol II, pp. 901–15. Warrendale, PA: AIME; 1983.

23. Lehner T, Carlson G, eds. *Proceedings Scaninject I, II, III, IV*. Lulea, Sweden: Jernkontoret, Mefos; 1977 (1980), (1983), (1986).

24. *Proceedings of Shanghai symposium on injection metallurgy, Nov. 1982, Shanghai, China*.

25. Engh TA, Larsen K, Venas K. Penetration of particle-gas jets into liquids. *Ironmaking and Steelmaking*, 1979; 6(6): 268–73.

26. Cordier J, Chaussy JP, Lhussier G, Bienvenu Y, and Dernier G. Desulfuration et traitement de la fonte par injection de magnesium: le procede USIRMAG 2. *Rev. Met. Paris.*, 1981; 78(3): 201–12.

27. Mehrotra SP, Chaklader ACD. Interfacial phenomena between molten metals and sapphire substrate. *Met. Trans B*, 1985; 16: 567–75.

28. Rohatgi PK. Interfacial phenomena in cast metal-ceramic particle composites, In: Dingra AK, Fishman SG, eds., *Interfaces in metal-matrix composites*, pp. 185–202. Warrendale, PA: Metallurgical Society of AIME; 1986.

29. Ohguchi S, Robertson DGC. Kinetic model for refining by submerged powder injection. Part 1 Transitory and permanent contact reactions. *Ironmaking and Steelmaking*, 1984; 11(5): 262–73.

30. Deo B, Boom R. *Fundamentals of steelmaking metallurgy*. Harlow, UK: Prentice Hall; 1993.

31. Sigworth GK, Engh TA. Chemical and kinetic factors related to hydrogen removal from aluminium. *Met. Trans B*, 1982; 13(3): 447–60.

32. Engh TA, Pedersen T. Removal of hydrogen from molten aluminium by gas purging. In: *Light metals*, pp. 1329–44. Warrendale, PA: Metallurgical Society of AIME; 1984.

33. Inouye M, ed. *Proceedings of the seventh international conference on vacuum metallurgy*, Keidanren Kaikan, Tokyo: ISIJ; 1982.

34. Turkdogan ET. *Physical chemistry of high temperature technology*. New York: Academic Press; 1980.

35. Ferguson FT, Nuth JA, Johnson NM. Thermogravimetric measurement of the vapor pressure of iron from 1573 to 1973K. *J. Chem, Eng. Data*, 2004; 49(3): 497–501.

36. Daehn KE, Cabrera Serrenho A, Allwood JM. How will copper contamination constrain future global steel recycling? *Environ. Sci. Technol.*, 2017; 51: 6599–606.

37. Savov L, Janke D. Evaporation of Cu and Sn from induction-stirred iron-based melts treated at reduced pressure. *ISIJ Int.*, 2000; 40(2): 95–104.

38. Duan YJ, Chen B, Ma YC, Gao M, Liu K. Determination of vapor pressure of liquid copper by carrier gas method. *J. Mater. Sci., Technol.*, 2013; 29(12): 1209–13.

39. Ward CP, Fang G. Expression for predicting liquid evaporation flux. Statistical rate theory approach. *Phys. Rev. E* 1999; 59(1): 429–40.

40. Bird, RB, Stewart, WE, Lightfoot, EN. *Transport phenomena*. New York: Wiley; 1960: p. 46.

41. ELIVAC: *Introduction to vacuum technology with reference to vacuum treatment of liquid steel*, pp. 1–24, p. 33; available at http://www.elivac.ch/download.

42. King CJ. *Separation processes*, 2nd edn. New York: McGraw-Hill; 1980.

43. Hultgren R, Desai PD, Hawkins DT, Gleiser M, Kelly KK, Wagman DD. *Selected values of the thermodynamic properties of the elements*. Metals Park, OH: American Society for Metals; 1973.

44. Richardson FD. *Physical chemistry of melts in metallurgy*, Vol. 2. New York: Academic Press; 1974: pp. 477–87.

45. Nisenfeld AE, Seeman RC. *Distillation columns* (Monograph Series 2). Research Triangle Park, NC: Instrument Soc. of America; 1981.

46. *Proc. third international symposium on distillation*, 2 vols. Rugby, UK: Institution of Chem. Engineers Symposium Series 56; 1979.

....

FURTHER READING

Harris R, Davenport WG. Vacuum distillation of liquid metals. Part 1 Theory and experimental study. *Met. Trans. B*, 1982; 13(4): 581–8.

Ozberk E, Guthrie RIL. A kinetic model for the vacuum refining of inductively stirred copper melts. *Metall. Mater. Trans. B*, 1986; 17(1): 87–103.

Geiger GH, Poirer DR. *Transport phenomena in metallurgy*. London: Addison-Wesley; 1973.

Labaj T, Oleksiak B, Siwiec G. Study of copper removal from liquid iron. *Metalurgija*, 2011; 50(4): 265–8.

ASM handbook: vacuum induction melting, Vol. 15. Casting Handbook Committee; 2008: pp. 1–8.

Sigworth GK. Gas fluxing of molten aluminum: Part 1, hydrogen removal. In: *Light metals*, pp. 641–8. Warrendale, PA: Metallurgical Society of AIME; 1999.

Sigworth GK, Engh TA. Refining of liquid aluminum–a review of important chemical factors. *Scand. J. Metallurgy*, 1982; 11(3): 143–9.

Senda Y. Theoretical Analysis of vacuum evacuation in viscous flow and its applications. *SEI Tech. Rev.*, 2010; No. 71: 1–10.

Daehn KE, Cabrera Serrenho A, Allwood JM. How will copper contamination constrain future global steel recycling? *Environ. Sci. Technol.*, 2017; 51: 6599–606.

5

Removal of Inclusions from Melts

5.1 Introduction

The solubility of dissolved impurities in molten metals is seen to decrease with decreasing temperature (see Chapter 2). During solidification, as the temperature drops, and the solubility decreases, impurities are precipitated as a second phase (non-metallic inclusions or intermetallic phases) in the metal. Inclusions come in all shapes and sizes. Similarly, they can be either solids or liquids, or even gases, within the melt, and either metal-phyllic or metal-phobic.

Inclusions can originate during the extraction of metals from their ores, and during the subsequent processing, alloying and refining, prior to casting. The inclusions are classified as being 'indigenous inclusions', or 'exogeneous inclusions', depending on whether their origin is part of the intrinsic processing steps prior to casting and solidification, or is the result of contamination by external sources, such as contacting gases, refractories, or slags, as shown in Fig. 5.1.

Inclusions can also arise from the mechanical entrapment of furnace and ladle dross or slag particles, and from the chemical and mechanical erosion of ladle and casting refractories. For steel, entrained air in the tundish and in the mould pool may be a source of inclusions. Air can be entrained even when submerged entry tundish nozzles are employed. Top-slag from the tundish and mould powder may also contribute to inclusions. The occurrence of such exogeneous inclusions is sporadic, but due to their large size they may have a deleterious effect on steel properties. Good handling practices and experience today make the avoidance of large exogeneous inclusions more of an art than a science. Exogeneous inclusions are therefore only briefly referred to in the following, but note that Chapter 7 includes some material concerning the content of inclusions in alloying additions.

Generally, inclusions are all harmful to the physical properties of the metal products manufactured following the casting of liquid metal. That aspect was covered in Chapter 1. As explained there, inclusions can harm tensile and ductile properties; i.e. they can result in a much shorter predicted fatigue life for a plane's landing gear, can ruin metal surface finishes, can act as sites for corrosion, etc. Similarly, they cause line breakages during copper or aluminium wire-drawing operations, or become nucleation sites for hydrogen gas formation within a metal. That, in turn, can lead to hydrogen-induced cracking (HIC) in pipeline steels, for example. In fact, the list of an inclusion's potential 'attributes' for a cast metal are

Principles of Metal Refining and Recycling. Thorvald Abel Engh, Geoffrey K. Sigworth, and Anne Kvithyld, Oxford University Press.
© Oxford University Press 2021.
DOI: 10.1093/oso/9780198811923.003.0005

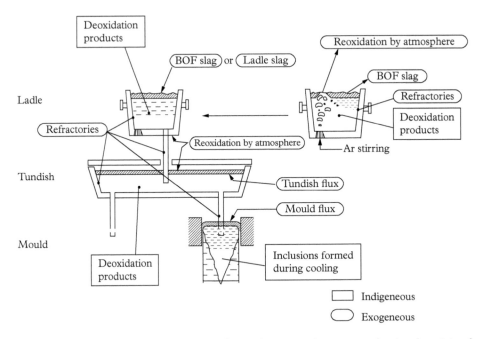

Fig. 5.1 *Potential sources of contamination in the continuous casting process, showing the origin of indigenous and exogenous inclusions in steelmaking.*

virtually endless. Their presence is usually highly detrimental, particularly so the larger the inclusion.

As such, the metals industry has largely been on an endless quest to make all commercial metals and alloys as 'clean as possible', prior to casting, but with varying degrees of success. Leaders in this effort have undoubtedly been member companies of the Iron and Steelmaking and Aluminium Industries. The approach to mitigate the presence of inclusions within a melt is to use stable refractories and a suitable protective slag. For instance, in steelmaking, a basic reducing slag prevents the pick-up of oxygen from the atmosphere.

Combined with the preventative methods mentioned above, dissolved impurity levels may be suppressed by the addition of elements with a high affinity for the impurity. A well-known example of this is the precipitation-deoxidation of liquid steel by additions of solid pieces of aluminium, which melt and dissolve into the molten steel, thereby precipitating billions of micron-sized alumina particles (inclusions).

Thus, the dissolved aluminium is convected and dispersed throughout the melt. Once the local solubility limit for Al_2O_3 formation is everywhere exceeded, solid alumina is precipitated, thereby removing dissolved oxygen. However, the first reaction products are fine particles, or films, of solid alumina, and these must then be removed, as far as possible, from the melt. Various methods for the removal of inclusions are discussed in this chapter. The methods are settling, flotation, filtration, and removal of inclusions to walls and top slag by stirring, although particle re-entrainment from overlaying slags must also be avoided.

5.2   Measurement of Inclusions

A number of methods for determining the content of inclusions within a melt exist: metal-lographic examination, centrifuge methods,[1] ultrasonic examination, the butanol method,[2] laser-induced breakdown spectroscopy (LIBS),[3] and finally the liquid metal cleanliness analyser (LiMCA) technique.[4]

In the metallographic techniques, samples are taken of the melt prior to casting or from the cast product. Alternatively, processed material (rolled or extruded product) may be examined. The volume of metal that may easily be analysed by metallographic examination is very small («1 g) so that a significant analysis of metal quality requires many hours of laboratory work.[5,6] The sensitivity or detection limit of this method is good, since it is limited only by the quality of metallurgical preparation and microscopy. The photographs in Fig. 5.2 show an electron

(a)   (b)
(c)   (d)

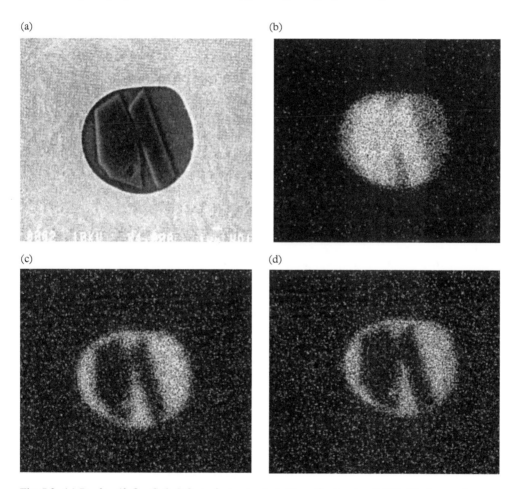

Fig. 5.2 *(a) Duplex Al_2O_3–CaO–SiO_2 inclusion in steel. Magnification is ×2 800. Electron probe microanalysis photographs are shown for (b) Al, (c) Ca, and (d) Si.*

microprobe analysis of an inclusion in steel. The two jagged particles are Al_2O_3 surrounded by a molten inclusion of SiO_2–CaO. Figure 5.3 shows an example of oxide films in an Al-Mg alloy.

Often in the description of inclusion removal, one resorts to the assumption that the inclusions are spherical. In fact, however, the inclusions may be elongated and irregular in shape. In liquid aluminium, one may find Al_2O_3 films torn off from the surface of the bath (see Fig. 5.3). A fundamental problem is that the metallographic picture is only two-dimensional. The three-dimensional shape must be derived from what is seen in two dimensions.

The volume of metal examined may be increased by using sampling filters.[7–9] Such filters operate by concentrating the inclusions either as a 'cake' on the inlet side of the filter or as a deposit inside the filter. More information concerning the distribution of inclusions is obtained when the particles are deposited inside the filter (in-depth filtration). On the other

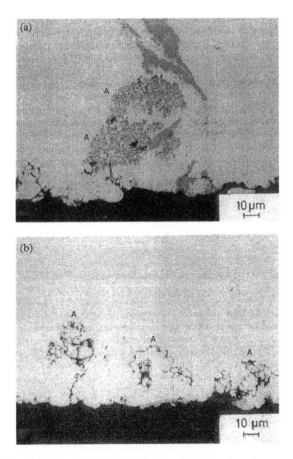

Fig. 5.3 *(a) Oxide films of sedimented Al_2MgO_4 clusters (A) in an aluminium–magnesium alloy, and (b) sedimented oxide films (A) in an aluminium–5 wt% magnesium alloy.*

hand, when samples are prepared from inside the filter, considerable care must be taken to conserve both the filter and metal material. When counting the inclusions, the filter material must be excluded. To speed up the counting of inclusions, automatic image analysis may be employed.[10]

The ultrasonic method[11] detects 'discontinuities' in the molten aluminium. In a commercial device,[12] a hollow titanium electrode is immersed into the melt, and an ultrasonic signal is transmitted and reflected within the probe. In practice, only particles larger than 100 μm are detected. Poor melt quality or detrimental handling practices are immediately apparent. The ultrasonic method can also be used in the control of melts of relatively good quality, especially if the inclusion size distributions have been determined by other methods. For very high-quality products, the probe may be too insensitive, even though a large amount of metal can be analysed, on the order of 5 weight per cent of the casting.

In the butanol method,[2] 0.3- to 5-g samples are dissolved in absolute 1-butanol. The content, the chemical composition, the structure, the shape, and the size distribution of the undissolved particles, for example intermetallic of Al–Fe–Si, have been determined by means of a series of methods including weight measurements, X-ray fluorescence, inductively coupled argon plasma spectrometry, Coulter Counter analyses, and X-ray diffraction. It has been shown that the butanol method can be used for quantitative analysis of α_1-AlFeXSi, ß-AlFeSi, α-Si, and $Al_{13}Fe_4$ particles in aluminium. It is worth mentioning that the volume size distribution can be determined by Coulter Counter analyses of the complex-shaped particles, and that the impurity concentration in the particles can be measured. This is not a routine method, however, as it is time consuming. The main reasons are that the dissolution rate must be continually controlled, the dissolution of 2-mm-thick samples takes about 2h, the filter must be washed for a long time with 1-butanol in order to avoid hydroxide formation, and the glassware in the apparatus must be thoroughly cleaned before each analysis. However, the method can be improved by increasing the dissolution rate of aluminium samples.

In metal processing, there has been significant interest in the use of laser-induced breakdown spectroscopy (LIBS),[2,3] as a tool for bulk chemistry measurement, as well as for the detection and analysis of individual inclusions in the size range $\sim$1–5 μm. Similar to conventional spark optical emission spectroscopy (OES), LIBS uses a short laser pulse to form a plasma on the metal surface. The elements present in the plasma emit characteristic EM radiation, which is collected and processed by a spectrograph. LIBS is technically a destructive test, since the sample studied is vaporized. However, the sample volume is on the order of 10^{-8} to 10^{-5} cm^3, which results in a very small sample size, even if thousands of measurements are taken. Other relevant advantages over other atomic emission techniques include: (1) LIBS can be applied to both conducting and non-conducting materials, (2) sample preparation is not necessary, (3) only an optical line of sight is required for measurement, and (4) measurements are taken in seconds.

The LiMCA technique instantaneously counts particles within a melt. The method has been applied to molten metals since 1985.[4,13] The probe detects particles by measuring the changes in electrical resistance of the liquid metal, when a suspended particle is carried with the melt, through an orifice, or through the electric sensing zone (ESZ). The main difficulty in the technique lies in selecting materials capable of withstanding the melt environment. The probe typically measures particles of size down to $\sim$10 μm using a probe orifice of approximately

250 μm diameter. Sample size is small, representing less than 0.01 per cent of the melt during casting. Similarly, the orifice through which the melt is sampled can become plugged by a single particle. As such, orifice size must align with the size distribution of inclusions. The technique has proven very useful for online measurements and is an important development and research tool.

No single technique is capable of describing all the information needed to assess inclusions—chemistry, concentration, and size distribution. It can be seen that an appropriate combination of methods is required to reach these goals. LiMCA is capable of generating the number per volume measurements instantaneously, which makes it popular in the cast house.

5.3   Removal of Inclusions Using 'Furniture' within a Tundish System

Figure 5.4 shows LiMCA data from a production plant of the former ALCAN company producing E.C. (electrical conductivity grade) wire by the Properzi Process. The standard practice at that installation was to allow for a period of quiescent settling as the last step in the batching procedure, prior to casting. The lower curve in Fig. 5.4 shows the results obtained using the LiMCA technique to monitor the metal cleanliness at the furnace outlet, following standard batching practice. There it is seen that the number of inclusions per kg of melt, greater than 20 microns, $N > 20$ μm per kg of aluminium, can run at about 550/kg, for a fully settled melt.

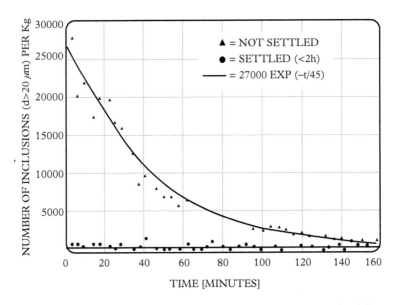

Fig. 5.4 *The settling of (Ti–V)B_2 particles from conductivity grade aluminum melts, following boron additions to an aluminium holding furnace.*[4, 13]

To investigate the kinetics of the settling process, a number of samples were taken from one batch in which the aluminium was cast immediately following mechanical stirring of the metal. This resulted in a high number of inclusions reporting within the metal initially, but gradually decreasing down to the baseline, as shown by the upper curve in Fig. 5.4. Careful analysis of the spatial variation of inclusion levels within holding furnaces revealed that inclusion counts were spatially uniform and independent of the location of LiMCA probes, at all times, during the observed decay in the settling of the $(Ti–V)B_2$ particles.[14]

Figure 5.5 shows equivalent TiB_2 results obtained as a function of time, starting near the end of one furnace batch, continuing for the entire duration of a second batch, and showing the first 60 minutes of a third batch. Parallel metal filtration (PoDFA) samples (Porous Disc Filtration Analyser) were also taken throughout the test period. The agreement matched, whilst LiMCA also had the advantage of being able to track changes, showing that the larger size particles were settling out more rapidly than the smaller ones, as expected.[14] This type of information later allowed ALCAN process engineers to proceed with optimizing their furnace-cleaning schedules and procedures, for jack-hammering out accumulated accretions of sintered aluminosilicate inclusions, etc., in the most efficient ways possible.

The 'furniture' technique for inclusion removal relies solely on fluid mechanics and is aimed at diminishing the disruptive effects on metal quality when a high speed, vertical turbulent jet of liquid steel enters the tundish, through a ladle shroud, from the emptying steelmaking ladle. On a positive note, the use of furniture can be quite effective, but on the negative side, it can be relatively expensive for the steel producer. Thus, Fig. 5.6 illustrates the use of a dam-and-weir system that was being researched for Broken Hill Proprietary

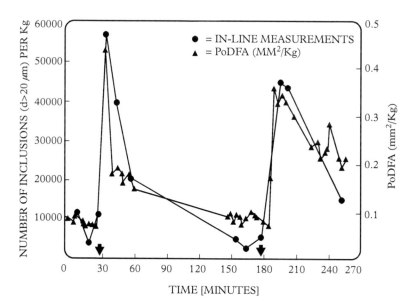

Fig. 5.5 *Simultaneous LiMCA and PODFA analyses for inclusions as a function of time into cast. Arrows indicate the times at which a furnace change took place.*[4,14]

(a)

(b)

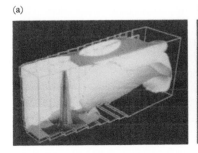
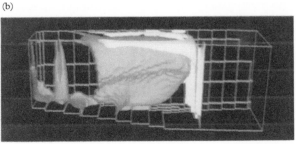

Fig. 5.6 *(a, b) Iso-concentration contours separating 120-micron-size inclusions from smaller ones. (b) shows that the placement of a dam and weir at 0.66L tundish is predicted to entirely block passage of the large inclusions into the exit ports of a slab/twin bloom tundish.*[15]

(BHP) Steel using a full-scale computational fluid dynamics (CFD) model of their proposed tundish from Voest-Alpine.[15] This revealed that a dam-and-weir arrangement, as illustrated in Figs. 5.6a and b, could be very effective in removing inclusions greater than about 100 microns, but would be relatively less effective in removing smaller inclusions.

Similar results were obtained experimentally, with an aqueous particle sensor (APS) at the Stelco research center, in 1982, using for a full-scale water model of their tundish system at their Lake Erie Works caster. The modelling work showed that there is a critical range of inclusion sizes that can be modified, by using dam-and-weir 'furniture'. Thanks to the development of the continuous APS system (equivalent to LiMCA for molten metals), the size distributions of inclusions entering and leaving could be directly measured, as a function of the dam-and-weir location within the tundish. Most revealingly, the residual ratio of inclusions (RRI) plot versus inclusion diameter illustrated that all inclusions greater than 110 microns would float to the surface of the tundish, whereas all inclusions smaller than about 50 microns would remain unaffected. Only the 50- to 110-micron inclusions could be diminished using a dam-and-weir system. The reason for this is that the floating times of inclusions smaller than 50 microns tend to be much greater than typical average residence times of the liquid steel within a tundish, of 7–14 minutes.[16]

Another of the early ways to remove inclusions from liquid steel was to build a weir enclosure to contain the incoming steel jet and have it fitted with adjacent outer dams, or with angulated holes, that directed outflows up towards the surface of the liquid steel. The purpose was to use fluid flows so as to force the churning liquid back up towards the surface. This again was successful in reducing larger inclusions entering the exit submerged entry nozzles (SENs) and passing them on into the continuous casting molds.

A later approach initiated at McGill was to build an 'impact pad', or 'pour box', which would reduce churning within the entering flow of steel so as to generate a fairly weak returning flow back up towards the upper surface. This up-flow does not disturb the protective cover slag too violently, which in turn acts as an absorber of inclusions. The McGill impact pad is shown located at the apex of a four-strand, delta-shaped, billet caster in Fig. 5.7a. This tundish model is a full-scale water model of the RTIT tundishes at Sorel–Tracy, Quebec. The

(a)   (b)   (c)

Fig. 5.7 *(a) The McGill impact pad used by QIT in their Delta-shaped, 4-strand tundish for many years. (b) is a model of an RHI impact pad, and (c) is a model 'Turbo-Stop' impact pad.*

device shown in Fig. 5.7b is an RHI pour box, located in the same model tundish, where one can see an overhanging lip constraining the outflow of liquid steel. Thus, the impacting flow of steel can be constrained and redirected using 'wall lips' so as to convert much of the kinetic energy contained in the incoming jet of steel into turbulence kinetic energy (TKE).

This potentially helps to cause individual inclusions to agglomerate together, thereby improving their collective chance of floating up to the surface of the tundish. Another pour box now widely used in North America, shown in Fig. 5.7c, is the so-called 'Turbo-Stop'. This box can also eliminate the splashing that occurs when the entering stream of liquid steel first impacts the bottom of an empty tundish. It is marketed by Vesuvius (formerly Foseco), and plant trials at AM-D, for instance, have proved the reliability of such a device. It is now a part of that company's standard casting practice, given its superior steel cleanliness performance during ladle changes.

5.4   Removal of Inclusions by Natural Flotation/Settling

Flotation or settling is the most common way inclusions can be removed, particularly in steelmaking, where all oxide-based inclusions are significantly less dense than the density of liquid steel (i.e.~3000 vs 7000 kg/m^3). For liquid aluminium, this is less easy, since the density of liquid aluminium (~2400 kg/m^3) is very similar to that of many of the inclusions found within it (e.g. Al_2O_3, Na_3AlF_6 (cryolite), TiB_2, etc.). Similar comments apply for inclusions formed in liquid magnesium. Furthermore, liquid aluminium forms a strongly adherent oxide film of alumina on the upper surface of the flowing metal, in contact with the atmosphere above it. Such a film, when disrupted by turbulence within the liquid, can become enfolded and entrained within the aluminium, thereby strongly compromising the quality and strength of the cast metal. For such reasons, flotation is less effective for many light metals, including molten aluminium and magnesium, but sedimentation nevertheless remains a possibility.

For liquid steel, flotation is a commonly used technique to rid liquid steel of larger inclusions (>100 microns). The rising velocity of inclusions can be readily estimated, by balancing the **buoyancy force,** $\rho_L g V$ on the particle, or inclusion, against the forces of gravity **(weight of inclusion)** $\rho_S g V$ and **drag,** $\frac{1}{2}\rho_L U_s^2 C_d A$, acting on an inclusion, as it rises to the surface. Thus,

$$\rho_L g V = \rho_S g V + 1/2\rho_L U_S^2 C_d A. \tag{5.1}$$

Here V is the volume of an inclusion, C_d is the drag coefficient for a spherical particle, A is the cross-sectional area of an inclusion perpendicular to the flow, and $\frac{1}{2}\rho_L U_s^2$ represents the dynamic pressure of the liquid acting on the spherical or particle. If we assume that all shapes of inclusions can be represented by equivalent diameter spheres, then $V = (\pi/6) D_e^3$, and $A = A_e = (\pi/4)D_e^2$, so that

$$U = (4\,(\rho_L - \rho_S)\,D_e g/3C_d\rho_L)^{1/2}. \tag{5.2}$$

For the special case of 'Stokesian' flow, where the flow around the inclusion can be assumed to be entirely laminar (Reynolds number less than 1), then the drag force, F_D, is analytically equal to $6\pi\mu R U_s$, or $3\pi\mu D U_s$, so that

$$U_S = (\rho_L - \rho_S)\,d^2/18\mu, \qquad \text{where } C_d = 24/\mathrm{Re}. \tag{5.3}$$

So, in steelmaking, one could, and did, just allow the liquid steel to 'rest' in a ladle for about 20 minutes, before teeming it out into ingot molds. This would allow most of the larger inclusions to rise to the steel's surface in the ladle. There they would join the upper insulating slag used to protect the freezing steel from re-oxidation and radiative heat losses. This step was practised in the 1920s–60s so as to allow the liquid steel within the ladle to become 'cleaner'. For instance, a 1000-micron, or 1-mm, inclusion can rise through a (stagnant) bath of steel at approximately $(7000 - 3000)\,[1000 \times 10^{-6}]^2/(18 \times 7 \times 10^{-3}) = {\sim}31.7$ mm/s, or 1.9 m in a minute. Of course, smaller inclusions would rise much more slowly.

Figure 5.8 shows a plot of the variation in the drag coefficient, C_d, or f, for a spherical particle versus the Reynolds number. There it shows the two distinct flow regimes, the Stokes and Newton, at $\mathrm{Re} \leq 1$, and $\mathrm{Re} \geq 1000$, respectively, together with an intermediate zone.

For the Stokes regime, the viscous drag coefficient diminishes rapidly with increasing Reynolds number, reaching a minimum at about $\mathrm{Re} = 1$. Beyond this Reynolds number, the fluid flow separates from the surface of the sphere, and turbulent flows set in at the back of the moving particle. At about $\mathrm{Re} = 1000$, the drag coefficient, C_d, becomes constant and equal to about 0.4. This is Newton's regime, where the drag force becomes proportional to the square of the relative velocity between the sphere and the liquid, U^2_∞, for which inertial forces are dominant.

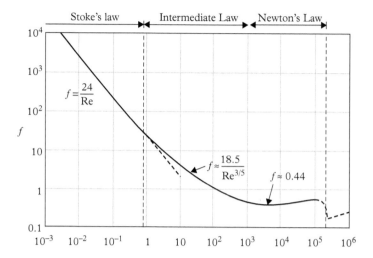

Fig. 5.8 *The friction factor f, or coefficient of drag, C_d ($f \equiv C_d$) of a spherical bubble or particle vs the Reynolds number. The friction factor, ϕ, or coefficient of drag, C_d, is defined as being equal to the drag force/$A_e(\frac{1}{2}\rho U^2)$. The Stokes regime, $f = 24/Re$, remains valid up to Re ~10, and Newton's regime between 1000 and 100000. In the intermediate region, $f \sim 18.5/Re^{3/5}$ applies.*

5.5 Introduction to Flotation by Bubbles

An important method for removing inclusions (particles) from molten metals (liquids) is by bubbling. Given the advent of continuous casting in the late 1960s, it was important to keep the liquid steel in the ladle at the correct superheat pouring temperature during the continuous casting process. This gradually led to the introduction of mandatory 'bottom gas bubbling' (argon or nitrogen) into a teeming ladle, as shown in Fig. 5.9. There the bubbles rose through the liquid steel using a novel (at that time) porous plug. This plug allowed gases to pass through its pores, but did not allow the liquid steel to pass through the other way, and so flood the shop floor!

The key 'invention' of the porous plug, credited to Dr. Robert Lee of Canadian Liquid Air (*Aire Liquide*), was first used at AM-D (DOFASCO) in ~1970, for electrical grades of steel high in silicon. These grades required a large amount of buoyant Fe–Si alloy lumps to be added to the bottom of the teeming ladle prior to tapping liquid steel from the BOF furnace. It required very strong stirring for homogenization of the melt within the ladle. These porous plugs have proved to be very effective for removing inclusions. The secret of its good performance is steel's high surface tension (1.85 N/m) and the non-wetting properties of the porous refractory in preventing the backflow of steel into the open pores of the 'plug'.

There has been a wealth of literature published by civil and chemical engineers over the past 75 years concerning the fundamentals of bubble formation and behaviour in water, and/or low-temperature organic liquids. One of the first most comprehensive studies was by Haberman and Morton.[18] A later book from Academic Press entitled *Bubbles, Drops and Particles* by Clift, Grace, and Weber[19] summarized somewhat more recent work on bubble formation in liquids. The various bubble regimes were categorized and are reproduced here as Fig. 5.10.

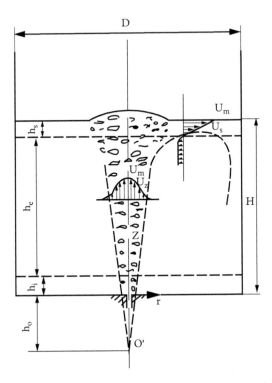

Fig. 5.9 *An early depiction of gas bubbling into a ladle of steel, which correctly shows spherical cap bubbles rising in a buoyancy-driven bubble plume, containing ~5% volume of spherical cap (SC) bubbles.*[17]

There they have plotted the Reynolds number (Re = $\rho U d/\mu$) as the ordinate, against the Eötvös (or Bond) number as the abscissa. The dimensionless Eötvös number contains the equivalent bubble diameter and the physical properties of the liquid ($g\Delta\rho D_e^2/\sigma$) for various Morten numbers, M (= $g\mu^4\Delta\rho/\rho^2\sigma^3$).

We see that very small bubbles rising in liquids are invariably purely spherical, such that their volumes are equal to $\{\pi D_e^3/6\}$, or $\{(4/3)\pi r_e^3\}$. Here D_e represents the equivalent spherical diameter, and r_e the equivalent spherical radius of a particle or bubble of any shape, and volume. For the case of a microbubble, $D_e = d$, and $r_e = r$. At Re $\geq$ 1, breakaway around the surface of a spherical bubble occurs at an angle of 135°, and a recirculating wake fills the low-pressure region behind it as it rises through the liquid. This leads to the formation of a turbulent wake which grows with increasing Reynolds number, interchanging turbulent kinetic energy with the bulk flow of water passing around the sphere. In the case of bubbles, the surfaces are not rigid, so the bubble's shape is governed by external events within the liquid such as the presence of a pressure gradient within the liquid. This leads to bubbles forming flattened, oblate spheroids. A little later, the growing bubble may increase in its velocity, or Re, and a rocking motion sets in owing to the periodic shedding of 'Karman vortices' from the object. As we move to ever larger bubble sizes, the bubbles become mushroom shaped,

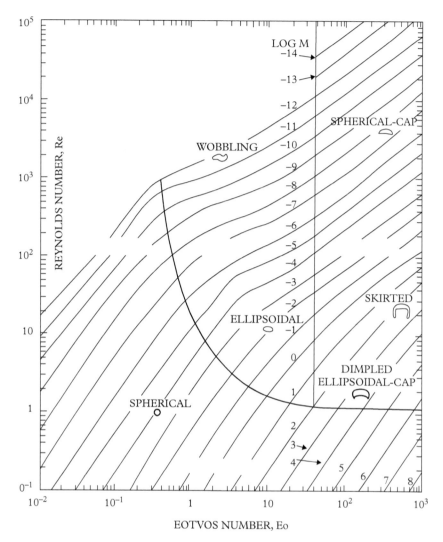

Fig. 5.10 *A plot of the Reynolds number for unhindered bubbles rising through a stagnant column of water, against the Eötvös number.*[19] *[Reprinted by permission from Dover Publications; copyright 1978.]*

commonly referred to as spherical cap (SC) bubbles. These bubbles are stabilized in stagnant, viscous liquids and can reach 200 cm^3 or more in volume.[20] Similarly, in still water, one can form SC bubbles with no satellite bubble formation up to volumes of about 45 cm^3.

Considerable work has also been invested in determining whether these trends also hold for liquid metals, given their much higher surface tensions, normally much higher densities, and their (usually) strong non-wetting characteristics with ceramic (brick) lined holding vessels. Irons and Guthrie,[21] for example, studied bubbles forming in molten iron, using graphite lances to bubble nitrogen, subsurface, into an inductively heated and stirred bath of molten iron. They were able to demonstrate that very large bubbles, of 60 cm³, would still be formed in molten iron, if leaking in gas through a small (1-mm) orifice, set in a horizontal plate of graphite, at 1 cc/s for ~60 s.

This is very different to the same experiment in water, for which tiny bubbles would be formed using a wetting orifice and stream towards the surface. This begs the question of what happens in liquid steel when purging gas enters through a submerged porous plug. The answer, of course, is 'coalescence' of forming gas bubbles, rendering them much larger, in some cases, than their (potential) birth size. These different phenomena can be explained by using a force balance. Thus, for gas injected into water through, say, a 0.2-mm diameter orifice of a wetting material (e.g. glass) of diameter d_i, we will have the attached rim of a forming bubble equal to the rim, or periphery, of the orifice, and we may write

Buoyancy force lifting bubble in water from 0.2-mm orifice = surface tension forces holding it to the orifice.

We have for the water system $V(\rho_L - \rho_G)g = \pi d_i \sigma$. So, ignoring ρ_G,

$$V_b = \pi d_i \sigma / \rho_L g. \tag{5.4}$$

Inserting appropriate values, we have $V_b = 3.14 \times 0.2 \times 10^{-3} \times (0.070)/1000 \times 9.81$, giving a bubble volume of bubble, $V_b = 0.00448$ cm³.

For steel, however, we have

$$V(\rho_L)g = \pi d \sigma. \tag{5.5}$$

Here d, the diameter, must be calculated. Assuming that the bubble forms a hemispherical bubble ($\Theta = 90°$) that spreads out across the surface of the porous plug until the forces of attraction to the non-wetting surface are overcome by the buoyancy force, we have $0.5[\pi d^3/6]\rho_L g = \pi d \sigma$, yielding $d = (12\sigma/\rho_L g)^{1/2}$, so $V = 0.5[\pi d^3/6]$, or

$$V_b = [\pi/12](12\sigma/\rho_L g)^{3/2}. \tag{5.6}$$

Inserting appropriate values of 1.85 N/m for σ_{steel}, and 7000 kg/m³ for ρ_L, yields $V_b = 1.5$ cm³. Thus, we see that a bubble forming in steel at a 20-micron 'non-wetting' orifice would be at least 335 times greater in volume (1.5/0.00448) than a bubble forming in water from the same (wetting) porous plug. Further, given that a porous plug will contain many small adjacent holes for allowing the passage of gas into the liquid steel, we see that one would expect the spreading and coalescence of forming bubbles, leading to much larger-sized bubbles versus those observed from a water model of such a system. This helps explain the formation of one large SC bubble, 60 cm³, every minute, when leaking in gas at 1 ml/s into the molten iron from a graphite surface, as previously mentioned.

This underscores the contention that the sizes of bubbles in liquid metals may be very different from those forming in aqueous solutions given the generally much higher surface tensions of liquid metals versus water, as well as typical contact angles between a metal and its container being greater than 90° (i.e. non-wetting).

These remarks have serious consequences as to what is possible in controlling bubble sizes and inclusion levels in liquid metals. For instance, if one wishes to remove inclusions using small bubbles, how can they possibly be formed for non-wetting ceramic–liquid metal systems? Many works have shown that much larger bubbles are the normal situation for liquid metals, including those formed on porous plugs in liquid steel.

We can look to the aluminium industry for a partial answer to this question. When using LiMCA for analysing liquid metal quality in an in-line ALPUR or similar degasser unit, this involved using a rapidly rotating hollow cylinder through which argon was being pumped into liquid aluminium passing through the launder system to the DC casters to produce cast ingots. It was surmised that microbubbles of argon must have been forming, since LiMCA monitors all inclusions in the size range 15–300 microns. So, if microbubbles are present, they will register as inclusions, exactly in the same way as the TiB_2 particles, etc., passing through the ESZ do. Process engineers at ALCAN Arvida were registering thousands of 'LiMCA events' after their new in-line gas injection degasser unit was introduced. These were far in excess of those being counted when the liquid metal entered their new in-line degasser unit. It was therefore making it impossible to register the cleaning efficiency of their new system, which was designed to replace their Alcan bed filter (ABF) unit. One proposed solution was to move the second LiMCA unit much further downstream so as to allow these microbubbles to float out. Unfortunately, that was impossible owing to the restricted length of the launder system. An alternate proposal was to develop a more sophisticated LiMCA detection system so as to distinguish these microbubbles from much denser particles, since the latter arrive at the event horizon for detection significantly more slowly.[22] As such, one can distinguish microbubbles from inclusions.

A further example of microbubbles being generated in liquid metals comes from TATA steel. Ijmuiden researchers some twenty years ago had analysed the quality of steel slabs coming off a new slab caster. The cast slabs revealed many perfectly spherical cavities (Fig. 5.11), as seen in the table included in the figure which shows a summary of their findings, indicating that a majority of the microbubbles detected were in the size range 50–100 μm.[23] However, these cavities were not present in their final strip products after rolling down to 1-mm sheet steel from the 250-mm-thick slab. As such, they were not deemed to be an important factor but were attributed to being the result of argon purging. This was used to prevent re-oxidation of liquid steel passing through their SENs into the oscillating moulds.[23]

A final instructive example on how to make microbubbles in liquid metals comes from the minerals processing industry, where froth flotation machines and, increasingly, column flotation machines are used to separate the metal sulfide particles from the silica gangue particles. There they use submerged gas sparger systems, together with frothing and collector agents, so as to produce microbubbles. The collector/frothing agents coat the bubble interfaces with hydrophobic ligands to attach mineral particles. However, equally important, though not perhaps as well known, is that bubble coalescence is hindered by these agents as coated bubbles repel each other, thereby making for a stable froth for ore collection. Perhaps there are equivalent systems for liquid metals?

Inclusions can be transferred to the bottom or walls, or to the top slag, by various effects such as gravity and turbulence. If gas is bubbled through the melt, the particles may come into contact with the bubbles by the same mechanisms. Furthermore, if these particles or inclusions

SECOND SERIES OF MICROFOCUS Ti-SULC
Sample size: 3mm * 40mm * 65mm
Bubble diameter

μm	Nr. 8568	Nr. 8573	Nr. 8574	Nr. 8578	Nr. 8595
50–100	60	142	78	9	62
100–150	18	37	31	4	17
150–200	16	31	10	8	9
200–250	10	25	3	4	4
250–300	8	12	6	9	3
300–350	1	0	1	1	0
350–400	3	2	0	1	0
400–450	0	0	0	0	0
450–500	0	0	1	0	0
500–550	0	0	0	0	0
550–600	0	0	0	0	0
600–650	0	0	0	0	0
650–700	0	0	0	0	0
700–750	0	0	0	0	0
>750	0	1	0	0	0

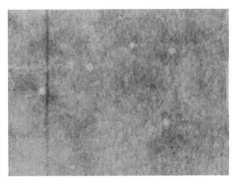

Fig. 5.11 *Photograph of just one surface of a continuously cast Ti-SULC steel slab, showing spherical cavities corresponding to microbubbles. Sample sizes 3 mm × 40 mm × 65 mm.*

are wetted by the bubbles but not by the melt, there is a high probability that particles remain trapped at the bubble–melt interface. They are then carried by the bubbles up to the dross or top-slag layer. The theory of removal of inclusions to bubbles (spheres) is treated in the following.

5.5.1   Attachment Mechanism to Bubbles

The mechanisms important for flotation are illustrated in Fig. 5.12. According to the interception mechanism, the centre of the particle follows the streamlines along the bubble, but the outer edge of the particle closest to the bubble intercepts the bubble–melt interface. Since velocities are high along a bubble, as opposed to along a solid wall, a large number of particles pass close to the bubble, so that removal in this manner is more important for bubbles than for a solid collector.

To develop a theory for flotation, we first assume that any particle is removed that 'stands in the path' of a rising bubble. Later, we will take into account that a particle is in fact partly swept sideways. The volume V_s that the bubbles sweep through per second is given by

$$\dot{V}_s = \text{cross-section of bubbles transverse to the direction of bubble flow}$$
$$\times \text{relative velocity between bubble and melt,}$$

or (for spherical bubbles of radius R)

$$\dot{V}_s = \pi R^2 N u_b,$$

where N is the number of bubbles present in the melt and u_b is the velocity of the bubble relative to the melt.

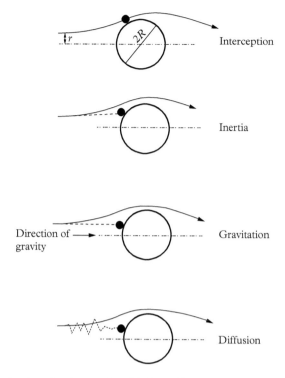

Fig. 5.12 *Various mechanisms for the transfer of (spherical) particles to a spherical collector (bubble).*

If the inclusions did not move sideways and stuck to the bubbles, once contact with the interface was established, the expression that would give the volume purified per second is

$$\dot{V}_s = \pi R^2 N u_{br},$$

where u_{br} is the relative velocity between bubbles and particles in the melt. However, Fig. 5.13 illustrates that only inclusions approaching the bubble within a cross-sectional area πr^2 attain contact. Thus the effective volume intercepted is only

$$\dot{V}_e = \pi r^2 N u_{br}. \tag{5.7}$$

The number of encounters in unit time between particles and bubbles is $c\dot{V}_e$. c is the number of particles in unit volume of melt.

As shown in Eq. (4.15), N is equal to

$$\frac{3\dot{Q}\tau_b}{4\pi R^3} = \frac{A}{4\pi R^2}.$$

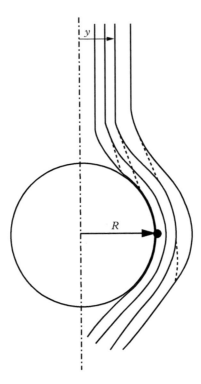

Fig. 5.13 *Collision cross-section πr^2 for particles removed by a spherical collector.*

In terms of the bubble-melt contact area A, Eq. (5.7) gives

$$\dot{V}_e = \frac{r^2}{R^2} \frac{A u_{br}}{4}.$$

(5.8)

Here $r^2/R^2 = \pi r^2/\pi R^2 = \eta$ is called the collision efficiency; η is the fraction of the cross-sectional area of the bubble where particles collide with the bubble. A more correct definition based on the particles instead of on the fluid is

$$\eta = \frac{\text{number of particles that collide with the bubble per unit time}}{\begin{array}{c}\text{number of particles per unit volume}\\ \times \text{ cross-sectional area of the bubble normal to the particle velocity}\\ \times \text{ velocity of particles relative to the bubble}\end{array}}$$

(5.9)

If the particles have the same velocity as the melt, the definition reduces to

$$\eta = \frac{\text{Volume flow of melt containing particles that will collide with the bubble}}{\text{Volume flow approaching the bubble}} \tag{5.10}$$

To illustrate the use of this equation, η is calculated for the interception mechanism. Since the flow around a bubble is only slightly perturbed from that of irrotational flow, we assume that we have potential flow. Furthermore, for small bubbles, we may presume that the form is spherical. Then the stream function is given by[24]

$$\psi = \frac{u_b r^2 \sin^2\theta}{2}\left(1 - \frac{R^3}{r^3}\right), \tag{5.11}$$

where r is the distance from the centre of the sphere and θ is the polar angle. The curves $\psi =$ constant are streamlines. The solution is chosen such that $\psi = 0$ for $r = R$. The flow of liquid in an annulus between r and R is equal to $2\pi\psi$.[24] Particles of radius a will collide with the bubble if they lie within a distance $r = R + a$ (see Fig. 5.12). The volume flow approaching the spherical bubble is $u_b \pi R^2$. Then according to the definition of η

$$\eta = \frac{\sin^2\theta}{R^2}\left\{(R+a)^2 - \frac{R^3}{(R+a)}\right\} \approx \frac{3a\sin^2\theta}{R}$$

for $a \ll R$. Inclusions collide with the bubbles from the 'pole' at $\theta = 0$ to the 'equator', $\theta = \pi/2$. This gives

$$\eta = \frac{3a}{R} \tag{5.12}$$

for the collision efficiency of interception. η is usually much smaller than 1. For instance, inclusions of diameter $2a = 100$ μm $= 10^{-4}$ m and bubbles of radius $R = 10^{-2}$ m give $\eta = 3 \times 10^{-4}/(2 \times 10^{-2}) = 0.015$.

Alternatively, Eq. (5.12) can be derived directly from the definition of η in Eq. (5.10):

$$\eta = \frac{2\pi R a u_\emptyset\left(\theta = \frac{\pi}{2}, r = R + a\right)}{\pi R^2 u_b},$$

where the tangential velocity is given by

$$u_\theta\left(\theta = \frac{\pi}{2}, r = R + a\right) = \frac{1}{r \sin\theta}\frac{\partial\psi}{\partial r} \approx \frac{3}{2}u_b \text{ for } a \ll R.$$

It is seen that the tangential velocity is higher than the approach velocity; the velocity must increase for the melt to get around the bubble:

$$\eta = \frac{2\pi R a_{\frac{3}{2}}^{\frac{3}{2}} u_b}{\pi R^2 v_b} = \frac{3a}{R}.$$

Thus, for the interception mechanism, the definition (5.10) may be replaced by

$$\eta = \frac{\begin{array}{c}\text{Circumference of the bubble normal to the direction of flow} \times \text{Particle radius} \\ \times \text{Velocity in the direction of flow along the circumference}\end{array}}{\text{Volume flow of melt approaching bubble}} \qquad (5.13)$$

Note that the efficiency is inversely proportional to bubble size, R. The interception mechanism is often augmented by other mechanisms such as the removal by gravity or buoyancy. This will be discussed later. For a filter, settling may even be a dominating effect.

The number of particles that collide with spherical bubbles per unit time is

$$\dot{V}_{e}c = \eta \frac{A}{4} u_{br} c,$$

where c is the number of particles in unit volume of the melt. $A/4$ is the projection of the bubble on the area normal to the relative velocity between bubble and inclusions. From this one can relate η to the coefficient for transfer of particles, k_t, introduced previously. k_t is defined so that the number of particles transferred to the bubble surface A is $k_t A c$. This is equal to $\dot{V}_e c$. Then by comparison with the previous expression, one finds that

$$k_t = \frac{\eta u_{br}}{4} \qquad (5.14)$$

Here u_{br} is the particle velocity relative to the bubbles. Particles only collide by the relative movement, μ_{br}, between particles and spherical bubbles. This is the reason for the factor $\frac{1}{4}$. If the bubble is not spherical, the factor $\frac{1}{4}$ in Eq. (5.14) must be replaced by the factor b equal to the ratio between the projected area normal to the relative velocity between bubbles and inclusion and the surface area at the bubble.

By analogy, relations derived for mass transfer of a dissolved component also apply to flotation if the mass transfer coefficient k_t is replaced by $\eta \mu_{br}/4$ and the concentration in moles/m^3 is replaced by the number of inclusions per m^3. The results are also utilized to describe what happens to inclusions in completely mixed melts.

The concept of collision efficiency is also very useful in order to describe filtration. There is considerable interest in comparing the relative merits of filtration and flotation. A discussion of various collision mechanisms is presented in Section 5.6 on filtration.

Returning to the results of using argon gas injection through porous plugs into the bottom of a teeming ladle by steelmakers so as to keep steel superheat temperatures roughly stable throughout emptying its contents into the tundish, they found that their total oxygen readings (which was, and still is, used as a quantitative measure for determining inclusion levels within gas-bubbled steels intended for teeming into continuous casting machines) were very much lower than in those steels intended for ingot casting! This meant that the steel was much cleaner with gas bubbling than with no gas bubbling, much to steelmakers' astonishment at that time.

5.5.2   Removal of Inclusions by Flotation (Bubbles)

A final way of removing the smaller solid inclusions from liquid steel is to make use of their non-wetting behaviour in liquid steel. Thus, many inclusions such as Al_2O_3, CaO, and MgO have contact angles greater than 90° in liquid steel. As such, they will potentially attach to rising bubbles and be carried up to join the surface slag. This explains why, for instance, an aluminium-killed steel contained in a teeming ladle destined for the continuous caster, using bottom gas bubbling through a porous plug, was able to report total oxygen levels significantly lower, as just mentioned, than oxygen levels in an Al-killed steel destined for teeming in an ingot aisle, for which bottom bubbling was not necessary.

The chances of an Al_2O_3 inclusion agglomerate attaching to a bubble's surface depends on a number of factors: (1) the local number concentration of inclusions within the liquid steel, (2) the relative velocity between bubbles and inclusions, (3) the probability of an inclusion becoming attached by adsorption to a bubble, and (4) the local number concentration of bubbles.

Our computational models using CFD clearly demonstrate that for a given volume of gas that is available, it is far better to use microbubbles rather than the larger bubbles normally present in steelmaking bubbling systems, in order to remove these inclusions (see Figs. 5.14 and 5.15). So the problem is therefore how to generate microbubbles in liquid metals in order to strip inclusions from a melt, or alternatively, how to generate much larger spherical cap bubbles for bulk stirring of a liquid metal in a ladle. Similarly, one must be aware of how to form intermediate sizes of bubbles and what the special properties of a liquid metal play

(a)   (b)

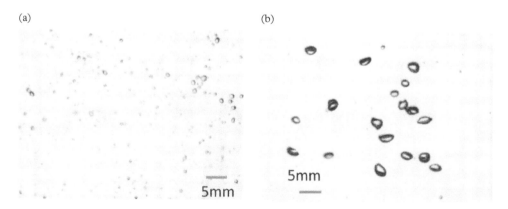

Fig. 5.14 *Two regimes of bubble formation: (a) micro-bubbles for U = 1.5m/s, (b) mm-sized bubbles for U = 0.5m/s.*[25] *(Reprinted by permission from Springer Nature: copyright 2017.)*

in determining bubble sizes. For these answers, one must study the voluminous literature on bubble formation in liquids.

Thus, two regimes of bubble formation are possible for steel flow transfer systems: (a) small bubbles less than 1 mm are formed when the shearing water speed is higher and the air injection rate is lower; (b) big bubbles greater than 3 mm are formed when the water speed is lower and gas injection rate is higher.[25]

5.5.3   Removal of Inclusions Using Microbubbles

Thus, the formation of microbubbles is indeed possible in liquid metal systems, provided one has the following elements in place: a *rapidly shearing flow system, strong convective flows, a dispersed flow of bubbles, plus an abundance of kinetic energy of turbulence.* The problem of how to produce a dispersed cloud of microbubbles within liquid steel or aluminium, such that they do not coalesce with one another, allows us to conclude that microbubbles are not possible using submerged gas injection through a porous plug placed at the bottom surface of a tundish. The most likely place for success is to inject microbubbles into the top of the ladle shroud sitting above a tundish, rather than into the SENs of individual moulds. That will allow us to make use of the tundish so as to collect all sizes of inclusions, down to ~20 microns, by allowing them to float up to the surface on inclusion-laden microbubbles. Residence times for liquid metals within the tundish are about 7 to 14 minutes. CFD computations have predicted this is the residence time for bubbles down to 500 μm, without being entrained in the exit flows to the casters.

So, two regimes of bubble formation were observed for flow through a small-scale simulation of our full-scale ladle shroud system:

(1)   When the water speed was higher (1.5 m/s) and the air injection rate was lower (0.05 L/min), then small bubbles were generated as shown in Fig. 5.14a;[25] most of these bubbles were less than 1 mm in diameter; and

(2)   When the water speed was low (0.5 m/s) and the air injection rate was higher (0.25 L/min), mostly big bubbles, larger than 2 mm, would be ejected into the water issuing from the ladle shroud, as shown in Fig. 5.14b.

We see from Fig. 5.15 that the consequences of this are profound in that we have only 100 bubbles entrained in the outflow at a velocity of 0.5m/s, but 1 800 bubbles in the same outflow of water from the ladle shroud, at 1.5m/s! Figure 5.16 plots the experimental data versus the predictions by Marshall et al.[26] for upward gas flow into a horizontal flow of water, according to the semi-empirical relation

$$R_b = 0.48 R^{0.826} \left(U_{air} / U_{liquid} \right)^{0.36} \quad \text{where } U_{air} = Q_{air} / \pi R^2_{orifice}. \quad (5.15)$$

As seen, bubble size from this relation agrees well with the ladle shroud result, but quite a number of the bubble sizes that were recorded were somewhat lower. We believe that these were due to a second mechanism at work. Thus, for the 23.8% slide gate opening (the

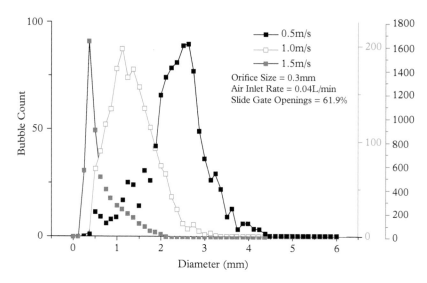

Fig. 5.15 *Number of bubbles vs the diameter of an equivalent volume spherical particle, D_e.*[25] *(Reprinted by permission from Springer Nature: copyright 2017.)*

black dots in Fig. 5.16), there were usually strong spatial velocity variations. This leads to the production of strong kinetic energy of turbulence (TKE) and an associated high rate of dissipation of the kinetic energy of turbulence. We can conclude that the smaller bubbles were the result of subsequent break-up processes after being formed. It also explains our findings that as the distance of gas injection below the slide gate nozzle increases, so are bubble sizes larger, since TKE values are lower.

These findings were then applied in the full-scale water model of our typical ladle–tundish–mould system used within the steel industry for the continuous casting of steel billets (a four-strand, 12-tonne, billet caster tundish, located at QIT, Sorel, Tracy, Quebec). By again using very small orifice diameters set just below the slide gate nozzle, and low gas flowrates, we were able to successfully reproduce the microbubbles of gas.[27]

Similarly, by using our ESZ system contained in the APS III (aqueous particle system) and monitoring both bubble size distributions and the residual ratios of inclusions in water, we were able to target smaller sized inclusions quite successfully. These findings are shown in Fig. 5.17b, where we show from more recent research that the removal of very small, hollow glass, microspheres in the sub-50-micron category by microbubbles in water is possible.

Further work in optimizing this system, plus new research using liquid metal experiments, is required to determine whether this approach can finally resolve the issue of removing the smaller inclusion agglomerates from liquid steel, prior to casting, and thereby allow us to usher in a new level of cleanliness for commercial quality steel.

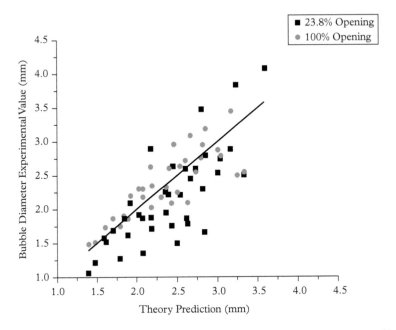

Fig. 5.16 *Experimental vs predicted spherical diameters of equivalent volume spheres, D_e.[25] (Reprinted by permission from Springer Nature: copyright 2017.)*

A practice similar to steel, based on the idea of stagnant melts, was also in place for liquid aluminium extracted from the Hall–Héroult aluminium cells and emptied into the top, gas-fired, aluminium holding furnaces in the Sagueney Smelter Plant of ALCAN in Quebec. These were, and still are, used for reheating, alloying, and cleaning the metal of inclusions. These holding furnaces are then slowly tilted so as to pour the molten aluminium metal carefully out from the so-called top layers of the supposedly 'cleanest' liquid aluminium into the launders connecting with the direct chill (DC) casters during continuous casting operations.

Given that many inclusions (Al_2O_3, Al_4C_3, $Al_2O_3 \cdot MgO$, TiB_2, etc.) are slightly denser while others are slightly more buoyant ($MgCl_2$, $NaCl$, $LiCl$, Na_3AlF_6,) or neutrally buoyant (e.g. Al_2O_3 films) compared to the liquid aluminium contained in these gas-fired, tilting, holding furnaces heated by flames from above, it was (and remains) common to have the free surface of aluminium in the holding furnace at a high temperature, say 750 °C, with metal at the bottom of the furnace registering about 660 °C. This was considered ideal since heavier inclusions were thought to gradually sink, leaving the cleanest metal near the surface and the dirtiest aluminium on the floor of the furnace. It was believed that strong thermal stratification was present within the furnace, rendering the melt immobile. In fact, it was argued that any stirring would prevent the aluminium metal from self-cleaning by preventing the majority of heavy inclusions from sinking to the bottom of the furnace. The fact that these furnaces had to be 'jack-hammered' out every month or so reinforced this widely held conviction.

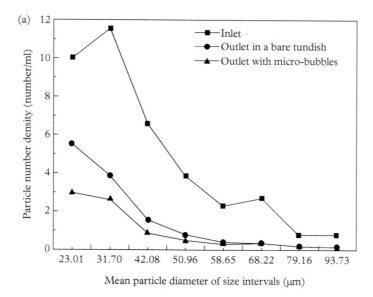

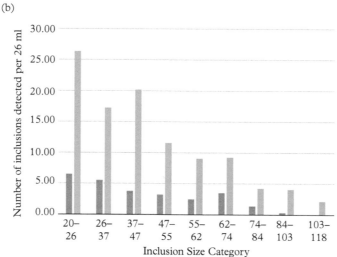

Fig. 5.17 *(a) The particle number densities for inlet, outlets in a bare tundish, and outlets with use of microbubbles. (b) Comparison of inclusions in the bottom water outlet using microbubbles, or not, to enhance removal of glass microspheres top the top outlet flow. Blue bars correspond to inclusion counts measured in the presence of microbubbles, and brown bars show bottom outlet particle flows without microbubbles. (Reprinted from Chang et al.[27] with permission from ISIJ.)*

However, following the development of the LiMCA,[4,13] which can analyse the local concentration of inclusions within a melt, it was conclusively demonstrated that turbulent natural convective heat losses within the liquid metal to the side walls of these holding furnaces lead to turbulent natural convection currents being generated within the melt; i.e. the melts were not stagnant as had been assumed. This, in turn, lead to the realization that inclusion levels within their holding furnaces were, in fact, 'well mixed' for slowly moving inclusions. As such, the top pouring concept for emptying these holding furnaces, where the 'cleanest metal' was located, is essentially incorrect.

It follows that the expensive tilting equipment for such holding vessels was, and remains, largely redundant in ensuring that clean aluminium is delivered to the DC continuous casters via the molten metal emptied into them via the launder system. It is more a consequence of good melt shop practices for maintaining metal cleanliness.

5.6   Introduction to Filtration

Filtration can be an effective means for removing inclusions, provided there are not too many of them to clog the filter devices. Unfortunately, commercial liquid steel is still relatively 'dirty', so that very fine filters whose pores are smaller than the inclusions themselves will rapidly saturate and cease to function. The solution is to rely on deep bed filtration, in which the inclusion particles become trapped within the pores, moving to more slowly moving parts of the filter bed. There they can accumulate and reduce the number of inclusions passing into the final product. This approach works well for liquid aluminium, where the number density of inclusions, and their sizes, are relatively small compared to equivalent parameters for liquid steel.

Figure 5.18a shows a plot of the filtration efficiency for a 30-ppi ceramic foam filter (CFF) as a function of casting time in an aluminium melt shop. LiMCA data revealed that the inclusions exiting are less numerous than those entering for much of the time. However, if the filter becomes saturated with inclusions, the number of inclusions per unit volume in the liquid metal exiting the filter, N"exit, can become greater than N" entry. Furthermore, in that case, the size distributions of inclusions released moved towards larger sizes, demonstrating that inclusions were agglomerating within the CCF filter before being released. This information was not possible before the LiMCA technology became available for aluminium producers.

Mention should also be made of the older ALCAN bed filter (ABF) system. This system passes liquid aluminium upwards through a packed bed of alumina spheres at a low speed ($\sim$1–5 cm/s), under Darcian flow conditions. The collisions and adherence of inclusions to the 'surfaces' of 'sticky' spheres, as the aluminium flows within the tortuous pathways, can produce removal efficiencies on the order of 95%, compared with 80–90% for a typical, new CFF system. Similarly, the ABF system has a much longer operating life than a CFF, but suffers from a large heel of liquid aluminium metal once an ingot 'drop' has been completed. This represents a large source of scrap that needs recycling. Some typical LiMCA readings for an ALCAN ABF system operating in Quebec in the 1980s is shown in Fig. 5.18b.

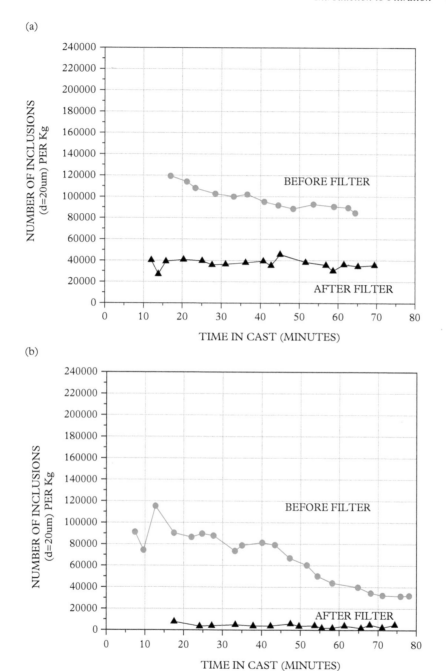

Fig. 5.18 *(a) Inclusion concentration in liquid aluminum on entry and exit into a single-use ceramic foam filter (CFF) versus time. (b) Inclusion concentration on entry and exit to the ALCAN bed filter (ABF) with direct chill (DC) casting time.*[13]

For liquid steels, many of the solid inclusions (e.g. Al_2O_3, CaO, MgO) are strongly ferro-phobic and will tend to cluster and sinter at $1600\,^{\circ}C$ to form much larger inclusions (50–350 microns) and even cause build-ups sufficient to block ladle and tundish nozzles, leading to aborted casting runs. Compared to the absolute cleanliness of liquid aluminium, liquid steel melts are far 'dirtier' and typically contain billions of inclusions per unit mass. For these reasons, the use of filters is generally impractical in continuously flowing liquid steel through ladle-tundish-mould systems, despite a number of reported attempts in the past. However, single-use filters have, and are, used in batch casting operations, such as for forming steel blanks in the production of railway wheels, or casting of 'fifth wheels' for transport trucks.

Here filtration is defined as the process of removing inclusions by forcing molten metal through a porous material. The most important characteristic of a filter is the filtration efficiency E defined by

$$E = 1 - \frac{c_0}{c_{in}},\qquad\qquad(5.16)$$

where c_{in} is the number of inclusions per unit volume that enters the filter, while c_0 is the number per unit volume that leaves.

There are at least three participants in the filtration process: the inclusions, the molten metal, and the filter material. The solid or liquid inclusions are described by their size distribution, composition, density, and shape. The properties of the fluid depend on alloy composition, kinematic viscosity, and density. The filter material is characterized by its geometrical dimensions: the size of the structural units, their distribution and arrangement in the filter, and their porosity.

The forces at the three interfaces (interfacial tensions) between the melt and filter material, melt and inclusions, and inclusions and filter play an important role. Also a gas or vacuum phase formed between the filter material and inclusion may influence the filtration properties.

The pressure drop, Δp, through a filter is important. In many applications in the aluminium industry today, the metallostatic head at our disposal is limited. Often 0.1 or 0.2 m of metal head must suffice to force the metal through the filter. Initially the pressure drop depends only on the properties of the fluid and the filter. As filtration proceeds the pressure drop becomes dependent also on the properties of particles deposited on or in the filter. If the particles are deposited on the filter it is said that we have 'cake mode' filtration. In this case, particles deposited previously act as a filter for the final particles approaching the filter. When the inclusions are deposited inside the filter we speak of deep bed filtration. If the size of the inclusions is comparable to the filter pore size, cake filtration is inevitable. Even if the inclusions are one or two orders of magnitude smaller than the pore size, the section of the filter closest to the inlet gradually fills up, so that the mechanism here more and more resembles cake mode filtration.

Filter capacity is defined as the quantity of deposited particles (usually expressed in grams or kilograms) which the filter is capable of treating before reaching a certain pressure drop (loss of metallostatic head). The filter capacity clearly depends on the type and size of the particles.

5.6.1 Cake Mode Filtration

Cake mode filtration as a 'quality assurance' is useful. For instance, in aluminium melt refining, should there be a surge in the level of inclusions this is helpful. This may be the case when metal is recycled or remelted when 'slurry' from the bottom section of a settling or holding furnace may suddenly appear downstream. Certain alloys may contain a large number of inclusions. Cake mode filtration results in the development of a superficial layer of inclusions separated on the substrate filter. Such a cake may lead to a high pressure drop. The cake is responsible for, and is also a consequence of, particle capture. The behaviour of the cake depends on the sizes and the size distribution of the inclusions, plus the pressure gradient experienced by the inclusions in the cake. Filter cakes can be broadly classified into two types: incompressible and compressible. An incompressible cake is one in which the void fraction, e, of the cake does not change within the pressure limits. Then an increase in pressure drop with time is essentially caused by increases in cake thickness. Compressible cakes typically develop when the separated solids are deformable and represent a broad size distribution.

It is important to be able to predict the pressure drop in cake mode filtration for a given alloy. Measurements of flow rates can be carried out in the laboratory at a given (constant) pressure.[28] This allows resistances for the cake, r_c, and for the filter medium, r_f, to be determined. From the Blake–Kozeny equation, valid for viscous laminar flows,[29] one obtains

$$A\frac{dp}{dz} = \dot{q}\rho v r_c \tag{5.17}$$

for the pressure gradient in the cake and similarly

$$A\frac{dp}{dz} = \dot{q}\rho v r_f. \tag{5.18}$$

for the pressure gradient in the filter. Initially the pressure drop at our disposal, Δp, is due to the filter medium only. Then the volume flow, $\dot{q}$, is given by

$$\frac{A\Delta P_f}{H} = \dot{q}\rho v r_f, \tag{5.19}$$

where H is the filter thickness.

As the cake builds up, an increasing part of the pressure drop is due to the cake. If the cake is incompressible so that r_c does not depend on dp/dz, Eq. (5.17) can be integrated to give a pressure drop over a thickness of cake z:

$$A\Delta p_c = \dot{q}\rho v r_c z. \tag{5.20}$$

Addition of the two resistances gives

$$A\Delta p = \dot{q}\rho v \left(r_f H + r_c z\right). \tag{5.21}$$

The growth rate of the cake dz/dt depends on the metal flow rate $\dot{q}$, on the content of inclusions in the melt, c (given in volume fraction), and on the volume fraction of inclusions constituting the cake, c_c. Then if it is assumed that all inclusions are captured, a volume balance gives

$$c_c A \frac{dz}{dt} = \dot{q} c. \tag{5.22}$$

From this equation it is seen that the cake thickness z for constants c and c_c is proportional to the volume of melt V that has passed through the filter:

$$z = \frac{c}{A c_c} \int_0^t \dot{q} dt = \frac{cV}{c_c A}. \tag{5.23}$$

If z is eliminated from Eq. (5.21), $\dot{q} = dV/dt$ is inserted, and the integration over time t is carried out, one obtains for V

$$A \Delta p t = V \rho v \left(r_f H + \frac{r_c c V}{2 c_c A} \right). \tag{5.24}$$

An example of the use of this equation is shown in Fig. 5.19. There the inverse average filtration rate t/V is given as a function of V. $\rho v r_f H/(A \Delta p)$ is obtained from the intercept of the straight lines with the y-axis, while the slope gives $\rho v r_c c/2A^2 \Delta p c_c$. The slope is seen to be much greater for the contaminated melt.

It may be illuminating to apply the above analysis to an industrial situation where the required head of aluminium is calculated.[28] r_f and r_c have been determined in the laboratory; for instance,

$$r_f H = 3 \times 10^7 \ m^{-1}$$
$$r_c = 2 \times 10^{11} \ m^{-2}.$$

The casting conditions are as follows: casting rate, $\dot{q}=4 \times 10^{-3} \ m^3/s$; drop size, $V = 33 \ m^3$; filter area (20" $\times$ 20") $= 0.2581 \ m^2$, $v = 0.4 \times 10^{-6} \ m^2/s$, $\rho = 2350 \ kg/m^3$, $c/c_o = 10^{-5}$.
    Then Eqs. (5.20) and (5.22) give

$$\Delta p = \frac{\dot{q} \rho v}{A} \left(r_f H + \frac{r_c V_c}{A c_c} \right) = 4.16 \times 10^3 N/m^2$$

or an aluminium head of $4.16 \times 10^3/(9.81 \times 2350) = 0.18$ m.
    This head represents a problem with the present design of cast houses. It may therefore be worthwhile taking a closer look at the resistances. The resistance in the cake, $r_c V c/(A c_c)$, dominates that in the filter, $r_f H$. This contrary to expectations, since the thickness of the cake $V c/(A c_c)$ is only about 1 mm compared to $H = 50$ mm. The explanation is that resistance in the cake is governed by inclusion size, while the resistance in the filter depends on the size of

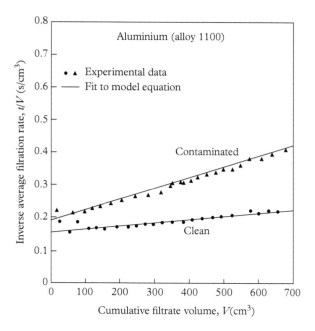

Fig. 5.19 *Inverse average filtration rate, t/V, as a function of the cumulative filtrate volume, V, for both clean and contaminated aluminium (1100 alloy) melts.*[28]

the pores. Thus, resistance in the cake must be high, as may be seen from the Blake–Kozeny[29] equation for a packed bed,

$$r_c \approx \frac{150(1-\varepsilon)^2}{d_p^2 \varepsilon^3}, \tag{5.25}$$

where d_p^2 is the average squared particle diameter and ε is the porosity. For instance, for a cake with inclusions of diameter $d_p = 30\mu m$ and a porosity of $\varepsilon = \frac{1}{2}$

$$r_c = \frac{150 \times 0.5^2}{\left(30 \times 10^{-6}\right)^2 0.5^3} = \frac{30}{9} \times 10^{11} \ m^{-2}.$$

This happens to be close to the value used in the previous calculation of pressure drop. In a ceramic foam filter, pore size is usually in the range 500–1000 μm. Thus, the pressure gradient is several orders of magnitude less than that for the cake.

This section is concluded by stating that cake filtration very effectively removes practically all inclusions since the inclusions are their own filter. The voids between the inclusions are small enough to trap all but the smallest particles in the size distribution. Some disadvantages are that particles may escape during the early stages, since it takes some time to build up a cake. Inclusions may be re-entrained in the melt from the 'back-side' of the cake, and the filter capacity is limited due to a rapid increase in the pressure drop with flowing time.

5.6.2   Deep Bed Filtration

In Section 5.6.1, the problems with cake mode filtration were outlined: a high pressure drop and a limited filter capacity. The reasons were that the pressure drop is caused essentially by the small inclusions, deposited only on the inlet surface of the filter.

If particles can be deposited on the walls inside a filter with collector dimensions much larger than the particles, the pressure drop can then be drastically reduced. Also, if the total surface area within the filter is much greater than the inlet surface, filter capacity is again much improved. To attain satisfactory deep bed filtration two or three points must be considered. First, cake mode filtration should only occur to a very limited extent. Second, the inclusions must in fact be deposited within the filter. Third, inclusions may protrude into the melt from the collector surface. Such 'dendrites' will give a significant increase in flow resistance.

To prevent cake mode filtration, the dimensions of the inclusions must be much smaller than the filter pore size. This means that ideally the melt should not contain large collections of inclusions or oxide films which may tend to choke up the filter inlet. Large inclusions and films should therefore be removed, ideally, in a previous separation stage.

The previous separation unit may be a settling furnace, a flotation unit, or a filter with large pores and a high filter capacity. When flotation is employed, it is possible to skim off the inclusions contained in a dross or slag layer. Both settling and flotation remove the larger inclusions. Gravity and inertia are the important mechanisms in this removal. When the density differences are small, it is important not to stir the melt too much. That would again disperse the inclusions after they had been removed to the bottom or top layers of the melt. In steel, where density differences are large, inclusions can be eliminated by the action of stirring and turbulence to be transferred to walls and top-slag (see Appendix 2).

It may be said that the larger inclusions (greater than 50 μm) are easily removed if reasonable measures are taken. Smaller inclusions (less than 1 μm) present no problem for metal quality currently. It is the intermediate size inclusions, say from 5 to 20 μm, that cause difficulty for aluminium melts. They are small enough to elude capture, but sufficiently large to damage the end metal product. In the rest of this chapter, priority is given to the study of the removal of this intermediate size of inclusions.

It will be seen that to assure the removal of intermediate size inclusions within a filter, a large inner surface area is required. To give a large surface area, the unit collector dimensions must be small. A balance must be struck, which depends upon the inclusion size: filter pore dimensions should be large enough to avoid cake build-up and small enough to capture inclusions down through the filter. Thus, in *sandwich* type filters, the inlet section has larger pores than the outlet part.

To describe the removal of inclusions in a filter, it is assumed initially that once the inclusions touch the collector, they adhere. A number balance over a differential slice of a filter (control volume) gives (see Fig. 5.20)

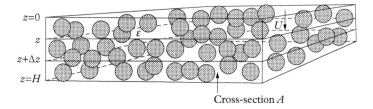

Cross-section A

Fig. 5.20 *A differential slice (control volume) in a filter comprising spheres (e.g. Al₂O₃ balls).*

flow of inclusions through the surface at position z = flow of inclusions through the surfaceat position $(z + \Delta z)$ + number of inclusions transferred to the collectors in unit time

or

$$c(z)Au\varepsilon = c(z+\Delta z)Au\varepsilon + k_t a_s(A\Delta z)\varepsilon c, \qquad (5.26)$$

where a_s is the surface area per unit volume of melt, and u is the mean particle velocity in the direction of z. $A\Delta z$ is the control volume and $A\Delta z\varepsilon$ is the melt volume. k_t is the coefficient for transfer of particles derived previously in Eq. (5.14) for the analogous case of transfer to bubbles,

$$k_t = \eta u b, \qquad (5.27)$$

where u is an average velocity of the particles in the voids between the collectors, and the approach velocities and tangential velocities to the collectors are taken to be the same. Finally, b, as defined previously, represents the ratio of the collector surface projected in the flow direction to the total collector surface area.

The collision efficiency for a collector in a filter is defined as previously in Eq. (5.9) for a bubble, except that the word collector replaces bubble in the definition. For instance, Eq. (5.13) for the interception mechanism gives

$$\eta = \frac{2\pi Rau}{\pi R^2 u} = \frac{2a}{R} \qquad (5.28)$$

for a packed bed. However, in fact, the velocity boundary layer around the solid collector must be taken into account. Velocities are much less than u in this layer. Therefore Eq. (5.28) strongly overestimates the value of η for inclusions smaller than the boundary layer.

If Eq. (5.27) is inserted into Eq. (5.26), and $\Delta z \to 0$, one obtains the differential equation

$$0 = \frac{dc}{dz} + \eta a_s bc \qquad (5.29)$$

or on integration

$$\frac{c(z)}{c_{in}} = \exp(-\eta a_s bz), \qquad (5.30)$$

where c_{in} is the number of inclusions per unit volume of melt that enters the filter.

The calculation has been carried out for the one-dimensional case. However, the derivation applies also in two or three dimensions if z is replaced by the arc length, s, of a stream-tube (a tube where the volumetric flow is the same throughout). Strictly, in this case η depends on u and thereby on s so that

$$\frac{c(s)}{c_{in}} = \exp\left(-a_s b \int_0^s \eta(s)ds\right).$$ (5.31)

In the one-dimensional case if the filter depth is H[30]

$$\frac{c}{c_{in}} = \exp(-\eta a_s bH).$$ (5.32)

This gives a filtration efficiency:

$$E = 1 - \frac{c}{c_{in}} = 1 - \exp(-\eta a_s bH).$$ (5.33)

For a void fraction ε and a spherical collector diameter of $2R$, the specific surface area becomes

$$a_s = \frac{\text{Surface area of collector}}{\text{Volume of melt}} = \frac{\text{Surface area of collector}}{\text{Volume of collector}}$$

$$\times \frac{\text{Volume of collector}}{\text{Volume of melt}} = \frac{4\pi R^2}{\frac{4\pi R^3}{3}} \times \frac{1-\varepsilon}{\varepsilon} = \frac{3(1-\varepsilon)}{R\varepsilon}$$ (5.34)

Then for spheres (where $b = \frac{1}{4}$)

$$E = 1 - \exp\left(-\frac{3\eta(1-\varepsilon)H}{4R\varepsilon}\right).$$ (5.35)

Previously, for interception on an isolated spherical bubble, we have found $\eta = 3a/R$. However, for spheres in a packed bed, we have set $\eta = 2a/R$:

$$E = 1 - \exp\left(-\frac{3a(1-\varepsilon)H}{2R^2\varepsilon}\right).$$ (5.36)

For instance; for $a = 5$ μm $= 5 \times 10^{-6}$ m, $R = 0.5$mm$= 0.5 \times 10^{-3}$ m, $H = 0.05$ m, $\varepsilon = 0.5$, these numbers give $E = 1 - 0.22 = 0.78$. As such, the concentration at the outlet is predicted to be reduced to 22 per cent of the inlet concentration.

It is seen that the exponent in Eq. (5.36) is proportional to the inclusion size a and inversely proportional to the square of the collector dimensions. Thus, the efficiency improves strongly with decreasing collector pore dimensions. On the other hand, the pressure drop also increases significantly.

5.6.3 Ceramic Foam Model

The performance of a CFF can be derived from models of the interaction between a particle in the liquid and a single collector. It is necessary to take the boundary layer into account. Most of the models assume that the metal flows through a tube, spheres, or in cells.[31,32] However, a closer look indicates that it is more realistic to regard the filter as a network of branches or struts (cylinders), with the metal flowing around the cylinders.

When considering the stream function for the boundary layer[35] on a cylinder axis, we obtain

$$\Psi(y,\theta) = \sqrt{\frac{v}{\frac{2u}{R}}} \left[\frac{2u}{R} (R\theta) f_1(\lambda) + 4\frac{-u}{3R^3} (R\theta)^3 f_3(\lambda) \right.$$

$$\left. + 6\frac{u}{60R^5} (R\theta)^5 f_5(\lambda) + 8\frac{-u}{2520R^7} (R\theta)^7 f_7(\lambda) \ldots \right]$$

$$= \sqrt{\frac{vuR}{2}} \left[2\theta f_1(y) - \frac{4}{3}\theta^3 f_3(y) + \frac{1}{10}\theta^5 f_5(y) - \frac{1}{315}\theta^7 f_7(y) \ldots \right],$$

where R is the radius of the cylinder, u is the approach velocity of the fluid normal to the axis, and v is the kinetic viscosity.

$$\lambda = \frac{y}{R}\sqrt{Re_c}$$

and

$$Re_c = \frac{2u_\infty R}{v}.$$

The functions $f_1, f_3, f_5,$ and f_7 from Schlichting[33] are given in Appendix 3 The collision efficiency due to interception according to Eq. (5.10) becomes

$$\eta_{in} = \frac{2L_c\Psi(y=R_P,\theta)}{2uRL_c} = \frac{\Psi(y=R_P,\theta)}{uR} = \frac{2\theta}{\sqrt{Re_c}}$$

$$\times \left\{ f_1\left(\frac{R_P}{R}\sqrt{Re_c}\right) - \frac{2}{3}\theta^2 f_3\left(\frac{R_P}{R}\sqrt{Re_c}\right) + \frac{\theta^4}{20}f_5\left(\frac{R_P}{R}\sqrt{Re_c}\right) - \frac{\theta^6}{630}f_7\left(\frac{R_P}{R}\sqrt{Re_c}\right) \ldots \right\},$$

$$(5.37)$$

where R_p is the radius of a spherical inclusion, L_c is the length of the cylinder, and θ is the collection angle, as illustrated in Fig. 5.21.

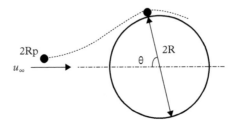

Fig. 5.21 *Two-dimensional view of a spherical inclusion in a stream approaching normal to a cylindrical collector.*

The collision efficiency increases with θ until the maximum collision efficiency is attained at $\theta = \theta_c$, given by

$$\frac{\partial \eta_{in}}{\partial \theta} = \frac{2}{\sqrt{Re_c}} \left\{ f_1\left(\lambda_c\right) - 2\theta^2 f_3\left(\lambda_c\right) + \frac{\theta^4}{4} f_5\left(\lambda_c\right) - \frac{\theta^6}{90} f_7\left(\lambda_c\right) \dots \right\} = 0 \qquad (5.38)$$

with

$$\lambda_c = \frac{R_P}{R}\sqrt{Re_c}$$

or

$$2f_7\theta^6 - 45f_5\theta^4 + 360f_3\theta^2 - 180f_1 = 0. \qquad (5.39)$$

This equation is solved graphically by equating the left- and right-hand sides of the equation.[34] The solution for Eq. (5.39) is given in Fig. 5.22. Inserting the collection angle θ_c in Eq. (5.37), we get the solution for Eq. (5.39)

$$\eta_{in} = 0.65\lambda_c^{1.73} / \left(Re_c^{0.5}\right). \qquad (5.40)$$

Equation (5.40) can also be written as

$$\eta_{in} = 0.65\left(R_p/R\right)^{1.73}\left(Re_c^{0.365}\right). \qquad (5.41)$$

It is seen that the collision efficiency η_{in} strongly depends on the ratio between particle size and cylinder size R_p/R. Also for a given particle size, η_{in} decreases strongly with cylinder radius R. There is a moderate increase of η_{in} with the cylinder Reynolds number, Re_c. When streamlines are pressed together in flow around the cylinder, particles in the liquid come closer to the walls and collide with the surface.

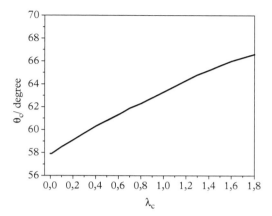

Fig. 5.22 *The maximum collection angle for collision by interception.*

The cylinders in the filter are generally not normal to the flow. It is assumed that the cylinders are oriented randomly relative to the velocity vector. For the flow normal to a cylinder, Eq. (5.41) gives the collision efficiency. β is the angle that the normal to the velocity makes with the cylinder axis. Following Schlichting,[35] we assume that when the cylinder axis makes an angle β with the normal to the flow, the velocity $u\cos\beta$ determines the boundary layer behaviour (independence principle):

$$\eta_{in}(\beta) = 0.65(R_p/R)^{1.73}(\mathrm{Re}_c\cos\beta)^{0.365}.\tag{5.42}$$

An average collision efficiency is determined:

$$\eta_{in-avg} = \frac{2}{\pi}\eta_{in}\int_0^{\frac{\pi}{2}}\cos^{0.365}\beta\,d\beta = 0.8109\eta_{in}.\tag{5.43}$$

Palmer et al.[36] have studied the capture of suspended particles with a silicone grease-coated single cylinder. Figure 5.23 shows the collison efficiency η_{in} calculated with Eq. (5.43) compared with Palmer's experimental data. The collison efficiency in our calculation is slightly higher than the experimental results, except in experiments 1 and 7. Two effects may explain: (1) not all the particles stick on the wall; and (2) the hydrodynamic effect, lift force normal to cylinder surface, may push particles away from the cylinder.

Although the collision efficiency is very small for a single cylinder, the resulting filtration efficiency may be high, for inclusion collision with a large number of branches (cylinders). A total of 2.5 mm of silicone grease coating was included in our calculation: $R' = R + 0.25$ cm. Gravitational collision is neglected due to the use of a vertical cylinder: $\rho_p = 1.03\rho_l$.

Another collision mechanism that is, or may become, important is gravity. Also hydro-dynamic forces may play a role in practice. For the removal of inclusions larger than 1 μm from metals, it seems that Brownian diffusion plays only a minor part.[37,38] Various

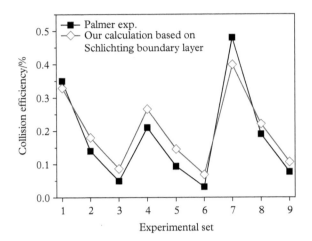

Fig. 5.23 *Collision efficiency on a single cylinder for Palmer's case for neutrally buoyant particles.*

collector mechanisms are illustrated in Fig. 5.12. Inertial forces are important for the removal of particles (aerosols) from gases, but usually not in the filtration of molten metals. The reason is essentially that the ratio of particle density to fluid density is a factor 1/1000 smaller for metal systems. The inertial force of the particles relative to the viscous force is small. Thus, they do not penetrate the boundary layer surrounding the collector. Only for velocities above 0.1 m/s and inclusion sizes of 100 μm or more do inertial forces become important.[38]

If the particle density is different from that of the metal, particles will settle out in the direction of the gravitational force. To determine the collision efficiency due to gravity or buoyancy, the velocity difference between fluid and particle (inclusion), u_r is introduced. In Fig. 5.24 u_r is taken in the downwards direction. The fluid velocity is in the horizontal direction. To discuss this case, Eq. (5.9) is employed, except that the word 'bubble' is replaced by the word 'collector'. Then, if the vertical projection of the collector is A_v and the projection on the surface normal to the mean particle flow direction is A,

$$\eta_r = \frac{A_v u_r}{Au}. \tag{5.44}$$

For a sphere $A_v = A$ and, the gravitational collision efficiency due to relative velocity becomes

$$\eta_r = \frac{u_r}{\left(v^2 + u_r^2\right)^{\frac{1}{2}}} \approx \frac{u_r}{v}. \tag{5.45}$$

This collection efficiency for the effect of gravity is often presented in the literature. Note that v is the relative velocity between the melt and collector, while u_r is the relative velocity between particles and the melt. Equation (5.45) applies only when v and u_r are in the horizontal and vertical directions, respectively. Other configurations are left as an exercise for the reader.

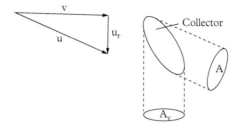

Fig. 5.24 *The vertical projection A_v and the projection on the surface normal to the path of particle A are needed to calculate the collision due to gravity or buoyancy.*

If particles are not spherical but elongated, irregular, or formed as plates, they will not follow the streamlines. They may wobble or tumble,[24,39] and this should give a higher collision efficiency. These phenomena are called hydrodynamic effects. They are difficult to describe and no method for the calculation of η is proposed for this case.

The terminal settling velocity of a particle is determined by the balance between the opposing forces of gravity and drag which give

$$U_S = \frac{2\left(\rho_p - \rho_l\right)gR_p^2}{9\rho_l \nu} \tag{5.46}$$

for particles with relatively small Reynolds number.[40] See also Eq. (5.3).

For instance, for alumina particles (ρ_p = 3.97 g/cm³, R_P = 10 μm) in aluminium metal (ρ_l = 2.37 g/cm³), the settling velocity is 0.28 mm/s. It takes 60 min to sink 1 m in a tank of stagnant aluminium.

To obtain the collision efficiency due to gravity η_g, it is assumed that the surfaces of the ceramic foam filters are randomly oriented relative to gravity. For inclusions heavier than the molten metal, only surfaces facing upwards capture inclusions. If γ is the angle between the normal to the surface and gravity, then the effective fraction of the surface in the gravity direction, f, is given by

$$f = \frac{1}{\pi} \int_0^{\pi/2} \cos\gamma \, d\gamma = \frac{1}{\pi}. \tag{5.47}$$

Although the velocity should be lower than the Stokes settling velocity U_s in Eq. (5.46) near a surface, we employ U_s to estimate the maximum removal due to gravity. Then

$$\eta_g = \frac{U_s}{U_\infty} \cdot \frac{1}{\pi}. \tag{5.48}$$

Removal from the eddies behind the cylinders is a complex problem and is not taken into account here. Gravitational collision dominates at the superficial metal velocity 11 to 12 mm/s for 40- to 60-μm inclusions studied in Tian and Guthrie's case.[41]

It is often necessary to include more than one mechanism in order to determine the total collision efficiency, η_t. For instance, often both interception and gravity should be taken into account. Then for $\eta = \ll 1$ we assume that the various mechanisms are independent of each other and regard the η as probabilities. The probability that a given particle approaching a collector does not collide with it is given by

$$1 - \eta = \left(1 - \eta_{in-avg}\right)\left(1 - \eta_g\right), \tag{5.49}$$

where $1 - \eta_{in-avg}$ is the probability that it does not collide due to interception. $1 - \eta_g$ is the probability that collision is not caused by gravity. The equation overestimates the total collision efficiency since sometimes the same particle may be removed by both mechanisms. For η_{in} and $\eta_g < 0.1$ Eq. (5.49) simplifies to[37]

$$\eta \approx \eta_{in-avg} + \eta_g. \tag{5.50}$$

The cylinders in the filter media are in general not normal to the flow direction. We consider cylinders which show an angle β, in the range of $-90°$ to $90°$, normal to the flow. The average cylinder surface area projected on the surface normal to the flow direction is

$$\frac{\int_{-\frac{\pi}{2}}^{\frac{\pi}{2}} 2RL_c \cdot \cos\beta d\beta}{\int_{-\frac{\pi}{2}}^{\frac{\pi}{2}} d\beta} = \frac{2}{\pi} \cdot 2RL_c \tag{5.51}$$

Then, the geometry factor b is equal to

$$b = \frac{\text{Collector surface are a projected normal to the flow direction}}{\text{Surface area of collector}} = \frac{2RL_c \cdot \frac{2}{\pi}}{2\pi RL_c} = \frac{2}{\pi^2} \tag{5.52}$$

There are various reported geometrical models, which describe ceramic foams as a collection of cells, polyhedral, etc. The specific surface area α_s is summarized in Table 5.1, where the structural parameter d_s is the strut (cylinder) diameter in the models.

The models from Innocentini et al.,[43] Richardson et al.,[44] Moreira et al.,[47,48] Lu et al.,[42] Bao et al.,[34] and Giani et al.[49] (in sequence from high to low specific surface area) give specific surface area higher than models by Lacroix et al.,[50] Fourie and Plessis,[45] and Buciuman and Kraushaar-Czarnetzki,[46] regarding the ceramic foams as polyhedral (12 or 14 faces). See Fig. 5.25 for open-cell ceramic foams.

The Buciuman and Kraushaar-Czarnetzki model[46] seems to give the best representation of the ceramic filter and is employed here. Of the various models, it gives the lowest specific surface area, as shown in Fig. 5.25.

Taking into account Eq. (5.52) and specific surface area from the Buciuman and Kraushaar-Czarnetzki model in Eq. (5.33), the filtration efficiency for a ceramic foam filter becomes

Table 5.1 *The ceramic foam models*

Reference	Model	Structural parameter	Specific surface area [m^2/m^3]
Lu et al. (1998)[42]	Open-celled metal foams are treated as cylinders.[a]	$d_s = d_{pore}\left(\frac{2}{\sqrt{3}\pi}\right)(1-\varepsilon)^{1/2}$	$a_s = \left(\frac{2\sqrt{3}\pi}{d_{pore}}\right)(1-\varepsilon)^{1/2}$
Innocentini et al. (1999)[43]	Ceramic foams are treated as cells. diameter = cylindrical form of the hydraulic cell diameter.	—	$a_s = \frac{4\varepsilon}{d_{pore}(1-\varepsilon)}$
Richardson et al. (2000)[44]	The pores are treated as uniform, parallel cylinders, each with a constant diameter.	$d_s = \frac{0.5338 d_{pore}(1-\varepsilon)^{1/2}}{1-0.971(1-\varepsilon)^{1/2}}$	$a_s = \frac{4\varepsilon}{d_{pore}(1-\varepsilon)}$
Fourie and Plessis (2002)[45]	Cellular metallic foams are treated as an array of tetrakaidekahedra.[b]	$d_s = d_{pore}\left(\frac{2}{3-\chi}-1\right)$	$a_s = \left(\frac{3}{d_{pore}+d_s}\right)(3-\chi)(\chi-1)$ with $\chi = 2 + 2\cos\left(\frac{4\pi}{3} + \frac{1}{3}\cos^{-1}(2\varepsilon-1)\right)$
Buciuman and Kraushaar-Czarnetzki (2003)[46]	Open-cell ceramic foams are treated as a package of tetrakaideka.[c]	$d_s = d_{pore}\left[\frac{1-\varepsilon}{2.59}\right]^{1/2}$	$a_s = 4.82\frac{(1-\varepsilon)^{1/2}}{d_{pore}+d_s}$

(Continued)

Table 5.1 *Continued*

Reference	Model	Structural parameter	Specific surface area [m²/m³]
Moreira et al. (2004)[47,48]	Ceramic foam is treated as cellular structure. The same as Richardson et al.[44]	—	$a_s = \dfrac{12.979[1-0.971(1-\varepsilon)^{1/2}]}{d_{pore}(1-\varepsilon)^{1/2}}$
Giani et al. (2005)[49]	Open-celled metal foams are treated as cubic cells. The same as Lu et al.[42]	$d_s = d_{pore}\left[\dfrac{4}{3\pi}(1-\varepsilon)\right]^{1/2}$	$a_s = \dfrac{4}{d_s}(1-\varepsilon)$
Lacroix et al. (2007)[50]	SiC foam bed is treated as the dodecahedra.[d]	$d_s = \dfrac{d_{pore}\left[\frac{4}{3\pi}(1-\varepsilon)\right]^{1/2}}{1-\left[\frac{4}{3\pi}(1-\varepsilon)\right]^{1/2}}$	$a_s = \dfrac{4}{d_s}(1-\varepsilon)$

[a] Reprinted from Lu et al.[42] with permission from Elsevier.
[b] Reprinted from Fourie and Plessis[45] with permission from Elsevier.
[c] Reprinted with permission from Buciuman and Kraushaar-Czarnetzki.[46] Copyright (2003) American Chemical Society.
[d] Reprinted from Lacroix et al.[50] Copyright (2007), with permission from Elsevier.

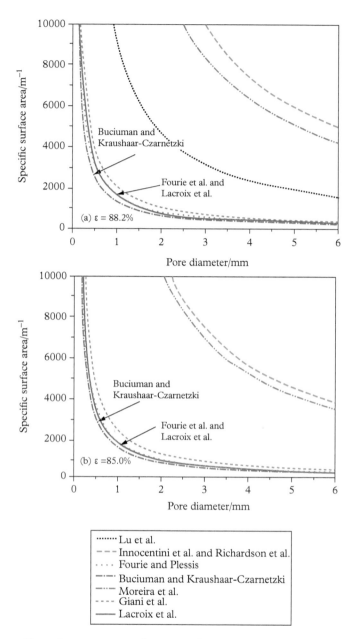

Fig. 5.25 *The specific surface area vs pore diameter with various models.*

$$E = 1 - \exp\left(-\eta H \cdot 4.82 \frac{(1-\varepsilon)^{\frac{1}{2}}}{d_{pore}\left(1+\left[\frac{1-\varepsilon}{2.59}\right]^{\frac{1}{2}}\right)} \cdot \frac{2}{\pi^2}\right), \tag{5.53}$$

where d_{pore} is the filter pore diameter and can be measured from the filter. ε is the filter porosity. Now, if the collision efficiency is known in Eq. (5.42) or (5.43) considering only interception, or in Eq. (5.50) taking into account interception and gravity, the filtration efficiency can be determined.

Bao et al. have carried out industrial scale Al filtration test with alumina and SiC ceramic foam filters. The filtration efficiencies calculated according to Eqs. (5.53) and (5.50) are shown in Figs. 5.26 to 5.29. The particle (inclusion) size is the initial size measured by LiMCA. Both collision efficiencies and filtration efficiency increase strongly with particle size. Note that it is assumed that inclusions stick to the walls on collision. However, probably adhesion effects should be considered.

The filtration efficiency will be a function of both transport efficiency (collision efficiency, η) and adhesion efficiency, q. We set $q = 1$. For particles less than 40 μm (45 μm for Exp. 4), the correspondence between the model and experiments is very good for Exps. 1 and 3 with Al_2O_3 filters. SiC filters give higher experimental results than calculated. This may indicate that SiC filters have better adhesion efficiency than Al_2O_3 filters. This also means that effects of the wettability (adhesion) of the filter and the metal should be considered. Few particles larger than 40 μm were measured by LiMCA in this pilot test.

The wettability of pure Al_2O_3–Al and pure SiC–Al has been systematically studied in the thesis of Bao.[51] It is concluded that the SiC–Al adhesion work reaches 60 per cent of the

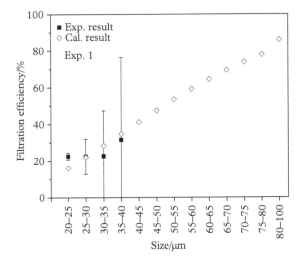

Fig. 5.26 *The filtration efficiency vs inclusion size in Exp. 1 with an Al_2O_3 filter.*

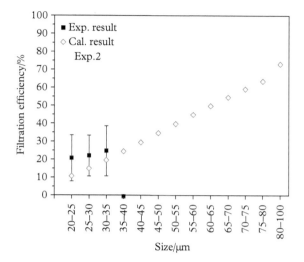

Fig. 5.27 *The filtration efficiency vs inclusion size in Exp. 2 with a SiC filter.*

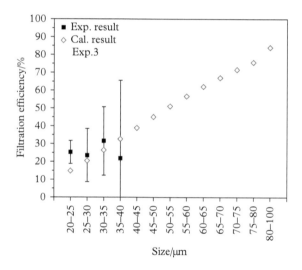

Fig. 5.28 *The filtration efficiency vs inclusion size in Exp. 3 with an Al$_2$O$_3$ filter.*

cohesion work of aluminium metal, and less than 44 per cent for the Al$_2$O$_3$–Al system. The adhesion work gives the attraction between unlike molecules (different phases), while the intermolecular attraction between like-molecules within a body is given by cohesion work. Thus, SiC has a greater tendency to attract aluminium metal than does Al$_2$O$_3$. As such, inclusions in the metal have more chances to collide with the walls in the SiC filter.

Figure 5.30 gives a sketch of filtration efficiency, including both settling and intercept collision as a function of flow rate. The filtration efficiency decreases with the flow rate until

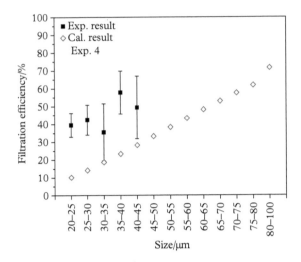

Fig. 5.29 *The filtration efficiency vs inclusion size in Exp. 4 with a SiC filter.*

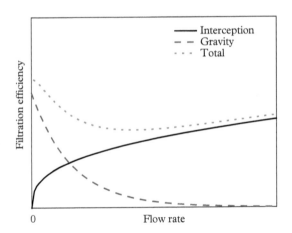

Fig. 5.30 *Sketch of collision efficiency vs flow rate.*

it reaches a minimum, and then increases. The greater the velocity, the less time particles have to settle. Gravitational collision must be taken into account at the lower flow rates. Intercept collision efficiency increases with the velocity, since more liquid and particles come into contact with the collector. Gravitational collision dominates at the flow rate 11 to 12 mm/s for 40- to 60-μm inclusions in Tian and Guthrie's case.[41] At the flow rate of 50–90 mm/s in the four tests, the settling contribution is very small and is neglected.

5.6.4 Re-entrainment of Inclusions

In Section 5.6.3, it was assumed that once an inclusion collided with the collector, it remained trapped at the point of collision. In such a case, the collection efficiency η in Eq. (5.33) is equal to the collision efficiency. Various experimental results indicate, however, that inclusions in aluminium can in fact be released from a filter.[4, 52] This may seem rather surprising, since the material used in the filters has a high interfacial tension relative to the melt. There is a net reduction of Gibbs energy when an inclusion resides at the interface between filter and melt. Since the interfacial tension is on the order of 1 N/m, the forces holding back an inclusion are of the order of magnitude a, where a is the inclusion radius. The viscous drag forces on the particle are roughly $av\rho v$. $v\rho v$ may be about 10^{-5} N/m. Thus, seemingly, the forces restraining the particles should be a factor 10^5 greater than the forces that re-entrain. An explanation is that both the particles and the filter have rough surfaces on a microscopic scale and that the true contact area between the particle and filter is only roughly 0.01 per cent of the apparent area.[53] One might believe that a remedy against re-entrainment would be to encourage the creation of liquid inclusions that could attain close contact with the collector or to cover the filter surface with a molten salt. However, a second liquid phase may tend, after a time, to run through the filter.[52] One recourse is to prevent fluctuations in the flow. Certainly, surges of metal through the filter are to be prevented. Also, if the filter depth is ample, re-entrained inclusions can be recaptured downstream.

It may be mentioned that there does not seem to be any evidence of re-entrainment of solid inclusions from alumina filters for steel. In contrast, such filters have tended to choke up, so that the pore sizes used must be larger than for aluminium.[54] This is because those inclusions that hit the filter medium sinter react (bond) chemically with the filter material at temperatures around 1600 °C.

The use of a greater pressure drop is obviously of interest in filtration, especially if 'dendrites' are formed. One problem is that the pressure gradients or velocity may become too high, so that particles are re-entrained as described above. Another difficulty-for flow upwards-with bed filters is that if the pressure gradient exceeds $(\rho_b - \rho)g(1 - \varepsilon)$, where ρ_b is the density of the collector medium, then fluidization or channeling may take place. This is obviously not a problem for rigid media filters. However, these are more expensive.

5.7 Rotational Forces for Removing Inclusions

A number of studies have been reported in which rotational forces are applied in tundish flow systems, so as to help separate inclusions from liquid metals.[55] The best examples of these are for entry flows from the ladle into the tundish, and from the SEN into the moulds. In these situations, produced by flow control devices (subject to erosion by steel), the lighter inclusions rotate towards the centre of the liquid stream. There they can agglomerate, become sufficiently buoyant, and rise to the protective top slag.

5.8 Electromagnetic Forces for Removing Inclusions

The use of electromagnetic forces to remove inclusions from liquid metals has been an important theme in the history of inclusion removal. When a strong electrical current is passed through a wire, it will generate a magnetic field around it. Similarly, if an external electromagnetic field is imposed on a liquid metal, then flow fields in the liquid can be strongly modified and can be used to enhance cast microstructures, as well as to promote the removal of inclusions.

For instance, Pechiney researchers developed a batch filter, tubular canister system comprising many small cylindrical pathways. All the treated liquid metal was then fed to the DC casting machines. By imposing a direct current on the liquid metal flowing through the filter system, inclusions which are non-conductive electrically were rejected to the outside circumference of the melt flow, due to the generation of a self-induced magnetic field within each of the circular passageways. In this way, they could potentially become entrapped, or bled off, as an inclusion-rich outer stream. This proved to be impractical, as the inclusion separation was a problem owing to the re-entrainment of inclusions from the sidewalls back into the bulk flow of liquid aluminium through the filter.

In the case of the LiMCA system, this rejection phenomenon of non-conducting particles being expelled from a melt in the presence of a high electrical current was circumvented by restricting the length of passage through the ESZ to about 1 mm. This prevents inclusions from colliding with the sidewalls of the ESZ. Thus, LiMiCA measurement provides a frequency size distribution of the total number of inclusions entrained within a flowing melt of an aluminium alloy. By taking advantage of an elliptically shaped ESZ, in the presence of a high current pulse (300 amps) at the end of each sampling, it was possible to break down any incipient formation of inclusion rafts at the entrance to the ESZ.

Building on this work, an alternative solution to the Pechiney approach was researched,[56,57] in which we mathematically modelled the concept of constructing a tube, in which metal would flow through a series of perforated discs, each containing a large number of small orifices. In the presence of a high DC, non-conducting inclusions could be deflected away from main flows passing into the next set of ESZ orifices and come to rest within the quieter regions setup within the system. Unfortunately, resistance to metal flow needs to be overcome, and the electrical power requirements would be significant.

Finally, in the steelmaking industry, magneto-hydrodynamics is commonly used to calm the entry flows of liquid steel entering a mould, and to re-direct incoming flows back up towards the meniscus. This improves flow control, allowing inclusions to be floated up towards the surface rather than becoming entrained within the body of liquid steel gradually freezing into a steel slab.

An early demonstration of the advantages of using an externally produced electromagnetic field can be inferred from Fig. 5.31, where it is seen that the numbers of entrained inclusions within the freezing steel can be reduced, according to liquid flow patterns. There, we see that flow patterns modified to produce flows directed towards the surface also produced enhanced removal of inclusions. We were able to confirm this using a water model of the flow system within a slab caster and comparing it with CFD predictions.[58]

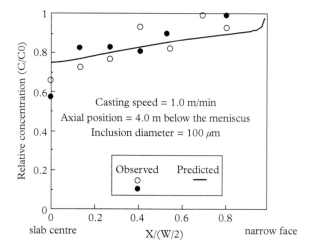

Fig. 5.31 *Relative concentration of inclusion glass microspheres in a water model of a continuous slab casting machine caused by up-flows at the narrow faces. These flows can be moderated by electromagnetic forces acting on the outlet jets of steel, leading to enhanced inclusion removal.*

5.9   The Number Size Distribution of Inclusions

Inclusions normally grow due to the diffusion of a supersaturated component to inclusions already present in the melt. Inclusions may also grow due to collisions between the inclusions.[59,60] For instance, for precipitation deoxidation, collisions may occur just after the addition of the deoxidizer before dissolution and mixing have properly taken place. Collisions should not normally be an important mechanism because the cleaner the melt, the less important the role of collisions. Collisions may occur due to the presence of large exogenous inclusions. However, this problem is not discussed here and the mechanism of collision is not included in the following.

A typical example of growth is the diffusion of oxygen in a steel melt to oxide inclusions.[61] In this case if we assume that the inclusions are spherical with radius a, that an inclusion contains ρ_o oxygen kmoles per unit volume inclusions, and that the mass transfer coefficient for oxygen is k_o then the growth rate of the inclusion, da/dt, is given by the kmole balance

$$\rho_o 4\pi a^2 \frac{da}{dt} = \frac{4\pi a^2 k_o \rho \Delta[\%O]}{16 \times 100},\tag{5.54}$$

where ρ is the density of the melt, and $\Delta[\%O]$ is the driving force for diffusion of oxygen (also called supersaturation). $\Delta[\%O] = [\%O] - [\%O]_{ie}$, where $[\%O]_{ie}$ is oxygen in the melt in equilibrium with oxygen in the inclusions (see also Section 4.4). Sixteen in the denominator is the atomic weight of oxygen. The right-hand side gives the number of kmoles of oxygen per unit time that is transferred to the inclusion. The left-hand side is the rate of accumulation of kmoles of oxygen in the inclusion since $4\pi a^2 da/dt$ is the volume increase per unit time. Equation (5.54) reduces to

$$\frac{da}{dt} = \frac{k_o \rho \Delta[\%O]}{\rho_o 100 \times 16}. \tag{5.55}$$

In Chapter 3 relations for the mass transfer coefficient k_o are given. ρ_o depends on the density and composition of the inclusions. It will be shown later that the supersaturation $\Delta[\%O]$ can be obtained from an oxygen balance for the melt.

Starting with Eq. (5.55), one finds that the number of inclusions per unit volume of melt of size a to $a + \Delta a\, f_N(a)\Delta a$ can be determined. The approach is to employ a number balance ('population balance') in the size interval from a to $a + \Delta a$: the rate of reduction of the number of inclusions of size a to $a + \Delta a$ is equal to the number of inclusions of size a to $a + \Delta a$ that are removed from the melt per second, or

$$-\frac{d(Vf_N(a)\Delta a)}{dt} = f_N(a)Ak_t\Delta a. \tag{5.56}$$

$Vf_N(a)\Delta a$ is the number of inclusions in the size interval a to $a + \Delta a$. k_t is the mass transfer coefficient for removal of an inclusion of size α; it has dimensions of velocity. A is the contact area between the melt, refractories, top slag, and bubbles. In Sections 5.10 and 5.11 relations for k_t will be derived.

The left-hand side of Eq. (5.56) may be split into two parts. One part takes into account the change of the distribution function with time. The other part is due to the growth of the inclusions with time. Thus, the left-hand side of Eq. (5.56) becomes

$$V\Delta a\left[\frac{\partial f_N\,(t,a)}{\partial t} + \frac{\partial f_N\,(t,a)}{\partial a}\frac{da}{dt}\right].$$

If there is little or no supersaturation of oxygen, that is, $\Delta[\%O] \approx 0$ so that $da/dt \approx 0$, only the time-dependent part remains.

As an example, we have evacuated the atmosphere above a steel melt initially containing 2×10^{-2} % oxygen in the form of inclusions, where very stable refractories have been used. We assume that we may neglect dissolved oxygen and that pick-up of oxygen from the refractories and atmosphere is small so the effect of supersaturation $\Delta[\%O]$ is negligible. The coefficient for transfer of the inclusions to the walls and top-slag is $k_t = 10^{-3}$ m/s, the wall and top-slag contact area is $A = 2$ m^2, and the melt volume is $V = 1$ m^3. One might then ask: How long time does it take for the oxygen (in the form of inclusions) to be reduced by 50 per cent to 1×10^{-2} %?

A mol balance for the oxygen content is then simply

reduction in oxygen in the melt = oxygen transferred to the walls and top slag

$$-\frac{\rho V d[\%O]_i}{M_O 100 dt} = \frac{\rho k_t A[\%O]_i}{M_O 100},$$

which after integration from $t = 0$ to t becomes

$$\frac{[\%O]_i}{[\%O]_i\,(t=0)} = \exp\left(-\frac{k_t A}{V}t\right).$$

Then by setting the left-hand side to 50 per cent and solving for the time, we find that

$$t_{1/2} = -\frac{V}{k_t A}\ln(0.5) = \frac{1}{10^{-3}\times 2}\times \ln(2) = 347\ \text{s} = 5\ \min 47\ \text{s}.$$

Now if we include some oxygen pick-up from walls and atmosphere and say that this has a mass transfer coefficient of $k = 10^{-4}$ m/s and a driving force of $\Delta[\%O] = 10^{-2}$, the mol balance gets one more term on right-hand side:

reduction in oxygen in the melt = oxygen transferred to the walls and top slag

$-$ oxygen in inclusions due to pick-up

$$-\frac{\rho V d[\%O]_i}{M_O 100 dt} = \frac{\rho k_t A[\%O]_i}{M_O 100} - \frac{\rho k A \Delta[\%O]}{M_O 100}$$

$$-V\frac{d[\%O]_i}{dt} = k_t A[\%O]_i - \rho k A \Delta[\%O]$$

$$\int_{[\%O]_i(t=0)}^{[\%O]_i(t)} \frac{d[\%O]_i}{[\%O]_i - \frac{k}{k_t}\Delta[\%O]} = -\int_{t=0}^{t} \frac{k_t A}{V}dt$$

$$\ln\left(\frac{[\%O]_i - \frac{k}{k_t}\Delta[\%O]}{[\%O]_i\,(t=0) - \frac{k}{k_t}\Delta[\%O]}\right) = -\frac{k_t A}{V}t.$$

With the given numbers, the time to reduce the concentration to 50 per cent when pick-up of oxygen is taken into consideration is then increased to

$$t_{1/2} = \frac{1}{10^{-3}\times 2}\times \ln\left(\frac{2\times 10^{-2} - \frac{10^{-4}}{10^{-3}}\times 10^{-2}}{1\times 10^{-2} - \frac{10^{-4}}{10^{-3}}\times 10^{-2}}\right) = 374\ s = 6\ \min 14\ s,$$

where $\partial f_N(t,a)/\partial t = 0$ under stationary conditions, so that $f_N(a)$ does not change with time.

An example of a near-stationary situation may be when an Al-deoxidized melt has been standing for a time without any additions or changes being made. Then dissolved oxygen and oxygen in the inclusions have reached steady state (see Section 5.10). The excess of aluminium keeps the oxygen supersaturation, $\Delta[\%O]$, at its constant low level. Then Eq. (5.56) simplifies to

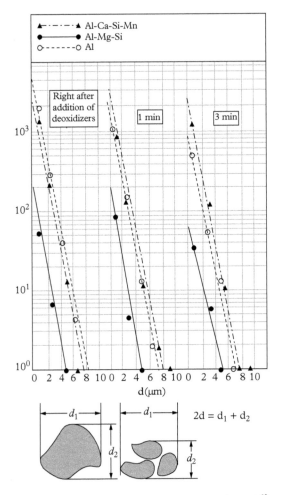

Fig. 5.32 *Number of inclusions in steel of size d. d is measured as indicated.*[63]

$$V \frac{df_N(a)}{da} \frac{da}{dt} = -f_N(a)Ak_t. \tag{5.57}$$

For a continuous steady-state back mixed reactor, V in Eq. (5.57) is replaced by $\dot{q}\tau$ where τ is the mean residence time in the reactor, $\tau = V/\dot{q}$, and $\dot{q}$ is the volume flow. When Eq. (5.55) is used to eliminate da/dt, one obtains

$$\frac{df_N(a)}{f_N(a)} = -\frac{da Ak_t \rho_o 16 \times 100}{\rho k_o \Delta [\%O] V} \tag{5.58}$$

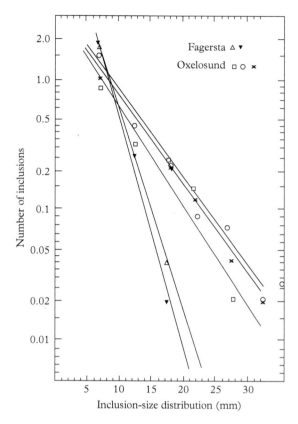

Fig. 5.33 *Number–size distributions of inclusions in stainless steel (Fagersta) and in carbon steel (Oxelosund).*[65]

If k_t/k_o does not depend on a (see Eqs. (3.56) and (3.67) and Sections 5.10 and 5.11 of this chapter), integration give

$$\frac{f_N(a)}{f_N(a_c)} = \exp\left(-(a-a_c)\frac{k_t A\rho_o 16 \times 100}{\rho k_o \Delta[\%O] V}\right), \tag{5.59}$$

where $f_N(a_c)$ is the number of particles of size a_c. It is seen that the number–size distribution of the inclusions is a decreasing exponential. Figures 5.32, 5.33, and 5.34 give examples from the literature[62–5] that, within the size range measured, show the negative exponential dependence of the number–size distribution. In this case, it is expected that removal of inclusion to the walls and top slag is due to turbulence with a very weak dependence on a. See Appendix 2. The exponential dependence found experimentally does not necessarily prove that k_t/k_o is independent of particle size, since often the particle size range studied is narrow. For filtration and settling, k_t increases with inclusion size.

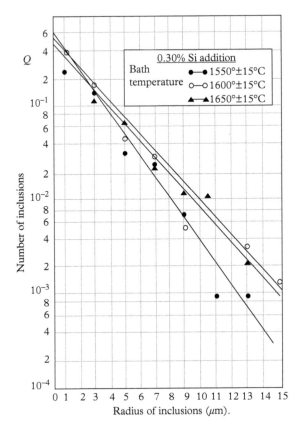

Fig. 5.34 *Number distribution of SiO₂ inclusions of size d in steel.[62] (This work is licensed under a Creative Commons Attribution 4.0 generic license.)*

The negative slope increases with the surface-to-volume ratio A/V and with the increasing ratio of the coefficients k_t/k_o. If the mass transfer coefficient, k_t, to the walls is large, the inclusions are essentially removed while they are still small. Methods of attaining this goal will be discussed in the following. If some of the inclusions are re-entrained, k_t is reduced accordingly. In the last section of Appendix 2 methods of increasing k_t are considered. It should be noted that the number of large inclusions can also be reduced by decreasing the supersaturation $\Delta[\%O]$.

If we wish to study removal of $\underline{N}$ instead of $\underline{O}$, we get

$$\frac{f_N(a)}{f_N(a_c)} = \exp\left(-\frac{(a-a_c)k_t A \rho_N 14 \times 100}{\rho k_N \Delta [\%N] V}\right), \tag{5.60}$$

where i is the diffusing component that limits growth of the inclusions, k_i is the mass transfer coefficient for dissolved component i to the inclusion, M_i, is the mole weight, and ρ_i is the number of moles of i in the inclusion per unit volume.

In general, $f_N(a_c)$ and $\Delta[\%i]$ depend on time. The interdependence of $f_N(a_c)$ and $\Delta[\%i]$ has been briefly addressed. In principle $\Delta[\%i]$ can be determined from a balance for component i for the melt. Strictly, the size distribution is not decreasing exponentially. However, even when k_t/k_i depends on a, Eq. (5.60) may be useful, over a limited size range, to discuss qualitatively the interaction between inclusions and dissolved elements.

5.10 Dissolved Elements and Inclusions

Since the deoxidation of steel is so well known, an oxygen balance is employed to illustrate how a balance for a melt component i is derived. It will be seen that the presence of component i also as a solute in the melt complicates matters somewhat. As already mentioned, oxygen is an example of a component present both as a dissolved element and as part of the inclusions. Oxygen is picked up by the melt from oxygen and moisture in the atmosphere and from the refractories (see Fig. 2.1). Sulfur is another example of a component present both in solution and in inclusions in steel. In aluminium melts, carbon is dissolved in the melt from carbon refractories and is found also in inclusions such as Al_4C_3.

Oxygen pick-up from slag or dross, refractories, and atmosphere by the melt may be described by

$$\dot{n}_j = \frac{\rho k_j}{100 \times 16} ([\%O]_{je} - [\%O]), \tag{5.61}$$

where $j = 1, 2, 3, 4$ refers to top slag, refractory, bubbles, and air interfaces, respectively, and k_j is the total two-film mass transfer coefficients introduced in Chapter 4. $[\%O]_{je}$ is the hypothetical dissolved oxygen concentration in the melt in equilibrium with the bulk oxygen concentration in phase j. $[\%O]$ is the dissolved bulk oxygen concentration in the steel. In terms of the number–size distribution function, $f_N(a)$, the moles of oxygen present as inclusions are

$$\frac{\rho[\%O]_i}{100 \times 16} = \rho_O \int_0^\infty f_N(a) \frac{4\pi a^3}{3} da. \tag{5.62}$$

ρ_O is the mass of oxygen per volume of inclusion. If aluminium has been added in order to deoxidize, oxygen is removed in two steps: by transfer of oxygen to the inclusions and by removal of the inclusions to wall, top-slag, and bubbles (see Fig. 3.4). A balance for moles of oxygen in a batch melt now gives

accumulation of oxygen in the melt (as dissolved oxygen and as inclusions)

= moles of oxygen picked up from atmosphere, walls and slag

− moles of oxygen transferred with the inclusions out of the melt.

Per unit time this becomes

$$\frac{d}{dt}\left(\frac{V_\rho\left([\%O]+[\%O]_i\right)}{100 \times 16}\right) = \dot{n}_1 A_1 + \dot{n}_2 A_2 + \dot{n}_3 A_3 + \dot{n}_4 A_4$$

$$-\frac{\rho[\%O]_i}{100 \times 16}\left(k_{t1} A_1 + k_{t2} A_2 + k_{t3} A_3\right) \tag{5.63}$$

The inclusion mass transfer coefficients are k_{t1}, k_{t2}, and k_{t3} for the top-slag, the wall, and bubbles, respectively. If Eq. (5.61) is introduced in Eq. (5.63), one obtains

$$V\frac{d}{dt}\left([\%O]+[\%O]_i\right) = \sum_{j=1}^{3} k_j A_j \left([\%O]_{je} - [\%O]\right)$$

$$- \left(k_{t1} A_1 + k_{t2} A_2 + k_{t3} A_3\right)[\%O]_i. \tag{5.64}$$

This equation contains two unknown variables, dissolved oxygen in the melt $[\%O]$ and oxygen contained in the inclusions $[\%O]_i$. However, $[\%O]_i$ depends on $[\%O]$ as may be seen from Eqs. (5.59) and (5.62). Thus, in principle Eq. (5.64) can be integrated to give $[\%O]$ as a function of time. Then also $[\%O]_i$ can be calculated from Eqs. (5.59) and (5.62).

It may perhaps be instructive to start with the special case that there are no inclusions, that is, $[\%O]_i = 0$. Then Eq. (5.64) reduces to

$$V\frac{d[\%O]}{dt} = \sum_{j=1}^{3} k_j A_j [\%O]_{je} - \sum_{j=1}^{3} k_j A_j [\%O] \tag{5.65}$$

This equation may be transformed to

$$V\frac{d[\%O]}{dt} = \sum_{j}^{3} k_j A_j \left([\%O]_\infty - [\%O]\right), \tag{5.66}$$

where

$$[\%O]_\infty = \frac{\sum_{j=1}^{3} k_j A_j [\%O]_{je}}{\sum_{j=1}^{3} k_j A_j}. \tag{5.67}$$

The $[\%O]_\infty$ is a mean value of the hypothetical oxygen concentration in steel in equilibrium with the oxygen activity of the four phases (corresponding to $j = 1, 2, 3,$ and 4) in contact with the melt. It is seen that the hypothetical concentrations $[\%O]_{je}$ are weighted with the kinetic factor

$$k_j A_j / \sum_{j=1}^{3} k_j A_j.$$

Since $[\%O]_\infty$ is constant, Eq. (5.66) integrates to give

$$\frac{[\%O] - [\%O]_\infty}{[\%O]_{in} - [\%O]_\infty} = \exp\left(-\frac{\sum_{j=1}^{3} k_j A_j t}{V}\right) \tag{5.68}$$

Here $[\%O]_{in}$ is the initial value at time $t = 0$, of the dissolved oxygen. For $t \to \infty$, the exponential approaches zero and $[\%O] \to [\%O]_\infty$ (see Fig. 4.9). $[\%O]_\infty$ is the final steady-state value and could have been obtained directly from Eq. (5.65) by noting that at steady state the left-hand side of the equation is zero. From the right-hand side one then obtains

$$[\%O] = \sum_{j=1}^{3} \frac{k_j A_j [\%O]_{je}}{\sum_{j=1}^{3} k_j A_j},$$

which is identical to Eq. (5.67).

In the general situation that we do in fact have inclusions, no attempt is made to derive a general relation for the time dependence of $[\%O]$ and $[\%O]_i$. If $[\% O]_i \gg [\%O]$, it is often a simple matter to calculate the time-dependent behaviour of $[\%O]_i$.

Dissolved oxygen often drops to the steady-state value in a very short time compared to the time it takes $[\%O]_i$ to reach the final value.

Here we look at the case when a steady state has been attained so that $[\%O]$ and $[\%O]_i$ no longer change with time. Then setting the right-hand side of Eq. (5.64) to zero gives

$$\sum_{j=1}^{3} k_j A_j [\%O]_{je} - [\%O] \sum_{j=1}^{3} k_j A_j = (k_{t1} A_1 + k_{t2} A_2 + k_{t3} A_3) [\%O]_i. \tag{5.69}$$

This equation can be solved for $[\%O]$ and $[\%O]_i$ employing Eqs. (5.59) and (5.62). Equation (5.69) says that oxygen pick-up from top-slag, walls, bubbles, and atmosphere is equal to oxygen removed in the form of inclusions to wall, top-slag, and bubbles. In terms of the supersaturation $\Delta[\%O] = [\%O] - [\%O]_{ie}$, Eq. (5.69) becomes

$$\sum_{j=1}^{3} k_j A_j \left([\%O]_{je} - [\%O]_{ie}\right) - \Delta[\%O] \sum_{j=1}^{3} k_j A_j = [\%O]_i (k_{t1} A_1 + k_{t2} A_2 + k_{t3} A_3). \tag{5.70}$$

Here $[\%O]_{je} - [\%O]_{ie}$ is the driving force for the transfer of oxygen from phase j to the inclusions. The first terms depend on the refractories, top-slag conditions, and atmosphere above the melt. For a given ladle and deoxidation practice, these terms may be taken to be constant. $\Delta[\%O] \sum_{j=1}^{3} k_j A_j$ gives the transfer of oxygen from the melt to the inclusions. The last terms in Eq. (5.70) take into account the transport of oxygen as inclusions out of the melt.

It is seen that the supersaturation, $\Delta[\%O]$, increases with decreasing content of inclusions, $[\%O]_i$. On the other hand, the number of nuclei increases with increasing supersaturation,

Δ[%O]. If there is a disturbance so that Δ[%O] is increased compared to the original steady-state value, then the number of nuclei and the resulting number of inclusions start to increase with time. However, in this situation, more oxygen is picked up by the inclusions and then removed by the inclusions as they are transferred to top-slag, wall, and bubbles. This in turn tends to reduce the supersaturation Δ[%O] and finally the original constant supersaturation level is recovered. If we start with a supersaturation that is lower than the original value, similar arguments return us to the steady-state situation.

The oxygen supersaturation in the example produces a large number of Al_2O_3 inclusions. However, [%O]$_i$ is very sensitive to the oxygen supersaturation. A slight decrease in the supersaturation gives a very strong reduction in the number of nuclei c^*. In the example, if it is possible to suppress oxygen down to [%O] = 0.003, then r^* = 10.2 Å and c^* = 2000 nuclei/m^3. In this case very few new Al_2O_3 inclusions are formed. Such a reduction may be attained by the addition of a deoxidizer stronger than aluminium, such as calcium. The addition of a stronger deoxidizer prevents nuclei of the weaker one from forming. After calcium is added, the melt still contains the inclusions present prior to the addition (plus the new inclusions formed by the stronger deoxidizer). These inclusions must be removed by stirring, that is, by increasing the transfer coefficient for the inclusions, k_t. Simultaneously, dissolved oxygen declines down to [%O]$_\infty$ even faster than the rate given by Eq. (5.68) since some of the dissolved oxygen in the melt is also transferred to the inclusions. From Eq. (5.67), it is seen that oxygen in equilibrium with the j interface is weighted by the kinetic factor

$$\frac{k_j A_j}{\sum_{j=1}^{4} k_j A_j},$$

where 4 refers to the inclusion phase. Only when the number of large inclusions is reduced to very low levels should casting be carried out. The times involved in inclusion removal for various types of stirring are discussed in Section 5.11.

To obtain a melt with a low inclusion content before casting, it is important that stable refractories and a reducing slag are employed so that the [%O]$_{je}$ values in Eq. (5.70) are low. This gives a low value of the supersaturation Δ[% O] and a correspondingly low number of large inclusions according to Eq. (5.59).

During casting care must be taken not to expose the melt to the atmosphere. Also entrapment of top-slag in the tundish must be prevented. If large exogenous inclusions are created during teeming, all our previous efforts may have been in vain. Equation (5.59) can only give the inclusion distribution if metal cleanliness can be retained during the casting operation.

5.11 Conclusions

We have seen how inclusions within a melt can be partially, or completely, eliminated from the liquid metal, using a variety of techniques. Even so, most liquid metal systems still contain many millions of smaller inclusions. Similarly, a few inclusions can appear in a much larger size range (i.e. >100 microns) in liquid steel and aluminium. These particles are often the

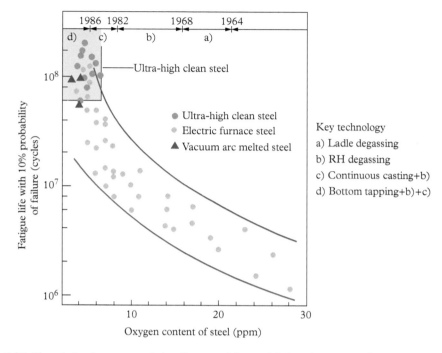

Fig. 5.35 *The relation between steel cleanliness and fatigue life for a bearing steel. [Reprinted from Morita and Emi[66] with permission from ISIJ.]*

result of various process upsets, or processing methods, used to produce the liquid metal from its concentrate and ore. As the inclusions, larger particles or agglomerated inclusions, can harmfully affect a metal's properties, the steel industry, particularly in Japan, has devoted much time and research to their removal.

For instance, Fig. 5.35, taken from the book on iron and steel by the Kawasaki Corporation, indicates that the reduction of total oxygen levels in steels can result in an exponential rise in fatigue life properties of ball bearings.[66]

Removal of inclusions by floatation, settling, and last but not least, filtration have been treated at some length. A branch ceramic foam model of filtration has been compared with plant experiments. The agreement is promising. However, there is a lack of understanding and measuring 'wetting' effects between inclusions and filter material. It is suggested that floatation removal of inclusions will become an important tool in the future. The wetting effect is expected to be important.

If we can eliminate macro-inclusions greater than 10 microns from a cast metal, we may be able to enter a new era of liquid metal cleanliness. So, much further research and development work remains, if we are to improve metals' intrinsic properties.

....

REFERENCES

1. Simensen CJ. Sedimentation and analysis of inclusions in aluminium and magnesium. *Metall. Mater. Trans. B*, 1981; 12(4): 733–43.
2. Simensen CJ, Fartum P, Andersen A. Analysis of intermetallic Particles in Aluminium by Dissolution of the Sample in Butanol. *Fresenius Z. Anal. Chemie*, 1984; 319: 286–92.
3. Hudson SW, Apelian D. Inclusion detection in molten aluminum: current art and new avenues for *in-situ* analysis. *Int. J. Metalcasting*, 2016; 10: 289–305.
4. Guthrie RIL, Doutre DA. Online measurement of inclusions in liquid melts. In: Engh TA, Lyng S, Øye HA, eds. *Proceedings of the international seminar on refining and alloying of liquid aluminium and ferro-alloys, Trondheim, Norway*, pp. 145–64. Düsseldorf: Aluminium Verlag; 1985.
5. Irwin DW. Sampling to detect inclusions in molten aluminium. In: Nilmani M, ed, *Aluminium melt refining and alloying—theory and practice*, pp. 51–8. Melbourne, Australia: University of Melbourne; 1989.
6. Gustavson S, Mellberg PO. Inclusion picture and inclusion origin in a calcium treated steel. In: Nordberg H, Sandstrom R, eds, *Swedish symposium on non-metallic inclusions in steel*. Stockholm, Sweden: Udeholms AB; 1981.
7. Simensen CJ. *Analysis of inclusions and hydrogen in aluminium and magnesium*. Trondheim, Norway: Dr. techn. Thesis, NTH; 1982.
8. Levy SA. Evaluation of molten metal quality for aluminium alloys. *Trans. American Foundryman*, 1985; 93: 889–94.
9. Bates DA, Hutter LC. Evaluation of aluminium filtering systems using a vacuum-filtration sampling device. In: *Light Metals*, pp. 707–21. Warrendale, PA: TMS; 1981.
10. Bathen E. Investigation of inclusions in aluminium melts by image analysis. In: Engh TA, Lyng S, Øye HA, eds, *Proceedings of the international seminar on refining and alloying of liquid aluminium and ferro-alloys, Trondheim, Norway*, pp. 173–91. Düsseldorf: Aluminium Verlag; 1985.
11. Mansfield TL. Ultrasonic technology for measuring molten aluminium quality. In: *Light Metals*, pp. 969–82. Warrendale, PA: TMS; 1982.
12. Mansfield TL. Molten aluminum quality measured with Reynolds 4M TM system. In: *Light Metals*, pp. 1305–28. Warrendale, PA: TMS; 1984.
13. Doutre DA, Guthrie RIL. *A method and apparatus for the detection and measurement of particulates in liquid metals (LiMCA)*. U.S. Patent 4,555,662, November 26, 1985. With counterpart patents, courtesy of Alcan, issued in Australia (564,184), Brazil, Canada (1203996), Austria, Belgium, Germany, France, Switzerland, Great Britain, Italy, Luxembourg, Netherlands, Sweden, Norway, South Africa (84/1303), Mexico, India, S. Korea, Japan, Spain (530 246).
14. Martin JP., Dubé G., Frayce D., Guthrie R. (2016) Settling Phenomena in Casting Furnaces: A Fundamental and Experimental Investigation. In: Grandfield J.F., Eskin D.G. (eds) Essential Readings in Light Metals. Springer, Cham., pp. 115–125. https://doi.org/10.1007/978-3-319-48228-6_15
15. Camplin JM, Herbertson J, Holl H, Whitehouse P, Guthrie RIL, Han JW, Hasan M. The application of mathematical and water modelling in the selection of tundish design for the proposed combicaster at BHP Whyalla. In: *Proc. sixth intl. iron and steel congress*, vol. 3, Steelmaking I, pp. 207–14. Nagoya: Iron and Steel Institute of Japan; 1990.

16. Nakajima H, Sebo F, Tanaka S, Dumitru L, Harris DJ, Guthrie RIL, On the separation of non-metallic inclusions from tundishes in continuous casting—a water model study. *Steelmaking Proc.*, 1986; 69: 705–16.
17. Hsiao T-C, Lehner T, Kjellberg B. Fluid flow in ladles—experimental results. *Scand. J. Metall.*, 1980; 9: 105–10.
18. Haberman WL, Morton RK. *An experimental investigation of the drag and shape of air bubbles rising in various liquids*. Washington, DC: David Taylor Model Basin Report 802; 1953.
19. Clift RC, Grace JR, Weber ME. *Bubbles, drops, and particles*. Mineola, New York: Dover Publications; 1978.
20. Guthrie RIL. *Spherical cap bubbles in liquid metal systems*. PhD thesis, London University; 1967.
21. Irons GA, Guthrie RIL. Bubble formation at submerged nozzles in molten iron. *Metall. Trans. B*, 1978; 9: 101–10.
22. Li M., Carozza C, Guthrie RIL. Particle discrimination in water based LiMCA (liquid metal cleanliness analyzer) system. *Canad. Metall. Q.*, 2000; 39(3): 325–37.
23. Abbel G. TATA Steel, Ijmuiden, Private Communication, July 4, 2017.
24. Lamb H. *Hydrodynamics*, 6th edn. Cambridge, UK: Cambridge University Press; 1932.
25. Guthrie RIL, Isac M, Ren XR. The fundamentals of forming micro-bubbling in liquid metal systems. In: Ratvik AP, ed., *Proc. the science of melt refining, Thorvald Abel Engh & Christian Simensen Honorary Symposium. Light Metals.* 2017; 1403–9.
26. Marshall SH, Chudacek MW, Bagster DF. A model for bubble formation from an orifice with liquid cross-flow. *Chem. Eng. Sci.*, 1993; 48(11): 2049–59.
27. Chang S, Cao X, Hsin C-H, Zou Z, Isac M, Guthrie RIL. Removal of inclusions using microbubble swarms in a four strand, full scale, water model tundish. *ISIJ Int.*, 2016; 56(7): 1188–97.
28. Apelian D, Eckert CE, Mutharasan R, Miller RE. Refining of molten aluminium by filtration technology. In: Engh TA, Lyng S, Øye HA, eds, *Proceedings of the international seminar on refining and alloying of liquid aluminium and ferro-alloys, Trondheim, Norway*, pp. 123–43. Düsseldorf: Aluminium Verlag; 1985.
29. Bird RB, Stewart WE, Lightfoot EN. *Transport phenomena.* New York: Wiley; 1960.
30. Frisvold F, Engh TA, Bathen E. Measurement of filtration efficiency of an Alcoa 528 filter. In: Bickert CM, ed., *Light metals*; 1990.
31. Acosta GF, Castillejos A, Almanza RJ, Flores VA. Analysis of liquid Flow through ceramic porous media used for molten metal filtration. *Metall. Mater. Trans. B*, 1995; 26(1): 159–71.
32. Tien C, Ramarao B. *Granular filtration of aerosols and hydrosols.* Oxford: Elsevier Science; 2007.
33. Schlichting H. *Boundary-layer theory*, 7th edn. New York: McGraw Hill; 1979.
34. Bao S, Engh TA, Syvertsen M, Kvithyld A, Tangstad M. Inclusion (particle) removal by interception and gravity in ceramic foam filters. *J. Mater. Sci.*, 2012; 47(23): 7986–98.
35. Schlichting H. *Boundary-layer theory*, 4th edn. New York: McGraw-Hill; 1960.
36. Palmer MR, Nepf HM, Pettersson TJR, Ackerman JD. Observations of particle capture on a cylindrical collector: implications for particle accumulation and removal in aquatic systems. *Am. Soc. Limnol. Oceanogr.*, 2004; 49(1):76–85.
37. Rajagopalan R, Tien C. Trajectory analysis of deep-bed filtration with the sphere-in-cell porous media model. *AIChE J.*, 1976; 22: 523–33.
38. Conti C, Jacob M. Depth filtration of liquid metals: application of the limiting trajectory method to the calculation of inclusion deposition in open cell structure filters. *Bull. Soc. Chim. Fr.*, 1984; 11–12: 297–313.
39. Cliff R, Grace JR, Weber ME. *Bubbles, drops and particles.* New York: Academic Press; 1978.
40. Engh TA. *Principles of Metal Refining.* Oxford: Oxford University Press; 1992.
41. Tian C, Guthrie RIL. Direct simulation of initial filtration phenomena within highly porous media. *Metall. Mater. Trans. B*, 1995; 26(3): 537–46.
42. Lu T, Stone H, Ashby M. Heat transfer in open-cell metal foams. *Acta Mater.*, 1998; 46(10): 3619–35.

43. Innocentini MDM, Salvini VR, Macedo A, Pandolfelli VC. Prediction of ceramic foams permeability using Ergun's equation. *Mater. Res.*, 1999; 2(4): 283–9.
44. Richardson J, Peng Y, Remue D. Properties of ceramic foam catalyst supports: pressure drop. *Appl. Catal. A: Gen.*, 2000; 204(1): 19–32.
45. Fourie JG, Du Plessis JP. Pressure drop modelling in cellular metallic foams. *Chem. Eng. Sci.*, 2002; 57(14): 2781–9.
46. Buciuman FC, Kraushaar-Czarnetzki B. Ceramic foam monoliths as catalyst carriers. 1. Adjustment and description of the morphology. *Ind. Eng. Chem. Res.*, 2003; 42(9): 1863–9.
47. Moreira E, Coury J. The influence of structural parameters on the permeability of ceramic foams. *Braz. J. Chem. Eng.*, 2004; 21(1): 23–33.
48. Moreira E, Innocentini M, Coury J. Permeability of ceramic foams to compressible and incompressible flow. *J. Eur. Ceram. Soc.*, 2004; 24(10-11): 3209–18.
49. Giani L, Groppi G, Tronconi E. Mass-transfer characterization of metallic foams as supports for structured catalysts. *Ind. Eng. Chem. Res.*, 2005; 44(14): 4993–5002.
50. Lacroix M, Nguyen P, Schweich D, Pham Huu C, Savin-Poncet S, Edouard D. Pressure drop measurements and modeling on SiC foams. *Chem. Eng. Sci.* 2007; 62(12): 3259–67.
51. Bao S. *Filtration of aluminium-experiments, wetting, and modelling.* PhD thesis, Trondheim, Norway: NTNU; 2011.
52. Apelian D, Luk S, Piccone T, Mutharasan R. Removal of inclusions from steel melts. *Steelmaking Proc.*, 1986; 69: 957–67.
53. Adamson AW. *Physical chemistry of surfaces*, 4th edn. New York: Wiley; 1982.
54. Saxena SK, Sandberg H, Waldenstrom T, Persson A, Steensen S. Mechanism of clogging of tundish nozzle during continuous casting of aluminium killed steel. *Scand. J. Metall.*, 1978; 7(3): 126–33.
55. Yokoya S, Takagi S, Iguchi M, Marukawa K, Yasugaira W, Hara S. Development of swirling flow generator in immersion nozzle. *ISIJ Int.*, 2000; 40(6): 584–8.
56. Afshar MR, Aboutalebi MR, Guthrie RIL, Isac M. Electromagnetic filtration of magnesium melts. MetSoc-COM 2008, August 24–27, Winnipeg: 229–38.
57. Afshar MR, Aboutalebi MR, Guthrie RIL, Isac M. Electromagnetic filtration of steel melts. In: *Proceedings of the 3rd International Congress on the Science and Technology of Steelmaking*, Charlotte, NC, pp. 661–71; 2005.
58. Aboutalebi MR, Hasan M, Guthrie RIL. Heat flow modelling for steel slab continuous casting systems. In: *Proc. 75th Steelmaking Conf., I.S.S. of AIME, Toronto, April 1992*, pp. 929–38.
59. Lindborg U, Torssell K. A collision model for the growth and separation of deoxidation products. *Trans. AIME*, 1968; 242: 94–102.
60. Torssell K. Removal of inclusions resulting from deoxidation of iron with silicon. *Jernkontorets Annaler*, 1967; 151: 890–949.
61. Turkdogan ET. *Physical chemistry of steelmaking*. New York: Academic Press; 1980.
62. Kawawa T, Okubo M. A kinetic study on the deoxidation of steel. *Trans. ISIJ*, 1968; 8: 203–19.
63. Asano K, Nakano T. Deoxidation of molten steel with deoxidizer containing aluminium. *Trans. ISIJ*, 1972; 12(5): 343–9.
64. Martin JP, Dube G, Frayce D, Guthrie RIL. Settling phenomena in casting furnaces: a fundamental and experimental investigation. In: Nilmani M, ed., *Aluminium melt refining and practice*, University of Melbourne, Australia; 1989.
65. Engh TA, Lindskog N. A fluid mechanical model of inclusion removal. *Scand. J. Metallurgy*, 1975; 4: 49–58.
66. Morita Z-I, Emi T. *An introduction to iron and steel processing*. Kawasaki Steel Techno-Research Corporation.

6

Solidification and Refining

6.1 Introduction

The solidification process involves a phase change from liquid to solid. The majority of impurities in metals have a lower solubility in solid than in the liquid phase. Consequently, impurities will usually be expelled at the solidification front. This presents an opportunity to refine or purify metals. The redistribution of impurities during freezing (called segregation) also has a considerable influence on the mechanical (and other) properties of the solidified casting.

In this chapter we will primarily deal with three subjects:

(a) The removal of impurities during solidification: There are several processes where a controlled solidification is used to produce a purified material. One example is the directional solidification of silicon for solar cells, or the crystal pulling of semiconductors, by the Czochralski process. Zone refining, also referred to as the float zone process, is another application of this principle.

(b) The transfer of unwanted elements to parts of the casting where they do not harm the finished product: this principle is applied in slab casting of steel, where the impurities and inclusions remain in the centre of the slab and later end up in the low stressed portions of the finished product. Another application is the Silgrain process[1] where the impurities remain at the grain boundaries.

(c) Grain refinement, where a nucleant is added to achieve a small (refined) grain structure.

In Fig. 6.1 segregation is illustrated for the case of a planar solidification front, where the solubility of impurities is higher in liquid than in solid. The molten metal in this figure solidifies from the bottom to the top. The impurity is progressively concentrated in the liquid phase, from the start to the end of the solidification. The solid is more pure than the original starting liquid. Figure 6.1 can be regarded as a micromodel for what happens at the solid–liquid interface in columnar, dendritic, or equiaxed solidification.

Principles of Metal Refining and Recycling. Thorvald Abel Engh, Geoffrey K. Sigworth, and Anne Kvithyld, Oxford University Press.
© Oxford University Press 2021.
DOI: 10.1093/oso/9780198811923.003.0006

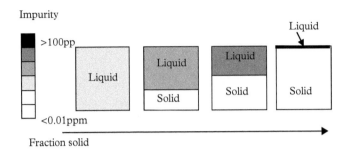

Fig. 6.1 *Principle for refining by solidification.*[2,3] *The dark colour indicates the level of impurity, where white is < 0.01 ppm and black is > 100 ppm*

6.2 Solute Distribution at the Solid–Liquid Interface

To describe segregation at the solid–liquid interface one usually assumes there is no diffusion in the solid. It is further assumed there is complete mixing in the liquid, and no mass flow in or out of the volume element. We then construct a simple conservation equation for the solute.

The amount of solute in the liquid is fc_l, where f is the fraction liquid and c_l is the concentration of the solute in the liquid. Later, when the fraction liquid is $f + \Delta f$ and the solute concentration is $c_l + \Delta c_l$, the amount of solute that remains in the liquid is $(f + \Delta f)(c_l + \Delta c_l)$. See Fig. 6.2. The concentration of solute in the solidified material is c_s so that the amount trapped in the solidified material is $-\Delta f c_s$. This is equal to the solute rejected by the liquid. Thus,

$$fc_l - (f + \Delta f)(c_l + \Delta c_l) = -\Delta f c_s. \tag{6.1}$$

On simplifying the equation and dividing by $\Delta c_l f$, we obtain

$$\frac{\Delta f (c_l - c_s)}{\Delta c_l f} + \frac{\Delta f}{f} = -1. \tag{6.2}$$

If Δc and Δf are infinitesimal, this becomes

$$\frac{df}{f} = -\frac{dc_l}{c_l - c_s}. \tag{6.3}$$

The ratio c_s/c_l is called the partition ratio or distribution coefficient, k. (See Fig. 6.5.) It can be obtained from a temperature versus concentration diagram (assuming no undercooling). For dilute solutions the temperature versus concentration lines may often be regarded as straight lines so that k is a constant. Then Eq. (6.3) becomes

$$(k-1)\frac{df}{f} = \frac{dc_l}{c_l}. \tag{6.4}$$

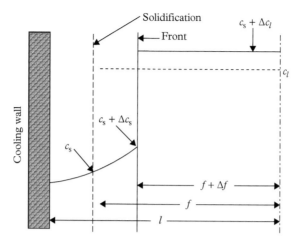

Fig. 6.2 *Concentration of solute in the melt, c_1, and in the solid, c_s, as a function of fraction melt, f.*

or, integrated,

$$(k-1)\int_1^f \frac{df}{f} = \int_{c_{in}}^c \frac{dc_l}{c_l}.$$ (6.4a)

This gives the Scheil equation

$$\frac{c_l}{c_{in}} = f^{k-1},$$ (6.5)

where c_{in} is the initial concentration of the solute when solidification begins. For the solid

$$\frac{c_s}{c_{in}} = kf^{k-1}.$$ (6.6)

Figure 6.3 shows the effect of a decreasing k, that is, the effect of the liquid rejecting the solute component. If $k = 1$, there is no segregation. If $k = 0.1$ the solute component is mainly found in the last fraction to solidify. There is a local variation of solute from the centre to the exterior portion of the solidified crystals or grains.

Equation (6.6) must be regarded as an extreme case of segregation. The mixing of the liquid may only be partial. There may also be rapid solid-state diffusion during solidification, particularly for carbon in cast iron and steel, but also for iron in silicon. These various effects tend to give a reduced 'effective' value of k, k_{eff}, often employed in an empirical version of the Scheil equation:

$$\frac{c_s}{c_{in}} = k_{eff} f^{k_{eff}-1}.$$ (6.7)

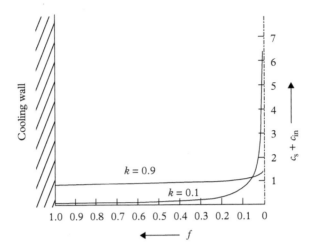

Fig. 6.3 *Ratio c_s/c_{in} between the concentration of solute in solidified material, c_s, and the concentration before solidification, c_{in}, as a function of fraction melt, f.*

We will find that the Scheil equation (6.7) can be expanded to include convection effects, evaporation from the melt surface, and solidification shrinkage. One example is the evaporation of oxygen from the melt surface that occurs during solidification of silicon. This will be treated in Section 6.9.

6.3   The Mass Transfer Coefficient k_t from Solid to Bulk Liquid

In Section 6.2 we looked at the special case with complete mixing in the liquid and no diffusion in the solid. In reality, there is always a viscous boundary layer at the solid–liquid interface where the transport of solute from the interface into the bulk is limited by diffusion.

Figure 6.4 sketches the solute concentration profile, c, from the solid to the bulk liquid. The solid is moving with velocity $v = dz/dt$ to the right. Relative to the solid the liquid moves to the left with velocity $-v$. To take mass transfer into account a turbulent diffusion coefficient, D_E, is introduced that is greater than the atomic diffusion coefficient. Then the flux of solute ($\dot{n}$) is given by

$$\dot{n} = -cv - D_E \frac{dc}{dz} \tag{6.8}$$

for $0 \leq z < \delta$. And in the solid

$$\dot{n} = -c_s v. \tag{6.9}$$

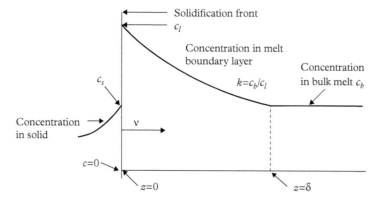

Concentration in solid

Concentration in melt boundary layer

$k = c_b/c_l$

Concentration in bulk melt c_b

Fig. 6.4 *Concentration of solute in the melt boundary layer at a solidification front moving with velocity v.*

$c = c_b$ at $z = \delta$ where δ is the thickness of the diffusion boundary layer. Since no solute is consumed in the boundary layer, this gives

$$-cv - D_E \frac{dc}{dz} = -c_s v$$

or

$$-D_E \frac{dc}{dz} = v(c - c_s)$$

or

$$\frac{dc}{c - c_s} = -\frac{v}{D_E} dz. \tag{6.10}$$

We integrate

$$\int_{c_l}^{c_b} \frac{dc}{c - c_s} = -\int_0^\delta \frac{v}{D_E} dz$$

and find

$$\frac{c_b - c_s}{c_l - c_s} = \exp\left(-\frac{v\delta}{D_E}\right). \tag{6.11}$$

Or if we introduce the mass transfer coefficient $k_t = D_E/\delta$ (see Chapter 3) to replace the quantities D_E and δ,

$$\frac{c_b - c_s}{c_l - c_s} = \exp\left(-\frac{v}{k_t}\right).\tag{6.12}$$

From this equation we see that if the mass transfer coefficient is small, that is, $k_t \ll v$, the solid concentration, c_s, is the same as the bulk melt concentration, c_b. If $k_t \gg v$ the liquid concentration at the interface, c_l, is the same as c_b.

To integrate the Scheil equation in the form

$$\frac{df}{f} = -\frac{dc_b}{c_b - c_s},$$

we must know the ratio between the concentration in the solid and that in the bulk melt:

$$k_{eff} = \frac{c_s}{c_b}.\tag{6.13}$$

Equations (6.12) and (6.13) give, upon inserting $c_l = c_s/k$,

$$k_{eff} = \frac{k}{k + (1-k)\exp\left(-\frac{v}{k_t}\right)}.\tag{6.14}$$

v/k_t may in general depend on time or on the fraction liquid, f. If $v/k_t \ll 1$, then $k_{eff} = k$. If $v/k_t \gg 1$, then $k_{eff} = 1$. If v/k_t does not depend on f and we insert Eq. (6.13) into the Scheil equation, we obtain Eq. (6.7). Stirring gives high values for D_E and k_t and a low value of k_{eff} and a reduced content of impurity in solid.

6.4   Constitutional Supercooling and Stirring

In Section 6.3 we have seen how a solute-rich layer normally builds up in front of a solidifying planar interface. This may lead to an unstable situation with breakdown of the planar front. In this section we will discuss this in more detail.

For most temperature versus composition systems, as illustrated in Fig. 6.5, the liquidus temperature drops with an increased content of solute. Therefore, there is a zone in front of the solidification front where the equilibrium freezing temperature is lowered. Since heat is being drawn away from the melt, the actual temperature is also lower at increasing distances from the solidification front.

This can result in a zone in front of the solid–liquid interface where the actual temperature is lower than the local equilibrium freezing temperature, yet where the liquid metal persists. In this case we say that the liquid is constitutionally supercooled. The term 'constitutional' reflects the importance of the solute composition change (see Fig. 6.5). In a supercooled zone nucleation may be enhanced, giving a smaller grain size. A criterion for the existence and extent of the supercooled zone is given in the following.

(a)

(b)

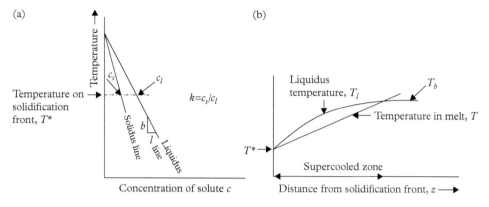

Fig. 6.5 *Constitutional supercooling in alloy solidification: (a) temperature versus composition diagram; (b) comparison of the liquidus temperature profile with the melt temperature.*

The concentration gradient, dc/dz, of the solute at the interface is found from Eqs. (6.8) and (6.9) and from the fact that $\frac{C_s}{C_l} = k$ at $z = 0$:

$$\frac{dc_l}{dz} = -\frac{v}{D_E} c_l (1 - k) \tag{6.15}$$

The relationship between temperature and composition is obtained from the slope of the liquidus line in Fig. 6.5:

$$\frac{\Delta T_l}{\Delta c} = -b. \tag{6.16}$$

Equations (6.15) and (6.16) then give, for the slope of the liquidus temperature at the solidification interface,

$$\left(\frac{dT_l}{dz}\right)_{z=0} = \frac{v}{D_E} b(1 - k) c_l. \tag{6.17}$$

If there is to be a supercooled zone, dT_l/dz must be greater than the temperature gradient $(dT/dz_{z=0})$:

$$\left(\frac{dT_l}{dz}\right)_{z=0} > \left(\frac{dT}{dz}\right)_{z=0} \tag{6.18}$$

or

$$\frac{vb}{D_E}(1 - k)c_l > \left(\frac{dT}{dz}\right)_{z=0}, \tag{6.19}$$

where c_l is obtained from Eq. (6.14). If there is little convection, that is, if the mass transfer coefficient k_t is small, then from $c_b = c_s$, Eq. (6.12), $c_l = c_b/k$, and Eq. (6.19) we find that

$$\frac{vb\,(1-k)\,c_b}{D_E k} > \left(\frac{\mathrm{d}T}{\mathrm{d}z}\right)_{z=0}. \tag{6.20}$$

When stirring is vigorous $c_l \approx c_b$ and the criterion for supercooling becomes

$$\frac{vb\,(1-k)\,c_b}{D_E} > \left(\frac{\mathrm{d}T}{\mathrm{d}z}\right)_{z=0}. \tag{6.21}$$

It is seen that the conditions for supercooling are favoured by a high velocity of solidification, v, and a large value of b (a steep equilibrium liquidus temperature versus composition curve). Furthermore, it is an advantage to minimize the temperature gradient, $\mathrm{d}T/\mathrm{d}z$. In practice this may be accomplished by reducing the pouring temperature and by electromagnetic stirring.

Supercooling results in the instability of the solidification front, since any protrusion forming on the interface finds itself in supercooled liquid and therefore will not disappear. This results in a cellular structure when inequalities (6.20) are fulfilled. This phenomenon can be seen in Fig. 6.6.

The intention of the experiment resulting in the microstructure of Fig 6.6 was to develop a new process for refining of silicon for solar cells.[3] There is a distinct boundary in the microstructure showing the transition from planar to columnar/equiaxed form. The concentration of the solutes followed the Scheil equation in the region with the planar front. Above the transition the concentration was equal to the average of the remaining liquid. The problem was solved by increasing the melt temperature and with electromagnetic stirring of the melt during solidification.

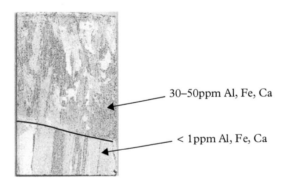

30–50ppm Al, Fe, Ca

< 1ppm Al, Fe, Ca

Fig. 6.6 *The microstructure of silicon solidified by directional solidification, from bottom to top. The initial values of Al, Fe, Ca in the silicon were in the range 30–50 ppm. The solidification rate was about 20 mm/hour. A change from planar to cellular growth can be seen from the change in microstructure. [Republished with permission of Springer from Crystal Growth of Silicon for Solar Cells, Nakajima and Usami, 2009; permission conveyed through Copyright Clearance Center, Inc.]*

With a planar front, impurities can be directly removed according to the principle illustrated in Fig. 6.1. For other types of solidification the refining must be accomplished by separating the solidified phase from liquid phase or the liquid phase from the solid, or by extracting wanted elements from the solid by crushing followed by leaching (solution refining).

At higher solidification velocities, the solute will pile up in the boundary layer. At low solidification velocities, there is complete mixing in the melt and sufficient time for the solute to diffuse through the boundary layer. For a planar front, as shown in Fig. 6.1, the segregation will be over the scale of the entire casting. For equiaxed or dendritic growth, the segregation will be on the length scale of a dendrite or a grain. In addition to solute distribution, segregation due to gravity, shrinkage, or other effects may occur. The length scale is often large and is therefore referred to as macrosegregation.

6.4.1 Macrosegregation

During solidification, segregation of the different elements usually occurs. The variation in concentration of an element due to solute distribution at the solid–liquid interface was discussed in Section 6.3. Variation in the concentration of an element due to other effects is treated here.

There may be several reasons for macrosegregation in a casting. It is also important to realize that macrosegregation may or may not be beneficial to the quality. Sometimes when some macrosegregation is unavoidable, a clever heat treatment of the casting may reduce the harm due to the segregation.

Many authors have discussed various mechanisms for macrosegregation.[4-9] Macrosegregation of an element in a casting is caused by a different content of solute in the solid and the liquid phases, combined with separation of the solid and the liquid phases.

A difference in the solubility of an element in the liquid and the solid phases is the normal situation during solidification when the equilibrium partition ratio (or the effective partition ratio, k_{eff}) is different from 1.

Table 6.1 gives a list of mechanisms that may separate liquid and solid. Normally in a casting the liquid phase may move while the solid phase stays fixed.

In the following a study of casting of FeSi is mentioned as an example of segregation during casting. Ferrosilicon is used as an alloy addition in steel-making (see Chapters 7 and 9), in which case it is important that the content of Si is the same throughout the charge

Table 6.1 *Mechanisms that cause separation of solid and liquid*

Liquid motion in mushy zone may be due to:
(1)   Temperature gradients
(2)   Concentration gradients
(3)   Thermal contractions and solidification shrinkage (volume change on freezing)
(4)   Convection in the bulk melt, which in turn may be induced by temperature or concentration gradients, thermal contractions, or electromagnetic stirring

material. The casting is a square block of iron-silicon with a content of 75 wt% Si. When a liquid of 75 wt% Si–25 wt% Fe is cooled down to 1315 °C the silicon starts to precipitate. The liquid will be enriched in iron until the eutectic composition is reached. Normally, with rapid solidification, the cast product consists of fine grains of silicon and $FeSi_2$ (actually non-stoichiometric $FeSi_{2.5}$). The melt has been cooled both from the bottom and from the top. When cooled from the bottom little vertical segregation seems to occur. The concentration in the solidified material is very nearly the same as in the melt, about 75% Si. This is shown in Fig. 6.7a. Figure 6.7b shows that when the block is unidirectionally cooled from the top with approximately the same solidification conditions, more segregation occurs and the concentration of Si is higher.

The reasons for the segregation with cooling from the top can easily be explained. As mentioned above the silicon will precipitate at the solidification front. The rejected melt will therefore be enriched in iron. This gives a heavier liquid than the bulk melt. This heavy liquid sinks downwards and induces a return convective flow of bulk liquid. Finally, the iron-enriched liquid will solidify in the bottom of the block. The result is a segregation profile as shown in Fig. 6.7b.

When the block is cast with unidirectional cooling from the bottom, a different segregation pattern occurs. The solidification front will consist of solid precipitated silicon and liquid iron–silicon. The rejected iron-rich liquid will be of higher density than the bulk density and will allow little bath movement. This results in less segregation as shown in Fig. 6.7a.

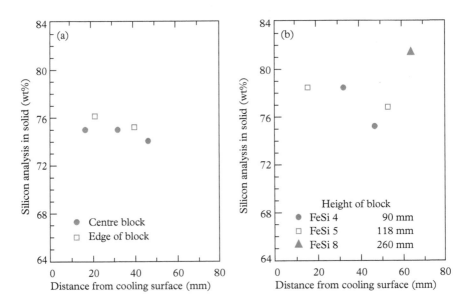

Fig. 6.7 *Si analysis in the solid as a function of the distance from the cooling surface.*[7] *(a) Cooling of the melt is from the bottom. (b) Cooling of the melt is from the top.*

In summary the following two important points should be taken into account for FeSi:

(1) The temperature–composition diagram (the partition ratio between solid and liquid) during solidification. Under thermodynamic equilibrium conditions only silicon should precipitate at temperatures above $1207\,°C$.

(2) The effect of liquid–solid separation during the solidification. The density of the melt depends on temperature and composition. Due to density differences, gravity creates bath movement and liquid–solid separation when cooled from the top. Gravity may prevent the bath movement (natural convection) when cooled from below, since the heavier iron-rich liquid phase tends to remain at the bottom. Thus, the direction of the solidification front is important.

6.4.2 Modelling of Macrosegregation

In general, macrosegregation is caused by a number of mechanisms and is a complex process. The Scheil equation (6.6) simply describes macrosegregation as complete mixing in the liquid zone and no diffusion in the solid.

A number of studies have been performed to describe the macrosegregation which results from mass flow of solute-rich liquid to feed solidification and thermal contractions. Flemings[4] gives an equation for the macrosegregation as a result of interdendritic flows:

$$\frac{\partial f}{\partial c_l} = -\frac{(1-\beta)}{(1-k)}\left[1 - \frac{v}{dz/dt}\right]\frac{f}{c_l}. \tag{6.22}$$

Equation (6.22) is for the one-dimensional case when the solidification front moves with velocity dz/dt in the z-direction. v is the local velocity of the interdendritic liquid relative to the solid. $\beta = (\rho_s - \rho)/\rho_s$ is the solidification shrinkage, and

$$1 - \beta = \frac{\rho}{\rho_s}.$$

If v is zero, Eq. (6.22) reduces to Eq. (6.3), except that now the difference in liquid and solid densities is taken into account. Often the solidification velocity, dz/dt, in Eq. (6.8) is given in terms of the rate of temperature change $\partial T/\partial t$ just after solidification (very close to the solidification front) and the corresponding temperature gradient $\partial T/\partial z$. Now, since temperature is a function of position z and time,

$$T = T(z,t)$$

and

$$dT = \left(\frac{\partial T}{\partial z}\right)_t dz + \left(\frac{\partial T}{\partial t}\right)_z dt. \tag{6.23}$$

On the solidification front the temperature is constant, so that

$$dT = 0.$$

Then

$$\frac{\partial z}{\partial t} = -\frac{\frac{\partial T}{\partial t}}{\frac{\partial T}{\partial z}}. \tag{6.24}$$

Now Eq. (6.22) becomes

$$\frac{\partial f}{\partial c_l} = -\frac{(1-\beta)}{(1-k)}\left(1 + \frac{v\partial T/\partial z}{\partial T/\partial t}\right)\frac{f}{c_l}. \tag{6.25}$$

Recently, several attempts have been made to calculate numerically the liquid fluid flow in the mushy zone during solidification. The procedure requires a calculation of the interdendritic fluid flow. Often this is performed by using Darcy's law.[8,9] Macrosegregation can be described using Darcy's law, together with mass balances, momentum, and energy balances. However, this procedure will not be examined here.

In the following it will be assumed, as in Section 6.3, that there is complete mixing in the bulk liquid with concentration c_b. Furthermore, we usually assume that there is no diffusion in the solid, so that Eq. (6.6) applies. Then c_l is replaced by c_b and the excess solute in the mushy zone is neglected. A total mass transfer resistance $1/k_t$ is introduced which sums the resistance to mass transfer in the mushy two-phase zone, and in the boundary layer between solid and bulk/melt. k_t must be regarded as empirical.

6.5 Segregation of Alloys Displaced from the Eutectic Composition

It will be seen that if the mass transfer coefficient, k_t, is too large, considerable segregation may occur during solidification of alloys near the eutectic composition. The results are of a somewhat general nature, although FeSi is used as an example.

If k_t is small and the ratio v/k_t is large, the iron that is rejected during precipitation of Si from the FeSi melt (with about 25% Fe) accumulates in the mushy zone. Then, after a short transient period, the mushy zone reaches the eutectic composition, c_E, about 42% Fe. Then Eq. (6.12) becomes

$$\frac{c_b - c_s}{c_E - c_s} = a, \tag{6.12a}$$

where, to simplify matters, $\exp(-v/k_t)$ has been replaced by a, and c_b can be integrated. The initial condition is $c_b = c_{in}$ for $f = 1$. Equation (6.12a) is employed to make use of this condition. Then rearranging Eq. (6.12a) slightly gives

$$c_E - c_s = \frac{c_E - c_b}{1 - a},$$

(6.11b)

and we set $c_1 = c_E$ also at $f = 1$.

In the Scheil equation (6.12a) we have from Eq. (6.11a)

$$c_a - c_s = a(c_E - c_s) = \frac{a}{1-a}(c_E - c_b).$$

Then

$$\frac{a}{1-a}\frac{df}{f} = \frac{dc_b}{c_E - c_b}.$$

If a is constant for all f, then we obtain from Eqs. (6.3a) and (6.11b),[2] integrating from $f = 1$ to f,

$$\frac{c_E - c_s(f)}{c_E - c_{in}} = \frac{f}{(1-a)}^{a/(1-a)}.$$

(6.26)

It is seen that when k_t (and therefore a) is small, there is little or no segregation. This is the situation in Fig. 6.7a. Figure 6.7b shows the segregation encountered when solidifying from the top. In experiments 4, 5, and 8 the solidification velocity v was estimated to be 4×10^{-5}, 3.4×10^{-5}, and 7×10^{-5} m/s, respectively.[7] From Eq. (6.26) it is found from Fig. 6.7 that $v/k_t = 1.3$, 1.9, and 2.7, respectively. This gives $k_t = 3 \times 10^{-5}$, 1.8×10^{-5}, and 2.6×10^{-5} m/s. Mass transfer in the melt boundary layer is caused by natural convection induced by temperature and concentration gradients in the melt. The effect of concentration gradients probably dominates.

Consider the mass transfer coefficient k_t for FeSi (75%). For natural convection assume that

$$Sh = 0.15 Ra^{1/3},$$

where

$$Ra = \frac{g\Delta\rho L^3}{vpD} \quad \text{and} \quad Sh = \frac{k_t L}{D}.$$

L is a characteristic dimension of the melt and $\Delta\rho$ is the density difference. On the basis of the three experiments, we take as a rough mean value that there is a 5 per cent difference in iron content between the melt and the solidified material. For 25% Fe the density of the melt is 3200 kg/m^3. For 42% Fe the melt density is 4400 kg/m^3. By interpolation this gives

$$\Delta\rho = (4400 - 3200)\,\frac{5}{42 - 25} = 353 \text{ kg/m}^3$$

$$D_{Fe} \sim 5 \times 10^{-9} \text{m}^2/\text{s}.$$

There are little or no experimental data for the kinematic viscosity v, but the result is not very sensitive to v. Assuming $v = 10^{-6}$ m^2/s, we have

$$k_t = 0.15 D_{Fe}^{2/3}\left(\frac{g\Delta\rho}{\rho v}\right)^{1/3}$$

$$= 0.15 \times 5^{2/3} \times 10^{-6}\left(\frac{9.81 \times 353}{3200}\right)^{1/3}\frac{1}{10^{-2}}$$

$$= 4.5 \times 10^{-5} \text{m/s}$$

In other words, the mass transfer coefficient caused by the concentration gradients is found to be $k_t \sim 4.5 \times 10^{-5}$ m/s. This value is roughly a factor of two too high. However, in the calculation, resistance to mass transfer in the mushy zone has not been taken into account. The admittedly rough correspondence with the calculated value for k_t indicates that k_t can indeed be obtained from computer calculations that should include both the temperature and the concentration effects causing natural convection in the mushy zone and in the melt.

If the ratio v/k_t is very small, there is no accumulation of iron in the mushy zone. The iron is instead transferred directly to the bulk melt. Silicon only is precipitated. Then the bulk melt should reach the eutectic composition c_E before precipitation of the eutectic can occur. Considering that we solidify FeSi (75% Si) very slowly with good mixing in the melt, the relative amounts of pure Si and eutectic (58% Si) is given by the mass balance for Si,

$$75 = 100x + 58\,(1 - x),$$

where x is the fraction that is pure Si. Then

$$x = \frac{17}{42} \text{ and } 1 - x = \frac{25}{42}.$$

That is, the ratio between the amounts of pure silicon and of eutectic precipitated from FeSi with 75% Si in this simple, idealized case is $\frac{17}{25} = 0.68$.

6.6   Refining Alloys by Partial Solidification

Refining by partial solidification or melting is based on the fact that when a melt containing one or more solutes is cooled, or conversely a solid is melted, the solid crystals have different impurity contents from the liquid. An example of industrial applications is the elimination of high melting-point impurities and dross from molten lead. Another process is the solvent refining of silicon. Usually silicon is melted together with another metal that has a low solubility in solid silicon. This process has been used by 6N Silicore and was extensively investigated by Morita for refining silicon for PV. The methods conventionally employed are skimming and sedimentation. At somewhat greater cost sedimentation can be enhanced by centrifuging. Filtration is employed to remove dross and intermetallics from molten metals such as aluminium and magnesium. The most commonly used method is to solidify everything and separate the pure crystals by crushing and etching/leaching. The main practical difficulty in all cases is usually an efficient and complete physical separation of solid and liquid. Traditionally, this has limited refining by fractional solidification to processes where only a small fraction of solid was removed. Techniques such as filtration under high pressures have improved the separation of solid and liquid. This is particularly the case if a non-dendritic structure is formed. This development allows refining by fractional solidification for higher solid fractions, for instance, up to ½.

Here we look at the application where a solidified fraction with a low content of solutes (impurities) is extracted from a molten alloy. This is possible in many cases, since very often the partition coefficient (see Fig. 6.3) for the solute is less than 1, for instance 0.1. Again the Scheil equation, Eq. (6.5), may be employed. It is integrated from liquid fraction $f = 1$ to f. Then the mean relative impurity content, $\bar{c}$, called the refining ratio, in the solidified part, $1 - f$, is

$$\bar{c}_s = \frac{1}{1-f} \int_f^1 c_s df$$

or

$$\frac{\bar{c}_s}{c_{in}} = \frac{k}{1-f} \int_f^1 f^{k-1} df = \frac{1-f^k}{1-f} \tag{6.27}$$

Equation (6.27) is shown in Fig. 6.8 for $k = 0.1$. If there is diffusion in the solid the ratio $\bar{c}_s/c_{in}$ is greater. Furthermore, the low refining ratios illustrated in the figure are achieved only if the liquid is completely removed from the solid.

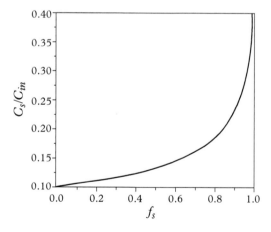

Fig. 6.8 *The refining ratio as a function of fraction solidified, $f_s = 1 - f$, for a partition ratio $k = 0.1$.*

Morita has derived the temperature dependence segregation coefficients for the most detrimental impurities for the Al–Si system. See Table 6.2.

High purity silicon for the chemical industry is produced by refining metallurgical grade silicon typically by Ca addition, solidification, and crushing followed by leaching.[12] The principle is that the impurities are transferred to the liquid phase during solidification and embedded in the CaSi$_2$ phase that is formed around the pure Si crystals. The leaching process will dissolve CaSi$_2$ and leave clean silicon particles. Elkem is marketing this product as Silgrain with a purity of > 99.99% Si; see Fig. 6.9. Research is ongoing to apply this method for solar grade silicon.[13]

Table 6.2 *Temperature dependence of segregation coefficients of metallic impurities in Al–Si Alloys*[10,11]

Element	1073 K	1273 K	1473 K	mp of Si
Fe	1.7×10^{-11}	5.9×10^{-9}	3.0×10^{-7}	6.0×10^{-6}
Ti	3.8×10^{-9}	1.6×10^{-7}	9.6×10^{-7}	2.0×10^{-6}
Cr	4.9×10^{-10}	2.5×10^{-8}	2.5×10^{-7}	1.1×10^{-5}
Ni	1.3×10^{-9}	1.6×10^{-7}	4.5×10^{-6}	1.3×10^{-4}
Cu	9.2×10^{-8}	1.2×10^{-7}	2.1×10^{-6}	1.0×10^{-5}
B	7.6×10^{-2}	2.2×10^{-1}	—	—
P	4.0×10^{-2}	8.5×10^{-2}	—	—

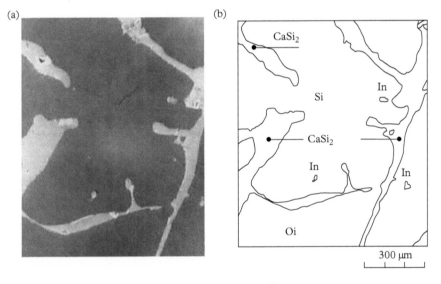

Fig. 6.9 *TEM Micrograph of silicon refined by addition of Ca.*[12] [*Reprinted with permission from Anders Schei.*]

6.7   Refining Alloys by Continuous Draining of Liquid

This process[14] involves heating an alloy within its liquid–solid region while simultaneously draining off the (interdendritic) liquid. The last to melt are the pure crystals that were the first to solidify. The liquid must be continuously expelled during heating, and heating must proceed at such a slow rate that diffusion in the solid has time to proceed. Filtration under high pressures is the method of removal. As shown theoretically in the following, remarkably low refining ratios may be attained. However, residual liquid increases the refining ratios significantly. The reason is good wetting between liquid and solid crystals.

Here we look at the ideal case where there is no residual liquid.

To describe the process we imagine that the temperature increases in small steps and that the temperature remains constant for a time sufficient for the mean solid composition $\bar{c}_s$ to come to equilibrium with the liquid, c_l,

$$\bar{c}_s < kc_l, \tag{6.28}$$

due to the gradient in a crystal as illustrated in Figs. 6.1 and 6.2.

We simplify and $\bar{c}_s$ now represents the solute composition throughout the solid particles. In our optimistic model the liquid with composition c_l is removed in each temperature interval and can no longer influence what happens later when the temperature is again increased slightly. Then we have

$$\frac{dc_l}{dT} = 0 \tag{6.29}$$

for the change of solute concentration with temperature T. The solute is in either the solid or liquid state so that

$$\bar{c}_s f_s + c_l f = \text{constant}, \tag{6.30}$$

or, in terms of temperature increase,

$$\frac{d}{dT} \left(\bar{c}_s f_s + c_l f \right) = 0, \tag{6.31}$$

where f_s is the solid fraction so that

$$f_s + f = 1. \tag{6.32}$$

Now Eqs. (6.29) and (6.31) give the following when the temperature is increased,

$$d \left(\bar{c}_s f_s \right) + c_l df = 0 \tag{6.33}$$

and employing Eqs. (6.28) and (6.32) gives

$$\frac{d\bar{c}_s}{\bar{c}_s} = \left(\frac{1}{k}-1\right)\frac{df_s}{f_s}.$$

Integration from $\bar{c}_s = c_{in}$ for $f_s = 1$ results in

$$\frac{\bar{c}_s}{c_{in}} = f_s^{1/k-1}. \tag{6.34}$$

For instance, if half the material is removed by partial melting, the refining ratio for the remainder is, for $k = 0.1$,

$$\frac{\bar{c}_s}{c_{in}} = 0.5^9 = 0.002.$$

This startling result depends, as previously pointed out, on an efficient continuous removal of interdendritic liquid and on heating so slowly that the solid is always in equilibrium with the liquid that is being melted off. The practical problem as mentioned before is to find a good way to get rid of the liquid. Centrifugation might be a solution.

6.8   Zone Refining

Zone melting, illustrated in Fig. 6.10, can avoid contamination from a container (crucible). Its potential for ultra purification of materials was first recognized when it was applied to germanium. The principle can also be used to grow single crystals and to obtain a controlled uniform or non-uniform composition within the crystal.

Only a small portion of the bar is melted and this molten zone is moved slowly from one end of the bar to the other. Movement is always in the same direction, for instance from bottom to top of the vertical bar. Figure 6.11 shows the solute distribution during and after one pass. The concentration in the solidified material is $c_s = c_l k$, or if the heater is moving too fast only $c_l k_{eff}$, where k_{eff} is given by Eq. (6.14). The rejected solute accumulates in the molten zone. When the heater has moved a distance dx, the accumulated solute is

$$l dc_1 = dx\,(c_{in} - c_s), \tag{6.35}$$

where c_{in} is the constant original composition of the bar. l is the length of the molten zone. At $x = 0$ there has been no accumulation of solute so that here $c_l = c_{in}$. Integration of Eq. (6.35), assuming the zone length l and k (or k_{eff}) are constants, gives

$$\frac{c_s}{c_{in}} = 1 - (1-k)\exp(-xk/l). \tag{6.36}$$

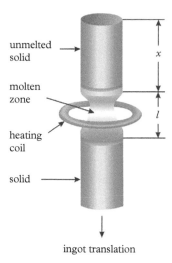

Fig. 6.10 *Molten zone of length l traversing a cylindrical ingot of length L. Based on Pfann.*[15]

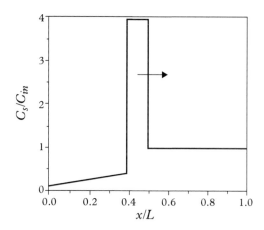

Fig. 6.11 *Solute distribution in zone refining during one pass. l/L = 0.1 and k = 0.1.*

There is no simple way to calculate the solute distribution after multipass zone melting, but a large number of computed curves are given by the inventor of this process, W. G. Pfann.[15] One set of these curves is shown in Fig. 6.12 for a single-phase alloy with a very high effective partition ratio, $k_{eff} = 0.9524$ and a ratio of zone length to ingot length of 1:100. n is the number of passes by the heater.

It is seen that the solute is slowly displaced to the right. For an infinite number of passes there is a limiting curve which may be calculated in a rather simple manner since it does not depend on time. When the heater moves a distance dx, the difference between the solute that

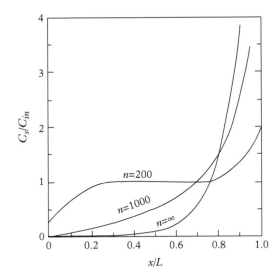

Fig. 6.12 *The refining ratio versus distance x/L with l/L =0.01 and a large k. k = 0.9524. n is the number of passes of the molten zone.*[15]

leaves the molten zone and the solute that enters is $dx(c(x+1)-c(x))$. This is equal to the accumulation, ldc_l.

Furthermore

$$c_1 k_{eff} = c_s(x)$$

so that

$$\frac{l}{k_{eff}} \frac{dc_s(x)}{dx} = c_s(x+1) - c_s(x) \qquad (6.37)$$

A solution of this equation is

$$\frac{c_s}{c_{in}} = A\exp(bx), \qquad (6.38)$$

where b is found by inserting Eq. (6.38) into Eq. (6.37):

$$\frac{lb}{k_{eff}} = \exp(lb) - 1. \qquad (6.39)$$

A is obtained from the condition that the solute remains in the ingot,

$$\int_0^L c_s(x)dx = c_{in}L,$$ (6.40)

where *L* is the ingot length. Thus

$$\frac{A}{b}(\exp(bL) - 1) = L.$$ (6.41)

Equation (6.39) gives, for $k_{eff} = 0.9524$,

$$\frac{bl}{0.9524} = \exp(bl) - 1.$$

With the solution $bl = 9.68/100$. Furthermore, from Eq. (6.41),

$$A = \frac{bL}{\exp(bL) - 1} = \frac{9.68}{\exp(9.68) - 1} = 0.0006.$$

This gives for Eq. (6.38)

$$\frac{c_s}{c_{in}} = 0.0006 \exp(9.68x/L).$$

At $x/L = 0.77$, c_s is equal to the original concentration, c_{in}. At the mid-point of the ingot, $x/L = 0.5$,

$$\frac{c_s}{c_{in}} = 0.076.$$

6.9 Refining Processes in Crystallization of Si for Solar Cells/Crystal Pulling/Directional Solidification

In the fabrication of materials for semiconductors and solar cells, high purity and a low amount of defects are essential for the performance of the devices. The two most important processes are directional solidification (DS) for solar cell silicon and crystal pulling (CZ and FZ) for solar cell silicon and other semiconductors. Common for the two processes is that the solidification is planar and grows in the opposite direction to the heat flow. Typically, a large ingot (as shown in Fig. 6.1) is produced and the solidification rate is controlled by the ratio between the thermal gradients, respectively, in the solid and liquid.

For solar cells the impurity of metals must be below 10^{-15} atoms/cm^3, resulting in processes with an extreme focus on impurity control. We will treat this in detail with other

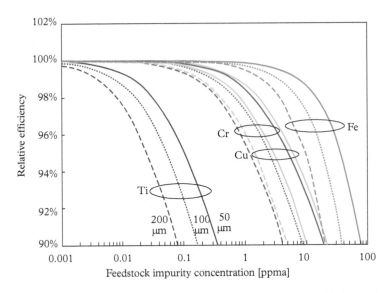

Fig. 6.13 *The effect of impurities on the relative efficiency of solar cells. [Republished with permission of John Wiley & Sons–Books from Di Sabatino[16]; permission conveyed through Copyright Clearance Center, Inc.]*

processes in mind. Coletti[16] (Fig. 6.13) has shown the relative solar cell efficiency that can be obtained for various concentrations of feedstock purity. The effect of impurities on the solar cell efficiency depends on the wafer thickness, as shown for Ti at 50, 100, and 200 µm. The impurities reduce the length that the electrons–holes separated by the energy from the sun can diffuse before they recombine. Higher diffusion lengths increase the probability for the carriers to diffuse to the surface of the wafers where the terminals are connected.

When the impurity levels are so low, one also has to be aware of the problem of contamination of the refined material during the refining process. In additions to metals, also the light elements N, C, and O can be detrimental for solar cell silicon due to the formation of particles. The reactions involved for the C and O circuit are shown for the Bridgeman process in Fig. 6.14.

The crucible and coating are the main source of impurities that will enter the silicon during melting and stabilization. Other sources of contamination are the feedstock, graphite dust in the furnace, impurities induced by handling, and those picked up from the atmosphere.

In the DS process we have a planar front where the impurities with lower solubility in the solid than in the liquid will be pushed to the top of the ingot. C and N will typically reach levels that are high enough to form SiC and Si_3N, respectively, as shown in Fig. 6.16. In Fig. 6.15, the C level measured in a 12-kg MC ingot is given as a function of distance from the bottom of the ingot.

The following reactions take place:

$$C + Si\,(l) = SiC\,(s)$$
$$4N + 3Si\,(l) = Si_3N_4\,(s).$$

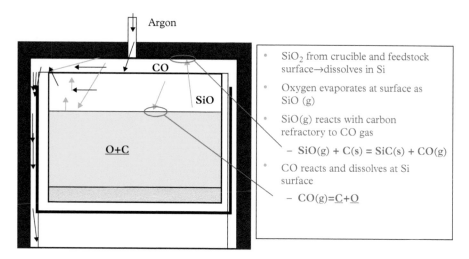

Argon

CO

SiO

O+C

- SiO$_2$ from crucible and feedstock surface→dissolves in Si

- Oxygen evaporates at surface as SiO (g)

- SiO(g) reacts with carbon refractory to CO gas

  - SiO(g) + C(s) = SiC(s) + CO(g)

- CO reacts and dissolves at Si surface

  - CO(g)=C+O

Fig. 6.14 *The C and O circuit in directional solidification of silicon by the Bridgeman process.*

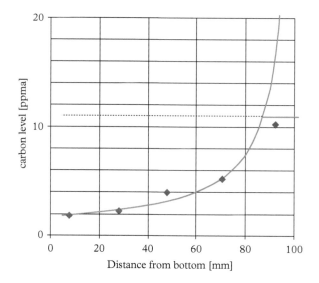

carbon level [ppma]

Distance from bottom [mm]

Fig. 6.15 *The C level presented as a function of distance from the bottom. The solid curve shows the theoretical segregation pattern (Scheil equation). The solubility concentration of C in equilibrium with SiC is indicated with a dotted line.*

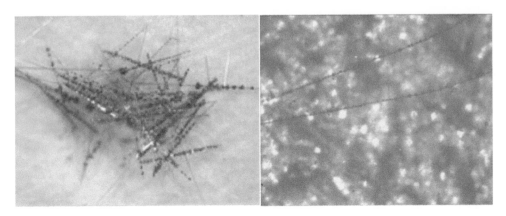

Fig. 6.16 *Nitride needles with SiC particles (left) extracted from the top of the ingot by dissolution of the silicon, and (right) visualized by a IR camera and backlight from a heat plate (silicon is transparent to IR radiation).*

The particles can create problems with instability during solidification and during cutting of material into thin wafers that are used as the basis for solar cells. The wafer thickness is typically 130 μm, so the particles are relatively large and can cause shunting of the solar cells, reducing the efficiency or creating hot spots resulting in the failure of the solar cell during operation.

The Czochralski process is one of the most important processes developed, making it possible to fabricate pure silicon single crystals used for electronic devices. In principle, a single crystal is pulled out of the melt using a seed to achieve a given crystal orientation. Dopants are added to give the wanted resistivity. Quartz crucibles are used. They are a source of oxygen, but other impurities also come as a result of dissolution of the crucible. For semiconductors oxygen clusters are used as gettering sites for impurities and to give additional mechanical strength to the substrates. The gettering is carried out by annealing of the silicon substrates.

During the crystallization, the melt is stirred by rotation and natural convection. Therefore, the Scheil equation can be used. It describes the segregation of the metallic impurities and carbon to a high degree of accuracy. For elements with high vapour pressure or elements dissolved from the crucible, the equation must be expanded.

6.10   The Czochralski Crystal Puller

A model for impurity pick-up from refractories in directional solidification has been developed as an extension of the model given by DiSabatino.[17,18] A simplified geometry of the Czochralski crystal puller is taken to be as shown in Fig. 6.17, which illustrates the situation for two different crucible materials. The model presented is essentially a mass balance where casting proceeds at a constant rate. Thermodynamics is taken into account with x_r and x_l being the mass fractions of the impurity and its solubility with the crucible, and the rate constants k_v, k_w the evaporation and transfer from the crucible.

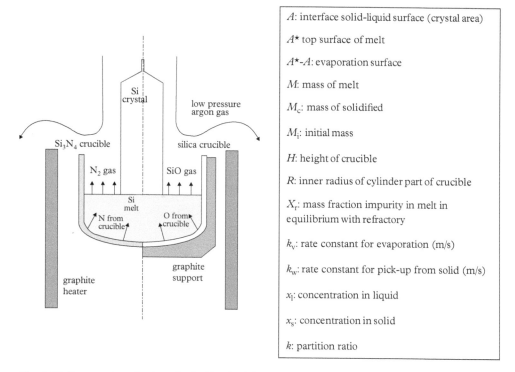

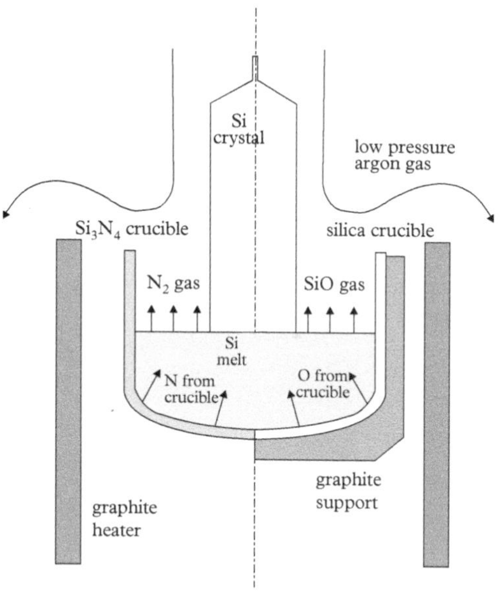
Si
crystal
low pressure
argon gas
Si3N4 crucible
silica crucible
N2 gas
SiO gas
Si
melt
N from
crucible
O from
crucible
graphite
support
graphite
heater

A: interface solid-liquid surface (crystal area)

$A^{\star}$ top surface of melt

$A^{\star}$-A: evaporation surface

M: mass of melt

M_c: mass of solidified

M_i: initial mass

H: height of crucible

R: inner radius of cylinder part of crucible

X_r: mass fraction impurity in melt in equilibrium with refractory

k_v: rate constant for evaporation (m/s)

k_w: rate constant for pick-up from solid (m/s)

x_l: concentration in liquid

x_s: concentration in solid

k: partition ratio

Fig. 6.17 *Transport mechanisms in the Czochralski process.*

The assumptions are:

1. complete mixing of impurity in the melt;
2. the evaporation of Si (as SiO) can be neglected;
3. k_V is the rate constant of evaporation;
4. k_W is the rate constant for transfer from the crucible to the melt at the crucible–melt interface;
5. the radial distribution for x_s of the concentration in the solid can be neglected;
6. the crucible is cylindrical; and
7. that the bottom of the crucible is not flat can be neglected.

A mass balance for impurity x becomes

$$\frac{d\,(Mx_l)}{dt} = -\dot{M}_c x_s - \rho k_v (A^* - A) x_l + 2\pi RHk_w \rho (x_r - x_l)\left(1 - \frac{M_c}{M_i}\right) + \pi R^2 k_w \rho (x_r - x_l), \quad (6.42)$$

$\uparrow (a) \qquad \uparrow (b) \quad \uparrow (c) \qquad\qquad \uparrow (d) \qquad\qquad \uparrow (e)$

where

(a) rate of accumulation in melt

(b) rate of transfer to solid

(c) evaporation rate

(d) net amount of x that enters the melt from the wall $\left[\frac{\text{kmol}}{\text{s}}\right]$

(e) net amount of x that enters the melt from the crucible bottom $\left[\frac{\text{kmol}}{\text{s}}\right]$

and x_r is the content of x in the melt in equilibrium with the refractory. Note that this equation only applies for $x_l > x_r$. This point will be treated in the discussion. Inserting $M_i = M + M_c$ gives

$$M_i \frac{dx_l}{dt} - \dot{M}_c(t)x_l - M_c \frac{dx_l}{dt}$$

$$= -\dot{M}_c x_s - \rho k_v \left(A^* - A\right) x_l + 2\pi RHk_w \rho \left(x_r - x_l\right)\left(1 - \frac{M_c}{M_i}\right) + \pi R^2 k_w \rho \left(x_r - x_l\right)$$

The partition coefficient obtained from the phase diagram is

$$k = \frac{x_s}{x_l}.$$

Thus, Eq. (6.42) becomes

$$(M_i - M_c)\frac{dx_l}{dt} = \dot{M}_c x_l \left(1 - k\right) - \rho k_v \left(A^* - A\right) x_l$$

$$+ 2\pi RHk_w \rho \left(1 - \frac{M_c}{M_i}\right)\left(x_r - x_l\right) + \pi R^2 k_w \rho \left(x_r - x_l\right).$$

Dividing by $(M_i - M_c)$ we get

$$\frac{dx_l}{dt} = \frac{\dot{M}_c x_l \left(1 - k\right)}{\left(M_i - M_c\right)} - \frac{\rho k_v \left(A^* - A\right) x_l}{\left(M_i - M_c\right)}$$

$$+ \frac{2\pi RHk_w \rho \left(1 - \frac{M_c}{M_i}\right)\left(x_r - x_l\right)}{\left(M_i - M_c\right)} + \frac{\pi R^2 k_w \rho \left(x_r - x_l\right)}{\left(M_i - M_c\right)}.$$

Introducing

$$\frac{dM_c}{dt} = \dot{M}_c \text{ or } dt = \frac{dM_c}{\dot{M}_c}$$

gives

$$\frac{\mathrm{d}x_l}{\mathrm{d}M_c} = \frac{x_l}{(M_i - M_c)}\left((1-k) - \frac{\rho k_v \left(A^* - A\right)}{\dot{M}_c} - \frac{\pi R^2 k_w \rho}{\dot{M}_c} - \frac{2\pi R H k_w \rho \left(1 - \frac{M_c}{M_i}\right)}{\dot{M}_c}\right)$$

$$+ x_r \left(\frac{2\pi R H k_w \rho \left(1 - \frac{M_c}{M_i}\right)}{\dot{M}_c \left(M_i - M_c\right)} + \frac{\rho \pi R^2 k_w}{(M_i - M_c)\dot{M}_c}\right).$$

Finally, when we introduce

$$B = \left((1-k) - \frac{\rho k_v (A^* - A)}{\dot{M}_c}\right) \text{ (melt surface/atmosphere)}$$

$$B2 = \frac{\rho \pi R^2 k_w}{\dot{M}_c} \text{ (bottom of crucible)}$$

$$B1 = \frac{2\pi R H k_w \rho}{\dot{M}_c} \text{ (side wall of crucible)}$$

we have

$$\frac{\mathrm{d}x_l}{\mathrm{d}M_c} = \frac{Bx_l}{(M_i - M_c)} - \frac{B2x_l}{(M_i - M_c)} - \frac{x_l B1}{M_i} + x_r \left(\frac{B1}{M_i} + \frac{B2}{(M_i - M_c)}\right).$$

When the mass fraction solid $f = \frac{M_c}{M_i}$ is introduced (note that now f refers to the solid; previously it referred to the fraction liquid, but we wish to avoid f_c), we get

$$\frac{\mathrm{d}x_l}{M_i \mathrm{d}f} = \frac{Bx_l}{M_i(1-f)} - \frac{B2x_l}{M_i(1-f)} - \frac{x_l B1}{M_i} + x_r \left(\frac{B1}{M_i} + \frac{B2}{M_i(1-f)}\right)$$

or

$$\frac{\mathrm{d}x_l}{\mathrm{d}f} = \frac{Bx_l}{(1-f)} - \frac{B2x_l}{(1-f)} - x_l B1 + x_r \left(B1 + \frac{B2}{(1-f)}\right).$$

We assume that B, B1, B2 are constant. This equation is solved in Appendix 4.
From Eq. (6.42) we get

$$\frac{x_l}{x_0 x_l^0}$$

$$= 1 + \frac{x_r}{x_0}\left[\frac{1}{x_l^0} - 1 + B \exp(B1)\left\{\frac{1-(1-f)^{1-\epsilon}}{1-\epsilon} - \frac{B1\left(1-(1-f)^{2-\epsilon}\right)}{1!\,(2-\epsilon)} + \frac{B1^2\left(1-(1-f)^{3-\epsilon}\right)}{2!\,(3-\epsilon)}\right\}\right].$$

$$(6.43)$$

Note that if $B_1 = 0$ and $B_2 = 0$, then

$$1 - \epsilon = B, \quad x_l^0 = (1-f)^{\epsilon-1}$$

$$\frac{1}{x_l^0} - 1 + B \left\{ \frac{1-(1-f)^{1-\epsilon}}{1-\epsilon} \right\} = \frac{1}{x_l^0} - 1 + \left\{ 1 - (1-f)^{1-\epsilon} \right\} = \frac{1}{x_l^0} - 1 + 1 - (1-f)^{1-\epsilon} = 0,$$

so that we are left with the Scheil equation:

$$\frac{x_l}{x_0 x_l^0} = 1, B_1 = 0, B_2 = 0,$$

if $f = 0$, $x_l^0 = 1$, and $x_1 = x_0$.
If $f \ll 1$, $B_1 \neq 0$, $B_2 \neq 0$, then

$$1 - (1-f)^{n-\epsilon} \approx f(n-\epsilon)$$

$$\frac{x_l}{x_0} = x_l^0 + \frac{x_r}{x_0} \left[1 - x_l^0 + Bx_l^0 \exp(B_r) \left(1 - \frac{B_1}{1!} + \frac{B_1^2}{2!} - \frac{B_1^3}{3!} + \ldots \right) f \right]$$

$$\frac{x_l}{x_0} = x_l^0 + \frac{x_r}{x_0} \left[1 - x_l^0 + x_l^0 Bf \right]. \tag{6.44}$$

An example of calculations of x_l using Eq. (6.43) follows:
$x_0 = 0.1$
$k = 0.1$

$$\frac{\rho k_v \left(A^* - A \right)}{\dot{M}_c} = 0.05$$

$B_1 = 0.1$
$B_2 = 0.05$
$f = 0.8$

$$\epsilon = k + \frac{k_v \rho \left(A^* - A \right)}{\dot{M}_c} + B_2 = 0.2$$

$$x_l^0 = (1-f)^{-1+\epsilon} \exp(-B_1 f) = 3.34$$

$$B = B_2 + (1-\epsilon) = 0.85$$

$$a = \frac{1-(1-f)^{1-\epsilon}}{1-\epsilon} - B_1 \frac{\left(1-(1-f)^{2-\epsilon}\right)}{2-\epsilon} + \frac{B_1^2 \left(1-(1-f)^{2-\epsilon}\right)}{2!} \frac{}{3-\epsilon} - \ldots = 0.854.$$

(a)   (b)

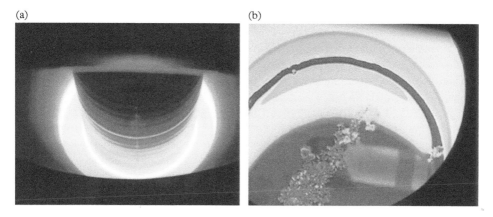

Fig. 6.18 *(a) View through the window of a Czochralski crystal puller showing the crystal being pulled out of the melt. (b) Silicon nitride particles flowing in the melt after removal of the ingot.*

We employ Eq. (6.43):

$$x_l = 0.1 \times 3.34 + x_r [1 - 3.34 + 0.85 \times 3.34 \times 1.105a]$$
$$x_l = 0.334 + x_r \cdot 0.34$$
$$x_r - x_l = 0.66 \times x_r - 0.334$$

$x_r - x_l$ is positive if $0.66 \times x_r - 0.334 > 0$. This means that the model with $f \leq 0.8$ is valid for $x_r > 0.51$.

The correct value of x_r is determined by the solubility in the metal of the impurity in the refractory. This must be determined experimentally and is shown in Fig. 6.15 for the solubility of C in silicon where the concentration of dissolved carbon reaches the limit where SiC is precipitated. In Fig. 6.18 we can see visually how Si_3N_4 is formed on the surface of liquid silicon in a Si_3N_4 crucible. The particles were formed during crystal growing due to an increase in the N level due to a higher solubility of N in liquid than in solid silicon.

6.11 Nucleation and Grain Refinement

Most properties of metals and alloys are affected by the as-cast structure, even after extensive deformation, heat treatment, and recrystallization have taken place. The reason is that these treatments cannot substantially change the distribution of many constituents and inclusions. Hence, it is important to achieve the best possible as-cast structure. This is often the responsibility of a process metallurgist working in the cast house.

Generally, the objective is to create a structure of fine, equiaxed grains having a random orientation. However, a fine grain structure is not always the most important condition for optimal properties. Especially in multiphase alloys, and in materials to be extensively worked, a small interdendritic spacing that produces a good distribution of phases and inclusions may be more important than grain refinement. Fortunately, grain refinement often reduces the variations of constituents in the microstructure.

One example is the formation of porosity in structural aluminium alloys. Figure 6.19 shows the grain structure observed in an A356 alloy casting. This casting was cut in half, polished on the cut surface, immersed briefly in Poulton's etch, and then illuminated from each side by two different coloured lights. When viewed in this lighting under a microscope, each individual grain has a slightly different colour. One can observe black areas between individual grains in this figure. These black areas are pores formed by segregation of dissolved hydrogen during solidification. These typically form during the last 5 to 10 per cent of solidification. Hence, pores need to somehow 'fit into' the spaces left in between already solidified grains. When the grain size is smaller, the pores also are smaller.[19] Smaller pores result in improved fatigue strengths.

Another example is the grain refinement of ingots used to make can stock. Figure 6.20 shows the grain structure in two castings made from 3004 alloy. These castings were made using the Alcoa test, which simulates conditions found in the DC casting of slab. These samples are 200 mm in length (from left to right in the figure). Both samples were held in a preheated, insulated mould and then cooled directionally by applying a water-cooler copper chill. The direction of solidification was from right to left as shown in the figure. In the absence of refinement, large elongated columnar grains are present in most of the top casting. Porosity and brittle intermetallic phases formed during solidification are located in between these large 'feather'-like grains. Hence, the mechanical properties in the casting (especially ductility or elongation) are highly directional. This can be a problem. Figure 6.21 shows ingots of this can stock alloy, which cracked during casting, owing to a loss of grain refinement.

The method of grain refinement most used industrially is the addition of nucleating agents, known as grain refiners or inoculants. This is especially true for the Al, Mg, and Cu alloys

Fig. 6.19 *Grain structure in an A356 (Al–7Si–0.3Mg) alloy casting (50×).*

chill

chill

Fig. 6.20 *Grain structure in 3004 alloy castings. (top: no refinement; bottom: 10 ppm B added as Al–5Ti–1B alloy).*

Fig. 6.21 *Cracks in 3004 alloy DC cast slabs.*

where the additions may be so small (< 0.10 per cent) that they do not appreciably affect other properties. For Ni alloys a substantial amount (> 1 per cent) must be added to get the refining effect.

Grain refinement in castings is produced by increasing the number of nucleating centres, so that a large number of crystals are formed. These nucleants must be solid in the melt.

They are sometimes already present in the melt, but are more commonly added purposely. A typical commercial refiner is the aluminium master alloy which contains 5% Ti and 1% B. This refiner was used to refine the casting in the bottom of Fig. 6.20. Figure 6.22 shows the typical microstructure of this master alloy. It is a mixture of $TiAl_3$ and TiB_2 particles suspended in a matrix of aluminium. The larger particles are $TiAl_3$. The smaller particles are only a few microns in size, and are TiB_2. Simple calculations will show that each cubic centimetre of this refiner contains about 500 million $TiAl_3$ and TiB_2 particles. Each has the potential to nucleate a solid grain of aluminium.

Many authors have given requirements for a good nucleant, but none of the criteria seem to be generally applicable.[20-2] The most widely accepted relationship between nucleant and nucleus is the similarity of lattice. It is an advantage that the nucleant has a crystal structure and lattice parameters very close to those of the nucleus. However, some nucleants are effective even though only a few atoms in the lattices coincide.

Furthermore, the nucleus must 'wet' the nucleant. Perhaps the best approach for the study of wetting is to look at the three-phase system: nucleant, solidified melt (nucleus), and melt. Such three-phase systems were discussed in Section 5.5. The interfacial tension between solid and liquid metal was examined in Section 2.2.8. If the nucleant is wetted by the nucleus, the growing nucleus can spread out over the nucleant. That is, when the interfacial tensions are favourable, wetting can occur. The problem is that very little reliable data are available for solid–liquid and solid–solid interfacial tensions.

In view of its commercial importance, the grain refinement of aluminium has been studied in greater detail than in other metals. Hence, it will be instructive to review our knowledge of grain refinement in Al-based alloys.

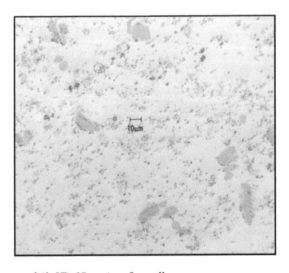

Fig. 6.22 *Microstructure of Al–5Ti–1B grain refiner alloy.*

6.11.1   Effectiveness of Nucleants

Arnberg, Bäckerud, and Klang[23] have shown that aluminium atoms on the (110) planes of TiAl$_3$ match well with the (112) planes of solid aluminium. Davies, Dennis, and Hellawell[24] found that the most common orientation relationship in their samples was Al(011)||TiAl$_3$(110). Marcantonio and Mondolfo[25] found in very slowly cooled samples that the (100) planes of TiAl$_3$ were active in nucleating solid aluminium. The most common orientation was Al(221)||TiAl$_3$(001). No measurable undercooling was found in this case. The orientation Al(001)||(001)TiAl$_3$ was also found at an undercooling of 3 to 5 °C. Kobayashi, Hashimoto, and Shinghu[26] examined the orientation relationship in thin film samples extracted from molten aluminium by dipping loops of molybdenum wire into the melt. The four relationships given above, as well as all others reported in the literature (there are 11 in all), were analysed in a subsequent review,[27] which showed that the simple model of lattice disregistry is not sufficient to explain the results. Two other theories resulting from models of grain boundary structure were proposed and found to have better predictive value. It is clear from these studies that the planes of TiAl$_3$ in contact with liquid metal—(001) and (011)—are effective nucleation substrates for solid aluminium.

6.11.1.1   The Peretectic Theory

In addition to the nucleating potential of TiAl$_3$, the presence of the peritectic reaction

$$L + TiAl_3 \rightarrow Solid$$

at temperatures above the melting point of pure aluminium has a profound influence on nucleation and growth of solid aluminium nuclei. (See the phase diagram shown in Fig. 6.23.) Consider a TiAl$_3$ particle in contact with pure aluminium, or a commercial alloy of low titanium content. Shortly after the addition of a master alloy, a diffusion field will be established in the vicinity of the TiAl$_3$ crystal. At the surface of TiAl$_3$, the titanium content in liquid metal is about 0.15 per cent. (Remember that the concentration away from the particle, in the bulk of the metal phase, is much lower.) Now, when the metal is cooled, because titanium raises the melting point of the metal, and because TiAl$_3$ is an effective nucleant, solid aluminium will form at the surface of the TiAl$_3$ particle at temperatures *above* the melting point of the bulk metal. Once solid has formed, however, the TiAl$_3$ particle becomes engulfed in the solid phase, and further growth is limited by the diffusion of titanium from TiAl$_3$ through the shell of solid aluminium. The combined effect of nucleating potency, and the ability to slow grain growth, makes a master alloy containing TiAl$_3$ particles an effective grain refiner.

Cooling curves from solidifying samples support the hypothetical sequence of nucleation and growth events described already. When no master alloy is present, an undercooling of a few degrees is observed, followed by recalescence and growth at a higher temperature. Coarse grains are found in the casting. When TiAl$_3$ particles from a master alloy are introduced in amounts large enough to produce a fine-grained casting (0.02 to 0.03% Ti in 99.7% Al), the undercooling disappears. Instead, the nucleation starts at a temperature slightly above the freezing point of the alloy, but still below the peritectic temperature. Fine

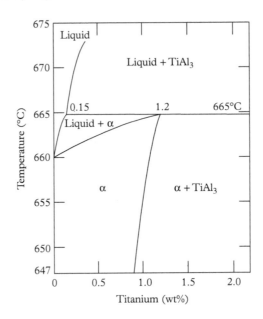

Fig. 6.23 *Aluminium-rich end of the Al–Ti composition versus temperature diagram[20] showing the peritectic at 665 °C and 0.15% Ti.*

grains are observed in the casting, and TiAl₃ particles are frequently observed in the grain centers.[20,24,25] However, after a long contact time—more than about 30 minutes—the grain size becomes coarse once again and the cooling curve reverts to the original form. Evidently, the TiAl₃ crystals have dissolved at longer holding times, and once they are gone, the entire sequence of events resulting in grain refinement no longer takes place.

6.11.1.2 The Role of Boron and Alloy Composition

Grain refinement was first accomplished in the 1930s by titanium additions. This was the established commercial practice until the 1960s, when the first Al–Ti–B master alloys were introduced. Rod feeders were introduced about the same time, so continuous additions of a grain refiner rod could be made into the launder or degasser during the length of a cast.

Boron-containing grain refiners are longer lasting. This is because TiB₂ particles are virtually insoluble in aluminium at normal casting temperatures. They are also more powerful, which means the addition level (and cost of refinement) can be reduced. The effect of adding boron to Al–Ti grain refiners can be seen in Fig. 6.24. Three Al–5% Ti alloys were prepared using a similar manufacturing process, only each alloy contained a different amount of boron. Their effectiveness was measured in 99.7% pure Al (P1020). Samples were taken at various times after the addition, between one-half and 500 minutes. (This is usually referred to as 'contact' time.) The beneficial effect of boron is obvious from these results: a better and longer lasting grain refinement results.

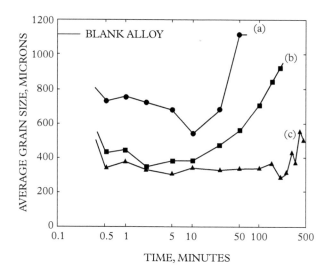

Fig. 6.24 *Influence of boron content on the grain-refining response of three master alloys in 99.7% Al (0.01% Ti was added at 700°C). (a) Al–5.35% Ti alloy (containing less than 20 ppm B), (b) Al–5.4% Ti–0.034% B alloy, and (c) Al–5.0% Ti–0.2% B alloy. (Plot produced from tabulated data by Guzowski and co-workers.[20])*

Numerous theories have been proposed for the beneficial effects of boron, and a review of the scientific literature will find conflicting, and sometimes controversial, theories. The subject is too complicated to discuss in detail here. The interested reader is instead referred to the paper by Guzowski et al.[20] This provides a good introduction to the subject.

A few simple observations can be made, however. First, it is clear that *in certain alloys* (as defined below) TiB_2 particles are excellent nucleants. Several researchers have observed TiB_2 particles in the centre of aluminium grains. Næss and Rønningen[28] examined samples carefully and established that the coincidence of lattice planes was $(201)TiB_2||(311)Al$. The effect of alloy composition was studied by Guzowski et al.[20] They produced master alloys containing only borides. Moreover, since AlB_2 and TiB_2 have the same crystal structure– and are mutually soluble in one another–they also produced $(Al,Ti)B_2$ borides having compositions intermediate between AlB_2 and TiB_2. None of these master alloys contained $TiAl_3$. When borides were added to relatively pure aluminium (99.7% Al containing 40 ppm dissolved Ti) there was only a slight grain refinement. However, when significant amounts of dissolved Ti were also present, the grain refinement was much better. Figure 6.25 shows a plot of their results. (A 10-g addition of each master alloy was made to a 14-kg melt held at 730°C, so the total amount of boron added was the same in each test.) These results point to an important factor which has not been considered yet: the role of alloy composition and its effect on grain growth.

Let us consider once again the redistribution of solute occurring on solidification, as illustrated schematically in Fig. 6.1. As each aluminium grain grows during solidification, it is surrounded by a boundary layer of solute-rich liquid. The situation at the solid–liquid

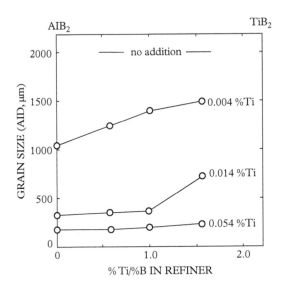

Fig. 6.25 *Grain sizes produced by boride additions to 99.7% Al.*

interface is illustrated in Fig. 6.4. The diffusion of solute away from the grain is necessary for the grain to grow. And because the diffusion-controlled solute transfer is rather slow, compared to thermal diffusivity in aluminium, this boundary layer restricts the growth rate of the newly formed grains.

The effect of alloy composition on grain growth is best considered by defining a growth restriction factor (GRF). This has been considered by several investigators, but the best study for our purposes is the paper by Easton and StJohn and colleagues.[29,30] The GRF is

$$GRF = \sum m_i c_i (k_i - 1),$$

where m_i is the slope of the liquidus in the binary Al–i system; or the freezing point depression which occurs when 1 wt.% of element i is added to pure aluminium, c_i is the concentration of element i in weight percent, and k_i is the distribution coefficient ($k_i = \%i_{solid}/\%i_{liquid}$).

Table 6.3 shows the values of these coefficients for a number of elements, calculated from the associated binary phase diagrams with aluminium.[29] Note that titanium has the most powerful effect on grain grown, on a weight percent basis, with a value of $m(k - 1)$ equal to 220. Of course the solubility of titanium is only 0.15% in pure aluminium at its melting point (%max in the table), so the maximum GRF possible by adding titanium is 0.15×220 or 33.

With the concept of the GRF in mind, it is possible to define an optimum grain refining practice for any particular alloy. Assuming that metal from the potrooms contains 50 ppm Ti and 20 ppm V (a typical value in many smelters), we may then calculate the GRF for any specific alloy, and this value can be used to estimate the grain size. This has been done for a number of alloys in Fig. 6.26. Alloys with a GRF less than a value of 10–15 will require

Table 6.3 *Alloy constants for several elements calculated from phase diagrams*[29]

Element	k	m	m(k − 1)	%max
Ni	0.007	−3.3	3.3	6
Fe	0.02	−3	2.9	1.8
Si	0.13	−6.6	5.7	12.6
Cu	0.17	−3.4	2.8	33.2
Zn	0.4	−1.6	1.0	50
Mg	0.51	−6.2	3.0	34
Mn	0.94	−1.6	0.1	1.9
Nb	1.5	13.3	6.7	≈0.15
Cr	2	3.5	3.5	0.4
Hf	2.4	8	11.2	≈0.5
Ta	2.5	70	105	0.10
Mo	2.5	5	7.5	≈0.1
Zr	2.5	4.5	6.8	0.11
V	4	10	30	≈0.1
Ti	≈9	30.7	≈220	0.15

Source: Reprinted by permission from Springer Nature: copyright 1999.

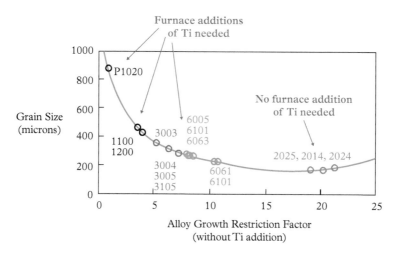

Fig. 6.26 *Predicted grain sizes for several Al alloys.*

furnace additions of Ti to increase GRF and produce an appropriate small grain size. Alloys with a higher value of GRF will not need furnace additions of titanium.

From the above analysis we can see that titanium has a dual function in the grain refining of aluminium. It can provide a nucleating substrate, when present in the form of $TiAl_3$ or TiB_2, and it also lowers grain size by restricting the rate of grain growth. Titanium can also be present in two forms in master alloys. The aluminide phase dissolves in aluminium, whereas the boride does not. Many have failed to realize these distinctions in the past, which has led to a good deal of confusion about the mechanisms of grain refinement in aluminium.

A number of other particles can be added to grain refine aluminium. However, only Al–Ti–C master alloys are used commercially. These are used only in a few aluminium alloys, where boron additions produce hot cracking during casting. This is because TiB_2 is generally a much better nucleant than TiC. Also, with longer contact times TiC particles are not stable. They react with aluminium to form Al_4C_3. Ultrasonic vibration can also produce some grain refinement, but this procedure is not used commercially.

A good deal of research has been done on the grain refining of magnesium alloys. This work was reviewed by StJohn and co-workers.[31] The grain refinement of magnesium alloys can be classified into two broad groups: alloys that contain aluminium, and those that do not. The alloys that are free of aluminium are generally very well refined by Zr-containing master alloys. The grain size is well characterized by the growth restriction factor (called Q in their paper). Evidently, the mechanisms for Zr grain refining of Mg are very similar to those for Ti-based grain refining of Al alloys. On the other hand, the grain refinement in Mg–Al alloys is more complicated, presumably owing to the interaction between impurity elements and dissolved aluminium. Balart and co-workers have conducted an extensive study of the grain refining of copper.[32] The situation appears to be rather complicated. Contrary to the results obtained in aluminium- and magnesium-based alloys, there is a poor correlation between grain size and the growth restriction factor in copper. Foundry studies of several copper-based alloys show that castability (especially resistance to hot cracking) is improved with grain refinement.[33,34] It was found that boron refined the yellow brasses, and zirconium refined other copper-based alloys. Boron refinement is long lasting, whereas zirconium refinement tends to 'fade' owing to its loss by oxidization.

An excellent review of grain refining has recently been completed by Greer.[35] This paper gives an extremely detailed explanation of the theories providing our current understanding of grain refinement. Liu has also given an extremely detailed review.[36] It is noteworthy for its coverage of grain refinement in zinc-based alloys.

....

REFERENCES

1. Aas H. The silgrain process: Silicon metal from 90% ferrosilicon. In: *Light Metals 1971*, 650–67; 1971.
2. Ceccaroli B, Øvrelid E, Pizzini S, eds. *Solar silicon processes: technologies, challenges, and opportunities*, 1st edn. Boca Raton, FL: CRC Press; 2017.
3. Øvrelid EJ, Geerligs B. Wærnes AN, et al. Solar grade silicon by a direct metallurgical process. In: Øye H, Brekken H, Foosnæs T, Nygaard L, eds, *Silicon for the chemical industry VIII*, pp. 223–34. Trondheim, Norway; 2006.

4. Flemings MC. Segregation in castings and ingots. In: *Proc.Elliott Symposium, June 10-13, Cambridge, Mass.*, pp. 253–72. Warrendale, PA: AIME; 1990.

5. Flemings MC. *Solidification processing*. New York: McGraw-Hill; 1974.

6. Minkoff I. *Solidification and cast structure*. New York: Wiley; 1986.

7. Tveit H. *Størkning av 75% ferrosilisium, forløp, struktur og styrke*. Dr.ing. Thesis, Metallurgisk Inst. (in Norwegian). Trondheim, Norway: NTH; 1988.

8. Mehrabian R, Keane M, Flemings MC. Interdendritic fluid flow and macrosegregation; influence of gravity. *Metall. Trans. B*, 1970; 1: 12097–20.

9. Ganesan S, Poirier DR. Conservation of mass and momentum for the flow of interdendritic liquid during solidification. *Metall. Trans. B*, 1990; 21: 173–81.

10. Yoshikawa T, Morita K. Refining of Si by the solidification of Si-Al melt with electromagnetic force. *Iron and Steel Inst. Japan Internat.*, 2005; 45(7): 967–71.

11. Yoshikawa T, Morita K. Refining of silicon during its solidification from a Si–Al melt. *J. Crystal Growth*, 2009; 311(3): 776–9).

12. Schei A. High purity silicon production. In: Engh TA, Lyng S, Øye HA, eds, *Proceedings of the international seminar on refining and alloying of liquid aluminium and ferro-alloys, Trondheim, Norway*, pp. 71–89. Düsseldorf: Aluminium Verlag; 1985.

13. Safarian J, Tranell G. Silicon purification through magnesium addition and acid leaching. In: *Proc. 32nd European Photovoltaic Solar Energy Conference and Exhibition*, pp. 1011–14; 2016.

14. Lux AL.Flemings MC. Refining by fractional melting. *Metall. Trans. B*, 1979; 10(1): 79–84.

15. Pfann WG. *Zone melting*, 2nd edn. New York: Wiley; 1966.

16. Coletti, G. Sensitivity of state-of-the-art and high efficiency crystalline silicon solar cells to metal impurities. *Prog. Photovoltaics*, 2013; 21(5): 1153–1170.

17. Di Sabatino M. Industrial production of silicon monocrystals doped with B, As and Sb. Master Thesis. Ancona, Italy: Marche Polytechnique University; 2002.

18. Di Sabatino M, Øvrelid EJ, Olsen E, Engh TA. Distribution of oxygen in mc-Si ingots for solar cell applications. In: *Proc. 136th TMS Annual Meeting. Orlando, FL*. TMS; 2007.

19. Fang QT, Granger DA. Porosity formation in modified and unmodified A356 alloy castings. *AFS Trans.*, 1989; 97: 989–1000.

20. Guzowski MM, Sigworth GK, Sentner DA. The role of boron in the grain refinement of aluminum with titanium. *Met. Trans. A*, 1987; 18: 603–19.

21. Quested TE. Understanding mechanisms of grain refinement of aluminum alloys by innoculation. *Mater. Sci. Technol.*, 2004; 20(11): 1357–69.

22. Perepezko JH, Uttormark MJ. Nucleation controlled solidification kinetics. *Metall. Mater. Trans. A*, 1996; 27A: 533–47.

23. Arnberg L, Bäckerud L, Klang H. Intermetallic particles in Al–Ti–B–type master alloys for grain refinement of aluminium. *Metals Technol.*, 1982; 9(1): 7–13.

24. Davies IG, Dennis JM, Hellawell A. The nucleation of aluminum grains in alloys of aluminum with titanium and boron. *Metall. Trans.*, 1970; 1(1): 275–80.

25. Marcantonio JA, Mondolfo LE. Nucleation of aluminium by several intermetallic compounds. *J. Inst. Metals*, 1970; 98: 23–7.

26. Kobayashi KE, Hashimoto S,Shinghu PH. Nucleation of aluminium by Al_3Ti in the Al-Ti system. *Z. Metallkunde*, 1983; 74(11): 751–4.

27. Hashimoto S, Kobayashi KE, Miura S. Roles of lattice coherency to the heterogeneous nucleation in the Al-Ti system. *Z. Metallkunde*, 1983; 74(12): 787–91.

28. Næss SE, Rønningen JA. TiB_2 particles as nucleants for aluminum. *Metallography*, 1975; 8(5): 391–400.

29. Easton M, StJohn D. Grain refinement of aluminum alloys: Part 1. The nucleant and solute paradigms—a review of the literature. *Metall. Mater. Trans. A*, 1999; 30: 1613–23.

30. StJohn DH, Prasad A, Easton MA, Qian M. The contribution of constitutional supercooling to nucleation and grain formation. *Metall. Mater. Trans. A*, 2015; 46: 4868–85.

31. StJohn DH, Qian M, Easton MA, Cao P, Hildebrand Z. Grain refinement of magnesium alloys. *Metall.Mater. Trans. A*, 2005; 36(7): 1669–79.

32. Balart MJ, Patel JB, Gao F, Fan Z. Grain refinement of deoxidized copper. *Metall. Mater. Trans. A*, 2016; 47: 4988–5011.

33. Sadayappan M, Thomson JP, Zavadil R, Sahoo M, Michels HT. Fading of grain refinement in permanent mold cast copper alloys. *AFS Trans.*, 2004; paper 04–12.

34. Sadayappan M, Cousineau D, Zavadil R, Sahoo M, Michels H. Grain refinement of permanent mold cast copper-base alloys. *AFS Trans.*, 2002; paper 02–108.

35. Greer AL. Application of heterogeneous nucleation in grain-refining of metals. *J. Chem. Phys.*, 2016; 145: 211704.

36. Liu Z. Review of grain refinement of cast metals through inoculation: theories and developments. *Metall. Mater. Trans. A*, 2017; 48(10): 4755–75.

7

Remelting and Addition of Alloy Components

7.1 Introduction

There are innumerable ways in which the properties of a metal can be modified and improved by the addition of alloy components. To produce alloys with specific engineering characteristics with regard to strength, fatigue, ductility, and other physical properties, the common practice is to start with a 'pure' solvent metal and introduce alloying elements at very precise, pre-designed levels. Thus, the addition of alloys, melting, and dissolution are necessary operations in the handling of melts before casting. The thermal requirements and the rate of solution needs to be understood to develop good alloying practices. Also, dissolution and melting rates may limit productivity in a number of processes.[1]

During recent years there has been an increasing focus on recycling metals, where metal is added to a molten pool of (more or less) the same composition. For instance, the aluminium industry uses more and more post-consumed or process-scrap in the production of new materials. There is a huge benefit if the cast-house can use the same alloy. This will reduce the need for more alloying elements or, in the worst case, require a large amount of primary metal to dilute elements present in the scrap metal not wanted in the final product. Furthermore, remelting aluminium requires only about 5% of the energy needed to produce it from bauxite ore.

Compared to solidification, the literature concerning the addition and melting of alloys is limited.[2-6] What happens to the additions proves to be surprisingly complex. For instance, a frozen shell may form for a transient period around the addition. Also, the addition may melt inside the frozen shell if the melting point of the addition is low compared to the melting point of the surrounding fluid. Seemingly, each combination of melt and alloy addition behaves in a complex and special manner. For a given combination it is possible to describe and model melting or dissolution with extensive numerical calculations.[7,8] However, a difficulty is always the lack of good data concerning physical quantities such as thermal conductivities and specific heats. Even worse is the problem of calculating the heat-transfer coefficient between melt and addition or frozen shell. On top of this, some alloys might form complex intermetallic phases during melting which the solute or solvent must diffuse through.

Principles of Metal Refining and Recycling. Thorvald Abel Engh, Geoffrey K. Sigworth, and Anne Kvithyld, Oxford University Press.
© Oxford University Press 2021.
DOI: 10.1093/oso/9780198811923.003.0007

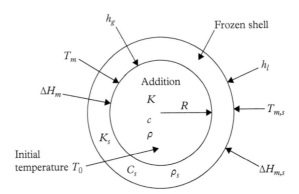

Fig. 7.1 *The various physical quantities needed to describe melting.*

Dissolution of an alloy addition in a melt may be divided into two main categories: class I alloys, where the melting range lies below the bath melting temperature, and class II alloys, where the melting range lies above the bath temperature. In the first case dissolution is usually limited by heat transfer from the melt to the lump or particle, whereas for class II alloys dissolution is controlled by diffusion through a melt boundary layer. This latter process is slow compared to melting rates for class I alloys. These two classes will be treated separately. Remelting aluminium scrap in an aluminium bath is regarded as a class I case since the metal bath is typically 50–100 °C higher than the melting point of the added scrap.

Figure 7.1 illustrates the various physical quantities needed to describe the problems and mechanisms associated with melting. These mechanisms or unit operations are described as simply as possible in terms of analytical solutions. The specific heats, c, densities, ρ, thermal conductivities, k, and thermal diffusivities, α, are assumed to be constants independent of temperature. Hopefully, the background given in this chapter will help the reader to formulate in quantitative mathematical terms the behaviour of the special combination of addition and melt. Then one must also be aware of the approximations and assumptions involved. A good physical feeling for the problem, consideration of the dimensionless groups that enter, and analytical solutions obtained in special cases are valuable as a starting point and a check for a numerical solution of the problem at hand.

Additions to steel are commonly in the form of ferroalloys, which often have melting ranges lower than those of pure elements and can be introduced more readily into the molten steel. Thus, ferroalloys often belong to class I. Most ferroalloys are produced in electric furnaces from the ores, usually with carbon as the reductant. Schematically, one may write

$$MO + Fe + C = FeM + CO.$$

The activity of the element, M, in the ferroalloy FeM is less than that for the pure element. This means in practice that FeM can be produced at lower temperatures and with less electrical

Table 7.1 *Thermodynamic and physical data for some solid and molten metals at the melting point*[7,11-14]

	T_m (°C)	Sensible heat (kJ/kg)	ΔH_m	ρ_s (kg/m^3)	ρ_l	c_s (J/kgK)	c_l	k_s (W/mK)	k_l
Fe	1538	1050	247	7270	7015	740	824	40	40
Al	660	672	397	2559	2385	1258	1177	210	95
Cu	1085	461	209	8320	8000	470	494	320	170
Mg	650	730	349	1640	1590	1326	1412	150	80
Zn	420	166	112	7140	6570	456	480	95	49.5
Cr	1859	1321	413	6950	6280	1136	956	36	—
Si	1412	1281	1787	2270	2510	1040	968	20	55
Mn	1246	910	235	7210	5730	842	874	—	95
FeCr (55% Cr)	1500	1100	325	7000	—	700	—	38	—
FeSi (75% Si)	1320	1000	300	3500	—	700	—	10	—
FeMn (80% Mn)	1266	900	—	7200	—	700	—	—	—

energy consumption than the pure element, M. The main requirements of ferroalloys are uniformity in size and composition, and these requirements are becoming ever more strict, reflecting steadily reduced tolerances on the composition of steels. Furthermore, the limits on impurities such as oxygen, nitrogen, carbon or carbides, and hydrogen are growing more severe.[9] Many ferroalloys have concentrations of oxygen and nitrogen up to about 0.2%[10] which is unacceptable when used for super clean steels with a total impurity level of maximum 0.005%. Attention needs to be paid to reducing nitrogen in ferroalloys. The steelmaking approach is to produce steel with a nitrogen concentration of 20–50 ppm, and then to shield the steel from the atmosphere during ladle treatment and continuous casting. Approximately 30% of N can be removed during vacuum treatment of low-carbon steels. However, in most steelmaking operations, any nitrogen introduced into the steels stays with the product. A problem with FeSi powder is that considerable amounts of nitrogen are located on or near the surface. Then the mean content of nitrogen becomes high.

Alloy additions to metals such as aluminium or magnesium often belong to the class II category. They dissolve into the melt by diffusion which is usually a slow process. Some melting points and other thermodynamic and physical data are given in Table 7.1. As stated in the table caption, the values are given at the melting point of its element or alloy. For instance, the average heat capacity for solid iron between 25 °C and the melting point at 1538 °C can be calculated by

$$\overline{C}_s = \frac{Sensible\ heat\ for\ iron}{T_m - 25°C} = \frac{1,050,000}{1513} = 694 \frac{J}{kgK},$$

which is a slightly lower value than the heat capacity of solid iron at 1538 °C. For aluminium, the average heat capacity is 1058 J/kgK and lower than the value at the melting point.

7.2   Change in Temperature from Alloying

One problem with alloying is that the enthalpy change due to melting and dilution (that is, not only the sensible heat) can have a large effect on the temperature of the mixture.

As an example, alloying aluminium with 1 wt% silicon is treated. The silicon is preheated to the temperature of the aluminium melt, which is at 750 °C before it is added to the melt. The heat required for melting and mixing into aluminium of silicon is, from Table 2.1, equal to 40.08 kJ/mol = 1427 kJ/kg. The heat capacity of the Al + 1% Si mixture is assumed to be equal to pure (liquid) aluminium, which is 1.177 kJ/kgK. Assuming an adiabatic system (no heat transport in or out), the result will be a drop in the temperature of the melt due to the enthalpy change. This temperature change is given by

$$\Delta T = -\frac{\Delta H}{mc_l} = -\frac{1.427\text{kJ}}{0.1\text{kg} \cdot 1.177\text{kJ/kgK}} = -12.1\text{K}.$$

There are two comments to this example.

1. It was assumed that the silicon had a temperature of 750 °C prior to addition. In practice, alloying elements are added at a lower temperature, although high enough to be completely dry, typically at 100–200 °C. Then the temperature drop will be larger due to the extra energy required to heat the silicon up to 750 °C.

2. The Gibbs energy for melting and mixing as given in Table 2.1 is given at the melting point of the elements. This means that since we consider dilution at 750 °C (which is lower than the melting point of silicon) one should take into account the heating of solid silicon to 1412 °C and cooling of liquid silicon to 750 °C by using the corresponding heat capacities for solid and liquid silicon, respectively. By doing this, the temperature drop in the example would be 12.5 °C.

In Table 7.2 the enthalpy and temperature change due to the addition of 1 mass% to aluminium for selected alloying elements are given. Note that some elements are already liquid at 750 °C.

There are several physical parameters which influence the dissolution rate of an alloying element.

- Density difference between alloy and melt: for instance, the density of Mn is much higher than for aluminium so that Mn has a tendency to sink to the bottom of a furnace, where both melt convection and temperature are lower than in bulk. This increases the dissolution time.

Table 7.2 *Calculated enthalpy change (ΔH) and temperature change (ΔT) when adding 1 mass% of an alloying element (which is at 750°C) to aluminium at 750°C*

	ΔH (kJ/g)	ΔT (°C)
Si(s)	1.43	−12.1
Zn(l)	0.16	−1.4
Cu(s)	−0.07	+0.6
Mg(l)	−0.36	+3.1
Mn(s)	−1.16	+9.9
Fe(s)	−1.50	+12.7
Li(l)	−3.89	+33.1

- Enthalpy change for dissolution: an exothermic dissolution (as Mg, Mn, etc.) may give a local temperature increase and thus increase the dissolution rate since the diffusion (and hence mass transfer) coefficient generally increase with temperature as given in Eq. (2.143).

In addition to these two parameters, the shape and number of possible intermetallic phases may influence the dissolution rate of alloying elements. This point will be addressed in Section 7.6.

In the following sections we study the dissolution times for class I and class II alloy additions. Initially when a cold body is immersed in a molten metal bath, the surface is heated up and a temperature gradient is set up in the body. Heat is extracted by the addition, and a shell of solidified metal may be formed even though heat enters the shell from the surrounding melt. Heat conduction to the body initially dominates heat transfer from the melt so that the shell grows. The temperature profile flattens out after a time; then heat transfer from the melt takes over and the frozen shell melts back.

The chapter starts with the case of remelting aluminium in an aluminium bath, employing a simple analytic model not considering formation of a shell. This model includes only sensible heat and latent heat for melting of pure metal. Then a complete model for the formation and melting of shell on a flat plate continuously fed into a melt is presented together with a simplified model for the shell. The simplified models are compared to measurements of plate feeding and to immersion of spheres into aluminium bath. These cases are relatively simple in the sense that the melting point is fixed and the same for the melt and added material.

The second part of this chapter concerns the addition of class II alloying elements to a molten bath, where the melting point of the added material is higher than the melt temperature. The impact of intermetallic phases formed during the melting of an alloy is also considered.

7.3 Models for Heating and Melting Pure Aluminium Metal in an Aluminium Bath

When describing melting and alloying in depth and in detail, one would like to know the temperature as a function of both time and position inside the added material. However, the most important question for the cast-house will be: how long time does it take for a metal part added to a bath to melt? This can be estimated if we consider the energy (that is, enthalpy) required for melting and have measurements for the rate of heat transfer in molten melts. And even better: the simplified model can be used to estimate a change in melting time due to changes in contact area or size, melt or preheating temperature, or even stirring (i.e. heat-transfer coefficient).

7.3.1 Energy Transport Model without Shell Formation

This model calculates the enthalpy required for heating an aluminium piece with mass m from a (preheating temperature) T_0 up to the melting temperature T_m with subsequent complete melting at T_m. The melt bath is considered to have a constant temperature of T_b (bulk temperature). The enthalpy required for this process is then the sensible heat plus the heat of fusion. From (published) values for heat-transfer coefficients h, the time required for energy transport is calculated.

The temperature in the addition is taken to be uniform before melting starts. This assumption is justified by the high heat conductivity of metals. And it allows that the two processes of heating and melting can be regarded independent of each other.

With no chemical reaction, energy is transported from a hot region to a cold region due to a temperature difference only. Fourier's law states the heat flux (energy per area and time) is proportional to the temperature gradient:

$$\vec{Q} = -k\nabla T \qquad (7.1)$$

In one dimension (that is, the x-direction) we get

$$\dot{Q}_x = \frac{dQ_x}{dt} = -k_x\frac{dT}{dx}. \qquad (7.2)$$

In analogy to mass transfer (see Section 3.1.1), a heat-transfer coefficient, h, is introduced (in order to avoid the problems with the actual shape of the temperature gradient). Thus:

$$\frac{dQ}{dt} = Ah(T_b - T), \qquad (7.3)$$

where T_b is the bulk temperature in the liquid metal and T is the temperature on the surface, A, of the added material.

During heating, dQ/dt is the increase in sensible heat per unit time. If we introduce mass and specific heat capacity, we get

$$\frac{mc\,dT}{dt} = Ah\,(T_b - T) \tag{7.4}$$

or

$$\frac{dT}{T_b - T} = \frac{Ah}{mc}\,dt. \tag{7.5}$$

Integration from $T = T_0$ (preheating temperature of the added part) to melting temperature T_m gives

$$\int_{T=T_0}^{T_m} \frac{dT}{T_b - T} = \frac{Ah}{mc} \int_{t=0}^{t_m} dt \tag{7.6}$$

$$\ln \frac{T_b - T_m}{T_b - T_0} = -\frac{Ah}{mc} t_m \tag{7.7}$$

or

$$t_m = \frac{mc}{Ah} \ln\left(\frac{T_b - T_0}{T_b - T_m}\right). \tag{7.8}$$

For the melting, we consider a solid volume that requires heat to melt. It is assumed that the temperature is constant throughout the volume during melting.

The enthalpy required for melting is given by the heat of fusion

$$Q_f = m\,\Delta H_f. \tag{7.9}$$

Employing the heat-transfer coefficient and the temperature difference between bulk melt and the added metal which has reached the melting temperature, we get

$$\frac{dQ_f}{dt} = hA\,(T_b - T_m), \tag{7.10}$$

which integrated gives

$$Q_f = hA\,(T_b - T_m)\,t_f. \tag{7.11}$$

Thus

$$t_f = \frac{m\,\Delta H_f}{Ah\,(T_b - T_m)}. \tag{7.12}$$

The total time for an addition to heat up and melt is then

$$t_{tot} = \frac{m}{Ah}\left[c\ln\left(\frac{T_b-T_0}{T_b-T_m}\right) + \frac{\Delta H_f}{T_b-T_m}\right].$$

(7.13)

As seen in Eq. (7.13), the time depends on both mass, m, and contact area, A, between the added part and melt.

7.3.2   Thin Flat Plate Continuously Fed into Melt

The flat plate has a thickness of $2l$ and a contact area between melt and solid of $A_{plate} = 2A$ (both sides of the plate count), whereas the sphere has a diameter d and a corresponding contact area of $A_{sphere} = \pi d^2$. One advantage by considering a flat plate is that even during melting, the contact area is nearly constant as long as $\sqrt{A} \gg 2l$. For such a thin plate:

$$\frac{m}{A} \rightarrow \frac{m}{2A} = \frac{\rho V}{2A} = \frac{\rho 2lA}{2A} = \rho l,$$

(7.14)

which gives a total time for heating and melting:

$$t_{tot} = \frac{\rho l}{h}\left[c\ln\left(\frac{T_b-T_0}{T_b-T_m}\right) + \frac{\Delta H_f}{T_b-T_m}\right].$$

(7.15)

Snorre Farner published in his Dr.ing. thesis[15] measurements from continuous feeding of a flat plate down into molten aluminium. Figure 7.2 shows measured penetration depth as a function of bulk melt temperature for six different feeding velocities. By dividing the measured penetration depth by the feeding velocity in Fig. 7.2, the melting time is found.

Farner calculated the heat-transfer coefficient in the liquid for these experiments to be 18.7 kW/m²K. With the use of this value and the plate thickness $2l = 0.54$ mm, Eq. (7.15) can be used to predict the heating and melting times. These measured times and model are presented in Fig. 7.3. From the figure, the model clearly underestimated the melting time for temperatures lower than about 780 °C. One must be aware that this model is very simplified. It assumes infinitely high heat conductivity in the solid material, and employs a constant heat-transfer coefficient, h. This model disregards formation and melting of a shell on the immersed plate.

7.3.3   Mass Transfer Coefficient Calculations

It is of interest to determine the mass transfer coefficient for the dissolution of aluminium plates in molten aluminium. As shown in Chapter 3, Eq. (3.49) gives the mass transfer of a solute element to a free surface

$$k = \left(\frac{4D}{\pi t}\right)^{\frac{1}{2}},$$

(7.16)

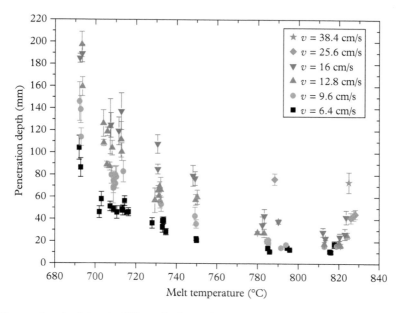

Fig. 7.2 *Penetration depth for six different feeding velocities versus melt temperature for the quiescent-melt experiments (AA1050 alloy). Error bars show the estimated standard deviation in each measurement.*[15]

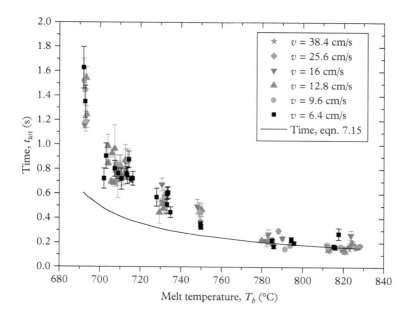

Fig. 7.3 *Calculated melting time for the measurements presented in Fig. 7.2. The calculation is by Eq. (7.15) and based on an enthalpy balance for the heating and melting of thin aluminium plates. (Based on data from Farner.*[15]*)*

Table 7.3 *Self diffusion coefficients for molten aluminium*[16]

T (K)	D (10^{-9} m²/s)
980 ± 2	7.2 ± 0.6
1020 ± 2	7.9 ± 0.7
1060 ± 2	8.8 ± 0.7

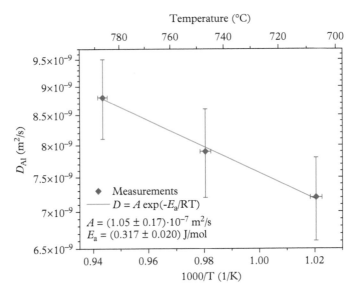

Fig. 7.4 *Diffusion coefficient for self-diffusion in molten aluminium. The fitted line represents an Arrhenius equation. (Based on data from Kargl et al.[16])*

where t is the time it takes for a plate to melt when dipped into the melt and D the (self-) diffusion coefficient of aluminium in molten aluminium. In 2012, Kargl et al. reported measurements of such self-diffusion coefficients in molten aluminium measured by neutron diffraction.[16] Table 7.3 shows the published self-diffusion coefficients for three different melt temperatures.

Figure 7.4 shows the diffusion coefficients plotted as function of the inverse temperature. An Arrhenius equation has been fitted to the measurements and the fitted parameters are given.

By using the fitted line in Fig. 7.4, Farner's melting time in Fig. 7.3, and Eq. (7.4), the mass transfer coefficient for the transport of aluminium from the plate can be calculated. These calculations are shown in Fig. 7.5.

In Fig. 7.22, in Section 7.6, mass transfer coefficients for dissolution of silicon in molten aluminium are shown. These are about one order of magnitude lower than the calculated values for dissolution of pure aluminium.

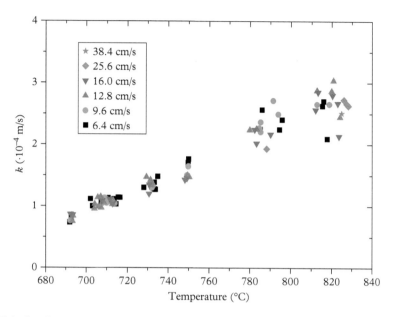

Fig. 7.5 *Calculated mass transfer coefficients during melting of an aluminium plate in aluminium. (Based on data from Farner.[15])*

7.3.4   Criterion for Shell Formation

Farner[15] gave a criterion for the temperature when shell formation occurs. If a shell exists, there will be two heat-transfer coefficients determining the heat flow into the plate.

1. The heat-transfer coefficient in the liquid, h_l, and
2. The heat-transfer coefficient for the boundary between the shell and plate, h_g.

A shell will form whenever the heat flow from the liquid into the shell is lower than the heat flow from the shell into the plate:

$$h_l (T_b - T_m) < h_g (T_m - T_0),$$ (7.17)

where T_0 is the preheating temperature in the plate before dipping. Introducing a dimensionless (superheat) temperature, θ_l, and the Biot and Nusselt numbers defined by

$$\theta_l = \frac{T_b - T_0}{T_m - T_0}$$ (7.18)

$$Bi = \frac{h_g b}{k}$$ (7.19)

$$Nu = \frac{h_l b}{k},$$ (7.20)

the criterium for shell formation can then be written as

$$\theta_l - 1 < \frac{Bi}{Nu}.$$ (7.21)

Farner calculated the Bi and Nu numbers for his measurements to be 0.010 ± 0.001 and 0.049 ± 0.001, respectively.[15] Using Eq. (7.21) with $T_m = 660\,°C$ and $T_0 = 25\,°C$, the critical temperature for formation of a shell becomes

$$T_{crit} < (790 \pm 10)\,°C,$$

which, as seen from Fig. 7.3, is about the temperature region where the model starts to underestimate the time for preheating and melting.

7.3.5   Melting of Spheres (with High Thermal Conductivity)

One difference between the flat plate and sphere is that the contact area of a sphere is reduced during melting.

For a sphere, the ratio between the mass (or volume) and surface area is given by

$$\frac{m}{A} = \frac{\rho V}{A} = \frac{\rho d}{6},$$ (7.22)

where ρ is the density and d the diameter of the sphere. Combining Eqs. (7.8) and (7.22) gives the time it takes to heat a spherical part from the preheating temperature to melting temperature:

$$t_m = \frac{\rho c d}{6h} \ln\left(\frac{T_b - T_0}{T_b - T_m}\right).$$ (7.23)

During melting, the size of the sphere decreases. A heat balance gives that the heat transported into the sphere per unit time is proportional to the volume removed after melting starts:

$$Ah(T_b - T_m) = \rho \Delta H_f \frac{dV}{dt},$$ (7.24)

which can be rewritten to

$$4\pi r^2 h(T_b - T_m) = \rho \Delta H_f \frac{4\pi r^2 dr}{dt}$$ (7.25)

or

$$\frac{dr}{dt} = \frac{h\,(T_b - T_m)}{\rho \Delta H_f}.$$ (7.26)

If we also assume that h does not depend on t or r, the radius is reduced at a constant rate; the time for melting a sphere with diameter d becomes

$$t_f = \frac{d\rho \Delta H_f}{2h\,(T_b - T_m)}.$$

Thus, the total time for a spherical part to heat up and melt is finally

$$t_{tot} = \frac{\rho d}{6h}\left[c\ln\left(\frac{T_b - T_0}{T_b - T_m}\right) + \frac{3\Delta H_f}{T_b - T_m}\right].$$ (7.27)

With the use of tabulated data for density, heat capacity, and heat of fusion as given in Table 7.1 in Section 7.1 together with a published heat-transfer coefficient of 25.1 kW/m^{2}K for a spherical addition to an aluminium bath,[17] the time for heating and melting a sphere of aluminium in aluminium bath can be calculated. Table 7.4 shows published data for melting aluminium spheres, together with the time calculated by Eq. (7.27). The simple model is in good agreement with experimental data.

Most interesting with the model presented in Eq. (7.27) is how the various parameters affect the time required for melting. By keeping various parameters constant, the effect on melting time can be visualized. From Eq. (7.27) it is clear that the total time is proportional to the sphere diameter. That is, doubling the particle size will increase the heating and melting times by a factor of 2. The heat-transfer coefficient can be increased by increasing the convection in the melt. This is possible by using an electromagnetic stirrer or performing gas purging. Both of these techniques will increase melt velocities close to the solid material and effectively decrease the thickness of a thermal boundary layer and in this way increase the thermal gradient. Both Farner[15] and Taniguchi et al.[17] showed that increasing the melt velocity, or stirring, in the liquid metal gave higher heat-transfer coefficients and thus shorter melting times. With gas purging, the heat-transfer coefficient increased from 25.1 to 68.2 kW/m^{2}K, which decreased the time from 35 to 13 s. Equation (7.27) predicts a melting time of 9.5 s,

Table 7.4 *Comparison between calculated heating and melting times from Eq. (7.27) and published data for the same metal system (pure aluminium spheres in aluminium)*

	Model in Eq. (7.27)	Taniguchi et al.[17]	Jiao and Themelis[18]
$T_b = 680\,^\circ$C	35 s	35 s	29 s
$T_b = 700\,^\circ$C	18 s	18 s	15 s
$T_b = 750\,^\circ$C	8.5 s	8 s	7 s

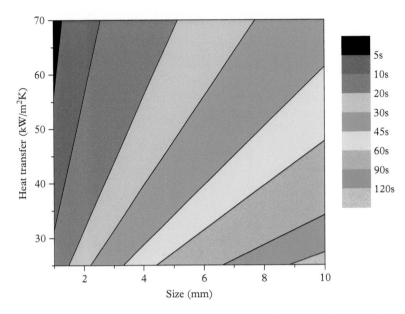

Fig. 7.6 *Colour contour plot of heating and melting times for an aluminium sphere added to an aluminium bath at 680 °C as a function of particle size and heat transfer coefficient.*

which is somewhat smaller than the measured melting time of 13 s. Figure 7.6 shows a contour plot where the total time for heating and melting time is given as function of the particle diameter and heat-transfer coefficient. As in Table 7.4, the bulk melt temperature is 680 °C.

There are two adjustable temperatures in the model in Eq. (7.27): preheating temperature T_0, and bulk metal temperature T_b. As expected, increasing the bulk temperature will decrease both the heating time from preheating temperature to the melting temperature T_m and the melting time, while increasing (or decreasing) the preheating time will only affect the heating time. Figure 7.7 show the total time as a function of both these temperatures. For the total time, increasing the bulk temperature is much more efficient than increasing the preheating.

The data and figures presented here are for pure aluminium. However, the model will be the same for other pure metals. For alloys, where there is a melting temperature range, the system will be somewhat more complicated.

7.3.6   Continuous Feeding and Melting of a Cylindrical Rod

A grain refiner is added to the melt in order to produce small grains in the final casting. This ensures uniform mechanical properties. The simple model in Eq. (7.13) can also be transformed into a cylindrical rod with heat transported into the cylinder side only and disregarding tip area. In this case

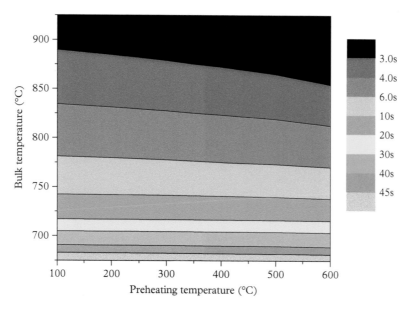

Fig. 7.7 *Heating and melting times as calculated from Eq. (7.27). The size of the sphere is set to be 23.8 mm when added to the melt, and the size constantly reduced during melting (dr/dt = const.). The heat transfer coefficient h = 25.1 kW/m²K.*

$$\frac{m}{A} = \frac{\rho V}{A} = \frac{\rho d}{4} \tag{7.28}$$

and

$$t_{tot} = \frac{\rho d}{4h} \left[c \ln \left(\frac{T_b - T_0}{T_b - T_m} \right) + \frac{\Delta H_f}{T_b - T_m} \right]. \tag{7.29}$$

Equation (7.29) can now be used to estimate the melting time of a rod continuously fed into the melt. Again, the heat-transfer coefficient must be determined from experiments. Also, if a melting time is measured, the dependency of this can be estimated with Eq. (7.29) if such a rod diameter or melt temperature is changed. Figure 7.8 shows the total time to transport enough heat to heat up and melt a rod of 10-mm diameter continuously fed into a melt as a function of melt temperature. The penetration depth shown on the right-hand axis is calculated by multiplying the total time on left-hand axis with a feeding velocity of 40 mm/s.

As seen in Section 7.2.4, a shell will form on the rod if the criterion in Eq. (7.21) is fulfilled. Forming and melting such a shell will increase the time even though the rod heats up under the shell.

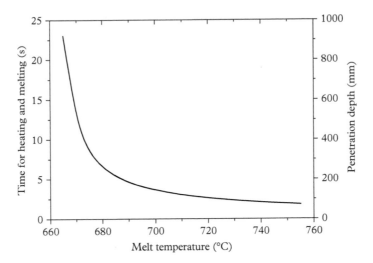

Fig. 7.8 *Heating and melting times for a 10-mm rod fed into a melt as a function of bulk melt temperature, as given by Eq. (7.29). The heat transfer coefficient used in the calculations was 25.1 kW/m²K. The right-hand axis gives the penetration depth with a rod-feeding velocity of 40 mm/s.*

7.3.7 Validity of the Energy Transport Model for Continuously Fed Material

It is interesting to consider for which size range the energy transport model for a plate or rod is valid. This can be estimated by looking at the dimensionless Biot number as in Eq. (7.19). This number depend on the thickness of the melting part, b or d, the thermal conductivity in solid metal, k, and the heat-transfer coefficients for the gap between shell and solid, h_g, and in the liquid, h_l. As long as the energy transport into the solid part dominates over the energy transport upwards in the plate or rod, that is,

$$h_g > \frac{k}{d} \qquad (7.30)$$

or

$$d > \frac{k}{h_g}, \qquad (7.31)$$

the model may be used.

The thermal conductivity of solid aluminium is given in Table 7.1 and is equal to 210 W/mK. As mentioned, the heat-transfer coefficients must be determined by experiments. Farner[15] carried out many experiments of melting thin plates both in quiescent melt and in launders where metal flows. He calculated the heat-transfer coefficient for the gap to be 4900 W/m²K for his experiments of melting plates into a launder with flowing metal. When this number is inserted into Eq. (7.31), the maximum diameter for the model to be valid is about 40 mm.

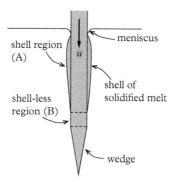

Fig. 7.9 *Continuous feeding of a metal plate into its own melt.*

7.4 Model Including Shell Growth and Melting

A metal plate ideally extending infinitely far upwards is fed vertically into a molten-metal bath at a steady velocity u as shown in Fig. 7.9. After a transient period, the system reaches a steady state. At the penetration point, i.e. where the plate penetrates the melt surface, the surface curves down and forms a meniscus. From this area and below, heat flows from the melt into the cold plate. If the supply of heat from the melt is lower than the heat flow into the plate, the melt closest to the plate solidifies and forms a shell on the plate. This is normally the case, and a criterion was presented in Section 7.2.4. As the plate moves downwards into the melt, the temperature in the plate increases, and the shell grows until the heat flows into and out of the shell balance each other. From this point on, the shell melts back and finally vanishes. Then heat flows directly from the melt into the plate, and after a distance the plate reaches its melting point and starts to melt. This final region is modelled as a wedge.

7.4.1 Dimensionless Groups

Figure 7.10 shows the coordinates, dimensions, and some other quantities representative of all models presented. The heat-balance equations and boundary conditions will be presented for each model. Dimensionless quantities will, in general, be used for simplicity and for generality.

The coordinate system has its origin in the centre of the thickness of the plate, on a level with the melt surface. The y-axis points downwards while the x-axis points horizontally out of the plate. The plate has width B (into the paper) and thickness $b \ll B$. It continues infinitely upwards but penetrates a finite distance P (the penetration depth) down into the melt. The shell region ends at P_A and the shell-less region at P_B. The temperature of the ambient air is T_0, which is also the temperature of the plate when it enters the melt. T_b is the temperature of the melt far from the plate, i.e. in the bulk melt. T_m is the melting point of the melt and the plate. We assume that both are of the same alloy. $T_p = T_p(y)$ and $T_s = T_s(y)$ denote the temperature of the plate and the shell, respectively, as functions of y.

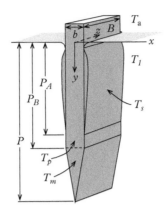

Fig. 7.10 *Three-dimensional sketch of the plate that is fed into the melt showing dimensions and temperatures.*

Four different dimensionless numbers will be used: Bi, Nu, Pe, and Sf. The Biot number, defined as

$$Bi = \frac{h_g b}{k},$$ (7.19)

is the ratio of heat transfer through the gap between the shell and the plate to the heat conduction in the plate. b is the thickness of the plate while k is its thermal conductivity. The heat-transfer coefficient h is defined by Newton's Law of Cooling, $q = h\Delta T$. The Biot number is a dimensionless version of the heat-transfer coefficient h_g. A Biot number much smaller than unity means that the dominating heat-transfer barrier is the gap and not in the plate.

The complex heat-transfer nature of liquid-flow boundary layers can also be described in terms of a heat-transfer coefficient, here denoted h_l (1 for liquid). The corresponding dimensionless number is the Nusselt number, defined as

$$Nu = \frac{h_l b}{k},$$ (7.20)

which may be understood as a ratio of the heat transfer through the boundary layer between the plate and the bulk melt and the heat conduction in and along the plate. If the Nusselt number is much less than unity, the boundary layer is the important heat-transfer resistance.

The dimensionless feeding velocity u is the Péclet number

$$Pe = \frac{\rho c u b}{k},$$ (7.32)

where ρ is the density of the solid plate and c its specific heat. The Péclet number is here the ratio of heat transport with the moving plate to heat conduction in and along it. For a value of 1, the heat transferred due to the motion of the plate is comparable to the heat conducted across it.

The last dimensionless number is the ratio between sensible and latent heat, called the Stefan number after Josef Stefan, who worked with free- and moving boundary problems already at the end of the nineteenth century. For convenience, we will use the inverse Stefan number, defined here as

$$Sf = \frac{\Delta H_m}{c\,(T_m - T_0)}. \tag{7.33}$$

This inverse Stefan number is the ratio between the latent heat of melting, ΔH_m, and the sensible heat of heating the plate from the ambient temperature T_0 to the melting point T_m.

In addition, dimensionless coordinates are introduced: $\xi = x/b$, $\eta = y/b$, and $\zeta = z/b$, the dimensionless penetration depth, $\Pi = P/b$, and the dimensionless temperature, $\theta = (T - T_0)/(T_m - T_0)$. If the temperature is equal to the ambient temperature $T = T_0$, then $\theta = 0$, and if the temperature is equal to the melting point, $T = T_m$, then $\theta = 1$.

7.4.2   General Assumptions

In order to derive models that can be treated analytically, a few basic assumptions have been made. Additional assumptions will be introduced when necessary.

1. The material properties are constant;
2. There is a well-defined melting point T_m, i.e. liquidus and solidus curves coincide;
3. The plate and the melt are of the same material and thus have the same melting point T_m;
4. The heat-transfer coefficients (h_g and h_l) are constant along the plate;
5. h_l from the bulk melt also accounts for the sensible heat released from the melt before solidification;
6. The liquid heat-transfer coefficient (h_l) from the melt is the same whether the heat flows to the shell or directly to the plate;
7. The shell is considered to be 'thin' and to have a constant temperature equal to the melting point T_m,
8. The temperature in the plate at the penetration point ($y = 0$) is constant and equal to T_0; and
9. A steady state is reached.

7.4.3   Main Model of the Plate with Shell Formation

The main model is a one-dimensional steady-state model consisting of three regions and a shell as shown in Fig. 7.11. Four connected heat equations can be established: one equation for each of the two regions A and B of the plate, one for the shell formed outside region A,

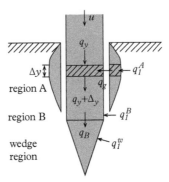

Fig. 7.11 *The heat flow in a cross-section of the plate.*

and the last for the wedge. In addition to the assumptions listed in Section 7.4.2, two more assumptions are necessary:

10. The thickness of the plate is so small that the isotherms inside the plate can be assumed to be horizontal; and
11. The plate ends in a wedge with constant temperature equal to the melting point.

7.4.3.1   Region A

In region A of the plate, the heat balance of a horizontal cross-section of thickness Δy of the plate (see Fig. 7.11) is governed by vertical conductive and convective heat flow in and along the plate,

$$q_y = -k\frac{\partial T_p}{\partial y}(y) + \rho c u T_p(y), \tag{7.34}$$

and heat transfer from the shell through a thin gap with heat-transfer resistance inversely proportional to h_g. The heat flow over the gap is expressed by Newton's Law of Cooling:

$$q_g = h_g\left(T_m - T_p(y)\right), 0 \le y \le P_A \tag{7.35}$$

While the plate temperature T_p depends on y, the shell temperature is constant and at the melting point T_m. P_A is the penetration depth of region A.

The resulting heat balance for the cross-section is thus

$$-\Delta\left(-kb\frac{dT_p}{dy} + \rho cubT_p\right) + 2h_g\left(T_m - T_p\right)\Delta y = 0. \tag{7.36}$$

Divide this by Δy, let Δy approach zero, and introduce dimensionless quantities (see Section 7.3.1), and we obtain the temperature profile along the plate, with primes denoting differentiation with respect to dimensionless distance η:

$$\theta_p'' - Pe\theta_p' - 2Bi\left(\theta_p - 1\right) = 0 \quad 0 \le \eta \le \Pi_A. \tag{7.37}$$

The general solution is

$$\theta_p^A(\eta) = 1 + Ae^{\lambda_1 \eta} + Be^{-\lambda_2 \eta} \quad 0 \le \eta \le \Pi_A, \tag{7.38}$$

where

$$\lambda_1 = \frac{1}{2}Pe\left(\sqrt{1 + \frac{8Bi}{Pe^2}} + 1\right) \tag{7.39}$$

$$\lambda_2 = \frac{1}{2}Pe\left(\sqrt{1 + \frac{8Bi}{Pe^2}} - 1\right)$$

The constants A and B will be determined from boundary conditions. The superscript A in Eqs. (7.37) and (7.38) means region A and is included to avoid confusion.

7.4.3.2 The Shell

The heat balance for a cross-section of the shell is reduced to a one-dimensional problem of the shell thickness due to Assumption 7 of the constant shell temperature. Figure 7.12 gives an overview of the heat-flux terms that must balance each other:

$$q_l^A = h_l\left(T_l - T_m\right) \tag{7.40}$$

$$q_g = h_g\left(T_m - T_p\right) \tag{7.41}$$

$$q_y = \rho c u T_m \tag{7.42}$$

$$q_{ly} = \rho c u T_m + \rho u \Delta H_m. \tag{7.43}$$

Thermal conductivities do not enter the expressions because the shell temperature is taken to be constant and equal to the melting point. The specific heat of liquid and solid aluminium differs insignificantly (less than 2 per cent), so only c is used in Eq. (7.43). Furthermore, we consequently use the solid density ρ in the heat balances since expansion and shrinkage due to melting and solidification do not affect the heat balance, only the convection in the liquid. Due

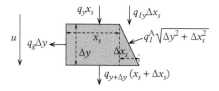

Fig. 7.12 *The heat balance of a section of the solidified shell.*

to assumption 5, that the sensible heat of the melt can be accounted for by the heat-transfer coefficient, T_l does not occur in q_{ly}. The heat balance becomes

$$q_{ly}\Delta x_s + q_y x_s - q_{y+\Delta y}(x_s + \Delta x_s) + q_I^A \sqrt{\Delta x_s^2 + \Delta y^2} - q_g \Delta y = 0$$

or, after dividing by Δy and letting it approach zero:

$$\rho u \Delta H_m \frac{dx_s}{dy} + q_I^A \sqrt{1 + \left(\frac{dx_s}{dy}\right)^2} - q_g = 0. \tag{7.44}$$

The square root appears because the outer surface of the shell curves and thereby slightly increases the surface area through which the heat flows from the melt.

Equation (7.44) is solved with respect to dx_s/dy, and the sign of the resulting square root is chosen such that $q_g = q_I^A$ when $dx_s/dy = 0$. We rewrite it in dimensionless quantities:

$$\frac{d\xi_s}{d\eta} = \frac{Bi(1-\theta_p)}{PeSf} \left(\frac{1 - \sqrt{1 + \left[1 - \left(\frac{Nu(\theta_l-1)}{PeSf}\right)^2\right]\left[\left(\frac{Nu(\theta_l-1)}{Bi(1-\theta_p)}\right)^2\right]}}{1 - \left(\frac{Nu(\theta_l-1)}{PeSf}\right)^2} \right). \tag{7.45}$$

If no approximations are introduced, this equation must be solved by numerical integration. A convenient approximation is presented and used in Section 7.4.5.

Note that $q_g = q_I^A$ is identical to $Bi(1-\theta_p) = Nu(\theta_l - 1)$ and that $\xi_s(\eta) = 0$ defines the length Π_A of region A.

7.4.3.3   Region B

There is no shell in region B, so heat flows directly from the bulk melt to the plate with the thermal boundary layer as the only thermal resistance,

$$q_I^B = h_l(T_b - T_p(y)) \, P_A \le y \le P_B, \tag{7.46}$$

where hl is the heat-transfer coefficient through the boundary layer of the liquid metal, and T_b is the temperature in the bulk melt. P_B is the penetration depth of region B. With dimensionless quantities we get

$$\theta_p'' - Pe\theta_p' - 2Nu(\theta_p - \theta_l) = 0 \quad \Pi_A \le \eta \le \Pi_B. \tag{7.47}$$

This equation is very similar to Eq. (7.37) except that now the temperature θ_l of the melt enters, and Nu replaces Bi. The general solution is thus

$$\theta_p^B(\eta) = \theta_l + A' e^{\lambda_1' \eta} + B' e^{-\lambda_2' \eta} \quad \Pi_A \le \eta \le \Pi_B, \tag{7.48}$$

where

$$\lambda_1' = \frac{1}{2} Pe \left(\sqrt{1 + \frac{8Nu}{Pe^2}} + 1 \right) \tag{7.49}$$

$$\lambda_2' = \frac{1}{2} Pe \left(\sqrt{1 + \frac{8Nu}{Pe^2}} - 1 \right).$$

The constants A' and B' are to be determined by the boundary conditions.

7.4.3.4 *The Wedge Region*

Finally, the heat flowing into the wedge from the melt must balance the latent heat of melting the plate material transported into the wedge region and the heat conduction $q_B = -k dT_p/dy$ between region B and the wedge. As a consequence of a constant wedge temperature, there are no heat gradients in the wedge, and in principle we cannot allow any heat conduction through the wedge or out of it. We therefore set $q_B = 0$, and obtain the steady-state heat balance for the wedge,

$$q_l^w \sqrt{1 + \left(\frac{2(P - P_B)}{b} \right)^2} = \rho u \Delta H_m, \tag{7.50}$$

where $P - P_B$ is the vertical length of the wedge (see Figure 7.10) and $q_l^w = h_l (T_b - T_m)$. Solving for $P - P_B$ and applying dimensionless quantities, we get

$$\Pi - \Pi_B = \frac{1}{2} \sqrt{\left(\frac{PeSf}{Nu(\theta_l - 1)} \right)^2 - 1}. \tag{7.51}$$

7.4.3.5 *Combined Solution*

We now have four unknown constants A, B, A', and B' in Eqs. (7.38) and (7.48), and two lengths Π_A and Π_B which must be determined.

The dimensionless penetration depth Π_A of the shell is determined by integrating Eq. (7.44) numerically and finding the solution of $\xi_s = 0$ for $\eta > 0$, or using the approximation presented in Section 7.4.5. Note that if $d\xi s/d\eta \leq 0$ at $\eta = 0$, there is no shell growth and $\Pi_A = 0$.

The remaining constants are determined with the aid of five boundary conditions. Four of them demand continuous temperature and derivative of the temperature between regions A and B and the wedge:

$$\theta_p^A (\eta = 0) = 0 \tag{7.52}$$

$$\theta_p^A \left(\eta = \Pi_A\right) = \theta_p^B \left(\Pi_A\right) \tag{7.53}$$

$$\frac{d\theta_p^A}{d\eta} \left(\eta = \Pi_A\right) = \frac{d\theta_p^B}{d\eta} \left(\Pi_A\right) \tag{7.54}$$

$$\theta_p^B \left(\eta = \Pi_B\right) = 1 \tag{7.55}$$

$$\frac{d\theta_p^B}{d\eta} \left(\eta = \Pi_B\right) = 0. \tag{7.56}$$

The constants A' and B' may be eliminated by starting with boundary conditions in Eqs. (7.55) and (7.56) to give

$$A' = -\frac{\lambda_2'}{\lambda_1' + \lambda_2'} \left(\theta_l - 1\right) e^{-\lambda_1' \Pi_B} \tag{7.57}$$

$$B' = -\frac{\lambda_1'}{\lambda_1' + \lambda_2'} \left(\theta_l - 1\right) e^{\lambda_2' \Pi_B}, \tag{7.58}$$

and the temperature profile in region B $(\Pi_A \le \eta \le \Pi_B)$ is thus

$$\theta_p^B \left(\eta\right) = \theta_l - \frac{\theta_l - 1}{\lambda_1' + \lambda_2'} \left(\lambda_2' e^{-\lambda_1' (\Pi_B - \eta)} + \lambda_1' e^{\lambda_2' (\Pi_B - \eta)}\right) \tag{7.59}$$

The boundary condition in Eq. (7.52) then gives

$$B = -(A+1), \tag{7.60}$$

and A is obtained from boundary condition (7.53):

$$A = \frac{\theta_l - 1 + e^{-\lambda_2 \Pi_A} - \frac{\theta_l - 1}{\lambda_1' + \lambda_2'} \left(\lambda_2' e^{-\lambda_1' (\Pi_B - \Pi_A)} + \lambda_1' e^{\lambda_2' (\Pi_B - \Pi_A)}\right)}{e^{\lambda_1 \Pi_A} - e^{-\lambda_2 \Pi_A}} \tag{7.61}$$

The temperature profile in region A $(0 \le \eta \le \Pi_A)$ becomes

$$\theta_p^A \left(\eta\right) = A \left(e^{\lambda_1 \eta} - e^{-\lambda_2 \eta}\right) + 1 - e^{-\lambda_2 \eta} \tag{7.62}$$

It is worth noticing that if $8Bi \ll Pe^2$, then $\lambda_1 \approx Pe$ and $\lambda_2 \approx 2Bi/Pe$, which is small. The same is the case in region B: when $8Nu \ll Pe^2$, $\lambda_1' \approx Pe$ while $\lambda_2' \approx 2Nu/Pe$.

Application of the remaining boundary condition (7.54) gives an implicit solution for Π_B,

$$e^{\lambda_2' (\Pi_B - \Pi_A)} \left(1 + \frac{\Lambda}{\lambda_2'}\right) = e^{-\lambda_1' (\Pi_B - \Pi_A)} \left(1 + \frac{\Lambda}{\lambda_1'}\right) + \frac{\Lambda}{\Lambda'} \Theta + \frac{\lambda_2}{\Lambda'} \frac{e^{-\lambda_2 \Pi_A}}{\theta_l - 1}, \tag{7.63}$$

where

$$\Lambda = \frac{\lambda_1 e^{\lambda_1 \Pi_A} + \lambda_2 e^{-\lambda_2 \Pi_A}}{e^{\lambda_1 \Pi_A} - e^{-\lambda_2 \Pi_A}}$$

$$\Lambda' = \frac{\lambda_1' \lambda_2'}{\lambda_1' + \lambda_2'}$$

$$\Theta = \frac{\theta_l - 1 + e^{-\lambda_2 \Pi_A}}{\theta_l - 1}.$$

A convenient way to find Π_B is to solve for the Π_B on the left-hand side and to iterate the resulting equation:

$$\Pi_B^{i+1} = \Pi_A + \frac{1}{\lambda_2'} \ln \left[\frac{\frac{\Lambda'}{\Lambda} \Theta + \frac{\lambda_2'}{\Lambda} \frac{e^{-\lambda_2 \Pi_A}}{\theta_l - 1} + e^{-\lambda_1'(\Pi_B^i - \Pi_A)} \left(1 - \frac{\Lambda}{\lambda_1'}\right)}{1 - \frac{\Lambda}{\lambda_2'}} \right] \tag{7.64}$$

With $\Pi_B^0 = \Pi_A$ as start value, rapid convergence is achieved.

7.4.4 Heat-Transfer Coefficient in Thermal Boundary Layer

As the plate slides into the melt, melt is dragged along with the plate surface, as shown in Fig. 7.13. We will use the boundary-layer approximation to describe the heat transfer from the melt to the plate or the shell. This approach allows us to find values for the melt heat-transfer coefficient h_l for a stagnant bath as well as for melt flow past the plate. We will not consider natural convection.

7.4.4.1 Boundary-Layer Theory for Molten Metals

When a fluid flows past a flat plate, heat and momentum are exchanged between the fluid and the plate. Boundary-layer theory considers the fluid outside a distance $\delta(x) \propto \sqrt{x}$ from the plate to be independent of the presence of the plate. The layer between the plate and the outer fluid is the boundary layer. In general, heat and momentum are transported by different

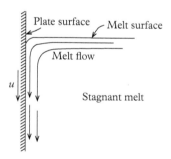

Fig. 7.13 *The flow of melt due to movement of the plate.*

mechanisms, so it is natural to distinguish between a viscous boundary layer with thickness $\delta_u \sim \sqrt{\nu x/u}$ and a thermal one with thickness $\delta_T \sim \sqrt{\alpha_l x/u}$). As indicated by the dependence on x, the thickness varies along the plate as shown in Fig. 7.14. Across the thermal boundary layer, the temperature changes from the plate temperature to the temperature of the bulk fluid, and similarly for the viscous boundary layer. The temperature and velocity profiles across their respective boundary layers are also shown in the figure for a distance x_0 from the start of the plate. A more detailed discussion of the boundary-layer theory can be found elsewhere.[19]

When a fluid has a large viscosity, momentum is easily transferred, the velocity gradients become small, and the boundary layer is thick. The same is the case for the thermal diffusivity $\alpha_l = k_l/\rho_l c_l$, which is relatively high for fluids that conduct heat well and have low heat capacity such as metals. The Prandtl number is the ratio of the kinematic viscosity ν and the thermal diffusivity: $Pr = \nu/\alpha_l$. For fluids with $Pr \sim 1$, like gases, the thermal boundary layer will be of similar thickness as the viscous boundary layer. If $Pr \gg 1$, as is the case for 'normal' liquids, especially oils, the thermal boundary layer will be insignificant compared to the viscous boundary layer. For metals, however, we have $Pr \ll 1$, as implied in Fig. 7.14, and the melt is at rest in most of the thermal boundary layer.

Aluminium has $\alpha_l = 34 \times 10^{-6}$ m²/s and $\nu = 0.50 \times 10^{-6}$ m²/s (see Table B.1 in Farner[15]), so Pr becomes 0.015. We can thus disregard the viscous boundary layer and assume that the melt velocity is zero throughout the thermal boundary layer.

For laminar flow, the heat equation in a coordinate system following the plate at negative velocity v along the x-axis, as shown in Figure 7.14, is

$$\rho_l c_l \left(u_x \frac{\partial T}{\partial x} + u_y \frac{\partial T}{\partial y} \right) + \rho_l c_l u \frac{\partial T}{\partial x} = k_l \left(\frac{\partial^2 T}{\partial x^2} + \frac{\partial^2 T}{\partial y^2} \right), \tag{7.65}$$

where u_x and u_y are the melt-velocity components, and the last term on the right-hand side enters due to the motion of the coordinate system. The flow is laminar[19] when $Re = ux/\nu < 10^5$.

Now, $u_x = 0$ because the viscous boundary layer is disregarded, and $u_y = 0$ as a consequence of the continuity constraint ($\nabla \cdot \vec{u} = 0$). Furthermore, the temperature inside the thermal boundary layer changes much more rapidly perpendicular to the plate than along it, so we can also neglect $\partial^2 T/\partial x^2$ compared to $\partial^2 T/\partial y^2$.

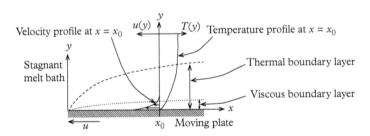

Fig. 7.14 *The velocity and temperature profiles close to a plate which moves to the left through a stagnant melt bath. The dashed lines mark the thickness of the boundary layers.*

This leaves only two terms:

$$u \frac{\partial T}{\partial x} = \alpha_l \frac{\partial^2 T}{\partial y^2}. \tag{7.66}$$

We use the similarity transformation

$$\phi = y \sqrt{\frac{u}{4\alpha_l x}} \tag{7.67}$$

and rewrite Eq. (7.66) to

$$T''(\phi) + 2\phi T'(\phi) = 0, \tag{7.68}$$

where primes denote differentiation with respect to ϕ.

Now, $T'(\phi) = Ce^{-\phi^2}$ is a solution of Eq. (7.68) where C is independent of ϕ. In normal coordinates, the horizontal temperature change is

$$\frac{\partial T}{\partial y} = Ce^{-\phi^2} \frac{\partial \phi}{\partial y} = Ce^{-\phi^2/4\alpha_l x} \sqrt{\frac{u}{4\alpha_l x}}. \tag{7.69}$$

We notice that the temperature slope and thereby the heat flow is infinite at $x = 0$, the first contact point between the plate and the melt, where the boundary-layer thickness is zero.

To obtain the temperature profile in the thermal boundary layer, Eq. (7.69) must be integrated across the boundary layer,

$$T_l - T_m = \int_0^\infty \frac{\partial T}{\partial y} dy = C \int_0^\infty e^{-\phi^2} d\phi = C\sqrt{\frac{\pi}{4}}, \tag{7.70}$$

where the error function erf(x) with relations

$$\mathrm{erf}(x) = \frac{2}{\sqrt{\pi}} \int_0^x e^{-x^2} dx$$

and $\mathrm{erf}(x \to \infty) = 1$ has been used. To calculate the *local* heat-transfer coefficient $h(x)$ at position x along the plate ($y = 0$), we use its definition

$$q_{y=0} = k_l \frac{\partial T}{\partial y} = h(x) \cdot (T_l - T_m). \tag{7.71}$$

Now, we apply Eqs. (7.69) and (7.70) to this to get

$$h(x) = \sqrt{\frac{k_l^2 u}{\pi \alpha_l x}}.$$ (7.72)

For a plate of length L, Eq. (7.73) gives the mean heat-transfer coefficient

$$\bar{h} = \frac{1}{L} \int_0^L h(x) dx = \sqrt{\frac{4 k_l^2 u}{\pi \alpha_l L}}.$$ (7.73)

Using dimensionless quantities, we obtain the well-known Nusselt relation for the mean heat transfer across a laminar boundary layer set up by a flat plate moving through a molten metal,

$$Nu = \sqrt{\frac{4}{\pi} PrRe},$$ (7.74)

where $Nu = hL/k_l$ as defined in Eq. (7.20), $Re = uL/\nu$ is the Reynolds number, and $Pr = \nu/\alpha_1$ is the Prandtl number discussed earlier in this section.

Due to the movement of the plate in the melt, there will be a boundary layer between the plate and the bulk melt.

In the present system, melt is drawn down from the melt surface with the plate as illustrated in Fig. 7.13. Hot melt flows from the bulk melt via the melt surface and meets the cold plate. We can expect that the thermal boundary layer is very thin here, giving a very high heat transfer, and that it grows thicker down along the plate. We apply the boundary-layer solution to our plate-feeding problem as a first-order approximation, as illustrated in Fig. 7.15. Note that the coordinates x and y are exchanged.

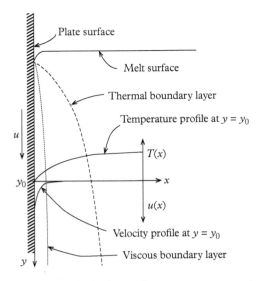

Fig. 7.15 *The heat-transfer coefficient for feeding a plate into a stagnant melt.*

To obtain an estimate for h_l, we introduce the penetration depth of the plate, P:

$$h_l = \sqrt{\frac{4k_l^2 u}{\pi \alpha_l P}} \tag{7.75}$$

7.4.5 Simplified Model with Shell Formation

Instead of integrating the complicated shell-growth (Eq. (7.45)) numerically, we can step back to the heat balance of a cross-section of the shell, (Eq. (7.40)), and assume that $(d\xi_s/d\eta)^2 \ll 1$ (that is, the shell is flat). Then the square root vanishes, and we find a much simpler formula for the shell growth:

$$\frac{d\xi_s}{d\eta} = \frac{Bi\left(1 - \theta_p\left(\eta\right)\right) - Nu\left(\theta_l - 1\right)}{PeSf}. \tag{7.76}$$

The constant A in (Eq. 7.62) is very small and its only significance is to provide a smooth temperature profile across $\eta = \Pi_A$. Therefore, we set $A = 0$ and substitute Eq. (7.62) for θ_p in Eq. (7.76) and integrate to obtain

$$\xi_s\left(\eta\right) = \frac{\frac{Bi}{\lambda_2}\left(1 - e^{-\lambda_2 \eta}\right) - Nu\left(\theta_l - 1\right)\eta}{PeSf}. \tag{7.77}$$

The first root of $\xi_s(\eta) = 0$ for $\eta > 0$ is Π_A, which can be found by rearranging the equation to

$$\eta_{i+1} = \frac{Bi}{Nu\lambda_2\left(\theta_l - 1\right)}\left(1 - e^{-\lambda_2 \eta_i}\right) \tag{7.78}$$

and iterating this until $|\eta_{i+1} - \eta_i| < \varepsilon$. A reasonable initial guess value is $\eta_0 = 100$, and $\varepsilon = 10^{-2}$ should be more than sufficient.

Figure 7.16 shows models with shell (Eq. (7.78)) and without shell (Eq. (7.15)) formation and subsequent melting compared to measurements of a flat plate submerged in stagnant aluminium melt. The heat-transfer coefficient according to Eq. (7.75) is used for the model with shell formation, whereas a constant value of $18.7\,\mathrm{kW/m^2K}$ is used for the model without shell formation.

It can be noted that an alternative to the iteration of Eq. (7.78) can be found if we realize that for enough numbers of iterations

$$|\eta_{i+1} - \eta_i| \to 0. \tag{7.79}$$

Thus

$$\lambda_2 \eta = \frac{Bi}{Nu\left(\theta_l - 1\right)}\left(1 - e^{-\lambda_2 \eta}\right) \tag{7.80}$$

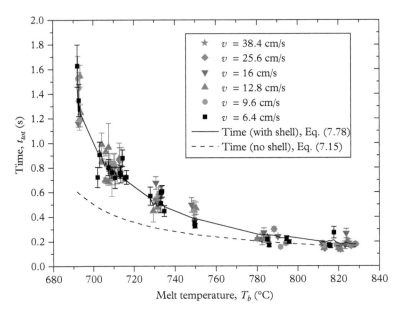

Fig. 7.16 *Simplified models with and without shell as calculated with Eqs. (7.78) and (7.15), respectively, compared to measurements. Equation (7.78) was iterated three times. (Based on data from Farner.[15])*

and

$$\frac{Nu\,(\theta_l - 1)}{Bi} = \frac{\left(1 - e^{-\lambda_2 \eta}\right)}{\lambda_2 \eta}. \tag{7.81}$$

Then Eq. (7.81) can now be solved implicitly by adjusting $\eta\lambda_2$ so that right-hand side equals the value of left-hand side.

The foregoing mathematical analysis shows how dimensionless groups often arise when analysing a practical problem. Their physical significance is also clearly illustrated. Chemical engineers often use a dimensional analysis when faced with a complicated process, which can involve a dozen or more seemingly unrelated variables. Metallurgists should be fully aware of this powerful tool. Dimensionless groups can help one to analyse complicated process data, to scale up a laboratory process, or to determine whether an analogous model (e.g. in water) can be applied to a metallurgical process. The latter point in particular is considered in some detail in the discussions which follow in this chapter.

7.5 Alloying

7.5.1 Diffusion-Limited Dissolution of Alloys

Alloys with a melting point higher than the bath temperature dissolve by mass transfer in the melt boundary layer. For metal systems mass transfer is a much slower process than heat transfer. Furthermore, convection and diffusion in the melt boundary layer proceed rapidly compared with solid state diffusion. Thus, once the solid shell that initially forms has melted, dissolution is determined by the mass transfer coefficient k as described in Chapter 3. Then the solid surface is heated up to the bath temperature while the surface composition may remain unchanged. Since viscosities and diffusion coefficients in molten metals are very similar to what we find in aqueous systems, often chemical engineering theory and correlations can be applied. There are two main problems. One is to obtain reliable data, especially for the diffusion coefficients, D (see also Section 2.2.9). The mass transfer coefficient k depends on D, for instance to the power $2/3$ (see also Chapter 3). The other problem is that dissolution sometimes does not proceed in a planar fashion.[20] However, in the following it is assumed that dissolution takes place on a well-defined planar front.

The rate of dissolution depends on the temperature of the melt, how much turbulence there is, particle size, physical properties, the presence of intermetallic phases or clean metal, and finally the molten metal composition. The part of the alloy elements that is lost mainly enters the top layer which can be covered by a slag or aluminium dross. This can happen with master alloys that have low density and with powder that floats up to the surface and in the oxide. It is therefore important that the alloys are added to the melt when there is a minimum of slag or dross on the surface. Magnesium and strontium may oxidize on the surface or even evaporate. Elements and intermetallic phases that have high relative density can settle down to the bottom of the furnace and result in a low dissolution rate due to both lower (local) temperature and less convection leading to a slower dissolution rate.

Once the mass transfer coefficient k is known one can calculate the dissolution time. A mass balance for the alloy component M gives:

number of moles of M removed from the alloy addition per unit time = flow of moles of M transferred through the diffusion boundary layer in the melt

or

$$-\frac{dr}{dt} A \rho_M = kA \left(c_M - c_{Mb} \right), \tag{7.82}$$

where it may be noted that $\frac{dr}{dt} A$ is the rate of change of a volume of a spherical shell and A is the melt–shell contact area, r is the radius, and ρ_M is the number of kmol/m^3 in the alloy addition. Here k is the mass transfer coefficient as described in Chapter 3. It should not be confused with the thermal conductivity. c_M is the concentration of M in the melt in equilibrium with the concentration in the solid in units of kmol/m^3. c_{Mb} is the concentration in the bulk melt. Thus $c_M - c_{Mb}$ is the driving force for diffusion through the melt boundary layer.

Integration of Eq. (7.82) gives, disregarding the dependence of k on r,

$$t = \frac{\rho_M \left(R - r \right)}{k \left(c_M - c_{Mb} \right)}. \tag{7.83}$$

Here R is the original radius of the addition, and t in Eq. (7.83) is the dissolution time from size R to r. The total dissolution time for the addition is

$$t_d = \frac{\rho_M R}{k \, (c_M - c_{Mb})}. \tag{7.84}$$

To obtain the total time, the shell time, t_s, must be added to t_d. We find that usually t_s is much smaller than t_d.

As an example of the calculation of t_d we take the addition of niobium particles of 4 mm diameter to steel. The melting point of Nb is high, about 2500 °C, so this should be a class II type of addition. The diffusion coefficient of Nb in Fe at 1570 °C is taken to be

$$D = 10^{-9} \frac{m^2}{s}$$

as a rough estimate.

Equation (5.2) with $\phi = 0.22$ for $\rho_M = 8570$ kg/m^3 now gives for the relative velocity between particle and melt

$$u_r^2 = \left(\frac{8570}{7030} - 1 \right) \times \frac{9.81 \times 4 \times 0.002}{0.22 \times 3}$$

or

$$u_r = 0.16 \text{ m/s}.$$

This gives

$$Re = \frac{0.16 \times 0.004}{0.8 \times 10^{-6}} = 800.$$

Then according to Eq. (3.60)

$$Sh = \frac{kd}{D} = 1 + 0.724 \times 800^{0.48} \left(\frac{0.8 \times 10^{-6}}{10^{-9}} \right)^{\frac{1}{3}}$$

or

$$k = 4.2 \times 10^{-5} m/s.$$

From Eq. (3.90) we see that we must know c_M, the concentration of Nb in the melt in equilibrium with pure Nb. In fact, from the Fe–Nb phase diagram[13] it is seen that a number of solid intermediate phases form. The result is that dissolution in fact does not proceed in a planar fashion. Rather, a hollow dendritic structure remains after most of the metal has diffused into the steel bath.[20] The melt in contact with the solid only contains about 20 mol% Nb. If c_{Nb} is negligible and Eq. (3.90) is employed despite the non-planar dissolution:

$$t_d \sim \frac{R}{k*0.20} = \frac{0.002}{4.2 \times 10^{-5}*0.20} \approx 240 \text{ s}.$$

In a stagnant bath the addition would settle to the bottom before dissolution was complete. Here the rate of dissolution would be very slow. Thus, stirring (see Section 5.8) is important in order to keep the additions suspended and to increase k. Alternatively, particles smaller than 4 mm can be injected into the melt in order to reduce the relative velocity, or FeNb may be added. Then the density difference is less so that the relative velocity u_r is reduced. This can be seen from Eq. (5.3), Stokes' equation for settling when the Reynolds number is low.

For other high-melting alloys and ferroalloy additions to steel such as FeMo, FeB, and FeZr the problems are similar to those sketched here for FeNb. Dissolved oxygen slows down the dissolution rate.[4] For instance, 700 ppm oxygen reduces the dissolution rate for FeNb by a factor of 2.[4]

7.5.2   The Heat-Transfer Coefficient for Molten Metals

In the calculation of the heat-transfer coefficient for molten metals the high value of the thermal conductivity must be taken into account. As a result, the thermal conductivity, k, and the thermal sublayer in the melt along the body play a much greater role than for normal liquids. Therefore, one should be wary of applying heat-transfer correlations derived for normal fluids to molten metals. For molten metals the ratio between kinematic viscosity and thermal diffusivity $Pr = \nu/\alpha$ is very small, in the range 0.003–0.1. By comparison Pr lies between 0.65 and 1.1 for gases, between 0.9 and 50 for 'common' liquids such as water, and beyond 20 for viscous liquids such as oils and slags. Due to the importance of thermal diffusion, α, in the wall boundary layer for metals, the kinematic viscosity, ν, and the roughness of the walls play only a minor role, except that they have an effect on the velocity distribution. In part this is accounted for by including the variables u and d. This leaves only one dimensionless (Péclet) group to characterize heat transfer,

$$Pe = \frac{ud}{\alpha} = \frac{ud}{\nu}\frac{\nu}{\alpha} = RePr, \tag{7.85}$$

where u is the (mean) velocity of the metal past the solid surface, d is a characteristic dimension: for flow around a sphere, d is the diameter; for pipe flow, d is the diameter of the pipe. The Péclet number compares heat transport by convection to thermal conduction.

For turbulent pipe flow at very high Re numbers it is found empirically that the following equation represents quite well the data for many liquid metal systems,[21]

$$Nu = 4 + 0.014 Pe^{0.8}, \tag{7.86}$$

where $Nu = hd/k$. The equation is valid when the length of the pipe is greater than 60 d. Also $Pe > 100$ and hence $Nu > 4.5$.

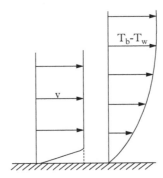

Fig. 7.17 *Comparison of the velocity and thermal boundary layers for flow of a liquid metal along a solid surface.*

For smooth plates, laminar flow persists up to Re numbers of roughly 5×10^5. For laminar-forced convective flow over a plate, with $Pr \ll 1$[6] theory gives a heat-transfer coefficient[22]

$$Nu = 1.13 Pe^{\frac{1}{2}}. \tag{7.87}$$

This is the same result as for transfer to a free surface (cf. Eq. (3.50)). Figure 7.17 compares the thermal and velocity boundary layers for $Pr \ll 1$. It is seen that the thermal boundary layer is much greater than the velocity boundary layer. Then compared to the free surface case there is only a slight reduction in the flow of melt that supplies heat to the surface. For $Pr < 0.4$ one may use a correction factor $(1 + 10Pr)^{-\frac{1}{6}}$.[23] For instance for $Pr \approx 0.1$ this gives 0.89 so that

$$Nu \approx Pe^{\frac{1}{2}}.$$

For flow past a sphere it is found experimentally that[24]

$$Nu = 2 + 0.38 Pe^{\frac{1}{2}}. \tag{7.88}$$

If the velocity boundary layer that surrounds the sphere is disregarded, one would obtain (see also Eq. (3.57))

$$Nu = 2 + 1.13 Pe^{\frac{1}{2}}. \tag{7.89}$$

The discrepancy between Eqs. (7.89) and (7.88) may be partly explained by the effect of the velocity boundary layer. Probably this effect is somewhat greater for a sphere than for a plate. However, of greater importance is the separation of the melt from the spherical surface at a polar angle of about $110°$.[22] In general the fluid particles near the surface beyond the point of separation move in directions that differ from the course of the external stream. This 'backwater' fluid has a long residence time and is of less use in transferring heat. Other causes for the lower value are discussed at the end of this section.

The constant 2 on the right-hand side of Eq. (7.89) takes pure conduction into account. This term is on the same order of magnitude as the convection term and even larger in some practical cases. The reason is the high thermal conductivity of metals. For instance, even if the Reynolds number for the addition is high, say 15 000, due to the low value of $Pr \sim 0.1$ the convection term is only

$$0.38 Pe^{\frac{1}{2}} = 0.38 \cdot (0.1 \cdot 15000)^{\frac{1}{2}} = 14.7.$$

Thus, even in this example with very high relative velocities convection-induced heat transfer is only around seven times greater than for pure heat conduction. A Reynolds numbers of 15000 corresponds to a diameter of 0.02 m and relative velocity in steel of 0.6 m/s. For smaller dimensions and lower velocities pure heat conduction plays a relatively greater role. Initially when the addition is made to the melt the temperature gradients are high. This causes the melt to freeze around the addition. Only when the internal temperature gradients have been substantially reduced, and the temperatures increased correspondingly, does the shell start to remelt. The addition of FeCr to steel is taken as an example. The melting-point of the steel is assumed to be $T_m^* = 1500\ °C$. If the bath temperature is $1700\ °C$ this gives $T_b - T_m^* \approx 200\ °C$. Then thermal conduction in the frozen shell and heat transfer in the melt are equal when

$$k_s \frac{dT}{dr} = h \left(T_b - T_m^* \right). \tag{7.90}$$

Here k_s is the thermal conductivity of the shell and h is the heat-transfer coefficient between melt and shell or, recalling that $Nu = hd/k_{Fe}$ (liquid),

$$\frac{dT}{dr} d = Nu \left(T_b - T_m^* \right) \frac{k_{Fe}\ (\text{liquid})}{k_{Fe}\ (\text{solid})} \tag{7.91}$$

or, disregarding the difference in the thermal conductivities,

$$\frac{\frac{dT}{dr} \frac{d}{2}}{T_m^* - T\,(r = 0, t = 0)} = \frac{Nu \left(T_b - T_m^* \right)}{2 \left\{ T_m^* - T\,(r = 0, t = 0) \right\}}. \tag{7.92}$$

If $Nu = 16.7$, as in the previous example,

$$\frac{\frac{dT}{dr} \frac{d}{2}|}{T_m^* - T\,(r = 0, t = 0)} = \frac{16.7 \cdot 200}{2 \cdot 1300} = 1.29,$$

where $T_m^* - T(r = 0,\ t = 0)$ is taken as $200\ °C$. Figure 7.18 sketches this situation for the temperature gradient and profile when the shell starts to remelt. Later when the shell has completely remelted, temperature gradients are even smaller and temperatures are higher.

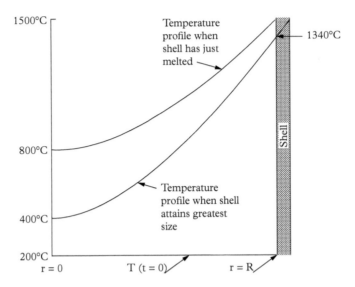

Fig. 7.18 *Temperature profiles when the shell attains its greatest size and when the shell has just remelted. Bi* = 1.29, Bi*b* = 2.0, k/k_s = 1, and α_s/α = 1.*

There is considerable scatter in the experimental results for metals where $Pr \ll 1$. Experiments also tend to give Nu numbers lower than the values obtained from the theoretical formulas. There are a number of possible causes:

(1) oxide formation on the wall that introduces an extra thermal resistance;

(2) failure of some liquid metals to wet the wall; and

(3) departures from a constant temperature along the wall.

Similar phenomena will also be present when additions are made to a molten bath, and this variability must be taken into consideration when calculations concerning melting are performed.

7.6   Dissolution Rate and Intermetallic Phases

For class II alloys, no analytical solution has been found for the complete melting process. Some work has been presented on numerical modelling of temperature evolution inside spherical additions where formation and remelting of shell are also discussed. However, the numerical procedures are not the scope of this section and readers interested in this may consult the references.[25–7]

As discussed earlier, if the temperature and heat-transfer conditions are such that a shell will form (see Section 7.3.4), this shell may also consist of an alloy with different thermal and physical properties than both the melt and the added alloy material, dependent on the relevant phase diagram. In addition, if there are any intermetallic phases in the Al–X phase diagram (where X is pure element or an AlX master alloy), these intermetallic phases will form and remelt during the melting process before the alloying element has dissolved completely. Three examples are treated here: Al + Mn, Al + Fe, and Al + Si.

Alloying elements with high relative density will have a tendency to sink to the bottom. This can lead to a slower dissolution process since this is (generally) a colder place and has lower melt convection. However, it seems that this is not the main problem with, for instance, manganese or iron. In these cases it has been shown that intermetallic phases play a significant role in the dissolution process.

Figures 7.19 and 7.20 show the binary phase diagrams for Al–Mn and Fe–Al, respectively.[14] Razaz and Carlberg have studied the dissolution of manganese and iron in pure liquid aluminium.[28] They took melt samples during the dissolution and in both cases found intermetallic phases according to the phase diagrams.

In one of the experiments performed with a melt temperature of 720 °C, examined samples showed a partially dissolve Al–90% Mn master alloy particle with layers of intermetallic phases in the following order outwards: Mn with dissolved Al, γ_2, Al$_{11}$Mn$_4$, μ, and finally aluminium with dissolved Mn.[28] They concluded that the γ_2 layer represented the largest diffusion barrier for diffusion of Al from outside of this layer into the remaining Mn particle. From this, it was also stated that using a master alloy with not more than 60% Mn content will increase the dissolution time since diffusion through the γ_2-phase is eliminated.

The situation with iron is similar. By adding pure Fe to a melt at 720 °C and examining samples taken from the melt during dissolution the authors found two intermetallic phases, namely the Al$_5$Fe$_2$ and Al$_3$Fe between the undissolved iron and aluminium metal.[28]

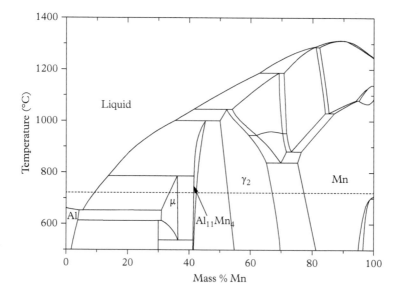

Fig. 7.19 *Simplified Al–Mn phase diagram.*[14]

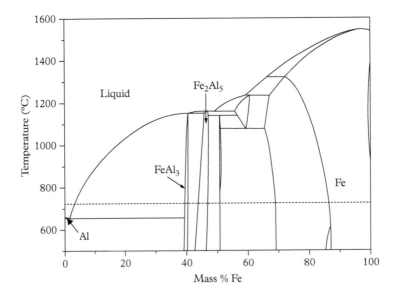

Fig. 7.20 *Simplified Fe–Al phase diagram.*[14]

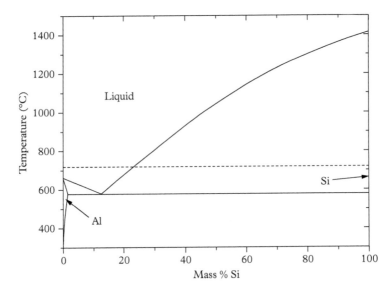

Fig. 7.21 *Al–Si phase diagram.*[14]

By comparison, the Al–Si phase diagram is shown in Fig. 7.21. Although the dissolution of silicon is an endothermic process (as shown in Table 7.2) that is locally giving a reduced temperature, the dissolution of silicon in molten aluminium has been shown to be relatively fast. As seen from the phase diagram there are no intermetallic phases forming during the

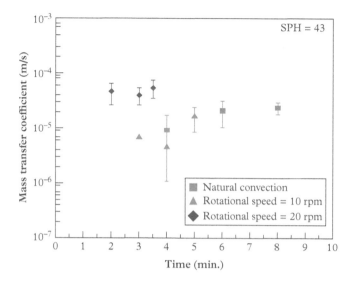

Fig. 7.22 *Mass transfer coefficients for dissolution of silicon in molten aluminium under different melt stirring conditions and temperatures. The melt superheat is shown in upper right corner.[29] [Reprinted with permission from Springer Nature.]*

dissolution. As long as the melt temperature is kept over 660 °C (locally) all silicon will dissolve directly into the aluminium liquid.

Ahmadi et al. studied the dissolution rate of silicon in molten aluminium at melt temperatures from 681 to 738 °C for three different melt convection velocities.[29] They also calculated and plotted mass transfer coefficients (as defined in Chapter 3) for the systems. The measurements done at 43 °C superheat (703 °C) are shown in Fig. 7.22.

Comparing the mass transfer coefficient of the dissolution of silicon in Fig. 7.22 with the mass transfer coefficient for the melting of pure aluminium in Fig. 7.5 it is seen that silicon dissolves at a rate of about one order of magnitude lower than aluminium in its own melt.

7.7 Practical Alloy Additions to a Melt

This section summarizes some of the typical practices for alloy additions. The choice of method depends on alloy type (i.e. element one wants to increase), bulk metal type and temperature, and furnace or equipment type available.

The bulk addition of coarse material is widely practiced. Provided good recovery can be achieved, this method dominates due to the favourable logistics and equipment costs. In the aluminium industry where gas-fired furnaces are common, it is common to have electromagnetic stirrers. These stirrers help transport heat downwards into the bulk metal and mix the alloying elements. Additions of bulk material to a tapping stream should be sized such that the material dissolves within the time of turbulent mixing.

Submerged powder injection has inherently good heat and mass transfer characteristics and can be used for virtually all alloying elements.[30-3] The buoyancy of the carrier gas bubbles and the momentum of the particles provide the stirring of the bath. Usually stirring is an advantage, but it may become a problem if top-slag or dross is entrained and an 'eye' exposer to the atmosphere is formed. Other considerations are the relatively high capital costs for reliable and flexible injection systems.

Injecting powder can give better contact surface area relative to the other methods. The contact area per kilo element increases with decreasing powder size. If the lance is moved, elements can be evenly distributed, avoiding local cooling of the melt. This allows for a higher temperature during dissolution, promotes stirring, and results in a more homogeneous melt. The concentration gradient between the alloy element and melt is thus larger. Thus, the driving force for the dissolution of the elements becomes larger.

Metal-encased alloy wires[34-6] are clearly more expensive than other forms of additions. Consequently, this method is only used for reactive elements such as calcium, titanium, or aluminium or for microalloying close to the casting operation. Wire feeding is attractive due to the simple equipment involved and the highly reproducible results. The metal sheath surrounding the alloys protects them from oxidation by the atmosphere and slag, and the wire penetrates some distance down into the metal before the sheath cover melts. This ensures a high yield.

Depending on their density relative to the melt, additions float up to the surface or settle to the bottom. In the first case, there may be considerable losses due to oxidation or evaporation. In the second case, dissolution rates may be prohibitively low, and the yields may vary from charge to charge. Therefore, the above methods can be supplemented by the use of ingots or waffles held in place in the melt with baskets or cages. However, such methods normally are time- and labour-consuming.

Another approach is to use master alloys as with ferroalloys in iron. For instance, FeMn and SiMn are produced directly in submerged arc furnaces and are used directly in the steel industry. For aluminium there is available on the market aluminium alloyed with Sb, Bi, B, Cr, Co, Cu, Fe, Mg, Mn, Ni, Si, Sr, Ti, V, Zn, and Zr. These master alloys contain at least 50% Al, often much more. Dissolution of the element from the master alloy is more rapid than for the alloy element alone. One reason is that the master alloy has a density close to the melt density. Thus, the tendency to float or settle is reduced. Secondly, as described in Section 7.5, by using a master alloy the slowest diffusion resistance can be avoided.

Briquettes and tablets are compressed powder additions. They are used for alloying element like chromium, iron, and manganese into aluminium. Briquettes and tablets have in many cases replaced master alloys. After addition, they disintegrate and dissolve faster than bulk addition due to the many small particles with large specific surface contact area towards the bulk metal. They are also easier to use and cheaper. Again. good stirring is required to get rapid dissolution. The downside with this method is that gas entrapped in the porous tablets creates turbulence on the metal surface, which promotes greater oxidation and hence metal loss to slag or dross.

A problem is the solubility of the alloy additions. For instance, if the molten master alloy is to be mixed with three times the amount of metal, the alloy element concentration of the master alloy must be four times that of the final product. If the aluminium is to contain 7% Si, the master alloy must dissolve 28% Si. From the Al–Si temperature–composition diagram[13] it is seen that this requires a temperature of 800 °C.

It is also important to consider the order in which the alloying elements are added to a charge. Razaz and Carlberg[28] studied the influence of Si and Ti on the dissolution rate of Mn and Fe in molten aluminium. They concluded that while Si inhibited the Fe dissolution, Ti affected the Mn dissolution. Therefore, Fe should be added before Si, and Ti before Mn.

7.8 Safety

Last but not least, it is important to always think about safety when working with molten metals and in particular the combination of solid and liquid metal together. Firstly, molten metal is hot and has a large amount of energy accumulated as heat. Secondly, molten metal can react chemically with, for instance, oxygen in air, or the solid may oxidize and release large amounts of chemical energy in an uncontrolled oxidation process. And thirdly, molten metal in contact with water may create steam which will expand in volume and may throw molten metal several meters.

Molten aluminium is generally much colder than molten silicon or iron. A typical melt temperature for aluminium is 720–750 °C when undergoing alloying or melt treatment. Molten silicon and iron have temperatures of more than 1410 and 1540 °C, respectively.

Aluminium is a strong oxidizing agent. Therefore, it is important to avoid any contact between molten aluminium and common oxides such as Fe_2O_3 (rust), Fe_3O_4, Cr_2O_3, CuO, and ZnO. A chemical reaction between Fe_2O_3 and aluminium is a well-known so-called thermite reaction which can release heat enough to heat the system to more than 3600 °C if ignited.[14] Thus, all tools must have a coating designed for its purpose before the tool comes in contact with the melt.

Water is particularly dangerous in contact with molten metal. For instance, one litre of water (or snow or ice) which is submerged in molten aluminium will instantaneously transform to steam and expand to 4.5 m^3 when evaporated and heated to 700 °C. Thus, all material or tools must be preheated properly before they are submerged in molten metal. Scrap material is often stored outside or in humid atmospheres and must be either charged in an empty furnace or dried completely before charging. Any closed containers must be opened before charging. This means that metal bales should be shredded or opened.

In order to learn from injuries and accidents related to the aluminium industry, the Aluminum Association (http://www.aluminum.org) present annual reports on accidents worldwide.[37] They characterize the incidents into three groups as shown in Table 7.5.

Figure 7.23 shows the reported worldwide incidents for melting and casting from 1990 to 2018, whereas Fig. 7.24 shows force 2 and 3 incidents for 1981–2018. Between 1981 and 2018 there has been on average 3.1 force 3 accidents per year. For the past ten years, the average number has dropped to 0.7 per year, showing that an increased awareness on safety is important.

Table 7.5 *Explosion rating force criteria as given by the Aluminum Association*

Guidelines	Force 1	Force 2	Force 3
Property damage	None	Minor	Considerable
Light	Minimal	Flash	Intense
Sound	Short cracking	Loud report	Painful
Vibration	Short and sharp	Brief rolling	Massive structural
Metal dispersion	< 5 m (15 feet)	5–15 m	> 15 m

Source: Reprinted from the Aluminum Association[37] with permission.

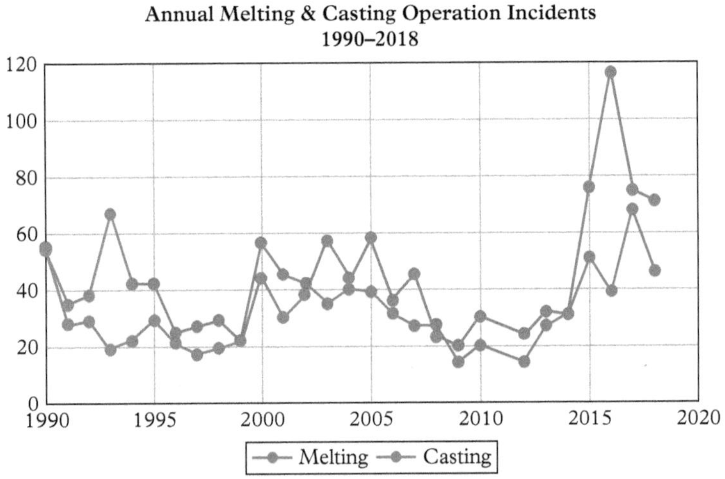

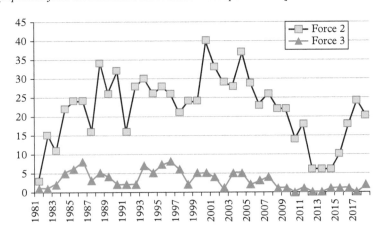

Fig. 7.23 *Annual worldwide reported incidents related to melting and casting in the aluminium industry. [Reprinted from the Aluminum Association[37] with permission.]*

Fig. 7.24 *Annual reported force 2 and 3 incidents worldwide in the aluminium industry. [Reprinted from the Aluminum Association[37] with permission.]*

7.9 Summary

It is difficult to describe alloy additions to a melt in a general manner. Complete analytical models do not exist. A simple energy transport model is presented for melting submerged spherical additions and continuously fed plates and rods to aluminium bath. This model fits reasonably well to situations without shell formation and where heat conductivity in the solid part is disregarded. A more sophisticated model with shell formation is also presented for plate feeding. In the simplified treatment given here energy required for heating and melting is considered.

Systems have not been considered where the heat of solution is important. For instance, the addition of FeSi to molten steel is not discussed. Ferrosilicon has an exothermic heat of solution in a steel melt and thus dissolves rapidly.[38] Calculating the shell time was illustrated using FeCr as an example.

Special problems are the lack of good data for the thermal conductivity and difficulties in determining both the heat-transfer and diffusion coefficients. Here it is assumed that the addition is removed as it melts.

It has been shown that the binary phase diagrams may give a hint of the dissolution rate. For instance, the Al–Si system, without any intermetallic phases, is much simpler than both the Al–Fe and Al–Mn systems where intermetallic phases plays an important role in the dissolution process.

This chapter has briefly discussed some aspects of safety concerns during alloying operations. Safety should always be our first concern. See, for example, the Aluminium Association's reports and guidelines.

....

REFERENCES

1. Wright JK. Steel dissolution in quiescent and gas stirred Fe/C melts. *Met. Trans.*, 1989; 20B: 363–74.
2. Guthrie RIL, Clift R, Henein H. Contacting problems associated with aluminium and ferro-alloy additions in steelmaking-hydrodynamic aspects. *Met. Trans.*, 1975; 6B: 321–9.
4. Argyropoulos SA. On the recovery and solution rate of ferroalloys. *Iron and Steelmaker*, 1990; May: 77–86.
3. Guthrie RIL. Addition kinetics in steelmaking. In: *Proc. 35th Electric Furnace Conference*, pp. 30–41. Warrendale, PA: Iron and Steel Soc. of AIME; 1977.
5. Joo S, Guthrie RIL. The role of physical and mathematical models for ladle metallurgy operations. Iron and Steel section. *International Symposium on Ladle Steelmaking and Furnaces, C.I.M. Montreal, Quebec*; 1988; pp. 1–28.
6. Mair H, Marsh DB. Advancement of cored wire applications within the steel and cast metal industries. In: *Proceedings INFACON'89, New Orleans, LA*, vol. 1, pp. 132–50; 1989.
7. Argyropoulus SA, Guthrie RIL. Dissolution kinetics of ferro-alloys in steelmaking. *Proceedings Steelmaking Conference,Pittsburgh*, pp. 156–67; 1982.
8. Argyropoulus SA, Guthrie RIL. The dissolution of titanium in liquid steel. *Metall. Trans.* B, 1984; 15: 47–58.

9. Aukrust E. Trends in steel developments and the impact on ferro-alloy production. In: *International seminar on refining and alloying of liquid aluminium and ferro-alloys*, pp. 37–48. Düsseldorf: Aluminium-Verlag; 1985.

10. Wada H, Lee SW, Pehlke RD. Nitrogen solubility in liquid Fe and Fe-Mn alloys. *Metall. Trans. B*, 1986; 17: 238–9.

11. Hultgren R, Desai PD, Hawkins DT, Gleiser M, Kelley KK, Wagman DD. *Selected values of the thermodynamic properties of binary alloys*. Metals Park, OH: American Soc. for Metals; 1973.

12. Barin I. *Thermochemical data of pure substances, parts I and II*. Weinheim, Germany: VCH Verlagsgesellschaft; 1989.

13. Brandes EA. *Smithells metals reference book*. London: Butterworths; 1983.

14. FactSage 7.3, FTlite database. FactSage.com.

15. Farner S. *Remelting of aluminium by continuous submersion of rolled scrap*. Dr.Ing. Thesis. Trondheim, Norway: NTNU; 2000: p. 127.

16. Kargl F, Weis H, Unruh T, Meyer A. Self-diffusion in liquid aluminium. *J. Phys. Conf. Ser.* 340 012077; 2012.

17. Taniguchi, S., Ohmi, M. and Ishiura, S. (1983), A hot model study on the effect of gas injection upon the melting rate of a solid sphere in a liquid bath. *Trans. Iron Steel Inst. Jpn.*, 1983; 23: 571–7.

18. Jiao Q, Themelis NJ. Mathematical modelling of heat transfer during the melting of solid particles in a liquid slag or metal bath. *Canad. Metall. Q.*, 1993; 32(1): 75–83.

19. Schlichting H. *Boundary-layer theory*, 7th edn. New York: McGraw-Hill; 1979.

20. Gourtsoyannis L, Guthrie RIL, Ratz GA. The dissolution of ferromolybdenum, ferroniobium and rare earth (lanthanide) silicide in cast iron and steel melts. In: *Proceedings of the electric furnace conference, AIME*, vol. 42, pp. 119–32; 1984.

21. Davies JT. *Turbulence phenomena*. New York: Academic Press; 1972.

22. Schlichting H. *Boundary-layer theory*, 6th edn. New York: McGraw-Hill; 1968.

23. Bennett CO, Meyers JE. *Momentum, heat, and mass transfer*. New York: McGraw-Hill; 1974.

24. Guthrie RIL. *Engineering in process metallurgy*. Oxford: Clarendon Press; 1989.

25. Røhmen E, Bergstrøm T, Engh TA. Thermal behaviour of a spherical addition to molten metals. In: Tuset JKr, Tveit H, Page IG, eds, *Proceedings of INFACON 7*. Trondheim Norway: FFF; 1995.

26. Jardón-Pérez IE, Ramírez-Argaez MA, Conejo AN. Melting rate of spherical metallic particles in its own melt: effect of particle temperature, bath temperature, particle size and stirring conditions. *Trans Indian Inst. Met.*, 2019; 72(9): 2365–73.

27. Røhmen E. *Thermal behaviour of a spherical addition to molten metals*. Dr.Ing Thesis. Trondheim, Norway: NTH; 1995: p. 55.

28. Razaz G, Carlberg T. On the dissolution process of manganese and iron in molten aluminum. *Met. Trans A*, 2019; 50: 1873–87.

29. Ahmadi MS, Argyropoulos SA, Bussmann M, Doutre D. Dissolution studies of Si metal in liquid Al under different forced convection conditions. In: Lindsay SJ, eds, *Light Metals*, 2011; 809–14. Springer, Cham.

30. Lehner T. *Reactor models for powder injection*. Luleå, Sweden: Scaninject (International Conference on Injection Metallurgy), 1977; 11: 1–48.

31. Engh TA, Larsen K, Venas K. Penetration of particle-gas jets into liquids. *Ironmaking and Steelmaking*, 1979; 6: 268–73.

32. Pedersen T, Myrbostad E. Alloying in aluminium by injection. In: *International Seminar on refining and alloying of liquid aluminium and ferro-alloys*, pp. 333–42. Düsseldorf: Aluminium-Verlag; 1985.

33. Irons GA. *Fundamental and practical aspects of lance design for powder injection processes.* Luleå, Sweden: Scaninject IV; 1980: 3.1–25.
34. Mucciardi FA, Guthrie RIL. Aluminium wire feeding in steelmaking. *Iron and Steelmaker Trans.,* 1983; 3: 53–9.
35. Herbert A, Notman GK, Morris D, Knowles S, Jemson C. *Experiences with powder and wire injection at British Steel Corp., Lackenby works, basic oxygen steelmaking plant.* Luleå, Sweden: Scaninject IV; 1986: 27.1–36.
36. Kulunk B, Guthrie RIL. Optimisation studies for aluminium wire feeding operations in steelmaking ladles. *Ironmaking and Steelmaking*, 1988; 15(6): 293–304.
37. The Aluminum Association: 2019 Molten Metal Incident Report. https://www.aluminum.org/resources/electrical-faqs-and-handbooks/safety, Annual summary report on molten metal incidents in 2018.
38. Argyropoulos SA Guthrie RIL. The exothermic dissolution of 50 wt.% ferro-silicon in molten steel. *Canad. Metall. Q.* 1979; 18(3): 267–81.

8

Metal Processes and Applications—An Overview

Christina Meskers

As background to the other chapters, a short overview of each metal's physical properties, production process, applications, and recycling is given. One way of approaching the elements is through their place in the periodic table of elements, which can be used to predict physical and chemical properties and behaviour. Each group in the periodic table corresponds to a section in this chapter. Still, how the metals relate to each other for a particular property is not directly apparent from the periodic table or from Table A5.1 in Appendix 5 where the key properties of metals are listed. They can be ranked based on density or melting point (Fig. 8.1) or strength, which are important for structural applications. But metals are not only used for that. Some are used because of their ability to form alloys and improve alloy properties, to catalyse reactions, or to conduct electricity or heat. The application also determines the purity of the metal. Whereas for ferroalloys impurity limits are typically in the range of 0.01 to 0.1 wt.%, impurities in semiconductor grade silicon or germanium are specified in atoms per cm^3. Extensive refining is necessary to achieve such purity. Another way to group the metals is by the method or equipment to produce or refine them. For example, molten salt electrolysis is used for reactive metals such as aluminium, lithium, magnesium, and rare-earth elements; metallothermic reduction is the method of choice for titanium, zirconium, magnesium, strontium, and tantalum, while carbothermic reduction is used for iron and steel, and copper, zinc and lead are typically produced via sulfide smelting.

This chapter utilizes a systematic approach. The linkages and similarities between different metals become apparent, providing valuable insights for primary production, recycling, residue treatment, technology development, alloy and product design, substitution, etc.

The metal wheel (Fig. 8.2) is a key tool for this.[1-3] It illustrates the interconnectedness between metals, which is particularly relevant for metallurgists in the context of metal refining and the circular economy. It applies to metals from primary resources, production scrap, end-of-life products, and residues generated during metal production and recycling processes. The metal wheel summarizes this chapter, and may provide guidance when designing products and assessing the recycling possibilities of metal-containing products.

The wheel is divided into segments that represent the main metal processing structure, as labelled in the inner circle. Three additional rings indicate the destination of the other elements that enter the main metal process route. From inside to outside: dissolution in the main metal (dark grey); accumulation in dust, slime, speiss, or other (high-value) by-product (middle

Principles of Metal Refining and Recycling. Thorvald Abel Engh, Geoffrey K. Sigworth, and Anne Kvithyld, Oxford University Press.
© Oxford University Press 2021.
DOI: 10.1093/oso/9780198811923.003.0008

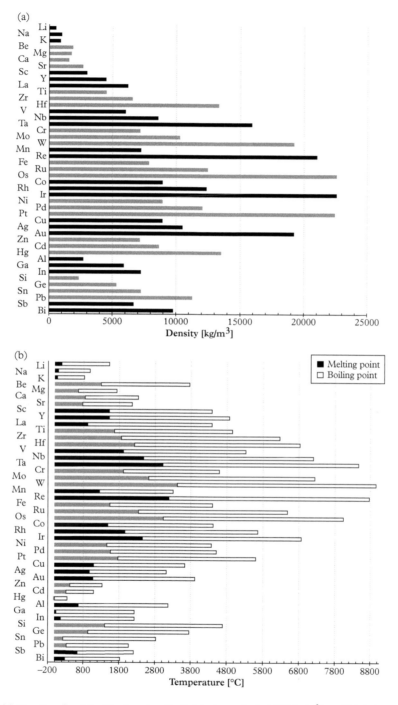

Fig. 8.1 *(a) Density of metals. The low density, lightest metals (< 3000 kg/m³) and high density, heaviest metals (> 15000 kg/m³) are easily identified. (b) Melting point (dark color) and boiling point of metals (light color). The groups of the periodic table are indicated by grey and black shading. In this way the trends within a group can be easily seen and compared to other groups and metals.*

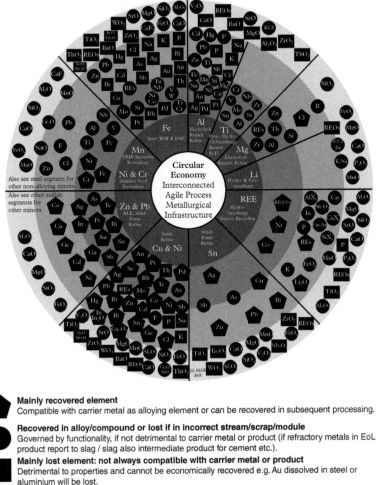

 Mainly recovered element
Compatible with carrier metal as alloying element or can be recovered in subsequent processing.

 Recovered in alloy/compound or lost if in incorrect stream/scrap/module
Governed by functionality, if not detrimental to carrier metal or product (if refractory metals in EoL product report to slag / slag also intermediate product for cement etc.).

Mainly lost element: not always compatible with carrier metal or product
Detrimental to properties and cannot be economically recovered e.g. Au dissolved in steel or aluminium will be lost.

Legend
Economically viable destinations of complex resources and materials, designed
functional material combinations, scrap, residues, etc. to metallurgical processing
infrastructures (each segment) to produce refined metal, high-quality compounds, and
alloys in best available technology

 Circular economy's agile carrier metals processing infrastructure
Extractive metallurgy's backbone, the enablers of a circular economy (CE) as it also recovers
technology elements used e.g. in renewable energy infrastructure, IoT, eMobility, etc.

 Dissolves mainly in metal if metallic (mainly pyrometallurgy—smelt)
Valuable elements recovered or (dissipative) lost (metallic, speiss, compounds, alloy in EoL also
determines destination). Linked hydro- & pyro-metallurgical infrastructure determines % recovery.

 Compounds mainly to dust, slime, speiss (mainly hydrometallurgy—refine)
Collector of valuable minor elements as oxides/sulfates/chlorides, etc. and mainly recovered in
appropriate mainly hydrometallurgical infrastructure if economical. Often separate infrastructure.

 Mainly to benign lower value building material products & dissipative loss
Relatively lower value but inevitable part of society and materials processing. A sink for metal and
loss from the CE system as oxides / compounds. Usually linked by separate infrastructure.

Fig. 8.2 *The metal wheel developed by MA Reuter and A Van Schaik, showing the interconnectedness between metals in the metallurgical infrastructure.*[1–3]

grey); and accumulation in residue e.g. slag and leach residue with low or no value or use (light grey). The shape of the symbol indicates whether the element can be recovered from the main metal or brings functionality as an alloying element (pentagons), provides functionality to an alloy or compound and otherwise is lost (circles), or is detrimental to the properties and/or cannot be recovered economically (squares).

The metal wheel can be used in different ways. It shows that many metals do not have their own production infrastructure, e.g. Se, Te, Ge, In, Re, Bi, or are produced as main metal and as a by-product, e.g. precious metals, Cu, Ni, Pb. This means that the availability (quantity produced) of these metals is fully dependent on the global production of the major metal, the price of the main and by-product metal, and the efficiency and costs of the by-product metal recovery processes.[4,5] For example, the copper segment shows how the elements present in a copper ore or a printed circuit board are distributed. Precious metals follow copper (dark grey circle) and can be recovered from the metal via the slimes produced in eletrolytic refining (pentagon). Iron, as sulfide in the ore and as metal in a circuit board, ends up in the slag (light grey circle) and provides functionality to the slag when it is used as construction material (sphere). Lastly, we consider a recycling example. Copper impurities in recycled steel scrap are detrimental to the steel quality, as indicated by the square symbol of Cu and its placement in the dark and middle grey circle. Thus copper should be removed prior to metallurgical recycling of steel. If the copper impurities instead are present in aluminium scrap, they will dissolve in the metal (dark grey circle). Hence, whether copper is detrimental depends on the alloy produced, as indicated by the sphere symbol. In this way valuable information is provided for the design of alloys, products, and recycling processes.

8.1   Alkali Metals (Na, K, Li)

The alkali metals were discovered in the early 1800s. Sodium and potassium were discovered by Davy in 1807, lithium in 1818 by Davy and Brandé. The alkali metals have all the characterstics of metals to a high degree. They are soft enough to be cut with a knife, are lighter than water, and have melting points ranging from 181 °C for lithium down to 28.5 °C for cesium. These properties make them unsuitable for structural applications. The alkali metals are the most electropositive elements. With a valence of 1 their physical and chemical properties are predictable and regular.

All alkali metals react vigorously with a range of non-metals. This includes oxygen—liquid metals can ignite in air—and hydrogen, nitrogen, and halides. These reactions are highly exothermic, and can be explosive in case hydrogen is one of the products. Sodium's ability to reduce metal halides is used in metallurgy for refining. Furthermore, they form alloys with a wide range of metals. Potassium is the most powerful industrially available reducing agent, as cesium and rubidium are too reactive for safe handling.

8.1.1   Sodium (Na)

8.1.1.1   *Production*

Sodium metal, due to its high reactivity, occurs in nature only in the form of salts such as chloride (NaCl, rock salt), carbonate (soda ash), nitrate, sulfate, and borate. Rock salt is the most widely used raw material in the inorganic chemical industry.[5] Sodium metal can be obtained from all sodium salts using high-temperature reduction. In the early days, since the 1850s, sodium carbonate and sodium hydroxide were used as feedstock. Since the 1920s the molten salt electrolysis of sodium chloride has been the industrial process for metal production. Natural rock salt contains impurities such as calcium sulfate ($CaSO_4$), which is removed by dissolving rock salt in water, then adding $BaCl_2$ and $NaCO_3$, and precipitating the impurities as $BaSO_4$ and $CaCO_3$. The clean brine is then dried till < 600 ppm of water is present.

Molten salt electrolysis is used to produce metal in a modified Downs cell.[6] The electrolyte consists of 28 wt.% NaCl, 47 wt.% $BaCl_2$, and 25 wt.% $CaCl_2$. Using this ternary mixture lowers the operating temperature to $600\,^\circ$C, compared to using pure NaCl ($T_{melt} = 808\,^\circ$C), and reduces corrosion. Each cell consists of four steel cathodes and four graphite anodes. Because graphite anodes are used, the electrolyte feedstock has to be oxide-free; otherwise, the graphite would react to form CO_2. The cell has a closed dome shape, so the generated chlorine gas can be easily collected and sold as by-product. The liquid sodium droplets rise to the top of the melt and are collected. Some metal is lost in the form of sodium chloride due to the reaction of sodium with the chlorine gas. Important for the functioning of this process are the wide liquidus temperature range of Na (98–$881\,^\circ$C), the low vapour pressure, the ability to use an abundant and cheap sodium feedstock, and a high current efficiency. Typically each cell contains about 8.4 t of electrolyte, produces $\pm$ 700 kg/day of sodium metal, operates about 3 years at 6.5–7V with a current efficiency of 80–90%, and requires 9.8–10.5 kWh of electricity per kg of sodium produced.[6] The entire production process is energy intensive; therefore, most of the production has moved from developed countries to low-cost countries such as China.[7]

The crude sodium metal contains about 1% calcium, calcium, and sodium oxides (0.3 wt.% each) and some chlorides. The impurities are removed by cooling the liquid sodium metal to $120\,^\circ$C, which is below the melting point of calcium and the chlorides, and filtering out the solid particles. The liquid sodium is then cast in blocks.

8.1.1.2   Applications

Until the early 1900s the use of sodium metal was mainly in metallurgical applications such as the reduction of aluminium trichloride to metal, and the reduction of titanium chloride to titanium using the Hunter process. Furthermore, it was used to produce cyanide for the extraction of gold. About 98.4 kt of sodium metal was produced in 2016, mainly in China;[7] about 53 kt of that is used in the Asia–Pacific region.

Today a large number of products derived from sodium metal are employed in the chemical industry. Of importantance is sodium amide used to synthesize indigo, which is applied in the denim (jeans) industry. Sodium methylate is a catalyst for the large-scale production of biodiesel. These two applications are considered growth areas for sodium. Organometallic compound production and the synthesis of vitamin A both use sodium metal. There are a range of metallic and metallurgical applications. It is used as a reducing agent in lead refining, to modify Al–Si alloy castings, and in potassium metal production. Liquid sodium is a coolant in nuclear power stations, and sodium–potassium alloys are used in heat exchangers because of their excellent heat-conducting properties.

A potential future use of sodium is in sodium-ion batteries,[8] metal/sulfur batteries that operate at 300 °C with molten sodium and molten sulfur electrodes and a solid electrolyte[9] and sodium-air batteries (for example[10]) which are a type of metal–air battery.

8.1.1.3   EHS

Special care needs to be taken when handling and working with sodium metal, and its burning behaviour needs to be studied. Combustion is strongly exothermic. In atmospheric air with normal humidity, molten sodium can *spontaneously* ignite at 115 °C and higher temperatures. Sodium reacts with water to form sodium hydroxide and hydrogen gas. If the hydrogen gas mixes with air the gas mixture can explode violently. Also, because of its low density sodium will float on water. Fire extinguishing should be with inert gases, *dry* salt, and *dry* sand. Avoid water, CO_2, or silicate-containing materials. Nitrogen is a suitable protective gas, also at high temperatures, for production, storage, and transport of sodium.[6]

8.1.2   Potassium (K)

8.1.2.1   Production

Potassium is geologically mainly found as silicate minerals, which are not suitable for the production of metal. The raw materials for metal production are mined from salt deposits, often in combination with sodium or magnesium salts. Typical examples are sylvinite, containing KCl and NaCl; carnalite, containing KCl and $MgCl_2$; and langbeinite, containing K_2SO_4 and $MgSO_4$.

In contrast to the other alkali (earth) metals, potassium cannot be produced using electrolysis, as it reacts with the graphite electrode used in this process. Instead, thermal processing is used where potassium chloride reacts with sodium metal to produce potassium vapour and sodium chloride. The thermodynamics for this reaction are unfavourable—the equilibrium lies on the KCl and Na-metal side of the reaction—unless the produced potassium vapour is removed continuously. Then chemical equilibrium is not reached. The vapour is condensed and solidified for use as pure metal or in alloys. Sodium chloride is drained off, and can be used for other applications.[11]

8.1.2.2   Applications

Potassium and sodium have similar characteristics, but are difficult to handle individually, while an alloy of the two is easier. The Na–K alloys are liquid at room temperatures over a wide composition range—from 40 to 90 wt.% K. Therefore the Na–K alloys have commercial importance, especially the liquid alloy at the eutectic composition of 78 wt.% K. These alloys are used in organic synthesis reactions, such as alkylation, isomerization, condensation, and transesterification, where potassium is a catalyst. Drug and chemical synthesis applications are based in the reaction of K metal with alcohols and amines. Furthermore, it is used as a cooling liquid in heat exchange applications, including nuclear reactors. Potassium has the ability to reduce oxides, hydroxides, and salts of many (base and technology) metals to the metallic state. It is used to create reactive metal powders, and can decompose silicates like glass, porcelain, and mica at high temperatures.[11]

8.1.2.3   EHS

Similar to sodium and other alkali (earth) metals, the main risks when handling potassium or its alloys is combustion and its liquidity at ambient temperatures. Combustion behaviour should be studied before handling. In dry air or oxygen metallic K burns to KO_2. This superoxide can explode in the presence of hydrocarbons. Storage of potassium and its alloys is best done under inert gas, as it will react with any oxygen or air entrapped in oils.

Potassium metal floats on water, while it reacts extremely fast with it. KOH solution and hydrogen gas are formed, the latter can explode violently in air, and the remainder of the metal melts and burns with air. Potassium also reacts violently with Teflon, even under inert atmospheres. As alloys are liquid they can flow and spread easily and react with other materials. Fires should be extinguished with dry sodium chloride, dry sand, or cement. Avoid water, carbon dioxide, or halogenated hydrocarbons.[11]

8.1.3   Lithium (Li)

8.1.3.1   Production

Lithium occurs in nature only in the form of salts and a silicate called spodumene ($LiAlSi_2O_6$). Both forms are mined.

Spodumene is found in crystallized magmatic fluids (pegmatites) in Canada, China, and Australia. $LiAlSi_2O_6$ theoretically contains about 7.9% Li_2O or 3.7% Li, while the richest deposit, the Greenbush deposit in Australia, contains about 3% Li_2O.[12] After separating the spodumene from the gangue minerals a concentrate with 4.8–7.5% Li_2O is obtained.[13] Subsequently the spodumene is treated in a kiln at 1150 °C and leached at 250 °C using sulfuric acid to produce lithium sulfate, which is concentrated and purified. Finally, lithium sulfate reacts with sodium carbonate (Na_2CO_3) to produce lithium carbonate (Li_2CO_3) and sodium sulfate.[12]

Lithium salts are mined from continental brines in the South American Andes, which contain in addition to lithium also potassium, boron, magnesium, and calcium salts. The Mg/Li ratio in the brine is important, as lithium and magnesium carbonates precipitate together. This ratio varies widely from 2 in Salar de Hombre Muerto up to 18 in Salar de Uyuni.[12] Typically the brine is concentrated stepwise in ponds using solar evaporation to precipitate NaCl, Ca salts, Mg salts, and sulfates. Co-precipitation of lithium in complex salts needs to be avoided. To the concentrated brine, ± 6% Li chloride, sodium carbonate is added to precipitate lithium carbonate. Alternatively an ion exchange process can be used.[12] Metal is produced by molten salt electrolysis using an electrolyte of potassium chloride (KCl) and 45–55 wt.% lithium chloride. This is close to the mixture's eutectic temperature (352 °C; 44.3% LiCl) and allows the cell to operate at 400–60 °C, well below T_{melt} of lithium chloride at 605 °C. The cell and cathode are made of steel; the anode is graphite. Lithium metal forms at the cathode, rises to the top of the cell, and is collected under an argon atmosphere. Each cell produces about 200–300 kg/day of Li metal, operates at 6–6.8 V with a current efficiency of 90%, and requires 28–32 kWh/kg Li.[13] The higher cell voltage, compared to the decomposition voltage of 3.68 V of LiCl, is due to the large anode-to-cathode distance. The metal has a purity of > 97% and can be remelted to reduce the sodium content to less than

100 ppm using the insolubility of the liquid phases. Standard grade lithium metal is > 99% pure with about 0.5% Na. Battery grade lithium contains less than 200 ppm sodium. Metal is available as ingots, rods, granules, foils, and powders.[13]

8.1.3.2   *Applications*

After World War I a market for lithium metal and compounds developed. Early use was in lead alloys (0.04% Li) for railway bearings until 1955 and in the chemical industry. About 197 000 tonnes Li-carbonate equivalent was produced in 2016.[14] Main markets are rechargeable batteries (43%) requiring 99.5% purity, greases manufacturing (7%), glass ceramics (12%), ceramics (12%), and metallurgical powders (4%) in 2016.[14] By 2019 the share of rechargeable batteries, driven by growth in e-mobility, increased to 54%.[15]

The battery application includes both lithium ion batteries, using carbonate or hydroxide,[15] and primary (non-rechargeable) batteries (2% of demand[14]) using the metal, because of its high electrochemical potential combined with low mass and long operating life. In glasses and ceramics lithium, added as spodumene, provides reduced melting temperatures and increased thermal shock resistance. Aluminium–lithium alloys are used in aerospace applications, as Li simultaneously reduces density (−10%), increases hardness (+10%), and improves strength through Al_3Li formation.[16,17] Typically the alloys are based on the Al–Cu group, e.g. alloy 2195, with 1–2% Li; the Al–Mg–Li group, e.g alloy 5091, with 4% Mg and 1% Li; or the high Li group with up to 4% Li. The addition of lithium exacerbates high-temperature oxidation at all stages of metal treatment, processing, and fabrication. Furthermore, the alloys contain higher amounts of hydrogen. Therefore, production of Al–Li alloys requires special adaptations and improverments in melting, melt treatment, and casting procedures. Lithium metal is also used for deoxidizing and desulfurizing copper, nickel, and steel alloys. Carbonate is sometimes added during aluminium electrolysis to form lithium fluoride (LiF) to improve current efficiency. Spodumene is used in continuous steel casting.[13] Future uses are lithium metal or alloy as anodes in rechargeable batteries, so-called solid-state batteries, and in next generation batteries such as lithium–sulfur and lithium–air batteries.

8.1.3.3   *Recycling*

Recycled lithium mainly comes from batteries used for e-mobility (vehicles, bikes, buses) and portable electronics (IT equipment). Collection of these is essential to get the batteries into a recycling chain. Recycling is driven by the value of the metals contained, and supporting legislation, for example, in the European Union.[18]

Two approaches are possible. The universal process, e.g. Umicore, treats all battery chemistries together, and focuses on raw material recovery, throughput, and minimal pretreatment. The pyrometallurgical process gives an alloy with Co, Ni, Cu, Mn, and a Li-rich slag. Both go into specific hydrometallurgical processes: the alloy is treated for recovery of Co, Ni, and other metals, while the slag is treated for the recovery of lithium salts.[19] The battery chemistry-specific processes uses dismantling, shredding, and sorting to create different fractions, e.g. aluminium foils and black mass, suitable for further (hydrometallurgical) processing and refining. This approach has high requirements for collection and sorting based on battery knowledge.[19] (See Section 10.5.3 for additional information.)

8.1.3.4   EHS

Special care needs to be taken because of the reactivity of lithium and its compounds. The metal can be processed in dry air. Long exposure of metal to air leads to the formation of lithium nitride inside the ingot and ignition. Noble gasses, sulfur hexafluoride (SF_6) up to 225 °C, and mineral oil can be used to prevent combustion. Burning molten metal with flame temperatures up to 1100 °C reacts vigoriously with silicates, carbon dioxide, and water. Li metal reacts with water to form hydrogen gas.[13]

8.2   Alkaline Earth Metals

The alkaline earth metals were discovered in the late eighteenth and early nineteenth centuries. Their properties are partly close to the alkaline metals and partly similar to aluminium. They are easily oxidized in air; only Be forms a protective oxide layer. Metallothermic reduction is used to produce the alkaline earth metals. Magnesium is also produced by molten salt electrolysis. A major application of these metals is as an alloying element. Only Mg is used as metal for structural applications.

8.2.1   Beryllium (Be)

Beryllium was discovered in 1797 by the Frenchman Nicolas-Louis Vauquelin. It took until 1828 before elemental Be was isolated by Friedrich Wölhler in Germany and Antoine Bussy in France through reduction of beryllium chloride. Commercial use began in 1926.[20]

8.2.1.1   Physical Properties

Beryllium is greyish and its chemical properties resemble these of aluminium.[21] Beryllium is resistant to air oxidation under normal conditions due to the formation of a thin protective layer of oxide. However, it is readily attacked by acids and alkalis. The applications of beryllium depend upon its unusual combination of physical properties. Its density (1846 kg/m^3) is about 30% less than aluminium, and its specific rigidity—the ratio of modulus to density—is about 50% greater than steel.[20] Furthermore, Be is nearly transparent to X-rays and sound travels fastest through beryllium. It has non-magnetic and non-sparking properties. When added to copper and other metals as an alloying element, it results in controllable and predictable strengthening mechanisms, while maintaining good machineability, and electrical and heat conductivity. This is caused by the formation of beryllides, intermetallics of beryllium with nickel or cobalt, that form precipitates located within the crystalline structure of the metal matrix. These irregularities in the structure enhance strength, stiffness, and hardness.[20]

8.2.1.2   Production

Beryllium is found in nature as silicates, sulfates, carbonates, hydrated minerals, and an oxide. The economically relevant minerals are beryl $Be_3Al_2(SiO_3)_6$ containing 3.85 wt.% Be and bertrandite $Be_4SiO_7(OH)_2$ containing 0.27 wt.% Be.[22] Most of the world production comes from the Spor Mountain deposit in Utah, USA.[20] Beryl concentrate is also a by-product from gemstone mining or feldspar quarrying. Typical impurities are alkali and alkali-earth metals, iron, manganese, phosphorus, and aluminium oxide and silicon dioxide.

Production of beryllium metal from concentrates requires four steps. First, the beryl concentrate is treated using the Kjellgren–Sawyer sulfate extraction method. Beryl is heated to 1700 °C and quenched in water to form a frit. The frit is heated to 1000 °C, ground to −75 microns, and leached in concentrated sulfuric acid at 250–300 °C. Several alternative processes exist.[20] Bertrandite is ground to a slurry and dissolved in sulfuric acid at 95 °C. The sulfate solution is subjected to solvent extraction to remove impurities, creating a clean Be hydroxide. About 80% of the Be is recovered from the concentrate.[20] Next the hydroxide is converted to fluoride by dissolving it in ammonium hydrogen fluoride (NH_4HF_2), heating it to remove impurities like Al, Cr, Mn, Cu, and Ni. Subsequently, the fluoride is reduced with magnesium metal between 900 and 1300 °C, producing a 96% pure metal and a beryllium and magnesium fluoride slag.[20] Lastly, the impurities—1.5 wt.% Mg, 2 wt.% F, Na, and Cl—present mainly as entrapped slag droplets, are removed by vacuum melting in a beryllium oxide or magnesium oxide crucible, and the metal is cast into ± 200 kg ingots.[22] The ingot is converted to chips and to metal powder to make parts. A major problem with the processes is the high content of salts in the raw metal produced. To produce Cu–Be alloys with 0.2–2 wt.% Be a master alloy with 3.5–10 wt.% Be is used, which is diluted with additional copper and alloying elements, such as nickel and cobalt. To obtain the master alloy an electric arc furnace (EAF) is used and calcined Be hydroxide, copper, and graphite (carbon) as reducing agent are added. The Cu–Be alloys are solution-annealed: heating to fully dissolve alloying elements softens the copper matrix and homogenizes and refines the size, shape, and location of beryllide precipitates. Afterwards work hardening or heat treatment gives the desired combination of alloy properties.[20]

8.2.1.3   Applications

The world production in 2015 was 465 t Be units[22] and refined beryllium about 220 t. USA produces about 65%.[23] Roughly 75% is used as an alloying element in copper: the 'gold alloys' contain 1.8–2% Be and are used where strength is key; the 'red alloys' contain 0.10–0.25% Be and are used where thermal and electrical conductivity is key.[20] These alloys are used in components in aerospace, military, electronic, and automotive hardware. Electronic applications are aircraft guidance systems, electromagnetic raditation shielding, airbag impact sensors, and medical diagnostic hardware. Small quantities of Be are sometimes used in magnesium, aluminium, and nickel alloys.

About 20% of the beryllium is produced in the form of > 99.5% pure metal and alloys with over 60% Be. It is used for critical components in high technology equipment, e.g. space applications. Thin foils are used as a window for X-ray tubes and detectors, for example in security devices and high-resolution imaging equipment for medical applications. It is also used in materials test reactors and fundamental particicle-physics research. An alloy with up to 62 wt.% Al is used for lightweight, high-strength structural components of aerospace systems. The remaining 5% is used as beryllia (BeO), which is an electrical insulator.[20]

8.2.1.4 Recycling

New scrap, especially from the aerospace industry, is commonly recycled by remelting to new alloys. Recycling saves about 70% energy compared to Be from ores.[15] Parts that can be recovered at end of life are recycled as well. Cu–Be alloys used on electronic components are usually not recycled for beryllium recovery as the Be(alloy) content in each device is small, and the Be content in the alloy is below 1.25 wt.%. The electronics end up in copper smelters for recovery of copper, where beryllium reports to the slag and is effectively lost.

8.2.1.5 EHS and Sustainability

Beryllium metal and beryllium oxide powders have to be handled with care. Soluble beryllium compounds in the form of solutions, dry dust, or fumes are toxic. When inhaled they may produce acute effects similar to those caused by phosgene. Some individuals develop an acute allergic lung disease called berylliosis, or chronic beryllium disease, after exposure to beryllium-containing dusts or fumes. This disease may sometimes develop years after exposure.

8.2.2 Magnesium (Mg)

Magnesium was first isolated as metal by Humphrey Davy in 1808 via electrolysis of magnesium sulfate. In 1833 Faraday electrolyzed $MgCl_2$ to $Mg(l)$ and $Cl_2(g)$. Robert Bunsen made first commercial quantities in 1852. Commercial, larger-scale production by electrolysis of carnallite started in 1886 in Germany,[24] about the same time as aluminium production.

8.2.2.1 Physical Properties

Magnesium is the lightest metal that can be commonly used for structural purposes. Its density is 65% of aluminium and 22% of iron. It is also the easiest structural material to machine. Both properties are widely exploited. Its thermal and electrical conductivity and its melting point are very similar those to aluminium. Whereas aluminium is attacked by alkalis but is resistant to most acids, magnesium is resistant to most alkalis but is readily attacked by most acids. Due to the formation of a thin but dense layer of oxide, it is stable in air and water at normal temperatures but burns when heated in air and is attacked by steam. Hydrogen is readily soluble in liquid magnesium; however, the reduction in solubility during solidification is relatively small. Ni, Co, and Cu create corrosion problems already in the part-per-million range and are nearly impossible to remove from liquid metal. Mg is a strong reducing agent and forms a wide range of stable compounds including oxides, sulfates, carbonates, nitrates, halides, etc.[25]

8.2.2.2 Production

Magnesium metal can be extracted from a wide range of minerals. Commonly used are dolomite ($CaCO_3 \cdot MgCO_3$), magnesite ($MgCO_3$), and to a lesser degree magnesium sulfates, chlorides, e.g. carnallite ($KCl.MgCl_2.6H_2O$), and silicates, e.g. olivine ($(Mg,Fe)_2SiO_4$).[25] Two production routes are used: electrolytic process and metallothermic reduction. Metallothermic reduction typically uses ground calcined dolomite—$MgO \cdot CaO$ with minimum 8 wt.% Mg—as feedstock, which is reduced with ferrosilicon lumps in the Pidgeon process, invented by Lloyd Pidgeon in Canada in 1944. This is a cyclic batch process. The dolomite and ferrosilicon are mixed in a 5:1 ratio, briquetted, placed in horizontal 3-m-long, 30- to 60-cm-diameter retorts and externally heated to 1200 °C under vacuum.[24,26] Silicon reduces MgO to Mg vapour, which flows into a condenser, solidifies, and is removed. The remaining calcium–iron–silicate can be used as a cement substitute. Alternatively the Bolzano process is used or Si is replaced by aluminium as reducing agent.[24] The Pidgeon process produces 70 kg Mg/day (per retort) and 6.2 t slag/t Mg while it consumes 0.9 MWh electricity, 10.5 t coal, 11 t dolomite, and 1.1 t of 75% ferrosilicon per t Mg metal produced. The Mg recovery is 82%.[24] Roughly, 80–5% of the global production, mainly in China, is produced this way.[27] The CO_2 footprint of this process route is 25–40 kg CO_2-eq./kg Mg.[27]

The electrolytic process starts with the production of dehydrated $MgCl_2$ by either chlorination of MgO or $MgCO_3$ or the dehydration of $MgCl_2$ brines or carnallite. This allows for coproduction of concentrated (> 98 wt.%) Cl_2 gas, high current efficiency in the cell, and lower anode carbon consumption as no water is present.[24] $MgCl_2$ is electrolysed using a molten alkali salt mixture containing 8–20 wt.% $MgCl_2$ at 660–800 °C.[24] Above 800 °C the vapour pressure increases rapidly. Electrolyte mixtures with different ratios of KCl, NaCl, $CaCl_2$, and LiCl are used. Typically, the density of the electrolyte is higher than Mg, so the metal floats on top. Some MgO is formed. The brick-lined or steel cell with steel cathodes and graphite anodes produces 200 kg/day Mg metal and 2.7–2.85 t Cl_2/t Mg, consumes 15–20 kg graphite/t Mg, operates at 5–7 V, 340–430 kA, with a current efficiency of 85–95%, and requires 10–20 kWh/kg Mg, depending on the cell design.[24] Electrolytic processes are used in USA, Russia, and Israel. The CO_2 footprint of this process route is about 19 kg CO_2-eq./kg Mg and about 7 kg CO_2-eq./kg Mg when renewable energy is used.[27]

The resulting > 99.8 wt.% Mg metal is molten, refined, and alloyed to the desired composition. Cast alloys are Al–Mn, Al–Si, Al–Zn, or Zn–rare earth containing. Special alloys can contain Ca and Sr. Wrought alloys are Al–Zn, Th–Zr, or Zn–Zr containing.[28] The metal refining possibilities of magnesium are limited and impurities lead to corrosion problems. Hence contamination is to be avoided. A continuous pick-up of soluble elements from the equipment or tools may occur if the melt is not saturated, especially iron. Pure Mg contains 0.02–0.04 wt.% Fe, equal to its solubility in molten Mg, and it can be lowered by adding Mn or Al to create solid intermetallics. Ni, Cu, and Co are highly detrimental and their content can only be lowered via dilution with virgin material or via vacuum distillation. Si is considered an impurity above 0.1 wt.% in silicon-free alloys. Chlorides from electrolysis or salt flux melting attach to oxides, MgO, and nitrides, Mg_3N_2, that form due to the reaction between magnesium and air. These are found to exist as thick films or flakes of loosely connected particles. Their high surface-to-volume ratio hinders their removal by settling; instead filtration or gas purging can be used.

To prevent oxidation or burning of molten metal, SF_6 is used at the present time as a protective gas cover. The SF_6 blend reacts with magnesium forming a dense oxide layer that contains small amounts of MgF_2. The MgF_2 crust may enter the melt. SF_6 has a 23900 CO_2-eq. impact and is more and more being replaced by SO_2 for example.[27] Beryllium is added to magnesium alloys (typically 5–15 ppm) to reduce the oxidation of the molten alloy. This limit is determined by the max solubility of Be in Mg. Be is gradually lost during the remelting of the alloys and accumulates in dross and sludge. When recycling Mg alloys, either in-house or external, the lost Be has to be replaced. Al–Be master alloys are normally used for this purpose.

8.2.2.3 Applications

Global primary production of magnesium is about 1100 kt/y in 2019, excluding production in the US.[23] The contribution from recycling is about 265 kt.[27] About $^1/_3$ of magnesium is used in casting alloys for lightweight structural applications, mainly in automotive applications ($\pm$ 75%), communication equipment ($\pm$ 10%), tools, and rail and air transport.[27] This application is driving magnesium demand. In China there are plans to increase the magnesium content in cars from 8.6 kg in 2017 to 45 kg by 2030.[29] Another $^1/_3$ is used as an alloying element for aluminium alloys used in packaging and automotive applications. Magnesium is a powerful reducing agent: about 13% is used in thermal reduction to produce metals such as titanium, hafnium, and zirconium from their compounds, while another 10% is used in the desulfurization of metals. Other applications are sacrificial anodes for cathodic protection of materials and pharmaceutical applications.[27] Wrought alloys exist, but the wrought alloy industry is still very small.

8.2.2.4 Recycling

See Chapter 10 for a more detailed discussion of Mg recycling.

Only casting alloys are recycled and need to meet the same specifications as primary alloys. Therefore, impurity management and alloy-specific sorting is key. Recycling requires less than $\frac{1}{2}$ of the energy necessary for primary production.[25] New clean scrap is recycled easily using remelting under protective gas (and minimal salt amount) in-house at the die-cast shop. Alternatively, flux-based recycling can be used. Flux-based recycling uses a salt flux to protect the magnesium from oxidation and to remove impurities. The flux has a near-eutectic composition in the $CaCl_2$–NaCl–KCl–$MgCl_2$ system and sufficiently different density from the Mg metal. Dross from die-casting with 70–90% metal can be recycled using a flux-based process. New scrap with inserts is feedstock for sacrificial anodes, steel desulfurization, or an alloying element for aluminium. Scrap from machining can go to desulfurization.

Old scrap, e.g. from vehicles, is usually insufficiently sorted and contaminated with paint, coatings, inserts, etc. The small quantity of Mg alloys in vehicles is mixed with aluminium alloys, and enters the aluminium recycling processes. There the Mg is likely lost through oxidation or removed during refining. If a separate Mg alloy fraction is obtained after sorting and recycled, it is highly contaminated and requires a refining/dilution with virgin magnesium.

Magnesium in aluminium alloys, e.g. in used beverage cans, can be recycled/recovered together with the aluminium in the aluminium recycling processes.[25]

8.2.2.5 EHS and Sustainability

Magnesium burns in air with an intense white flame. The ignition temperature is 645 °C in dry air but decreases with increasing moisture content. Burning magnesium reacts violently with water. Magnesium fires are extinguished with dry salts or dry salt fluxes.[24]

Recycling requires less than $\frac{1}{2}$ of the energy necessary for primary production,[25] and the low density of Mg is an advantage when comparing the energy consumption to other materials on a per cm^3 basis. Several environmental factors will play a role in the future of magnesium production.[27,29] Changes in the production process are likely. The Pidgeon process is quite polluting and energy intensive compared to alternatives developed in Australia,[30] Canada[31,32] and Norway, as mentioned in the production section (Section 8.2.2.2).

8.2.3 Calcium (Ca)

8.2.3.1 Physical Properties

Calcium is a highly reactive element, and never found in nature in its metallic state. It was discovered in 1808 simultaneously and independently by Sir Humphrey Davy and by Jöns Jacob Berzelius and Magnus Martin de Pontin. The metal is silvery-white, very ductile, oxidizes easily, and forms a white coating in air. The metal burns with a white light and yellow-red flame. It reacts sponteanously with water to form hydrogen gas and $Ca(OH)_2$. Its high reactivity and the heat of formation of some of its compounds enables its use as a reducing agent for the production of metals such as chromium, thorium, zirconium, and uranium. It is one of the most efficient electrical conductors: It conducts 150% higher current density than copper at room temperature, a property that cannot be used on earth, but can be of use in the vacuum enviroment of space.[33] Economically important compounds are carbonate limestone $CaCO_3$, gypsum $CaSO_4 \cdot H_2O$, fluorite CaF_2, and complex silicates.

8.2.3.2 Production

The major raw materials for metal production are limestone ($CaCO_3$), dolomite ($CaCO_3 \cdot MgCO_3$), gypsum $CaSO_4$, and wollastonite $Ca_3Si_3O_9$.[29] Two process routes are in use: molten salt electrolysis and aluminothermic reduction. The latter is the standard production process, as it is simple and uses cheaper raw materials.[34]

Molten salt electrolysis uses anhydrous $CaCl_2$, which is difficult to prepare and keep water-free, or a 15% KCl mixed salt, with T_{melt} of 772 and 630 °C, respectively. As the formed liquid metal is highly combustible, contact electrolysis is used to recover calcium in solid form. The closed cell is operated at a temperature slightly lower than T_{melt} (839 °C) of calcium. The steel cathode is in contact with the bath, so calcium solidifies on the cathode. By continuously pulling out the solid calcium a 'carrot'-shaped stick is formed.[34] Chlorine gas is formed at the graphite anode. The cell operates at 25 V with a current efficiency of 65–82% and requires 40–50 kWh/kg Ca. The metal has a purity of 98%.[34]

Thermal reduction for calcium metal is similar to the metallothermic Pidgeon process used for magnesium. In this case CaO, obtained by calcining limestone, reacts with aluminium metal. A MgO content below 0.5% in the feed is desired to avoid the thermodynamically preferred reaction between aluminium and MgO, which lowers the reaction efficiency for calcium metal. The finely ground materials are mixed in a molar ratio of CaO:Al of 3:2,[34] briquetted, and charged in a retort. Vacuum is applied and the retort is heated to 1200 °C. Calcium metal vapour condenses at the cool end of the retort. After about 8 hours the 23- to 34-kg pieces of Ca metal are removed[33] as well as the calcium aluminate residue. The metal is 99.5% pure commercial grade. Magnesium (0.5 wt.%), aluminium (0.3 wt.%), iron (0.008 wt.%), and manganese (0.01 wt.%) are the main impurities.[33] Further vacuum distillation at 900–25 °C is necessary to obtain high-purity, 99.9% calcium metal. Distillation reduces impurities that do not volatilize, such as Al, Fe, and Mn, while the magnesium content is more difficult to lower.

8.2.3.3 Applications

Major producers are China, Russia, the USA, Canada, and France. The global production capacity is about 40 000 t.[35]

Most of the calcium metal, as commercial grade, is used in steelmaking to improve the quality of steel. For many years this has been in the form of calcium–ferroalloys. In the 1970s wire-feeding technology was developed, where steel-clad calcium metal wire is inserted into the melt. This avoids the problems resulting from adding calcium metal directly, due to its high vapour pressure and low melting point, as discussed in Chapter 2. Calcium modifies the melting point of inclusions, making them float out rapidly. The shape and size of the remaining inclusions is changed to spherical, very small and finely dispersed in the metal. This results in major quality improvements in castability, mechanical properties, and hydrogen-induced cracking resistance. Calcium is applied in ladle metallurgy, tundish and mold injection for continiuous casting, especially thin-slab casting, and manufacturing of ultraclean steels.[33] Furthermore, Ca is used in ironmaking.

Calcium is used as an alloying element in 0.1 wt.% Ca–Pb grid alloy that is used in maintenance-free lead acid batteries. Addition of Ca improves the conductivity and current capacity of the cell, and reduces gassing so the cell can be closed. This prevents water loss and extends battery life.[33] It is also used as an alloying element in special aluminium alloys. High-purity calcium metal can also be used as a reductant in the production of high-energy-density magnetic materials such as samarium–cobalt, and neodymium–iron–boron. The Nd–Fe alloy is produced by calcium thermal reduction of Nd fluoride, Nd chloride, or Nd oxide.[33] It is also used as a reductant to refine zirconium from Zr fluoride, to deoxidize magnesium, to debismuthize lead by the Kroll–Betterton process, and to reduce uranium fluoride or oxide to metal.[33]

8.2.3.4 Recycling

Calcium is not recycled, as it is mainly used as a reducing agent and thus converted to its oxide. It ends up in slags and residues that may be used in other applications such as construction materials.

8.2.4 Strontium (Sr)

The element strontium was discovered in 1787 by Adair Crawford in the form of strontium carbonate from a mine close to Strontian, Scotland, UK. It took 20 years before strontium metal itself was isolated by Humphrey Davy using electrolysis, the process he also used to isolate sodium and potassium.[36]

8.2.4.1 Physical Properties

Strontium is a lustrous, silvery-white metal. Located between Ca and Ba in the periodeic table means its reactivity is between these metals. A white oxide is formed on its surface when exposed to water and air. It reacts with water, oxygen, nitrogen, sulfur, and fluoride to form compounds. The sulfate and carbonate have low solubility. The sulfide reacts with O_2 and CO_2, while the chloride does not. Strontium compounds give a carmine red colour to a flame or fireworks.[36]

8.2.4.2 Production

Of economic importance are the carbonate strontianite, $SrCO_3$, and the sulfate celestite, $SrSO_4$. The feedstock for producing strontium metal, and other Sr compounds, is celestite. It is mined from deposits in China, Iran, Spain, Mexico, Turkey, and Argentina, resulting in a global mine production of 220 000 t/y.[23] Celestite (3.8–3.9 g/cm^3) is concentrated using density separation to remove lighter impurity minerals such as limestone, quartz, gypsum, and dolomite. A > 95% $SrSO_4$ concentrate is obtained. Conversion of the sulfate into metal consists of three steps. First, it is converted into carbonate by reduction with carbon in a countercurrent rotary kiln at 1100–1200 °C to form strontium sulfide, SrS, which is extracted using hot water. Then it is precipitated as carbonate using CO_2 gas. An alternative process is the reaction with hot sodium carbonate solution followed by several precipitation and solution steps. Its main disadvantage is the high use of reagents such as $NaCO_3$ and NaOH.[36] The obtained strontium carbonate is 98–98.5 wt.% pure with 1–2.5 wt.% barium carbonate as the main impurity. Next, the carbonate is converted to oxide by heating it with carbon in an electric furnace. Lastly, the oxide is converted to metal by aluminothermic reduction, a variant of the Pidgeon process. Under vacuum the strontium metal is formed as gas and solidifies at the cooler end of the retort. Part of the strontium reacts to $SrO\cdot Al_2O_3$, reducing the efficiency of the reaction. The 99% pure metal contains Mg, Ca, and Ba as main impurities. It can be further alloyed to aluminium and aluminium–magnesium master alloys with up to 10% Sr.[36]

8.2.4.3 Major Applications

Strontium metal has one main application: improvement of the mechanical properties and machineability of aluminium alloys. It is added in the form of pure metal or as Sr–Al master alloys. In Al–Si casting alloys an addition of 0.005–0.015 wt.% Sr to hypoeutectic Al–Si alloys improves the tensile strength and ductility. It changes the morphology of the eutectic silicon phase to a fine, interconnected fibrous morphology. The impact is also influenced by the solidification rate, with faster rates giving better mechanical performance. Excess Sr will deteriorate the alloy's properties.[36,37] In wrought aluminium alloys strontium improves the mechanical and forming properties. In 6xxx series alloys additions of Sr promote formation of a-AlFeSi, leading to better surface quality and faster extrusion speeds. In the 7xxx series alloys the presence of 0.005 wt.% Sr or more refines intermetallic phases such as Mg_2Si, Al–Cu–Mg, and Al–Cu–Fe, resulting in better toughness, finer grain size, and faster processing due to a shorter homogenizing time.[36] In sheet alloys strontium reduces the number of intermetallic precipitates, enhancing the formability and reducing edge cracking. Furthermore strontium metal is applied in gray iron casting, as part of the ferrosilicon additions.[36] It is also used in the heat-resistant magnesium AJ and AXJ alloys.

Strontium carbonate is used for the manufacture of permanent ceramic ferrite magnets and the production of electrolytic zinc. Adding strontium carbonate dissolved in sulfuric acid reduces the lead content of the electrolyte, and of the deposited zinc metal.[38]

The radioactive isotope strontium-90, generated in nuclear reactors, can be used to generate electricity for space vehicles, remote weather stations, and navigation buoys, as well as for thickness gauges.[39]

8.2.4.4   *Recycling*

Strontium is used as alloying elements and will enter a recycling process if the corresponding alloy is recycled. Sr oxidizes slowly during remelting and is partially lost from aluminium alloys during remelting.

8.2.4.5   *EHS and Sustainability*

Strontium metal and the 90% Sr–10% Al alloy need to be packed under inert atmosphere to prevent oxidation. The aluminium-containing master alloys do not require this. Strontium is a strong reducing agent and is to be handled appropriately. Strontium behaves similarly to calcium both chemically and biologically, depositing in bones and teeth of humans. The radioactive Sr isotopes have to be carefully managed as the Sr isotopes from H-bomb tests have posed a serious environmental problem.

8.3   Rare Earths: Scandium, Yttrium, and Lanthanides

The rare-earth elements (REE) are defined as the 15 lanthanides, yttrium, and scandium.[40] Scandium behaves quite different from the other elements, and will be covered separately in this chapter. The remaining elements are divided into the light REE—lanthanum (La), cerium (Ce), praseodymium (Pr), neodymium (Nd), promethium (Pm; rarely found in nature due to its short half-lifetime), and samarium (Sm)—and the heavy REE—europium (Eu), gadolinium (Gd), terbium (Tb), dysprosium (Dy), holmium (Ho), erbium (Er), thulium (Tm), ytterbium (Yb), and lutetium (Lu). Yttrium behaves as a heavy REE.

Rare earths were isolated as a mixture from a mineral found at Ytterby, Sweden in 1794 by the Finnish chemist Johan Gadolin. Scandium was discovered at the same location by Lars Frederik Nilson in 1879. It took over 150 years until the last rare earth, promethium, was found in 1947. The rare earths are not rare. They are as abundant as copper, lead, or bismuth. The light REE are 200× more abundant than heavy REE, with the even atomic number REEs (Ce, Nd, Sm, Gd, Dy, Er, Yb) being more abundant than the neighbouring REEs with odd atomic number (La, Pr, Eu, Tb, Ho, Tm, Lu). This is called the Oddo Harkins effect.

8.3.1   Scandium (Sc)

8.3.1.1   *Physical Properties*

Scandium is a silvery-white metal and can be considered a light metal. Its density, 2.99 g/cm^3, is close to aluminium, but its melting point, $1541\,^{\circ}\text{C}$, is considerably higher. The chemical properties of scandium differ considerably from the other rare-earth elements. During the geological processes scandium has segregated from the REE. Thus, Sc is seldom associated with them. Due to the small size of the ion it does not selectively concentrate in nature. Therefeore scandium deposits are rare.

8.3.1.2 Production

Scandium is found in over 100 minerals in small concentrations. The minerals with considerable scandium content thortveitite ($(Sc,Y)_2SiO_7$), euxenite ($(Y,Ca,Ce,U,Th)(Nb,Ta,Ti)_2O_6$), and gadolinite ($(Ce,La,Nd,Y)_2FeBe_2Si_2O_{10}$) are rare. Therefore, scandium mines do not exist; all is won as by-product from other metals. Scandium is concentrated during the processing of various ores, specifically uranium, thorium, aluminium (in red mud), tungsten (from wolframite), tin, tantalum, zirconium, titanium (acid waste stream), nickel–cobalt laterites, and REEs. Identified resources containing 0.002–0.005 wt.% Sc are found in Australia, Canada, China, Kazakhstan, Madagascar, Norway, Philippines, Russia, Ukraine, and the USA.[41]

The extraction processes for the recovery from residues generally consist of leaching to remove Sc together with Y, La, and impurity elements from the residue. This is followed by solvent extraction and precipitation to obtain scandium hydroxide or oxalate. Lastly, calcination produces Sc_2O_3 in 99 or 99.9% purity.[42] Recovery from uranium processing and rare-earth processing is currently done on an industrial scale. The recovery from other residues is in a research and pilot stage. Detailed process descriptions are available in the literature.[41]

In case thortveitite is available it can be subjected to fractional sublimation to recover Sc as chloride by heating it to 900–1000 °C with coal and chlorine gas. The impurities Si, Ti, Al, Fe, and Zr form chlorides with a sublimation point below 350 °C and Y chloride remains in the residue. Sc chloride sublimates around 967 °C and is deposited in a cool zone of the furnace.[41]

8.3.1.3 Major Applications

The global production and consumption of scandium is estimated to be 15–20 t/a.[22] The market and number of producers is very small; hence, data are scarce and less reliable. It is available as 99.99% Sc_2O_3 and a 2% Sc–Al master alloy. China, Russia, and the Philippines are the main producers.[22] New projects are in the feasibility or development stage in the USA, Australia, and Russia.[22]

Globally the vast majority of scandium is used in aluminium alloys for specialized automotive and aerospace applications, e.g. in the Sc–Mg–Al family. Sc–Al alloys with 0.25–0.4 wt.% Sc are very light and contribute to fuel effiency.[41] Small additions of Sc to aluminium refine the crystal structure and reduce hot cracking, allowing welding without strength loss, and increase plasticity, corrosion resistance, and thermal conductivity.[43] It can be substituted by titanium, lithium, or carbon fibre materials. Sc–Al alloys are also used in high-quality sports gear like bikes, baseball bats, and golf clubs.[42]

The remainder is used in metal halide lamps (10%) and in solid oxide fuel cells (SOFC) (5%). Sc_2O_3 stabilizes the zirconia-based solid electrolyte and can be substituted by yttrium. Compared to Y, scandium is a better ionic electrical conductor, lowering the temperature of the electrolyte from 1000 °C to 750–800 °C, and increasing the lifespan and efficiency of the cell.[42] Minor applications are in Eu:YSGG lasers, phosphors for lighting, 80% titanium–20% scandium carbides, and dopants in glasses, glazes, and ceramics.[43]

8.3.1.4 Recycling

Scandium is currently not recycled. During Al–Sc and Mg–Sc alloy production scandium is partly lost to dross because of its high reactivity. Dross from Mg–Sc alloy production can contain 12–23% Sc. This can be recovered by leaching the dross in HCl followed by solvent extraction, precipitation, and calcining. This process is not applied industrially as the quantity of scandium-containing drosses is too small for an economically feasible process.[41]

8.3.2   Rare-Earth Elements & Yttrium (Y)

8.3.2.1   Physical Properties

The REEs are much alike because the differences in their electronic structures chiefly involve the inner 4f electrons, whereas the outer s and p (sometimes d) electrons determine their chemical behaviour. The variation in the 4f electrons gives the strongly paramagnetic and spectroscopic properties used in many REE applications.[44] Y and La do not have 4f electrons.

The aqueous chemistry of all rare earths is very similar and changes only slightly in progressing along the lanthanide series. Therefore, it is difficult to separate individual rare earths. All of the elements form trivalent compounds, and in the crystal lattices of such compounds one rare-earth ion readily replaces another. The rare-earth elements when heated react strongly with non-metallic elements to form very stable compounds. They are never found as free metals in the earth's crust.

Physical properties, however, differ as much across the lanthanum series as they do for most other series in the periodic table. The melting point of Lu, for example, is almost twice that of La, and the vapour pressures of Yb and Eu at 1000 °C are millions of times greater than those of La and Ce. Each of the rare-earth metals readily combine with almost any other metallic element to form alloys that can be hard or soft, brittle or ductile, and they can have high or low melting points. The lanthanides are invaluable in the study of the metallic state, properties of alloys, the existence of intermetallic compounds, and the effect of impurities. The pure rare-earth metals are bright and silvery. Lanthanum, Ce, and Nd corrode readily in air. Metallic Y, on the other hand, remains bright and shiny for years. Properties are frequently sensitive to impurities. For example, the light lanthanide metals will corrode much more rapidly if small amountsof Ca or Mg, or rare-earth oxides, are present in the metal.

8.3.2.2   Production

Rare-earth elements are difficult to separate from each other, prefer to stay in the oxide state, and are nearly universal solvents for other elements. This makes their production highly complex, where the elaborate processing routes depend strongly on the specifics of the ores and metal. Production of rare earths has been extensively discussed.[45] The main, economically relevant minerals are the phosphates: monazite, $CePO_4$, with 70 wt.% light REO and the balance heavy REO; and xenotime, YPO_4, with 61 wt.% heavy REO and the balance light REO; the carbonate bastnäsite, $CeCO_3F$, with 75 wt.% REO; and the oxide loparite, $Ce(Ti,Nb)O_3$, with 30 wt.% REO.[44] To include also the other rare-earth elements present in the mineral, Y and Ce are often replaced by REE in the chemical formula. REE can also be substituted in apatite ($CaPO_4$), zircon ($ZrSiO_4$), and fluorite (CaF_2). Typical impurity elements in the minerals are the elements from group 2 and group 4, as well as uranium and thorium.

REE have dedicated mines and are also produced as by-products from Nb, Th, and U, or zircon, rutile, and ilmenite (beach sand placer/mineral sands) mining. REE from clay deposits are excavated and leached, or leached *in situ*, with neutral or acid solutions with an ion-exchange agent such as ammonium sulfate or EDTA. REO-bearing minerals are separated from the other minerals using gravity, magnetic, and electrostatic separation methods. From (underground) mines the ore is crushed, ground, and concentrated by flotation to obtain a bastnäsite or monazite concentrate.[44] The REE process from concentrate to metal consists of four steps.

The first step is converting the REE mineral concentrate into carbonates, chlorides, or oxalates. This is done by acid roasting followed by multistep leaching. The impurities are removed and the concentration of REO is increased to $\pm$ 90% after leaching.[46] High-temperature acid roasting, for example with 98% H_2SO_4, in a rotary kiln heated to about 200 °C for monazite and about 500 °C for bastnaesite, is a common practice in China. The emitted exhaust gasses HF, SO_2, SO_3, and SiF_4 are captured in water scrubbers.[46] The obtained RE sulfates, $RE_2(SO_4)_3$,[47] are leached with water from the rest of the material and precipitated as double sulfates, $Na_2SO_4.REE_2(SO_4)_3.nH_2O$, by adding excess NaOH[47,48] or carbonates when NH_4HCO_3 is used.[46] Finally, leaching by HCl is done to convert the sulfates to chlorides, 92% $REECl_3$.

In the second step the REE are separated from each other by solvent extraction commonly using organic solvents. The pH is varied to selectively extract the REE from light to heavy. Each solvent extraction step is repeated a minimum of 12 times for each REE,[47] making this a time- and cost-intensive process. The rare earths are precipitated as $RE_2(C_2O_4)_3$ or $RE_2(CO_3)_3$ from the solvent using an inorganic salt, e.g. ammonium bicarbonate,[46,47] and calcined to form up to 99.99% pure REO. Typically, this yields lanthanum oxide, cerium oxide, neodymium and praseodymium oxide, and other heavy REOs.[48]

Step three is the reduction of the REO to metal. This is done by molten salt electrolysis (similar to aluminium, magnesium, and other alkaline metals). For example, Nd_2O_3 is dissolved in a 85–90% NdF_3-10–15% LiF molten salt to obtain Nd metal, using a cell of 4–10 kA operating at 1050 °C. Nd reduces at the tungsten or molybdenum cathode, sinks, and is collected in a Mo crucible; CO and CO_2 are formed at the anode.[49] The process emits large quantities of CF_4 (7%) and C_2F_6 (0.7%) in the offgas, which are perfluocarbons with high greenhouse gas impact. The use of automated process control, as in aluminium production, can keep the process within a near PFC-free operating window.[49] Alternatively, calcium thermal reduction can be used to directly produce a Nd–Fe alloy by mixing Nd_2O_3 or Nd fluoride with Ca metal and Fe metal, which is less capital intensive and produces non-toxic by-products.[33]

Lastly, refining and alloying employs techniques, such as fused salt electrolysis, molten salt oxide reduction, metallothermic reduction, zone melting, and solid-state electrolysis. The molten metal attacks any crucible in which it is melted, and the final product generally is a rare-earth-rich alloy of the crucible elements. Tungsten and tantalum make the best crucibles. They are only weakly attacked by the molten metals below 1000 °C. Impurities from the atmosphere can be largely eliminated by carrying out all operations in an environment of helium or argon and by the use of high-vacuum ion pumps.

8.3.2.3   Applications

The global production of REO was 210 kt in 2019.[23] The metal production comes largely (90%) from China, with Vietnam, Thailand, and Japan as smaller metal producers.[42] In the mine production China plays a smaller role (68%), together with Myanmar, Australia, and USA (each $\pm$ 10%).[50] About 30 kt is Nd.[49] The mine production of the individual REE is determined by the ratio in which they are present in the ores. Hence the majority of REE produced is La and Ce, while the majority of REE demand is the magnet elements Nd, Pr, and Dy. This can lead to overproduction of some REEs and a need for recycling of other REEs to meet demand.[51]

The main applications for REE are magnets (29%), catalysts (21%), glass polishing powders (13%), metallurgy (8%), glass (8%), batteries (7%), ceramics (4%), and phosphors (1%) in 2019, amounting to 139 kt of REO consumed globally.[42] They are often used in small quantities, and are key in digital and energy efficiency technologies. REEs have two applications in which they are used as metals: permanent magnets and metal alloys/metallurgy. The three applications where they are used as oxides are catalysis, ceramics and glass, and phosphors. Neodymium–iron–boron ($Nd_2Fe_{14}B$) is the strongest permanent magnetic material, and is used in a wide range of applications including electromotors, frictionless bearings, acoustic transducers (microphones and speakers), and power generators. These are used in wind turbines (9%), electric vehicles (9%), automotives (15%), consumer electronics (21%), and other applications. The main use of Dy and its substitute, Tb, is as an addition to the magnet alloy to improve performance at higher temperatures.[42,44,52] Samarium–cobalt ($SmCo_5$ or Sm_2Co_{17}) magnets are the main application of Sm used in specific applications such as aircrafts.[52]

Typical metallurgical applications are in the form of mischmetal, a mixture of light REE. Mischmetal is used during the production of specific steel alloys (to desulfurize and deoxidize the metal), and in nickel-metal hydride batteries. Y, Ce, Pr, and Nd are constituents of numerous ferrous and non-ferrous alloys to improve properties. An addition of up to 1% Gd significantly improves the workability and high-temperature oxidation resistance of iron, chromium, and their alloys,[42] while REE additions to Fe–Cr and Fe–Cr–Ni alloys increase their resistance to corrosion.

Catalysts in fluid catalytic cracking for petroleum cracking use La and Ce halides to stabilize the zeolite structure of the catalysts. In car exhaust catalytic converters cesium's ability to form non-stoichiometeric CeO_2 improves the oxidation of unburned hydrocarbons.[44] Powders for glass polishing are a main application for cerium oxide.[44] Phosphors of Y, Gd, and Eu are used in low-energy fluorescent lighting (Eu, Tb, Y) and LEDs for lighting and display applications. Laser crystals used in 5G frequency generators for telecommunications are doped with Y, Nd, Er, Yb, and Tm.

8.3.2.4   Recycling

REE recycling possibilities and technologies for different applications, such as magnets, batteries, and lamp phosphors, have been extensively discussed.[52] Less than 1% of REE are currently recycled from end-of-life products. Generic challenges are that the REE are often in small quantities in highly complex products with varying lifetimes, ranging from several years for mobile phones to over a decade for wind turbines and vehicles. Collection of the end-of-life products from consumers and separation of the REE components are major challenges, in addition to the recycling process itself. Just like the primary production process, these are energy intensive and complex. When these challenges are overcome recycling can play an important role in the future supply of REE.[53,54] Recycling of industrial magnets, e.g. from windmills, is already taking place as collection is much easier.

8.3.2.5   EHS and Sustainability

REEs have low toxicity. Long exposure to Ce-oxide polishing powders can lead to lung problems. The main environmental issues are linked to the presence of radioactive elements Th and U and to the energy consumption, chemicals usage, and emissions from the extraction and separation of REE.[44,47–9] Alternative sources of REE could be from the processing of industrial by-products and residues of other metals such as tin, niobium, iron, apatite, mineral sands, bauxite, and red mud. Lastly, deep ocean manganese nodules, crust, and muds are potential future sources.[44]

8.4   Titanium, Zirconium, and Hafnium

In the metallic form all three elements have excellent corrosion resistance. Titanium has the highest strength-to-density ratio of any metallic element. In its unalloyed condition, titanium is as strong as some steels, but less dense. Titanium also has excellent high-temperature strength and is used for turbochargers and compressors, and for numerous aerospace components. Zr and Hf are used for nuclear applications for their neutron capture cross-section, which are completely opposite. Otherwise, they behave very similar. Ti and Zr are also used extensively as alloying elements in aluminium and steels, and for pigments and numerous chemical applications. Strikingly, all three elements are produced by magnesiothermic reduction via the Kroll process.

8.4.1   Titanium (Ti)

Titanium's discovery began with the isolation of its oxide from a black iron sand in the UK by William Gregor in 1791, who concluded it contained an unknown metal. In 1795 Martin Heinrich Klaproth named it titanium. The fundamental reactions for producing titanium metal were already known in the 1800s. It took till 1908 before pure titanium oxide was prepared and titanium metal was made in 1910 by Matthew Hunter. Its high affinity for O, N, and C made extraction difficult. Anton Eduard van Arkel developed a larger scale process in 1922, based on the reduction of potassium hexafluorotitanate with sodium.[55] The breakthrough for industrial scale production was William J. Kroll's discovery of the reduction of titanium tetrachloride by magnesium,[56] and FS Wartman's method for melting titanium sponge in an EAF.[57] Commercial production started in the 1940s driven by the aircraft industry.[58]

8.4.1.1   Physical Properties

Titanium is a silvery-white, ductile metal. It is a quite light metal (4.51 g/cm^3), about 60% the weight of 12% Cr stainless steel,[42] with a much higher melting point than Al and Mg, at 1668 °C. It has two crystal structures: α-Ti (hcp) below 882.5 °C, which is similar to the structure of magnesium, and β-Ti (bcc) above 882.5 °C.[57] The strength, hardness, and electrical resistivity are strongly influenced by O, N, C, H, and other impurity contents in the metal. Even a very small amount of dissolved oxygen makes the metal brittle and hard. Corrosion resistance is due to the thin, dense, stable adherent oxide layer, which increases thickness with temperature. This layer is destroyed in water-free environments, e.g. dry chlorine or oxygen, in reducing corrosive media, and in water or humid environment with chloride present. Corrosion resistance can be improved by adding noble metals. Titanium is stable in molten alkali metals up to 600 °C, magnesium up to 650 °C, and up to 300 °C in Ga, Sn, Pb, and Pb–Bi–Sn alloys.[57] Ti forms alloys with most of the other metals. Ti and its alloys have good cold working properties.

8.4.1.2 Production

Because of its high oxygen affinity, titanium is found in nature as an oxide. The industrially relevant minerals for metal production are rutile, ilmenite, and titanomagnetites and hematites. The magnetites can be vanadium-bearing.[59] Rutile, 90–7% TiO_2 with silica, iron, oxide, and V, Nb, and Mo compounds as impurities, is mined from heavy mineral beach sands and placer deposits. Ilmenite, $FeTiO_3$, is magnetic and mined from beach sands and placer deposits if TiO_2 content > 0.3 wt.%. In addition, it is obtained from titanomagnetites (13–35 wt.% TiO_2) in the form of Ti slags. The minerals are mainly used for TiO_2 pigments; only a small share ($\pm$ 4%[57]) is employed for metal production. Pigment-quality TiO_2 is unsuitable for metal production due to the impurity levels.[57]

Direct reduction of TiO_2 to metal is not economically possible because of the large heat of formation of TiO_2 (945.4 kJ/mol) and the high solubility of oxygen in Ti at high temperatures. Reduction by carbon is possible only above 6000 °C. Therefore, a chloride route is used.[57] Metal production consists of upgrading of the concentrate, chlorination to $TiCl_4$, reduction to Ti metal, and melting/alloying.

The first step is to produce sufficiently pure TiO_2 (rutile) from the different feedstocks. Physical separation methods are used for natural rutile to remove zircon and ilmenite to obtain 95% TiO_2 concentrate. Ilmenite is converted to synthetic rutile (92% TiO_2) by reductive roasting in a rotary kiln at 1200 °C, followed by leaching to remove the iron.[58] Titanomagnetite (and ilmenite) is processed in a pig iron process, similar to vanadium. It is smelted in an EAF into a 85–90 wt.% TiO_2 slag and pig iron.[58]

Chlorination of rutile to $TiCl_4$ is done in a fluidized bed reactor at 900–50 °C. The reaction is exothermic above 600 °C, and no further heating is necessary. About 90–5% of rutile is converted.[57] Distillation and condensation removes $SiCl_4$ and $SnCl_4$ below 136 °C, and $AlCl_3$ and $FeCl_3$ above 136 °C. Vanadium is difficult to remove from Ti as their boiling points are close together. H_2S gas and copper metal are added to reduce $VOCl_2$ to $VOCl_3$ to remove it. The Ti slags used in the process contain more Fe, as well as Ca and Mg, causing problems that can be managed by adding sand to the bed.[57]

Reduction to metal is done using the Kroll process, consisting of a reduction and distillation step. The process is based on magnesiothermic reduction of $TiCl_4$ to metal and the difference between T_{boil} of Mg and T_{melt} of $MgCl_2$, 1120 and 711 °C, respectively. The batch process takes place in a carbon steel or Cr–Ni steel reactor, with a sponge capacity of 1.5–10 t. The inside walls are brushed or titanium coated. The reactor is filled with oxygen-free Mg lumps that are melted under argon atmosphere. Around $700\,^\circ$C the $> 99.9\%$ pure $TiCl_4$ is added or blown in and reacts to Ti and $MgCl_2$. The temperature is kept at 850–$950\,^\circ$C to avoid a reaction between titanium and the reactor wall. Titanium deposits on the walls as sponge, forming a porous cake above the Mg metal. Mg metal moves up through the sponge by capillary forces to react with $TiCl_4$. The molten $MgCl_2$ collects below the Mg metal at the bottom of the reactor, and is drawn off and recycled to Mg by electrolysis. When the reactor has cooled to $200\,^\circ$C, it is opened in a dry room as the salt residues are hygroscopic. The sponge is removed from the reactor, crushed, and purified by vacuum distillation at 0.1 Pa and 900–$1000\,^\circ$C to remove Mg and chloride impurities. The Ti yield is about 98%.[57] The process takes roughly 95 h for reduction and 85 h for distillation,[57] making it time, energy, and thus cost intensive.[58] Alternative processes such as direct reduction and electrolysis have been developed.[58]

Melting of the metal requires appropriate crucibles as the group 4 metals react with common refractory mould materials. Copper moulds cooled with water or molten sodium–potassium (Na–K) alloy are used. Melting is done in a vacuum arc furnace, and a consumable electrode of Ti sponge, scrap, and alloying elements is used. The molten metal drips in the copper mould or crucible forming a crude ingot of 3–8 t and 50–110 cm diameter. Depending on the application the metal is melted two or three times. Typically the ingots contain 0.12–0.35% O, 0.05% N, 0.06% C, and 0.013% H.[57] Ti castings use tiltable crucibles for melting, followed by casting in graphite, ceramic, or metallic moulds. It account for less than 1% of the output of semi-finished productions.[57] Powders can be obtained by atomization.

Ferrotitanium, 28–50 wt.% Ti and 4.5–7.5 wt.% Al, is made using aluminothermic reduction, as carbothermic reduction results in a too high-carbon content. In an EAF TiO_2 is reduced to TiO and then to Ti metal using Al and generating Al_2O_3. The yield is about 50%. The exothermic reaction consumes 1.89 kg Al/kg Ti plus electricity.[57] Alternatively ferrotitanium, particularly 65–75 wt.% Ti, can be produced by melting scrap Ti (consumable electrode) with iron in an induction furnace.[57]

8.4.1.3 *Major Applications*

The majority of the mined titanium is ilmenite used as TiO_2 pigments. About 10% or 600 kt/y of the mined Ti is rutile from Australia, Sierra Leone, and South Africa, about 20% each.[23] Furthermore, the Ti slags contribute to the available TiO_2. Main sponge producers are China, Japan, Russia, and Kazakhstan; Saudi Arabia is a new producer. Global sponge production was 210–23 kt in 2019.[23] Ingot production is located in China, the USA, Russia, and Japan.[60]

Ti alloys can be divided into 3 types: α alloys, β alloys, and $\alpha + \beta$ alloys. The α alloys, e.g. Ti–5Al–2.5Sn, contain alloying elements, mainly Al, which stabilize the α phase by increasing the transformation temperature. This results in a higher tensile strength.[61] Sn and Zr also stabilize the α phase. β alloys contain alloying elements stabilizing the β phase, such as Mo, V, Nb, Cu, and Si.[62] These alloys have extraordinary fatigue resistance, and high tensile strength for heavily loaded structural parts. The $\alpha + \beta$ alloys, e.g. Ti–6Al–4V, contain 2–6% α stabilizers and 6–10% β stabilizers and are most commonly used. They combine the best of both: low density; fracture toughness, creep strength, and ductility are better than those of α alloys; tensile strength and fatigue resistance are better than those of β alloys.[61] Corrosion properties can be improved by Mo, Zr, Hf, Ni, Ta, Nb, or Pd additions.[57] Many alloys are biocompatible and can be used as implants.

Ti alloys and metal are used in aerospace applications ($\pm$ 50%), automotive applications (20%), and hand-held equipment, medical equipment, e.g. dental and orthopedic implants, and alloys about 13% each.[42] Metallurgical (alloy) uses are in steelmaking as a deoxidizer, grain refiner, carbide former (TiC), nitrogen remover, and bonding with sulfur.[57] In aluminium alloys it is used as a grain refiner; see Section 6.11. Titanium is used to a great extent for electrochemical process equipment and for electrodes, e.g. anode material in the electrolysis of chlorine, chlorate, and manganese dioxide. Future applications can be shape memory alloys with 50 at.% Ni, superconductors with 50 at.% Nb, and hydrogen storage alloys with high Fe, Cr, Mn, and V contents.[57]

8.4.1.4   Recycling

Increased use of titanium increases the availability of new and old scrap. For old scrap sorting of Ti alloys from the other materials is a prerequisite for dedicated recycling. Before recycling, the various alloys need to be sorted, and scale and residues need to be removed to not contaminate the melt. The scrap is mixed with sponge or added as pieces. Untreated scrap can be used as an additive to steel, nickel, aluminium, copper, and zinc alloys; and processed into ferrotitanium or other master alloys. In the USA about 40% of Ti consumption is from processed scrap.[23]

8.4.1.5   EHS and Sustainability

Pulverized titanium metal that is created during cutting processes is pyrophoric and can spontaneously ignite below 55 °C in air.[42]

8.4.2   Zirconium (Zr)

Zirconium was discovered by Martin H. Klaproth in Germany in 1789, and Jöns Jacob Berzelius isolated the metal in 1824. It took another 100 years before commercialization of the metal took place, using the Van Arkel–De Boer process developed in 1925.[55] William J. Kroll developed an alternative, more cost-efficient process using magnesium as reducing agent, first in 1938, and in 1947 a pilot plant was running.[56] Oak Ridge National Laboratory, USA developed a Zr–Hf separation process in 1949–50, enabling zirconium as key material for the emerging nuclear power industry.[63]

8.4.2.1   Physical Properties

Zirconium is a shiny, strong, ductile metal which looks similar to stainless steel. It has three solid phases: ω-Zr stable below –73 °C, α-Zr stable between –73 and 852 °C, and β-Zr stable stable above 852 °C to T_{melt} of 1852 °C. Despite the high melting point the mechanical properties are more similar to those of metals with a much lower melting point, like the alkali earth metals (group 2) and aluminium (group 13): a low elastic modulus, strength decreasing with temperature, and hydrogen embrittlement. The mechanical properties of zirconium strongly depend on purity—N, O, and C decrease ductility—the degree of cold work, and the crystallography texture. With regard to the chemical properties, Zr is very reactive and develops in air or aqeous solution, a stable surface oxide film providing corrosion resistance, similar to aluminium. The oxidation reaction is exothermic and in sponge metal or powder form the reaction may cause ignition.[63] In most media Zr is more corrosion resistant than titanium or stainless steel. Its acid resistance is similar to that of tantalum, and in addition Zr is resistant to caustic media. Fluorine bonds strongly with Zr and trace quantities of fluoride will drastically reduce the corrosion resistance.[63] Its most important property for industrial use as metal is its low neutron capture cross-section,[64] while its oxide has low solubility in silica or silicates and can be used at high temperatures (1700–2000 °C) in refractories.[64]

8.4.2.2   Production

Zirconium is only found in an oxidized state in nature. It is mostly found as zircon, $ZrSiO_4$, consisting of 67 wt.% ZrO_2, including 0.5–2 wt.% Hf built into the ZrO_2 lattice, 33 wt.% SiO_2, and as impurities 0.2 wt.% Al_2O_3, 0.15 wt.% TiO_2, 0.1 wt.% Fe_2O_3, 0.1 wt.% P_2O_5, 0.025 wt.% U, and 0.020 wt.% Th.[63] It can be radioactive if the content of uranium and thorium are sufficiently high. Zircon can also have increased Y, Nb, and Ta contents, which are neighbouring elements in the periodic table.[64] Zircon is found as part of heavy mineral beach sands. It is separated from the other minerals using gravity and electrostatic and magnetic separation methods.[64,65] The industrial production process from zircon concentrate to nuclear grade Zr metal consists of five steps: ore cracking; hafnium (Hf) separation; reduction, refining, and melting.[66]

Ore cracking is done via carbochlorination and distillation of zircon in a fluidized bed reactor, where zircon, carbon, and chlorine gas are heated to 1200 °C and react to form $ZrCl_4$ (and $HfCl_4$) according to the endothermic reaction $ZrSiO_4 + 2C + 4Cl_2 \rightarrow ZrCl_4 + SiCl_4 + 2CO_2$.[64] $ZrCl_4(g)$ with less than 50 ppm Hf[63] is distilled off and precipitated as powder at 200 °C, while $SiCl_4(l)$ is collected at –20 °C. The $SiCl_4$ by-product is used for fibre optic cable, fumed silica, and silicon metal for solar cells.[66]

Separation of Hf from Zr is mainly done via solvent extraction and calcination to obtain HfO_2 and ZrO_2, followed by chlorination to $ZrCl_{4(s)}$. The alternative, pyrometallurgical method also used is extractive distillation.[55] It uses the difference in volatility between $HfCl_4$ and $ZrCl_4$, and their ability to dissolve in a molten salt of $AlCl_3$–KCl at a molar ratio of 1.04, at 350 °C.[66] This method has high maintenance cost. Metal is created via magnesiothermic reduction using the Kroll process to avoid contamination with C, N, and O. The crude sponge metal still contains some Mg and $MgCl_2$.

It is refined using distillation to obtain sintered zirconium sponge metal. Vacuum arc melting is used to obtain ingots of 75 cm diameter and weight up to 9 t.[63] The combination of pyro-hydro-pyro metallurgy operations leads to high production costs, intensive labour, and a heavy environmental burden.[66] Xiao et al. proposed an alternative 3-step process based on molten salt electrolysis of $(Zr,Hf)O_2$ to alloy, purification using molten salt and electrorefining to pure Zr-metal.[67]

Metal forming is done by hot working, followed by cold working to make sheets and strips. Welding is done using electron beam welding, while friction welding or explosive cladding is used to apply zirconium lining to ferrous alloys.[63]

8.4.2.3   Applications

Zirconium ores mine production was 1400 kt in 2019, with about 40% from Australia, 25% from South Africa, and the remainder from Kenya, Mozambique, Senegal, and China.[23] The applications of zircon relevant for the metallurgical industry are foundry sands for casting (15%), and refractories for glass, cement, and steel production (11–14%). Only 2–3% of zircon is converted to metal for super alloys and nuclear control rods.[64] The global Zr–metal production capacity is estimated at 8500 t.[66]

The key application of Zr–metal (±85%) is in the nuclear industry for the lining of reactors and cladding fuel rods. Zr has to be free of Hf and is alloyed with Sn and Nb, as well as Fe, Cr, and Ni. Alloys for piping and equipment in the chemical industry are Zr_7O_4 with Fe and Sn, or Zr_7O_5 with 2.5 wt.% Nb.[63] Master alloys with Zr are added to steel melts to control sulfide and oxysulfide in inclusions, to bind N_2, and to stop crystal growth and strain aging.[64] It is also added to superalloys and non-ferrous metals such as aluminium and wrought magnesium alloys.

8.4.2.4   Recycling

Foundy sands are commonly recycled; refractories are partly recycled. Zr in superalloys is recycled as new and old scrap back to superalloys. Zr metal in the form of machining chips and chops, crushed sponge, or Mg–Zr rests from the Kroll process is used as an alternative to master alloys.[63]

8.4.2.5   EHS and Sustainability

Zr is very reactive and develops in air or aqeous solution a stable surface oxide film providing corrosion resistance, similar to aluminium. The oxidation reaction is exothermic and in sponge metal or powder the reaction may cause ignition.[63]

8.4.3   Hafnium (Hf)

Hafnium was discovered in 1923, a century after the first zirconium metal was made. Hf and Zr are basically twins in their chemical properties and share a large part of the production process.

8.4.3.1   Physical properties

In appearance similar to stainless steel, hafnium is a heavy, hard, ductile metal. Its density, 13.31 g/cm^3, is double that of zirconium, though it is still lighter than gold, rhenium, and platinum, among others. The chemical properties are similar to Zr; hence, precautions need to be taken when analysing material to ensure that only hafnium is determined, instead of a combination of hafnium and zirconium. The differences are in the physical properties. Hf has two phases, α and β, a transition from α to β at 1760 °C, and T_{melt} of 2227 °C, which is higher than zirconium. Furthermore, Hf is known for its high thermal neutron absorption coefficient—1000× higher than zirconium.[68]

8.4.3.2   *Production*

Hafnium is always found together with zirconium. A typical zircon mineral, $ZrSiO_4$, contains about 0.5–2 wt.% Hf built into the ZrO_2 lattice, 67 wt.% ZrO_2, and 33 wt.% SiO_2. Hf is a by-product of the Zr production process, created as HfO_2 or $HfCl_4$ in the hafnium removal step, and follows the same process steps as for zirconium to obtain metal. In its oxide form it is converted to $HfCl_4$ by carbochlorination at 950 °C with maximum 1 wt.% $ZrCl_4$.[68] Next, $HfCl_4$ is purified to lower levels of Al, Fe, and U by sublimation under a nitrogen–hydrogen atmosphere.

Metal can be produced in two ways. It can be done by magnesiothermic reduction using the Kroll process operated at 850 °C, followed by vacuum distillation. The sponge is crushed into 2-cm pieces. Alternatively, molten salt electrolysis using a molten equimolar KCl–NaCl salt with a bit of NaF can be done. The $HfCl_4$ dissolves in the salt and reacts with NaF to form a hexafluorohafnate complex. After leaching out the salt, Hf metal crystals are obtained. Both sponge and crystals need to be further refined, as their impurity levels are too high: 670–875 ppm oxygen, 15–35 ppm nitrogen, up to 40 ppm carbon, 50–100 ppm chlorine, and hundreds of ppms of Al, Fe, and Mg.[68] The Van Arkel–De Boer process,[55] or iodide-bar or crystal-bar process, lowers all impurities to below 50 ppm. Hf sponge reacts with iodine vapour to form hafnium iodide vapour that diffuses to a hafnium filament heated to 1400 °C. Hf iodide dissociates and deposits on the filament. The filament grows from a 3-mm wire to a 40-mm crystal-faceted bar. O, C, and N are removed, and metallic impurities are reduced as their iodides diffuse less well. Electron-beam melting can be used as well, though it is less effective in removing impurities, especially O and N, rendering the metal unsuitable for nuclear applications.[68] Melting and alloying is done using vacuum arc melting.

8.4.3.3   *Applications*

The production of hafnium is fully coupled to the demand and production of zirconium. Global production of Hf metal is dominated by France and the USA, each counting for nearly half, and amounts to about 50–60 t/y in 2016.[42] Its main use is in nickel-based superalloys (61%)[42] for jet engines and gas turbines. An addition of 1–2 wt.% hafnium increases the high-temperature creep strength by forming stable precipitates at the grain boundaries.[68] Furthermore, hafnium is used to make Hf-carbide–Nb-carbide solid solutions for cutting,[68] nozzles for plasma cutting (15%),[42] in nuclear control rods (11%),[42] because of its ability to strongly absorb neutrons, and as an alloying element in Nb, Ta, Ti, Mo, and W alloys.[68]

8.4.3.4   *Recycling*

There is no information available about the recycling of hafnium.[42] It is probable that hafnium is recycled as part of the recycling of superalloys and turbine blade alloys, as production waste (new scrap) is recycled in-house.

8.5   Vanadium, Niobium, and Tantalum

These metals are characterized by high melting points and especially tantalum has a very high density (16 690 kg/m^3) and melting point (3017 °C). Their refining challenges are similar to the elements in group 4: high melting point and very stable oxides. Aluminothermic reduction is used to produce all three metals. Furthermore, Nb and Ta can be considered twins due to similarities in properties. Vanadium and niobium are mainly used as alloying elements in steel alloys to improve strength, toughness, and corrosion resistance. Tantalum's main use is in capacitors.

8.5.1   Vanadium (V)

Vanadium was discovered in 1801 by Andres Manuel del Rio in a Mexican lead vanadate ore. In 1831 Nils Gabriel Sefström discovered the element in iron converter slags and named it vanadium, after the Norse god of beauty, Vanadis.[59] Metal was first produced in powder form by Roscoe around 1867. Its first application was in alloyed steel in 1903, and Henry Ford started to use vanadium alloyed steel on an industrial scale in automobile construction in 1905, which is still a main application.[59]

8.5.1.1   Physical properties

Vanadium is a grey-white to silvery-blue-grey transition metal. It is hard but ductile and can be rolled and forged. The presence of impurities has a high impact on strength. O, H, N, and C decrease ductility and increase hardness and tensile strength. Oxidation takes place in air above 300 °C. At temperatures up to 500 °C vanadium absorbs hydrogen and becomes brittle and can be powdered. Hydrogen can be liberated by heating to 600–700 °C in vacuum. Nitrides are formed above 800 °C and carbides are formed at 800–1000 °C.[59] V and some of its alloys have good corrosion resistance towards molten low-melting metals and alloys, especially alkali metals, and towards tap and seawater due to its protective oxide film. It is quite stable in dilute sulfuric, hydrochloric, and phosphoric acid, and dissolves in nitric and hydrofluoric acid. Vanadium has several valence states: 2+ (VO), 3+ (V_2O_3), 4+ (VO_2), and 5+ (V_2O_5).[59]

8.5.1.2   Production

Vanadium is found in nature in about 152 minerals. The economic relevant minerals are solid solution in magnetite (Fe_2O_3), patronite (VS_4), vanadate ($Pb_5(VO_4)_3Cl$), and carnotite $K_2(UO_2)_2(VO_4)_2 \cdot 3H_2O$.[69] On average the V_2O_5 content ranges between 0.2 and 1.6 wt.%. The most important deposits are V-bearing titaniferous magnetites. Also from iron sands, phosphate and uranium ores vanadium is extracted.[69] After separating the V-rich magnetites from the gangue using magnetic separation, it is either processed in a pig iron flowsheet or a salt roast flowsheet.

In the pig iron flowsheet vanadium is a by-product from steelmaking. First, the magnetites are mixed with coal, silica, and limestone and partially pre-reduced, 37–45%, in rotary kilns. Then it is smelted in a submerged arc furnace where first Ti is slagged off as ferrotitanium melts at 1070–1135 °C, while ferrovanadium melts at 1699.5–1770 °C. The hot metal is tapped into ladles and top blown with oxygen to oxidize V, which forms a slag.[69] The slag contains about 25% V_2O_5 in the form of $FeO \cdot V_2O_3$, 1.5% Ca, 2% Mg, 2% Al, 9% Fe, 7% Si, 3.5% Mn, and 3.5% Ti.[59] The pig iron follows the steelmaking process. The vanadium slag goes to the salt roast flowsheet; any iron metal is removed prior to roasting.

The salt roast flowsheet processes V-rich magnetites, V-slags from iron making, and V-residues. The feed is roasted in a rotary kiln together with sodium carbonate and sodium sulfate, at about 1150 °C, to create water-soluble vanadates (vanadium oxides). Impurities such as Ca, Mg, and Al form water-insoluble vanadates and lower the vanadium yield. Si causes filtration problems and P goes with the vanadate and lowers the yield in the precipitation step. After cooling the calcine is ground in a wet ball mill, and the slurry is thickened to concentrate the V-bearing leach liquor. Ammonium sulfate is added; vanadium precipitates and is calcined to obtain V_2O_5. This is melted to produce the final V_2O_5 product.[69]

To produce ferrovanadium (FeV), V_2O_5 is reduced with aluminium granules (aluminothermic reduction or thermite process) together with steel shot. After ignition the reaction is self-sustaining. To ensure that equilibrium is reached, the melt is heated with an electric arc and then cooled for 2–3 days. A batch is about 0.5–1 t metal. As iron and vanadium are almost completely mutually soluble, a range of ferrovanadium compositions is possible: FeV 60, FeV 80, and FeV 90, with 90% V[59] with C, Si, Al, P, S, and Ti as main impurities.[69]

8.5.1.3 *Major Applications*

The global production of vanadium is about 73 kt/year (2019) mined mainly in China (55%), Russia (25%), and Brazil and South Africa (10%, each).[23]

About 90% is used as FeV in steel production. The remainder is used in non-ferrous alloys, in Ni–Co and Co–Mo catalysts and batteries.[69] In steel production about $^2/_3$ is used for high strength low alloy (HSLA) steel, nearly $^1/_3$ for special steels, and the little that remains for superalloys, cast iron, and stainless steel.[42,69] Vanadium (up to 0.05%) increases hardenability because it reduces the rate of grain growth during heat treatment and increases the temperature at which grain coarsening starts. Too much V leads to carbide formation and reduces hardenability.[69] The increase in strength can be up to 100% and reduces part weight by up to 30%.[42] Vanadium–aluminium master alloys are added to $TiAl_6V_4$ metal matrix alloys for strength and creep resistance, and used in jet engines and airframe construction.[59] Vanadium is used in alloys for metal hydride batteries. Vanadium redox (flow) batteries (VRB) used for grid storage are considered a potential growth market.[70] Vanadium's ability to move between its four oxidation states when dissolved in sulfuric acid is key to making the battery function.[71]

8.5.1.4 *Recycling*

Spent catalysts are recycled via acid leaching with SO_2 additions followed by oxidation and precipitation to produce V_2O_3 or V_2O_5 or treated via the salt roasting process.[59] Recycling from the catalysts can be up to 40% of catalyst demand.[23] The vanadium used in steel and alloys is recycled as part of the alloy.

8.5.2 Niobium (Nb)

Niobium was discovered in 1801 by Charles Hatchett in the mineral columbite and named columbium. In 1844 the name niobium was proposed by Heinrich Rose, who demonstrated how to distinguish Nb from Ta. Niobium in the USA is still referred to as columbium.[72]

8.5.2.1 Physical properties

Niobium is a grey, hard, paramagnetic refractory metal. Its melting temperature, 2468 °C, and boiling point, 4930 °C, are very high. It is highly resistant to chemical attack due to the formation of a protective oxide layer[73] and behaves as a superconductor at very low temperature.[42] Niobium's mechanical properties, like most refactory metals, are strongly influenced by metal purity, production method, and mechanical treatment. Small amoutns of impurities increase the hardness and strength, while reducing ductility, especially Nb carbide, which is extremely hard. Nb chemicals have unique optical, piezoelectric (electric charge generated by mechanical stress), and pyroelectric (electric charge generated by heating or cooling) properties.[42] It can be considered a twin of tantalum due to their similarities in properties.

8.5.2.2 Production

Niobium is found in nature only in its oxide form, never as pure metal. The main minerals are pyrochlore $(Na,Ca)_2Nb_2O_6(OH,F)$ and together with tantalum in columbite (Nb)–tantalite (Ta): $(Fe,Mn)(Nb,Ta)_2O_6$. It is also found with tin and titanium minerals. Niobium mines produce Nb_2O_5 concentrates from pyrochlore deposits in Brazil and Canada. The ore in Brazil contains about 1–7 wt.% Nb_2O_5 with iron oxide, phosphate, barite, zircon, and quartz as impurities. The Canadian ore contains 10× less, 0.1–0.7 wt.% Nb_2O_5 with 65% carbonate, and 10 wt.% each of iron oxide, silicate, and phosphate. After flotation, about 60–80% of the Nb_2O_5 is recovered in a 55–60 wt.% Nb_2O_5 concentrate. Further details on flotation can be found in Gibson et al.[74]

The oxide is reduced to ferroniobium metal by aluminothermic reduction and addition of iron oxide, if the concentrate is low in iron. The enthalpy of the reaction between Nb_2O_5 and Al (−276.1 kJ/mol Al) is below the threshold for self-sustaining reactions; therefore, pre-heating or mixing with compounds such as BaO_2, CaO_2, $KClO_4$, $KClO_3$, or $NaNO_3$ is necessary to provide oxygen.[72] Although thermodynamically possible, carbothermic and silicothermic reduction are unfavourable and not used. A single batch can produce Fe–Nb blocks of 11 t, starting from 18 t pyrochlore concentrate, 4 t iron oxide, 6 t aluminium powder, 0.75 t fluorspar, and 0.5 t lime. Also 20 t slag, consisting of 48 wt.% Al_2O_3–25% CaO and impurities, is produced.[72] The amount of Al determines the Nb yield and the oxygen content.

The alloy is then refined in a vacuum, EAF, or electron beam furnace to pure, low-oxygen, carbon-free alloy.[72] At the T_{melt} of Nb most other elements can be removed easily by vaporization. Special Nb alloys, e.g. Ni–Nb, Co–Nb, Al–Nb, and Cr–Nb, are also produced using the aluminothermic process, using Nb_2O_5 as feed material.[72] Production from columbite–tantalite is covered in Section 8.5.3.2.

8.5.2.3 Major Applications

The global production of Nb_2O_5 concentrates is about 100 kt/y.[75] About 90% is produced from mines in Brazil, 5–10% from Canada, and the balance from Russia.[75] Nearly 90% of niobium produced in 2017 was used as ferroniobium and is added to manufacture HSLA steels for oil and gas pipelines (17%), car and truck bodies (28%), stainless steel (14%), and construction steel (32%). Fe–Nb increases strength and toughness by refining the microstructure and forming nanoparticles,[42] consequentially reducing the weight of the part.[75]

Nb–Ti and Nb–Sn alloys are used in superconducting magnetic coils in magnetic resonance imagery (MRI) equipment, magnetic levitation transport systems, amd particle physics experiments. The resistance of these alloys drops to near zero at or below −268 °C (liquid He).[75] Aimone and Yang[73] have investigated new alloys with higher corrosion resistance and mechanical properties for application in the chemical process industry. Nb metal alloyed with Ta, Mo, W, Ru, and Pd was investigated as an alternative to tantalum (Ta-W$_3$) and zirconium (Zr$_7$0$_2$) alloys.

Nb carbide is used for cutting tools, and Nb oxide is used as a coating on glass for screens and camera lenses, and ceramic capacitors and to make lithium niobit (LiNbO$_3$) for surface acoustic wave filters.[75]

8.5.2.4   *Recycling*

Little has been reported about Nb recycling, and recycling back to Nb metal is not reported. When recycled it is recycled as part of an alloy. New scrap from steel manufacturing and part manufacturing, as well as specialty alloys, can be recycled. Old scrap could only be recycled when it is collected, and kept separate from other (steel) alloys.

8.5.3   Tantalum (Ta)

Tantalum can be considered a twin of niobium (Nb). Tantalum was discovered in 1802 by Anders Gustav Ekeberg. From 1900s it was used in light bulb filaments,[75] and since 1950s used to improve the performance of optical glasses. The former has lost its importance with the transition to energy-saving lightbulbs.

8.5.3.1   *Physical Properties*

Tantalum is an extremely hard, heavy, graphite-grey, shiny metal. It is paramagnetic and has the fourth highest T_{melt} of all metals: 3017 °C. Only W, Re, and Os have a higher melting point.[76] Its atomic weight and density (16.6 g/cm^3) are double those of Nb.[75] The characteristics of tantalum are very similar to those of niobium. It is also protected by a thin, dense, and stable oxide layer (Ta$_2$O$_5$) against corrosion by most acids, and the mechanical properties are strongly dependent on purity, structure, and crystal defects. Above 300 °C Ta forms oxides, nitrides, hydrides, and carbides.

8.5.3.2   *Production*

Tantalum, due to its strong affection for oxygen, is not found as metal in nature. The main mineral is columbite–tantalite, (Fe,Mn)(Nb,Ta)$_2$O$_6$, also dubbed 'coltan' where it is always found in combination with niobium. It can be found in complex ores together with Nb, S, W, Zr, Y, rare earths, and Li. Furthermore, Ta is found in tin-bearing minerals such as cassiterite with wt.% levels of Ta and Nb, and titanium-bearing minerals like rutile and titanate. Tantalum is produced from columbite–tantalite with Nb as by-product and as by-product from tin smelting.

Columbite–tantalite is often found as part of a heavy mineral beach sand or in a complex ore. Using a combination of gravity, magnetic, and electrostatic separation, or flotation, a concentrate with about 20–40% Ta_2O_5 is obtained. The Nb_2O_5 concentration varies with geography. Lower grade Ta concentrates can technically be further processed by hydrometallurgy, but the processing costs make this uneconomic.[75] Tin-smelting slags containing > 2% Ta_2O_5 can go directly to hydrometallurgical processing, while slags with 1–1.5% Ta_2O_5 need treatment. Processing in an EAF to a ferroalloy and further process steps increases the Ta level.[77] A synthetic concentrate with about 50% $(Ta,Nb)_2O_5$ is obtained[76,77] that can go to hydrometallurgical treatment.

The conversion from concentrate to metal is a hydrometallurgical process, often batch or semi-batch production with many products made to order.[75] The concentrate is digested at high temperature in a mixture of H_2SO_4 and HF, sulfuric acid and hydrofluoric acid, to obtain a K_2TaF_7 solution, or Ta_2O_5. The solution is treated for removal of As, Bi, and Sn.[75] Next solvent extraction is used to obtain a Nb- and a Ta loaded solution. A detailed overview has been provided.[78] The Ta (and Nb) is removed from the loaded solution as K_2TaF_7 through reaction with potassium fluoride, or removed as oxide by neutralization, precipitation as hydroxide, followed by calcination to oxide with 99.9% purity.[75,76]

K_2TaF_7 is reduced with sodium metal to obtain Ta metal. To control the highly exothermic reaction inert salts like KCl, NaCl, KF, and NaF are added. Afterwards the salts are washed out.[77] Oxides are subjected to aluminothermic reduction.[75] Lastly, the metal powder is sintered, resulting in 99.9% purity, or melted using vacuum arc or electron beam, and deoxidized with magnesium to obtain 99.98% purity metal or alloys.

8.5.3.3   Applications

Tantalum's global production was about 2250 t in 2018.[79] Primary mine production is mainly from Congo (40%), Rwanda (20%), Nigeria (12%), and Australia (14%).[23] Primary mining makes up about 60% of the global supply.[77] The remainder comes from secondary concentrates (10%), tin slags (10%), and scrap recycling (20%).[77] Metal production is done in China, Germany, Japan, the USA, and Brazil.[42]

Tantalum is used in capacitors (32%), superalloys (23%), sputtering targets (17%), chemicals (11%), semiproducts (10%), and Ta-carbide cutting tools (7%).[76] Ta capacitors work uniquely equally well at high temperatures and subzero temperatures. They are used in IT equipment, automotive airbag activators, GPS, aircraft, and medical appliances such as pacemakes and defibrillators.[75] In single-crystal superalloys for turbines, 4–7 wt.% Ta supports high-temperature strength, and reduces fatigue at high temperature and corrosion by sea air through the formation of precipitates.[77] Ta_2O_5 coatings have a high refractive index, making camera and spectacle lenses lighter and brighter.

8.5.3.4   Recycling

Scrap recycling is about 20% of the global production, mainly made up of new scrap. From superalloys it is recycled as part of the appropriately sorted superalloy. Hard metal scrap (carbides) are recycled by roasting and treating with caustic soda; for example, to obtain Ta oxide for further recovery. Clean capacitor anode scrap metal, 1.5–3% oxygen, can be melted again in an electron beam furnace. Oxidized metal can be reduced by magnesium or calcium addition. If impurities such as silver and manganese are present, treatment with nitric and hydrochloric acid is possible.[77] Ta capacitors from post-consumer scrap, e.g. mobile phones and printed circuit boards, are mostly not recycled as they have a low concentration, and end up in the slag of copper smelters. Physical separation from the circuit boards prior to smelting is labour and cost intensive, though robotization may be a solution to overcome this.

8.5.3.5 *EHS and sustainability*

Tantalum capacitor powders can ignite at higher temperatures, above $100\,°C$ in air, forming Ta_2O_5 and are classified as flammable solids.[77] To ensure that the proceeds of mining of Ta minerals in Africa or other countries is not used for financing civil war, OECD guidelines are in place and a conflict-free smelter programme has been set up by the Responsible Minerals Initiative since 2010.[80]

8.6 Chromium, Molybdenum, and Tungsten

These metals have extremely high melting points and are stable in a number of corrosive or oxidizing environments. This makes them suitable for special applications, such as filaments in incandescent light bulbs (W) or as a coating for gun barrels (Cr). They are also widely used to impart strength or corrosion resistance to specialty steel alloys. Tungsten carbide is essential for cutting tools.

8.6.1 Chromium (Cr)

Chromium was discovered in 1797 as part of the mineral $PbCrO_4$ by the French chemist Nicolas-Louis Vauquelin, and isolated as metal in the following year. Its first use was as chromates, followed by use in low-chromium alloy steels around 1860–70 in France. A process for ferrochromium production using an electric furnace and carbonthermic reduction was developed in 1893 by Henri Moissan and commercialized by Paul Héroult in 1899, which is the basis for modern production. Aluminothermic reduction was developed in 1898 by Goldschmidt, though not in use today. Electrolytic chromium metal production was first done by Robert Bunsen in 1854, and was commercialized in 1954.[81]

8.6.1.1 *Physical Properties*

Chromium (Cr) is white, hard, lustrous, and brittle; and is extremely resistant to ordinary corrosive reagents. This resistance, combined with its hardness, mirror-like brightness, and low coefficient of friction, accounts for its widespread use as an electroplated protective coating. Chromium has a high melting point, $1907\,°C$, very high thermal conductivity, and low coefficient of expansion. At elevated temperatures chromium combines directly with O, S, Si, B, N, C, and the halogens. It has 2, 3, and 6 as valence states, allowing for its use as a catalyst. The different oxidation states result in a range of colours. Chromium and its Cr-rich alloys are brittle at room temperature; hence, there is limited use in structural applications.[81]

8.6.1.2 Production

Chromium is found in nature as a solid solution between $FeCr_2O_4$ (chromite) on the one end and $MgCr_2O_4$ magnesio-chromite on the other end, and has as general formula $[Mg, Fe^{2+}][Cr, Al, Fe^{3+}]_2O_4$.[82] It is classified based on its Cr content, the Cr/Fe ratio, and the alumina content, which also determines its use. Typically Ti, V, Mn, Ni, Co, and Zn are found in the mineral. Chromite is obtained from chromite mines and as a by-product from platinum group metal mining in South Africa.[82]

To manufacture ferrochromium, high-Cr chromite, 45–55% Cr_2O_3, is used. For high-carbon ferrochromium with 4–12% C and 45–75 wt.% Cr, chromite is fed together with coke into a three-phase submerged arc furnace for carbothermic reduction. The resulting alloy is tapped from the furnace, together with a slag containing generally 30% SiO_2, 30% MgO, and 30% Al_2O_3 with some Cr_2O_3, CaO, MnO, and FeO. The furnace operates at 10–50 MVA and produces 15–60 kt/y of ferrochrome.[81] Nelson provides a detailed overview of industrially used submerged arc furnaces.[83] To produce medium- (0.5–4% C) and low-carbon (0.01–0.5% C) ferrochromium, a silicothermic reduction process is used, for example the Perrin process.[81] It uses a submerged EAF to produce ferro-silicochromium, 40–45% Si, 40–45% Cr, and little carbon, and a final slag. Reduction takes place analogously to the production of FeSi. In a second furnace this alloy is mixed with a slag rich in Cr_2O_3 so the silicon in the metal reduces the chromite. An effort is made to obtain a final alloy low in Si and a waste slag low in Cr.

Chromium metal is produced from high-iron chemical-grade chromite ore. First, chromium is extracted from the ore via alkaline or acidic dissolution,[81] followed by the production of metal via aluminothermic reduction of Cr_2O_3 or electrolysis of Cr(III) solutions. Extraction by alkali dissolution takes place in a rotary kiln where the ore is roasted with $NaCO_3$ at about 1100 °C to form sodium chromate, $Na_2Cr_2O_7$, which is then leached from the calcine. For electrolysis the 6-valent Cr in solution is reduced to 3-valent state using SO_2. Alternatively, $Na_2Cr_2O_7$ can be crystallized and converted to Cr_2O_3. Acid dissolution entails reduction of chromite with carbon to ferrochromium, which is ground and then dissolved in H_2SO_4 followed by purification before electrolysis takes place. Electrolysis for metal recovery has been described in detail.[81] Cr is plated from solution on stainess steel cathodes. The metal is stripped and deoxidized using hydrogen or a vacuum furnace. Metal production by aluminothermic reduction starts with Cr_2O_3 mixed with aluminium metal powder. This exothermic reaction generates temperatures above 2000 °C, resulting in a good separation of metal and slag. Alternatively carbon reduction can be used in a furnace at 1400 °C and over 40 Pa pressure.[81] Standard Cr-metal grades range from 99 to 99.4%.[84]

8.6.1.3 Applications

The global chromite production is about 44 Mt/y from South Africa (39%), Turkey (23%), Kazakhstan (15%), India (9%), and Finland (5%).[23] Chromite is divided based on its Cr_2O_3 and Al_2O_3 content, and the Cr/Fe ratio. High-chromium chromite, 46–55% Cr, is used for metallurgical applications. High-iron chromite, 40–46% Cr_2O_3 and Cr/Fe of 1.5–2, is used to produce metal and for the chemical industry. High-aluminia chromite, 33–38% Cr_2O_3 and

22–34% Al_2O_3, can be combined with magnesite and used for high-temperature refractories[82] in furnaces for Cu, Ni, and Pt production, and can be used as mould and foundry sands.[84]

About 93% of the chromite production is used for ferrochrome. The global ferrochrome production is about 6.2-Mt Cr content with about 37% from China, 28% from South Africa, 14% from Kazakhstan, and the rest from India, Finland, and Russia.[84] Roughly 73% of ferrochrome is used for stainless steel; the remainder is used for specialty steels.[84] An addition of 12.5–26% Cr results in the formation of a shiny, stable Cr-oxide layer that protects against corrosion and oxidation.[84] Up to 3% Cr is added to improve the physical properties and increase heat treatability. Steels with 5–6% Cr have increased corrosion and oxidation resistance and are employed in the form of tubes in the oil industry.

Chromium metal is used for alloying non-ferrous alloys, e.g. with nickel and cobalt-based super alloys.[81] It is also used as a surface coating in a range of applications.

8.6.1.4   Recycling

Chromium in stainless steel is recycled via the stainless steel cycle, as it remains part of the stainless steel alloys and keeps its functionality.[84] In the old scrap of other steel types it is not functionally recycled, as it ends up in carbon steel. New scrap and home scrap is being recycled.[84]

8.6.1.5   EHS and Sustainability

Many Cr(VI) compounds used in surface treatment processes have carcinogenic and muta-genic properties and are harmful to health and environment. Since September 2017 the EU's REACH regulation has put strict requirements on their use, basically banning their use unless authorization is granted.[84]

8.6.2   Molybdenum (Mo)

Molybdenum was discovered in 1778 by Carl Wilhelm Scheele, though not named, and isolated and named as metal by Peter Jacob Hjelm in 1781. It took until World War I before its first major use was found as a steel addition to improve strength and toughness in tank armor and aircraft engines, followed by use in automotive applications from the 1950s onwards.[85] In the nineteenth century Norway was the global main supplier. In the twentieth century the USA, Chile, and Mexico became additional suppliers. Development of selective flotation of molybdenite (MoS_2) from copper ores in 1933 resulted in production of Mo as a by-product from copper,[85] which still accounts for about 60% of production today.[84]

8.6.2.1   Physical Properties

Molybdenum (Mo) is a shiny silvery metal with a high T_{melt} of 2624 °C and a density, 10.22 g/cm^3, much lower than that of other metals with a high T_{melt}. The thermal expansion coefficient is the lowest of all metals, and it has high corrosion resistance.[84] It is inert to oxygen at normal temperatures, but combines with it when heated, forming Mo oxide. At 600 °C the oxide sublimes, and the exposed metal oxidizes quickly and can burn between 500 and 600 °C in oxygen.[85] Furthermore like Cr, Mo reacts with B, C, N_2, and Si at high temperatures to form hard, refractory, and chemically inert compounds. Mo can have a valence of 2, 3, 4, 5, or 6. Its physical properties depend on the method used to produce the metal and the degree of treatment.[85] Molybdenum is inert to dilute acids, concentrated HCl, and near-inert to alkaline solutions and fused alkali metal hydroxides among others. Exposure to concentrated nitric acid leads to passivation.[85]

8.6.2.2   *Production*

Molybdenum mainly occurs in nature as molybdenite, MoS_2, though it is also found as $PbMoO_4$ and $Ca(Mo,W)O_4$. Molybdenite is the the the only economically relevant mineral.[85] It is mined from molybdenite deposits with Mo content between 0.05 and 0.25 wt.%, and a Mo/Cu ratio > 1. The main source of MoS_2, 60% of world production, is as a by-product from copper deposits. These deposits contain < 0.05 wt.% Mo.[84] In both cases the ore is subjected to flotation to produce a concentrate with 85–93 wt.% MoS_2,[84] up to 15% gangue minerals (SiO_2, Al_2O_3), and up to 2.5% Cu, 0.9% Pb, 3% Fe.[85] If the impurity content is too high for steelmaking, these are removed by sodium cyanide leach (Cu and Au), iron (III), chloride leach (Cu, Pb, and Ca), or hydrochloric acid leach (Pb and Bi).[85]

The first step in the process to obtain Mo metal or ferromolybdenum is roasting the sulfide in air between 500 and 650 °C to technical grade MoO_3. This is done in a multiple-hearth furnace in which the hearths are stacked vertically, and in each hearth a different part of the process takes place. At the top Mo concentrate is fed, and SO_2 gas is captured and converted to acid. Any rhenium oxide (Re_2O_7) present in the concentrate reports to the off-gas, is removed by scrubbing, and is further recovered. At the bottom the oxide leaves and air passes upwards.[85] Water evaporates in the upper hearths followed by the highly exothermic reaction to oxide in the intermediate hearths. During this reaction at first mainly MoO_2 forms and only small amounts of MoO_3 until the quantity of MoS_2 remaining is equal to the amount of MoO_3 formed. Then MoO_2 is converted to MoO_3. Technical grade oxide contains 85–90 wt.% MoO_3, the balance being silicate with Fe_2O_3 and Al_2O_3.

The second step is the purification of trioxide to > 99.9% MoO_3 either by sublimation at 1100–1200 °C or by calcination of ammonium dimolybdate (ADM) at 420 °C. ADM is manufactured by leaching technical grade oxide, dissolving it in 10–20% ammonium hydroxide solution at 40–80 °C to form $(NH_4)_2MoO_4$. After purification and evaporative crystallization $(NH_4)_2Mo_2O_7$ crystals (ADM) are obtained.[85] Mo can also be recovered from the residue created during scheelite ($CaWO_4$) ore leaching in tungsten production. The residue contains MoS_3 and WS_3. It can be calcined and leached with strong HCl to dissolve Mo and leave W in the residue.[85]

Lastly, metal is produced by reduction of MoO_3 with hydrogen in two steps using a walking beam or roatary kiln furnace with a countercurrent hydrogen flow. First, the exothermic reaction of MoO_3 to MoO_2 takes place at 600 °C, to prevent melting of MoO_3. Next, the metal powder is formed at 1050 °C, consisting of 2- to 10-micron particles containing 100–500 ppm oxygen.[85] High purity, 5N5, metal can be produced through multiple steps of recrystallization and solvent extraction. Special linings and boats are used to prevent contamination with iron and ceramic particles. Alloys and metal parts are produced using mainly powder metallurgy, or vacuum arc remelting and electron beam melting.[85] Mo alloys are strengthened by solution hardening with Ti and Zr, Hf, Hf and Zr, or W and Re, and by particle hardening using carbides.[85]

Ferromolybdenum (FeMo), containing 60–75 wt.% Mo, can be produced by carbothermic reduction as well as silicothermic reduction. The latter is the commonly used process.[86] The single-step process uses ferrosilicon as the Si source, mixed with Al and Ca metal to self-sustain the reaction. Using only Al metal (aluminothermic reduction) provides too much exothermal heat, resulting in excessive Mo-oxide evaporation and an explosion danger. The slag containing lime, fluorspar, FeO, and < 1% Mo is tapped, and the red hot (solid) metal is removed from the furnace.[85] Dust is recycled because of its Mo content. The overall yield is 97–99%, and the reaction time is 30–60 min for 1–4 t of metal, followed by 30–60 min settling time.[85]

8.6.2.3   *Major Applications*

World mine production is about 290 kt Mo content per year mainly from China (44%), Chile (19%), the USA (15%), Peru (10%), and Mexico and Canada, among others.[23] Metallurgical applications, such as FeMo to increase corrosion resistance and high-temperature strength, is about 87% of its use. The remainder is in chemicals/catalysis.[87] Catalysts with Mo oxide or sulfide, or mixed with Co, Ni, Fe, V, or W and Bi, on an alumina substrate, are used in the petroleum refining industry for desulfurization.[85]

Metallurgical applications can be divided into engineering (carbon) steels (42%) with maximum 1% Mo; stainless steels (23%) with 2–3% Mo; tool steels (7%) with 3% Mo for regular and 5–10% Mo in high-speed steels; nickel-based superalloys (3%) with up to 28.5% Mo for jet engines and turbines; and Mo metal and alloys (12%) such as Mo–W alloys used in zinc pyrometallurgy.[84,87]

8.6.2.4   *Recycling*

Mo contained in ferrous and superalloy scrap is recycled as part of the ferrous and superalloy stream and functionally reused. The amount may be about 30% of the apparent supply of Mo.[23] Catalysts are recycled to recover the metal values and used again in catalysts. A typical composition of spent catalysts is 2–10% Mo, 0–12% V, 0.5–4% Co, 0.5–10% Ni, 10% S, 10% C, and the rest Al_2O_3. First, the material is heat treated to remove residual C, S, and hydrocarbons and to oxidize Mo and V to soluble oxides. Using selective leaching and precipitation Mo, V, Co, and Ni are separated and recovered in a form that can be processed in the typical metal recovery processes.[85]

8.6.3   Tungsten (W)

Tungsten pigments were used in China for porcelain in the sixteenth century, unknown to Europeans, although the mineral wolframite was known from tin production. The mineral $CaWO_4$ was found in Sweden around 1750 and named tungsten, based on the Nordic word for heavy stone. In 1781 Carl Wilhelm Scheele identified it as a new element. Two years later it was isolated as metal by Juan Jose and Fausto de Elhuyar in Spain and named wolfram. Robert Oxland in the UK patented processes for the production of tungstate oxide and metal in 1874, and Robert Mushet identified tungsten's role in steel hardening. Industrial use started around 1890 with the production of carbide (WC) by Henri Moissan followed by use of tungsten in high-speed, high-temperature steel (1900), in incandescent lamp filaments (1903), in powder metallurgy for wire production invented by William D. Coolidge (1909), and in the first cemented carbide cutting tools in 1925. In the 1930s and 1960s catalysts based on W for treatment of oil and exhaust gases were developed.[88,89]

8.6.3.1   *Physical Properties*

Tungsten is a metal with extreme properties. It has a very high density, 19.3 g/cm^3, comparable to gold, the highest T_{melt} of metals (3410 °C) behind carbon and carbides of Nb, Zr, Ta, and Hf, and high thermal conductivity, elasticity, and hardness. It also has the lowest vapour pressure of all metals, lowest compressibility, and low thermal expansion.[89] Tungsten is brittle at room temperature, and becomes ductile at higher temperatures (100–500 °C), depending on the presence of dissolved O, C, and N and degree of deformation.

In air tungsten is stable until about 350 °C, then starts to form stable blue tungsten oxides, which start to crack at higher temperatures and oxidation continues. Above 800 °C rapid oxidation results in WO_3 formation and sublimation, while in oxygen the metal burns at 500–600 °C. Carbides are formed by reaction with CO above 800 °C. Above 1400 °C tungsten reacts with B, C, N_2, and Si to form hard, refractory, and chemically inert compounds.[89] Its corrosion resistance is excellent, and it does not react with inorganic acids, alkaline solutions, molten alkalis, glass, or silica. Tungsten can be dissolved in hydrogen peroxide and mixtures of HF and HNO_3.[88,89]

8.6.3.2   *Production*

Tungsten is found in nature only as minerals with other elements. The economically important minerals used for metal production are scheelite, $CaWO_4$, containing 80.6 wt.% WO_3 and wolframite, $(Fe,Mn)WO_4$, containing about 76.4 wt.% WO_3. Scheelite can contain Mo as an impurity. Tungsten hydroxides, e.g. $H_2WO_4 \cdot H_2O$, can be present and negatively influence tungsten recovery. Typical ores contain 0.1–1 wt.% WO_3, which is processed to a concentrate with 65–75 wt.% WO_3.[88] Metal is produced via a combined hydrometallurgical and pyrometallurgical process route, and ferrotungsten can be produced directly from concentrate. The different processes for metal recovery have been described in detail.[90] Synthetic scheelite can be obtained as a by-product from tin concentrate processing. During leaching of tin concentrate with sodium carbonate or ammonia, the tungsten dissolves and can be precipitated as scheelite.[91]

Ferrotungsten containing 70–75 wt.% W or 75–82 wt.% W is produced by smelting concentrate, synthetic scheelite, or soft scrap with charcoal and coke in an EAF. Tungsten oxide is reduced carbothermically or silico-carbothermically. Metallothermic reduction with Al or Si is possible, though not preferred.[86,89] Formation of W carbide is prevented by controlling the CO/CO_2 ratio during the process, so a maximum of 1% carbon is present in the alloy. The liquidus temperature of ferrotungsten (80 wt.% W) is over 2500 °C, while the solidus temperature is about 1640 °C. Hence, a solid ferrotungsten ingot and a liquid slag are formed. First, in the refining stage high-grade ferrotungsten and a WO_3-rich slag are produced. Next, in the reduction stage the slag is processed to produce ferrotungsten with 50% W, and a slag with less than 1% WO_3. The ferrotungsten is recycled to the reduction stage.[89] Typical yield is 97–8%. The process requires per ton of 80% FeW: 1.6 t W concentrate, 150 kg coal, 80 kg fluorspar, 50 kg quarzite, 75 kg steel scrap, 80 kg bauxite, 120 kg electrodes, and 7500–8500 kWh electricity.[89] If a silico-carbothermical process is used, then the process has three stages.[89]

Tungsten metal is produced from concentrate, synthetic scheelite, and scrap containing 40–95% W. Typically hydro-metallurgy is used to convert the feed into ammonium paratungstate (APT) by alkaline digesting or pressure leaching the feed materials to create a sodium tungstate solution, followed by conversion to APT using solvent extraction or ion exchange. The APT, $(NH_4)_{10}(H_2W_{12}O_{42})_4 \cdot H_2O$, has 90–9% purity. Pyrometallurgy is used to calcine APT to (blue) tungsten trioxide in a rotary furnace at 500–700 °C, under reducing conditions. The duration and temperature of heating determines the micromorphology of the particles, which is important for W-powder production.[89] The conversion to metal is done under reducing conditions in push-type or rotary furnaces between 600 and 1100 °C, where hydrogen is flowing through countercurrent.[89] A metal powder with particle sizes in the range of 0.1–60 microns, depending on process conditions, is obtained. The powder is processed using powder metallurgy: sintering and compacting at 1800–2500 °C followed by working and rolling. For the production of massive metal (sheet, rod, and wire) a wide particle size range is required to give maximum compaction. A narrow size range is necessary when the particles are to be converted to carbides for hard metal production. The carbides W_2C and WC are produced by reacting metal powder with carbon powder and creating a eutectic WC and W_2C melt. After rapid quenching extremely hard particles are formed.[88] For increased shock resistance Co is used as a binder.

8.6.3.3 Applications

The world mine production is about 85 kt W content of which about 80% is mined in China. Other producers are Vietnam, Rwanda, Russia, Mongolia, Boliva, Austria, and North Korea, among others.[23]

Key applications for tungsten are in hard metals—as carbides—used for cutting and drilling (58%), steel and other alloys (25%), and tungsten mill products, e.g. wire, sheet, and rods (8%). The remainder is used as W compounds in the chemical industry in applications such as catalysts, pigments, and thin film deposition. The hard metals are used in the mining, oil and gas, and construction sector and general industrial use. When W is added to steel and other alloys, the physical and chemical wear and heat resistsance increases. Tungsten is a key ingredient in high-speed steel and superhigh-speed steel for specialist engineering applications; in superalloys used for flue gas desulfurization systems, heat exchangers, industrial furnaces, and jet engines; and in tungsten heavy metal alloy (WHA) containing 90–8% W alloyed with Ni, Fe, Cu, or Co for use in X-ray shields and counterweights. Non-ferrous tungsten alloys are typically with copper (WCu), and nickel (WNi), and for specialist applications MoW, WRe, TaW, and NbW.[88]

8.6.3.4 Recycling

Recycling plays an important role in the global tungsten supply. Over 30% of the world supply is estimated to be from recycled materials, be it old or new scrap. This includes carbides, metals and alloys, and catalysts. The main barriers to recycling are lack of collection, cost of impurity removal, dissipative use (e.g. pigments), and wear of carbide parts. As described in the production section (Section 8.6.3.2), scrap can be processed to APT and metal. High-speed steel scrap is recycled separately and and a typical steel scrap melt contains about 60–70% scrap.[88]

8.6.3.5   EHS and Sustainability

Tungsten, as wolframite, is considered a conflict mineral and subjected to conflict mineral regulations in the USA and EU. OECD guidelines are in place and a conflict-free smelter programme has been organized by the Responsible Minerals Initiative since 2010.[80] Tungsten is much less toxic than lead; hence, it is used as a substitute in ammunition, for example. However, it is not completely non-toxic and the environmental and health impacts are not fully understood.[88]

8.7   Manganese and Rhenium

Manganese is used primarily as an alloying element in steel alloys to improve strength, hardenability, and ductility; and in aluminium alloys to improve corrosion resistance. It is also important for several new battery chemistries. Rhenium is used in nickel-based superalloys, in catalysts, and in a number of special applications. Technetium (Tc) has essentially no application.

8.7.1   Manganese (Mn)

Manganese oxide rock has been used as pigment since the prehistoric era 30 000 years ago in painting, ceramics, glass, etc. around the world.[39] Manganese as an element was found in 1774 by Carl Wilhelm Scheele, and isolated as metal by the Swedish chemist Johan Gahn. Mn probably was used in iron making for many years, but formal usage started in the nineteenth century. Industrial scale production of ferromanganese (FeMn) was started in 1841 by Alexandre Pourcel in France, Siemens patented the use of FeMn as a steel additive in 1866, and FeMn production using an EAF began during 1890. Manganese metal was first industrially produced in 1898 by aluminothermic processing. The first patent for Mn in a battery application was filed in 1866.[92]

8.7.1.1   Physical Properties

Manganese is a grey, hard, and brittle metal which resembles iron (group 8) in properties.[69] It has six oxidation states: 2, 3, 4, 6, 7, and −1. The 2 and 4 states are industrially most important.[92] It has four crystal forms: α below 727 °C, β between 727 and 1095 °C, γ between 1095 and 1104 °C, and δ between 1104 and T_{melt} at 1244 °C.[92] Due to its brittleness manganese is not used in structural applications.[69] At room temperature Mn is inert to O, N, or H but at high temperature it reacts violently with O, S, and P, making it very useful for deoxidizing and desulfuring metals. Mn has a high tendency to oxidize at elevated temperatures, leading to oxides of different colours on steel, for example. It dissolves in acids, forming hydrogen and Mn (II) salts, and SO_2 gas in H_2SO_4 and H_2, N_2, and N_2O in nitric acid.

8.7.1.2 Production

Over 250 minerals with Mn exist. Only the oxides pyrolysite MnO_2, braunite $Mn^{2+}Mn^{3+}_6(O_8/SiO_4)$, and manganite $Mn_2O_3 \cdot H_2O$ are considered ores, containing 60–3 wt.% Mn.[69] The ores are classified based on their application area. Battery grade is used for producing Mn compounds for batteries and contains 44–54 wt.% Mn, less then 0.05 wt.% of Cu, Ni, Co, or As, and has specifications with regard to physical properties. Chemical grade ore contains 35–43 wt.% Mn. These two grades are processed hydrometallurgically. Metallurgical grade is used for producing ferromanganese alloys, contains 28–48 wt.% Mn and less than 0.1 wt.% P, and has a Mn/Fe ratio of 7.5 and specifications for Al_2O_3, SiO_2, CaO, MgO, and S contents, as these can end up in the alloy and negatively impact steelmaking processes.[69] A possible future source of Mn are ocean floor nodules, and processes for the mining and recovery of Mn from them are being developed.

Ferromanganese (FeMn) alloys (75–92 wt.% Mn) are available with different carbon contents: high 7.5%, medium < 2%, and low 0.5–0.75%, where high is most used. Furthermore, silicomanganese (SiMn) alloys, 65–75 wt.% Mn, 15–25 wt.% Si, < 2.5 wt.% C, and ferrosilicomanganese (FeSiMn), 58–72 wt.% Mn and 23–35 wt.% Si, are produced.[92]

In the pyrometallurgical production processes, similar to iron, the reduction of MnO_2 to Mn takes place by first forming Mn_2O_3, which is then reduced to Mn_3O_4, which in turn is reduced to MnO, and then metallic Mn is formed. The first two reactions are exothermic; the third reaction is exothermic when CO is used and endothermic when C or methane (CH_4) is used. The last step requires strongly reducing conditions. Also FeO and P_2O_5 are reduced and dissolve in the metal. SiO_2 will be reduced if the temperature is increased. Besides the CO/CO_2 ratio in the process gas, slag chemistry plays an important role. The MnO content of the slag is influenced by temperature and the basicity ratio, $(CaO + MgO)/SiO_2$. When (ferro)silicon is used as a reducing agent the slag composition impacts its effectiveness. Sufficient lime needs to be present in the slag to ensure the activity of SiO_2 in the melt is low, forcing the reduction of Mn oxide to Mn metal to take place.[92]

In the past blast furnaces, similar to iron making, were used to produce ferromanganese. Today only the submerged arc furnace is used for both carbothermic and silicothermic reduction. To produce high-carbon FeMn the feed materials—manganese ore, limestone, dolomite, or silica—are fed from the top into the submerged EAF. The electrodes are Søderberg electrodes, which are also used in the production of primary aluminium. The furnace is lined with carbon blocks. The feed is reduced by the gas travelling up through the burden. The CO-rich gas, leaving the furnace at 290 °C, is used to generate electricity, or can be used to pretreat the ores.[93] The molten high-carbon FeMn and slag are tapped separately. The slag, with about 30 wt.% MnO, is used to produce FeSiMn by silicothermic reduction using quartz in another EAF. FeSiMn can also be produced by directly reducing metallurgical grade ore with high-silica ore and coke in an EAF.[92] Nelson provides a detailed overview of industrially used submerged arc furnaces.[83]

FeSiMn together with metallurgical grade Mn ore is the feedstock for medium- and low-carbon FeMn, or industrial Mn metal. It is obtained by silicothermal reduction in an arc or ladle furnace. To produce medium-carbon FeMn also high-carbon FeMn can be used and decarburized by blowing in oxygen in an oxygen furnace. This method is not suitable for low-carbon FeMn due to high losses of Mn as MnO in the off gas.[92]

Electrolytic manganese metal (EMM) and dioxide (EMD) are produced from battery and chemical grade ores in three stages. First, the ore is dry-milled to obtain sufficient reactivity. After milling, 80–90% of the particles are smaller than 75 μm and the surface area is 2–5 m^2/g. Then Mn in the oxide is converted from 4-valent to 2-valent (MnO), which is soluble in acid, by calcination under reducing conditions with a carbon source at 850–1000 °C. Iron and other elements are reduced as well. Next, the reduced ore is dissolved/leached in spent electrolyte, containing 2–5% H_2SO_4, where both Mn and Fe(II) as well as other compounds dissolve. These impurities are removed by purification using air oxidation to remove Fe, pH adjustment to precipitate other metals as sulfides, although some Mn sulfide coprecipitates, and aluminium hydroxide, $Al(OH)_3$, to remove soluble silica. The electrolyte contains about 10% manganese sulfate, and ammonium sulfate is used as a supporting electrolyte. Lastly, electrowinning of the metal takes place in a diaphragm cell with cathodes made of Ti, stainless steel, or proprietary alloys, and anodes of Pb–Ca or Pb–Ag alloy or coated with platinum group metals. The addition of SO_2 to the electrolyte ensures deposition as α-Mn, which is hard and brittle. SeO_2 can be used as well, though this results in a few hundred ppm of Se in the product. At the cathode, Mn metal is deposited and formation of H_2 from water takes place. At the anode water is electrolysed to O_2 and H^+, and MnO_2 is formed. At the cathode the current density is 500 A/m^2, 4% $MnSO_4$ in solution, pH is about 7, and the cell potential is 5–7 V. The current efficiency is 42–62% and the power consumption 9 000–12 000 kWh/t Mn. After about 12–30 h a layer of 1.5–2 mm metal, > 99.5% purity, is obtained and the deposit is stripped. After washing, drying, and lowering absorbed hydrogen by heating, the metal (EMM) is sold as flake, powder, or grains, or sometimes as briquettes.[92]

EMD is also produced by electrolysis. After milling, calcining, and dissolution the 75- to 160-g/L $MnSO_4$ sulfate solution is electrolysed. MnO_2 forms at the titanium anode, current density > 150 A/m^2, > 90% yield, and is stripped after 14–20 days when the deposit is 20–30 mm. EMD contains 91 wt.% MnO_2, 3–5 wt.% moisture, < 0.02 wt.% Fe, and < 0.001 wt.% of Pb, Cu, and Co.

8.7.1.3   Applications

Manganese world mine production in 2019 was 19 Mt Mn-contained, mainly from South Africa (29%), Australia (17%), Gabon (13%), and China, Brazil, and Ghana—about 7% each.[23] China is the leading consumer of concentrates, about 2/3, and produces about 50% of the Mn alloys, followed by India, South Africa, Ukraine, and Norway.[84]

About 90% of all Mn is used in steel manufacturing and added as FeMn, and more and more as Mn metal,[92] to fix sulfur, deoxidize, and improve mechanical properties. Mn additions increase hardenability, and lower the martensitic start temperature, for example with 200 °C with 5% Mn addition.[92] Mn is found in nearly every kind of steel, as the Mn consumption in Europe shows. About 25% is used in construction steel, about 5% is domestic appliances steel, and about 10–15% each in automotive, mechanical engineering, structural works, metalware, and tubes. Mn is also used as an alloying element in Al–Mn alloys for cans and food packaging to improve corrosion resistance, and in copper alloys to improve strength and hot workability.[84]

Besides chemical applications Mn is used as cathode material in alkaline batteries, and as a cathode material constituent in lithium ion batteries of the nickel–manganese–cobalt (NMC) type and the lithium–manganese–oxide (LMO) type used in electrical mobility and energy storage applications. Demand in this segment is expected to increase strongly.

8.7.1.4   Recycling

Manganese is recycled as part of the steel alloys it is used in, and remains part of the steel alloys. Hence, specific recovery of Mn is not taking place. Mn is separately recovered during the recycling of Li-ion batteries from (hybrid) electric vehicles.

8.7.1.5   EHS and Sustainability

Electrolytic manganese metal (EMM) that will be milled to fine powders must have an H content below 0.0005% to minimize explosion risk. A hydrogen explosion can initiate a manganese dust explosion.[92]

8.7.2   Rhenium (Re)

Rhenium was discovered in 1925 by Ida Tacke, Walter Noddack, and Professor Otto Berg and first isolated from Norwegian molybdenite. It was the last refractory metal to be discovered. Industrial production started in Germany in the 1930s. Technology advanced when the production from flue dusts and gas from molybdenite roasting was improved by Kennecott in the USA.[94] Rhenium production started in 1960s in the USA.[95]

8.7.2.1   Physical Properties

Rhenium is a ductile, highly corrosion-resistant silvery-white metal. It has a high density, 21.04 g/cm^3, and has the second highest melting point of all metals: 3180 °C. It is a high-melting metal together with metals from groups 5 and 6: Mo, W, Nb, and Ta with similar properties. In contrast to other metals Re does not have a ductile to brittle transition temperature; it is ductile from subzero to very high temperatures. Rhenium's valences are commonly 7+, 6+, 5+, and 4+ and it can change valence easily. The chemical properties are similar to those of manganese. It also shares properties with the platinum group metals. It is nearly insoluble in hydrochloric and hydrofluoric acid, while dissolving in oxidizing acids to form perrhenic acid. Rhenium does not form carbides, though forms volatile oxides in oxygen at high temperatures. It can form fluorides, chlorides, silicides, borides, and phosphides.[94]

When alloyed with Mo or W the alloy has the best properties of both and none of the bad. It reduces brittleness, increases the recrystallisation temperature, ductility, and ultimate tensile strength[95] caused by alloy softening. W–Re alloys with 25–30 wt.% Re have good cold ductility. Re is very soluble in tungsten, 28 wt.% at 1600 °C, and vice versa. Tungsten's solubility is 11 wt.% at 1600 °C. Mo–Re with up to 50 wt.% Re has high tensile strength and good elongation over a wide temperature range from 0 to 1800 °C.[94]

8.7.2.2   Production

Rhenium is a very rare element. Its mineral is found in only a few places in small, not commercially viable quantities. Rheniite (ReS_2) is found at active volcanos in Japan and the western Pacific, for example. Industrially relevant minerals are molybdenite (MoS_2), which can contain 10–1000s ppm Re, and castaingite ($CuMo_2S_5$), which can contain up to 1 wt.%

Re.[95] Most rhenium is produced as by-product from molybdenum production, which is also a by-product from copper smelting. A currenly unused source is the flue dust of Ni–PGM smelters, as Re is present in Ni–Cu–PGM deposits.[95]

The production process of Re metal consists of three main steps. During the roasting of molybdenite (500–700 °C) or the smelting of copper ore, rhenium is oxidized to volatile Re_2O_7 due its high vapour pressure, and exits with the SO_2 containing off-gasses. Scrubbing dissolves over 90%[94] of the Re_2O_7 as crude perrhenic acid, $HReO_4$, and solid impurities such as Mo oxides, sulfides, and selenium are removed. Part of the Re forms non-volatile, stable perrhenates with alkali and alkaline-earth metals present in the feed material, lowering the yield.[94]

Next, rhenium is recovered via solvent extraction or solid bed ion exchange. It is stripped from the solution, crystallized as ammonium perrhenate (APR), NH_4ReO_4, to obtain 99.95% purity. APR is a salt traded as a commodity in basic grade or catalyst grade. The latter contains minimum 68.5 wt.% Re and has stricter impurity limitations to avoid poisoning of catalysts.[96] An alternative process is the Kennecott Molybdenite Autoclave Process (MAP) in use since 2010. The high-pressure oxidation of molybdenite gives a higher recovery and purity of molybdic oxide and ammonium perrhenate.[95]

Lastly, APR is converted to metal by reducing with hydrogen gas in a counter current flow tube furnace. Reduction to ReO_2 occurs at 300–50 °C, followed by reduction to metal at 800 °C. The powder is pressed into pellets, sintered at 1000–1500 °C under hydrogen atmosphere to remove gasses like oxygen and improve physical integrity, and shipped to alloy producers. Re-metal products are made by pressing the powder into bars or rods, followed by sintering followed by wire drawing or rolling into sheets and plates.[95] It can be produced at almost its theoretical density by hot isostatic pressing (HIP) and electron beam or electron arc melting. Re can also be deposited on parts via chemical vapour deposition.[94]

8.7.2.3   *Applications*

Global Re-metal production was 49 t in 2019 from Chile (55%), Poland (19%), the USA (17%), China (6%), and Kazakhstan (2.5%).[23] Recycling contributes about 20 t/y[42] to meet a global demand of 60–5 t/y.

The main use of Re (83%)[42] is as an alloying element in superalloys for high-pressure gas turbines (HPT) in aerospace and land-based power generation. As much as 3–6% Re is added to nickel-based superalloys, which are casted into single-crystal turbine blades. Re increases the creep resistance through formation of the gamma prime (γ') phase and increases blade longevity. This enables higher operating temperatures, increased fuel efficiency, and reduction of nitrous oxide emissions to the atmosphere.[95] Furthermore, W–Re and Mo–Re alloys are used in the manufacturing of thermoelements. W–Re alloys (W5Re, W10Re) are used to manufacture rotating X-ray anodes for X-ray diagnosis and computer tomography. The alloys are applied to the molybdenum anode base as a coating.[94]

Re is also used in catalysts for the petroleum industry (17%), because of its range of valencies and its ability to change valency easily. Bimetallic reforming catalysts contain 0.3 wt.% Re and 0.3 wt.% Pt. Possible other applications are in silver catalysts for ethylene oxide production and in gas-to-liquid (GTL) catalysts based on Fe, Co, or Ni.[95]

8.7.2.4 Recycling

Recycling of Re-containing scrap and products is common practice, driven by the value of the material. Spent catalysts are recycled mainly for the platinum value. The Al_2O_3 catalyst carrier and Re are dissolved in sodium hydroxide or sulfuric acid. Pt remains in the residue. Ion exchange is used to recover Re.[94] Companies in the USA and Germany recover about 15 t/y, which is in a closed, continuous loop with catalyst manufacturers.[95] Alternatively, a pyrometallurgical fusion process can be used. The obtained alloys are treated by dissolution and separation.

Superalloy scrap, such as risers, grindings, and old scrap, are recycled by removing Re from the alloy and converting it to APR. New scrap and clean alloys are directly remelted.[95]

8.7.2.5 EHS and Sustainability

Rhenium has two isotopes. About 60% of rhenium consists of the ^{187}Re isotope. It is radioactive with a half-life of 4.3×10^{10} years; however, the radiation cannot penetrate the human skin, hence poses little risk.[95]

8.8 Iron

Iron and the iron group play the dominant role for a number of reasons. Iron is number four in relative abundance in the Earth's solid crust and has important unique chemical and metallurgical properties. The iron group elements can be melted by standard metallurgical techniques, their densities are lower than for the two other subgroups, and their relative abundance is much higher (and the price much lower).

We give here only a very brief overview of iron and steel production methods and uses. The fundmentals of steel refining are also illustrated in chapters in this book. A detailed treatment would require a separate chapter, or perhaps a separate book. The reader seeking additional information can easily find excellent reviews.[97,98] For the reader looking for a 'deep dive' into iron and steel technology, consulting *The Making, Shaping and Treating of Steel*[99] is recommended.

8.8.1 Iron (Fe)

8.8.1.1 Physical Properties

The main impurities in iron and steel are the non-metallic elements O, H, N, S, and P. These are 'surface active' and tend to accumulate in the interdendritic liquid during solidification. This leads to the formation of non-metallic inclusions, oxides, sulfides, oxysulfides, and so on, with iron or with the various alloy components. Porosity may be caused by the nucleation and growth of H_2 or CO bubbles.

Carbon is also another element that is always present, but its content is controlled carefully to create a wide range of mechanical properties. Pure iron is soft. Adding carbon increases strength by forming carbides, usually Fe_3C. Carbon also dissolves in liquid iron, forming a eutectic at 4.3% C and $1148\,°C$. This temperature is nearly $400\,°C$ less than the melting point of pure iron ($T_{melt} = 1539\,°C$), which makes it much easier to makes net-shaped castings.

High-carbon iron materials are usually called cast iron. They are strong and have excellent heat resistance, but somewhat low ductility. The low carbon alloys (less than about 0.5% C) are called steels, and have much better ductility and good fatigue resistance.

The wide range of material properties available by varying carbon content, together with the abundance of iron ores and the relatively low cost of extraction, have made iron and steel an extremely important part of our modern technological society.

8.8.1.2   Production

The main minerals in iron ore are hematite (Fe_2O_3), magnetite (Fe_3O_4), and goethite (FeOOH). The process from ore to steel consists of four main steps: iron making, steelmaking, refining, and casting.

Iron ore concentrates that are not in the form of lumps must be agglomerated before they can be used in the blast furnace. The concentrate is either sintered or converted into pellets by adding slag formers and carbon. Besides chemical composition requirements, the sinter and pellets have to be strong enough to withstand handling, e.g. feeding into the blast furnace, as well as maintain their strength while descending down the blast furnace with the weight of the rest of the burden on top, so the pellets will only melt when reaching the melting zone in the blast furnace. The blast furnace is filled with alternating layers of pellets/sinter and coke lumps. Preheated air with pulverized coal (or sometimes oil) at 1000–1100 °C is blown in at the bottom of the blast furnace, and reacts with coke to form a CO/CO_2 gas mixture, creating a hot zone of 1900–2000 °C where it enters. The gas travels countercurrently from the bottom of the furnace up through the burden, reducing the iron ore to iron along the way. After leaving the furnace the gas is cleaned and can then be used for pre-heating, as it contains combustible CO. The porosity of the burden is a key process parameter. Too much porosity means there is not enough time for the reaction to take place. Too little porosity results in the gas mixture not travelling up through the burden.

The reduction of iron ore to Fe takes place by first forming Fe_2O_3, which is then reduced to Fe_3O_4, wich in turn is reduced to FeO and then metallic Fe is formed. The first two reactions are exothermic; the third reaction is endothermic and requires strongly reducing conditions. The molten iron (pig iron) with slag on top accumulates at the botton of the furnace, and are tapped separately. After crushing and removal of iron droplets, the slag can be used as construction material. The pig iron, containing among other elements carbon, silicon, sulfur, and oxygen, is sent to the basic oxygen furnace (BOF), where the pig iron is converted into steel. Alloying elements are added and impurities removed. Up to 30% of steel scrap can be added as cooling agent to the BOF furnace. Final adjustments are made using ladle metallurgy before casting the metal. The slabs or billets are converted to the desired shape and thickness by hot rolling, cold rolling, and coating with tin (tin plate), zinc (galvanizing), or other coatings.

8.8.1.3   Applications

Global crude steel production in 2019 was about 1869 million tons, with China, India, Japan, the USA, Russia, and South Korea as major producers. Steel is produced in tens of other countries all around the world. Nearly three-quarters is produced via the BOF route; most of the remainder is via EAF furnaces.[100] About half of this production is used in building and infrastructure applications. About 15% is used in mechanical equipment, while the automotive industry accounts for about 12% of global consumption. Metal products (10%), other transport applications (5%), domestic appliances (2%), and electrical equipment (3%) make up the remainder.[100]

There are currently over 3500 steel alloys. Many of them have been developed in the past 20 years. Here no attempt is made to review the various Fe-based alloys, and reference must be made to the specialized literature.

8.8.1.4   *Recycling*

Recycling of iron and steel has been done for centuries. The large variation of alloys and variation in length of the use influence the scrap mix available for recycling. There is limited alloy-specific sorting, so many alloying elements are not functionally recycled. Steel alloys with a high content of alloying elements can be used instead of ferroalloys. Steel scrap can be added to the BOF as coolant. Most of the steel scrap is recycled via the EAF route, which can handle a 100% solid charge. Recycling/remelting or iron and steel has a much lower energy consumption and CO_2 emission than primary production.

After end-of-life products (e.g. cars, white goods, or cans) are collected, shredding and sorting processes take place. It is vital to remove the metals that are detrimental to the steel quality, as refining possibilities in the BOF and EAF are limited. In particular, copper needs to be removed. In many alloys the Cu content needs to be below 0.1%. Typically the copper content in scrap from shredder plants is about 0.3%.[101] Galvanised and tinned steel can be treated to remove zinc and tin before melting. The scrap is usually blended before it is fed to the EAF. The scrap is melted together with flux, to form a basic slag. The slag collects some of the impurities coming along with the scrap such as Al, Ca, Mg, Si, Ti, and Zr, as well as alloying elements such as Cr, S, and P that go partly to the slag. Elements such as Co, Cu, Mo, Ni, Sn, Ta, and W dissolve in the liquid metal. Zinc and lead go to the off-gas. Direct reduced iron (DRI) or hot briquetted iron (HBI) is added, among others, to dilute impurities. After tapping, the metal composition may be adjusted in the ladle before casting into products.

8.9   Cobalt

8.9.1   Cobalt (Co)

Cobalt compounds have been known and used since the Bronze Age around the world, to give a blue colour to glass, glazes, and ceramics. Cobalt metal was first isolated by Georg Brandt in 1735, and in 1780 Torbern Bergman showed the metal was a new element. Its first use as metal occurred in 1907 in Co–Cr alloys named stellites. It was later used as an alloying element in stainless steel. In 1930 it was discovered that a cobalt addition to alloys of nickel, iron, and aluminium enhanced their permanent magnetic properties. Today cobalt is used in a wide range of alloys—steel, cemented carbides, superalloys—and compounds for pigments, catalysts, and rechargeable batteries, among others.[102,103]

8.9.1.1   *Physical Properties*

Polished cobalt is silvery-white with a faint bluish tinge. The close-packed hexagonal ε-form is stable below 427 °C while the face-centred cubic α-form is stable at higher temperatures.

It is ferromagnetic up to 1121 °C, the highest known Curie point of any metal or alloy. Cobalt has 2+ and 3+ as oxidation states and a high melting point of 1439 °C.[102,103] It is rarely used as a pure metal. Most commonly it is used as an alloying element or a compound, for its chemical or wear resistance, magnetic properties, or high-temperature strength.[102]

8.9.1.2 Production

Even the best cobalt ores contain only low concentrations of the element. Cobalt bonds with Fe, Ni, Cu, and S. Cobalt is found as colbaltite $(Co,Fe)AsS$, carrolite $CuCo_2S_4$, linnaeite $(Co,Ni)_3S_4$, and skutterudite $(Co,Fe,Ni)As_2$. Cobalt carbonate $(CoCO_3)$ and hydroxide are found as well.[102] Most cobalt is recovered as a by-product from the ores of other metals, particularly nickel (50%) and copper (35%), and the remainder from arsenide ores.[104] Ocean floor nodules and crusts may be a future source of cobalt, containing about 0.74 wt.%.

Nickel sulfide ores typically contain about 0.05 wt.% Co.[104] Co follows nickel in the pyrometallurgical smelting–converting flowsheet of nickel, until the nickel refinery. There cobalt and nickel are dissolved in chloride or sulfate solutions and separated by solvent extraction. The cobalt chloride or sulfate solution is sent to the cobalt refinery. Nickel laterite ores contain about 0.025–0.18 wt.% Co[102] and only the limonite (goethite) section is processed to recover cobalt. Cobalt follows nickel in the hydrometallurgical processing of limonite, and is separated from nickel in the nickel refinery.

Copper–cobalt ores are mainly found in Central Africa, and contain dolomite and quartz as gangue minerals. The ores have an 'oxide' layer of $CoO \cdot OH$ and $CoCO_3$ with $CuCO_3 \cdot Cu(OH)_2$ and $CuSiO_2 \cdot H_2O$ that typically contains 0.3 wt.% Co and 3 wt.% Cu, and a sulfide layer containing mainly Co_2CuS_4.[102] The oxide layer is processed in a whole ore leaching (WOL) process. After grinding, the ore is leached under reducing conditions with sulfuric acid and SO_2 gas to dissolve cobalt and copper. The solution then goes to a solvent extraction process to remove copper, which is recovered by electrowinning, and the solution is further purified. By bubbling SO_2 and air into the solution and increasing the pH, Fe, Al, and Mn are removed as hydroxide precipitates. Any remaining Cu and Mn is removed by forming precipitates. From the purified solution cobalt is precipitated as hydroxide, $Co(OH)_2$, and then sold or refined to metal.[102,104]

Cobalt recovery from arsenide ores is under development, e.g. Co arsenide in Morocco, Co–Au–Bi arsenide in the Northwest Territories, Canada, or Co–Cu–Au sulfide/arsenide deposit in Idaho, USA.[105,106]

Cobalt refining is similar to nickel refining. Electrowinning of 99.95% pure cobalt is done from chloride or sulfate solutions. The main difference is that for Co-sulfate electrowinning no diaphragm between the anode and cathode is required. The current efficiency is about 90%, cell voltage is about 3.7 V, and the current density is 200–50 A/m^2. Hydrogen reduction of cobalt from ammoniacal sulfate solutions is similar to the process used for nickel, although cobalt reduction goes a bit faster. The recovery efficiency is about 99%. Detailed descriptions of the processes are available.[103,104,107]

8.9.1.3 Applications

The world mine production of cobalt is about 140 000 t/y, mainly produced in Congo (70%), and Russia, Australia, the Phillipines, Papua New Guinea, New Caledonia, Canada, Cuba,

Madagascar, Morocco, and South Africa with each less than 5%.[23] China is the largest producer of refined cobalt (50%), and Finland, Canada, Belgium, Japan, Norway, and France are major refining countries.[102]

The major application of cobalt in 2016,[42] about 50% of total use, was in rechargeable batteries for portable electronics, handheld equipment, electrical vehicles, and energy storage. Depending on the battery composition, the cobalt content varies. It is 60% for NMC Li-ion batteries, 9% for NCA Li-ion batteries, and up to 15% for nickel metal hydride batteries. Nickel–cadmium batteries contain 1–5% Co.[102] Furthermore, cobalt is used in superalloys (18%), carbides and diamond tools (8%), ceramics and pigments (6%), catalysts (5%), other alloys (4%), and magnets (3%).[42] In superalloys, high-speed steels and high-strength alloys and medical alloys (e.g. Co–Cr–Mo–C), cobalt is used as an alloying element. It is a key ingredient for samarium–cobalt high strength magnets and AlNiCo magnets. Typical catalyst applications are GTL catalysts, polymerization of plastic resins, and desulfurization of crude oil.[102]

8.9.1.4 *Recycling*

Recycling of cobalt contributes to about 20% of consumption.[104] Scrap metal, spent catalysts, and batteries are typical sources of recycled material, as use as pigment is dispersive. Superalloys and hard metals (carbides) are recycled by those sectors. Catalyst and battery recycling generally uses the cobalt industry infrastructure.

The increasing demand for cobalt also drives the recycling of rechargeable batteries, as this is a major application of cobalt. In Europe the collection and recycling of batteries is governed by the Battery Directive.[18] Achieving a high collection rate of portable devices, including the battery, is challenging. Furthermore, there also remain business, technical, and upscaling challenges to be solved, especially in the recycling of batteries from electrical vehicles.[108]

8.9.1.5 *EHS and Sustainability*

Overexposure to cobalt can be avoided by practicing good workplace hygiene measures and wearing the appropriate personal protection equipment. Inhalation of high concentrations of cobalt is linked to lung disease.[102] Responsible sourcing of cobalt is emphasized by the Responsible Minerals Initative.[80]

8.10   Nickel and Platinum Group Metals

Nickel is mainly used as an alloying element in steel. The platinum group metals (PGMs) have catalysis as their main application area. Both metals are essential in the energy transition: nickel is used in rechargeable batteries and PGMs are applied in fuel cells. Nickel and PGMs are often found together in ores and processed together.

8.10.1   Nickel (Ni)

Nickel has been found in metallic artefacts from over 2000 years ago. Nickel-containing alloys, such as 'white copper' with 40% Cu, 32% Ni, 25% Zn, and 3% Fe were made in China for centuries. It was identified and isolated as an element in 1751 by the Swedish chemist Cronstedt, though its properties were not described until 1804 by Richter.[109] In the nineteenth century nickel started to be used for plating and in coins, alloyed with copper, from 1857 in the USA. Stainless steel, discovered in the early twentieth century, and the benefical role of Ni in it became a key application for nickel. Today, nickel is a key ingredient of rechargeable batteries used in electric vehicles for clean mobility, in storage of renewable energy, catalysis and high-corrosion resistance, and high creep strength alloys. Together with its applications in a wide range of other sectors, nickel is a key element for a sustainable future.

8.10.1.1   Physical Properties

Nickel is a silvery-white metal with high electrical and thermal conductivity, and its alloys with copper are substantially white. The ability of nickel to form strong ductile alloys with many metals, including iron, chromium, cobalt, copper, and gold, is widely utilized in industry. It has a high melting point, 1453 °C, is magnetic at room temperature albeit less than iron, and has catalytic properties. The Ni^{2+} ion forms complexes easily, in particular with ammonia. Nickel is resistant to corrosion by fluorine, alkalis, non-oxidizing acids, air, sea water, and a variety of organic materials. It also easily absorbs hydrogen, especially as fine powder.[109]

8.10.1.2   Production

Nickel is found as sulfides and oxides (or silicates), or arsenide, never as native metal. Of economic relevance is the sulfide pentlandite $(Ni,Fe)_9S_8$ found together with nickeliferous pyrrhotite, Fe_7S_8 with 0.2–0.5% Ni in solid solution, copper sulfide, and ilmenite $(FeTiO_3)$. Cobalt, PGMs, gold, silver, selenium, and tellurium are found as solid solutions in the sulfides and can be recovered as by-products.[109] A typical ore contains 1–2% Ni, 1–2% Cu, 0.01–0.05% Co, and 0.03–0.06% PGM. The other main source of nickel are laterite deposits, which are located close to the surface, contain typically 1–2% Ni and 0.1% Co,[104] and consist of minerals formed after weathering of peridotite; thus, the deposit composition changes with depth. The upper layer, called limonite, contains goethite mixed with nickel hydroxide $(Fe,Ni)O(OH)\cdot n\,H_2O$, and is used to recover nickel metal. The lower layer, called saprolite, consists of garnierite $(Ni,Mg)_6Si_4O_{10}(OH)_8$ and is used to produce ferronickel.[104] Nickel extractive metallurgy has been extensively discussed.[107]

Nickel sulfide ore is concentrated by grinding and flotation to produce a concentrate with 15% Ni, 0.2% Co, and PGMs, if present. A copper sulfide concentrate is obtained as by-product for processing in a copper smelting flowsheet. The Ni–Co concentrate is smelted into a matte containing about 40% Ni, 0.5% Co, 25% Fe, PGM, and the remainder sulfur. About 75% of Ni-sulfide smelting is done using a flash smelting furnace. The other quarter is processed using a fluidized bed roaster combined with an electric furnace.[104,109] Flash smelting combines roasting and smelting at about 1300 °C and uses the heat from the exothermic reaction of the feed with oxygen from enriched air to smelt the particles. This creates a nearly autogeneous process. Dry concentrate (2000 t/d)[104] and silicate flux are blown into the vertical reaction shaft and reacts in a few seconds, before reaching the settling area. Here the fayalite slag and matte are separated and tapped through separate tapholes. The slag contains nickel matte droplets, which are removed in an electric settling furnace. Off-gas, containing 20–50 vol.% SO_2 is treated to remove dust before sulfuric acid is produced as by-product. The matte is converted at 1200–50 °C in a Pierce-Smith converter, also used in copper converting, to reduce the iron content to about 1 wt.% by oxidation and collecting it in a iron-silicate slag. The Ni content and Co content in the matte increases to 50–60 wt.% and 1 wt.%, respectively.[104] Furthermore, the matte contains the PGMs. The slag is processed in an electric settling furnace to remove matte droplets and then discarded. Alternatively, a direct nickel flash smelting process can be used, which produces a high-grade nickel matte.[110]

Limonite concentrate from laterite deposits, 1.2 wt.% Ni and 0.1 wt.% Co, is treated by hydrometallurgical methods. First, it is treated in a high-pressure autoclave leaching (HPAL) process, using sulfuric acid at 40 bar and 250 °C to dissolve Co and Ni. The autoclave typically is 20 m long, 2 m in diameter. The solution is purified, removing Al, Cr, Si, Fe, Cu, and Zn, and then precipitated in an autoclave using H_2S gas as Ni–Co sulfide containing $\pm$ 55 wt.% Ni, $\pm$ 5 wt.% Co, $\pm$ 1 wt.% Fe.[104] The sulfide is treated in the nickel refinery.

Refining to metal from sulfide feeds such as matte and Ni–Co sulfide from laterite processing consists of four steps: (re)leaching in chlorine, ammonia/oxygen, or sulfuric acid/oxygen; followed by purification to remove Fe, As, Pb, Mn, Ca, and Zn; solvent extraction to separate Ni and Co in separate sulfate or chloride solutions; and lastly electrowinning or hydrogen reduction of those solutions to obtain 99.9% pure Ni metal. Note that during chlorine leaching, a copper (CuS) precipate, including PGMs, is formed. This is further processed in a copper/precious metals plant. Electrowinning takes place in separate cells, which are created by placing a diaphragm, usually a bag around the cathode (sulfate electrolyte) or anode (chloride electrolyte). The electrodes are 0.8 m wide and 1 m deep. Plating occurs for 7 days, 99% current efficiency, 3 V cell voltage, and 240 A/m^2 current density. Chloride electrowinning uses as a cathode stainless steel blanks or starter sheets of nickel on titanium, and titanium anodes coated with Ru or Ir oxide. Sulfate electrowinning uses lead alloy anodes instead of titanium.[104] Hydrogen reduction requires air/ammonia leaching or the addition of ammonia to sulfate solutions. In an autoclave batch process nickel is precipitated on nickel metal seed particles by bubbling in hydrogen at 180 °C, 30 bar. The process is repeated 60–80 times till about 20–50 ton metal is produced. Ammonium sulfate fertilizer is produced as a by-product.[104] Electrorefining of nickel matte anodes is possible and still practiced, though mainly replaced by electrowinning, as well as carbonyl nickel refining.[104]

Ferronickel, 20–40 wt.% Ni and 80–60 wt.% Fe, is produced using garnierite ore from laterite deposits by first dewatering and then calcining and partly reducing the concentrate together with about 5% coal in a continuously operated rotary kiln, 100 m long and 5 m diameter. The burner is located at the bottom end of the kiln, so reducing gas travels counter current to the feed. It leaves the kiln at the top and dust, $\pm$15% of feed, is captured and recycled into the kiln. During the process the H_2O in the minerals is removed, about $\frac{1}{4}$ of the nickel and 5% of iron is reduced to metal, and most Fe^{3+} is reduced to Fe^{2+}. The calcine leaves the kiln at 900 °C, containing 1.5–3 wt.% Ni, 15 wt.% Fe, 1–2 wt.% C, and Mg and Si oxides.[104] It is directly fed from the top into the electric furnace. Nelson provides a detailed overview of industrially used electrical furnaces.[83] As nickel and iron oxide are almost equally easily reduced, the result of melting is always a nickel–iron mixed alloy. Some of the silica and Cr_2O_3 are reduced as well and dissolve in the ferronickel.[104] Any Co present has no economic value. In the furnace molten ferronickel collects on the bottom of the furnace, the lighter SiO_2–MgO–FeO slag with 0.1–0.2% NiO, 3 kg/m^3 density, and T_{slag} of 1550 °C floats on top. CO-rich gas leaves the furnace at 900 °C and enters the afterburner and off-gas treatment system. The crude FeNi and slag are tapped separately. The slag may be used as building material or is landfilled.[104] Crude ferronickel is refined similar to iron. First, under reducing conditions soda ash, lime, or calcium carbide is added to lower S from 0.4 to < 0.03 wt.%. Then under oxidizing condition in a oxygen-blown converter or ladle with oxygen lance, the P content is reduced from 0.06 to < 0.02 wt.% and C and Si are removed. It is tapped and granulated or cast in pieces.[104,109] In China also nickel pig iron is produced by adding saprolite ore to iron-making furnaces.[104]

Nickel metal, crude nickel sulfate, and crude nickel chloride form the starting materials for the production of specialty nickel compound manufacturing including nickel sulfate, chloride, hydroxycarbonate, nitrate, oxide, hydroxide, sulfide, etc.,[111] which can have application-specific requirements such as impurity limits and particle size and shape, for example for battery materials.

8.10.1.3   Applications

World mining production of nickel is about 2.4 Mt per year,[23] of which about 60% is from sulfide ores.[104] The major mine production countries are Indonesia (30%), the Philipines (15.5%), Russia (10%), New Caledonia (8%), Canada and Australia (6–7% each), and China, Brazil, Cuba, and the USA, each less than 5%.[23] Nickel refinery production mainly takes place in China, Russia, Japan, Canada, Australia, Norway, Brazil, New Caledonia, Finland, Colombia, France, and others.[84]

Nickel is used mainly as an alloying element in stainless steel (70%), nickel- and copper-based alloys (8%), alloy steels and castings (8%), plating (8%), and batteries (5%).[111] Additions of 8–10% Ni to stainless steel as an alloying element (together with Cr) improves formability at low temperature, weldability, ductiliuty, and corrosion resistance while maintaining high-temperature strength. Furthermore, Ni-containing stainless steels are non-magnetic and make up 75% of all stainless steel produced.[111] Nickel alloys include wrought nickel, Fe–Ni alloys used as soft magnetic materials, and defined thermal expansion materials. The Ni–Cu, Ni–Mo, Ni–Cr, Ni–Cr–Fe, Ni–Cr–Mo, and Ni–Cr–Co alloys are known for their high-corrosion and high-temperature resistance in a wide range of environments. Ni–Ti alloys with 45 wt.% Ti have shape memory properties. Copper–nickel alloys, most commonly with 10 or 30 wt.% Ni, are extremely resistant to corrosion by seawater, and have good thermal conductivity and low macro-organism attachment in marine environments. These alloys are used in offshore oil and gas, shipbuilding, desalination, and power generation.[111] Nickel plating or coatings are applied to metals and plastics to provide corrosion and wear resistance; brightness, lustre, and appeal; and adhesion for other coatings like chromium.

Rechargeable batteries contain nickel in the cathode material. It started with the Ni–Cd batteries, followed by Ni-metal hydride batteries in the1980s in electronics and power tools followed by use in vehicles in the mid-1990s. Both battery types contain NiO·OH as cathode. Li-ion batteries started to be used in electronics in the late 1990s followed by use in vehicles, with Ni–Mn–Co (NMC) type or in the Ni–Co–Al (NCA) type cathodes.The Li-ion battery segment is growing strongly and by 2025 it is expected that 58% of the Li-ion batteries will contain nickel.[112] The use of nickel in the cathode material increases energy density and storage at lower cost, and makes batteries smaller and lighter.[112]

Catalysts are a specific application of nickel where nickel is used together with molybdenum (Ni–Mo) or tungsten (Ni–W). The catalysts are used in oil refineries to remove sulfur from fuel, reducing sulfur emissions when the fuel is used.

Nickel-based superalloys are designed for performance under highly stressed conditions at high temperature. This requires high creep resistance or low thermally activated movement of dislocations through the crystal lattice. It can be achieved by solid solution strengthening, and nickel forms solid solutions with a wide range of elements: Cr, Co, Mo, and W. Even higher creep resistance is achieved with precipitation strengthening through a stable finely dispersed second phase such as the intermetallic compound $Ni_3(Al,Ti)$ or with Nb, Ta, and Hf, and carbides, formed by Ti, Ta, Hf, Nb, or W, Mo, Cr, can contribute to lattice strengthening. Re is added to alloys for high-pressure turbines, where control of the additions and contaminants is critical. For example the CMSX-4 alloy consists of: 6.5% Cr, 9% Co, 0.6% Mo, 6% W, 6.5% Ta, 3.0% Re, 5.6% Al, 1% Ti, 0.1% Hf, 61.7% Ni.[94] Superalloys with over 50% Ni can solidify as a single crystal.

8.10.1.4 Recycling

Nickel is used in durable products such as stainless steel, which has a long lifetime. About 68% of the nickel available from consumer products is recycled. In addition 15% nickel enters the carbon steel recycling process.[111] Nickel recycling accounts for about 40% of nickel consumption.[104]

Nickel can be recovered from flue dust, grinding swarf, mill scale, and shot blast generated during the manufacture of stainless steel, as well as filter cakes, plating solutions, spent catalysts, spent pickle liquor, sludges, and all types of spent nickel-containing batteries.[23] Superalloys, both old and new scrap, are recycled at superalloy manufacturers, and keeping the different alloys separate is especially important because of the strict composition limits.

8.10.1.5 EHS and Sustainability

There are indications that inhalation of nickel or dust fumes over long periods of time may lead to some forms of cancer. Direct skin contact with alloys or jewellery that contains Ni can give allergic reactions.

8.10.2 Platinum Group Metals (Pt, Pd, Rh, Ru)

Platinum group metals have been known to man since ancient times. Archeologic evidence of use of platinum metal in jewellery was found in Egypt, around 1000 BC, and in the Inca empire in South America around 500 AD. Platinum occurs as metal in nature and can be found in river deposits, similar to gold. Platinum arrived in Europe around 1750. Palladium was discovered in 1803 by William Hyde Wollaston, followed by rhodium, iridium, and osmium in 1804, and ruthenium in 1844. Wollaston developed around 1810 the process to make platinum sponge metal the foundation of platinum technology. Initially, all platinum came from South America. Russia became the main producer in 1825, and in the early 1900s was joined by Canada and South Africa. From the 1920s onwards the PGMs became essential in technological development in catalysis, electric applications, emission exhaust catalysts, and fuel cells. Their ability to store large amounts of hydrogen makes them important for the hydrogen economy.[113,114]

8.10.2.1   *Physical Properties*

The PGMs consist of ruthenium (Ru), rhodium (Rh), palladium (Pd), osmium (Os), iridium (Ir), and platinum (Pt). They are part of groups 8, 9, and 10 in the periodic table. They are found together with iron, cobalt, and nickel, and neighbours of copper, silver, and gold. Pt, Os, and Ir are the densest known metals (21.45–22.61 g/cm^3), denser than gold. Pd, Rh, and Ru can be considered 'light' PGMs (12.02–12.45 g/cm^3), though denser than silver and copper.

These metals are notable for their wide range of applications, many of which are based on their high melting points, resistance to corrosion—Renner gives an excellent overview[113]— and catalytic properties. Platinum and Pd are highly resistant to heat and corrosion, are soft and ductile, and can be worked hot or cold. Pure platinum and Pd are too soft for many applications and are hardened by alloying with other metals, usually from the platinum group (for example, Ir or Ru). Pt and Rh are able to withstand and do not react with molten glass at high temperatures. Rh and Ir are more difficult to work, while Ru and Os are hard and brittle.

All PGMs have a wide range of oxidation states, making them good catalysts, although Ru, Rh, and Pd (light PGMs) behave differently in catalysis compared to Pt, Os, and Ir (heavy PGMs). Many physical properties are similar within the light PGM and heavy PGM group, and distinctly different between light and heavy.[114]

8.10.2.2   *Production*

PGMs often occur in nature in the free state, but not pure. In working secondary (alluvial) deposits, the high density of the platinum metals is relied upon for separation, as is the case with gold. PGMs form alloys with each other and with Fe, Sn, Cu, Pb, Hg, and Ag. In minerals they are bonded to S, As, Sb, Te, Bi, and Se. PGMs are produced from PGM-dominant deposits where PGMs provide the economic driver for mining and Ni and Cu are by-products, for example in South Africa or the USA. Furthermore, Ni–Cu-dominant ores with PGM, Co, Ag, and Au as by-products are used for PGM production. For example, in Norilsk-Talnakh, Russia, the ore averages 1.77% Ni, 3.57% Cu, and 0.06% Co; and 1.84 g/t Pt and 7.31 g/t Pd.[113] In both cases the processing is similar. A roasting–smelting–converting process is used to obtain a PGM-bearing nickel–copper matte in which the metals are present as sulfides. Iron, cobalt, and other metals can be present as well. The matte is leached to generate a PGM concentrate and copper, nickel, and cobalt in solution; or a Cu–PGM concentrate with nickel and cobalt in solution. The PGM concentrate is sent to the PGM refinery. The Cu–PGM concentrate is processed in a pyrometallurgical copper flowsheet together with other copper ores having very small quantities of PGMs. The PGMs collect in the anode slimes together with with Ag and Au. The slimes are smelted into a PGM-bearing matte and sent to the PGM refinery.

In the refinery the PGM-containing materials are dissolved in aqua regia or HCl/chlorine. The solution is then treated to separate the individual metals as metals or salts. Refineries with a stable feed that has a consistent metal ratio can first remove silver by precipitation as AgCl$_2$, followed by solvent extraction of Au, Pd, and Pt and removal of Ru by distillation. Ir is removed by solvent extraction and Rh is obtained by removing the non-preicous metals with cation exchange. For a feed that has varying metal ratios and impurities, a flexible process using oxidation, precipitation, reduction, crystallization, and distillation processes is more common.

The obtained crude salts can be converted to metal by reduction or calcination, and redissolved and precipitated to obtain the desired quality. The PGMs can be traded as sponge and powders, and semi-manufactured as ingots, wire, and mesh. PGM metallurgy has been described extensively.[107,114]

8.10.2.3 Applications

The world production of PGM metals is expressed in troy ounces (tr.oz), which is equivalent to 31.10 g. Primary production of Pt was 6 million tr.oz. in 2019, mainly from South Africa ($\pm$ 75%) and Russia ($\pm$ 10%), and about 2.2 million tr.oz. from recycling. Palladium primary production was about 6.8 million tr.oz. from South Africa (38%) and Russia (40%), and about 3.4 million tr.oz. were supplied through recycling. Primary production of Rh is about 746 000 tr.oz., mainly from South Africa (83%), complemented by 372 000 tr.oz. from recycling. The annual demand of Ru is about 1.2. million tr.oz. and the demand of Ir is about 265 000 tr.oz.

Because of their similar properties the PGMs can be exchanged among each other, depending on the price of each. Platinum and palladium are regularly replaced by each other in automotive catalysts. Furthermore, due to their high price, there is strong interest in reducing the amount of PGM used in an application while still achieving the same performance. This is done in automotive and chemical catalysts and fuel cells, among others.

Automotive catalysts accounted for about 34% of Pt, 84% of Pd, and 88% of Rh use. The PGMs are used to convert noxious emissions from exhaust systems to harmless, non-toxic products. Pd is also used in catalysts for stationairy applications such as diesel generators, gas turbines, and manufacturing processes. Increasingly stringent emission regulations drive the demand for PGM.[113,115] Jewellery applications used about 25% of Pt and about 1% of Pd. PGMs, in particular platinum, are considered a safe investment similar to gold and silver. This accounts for 13% of platinum demand. Platinum (28%) and palladium (15%) are used in a wide range of industrial applications including Pt or Pt–Ph alloys for glass manufacturing vessels and bushings for fibre glass production; catalysis of bulk and fine chemical and pharmaceutical processes as well as in petroleum refining and production of petrochemicals. In electronics Pt is used in hard disk coatings, Pd is used in multilayer ceramic chip capacitors (MLCC), and both are used for plating of integrated circuits.[113,115]

Ruthenium and iridium are mainly used in chemical, electrical, and electrochemical applications,[115] such as Ru in perpendicular magnetic recording used in hard disks. Rhodium is alloyed with platinum in Pt–10% Rh thermocouples, which are routinely used for the measurement of temperatures up to about 1500 °C. For higher temperatures, Pt–7% Rh and Pt–30% Rh thermocouples are available. Platinum alloyed with 3.5% Rh is the standard material for crucibles in the laboratory where they are exposed to a variety of corrosive melts, including molten slags and salts. Pd and Pt are capable of absorbing large volumes of hydrogen. At 80 °C and one atmosphere Pd will absorb up to 900 times its own volume. Hydrogen diffuses rapidly through Pd. It is used industrially to separate high-purity hydrogen from mixed gases.

8.10.2.4   Recycling

Recycling of PGMs makes a key contribution to the global supply. About ¼ of Pt demand and about ⅓ of Pd demand was met by recycled material in 2019.[115] Recycling is driven by the value of the metals. The lifetime of PGM-containing products can be years or even decades. This has to be taken into account when evaluating the recycling system performance. Closed-loop recycling takes place for the recycling of PGM from industrial catalysts. Highly efficient recycling processes require only a minimal amount of primary metal to make up for losses. The recycling loop is more open for automotive catalysts, PGM from electronics used in consumer electronics, and other applications. Collection is less effective, and in particular for electronics the PGM may not end up in a recycling stream from which they can be recovered. This is discussed in detail by Hagelueken and Meskers.[4,115]

8.10.2.5   EHS and Sustainability

The PGM metals in their metallic form are inert, non-toxic, and non-allergenic. Their compounds, in particular some chlorinated salts, are highly toxic and allergenic. Platinosis, the allergic effects of platinum salts, is known to PGM refinery workers.[113]

8.11   Copper and Precious Metals

Group 11 is composed of the metals copper, silver, and gold. They were the first metals found by Neolithic man and have been smelted and worked for millenia, creating a vast body of metallurgical and materials knowledge. The characteristics of these metals are that they are chemically inert, have a high electrical conductivity, and are extremely malleable and ductile. A gram of gold (Au) can be hammered out to more than one square metre.

8.11.1   Copper (Cu)

Copper was one of the first metals man found and has been in use since about 7000 BC, first as copper, and later as bronze (alloyed with Al and Sn) and brass (alloyed with Zn), leading to a tremendous knowledge about copper metallurgy. It is named after the island Cyprus by the Romans, where it was produced since millenia BC, and the word cuprum has corresponding words in many European languages. Copper processing and metallurgy was first noted in *De Re Metallica* by Agricola in 1556. The inventions in electricity and magnetism in the late eighteenth century using copper products started a new era for copper and the Industrial Revolution. Copper-processing technology has made great progress in the past 120 years. Copper electrolysis was invented in 1865 by JB Elkington (UK) and in 1876 by E Wohlwill (Germany). Manhès and David in 1880 (France) developed copper converting as it is practiced today by applying the Bessemer converter to copper. This process was replaced by reverbatory furnaces in 1920, followed by flash smelting in the 1950s; and a variety of new smelting furnaces have been developed since then.[117] An estimated 80% of all copper ever mined is still in use today. About 95% of all copper mined has been mined since 1900.

8.11.1.1 Physical Properties

Copper is known as the red metal. Besides gold it is the only element that does not have a grey shade.[117] It is one of the most ductile metals, and is not especially strong or hard. Strength and hardness are appreciably increased by cold working due to the formation of deformed crystals of the same face-centred cubic (fcc) structure present in the softer annealed copper. Strength-lowering impurities are Pb, Bi, Sb, Se, Te, and S.[117] Its electrical and thermal conductivity is only exceeded by silver, and its corrosion resistance makes it a half-noble metal.[117] It also has antimicrobial properties similar to those of silver.[118]

The solubility of gasses in molten copper follows Henry's Law for O_2 and SO_2, H_2 follows Sievert's Law, steam is not soluble, and N, CO, and CO_2 are practically insoluble in liquid or solid copper. Dissolved gases, especially O, greatly affect the mechanical and electrical properties of the solidified metal. Hydrocarbons do not react with copper, except acetylene, which forms highly explosive acetylides Cu_2C_2 and CuC_2. Copper has a high positive reduction potential in water; it is therefore not soluble in acids with the evolution of hydrogen. Copper resists attack by the atmosphere and sea water. Exposure for long periods to air, however, results in the formation of a thin protective coating of green, basic copper sulfate, and carbonate (as a patina).

8.11.1.2 Production

The main copper sulfide minerals are chalcopyrite ($CuFeS_2$), chalcocite (Cu_2S), covellite (CuS), bornite (Cu_5FeS_4), tetrahedrite ((Cu,Fe)$_{12}Sb_4S_{13}$), and enargite (Cu_3AsS_4). The largest source of copper is from porphyry ore deposits which contain one or more of these minerals. A typical copper sulfide ore contains various levels of iron sulfides that generally include pyrite (FeS_2) and pyrrhotite ($Fe_{1-x}S$). Small amounts of gold and silver are often present, which may be associated either with the sulfides or as free metals. The sulfides are readily liberated from gangue compenents by flotation.

Copper ores usually contain about 0.5% Cu, although there may be 2% or more Cu in new underground mines. Copper metal is produced from these ores through smelting, converting, and refining, i.e. by pyrometallurgical processes. It is the major route for primary production. Hydrometallurgical processes are used to produce copper from copper oxides, carbonates, and sulfates; and to recover copper from lower grade ore deposits, such as tailings from earlier mining operations. This includes (heap) leaching with sulfuric acid, solvent extraction (SX) to separate copper from the other elements (Fe, Zn, ...) that are dissolved during leaching, and finally electrowinning (EW) to produce copper metal cathodes.[119] During the leaching using the SX-EW process, precious metals are not recovered.

Prior to pyrometallurgical processing of the sulfide minerals a concentrate is obtained by beneficiation. The ore is ground to particles smaller than 100 μm to liberate the Cu-containing minerals, followed by flotation to obtain a concentrate with about 30% copper.[119] The gangue minerals are landfilled in tailings ponds. If necessary a roasting step, e.g. to remove arsenic, can be included prior to smelting.[117]

Next the concentrate is smelted at 1225–1375 °C to produce a sulfide melt, known as matte, containing Cu_2S (and some Fe sulfide), and a slag is produced by using oxygen and adding quartz as flux. The Cu–Fe–S minerals oxidize to iron oxides (FeO, Fe_2O_3, and Fe_3O_4), which form an iron silicate, liquid fayalite (Fe_2SiO_4) based slag. The slag composition is selected to minimize the amount of copper in the slag. It has a low density (3.1–3.6 g/cm^3), a low viscosity and high surface tension, and a low, narrow melting range. Copper can be present in the slag as matte droplets, dissolved copper (I) sulfide, and slagged copper (I) oxide.[117] The liquid matte, containing 25–70% Cu, has a higher density (4.1–5.2 g/cm^3)[117] than the slag, so the slag floats on top. The generated sulfur dioxide (SO_2) is captured in the off-gas treatment system. Several smelter processes and furnace designs are used, grouped as bath smelting and autogenous (flash) smelting processes.[117,119] Flash smelting was used to produce 69% of copper in 2018,[118] mostly using the Outotec process. The process uses enriched air to oxidize the sulfide feed. The resulting heat smelts the particles, creating an autogenous and economical process. Dry concentrate and flux is blown into a vertical reaction shaft and has only a few seconds to react before it reaches the rectangular settling area (8 m wide, 26 m long), where matte and slag are separated and tapped through separate tapholes. The fast reaction times mean that thermodynamic equilibrium is not reached—kinetic conditions prevail. A small additional burner is used to provide a good heat balance. Typically an Outotec flash furnace smelts up to 4500 t/d concentrate into a 65% Cu-containing matte, a slag (1–2% Cu), and a 30–70 vol.% SO_2-containing off-gas.[119] Bath smelting technologies include submerged tuyere smelting, with horizontal cylindrical Teniente and Noranda furnaces and top submerged lance (TSL) smelting using vertical cylindrical ISA and Ausmelt furnaces.[119] The impurity and accompanying elements divide themselves between the matte, slag, and flue dust phases. Over 90% of Ni, Co, Ag, Au, and PGMs go to the matte; Se and Te divide between matte and flue dust, while the majority of Bi goes to flue dust. The other elements (As, Sb, Pb, Zn, Sn) distribute over the three fractions, as can also be seen in Fig. 8.2.[117] If the slag contains too much copper, a slag cleaning step is added.

In the 2-stage converting process the copper matte is further oxidized to produce blister copper. This is mostly done in Peirce-Smith converters, which are cylindrical, 4.5 m wide, and 12 m long, furnaces with a capacity of 600–1000 t/d matte. Oxygen or oxygen-enriched air is blown into the molten matte through submerged tuyeres. In the first, slag-forming stage, FeS, due to its higher affinity for oxygen, is oxidized to produce FeO, Fe_3O_4, and SO_2. Less noble impurities like Pb and Zn are removed too. In this slag-forming stage the FeO is fluxed with SiO_2 and tapped from the converter. During the second copper-making stage, the sulfur in Cu_2S is oxidized to SO_2, until the first appearance of Cu_2O, while metallic copper is produced. The so-called blister copper contains about 98–99% Cu, 0.6% O, 0.01–0.03% S, and impurities such as Ag, Au, Se, Te, and As that have not reacted with oxygen. This is clearly too impure to cast into slabs for working or even to cast into anodes for purification by electrolysis. Besides the two-step smelting–converting approach, direct flash smelting to blister copper is also practiced.[119]

Refining consists of fire refining, followed by eletrorefining. During fire refining in a rotary furnace, dissolved sulfur is oxidized to SO_2-gas to obtain 0.003% S in the metal. Next, oxygen is reduced to 0.16% O in the melt through the use of injected hydrocarbons, such as natural gas. The removal of impurities will be limited by mass transfer (diffusion) of the impurity element to the melt–gas interface. Therefore, it is important that the contact area between melt and gas is large. The molten copper is then cast as anodes. These weigh about 350 kg, and are typically 1 m square and 5 cm thick. In the subsequent electrorefining process the anodes are electrochemically dissolved into an electrolyte containing copper sulfate ($CuSO_4$) and sulfuric acid (H_2SO_4). Copper is electroplated onto cathodes of copper or stainless steel,[117] producing 99.995% pure Cu cathodes of 60–160 kg.[119] The behaviour of the impurities (inclusions) in the anodes depends on their solubility in the electrolyte and the solubility of their compounds, as well as their reduction potential. Precious metals selenium and tellurium, which enter as Cu_2Se, Cu_2Te, Ag_2S, Ag_2Se, and Ag_2Te inclusions, end up in the anode slimes at the bottom of the electrorefining cells together with Pb and Sn as insoluble $PbSO_4$ and $Sn(OH)_2SO_4$. These slimes are further treated to produce Ag and Au as by-products, while Se and Te are extracted as well. Any As, Bi, Sb, Co, Fe, and Ni dissolve into the electrolyte. They are prevented from accumulating in the electrolyte by bleeding off part of the circulating electrolyte to a purification circuit. Firstly, arsenic, bismuth, and antimony are electrowon from the copper sulfate electrolyte along with copper for impure Cu–As–Bi–Sb cathodes or sludge, which can be further processed for refining. Then, nickel is precipitated as nickel sulfate. The cathodes are stripped from the steel starter sheet, melted down, and cast into slabs for rolling or extrusion.

8.11.1.3 Applications

World mining production of copper is about 20 million tons. The largest mines are located in South America, e.g. Escondida (1400 kt/y), Collahuasi (570 kt/y), and Cerro Verde II (500 kt/y) in Chile.[118] Roughly one-third of the world mine production comes from Chile, while Peru accounts for about 10%.[118] China, the USA, and Congo represent about 5–10% each. The remaining 25% is mined in a number of other countries.[23] The concentrates are shipped across the globe. China is the largest producer of blister and anode copper accounting for about 9.2 Mt/y, followed by Chile (2.5 Mt/y), Japan (1.7 Mt/y), and the USA, Russia, and Congo each with about 1 Mt/y. Global refinery production is about 24 Mt, including 4 Mt secondary refined copper and 6 Mt scrap direct melt material (directly inserted in fabrication processes).[118] Copper is shipped to fabricators mainly as cathode, wire rod, billet, cake (slab), or ingot. Through extrusion, drawing, rolling, forging, melting, electrolysis, or atomization, fabricators form wire (down to 0.004 mm diameter), rod, tube, sheet, plate, strip, foil (down to 0.002 mm thick), castings, powder, and other shapes.[117] The fabricators of these shapes are called the first users of copper.[118]

Major uses of copper and its alloys are electrical infrastructure, electronics and telecom (16%); construction, plumbing, fittings, and roofing (28%); industrial machines and equipment (31%); industrial use (12%); and transportation use in motors, radiators, wiring, etc. (13%). A typical combustion engine car contains about 22.5 kg copper, while an electrical vehicle contains about 80–100 kg.[118]

Important copper alloy types are wrought copper, wrought brasses, wrought bronzes, wrought Cu–Ni, casting alloys, and master alloys.[120] Wrought coppers contain over 99.3 or 96.0 wt.% Cu and can have Ag, As, Cd, Te, Be, Co, Zr, Cr and Fe, Pb, and Sn as alloying elements. The main grades are tough-pitch copper, deoxidized copper, and oxygen-free copper. The latter two are highly suitable for welding and brazing, as hydrogen embrittlement does not occur. Tough-pitch copper, with 0.02–0.04 wt.% O as Cu_2O, represents about 80% of refined wrought coppers.[117] At these oxygen concentrations the formation of internal H_2O pores can compensate for the 4.7% shrinkage during solidification so the surface of the casting remains level and flat. It is not suitable for welding because hydrogen embrittlement may occur. The relative electrical conductivity of alloys is given in % IACS—the International Annealed Copper Standard—based on a 1-g, 1-m-long copper wire with an electrical resistance of 0.15328 Ω.[120]

Wrought brasses or Cu–Zn alloys (typically 30% Zn), where zinc impacts also the colour of brass, are subdivided into lead (max 3% Pb) and tin (up to 3% Sn) brasses. The bronzes or Cu–Sn alloys contain about 0.01–0.35 wt.% P, up to 10% Sn, and up to 8% Pb for leaded bronzes. Furthermore, aluminium bronzes (2–15 wt.% Al) and silicon bronzes (max. 4% Si) are available. Wrought Cu–Ni alloys (up to 30% Ni) have Fe, Mn, and Co as main alloying elements. Cast alloys are not worked easily and need to be near final shape. Many alloying elements are the same as for wrought alloys, e.g. Pb, Sn, Fe, Al, Zn, Mn; only they are present in higher quantities.[120] Master alloys are used to alloy elements that are difficult to dissolve, or which develop fumes when added directly to liquid copper.

8.11.1.4   Recycling

Copper is one of the most recycled metals. Recycled copper makes up roughly $^1/_3$–$^1/_2$ of global copper use.[118] The copper production process depends on the type of scrap or end-of-life copper product and the alloy. Home scrap is produced in the primary smelters and includes anodes, cathodes, or rods that do not meet the required specification. It is put back into the converter or anode furnace, and does not appear in statistics. New scrap is produced during the manufacture of copper products. It may have been alloyed, coated, or covered with other materials. The treatment of this scrap depends on its chemical composition and the treatments the copper had, e.g. presence of coatings or insulation. If clean it can be directly remelted by fabricators, and when contaminated it is processed by smelters and refiners using the technologies described in the production section (Section 8.11.1.2).

Old, post-consumer scrap is divided into several categories.[101] Pure scrap > 99% Cu is directly remelted, lightly contaminated scrap with > 88% to be refined. Shredder material containing $\pm$ 60–5% Cu, e.g. from vehicles or white goods, is treated in primary copper smelters, since it needs to be smelted, converted, and refined to remove impurities. Electronic scrap from waste electric and electronic equipment (WEEE) contains 5–30% copper, is very complex in composition, and contains organic (plastic) material with low Cu-metal content. Printed circuit boards are a typical example, or shredded fractions that include circuit board pieces, electrical wire pieces, other metal, non-metal, and plastic housing pieces. These fractions are recycled via the primary copper route where copper and a wide range of other metals can be recovered. Because the WEEE-scrap composition is different from concentrates, impurity and slag management is key: aluminium impacts slag properties, whereas increased quantities of Sb, Sn, and Pb from flame retardants and solders impact refining, for example. The economic value is determined by the content of copper and other (precious) metals. Sampling feedstock is extremely important, as the composition of end-of-life materials changes continuously.[116]

8.11.1.5   EHS and Sustainability

Hydrocarbons do not react with copper, except acetylene, which forms highly explosive acetylides Cu_2C_2 and CuC_2. so acetylene gas cylinders should not be equipped with copper fittings.[117]

8.11.2   Silver (Ag)

Silver has been used by man since prehistoric times across the globe; named argentum in Latin, arjuna in Sanskrit, and airgead in Gaelic. It was highly valued by ancient Egyptians (2600 BC). It was first used as coins around 560 BC and continued to be used as currency and for decorative purposes throughout the centuries. Silver photography was discovered in the nineteenth century, while its use in catalysis, brazing alloys, and microelectronics developed after World War II. Major silver producers were Greece and Asia Minor around 500 BC; Central Europe (Bohemia) around AD 1400; Mexico, Peru, and Bolivia in the fifteenth to eighteenth centuries; and the USA and Mexico in the nineteenth and twentieth centuries, with a global distribution today. During the second half of the nineteenth century most of the processes used today for silver refining were discovered.

8.11.2.1   *Physical Properties*

Silver is a highly lustrous, ductile metal with a warm-white lustre. It is a typical transition metal, able to form complexes and act as a catalyst, like its neighbor Pd. The physical properties are between copper and gold. Silver has the highest electrical conductivity, thermal conductivity, and lowest electrical contact resistance of all metals. It also has anti-microbial properties. Silver alloys well with the elements from groups 10–15, except for Ni, forming industrially important binary alloys; it does not alloy well with elements from group 4–9 due to poor miscibility. Silver can be cold-worked to more than 90% by rolling and drawing.

Silver and halogens react violently at red heat and humid chlorine gas corrodes silver at low tempreature. H_2S and sulfide aqueous solutions result in a black Ag_2S coating on the metal surface. Molten silver dissolves oxygen easily, while H is slightly soluble and N, CO, and CO_2 are insoluble. The dissolved oxygen escapes during solidification, leading to bubbling of the metal surface. Fused salts of alkali metals attack silver. Ag can be dissolved in hot concentrated nitric or sulfuric acid, and in dilute sulfuric acid with oxygen or hydrogen peroxide, and is attacked by oxidizing aqueous media. It is resistant to aqueous alkali and organic acid solutions.

8.11.2.2   *Production*

Silver is mainly a by-product from base metals and gold mining, and true silver mines are a minority.[23,121] Silver typically forms sulfides, and can include compounds with Se, Te, As, Sb, and Bi. The most common mineral is argentite Ag_2S, found with lead and copper sulfides. Pyrargyrite Ag_3SbS_3 and proustite Ag_3AsS_3 are typical constitutents of true silver ores. True silver mines were found and mined to exhaustion in many places. Today true silver mines are found in Mexico, Bolivia, the USA, and Canada. Silver is also found as a native metal. Oxides, silicates, carbonates, etc. are not found.

About 32% of primary silver is produced as by-product from lead and lead/zinc ores.[121] In the zinc roast-leach-electrowinning (RLE) flowsheet silver is concentrated in the Pb–Ag residue generated during hot acid leaching. The residue is treated with lead and lead/zinc ores in the lead flowsheet. Silver leaves the lead flowsheet in the rich crust generated by the Parkes process. The Parkes process uses two key properties of zinc for the extraction of silver:

zinc is immiscible with lead and zinc dissolves 300× more silver than lead.[122] The silver in the Zn–Ag-rich crust is concentrated up to 10%. The rich crust is hot pressed to remove liquid lead, which is returned to the lead flowsheet. Then zinc is removed by distillation in a retort or vacuum furnace. The resulting lead–precious metals alloy contains about 50% Ag, 1% Zn, and Cu, Bi, As, Sb, and Au if present. Using cupellation the silver (and gold, PGM) are separated from the base metals in the crust. The crust is molten at 900–1000 °C. If the operating temperature is too high silver is lost by vaporization. Air is blown into the melt to oxidizes Pb to PbO. Cu, Zn, Sb, and Bi also report to the oxide phase (litharge), which is continuously removed during the process. When cupellation is complete the obtained crude silver should have less than 1% Cu and 0.1% Pb. Typically, the capacity of a cupellation furnace is 1000–2000 kg silver.[123]

Roughly 20% of primary silver is a by-product from Cu, Cu–Ni, and Cu–Ni–Co ore processing in smelters.[121] Silver (and gold) follows the copper and—together with Au, PGM, Se, Te, Pb, and Sn—concentrates in the anode slimes. The complex and varying composition of the slimes requires a multistep treatment process. First, large copper metal particles are removed by sieving. Then the slimes can be treated by roasting, Doré process, cupellation (see above), or sulfation to obtain crude silver, or hydrometallurgical processes are employed. Oxidizing roasting at 300–400 °C or 600–800 °C converts Ag and Cu tellurides and selenides into soluble tellurites and selenites. After dissolution in sodium hydroxide to dissolve Se, and dissolution with sulfuric acid to dissolve Te and some Ag, a silver-rich residue remains.[123] The Doré process is a batch process during which about 10 t of anode slimes are treated for 3–6 days in a Doré furnace.[123] The slimes undergo three slagging steps. First, silica and lime are added to form a silicate slag that removes Fe, As, Sb, Pb, Ni, and Sn from the melt. A selenium matte, consisting of Cu_2Se, Cu_2Te, Ag_2Se, and Ag_2Te, forms between the metal and slag phases. The silicate slag is removed, sodium carbonate and nitrate are added, and air is blown in the melt. A slag containing sodium selenite, selenate, tellurite, and tellurate, and about 1% Ag is tapped off. Lastly, sodium nitrate is added to remove the last Cu, Pb, Se, and Te impurities in a nitrate slag. The Doré silver contains at least 95% Ag, up to 4% Au, Pt, and Pd, and 0.1–0.5 wt.% Cu, while the flue dust is a major source of Se and Te.[122] Cupellation can be used if the anode slimes contain moderate quantities of impurities.[123]

Silver mines account for about 29% of silver production.[121] Silver ores are not directly smelted anymore. They can be directly added to lead and copper smelting processes if their gangue minerals can be used as slag-forming agents in the smelting process.[123] Cyanide leaching is the most commonly used process for silver[122] and is basically the same as used for gold. The ore is finely ground and treated with a 5–10 g/l NaCN solution and blowing air into the solution, in a so-called Pachuca tank to dissolve silver as $NaAg(CN)_2$. Silver is precipitated by cementation on zinc dust. Though removal by carbon-in-pulp processes is possible, this is not often used. The reaction time of sulfides is longer than elemental silver. If insoluble minerals such as Ag_3AsS_3 and Ag_3SbS_3 are present, these need to be converted to chlorides by a chloride roasting process prior to cyanidation. Yields range from 80 to 98%, depending on the ore mineralogy.[123] The silver cyanide leaching process is much slower than for gold.

An additional 15% of silver is a by-product from gold.[121] Silver follows gold in the processes and is separated from crude gold in a borax slag during the Miller process. The slag is leached with sodium chlorate in hydrochloric acid and a $AgCl_2$ residue remains. This is reduced with zinc dust to crude silver.

Silver refining produces 99.9, 99.97, or > 99.99% pure silver. The latter is more and more used for industrial applications. Electrolysis, either in a Möbius (Moebius) cell or Balbach-Thum cell, is mostly used. Crude silver is cast into anodes. The anodes are placed in bags in which the anode slimes accumulate. The slimes consist of Au as metal, selenide or telluride, PGM, PbO, and silver metal particles. Ag dissolves in the electrolyte a silver nitrate, sodium nitrate, and nitric acid solution with about 50 g/l Ag, and precipitates on stainless steel cathodes as crystals. The current density at the anode is 400–500 A/m^2, the cell voltage is 2.5 V, and electricity consumption is about 0.6 kWh/kg silver.[122,123] The crystals of 99.95–99.99 purity are washed and melted into ingots or granules. To obtain 99.995–99.999% purity the electrolysis process is repeated.

If the crude silver contains very little Au and PGM a smelting process is used. Here silver is refined by adding alkali metal nitrates or carbonates, or silver sulfate, and blowing oxygen.

8.11.2.3 *Applications*

Global primary production of silver is about 26 000 t/y or 836.6 million ounces.[121] Silver is found and mined in many countries around the world. It comes from silver mines, as well as a by-product of lead, zinc, copper, and gold mining and refining. The supply as by-product accounts for about 70% of silver production. Main producers are Mexico 23%, Peru 14%, China 13%, and Russia 8%, as well as Poland, Australia, Chile, Argentina, and Boliva with about 5% each. Primary production is complemented by about 5284 t/y from recycling, so overall world production is nearly 32 kt/y.[121]

Silver is used in a wide range of applications which can be divided into industrial (51%), jewelry (20%), investment (19%), silverware (6%), and photography (3%).[121] Industrial uses include use as electric contacts and conductors in industrial and consumer electrical and electronic applications; in automotive applications especially in hybrid, electric, and autonomous vehicles; and in photovoltaic energy generation, which accounts for about 3% of global silver demand.[121] In these applications thrifting, or reducing the amount of silver used for the same functional performance, takes place. Another key industrial application is its use as Ag-α-Al_2O_3 catalyst in the production of ethylene oxide (30 t silver per plant) and as silver wire mesh or crystals in formaldehyde production.[123]

The most important silver alloys are Ag–Cu and Ag–Au–Cu (–Pt,Pd) used for coinage and jewellery, e.g. sterling silver Ag 925 Cu 75; Ag–Pd, e.g. Ag 23 Pd 77 used for hydrogen diffusion membranes; Ag–Ni with 0.15 wt.% Ni is very fine grained; Ag–Sn–Hg as dental amalgams; Ag–Cu–Zn as brazing alloys; and Ag 80 Cd 5 In 15 alloy in nuclear reactors.[123]

8.11.2.4 *Recycling*

The high value of silver drives recycling of the metal. This means that recycling has taken place for centuries. From a business economics standpoint it is important to recover the silver as fast as possible with high yields. Global recycling accounts for 5284 t/y Ag from industrial (60%) and photographic (13%) applications, jewelry (14%), silverware (12%), and coins.[121] Different recycling processes are used, depending on the application, similar to those described for primary silver. Jewelry scrap, silverware, and metallic silver can be processed by precious metals processors. The copper smelting flowsheet is typically used for scrap and waste from the electronics industry, such as printed circuit boards.

8.11.2.5   *EHS and Sustainability*

Poor working place hygiene in silver nitrate and silver-plating industries can result in argyria, a discoloration of the skin and accumulation of silver in organs. Colloidal silver, as well as massive metal silver particles, has bactericidal properties. Silver can form unstable compounds that can explode spontaneously, under the influence of light, drying, thermal, or mechanical stress. These compounds include AgN_3, $AgNH_2$, $AgCNO$, silver acetylide and derivatives, and silver oxalate. Silver or silver alloys should not come in contact with acetylene and ammonia, either in a gas or a liquid state.[123]

8.11.3   Gold

Gold was the first metal element humans found. The oldest archeological findings date from about 4000–3000 BC and were found in Egypt, in Mesopotamia, and on the shores of the Black Sea.[124] A ceremonial gold cape from 1900–1600 BC was found in Wales, UK.[125] Gold was mentioned in literature such as the Vedanta in India, and the Old Testament. The ancient Egyptians could separate gold from silver and copper, for example.[124] Gold metallurgy thus has a long history. Gold was used both as a means of payment and for decorative puposes. The first gold coins appeared about 560 BC. The gold standard was used between 1854 and 1914 by countries to fix the value of their currency.

The Miller chlorine refining process was developed in 1867, electrorefining was developed by Wohlwill in 1878, and cyanide leaching has been in use since 1888. These processes are still used today.[124]

8.11.3.1   *Physical Properties*

Gold is a very soft, yellow-coloured metal. Gold has a high conductivity of heat and thus reaches body temperature fast, making it pleasant to wear as jewelry. Its cold workability—malleabilility and ductility—is exceptional. One ounce of gold can be pulled to a wire 5 microns wide and 80 km long, or hammered in a sheet of 9 m^2 and 0.2 micron thick. Gold metal is chemically inert to water, air, oxygen, hydrogen, sulfur, fluorine, and hydrogen sulfide at room temperature. It is also resistant to dilute and concentrated anorganic and organic acids. It can be dissolved in aqua regia, acids, or alkali cyanides combined with an oxidizing agent, water and a halogen, or selenic acid. Gold reacts with chlorine gas.

It alloys with many metals. Zinc, lead, and copper can be used as collector metals for gold. Pb, Te, Se, Ab, and Bi are easily taken up as impurities.[124]

8.11.3.2   *Production*

Gold is found almost only as metal particles in placer or secondary deposits, in quartz veins or primary deposits, or in sandstone deposits called conglomerate deposits. Furthermore gold is found as small inclusions in sulfidic copper ores (< 1 ppm Au), in silver ores, and in other sulfides such as pyrite.

Depdening on the composition of the ore, gravity concentration, froth flotation, or cyanide leaching is used. Gravity concentration is used for gold present as small particles that are already liberated or can be liberated by grinding. This process uses the density difference between gold and gangue minerals for the separation. The resulting concentrate contains considerable amounts of iron pyrite and iron, and is amalgated to separate gold and silver as sponge gold metal. Amalgation is today not used on an industrial large scale anymore because of the health and environmental issues associated with the use of mercury. Amalgation is still used by small-scale operations in developing countries.

Cyanide leaching followed by carbon-in-pulp processing is currently the industry standard process. It is used when the gold particles are too small, e.g. in conglomerate deposits. The process consists of a series of 6–8 tanks where the leach slurry and carbon particles flow counter current. After milling, the ore, containing 5–10 ppm gold, is leached with 0.03% NaCN solution while bubbling in air, adding lime to control the pH of the process, and adding activated carbon particles. Gold forms a $NaAu(CN)_2$ complex and is adsorbed on the carbon. The loaded carbon particles with about 4–8 kg Au/t carbon are sieved off, and the gold is removed from the carbon using a concentrated cyanide solution. The carbon is reused. The concentrated cyanide solution is subjected to electrowinning. Gold and other metals deposit on steel wool cathodes.

If an ore contains sulfide minerals and/or organic carbon, it cannot be processed by cyanide leaching directly. The sulfides hinder the leaching of gold, and organic carbon absorbs gold–cynanide complexes. The ore is then pretreated by roasting to oxides, or dissolution of sulfides by bio-oxidation or pressure oxidation, or ultra-fine grinding. Heap leaching is used for gold ores with less than 1 ppm of gold. The dissolved gold is precipitated from the solution by cementation with zinc metal powder.

Gold follows copper and silver in the copper smelting flowsheet, and in the silver flowsheet. Gold refining is done using the Miller process or the Wohlwill eletrorefining process. The Miller process uses chlorine to separate gold from silver, copper, iron, zinc, and lead impurities. The raw gold is melted in an induction furnace (500 kg batches) at about 1100 °C and chlorine is blown into the melt. Gold chlorides are unstable above 400 °C, while the impurities form stable chlorides. First, volatile Fe, Zn, and Pb chlorides are formed. Then molten Cu and Ag chlorides are formed, which are removed as a slag created by adding borax ($Na_2H_4B_4O_9 \cdot nH_2O$). The gold metal contains less than 0.4 wt.% Ag. The chloride slag is ground and screened to remove entrained gold droplets; copper is recovered by oxidation–dissolution–cementation. Silver is removed and sent to silver refining. PGMs cannot be separated from gold in this process, as their chlorides are also unstable at the process temperature.

In the Wohlwill process raw gold is first cast as 12-kg anode plates. These are placed in a electrolyte containing hydrochloric acid and tetrachloroauric ($[AuCl_4]^-$) acid. Gold is plated for about 2 days on titanium or tantalum cathodes, from which it can be peeled off easily. Ag chloride, Rh, Ir, Ru, and Os accumulate as anode slimes. Pt, Pd, Cu, Ni, and Fe dissolve in the electrolyte. The process operates at 65–75 °C, 1.5 V cell voltage, and 1500 A/m^2 current density. The precious metals are recovered from the slimes and electrolyte. The 99.95 or 99.99% pure gold is melted and cast.

8.11.3.3   Applications

World primary (mining) production of gold is about 3500 t/y.[126] Gold is found and mined in a large number of countries around the world. Major producers are China (12%), Russia (9%), Australia (9%), the USA (6%), and Canada (5%), which together account for about 44% of global primary production.[23] About 17% of production comes from South and Latin America, while 19% comes from Africa.[126] Recycling contributes about 1280 t/y to the gold supply.[126]

The main applications of gold are still the same as centuries ago: nearly 50% as jewellery, nearly 30% as bars, coins, and ETFs for investment, and about 15% to central banks. Only 7.5% of the gold production is used in industrial applications. The small quantity of gold used in each device or application means that gold is all around. The electronics industry uses about 6% of the gold production for plating contacts and printed circuit boards. The remainder is used as dental alloys, catalyst nanoparticles for the production of vinyl acetate and pharmaceuticals.

Typical jewellery alloys are yellow gold, white gold (Au–Ag or Au–Pd (Pt)), and rose gold (Au–Cu or Au–Ag–Cu), where the alloying elements depend also on the purity of the gold jewellery.

8.11.3.4   Recycling

Recycling makes up about one-third of the gold supply. The high value of gold drives recycling, even when gold is present in small quantities. Gold alloys can be recycled using the Miller process, or specific dissolution processes to separate base and precious metals. Surface coatings of gold on a base metal are removed by dissolution and the base metal is sent to its own recycling process. Printed circuit boards can be deplated or treated in copper smelters.[124] The obtained crude gold is further refined using the methods described above.

8.11.3.5   EHS and Sustainability

Gold is subjected to a number of voluntary and mandatory standards and certification programmes. Gold is considered a conflict mineral and subjected to the OECD due diligence guidance.[80]

The use and handling of cyanide is governed by the International Cyanide Management Code and the phase out of mercury used in amalgation is globally agreed on in the Minamata convention on mercury, managed by the UN Environmental Program.

8.12   Zinc and Cadmium

The elements in this group have comparatively low melting and boiling-points and correspondingly high vapour pressures. The boiling point of mercury—the only common metal that is liquid at room temperature—is lower than that of any other metal. The pyrometallurgical extraction and refining of metals with high vapour pressures present special problems in pyrometallurgy. For the production of highly pure zinc it is in generally more economical to use an electrolytic process.

8.12.1   Zinc (Zn)

Zinc metal was produced in India and China before 1300 for coins, and known in Europe from late sixteenth century onwards through trade with the Far East. This is much later than the other common metals Cu, Pb, Sn, and Fe, because ZnO cannot be reduced to Zn by carbon at temperatures below its T_{boil} (907 °C). Brass was produced and has been used since 8000 BC, but it was not until the early 1700s that Europeans realized the mineral smithsonite $ZnCO_3$ added a metallic component to copper to form brass. Large-scale zinc metal production started in 1720 in the UK, using methods from the Far East. In 1746 in Berlin Marggraf produced zinc by distillation, and industrial zinc smelting works were built in the UK, Germany, and Belgium, followed by the USA, producing Zn around 1840. By 1907 the USA replaced Germany as the leading producer. Electrolytic production from sulfuric acid solutions was developed in the early 1920s in North America. Global production in 1871 was 121 kt/y, increased to 472 kt/y around 1900, and was 1 million t/y in 1913. Zinc rolling was discovered in 1805 and hot-dip galvanizing to prevent corrosion was discovered in 1836.[127] Today it is the fourth most used metal after iron, aluminium, and copper.[84]

8.12.1.1   Physical Properties

Freshly cast zinc has a bluish-silver surface, but slowly oxidizes in air to form a greyish protective oxide film. Highly pure zinc (99.99%) is ductile; the so-called prime western grade (99.98%) is brittle when cold but above 150 °C can be rolled into sheets that remain flexible at lower temperatures. Zinc is a metal with quite low melting and boiling points, 419 and 906 °C, respectively. The low T_{boil} is used in pyrometallurgical zinc production, as well as in other metal production processes for zinc removal. The low T_{melt} is useful for galvanization processes. When iron and zinc together are exposed to a corrosive medium, they constitute an electrolytic cell, and the zinc is attacked (oxidized to the Zn^{2+} ion) preferentially and sacrificially, since its reduction potential of –0.76 is lower than that of iron, which is –0.41. This reaction, coupled with the much greater corrosion resistance of zinc under atmospheric conditions, is the basis for galvanizing. Zn dissolves in mineral acids with the evolution of hydrogen, and evolves NO_x when dissolved in nitric acid. Zn chloride and sulfate are very soluble in water; the other compounds and organic complexes are barely soluble. It can form complexes with ammonia, amines, cyanide, and halide ions. It is used as a strong reducing agent, and zinc dust can be highly reactive and pyrophoric with O, Cl, and S.[127]

8.12.1.2   Production

The major raw material for zinc production is sphalerite (zinc blende) (ZnS). Commercially viable zinc ores usually range from 4 to 20% zinc. Typical by-product elements found with zinc are Cd, Ag, Ge, In, Pb, and Cu. The main impurity is FeS, ranging from 0.3 to over 20 wt.%, while minor amounts of As, Sn, Bi, Co, Ni, Hg, and Tl can be present too. Zinc carbonate and zinc silicate are not used for zinc metal production.[127] The minerals in the ores are highly interlinked, so the ore is ground to particles <50 µm and a zinc sulfide concentrate with 48–56% Zn is obtained by flotation.[128] Primary zinc is almost exclusively (90%) produced hydrometallurgicaly through the roast-leach-electrowinning (RLE) process.

The remainder is produced pyrometallurgically using the Imperial Smelting Furnace (ISF) route.

The RLE process starts with roasting, as ZnS cannot be leached in sulfuric acid. Zinc sulfide is converted into oxides between 900 and 980 °C, while the FeS reacts with ZnS and oxygen to zinc ferrite, $ZnOFe_2O_3$. The generated SO_2 is captured to produce sulfuric acid. Volatile elements such as Hg, Cl, F, and Se are removed.[127] Fluid bed roasters are commonly used with a capacity of 250–900 t/d concentrate, producing a fine-grained material.[128]

In the subsequent leaching stages zinc oxide reacts with sulfuric acid to form a zinc sulfate solution, and elements that are more electropositive than zinc are removed to not disturb the electrolysis process. Four different leaching conditions are used, ranging from pH 5.5 and 60 °C in the first step to 120 g/L of H_2SO_4 and over 90 °C in hot acid leaching. In the first, neutral leach step, neutral liquor (Zn sulfate solution) is created together with a residue with Zn ferrites, silica, gypsum, and Pb and Ag compounds that contains about 22% Zn. The residue undergoes hot acid leaching, dissolving both Zn and Fe, leaving lead and silver in the final residue. Pb and Ag can be recovered in a lead smelter or ISF. Iron is removed from the solution by precipitation as jarosite $K[Fe_3(SO_4)_2(OH)_6]$ or goethite ($FeO(OH)$) by adding K^+, Na^+, NH^+_4 ions or air and some roasted concentrate, leaving $\pm$ 6 wt.% Zn in the iron residue, which is currently landfilled.[127] In and Ge coprecipitate in the iron residue, and can be recovered if present in significant amounts. The solution is returned to neutral leach. The neutral liquor is further purified in a multistep process so the electrolysis process is not affected negatively. First, Cu, Co, Ni, and a bit of Cd are precipitated by adding zinc dust and arsenic trioxide. Next, the remaining Cd is removed as a cement with Zn dust, followed by a final 'polishing' stage using Zn dust. The precipitated metals can be recovered.[127,128]

The cleaned liquor is transferred to the electrowinning process where Zn is plated on aluminium (99.5%) cathode sheets. The anode is lead alloyed with silver.[127] With a standard electrode potential of -0.763 V it appears impossible to electrodeposit Zn, since H_2 should be formed in preference to Zn. However, the overvoltage for H_2 formation on pure Zn increases so rapidly with current density that Zn is deposited instead of H_2 at high current densities. In practice about 400–600 A/m^2 is used at 40 °C, the current efficiency is 93%, the electricity consumption is 2900–3300 kWh/t Zn, and the cathodes are stripped after 24–72 h. The zinc content in the electrolyte is reduced from 150 to 60 g/L, with a final H_2SO_4 concentration of 165 g/L and recycled to neutral leach.[128]

Pyrometallurgical production relies on the carbothermic reduction of zinc oxide to produce volatile zinc metal, which is collected in the off-gas system, as in the Imperial Smelting Process. First, Zn concentrates mixed with other feed is roasted and sintered to a lumpy porous cake, while sulfur is eliminated. Then it is fed to the ISF, a cylindrical shaft of 6.3 m × 3.2 m, together with coke and air. At a (fayalite) slag melting temperature around 1200 °C the zinc oxide is reduced to metal. It leaves the shaft as volatile Zn and enters the lead splash condenser, where molten lead is used as a cooling medium and solvent for zinc. By decreasing the temperature of the molten lead to 450 °C, the solubility of zinc in lead is reduced. Zn floats to the surface and is removed for further refining (98.6% Zn) by distillation, while the cooled lead is circulated back to the condenser.[128] Due to its lower vapour pressure, lead, together with dissolved copper and precious metals from the feed, collects at the bottom of the shaft together with the slag. They are tapped regularly and separated. Lead is further refined.The retorting process uses the same principle. Only one vertical retort is still operational in Huludao, China.

Fuming processes also perform a carbothermic reduction of zinc oxide to zinc metal vapour. However, it is then reoxidized to a fine zinc oxide or fume, which is separated from the off-gas. This process is mainly used with low-grade and secondary materials, because of the difficulties of performing a direct condensation of zinc metal from vapour with the resulting gas composition and dust loadings. The Waelz kiln process is one of the most common fuming processes.[128] The kiln, which can be over 100 m long, is on a slight incline so material moves slowly through while being heated up by the counter current flowing gas, reaching temperatures of minimum 1000 °C. The feed is agglomerated with reduction agents and fluxes before entering the kiln. The Waelz slag, containing 30–50 wt.% FeO, leaves at the end of the kiln, while the Waelz oxide (55–8 wt.% ZnO, 7–10 wt.% Pb, 3–5 wt.% FeO) leaves at the beginning of the kiln and is washed to remove halogens, before being further refined in a zinc plant.[129]

8.12.1.3   *Applications*

The zinc world mining production was 13 Mt zinc-units in 2019, of which about $^1/_3$ came from China, and 10% from Australia as well as Peru. India, Russia, the USA, Bolivia, Canada, Mexico, Kazakhstan, and Sweden are additional producers.[23] World metal production is about 13.5 Mt zinc. About 42% is produced in China, followed by South Korea (7%), and India, Canada, Japan, Spain, and Australia with about 5% each.[84] Recycled zinc production is about 820 kt/y.[84]

Over half of the zinc consumption is used for galvanizing. It is also used for alloying (17%), brass and bronze (17%), zinc part manufacturing (6%), and chemicals (6%).[84]

Zinc can only be used as structural material when alloyed, as it has low creep resistance. Casting alloys include Zn(Al)Cu alloy series with Mg as an additional alloying element. Ductile alloys include the ZnCuTi alloy series with 0.08–1.0 wt.% Cu, 0.06–0.2 wt.% Ti, and max 0.015 wt.% Al. Zn-based alloys are used extensively to make complex shaped components by high-pressure die casting. Alloys for hot-dip galvanizing of steel can contain 0.005–0.3 wt.% Pb and 0.15–0.35 wt.% Al (Sendzimir galvanizing); 55 wt.% Al and 1.0 wt.% Si (galvalume, zincalume); or 5 wt.% Al and 0.1 wt.% SE (galfan).[130]

8.12.1.4   *Recycling*

The major secondary materials are ashes and dross from the galvanizing process and EAF dust created during steel recycling. (See Chapter 10.) The galvanizing ashes contain 65–75% Zn (as oxide, metal, and chloride), while the dross contains around 96% zinc as a zinc–iron alloy. The zinc of the galvanizing dross is recovered through distillation, being a raw material for the production of zinc oxide. Furthermore, the galvanizing ashes have different possibilities for their treatment. They can be milled, screened, and fed to the roasting process of a RLE plant. There, the chloride of the ashes is eliminated. The ashes can also be washed with a sodium carbonate solution to extract chloride. Alternatively, the ashes can be fed to an ISF. However, the chlorine load on the condenser must be controlled to not become excessive.[128]

EAF dust contains 10–35% Zn. Half is present as zinc oxide (ZnO); the rest is zinc ferrite ($ZnFe_2O_4$) and zinc chloride ($ZnCl_2$). It is treated mainly in the Waelz kiln fuming process, where zinc is recovered as zinc oxide fume. The zinc oxide fume is fed into an ISF or a RLE plant.[128]

8.12.1.5 EHS and Sustainability

Zinc metal is slightly hazardous (a possible irritant) in case of skin or eye contact, or ingestion. Powdered metal should be handled carefully, as it can be extremely reactive. Breathing large amounts of zinc dust or zinc fumes can cause a condition called metal fume fever. This is usually a flu-like illness with a metallic taste in the mouth, throat irritation, and dry cough. This usually goes away when the zinc exposure stops.

8.12.2 Cadmium (Cd)

Cadmium was discovered in 1817 by Strohmeyer, while investigating zinc carbonate. Cadmium started to be used commercially from the 1940s onwards.[131]

8.12.2.1 Physical Properties

Silver-white and capable of taking a high polish, cadmium (Cd) is nearly as soft as tin. When bent it makes a 'cry' just like tin. Cadmium melts and boils at low temperatures, 320.9 and 767 °C, respectively, and the temperature difference between the two is quite small. The metal is stable in dry air but becomes coated with oxide in moist air. If the metal is heated to volatilization, it burns with a red-yellow flame and forms poisonous cadmium oxide. Cd dissolves fast in nitric acid, but slowly in hydrochloric and sulfuric acid. The addition of zinc displaces it from solution—a property that is used industrially. Cadmium does not dissolve in bases.[131]

8.12.2.2 Production

Cadmium is found in several minerals: greenockite Cd sulfide, carbonate, and oxide. None are economically relevant as the deposits are too small. It is found in zinc minerals with a concentration between 0.05 and 0.8 wt.%. Lead and copper ores also contain small amount of cadmium. All cadmium is produced as by-product from zinc, lead, and copper.[131]

During the hydrometallurgical processing of zinc, cadmium is removed in the leaching step of the RLE flowsheet. In the third leaching-purification step cadmium is precipitated/cemented by adding zinc metal powder. The precipitate is sent to the Cd refining step. In lead and copper smelting cadmium winds up in the flue dusts. The cadmium in the dusts is concentrated and further processed in the Cd refining step.[131]

Refining via electrolysis is most used industrially. Before processing the feedstock is leached to remove impurities: Co, Cu, Pb, As, and Ni. Zinc is added to reduce Cd and precipitate a sponge metal with 90% Cd and up to 2.5% Zn.[131] Electrolysis in sulfuric acid uses lead anodes and aluminium cathodes, which are stripped every 12 hours. The process consumes 1250 kWh/t Cd electricity and the current efficiency is 93%.[131] Electrolytic cadmium sheets contain 5–10 ppm Cu, 15–20 ppm Pb, 5–10 ppm Tl, and 5 ppm Zn.[131] The sheets are used directly or melted and cast into ingots under a sodium hydroxide cover.[131] The systemic linkages and interactions between the Zn, Pb, Cu, and Cd primary and secondary processes have been discussed in detail.[132,133]

8.12.2.3 *Applications*

Cadmium is produced in about 20 countries;[84] global production is about 23–25 kt/y.[23,84] The main producers are China (35%), South Korea (22%), Japan (8%), Canada (7%), and Mexico, Netherlands, Kazakhstan, and Russia with about 5% each.[23]

The metal is used in NiCd batteries for industrial energy storage and in consumer goods (80%), pigments (11%), anti-corrosion coatings plated on steel, iron, copper, and brass used in military and aerospace applications (7%), and plastic stabilizers, mainly for PVC (polyvinylchloride) (2%).[84] A minor but important use is in cadmium telluride (CdTe) thin film photovoltaics. About 25 GW of capacity has been installed up to 2020.[134] The use of cadmium is regulated. In the European Union the use of NiCd batteries has been limited to industrial use since 2017. The use of cadmium as PVC stabilizers was voluntarily abandonded by the European Union.[84]

8.12.2.4 *Recycling*

Recycling is driven by the potential environmental impact of cadmium. NiCd batteries, and their production scrap, are recycled. From industrial uses the collection and recycling rate is high; from consumer use, collection of waste batteries is a challenge. The collected batteries are dissolved in 2 wt.% sulfuric acid and cadmium is selectively precipitated from nickel.[131]

CdTe photovoltaic panels have a service life of about 25 years, so many are still in use by industrial users and consumers. A recycling process is in place for scrap modules.[134] The life cycle including recycling has been detailed.[132,133] The aluminium frame is removed; the panel is shredded and leached with a sulfuric acid and hydrogen peroxide solution to dissolve both Cd and Te. After removing the (solid) glass the metals are precipitated. The sludge is recycled in the lead smelting process, where cadmium is captured in the flue dust and tellurium in the lead bullion.

8.12.2.5 *EHS and Sustainability*

Vapour or dust of cadmium is carcinogenic. Inhalation of fumes under industrial conditions can produce an acute, extremely severe inflammation of the lungs that may be fatal. Chronic poisoning from inhalation may injure the liver and kidneys. Because of its hazardousness and impact on environment the EU government has put restrictions on its use; see the applications section (Section 8.12.2.3).

8.13 Aluminium, Gallium, and Indium

8.13.1 Aluminium (Al)

Bauxite, the raw material for making aluminium, was discovered in 1821 in France near Les Beaux by Pierre Berthier. Aluminium metal was discovered after Na, K, Ca, Ba, and Mg, in 1827 by Friedrich Wöhler. Sainte-Claire Deville discovered how to separate aluminium from alumina in 1854, using cryolite as solvent. It was not until the 1880s that industrial scale processes for aluminium production were discovered. In 1886 Charles Martin Hall (USA) and Paul Hérault (France) develop independently the electrolysis process for making aluminium from alumina—the Hall–Héroult process. Karl Josef Bayer (Austria) patented the process to extract alumina from bauxite in 1887.[135] Within a few years these processes became industrial practice. Both processes are still in use today for commercial production of aluminium.

8.13.1.1 Physical Properties

Aluminium has a value of 1.43 on the electronegative scale, so it is not as strongly attracted to the halide elements, F_2, Cl_2, and Br (group 17) as the alkali and alkaline earth elements. This property makes it possible to remove the alkali and alkaline earth elements from aluminium. Aluminium has a density of about 2700 kg/m^3, less than one-third that of steel. The melting-point is 660 °C and pure aluminium (99.996%) is quite soft and weak. Small amounts of impurities increase the tensile strength and hardness considerably.[135] Aluminium is an excellent conductor of heat and electricity. Its thermal conductivity is about one-half that of copper, its electrical conductivity about two-thirds. Ductile and malleable, aluminium can be drawn into wire or rolled into thin foil.

Aluminium is protected by a thin, dense, transparent film that prevents further oxidation in air (or even by concentrated nitric acid). The oxide layer has a molecular volume 1.3 times greater than alumiunium, is under compressive stress, and heals quickly when damaged. The maximum oxide layer thickness is about 2.5–3 nm at room temperature in dry air. When temperature or the amount of moisture increases, so does the thickness of the oxide layer.[135] However, aluminium is attacked by other acids and by alkalis. Due to its strong attraction to oxygen, Al is often used in the production of various metals by reduction of the oxides (aluminothermal reduction).

8.13.1.2 Production

Aluminium is found in nature as bauxite ore. Bauxite is a weathered rock typically found in tropical and subtropical regions, as flat, extensive layers close to the surface. It consists of a mixture of the aluminium-containing minerals gibbsite ($Al(OH)_3$) and böhmite and disapore (both $AlO(OH)$), which have different crystalline structures. Typically, it consists of 80% Al-containing minerals, 5–10% goethite ($Fe(OH)_3$), 2–3% TiO_2, and 5% SiO_2. Small amounts of sodium, magnesium, and calcium chlorides are also commonly present in the ore.

The production process of aluminium consists of two processes: selective dissolution of Al minerals in caustic soda (NaOH) to obtain Al_2O_3 using the Bayer process, followed by electrolysis of alumina to aluminium using the Hall–Héroult process.

The Bayer process starts with the digestion of ground bauxite, which contains about 25–30% Al_2O_3, in a pressurized vessel at 140 °C for a high-gibbsite bauxite and 200–80 °C for a high-böhmite bauxite, and 3.5 Mpa pressure at 240 °C.[136] Excess silica is removed beforehand. The Al minerals dissolve as $Al(OH)^-_4$ and form a supersaturated solution or 'pregnant liquor'. The solution is cooled to about 106 °C followed by separation of the solids (bauxite residue; red mud) in a sedimentation process. The residue is washed to recover the caustic soda. After final filtration the solution is ready for precipitation. Precipitation takes place by progressive cooling of the liquor: small crystals of $Al(OH)_3$ are formed on fine seed crystals, which grow and agglomerate to large crystals. The spent liquor is evaporated to recover NaOH and water for reuse in the process. The gibbsite crystals are classified; the fine crystals are returned as seed crystals. Lastly, calcination is done at about 1100 °C in a gas suspension calciner, fluidized bed, or rotatry kiln. The obtained alumina Al_2O_3 can be used in the Hall–Héroult process or directly in the chemical industry.[135,136]

To produce 1 t Al_2O_3 requires 3 t bauxite, 0.1 t caustic soda, and 1.4 m^3 water, and produces ±1 t bauxite residue.[136] The average energy consumption is around 14.5 GJ per tonne of alumina, including electrical energy of about 150 kWh/t Al_2O_3. The bauxite residue typically contains iron oxides (10–30%), titanium dioxide (2–15%), silicon oxide (5–20%), and undissolved alumina. It can contain trace quantities of metals such as arsenic, beryllium, cadmium, chromium, lead, manganese, mercury, nickel, and naturally occurring radioactive materials, such as thorium and uranium.[136] If present in sufficient quantities recovery of e.g. iron, aluminium, titanium, lanthanides, scandium, yttrium, and gallium can be pursued.

The Hall–Héroult process electrolyses Al_2O_3 using a molten cryolite salt mixture containing over 75% cryolite (Na_3AlF_6), 4–8% calcium fluoride (CaF_2), 5–13% aluminium fluoride (AlF_3), 1–5% alumina, and sometimes LiF and MgF_2 to lower operating temperatures and increase current efficiency. The operating temperature is 940–70 °C. Control of the NaF-to-AlF_3 ratio is important, as too much AlF_3 decreases the solubility of alumina and electrical conductivity while increasing vapour pressure.[135]

The cell, 9–15 m long, 3–4 m wide, and 1–1.2m high, is completely closed to avoid emissions of HF, CF_4, C_2F_6, and SiF_4. It is lined with carbon, forming the cathode together with a steel sheet liner to supply the electricty and has carbon anodes. The anodes are consumed by the reaction with oxygen and are continuously slowly lowered into the salt mixture. The alumina is fed from the top. When enough aluminium has formed it is siphoned from the cell into a transfer ladle. The cells are connected in potlines with more than about 100 cells per line. Cell life is typically 4–6 years.

The production of a cell depends on its size and the current applied to the cell. Most modern cells operate at about 300 000 amperes and produce 2300 kg /day of aluminium metal; although larger cells have been developed, and many smaller (older) cells are still in operation. The off-gas is a mixture of carbon monoxide and carbon dioxide, which results in a consumption of anode which is 0.41–0.48 kg of carbon/kg aluminium. The cell operates at 4.2 V with a current efficiency of 93 to 95% and requires 12.5–16 kWh/kg Al. The 99.95% pure aluminium contains impurities such as 0.06–0.35% iron, 0.05–0.15% silicon, 3×10^{-5}% hydrogen, 0.002–0.004% of Na, Ca, and Li, and 0.005–0.020% of Ti, Zr, V, Cr, and variable amounts of oxides, carbides, and borides.[137] Hydrogen is the only gas significantly soluble in aluminium and can create porosity in castings. Its solubility in liquid aluminium is 20× the solubility in solid Al. This means that even when the hydrogen level in the molten metal is reduced below the equilibrium concentration by fluxing, during solidification the hydrogen concentrates in the remaining liquid phase.[137] This is mitigated by degassing using Ar, N_2, or Cl_2 prior to the casting process. Non-metallic inclusions are either oxides, borides, carbides, or salts (chlorides). Oxides have a film-like structure and form when the melt surface is disturbed, e.g. during tapping and alloying. An exception is spinel, which forms as particles on Al–Mg melts. Borides come from Zr, V, Ti, and B present in the alumina feed and anode or from the Al–Ti–B master alloys. Carbides, Al_4C_3, originate from the carbon lining in the reduction cell. The solubility of Al_4C_3 is about 50 ppm at the high temperature of molten aluminium coming from the pots, but the solubility is 1 ppm or less at normal alloying and casting temperatures (700–25 °C).[137] Chlorides originate from cryolite used during primary production, from salt fluxes used during recycling and for bath surface protection, and from the use of Cl_2 to remove alkaline and alkaline earth metals from aluminium. Higher purity aluminium can be produced using electrolytic refining or fractional crystallization.

The 99.95 wt.% Al obtained from the primary Hall–Héroult process is melted together with secondary aluminium ingots and in-house scrap, refined and alloyed to the desired composition. Then it is manufactured into the cast or wrought product. The properties of aluminium alloys depend on the metallurgical structure, which depends on composition, solidification process, and thermal and deformation processesing.[137] Alloying is usually done between 700 and 730 °C using pure metal for Zn, Mg, Cu, and Si that have lower melting points, while Mn, Cr, Fe, Zr, and B with higher melting points are added as master alloys, as briquettes, or injected as powders. Cu, Mg, Mn, Si, and Zn additions increase the alloy's strength and have solid solubility limits > 1.5%. Low solubility elements such as Cr, Zr, Ti, and B are used to control specific aspects, e.g. grain size or recrystallization.

There are two major groups of alloys: cast alloys and wrought alloys. Their composition is standardized and registered. About 230 cast alloy compositions and 400 wrought alloy compositions have been registered. The alloys are used in semi-finished products such as plates, sheets, and strips; bars, rods, and profiles; and foil, wire, tubes, and pipes.

8.13.1.3   Applications

The global production of primary aluminium was about 64.9 million ton in 2019. About half is produced in China. The rest of the production is spread around the world. GCC (Middle East) accounts for 9%; the rest of Asia, Eastern Europe and Russia, Western Europe, and North America each account for 5–7%; and Oceania, Africa, and South America account for less than 3% each.[138] This is complemented by nearly 33 million tons of secondary aluminium.[139] Main alumina producing countries are China, Australia, and Brazil.[84]

In the wrought or cast form the metal and its alloys are used extensively for aircraft and automotive (12%), building materials (34%), consumer durables (4%), electrical conductors (17%), chemical and food processing equipment (9%), beverage cans and packaging (10%), and foil of thickness down to 5 μm (10%).[84] Its light weight combined with high strength, the many ways in which it can be worked, and its good ability to be recycled make it a very attractive material, especially with regard to fuel efficiency in automotive and aerospace applications. In automotive applications Al is used for body panels as well as massive parts like engine blocks, transmission housings, and wheels. Also, many bike frames and wheels are aluminium. Common airplanes are made of 70–80% aluminium. Typical applications in construction are doors, window frames, cladding, and roofing. Packaging applications benefit from aluminium being non-toxic, and can be divided into flexible (foil), semi-rigid (trays), and rigid (cans) packaging. Consumer durables include cooking pots and the casings of IT equipment.[84]

Besides these uses, aluminium is also used as an alloying element in magnesium, copper, and zinc, and for aluminothermic reduction of many metals, especially calcium, strontium, and niobium.

8.13.1.4   Recycling

Recycling of aluminium is common practice around the world, driven by the value of the material. Recycling typically consumes 0.7 kWh energy/kg. Scrap is divided in new scrap—originating at the cast house and manufacturing of semi- and final products—and old scrap or post-consumer scrap. Most scrap for recycling comes from transport and packaging, due to the short (several weeks–years) to medium (1–2 decades) lifetime of the products. The share of building scrap is lower as these products have a lifetime of several decades. Also important are the collection rates (quantity available), the sorting effectiveness (contamination with other metals, alloy-specifc), the shape and thickness (amount of oxide/oxidation) which determines the type of pretreatment e.g. decoating, the recycling process, and the aluminium recovery rate.

The recycling process is a remelting process. This means it is not possible to reduce oxides (Al_2O_3) back to metal, and the possibility of removing impurities and contaminations, e.g. copper metal or sand, from the melt is limited. Still, the standard alloy compositions need to be achieved. Therefore wrought alloys may be recycled (downcycled) into cast alloys, and primary metal or new scrap is added to dilute the impurities to acceptable levels. Oxides come in via the scrap, the amount of oxide increases with the surface area of the scrap and thus is inversely related to the scrap particle size, or the oxides are formed during the remelting process by reaction with air or through the reduction of impurities such as sand (SiO_2) and rust (Fe_2O_3). Clean scrap is treated at remelters using a box-type furnace with a minium amount of salt flux. A dross, containing about 40% oxides and 60% metal, is skimmed from the surface and sent to a refiner for recycling.

At secondary refiners all types of scrap are usually treated in a rotary furnace with an excess amount of salt flux operated at 700 °C. The salt flux is a mixture of 30% KCl–70% NaCl and max. 5% NaF, AlF_3, or Na_3AlF_6, with $T_{melt} = 645$ °C and density 1.57 g/cm^3. The flux has the following functions/requirements: protect the metal against oxidation, enhance coalescence of metal droplets, suspend or dissolve alumina, have T_{melt} and density lower than aluminium, have low viscosity and surface tension, be inert to aluminium and furnace lining, and be recyclable. When the scrap is molten, the furnace stops rotating and the metal collects in the bottom and is tapped, followed by tapping of the salt slag. The slag contains about 8.5% metal, 59% salt, and 32.5% non-metallic components (alumina and others). To recover the remaining metal in the salt slag it is recycled and treated. Granulation and sieving produces Al granules and a salt–oxide mixture. This mixture is dissolved in water, filtered, and crystallized to obtain the salt flux. The filter residue contains the non-metallic components mixture, which goes to disposal.

8.13.1.5 EHS and Sustainability

The aluminium life cycle with its high degree of aluminium recycling is already quite sustainable. The industry has made a lot of progress in reducing the energy consumption, PFC emissions, and CO_2 emissions by using renewable energy, and the water consumption in the Bayer process over the past years. Some challenges remain. New applications for bauxite residue have been found,[140] and additional ones are being researched. Furthermore, the search for alternatives to the Hall–Héroult process and nonconsumable anodes continues. For example, Elysis is developing a carbon-free smelting process, producing oxygen as by product.[141]

Spent potlining remains a problem. During the operation of the cell, sodium from the electrolyte, aluminium, and fluorides are absorbed into the carbon lining. After the pot fails the lining is removed. This is called spent potlining (SPL), which is considered to be a hazardous waste since it contains toxic fluoride and cyanide compounds that are leachable in water, and it forms a high pH with water, owing to contained alkali metals. Lastly, it may react with water—producing inflammable and explosive hydrogen gas, and other toxic gases. The problem is discussed in detail by Sørlie and Øye.[142] New processes under development to recycle SPL have been reviewed by Holywell and Breault.[143]

8.13.2   Gallium (Ga)

Gallium's existence was predicted by Mendeleev and discovered by Lecoq de Boisbaudran in 1875. The metal had no application until the discovery of the semiconducting properties of gallium compounds in 1970, when industrial use and production began.[144]

8.13.2.1   Physical Properties

Gallium is a silvery-white metal with some special characteristics. It has the longest liquid range of any metal—2373°—caused by the very low T_{melt} of 29.8 °C and T_{boil} of 2403 °C.[145] Ga also expands 3.2% when it solidifies. Furthermore, it is magnetic and a good conductor of heat and electricity. The latter is strongly influenced by purity and temperature. Gallium, just like aluminium, has valence of 3, a low- and high-temperature form of Ga_2O_3, and two hydroxides, $Ga(OH)_3$ and $GaO.OH$.[144] In dilute mineral acides and solutions of hydrogen halides Ga dissolves slowly, but rapidly in aqua regia and concentrated sodium hydroxide solutions. Gallium diffuses easily in other metals, making it difficult to remove.

8.13.2.2   Production

Gallium is found only as trace quantities in other minerals. It does not exist as its own mineral. It is found in certain bauxite deposits (0.003–0.008 wt.% Ga), in sphalerite (ZnS), in phosphate ores, and in coal. Gallium concenctrates (0.01–0.1 wt.%) in the fly ash of coal combustion and in the flue dust from electrothermal phosphorus production, from which it is uneconomical to be recovered.[144] Gallium can be recovered from sphalerite when the RLE process is used to produce zinc. Gallium can be precipitated from the zinc sulfate solution during purification and is extracted from the solids or cemented product.[145]

About 90% of the primary gallium is a by-product from bauxite processing.[146] In the Bayer process the pregnant liquor can contain 100–25 ppm Ga.[145] Part of it is diverted for Ga recovery by fractional precipitation using CO_2, extraction with chelating agents, or electrolytic processes.[124,125] The crude 99–99.9 wt.% gallium metal is purified to reach 6N–8N (99.999999%) necessary for semiconductor applications. Vacuum distillation is used to remove Zn and Hg (mercury), followed by washing with hydrochloric acid or alkalies or electrolysis. Next, fractional crystallization, zone melting, or single-crystal growth is used to obtain the highest purity.[144] Impurites such as Ca, Cu, C Fe, Mg, Ni, Se, Si, Sn, and Te have to be < 1 ppb.[145] Gallium arsenide (GaAs) is made by melting high-purity Ga and As together and growing single crystals from the melt using the Czochralski method or the vertical gradient freeze method. After cutting, the epitaxial layers are grown using molecular beam epitaxy, metal organic vapour phase epitaxy, or liquid phase epitaxy.[145]

8.13.2.3 Applications

The quantity of primary gallium production is determined by the limited installed recovery capacity at alumina plants. Not every plant has technology installed due to low Ga concentrations and high production cost (labour, energy).[144] Global primary production of crude Ga is about 320 t (2019).[23] Eighty per cent is produced in China and the rest in South Korea, Russia, Japan, Ukraine, and Germany.[146] Purification to 6N Ga is also done in the UK, USA, and Slovakia.[146]

Roughly, 2/3 is used as GaAs, another 1/3 as GaN, and about 5% as GaP.[146] The major end use (50%)[23] is as integrated circuits (ICs) and microwave devices in cell phones, military applications, and wireless communications (e.g. satellites, GPS, WiFi, Bluetooth). GaAs has 5× higher electron speed and 6× higher electron mobility than silicon.[145] Laser emitting diode (LED) lighting (38%), replacing incandescent lighting, has lead to large reductions in energy usage for lighting. LEDs use GaAsP for orange, yellow, and red light; GaAlAs for infrared (used for touch screens) and ultra-red; GaN for ultraviolet and blue light; and GaP for green light. Smaller metallurgical end uses are CIGS photovoltaics (4%) for renewable energy generation, NdFeB magnets (3%) for increasing alloy fluidity and magnet efficiency, eutectic alloys with In, Se, or Zn for heavy current switches, and Ga–In–Sn alloys for replacing mercury in thermometers.[145]

8.13.2.4 Recycling

About 60% of production scrap arises during the fabrication of wafers. Its recycling is common practice; the recovery rate is 90%, and makes up about 40% of the Ga used in the GaAs sector.[146] Gallium can be recovered from end-of-life CIGS photovoltaic modules. Due to its long lifetime this material is not yet available. Post-consumer scrap such as LED lighting is not used to recover gallium as the metal is too dispersed. Ga is also found in bauxite residue if it is not recovered during the Bayer process. Research in recovery of gallium from bauxite residue is ongoing[145,146]

8.13.2.5 EHS and Sustainability

Gallium is essential for the transition to a low carbon society because of its use in energy efficient lighting (LEDs) and in photovoltaics for renewable energy generation. The main impacts of its production process are from the reagents and energy used during the Bayer process and the extraction process. Gallium should not be stored in a restricted container as it will expand when it solidifies.

8.13.3 Indium (In)

In 1863 Indium was discovered in Germany at the Freiberg School of Mines, during spectrometric analysis of sphalerite (ZnS) ore. Its name comes from its indigo blue spectral lines. Its first industrial application was found in the 1920s to stabilize alloys; its second main application since the 1990s has been in semiconductor compounds with elements from groups 14–16.[147,148]

8.13.3.1 Physical Properties

Indium is a white, soft, ductile, and mallable element—softer than lead—even at near absolute zero temperatures. It emits a 'cry' on bending and can be deformed almost infinitely under pressure. Similar to gallium, indium has a long liquid range from $T_{melt} = 156\,°C$ to $T_{boil} = 2080\,°C$, or $1924°$.[147] This property is extensively used for removal of metal impurities. Indium forms a mirror surface on glass with reflective properties similar to those of silver but with better corrosion resistance, forms alloys with many other metals, and forms transparent conductive coatings when alloyed.[148] Indium is not oxidized in air at room temperature, and reacts directly with As, S, P, S, Se, and Te, and with halogens and oxygen when heated. It is inert to alkali, boiling water, and organic acids, while it reacts with hot dilure or concentrated acids.[147]

8.13.3.2 Production

In nature indium occurs with elements from groups 11–15: Cu, Ag, Zn, Cd, Sn, Pb, and Bi. Sphalerite (ZnS) contains typically 10–20 ppm, exceptionally 200 ppm, and is the main raw material for 95% of the primary indium production.[148] The balance is produced from lead and copper sulfides and tin minerals (stannite, cassiterite), which can contain up to 0.21 wt.% indium.[148] Similar to gallium, it has no own minerals, and is solely produced as a by-product, in this case mainly from zinc. Indium concentrates in residues, flue dust, slags, and metallic intermediates.

In the zinc RLE flowsheet indium goes with zinc to the purification stage, where it can be precipitated in a solid residue 0.4% In) and partly as leach residue (jarosite).[147] Indium is recovered from the solid residue via several leaching–precipitation steps to remove Zn, Sn, As, and Cd and obtain crude indium hydroxide. The hydroxide is dissolved in hydrochloric acid and cemented with aluminiun to obtain indium metal sponge.[148,149] If a dedicated indium recovery process is not present, all indium ends up in the jarosite residue. Indium can also collect in the anode slime residue during electrowinning. The residue can be treated at the zinc operation, or sent to a lead imperial smelter where indium can be recovered. If zinc is produced pyrometallurgically, for example via direct zinc smelting,[150] indium will concentrate in the ZnO fume as it volatilizes easily and can then be recovered from the ZnO fume during hydrometallurgical processing. Pyrometallurgical leach residue treatment is industrially implemented, where the residue is fumed to obtain a ZnO fume including indium for further recovery.[150]

In lead flowsheets indium accumulates in fumes of the lead drossing furnace, which can be further processed to recover In in the form of sponge, Sn, Sb, and lead using a combination of pyrometallurgical and hydrometallurgical processes.[147] Indium can also concentrate in a residue from the Harris process used for lead refining.[147,151]

Sponge indium, 99.5% pure, is further purified/refined to obtain 99.99% (4N) to 7N indium for semiconductor applications. Multistage electrorefining can be used to obtain 5N purity. The sponge is cast into anodes, which are put in a cotton bag into the chloride solution electrolyte. The indium deposits on the pure indium cathodes, the anode slimes remain in the bag so contamination of the deposited indium and electrolyte is avoided. Further purification is done using zone refining, removing Cd, Cu, Zn, Au, Ag, and Ni, or by vacuum distillation, removing Cd, S, Se, Te, Zn, Pb, and Th followed by intensive cooling.[147]

8.13.3.3 Applications

Global indium refinery production in 2018 was about 741 t, mainly in China (40%), South Korea (31%), Japan (9.4%), Canada (8%), Belgium, France, and Peru.[23]

More than half of indium (56% in 2010) is used as thin films of 90% indium oxide–10% tin oxide (ITO) on glass. These films are transparent, electricity conducting, and heat reflective, and can transfer electricity to light-emitting or light-reflecting elements. ITO coatings are used in LCD displays, touch screen displays, architectural and windscreen glass, and solar collectors.[148] Substitutes are Sb–Sn oxide or Zn oxide doped with Al for coatings, carbon nanotubes, or PEDOT (graphene quantum dots). Indium semiconductors are used in photovoltaics (8%) such as CIGS, $CuInGaSe_2$, CIS, and $CuInS_2$, and in the production of LEDs in displays and laser diodes in fibre optics, such as InP. InP is key in telecommunications, as it is used in lasers and recievers that enable 3G, 4G, and 5G wireless networks with high-speed data transmission and large transmission distances.[147]

Indium metal is used in solders (10%) based on Sn, In, and Ag for electronics, and alloys (4%). Fusible indium alloys used for bending thin-walled tubes are based on bismuth with In, Pb, Cd, and Sn. Nuclear reactor control rod alloys contain 80% Ag, 15% In, and 5% Cd. In improves the machinability of gold alloys, and the resistance to fatigue and seizure of Pb–Sn bearings for heavy duty and high-speed applications, e.g. Formula 1 engines.[147]

8.13.3.4 Recycling

Recycling of indium production scrap is common practice and exceeds primary production by at least 250%.[148] During sputtering, over 70% of the ITO is not deposited on the part. Instead it is on the walls of the deposition chamber and in the spent sputtering target. Recycling is done by indium refineries or ITO manufacturers. Due to the large quantities and high purity of the material, recycling is economic and efficient.[148] Also alloy and solder scrap arising during manufacturing and industrial use is recycled. Indium recycling from electronic scrap is difficult and uneconomic, due to the low indium content in displays. Processes for recovery from displays continue to be investigated.[148] Indium in solders used on printed circuit boards is partially recovered, if the copper–lead smelter treating the board is equipped for indium recovery. Besides technical factors, the relatively low price of indium combined with the low concentrations and high cost for recovery and refining are barriers to indium recycling.

8.13.3.5 EHS and Sustainability

When handling ITO proper ventilation and wearing masks is sufficient to prevent health hazards in production.

8.14 Germanium, Tin, and Lead

Carbon and silicon are the first two of the group 14 elements. They play a very important role in reducing oxides to their metals and as alloying elements.

8.14.1   Germanium (Ge)

Germanium's existence was predicted by Mendeleev in 1871, and discovered in 1886 by the German Clemens Winckler in silver ore as Ag_8GeS_6. First industrial use started in 1947 when its semiconductor properties were discovered.[152]

8.14.1.1   *Physical Properties*

Germanium is a greyish-white brittle semi-metal with a bright lustre and is sensitive to thermal shock. Just like gallium, germanium expands considerably during solidification: 6.6%.[152] Metallic germanium is stable in air, while above 400 °C in oxygen a passivating oxide layer is formed, which is destroyed by water vapour. Germanium does not react with concentrated hydrochloric and hydrofluoric acids even at boiling points, but reacts slowly with hot sulfuric acid. It dissolves faster in nitric acid.[152] Ge reacts easily with halogens to tetrahalides; $GeCl_4$ boils at 85 °C, and sublimation of GeO_2 and GeS happens at 700 °C. These properties are used in its production.[153] Furthermore, it is completely miscible with Si, forming a continuous range of alloys, and the Ge–Sn system has a eutectic.

In applications the following characteristics of Ge are key: intrinsic semiconductor, transparent to infrared light, glass-former, high refractive index, low chromatic dispersion.[153]

8.14.1.2   *Production*

Germanium is not found in its metallic state. It forms 30 different minerals, up to GeO_2 with 70 wt.% Ge. It is found in many minerals such as zinc sulfide, copper sulfide, cassiterite (tin), and the iron minerals hematite and goethite. Commercially, Ge is recovered as a by-product from zinc (60%) and from coal fly ashes (40%). Lignite deposits in China and Russia contain sufficient Ge (up to 3000 ppm) for it to be recovered from the fly ash of lignite/coal-fired power plants. The ashes with 0.1–1% Ge as well as As, Zn, and Pb among others are filtered, pyrometallurgically treated, and leached with sulfuric acid. To obtain metal the process described below is used.

Zinc sulfide can contain up to 3000 ppm Ge. When it is processed in a pyrometallurgical process, e.g. imperial smelting process, it concentrates in the flue dust with zinc. After distillation and volatilization of zinc a Ge residue is obtained. If a RLE process is used, Ge is found in the leach residue after acid leaching. The Ge-containing residue is leached with HCl to obtain $GeCl_4$. It is purified by a series of fractional distillation steps to remove As, the main impurity, followed by its hydrolysis to GeO_2. GeO_2 is filtered and vacuum baked. Metal, 5N purity, is produced through reduction with hydrogen at temperatures below 700 °C in ultra-clean graphite boats. After reduction is completed, the temperature is raised to above the melting point and a polycrystalline bar is solidified from one end to the other. Impurities concentrate through segregation in the beginning or end of the bar. Hydrogen is highly soluble in Ge, so the crystallization speed is several cm/h to avoid porosity.[152]

High-purity metal (4N–13N) suitable for single crystals is obtained by zone refining to lower all impurities to below 5 ppb, followed by crystal pulling using the Czochralski vertical pulling method to grow crystals up to 30 cm diameter using small seed crystals.

8.14.1.3 *Applications*

The global refinery production of Ge was about 130 t in 2019,[23] of which about 65% is in China and the remainder in Belgium, Canada, Germany, Japan, and Russia. In 2016 about 30% of germanium[42] was used in fibre optics for telecommunications in the form of GeO_2 as dopant for the core of the optical glass fibre.[153] About 23% is used as single crystals in infrared optical systems such as infrared lenses for nightvision systems, satellite systems, and car safety systems.[42] Ge wafers are also used for space-based photovoltaic cells, for example on the Mars lander, and as a substrate for high brightness LEDs in displays and lighting. The last major application (31%) is as catalysts for polymerization to produce high-quality PET plastic for bottles, without any decoloring of the PET.[42]

8.14.1.4 *Recycling*

Recycling from new scrap is common practice. This includes wafer scrap, solutions, fibres, dust, and filters that contain over 2% of Ge. At refiners germanium is dissolved to $GeCl_4$ and follows the normal production process to produce the usual quality. Globally, about 30% of germanium used is from recycling. In infrared optics about 30% is from new scrap; in fibre optic cables this is 60%, and in electronics and photovoltaics it is about 50%. The recovery rates of germanium are very high: 80% and higher.[153] Germanium used as catalyst in PET is not recovered as it is included in the PET bottle. Old scrap mainly comes from decommissioned military equipment.[23] In other applications the quantities are low and difficult to collect—in particular, when used in photovoltaics in space applications.

8.14.1.5 *EHS and Sustainability*

Germanium metal is essentially non-toxic. A future source of Ge may be from the treatment of zinc residues, such as jarosite, which is currently being researched. Also alternative, more energy-efficient methods for recovery from coal fly ash are being investigated.

8.14.2 Tin (Sn)

Tin has been known for over 5000 year as a component of bronze (8–10% Sn with copper). The Bronze Age lasted roughly from 3300 BC to 1200 BC. Its name comes from the Norse 'tin' and the Old High German 'zin'. Berzelius proposed the symbol 'Sn' as used in the periodic table. Tin metal was first produced in China and Japan around 1800 BC; artefacts were found in Egyptian tombs from around 600 BC, and Cornwall, UK, provided Europe's tin between 100 BC and the thirteenth century AD to make plates, jugs, and organ pipes; tin mines were opened in Saxony, Germany, around 1150. It has been essential for economic development in many regions and is a key ingredient for solder (alloyed with lead). The tin reduction process used today is principally the same as described in *De Re Metallica* in 1556. Today, most tin comes from Asia and South America. Solder accounts for over 50% of tin's usage.[91]

8.14.2.1 *Physical Properties*

Tin (Sn) is a soft, silvery-white metal with a bluish tinge. In the periodic table it is located on the boundary with semi-metals like Sb, Ge, Si, and non-metals. Tin is non-toxic, ductile, malleable, and adapted to all kinds of cold working, such as rolling, spinning, and extrusion. Similar to Cd and In, pure tin gives a 'cry' when bent. The low melting point of tin (232 °C) and its firm adhesion to clean surfaces of iron, steel, and copper facilitate its use as an oxidation-resistant coating. It is also resistant to many organic acids in food, hydrocarbons, and alcohols.[91] Tin exists in two forms or allotropes: white or β-tin which is stable above 13 °C, and a grey or α form which is powdery. The transition from β to α has a high activation energy, results in a volume increase of 21%, and can be prevented by small amounts of impurities or deformations.[91] A thin invisible film of tin (IV) oxide is formed spontaneously by reaction with oxygen in air. Impurities such as Sb, Ti, Bi, and Fe promote oxidation. Solid tin does not dissolve N, H_2O, or F at room temperature and vigorous reactions occur with chlorine, bromine, and iodine. SO_2 reacts with molten tin to form SnO_2 and S, and SO_2 reacts with a copper–tin melt to form SnO_2 and Cu_2S. These are important reactions for metallurgy.[91]

8.14.2.2 Production

Tin is found mainly as tin oxide cassiterite, SnO_2, in both rock and sedimentary deposits, such as heavy mineral (beach) sands. It can take up a wide range of elements: Fe, Nb, Ta and can replace Sn in cassiterite; common trace elements are Ti, Li, Sc, Zn, W, and Mn, and to a lesser degree Zr, Hf, V, In, and Ga.[154] In ore deposits minerals of these elements are the typical accompanying minerals. Stannite Cu_2FeSnS_4 is only in South America relevant as ore mineral in itself.[154] Tin concentrates are obtained by separating cassiterite from the gangue by sulfide flotation for hard rock and physical separation for sedimentary deposits[154] and can contain 8–60% Sn.[91]

Iron is the most troublesome impurity, as can be seen from the Sn–O–C, Fe–O–C, and Sn–Fe phase diagrams. Both iron and tin are reduced by carbon above 1100 °C, up to 20 wt.% Fe dissolves in liquid tin, and during solification iron–tin compounds are formed; so a pure metal cannot be obtained this way. Furthermore, the iron content in tin metal depends on the Fe/Sn ratio in the slag. To obtain a high tin yield the tin content of the slag has to be low, resulting in high iron contents in tin. To obtain pure tin metal the iron content of the slag has to be high, resulting in high tin losses to the slag.[91] Hence, tin can be reduced only if all the iron is reduced. Tin reduction and refining metallurgy has been extensively described.[91]

The tin concentrate first undergoes metallurgical pretreatment to increase the tin content, and to remove iron and sulfur by pyrometallurgical enrichment or roasting. Tin is reduced to volatile tin sulfide in the furnace and oxidizes to SnO_2 in the off-gas treatment with a yield of 90–5%, to give a 40–60 wt.% tin concentrate.[91] Leaching with concentrated hydrochloric acid removes Fe, Pb, Cu, Sb, or As as soluble chlorides. It is also important for treatment of tungsten-containing concentrates.

Next it is reduced in a two-stage, counter-flow process. SnO_2 is reduced with carbon to crude tin (97–9% Sn)[154] and a tin-rich calcium–iron–tin slag (8–35% Sn) in the first stage at 1200–1400 °C. Various types of furnaces can be used, lined with high-grade, high-purity chrome-magnesite bricks to avoid reactions between the charge and iron oxide or silica in the bricks.[91] The flue dust is recycled for its SnO content. In the second stage the slag is processed under strongly reducing conditions into a discardable slag (< 1% Sn) and an impure iron–tin alloy with > 20% Fe called 'hard head'. The iron–tin alloy is returned to the first stage.

Alternatively a blowing process is used where tin is volatilized from the slag into flue dust and returned to the first stage, and a slag high in iron is discarded.[91]

The refining of crude tin to 99.99% Sn is similar to that of lead: in highly selective, time-consuming batch processes creating intermetallics. First, Fe is removed by liquation, as its solubility at 250 °C is 0.0058 wt.%. During cooling Fe and Fe–Sn intermetallics precipitate. Their density is similar to molten tin so 'poling', i.e. passing steam or air through the melt to coagulate the particles, is used to remove them as dross.[91] Ni, Co, Cu, As, and Sb are also partially removed. In subsequent steps Cu is removed using sulfur; As is removed using Al or Al–Sn master alloy to create intermetallics at 350–400 °C, which are removed by poling. The Al arsenide needs to be stored properly and processed quickly to inert oxides in a separate process, as reaction with water or moisture creates toxic arsine.[91] Pb (and Zn, Al) is removed using chlorine as $PbCl_2$ just above T_{melt} of tin to obtain 0.008 wt.% Pb in tin. Bi is removed to 0.06–0.003 wt.% Bi content by adding Mg metal scrap or Ca. Excess Mg is removed as chloride.[91]

Secondary raw materials, like ashes, residues, and slags, are reduced to crude tin, similar to the above process.[154]

8.14.2.3   Applications

Global primary production of tin was 310 kt in 2019, about ¼ each in Indonesia and China, about 17% in Burma, and about 5% each in Peru, Bolivia, Brazil, Australia, and Congo.[23] Global refined tin usage is about 371 kt.[84] Nearly half is used in solders, of which 85% is used in electronics, and 70% of solders are lead-free tin alloys driven by the EU and Japan RoHS regulations.[155] For connecting copper tubes, lead cables, and zinc roofing, 20–40% Sn solder is used, while for electrical cables and printed circuit boards 60–3% Sn solder is used.

Tin metal is also is used as tinplate layers of 0.3 micron thickness[154] (13%), in lead acid batteries (6%), copper alloys, e.g. bronze (6%), and in float glass production (2%). The liquid glass is poured onto a liquid tin bath containing about 200 t tin, to form a very thin, floating layer due to the surface tension and density difference between glass and tin.[154] The rest is used as chemicals (16%), among other applications as indium–tin oxide (ITO) on glass,[154] and as ornaments and organ pipes, as well as Pb–Sn bearings.[147] Potential future applications are as anodes in Li-ion batteries or fuel cells.[154]

8.14.2.4   Recycling

About 63 kt tin is recycled annually.[154] Alloys (e.g. bronze) and solders are about 70% recycled. New scrap needs to be sorted into lead-free and lead-containing. Electronics scrap, such as printed circuit boards, both new and end of life, is recycled via the recycling process of the boards, usually in copper and copper–lead flowsheets. The recycling rate of tin depends on the collection rate of the electronic scrap and the recovery efficiency in the metallurgical processes. Tin recycling from cans is increasingly difficult due to the thin layer that is not easily removed in de-tinning processes. Tin burns off in municipal waste incinerators. In steel recycling processes it can be remelted into tin-containing steels; othterwise, it is an impurity.[154]

8.14.2.5 EHS and Sustainability

Metallic tin is non-toxic. Massive inhalation of tin or tin oxide dust by industrial workers can lead to irritation of the respiratory tract. Tin hydride is very toxic, similar to arsine, and organotin compounds are also very toxic.[91]

8.14.3 Lead (Pb)

Lead has been used since 5000 BC in Egypt, India, China, Phoenicia, and ancient Rome, where it was used for water piping. Other uses over the centuries have been roofing, stained glass windows, coins, weights, projectiles, and seals. The alchemists also had a lot of interest in lead. In the twentieth century the introduction of the car resulted in a lot of new applications for lead: batteries, fuel additives, and bearings. There is an extensive body of knowledge about lead metallurgy, similar to copper.[151] Lead is an essential element in the recovery of other minor elements such as Bi, Sb, Ag, and an enabler of the circular economy.

8.14.3.1 Physical Properties

Like tin, lead (Pb) is lustrous, soft, and silvery white. When freshly cut, a dull grey lead carbonate coating is formed which protects the metal from further corrosion. Its melting point is low, 327 °C, and it is very malleable, ductile, and dense. The main alloying elements for strengthening lead are Sb, Ca, Sn, Cu, Te, As, and Ag. It can shield against gamma and X-rays. It has poor electrical conductivity and is extremely durable due to its resistance against corrosion in air, water, and soil. It resists dissolution in hydrochloric acid, sulfuric acid, and hydrofluoric acid, but dissolves in warm dilute nitric acid. Alkali remove the protective oxide coating and plumbites and plumbates are formed. During melting in air it forms PbO readily, the oxidation (drossing) is promoted or inhibited by dissolved impurities.[151]

8.14.3.2 Production

Galena (PbS) is the main mineral used in primary lead production, which is a lead sulfide mineral. It occurs often together with Ag, Zn, Cu, and Au and is a co-product of zinc copper and silver production. Lead ores are physically processed to concentrate the lead sulfide minerals by grinding and flotation, achieving a lead content between 55 and 75%.[156]

The extraction and recovery of lead from sulfide ores is performed through pyrometallurgical processes to obtain a lead bullion for further refining. Several technologies are available, e.g. sinter plant-blast furnace, ISF, and direct smelting (e.g Kivcet, TSL). Direct smelting technologies are now the most common process for primary smelting.[151]

A direct smelting process is a two-stage process. Firstly, the lead sulfide concentrate is oxidized to eliminate sulfur. Then, a second stage under reducing conditions is performed to lower the lead content in slag. As the oxidation of sulfur is exothermic, the fuel consumption in the process is reduced.[156] There are different technologies, which are bath smelting and oxygen flash smelting.[151] The Kivcet furnace combines an oxygen flash smelting shaft, where the oxidation of sulfur occurs, and an electric furnace, where the slag is cleaned and settled under a reducing atmosphere. These furnaces can be up to 24 m long and treat 1.400 tonnes of concentrate per day.[156] The Queneau–Schuhmann–Lurgi (QSL) furnace is a bottom-blown furnace. In this technology, the oxidation stage occurs with a layer of molten slag over the molten lead that avoids the lead sulfide volatilization. The reduction stage occurs in the same furnace and it is performed using coal, which also supplies the heat required. The length of QSL reactors can be up to 41 m.[156] The SKS process uses the same bottom-blown approach as the QSL furnace for the oxidation and reduction stages of the lead direct smelting process. The SKS process is currently substituting the QSL furnaces in China. Also, the top submerged lance (TSL) technology is also used for direct lead smelting. This process uses a top submerged lance to inject the gases and the fuels into a slag bath.[156]

The sintering-blast furnace route and the sintering-ISF are slowly disappearing. For both processes the feed is first sintered by roasting the concentrate to remove sulfur, and partially melted so that a porous cake (sinter) is formed.[156] In the rectangular blast furnace the feed is fed from the top, while the necessary heat and carbon monoxide (CO) for the lead oxide reduction are generated through the combustion of coke with an air blast from the bottom. A molten lead bullion and slag containing 1 to 4% lead oxide are generated, which are tapped from the furnace. Lead blast furnaces are 5–10 m high and can produce 220–650 t/d bullion.[156] The ISF was developed for the production of zinc. However, lead is also reduced in the ISF and recovered as bullion on the bottom of the furnace. Thus, it can be used to produce zinc and lead simultaneously.[156]

The lead bullion contains impurities such as copper, arsenic, silver, and tin, among others, which are removed through thermal refining processes—chemical, precipitation, and distillation—performed in (open) kettles with a diameter varying from 2.5 to 4 m, and containing between 100 and 300 t Pb. This is a time-consuming, mainly batch process, although some steps are operated continuously, similar to tin refining. First, copper is removed to < 0.01% using two steps: liquation or copper drossing followed by sulfur drossing. The Cu solubility at the bullion freezing point is 0.06%, and thus can be removed by cooling the bullion from about 950 °C to about 320 °C, close to the freezing point. Dissolved Fe and Zn precipitate first as oxide, magnetite, or spinel, followed by precipitation of Cu, and Ni and Co, if present, and partly As, Sb, and Sn. The dross contains considerable Pb and is further treated. During sulfur drossing, or Colcord process, sulfur is mixed in at 320 °C to remove copper as Cu_2S. Sulfur reacts also to PbS, but much more slowly than with copper. When silver and tin are present the speed is reduced even further. This difference in kinetics allows one to achieve 0.001–0.002% Cu after drossing, instead of 0.05% Cu under equilibrium conditions.[151,156]

As, Sn, and Sb are removed by softening, using either oxygen or caustic (hydroxides). During oxygen softening, oxygen is blown into the bullion at 500–800 °C and Sn, followed by As, Sb, and then Pb, is oxidized to form a slag with about 50–60% PbO.[151] The slag can be used to make hard lead.[151] Caustic softening, or the Harris process, is used for higher impurity levels. Sodium hydroxide, molten above 450 °C, and sodium nitrate react with As, Sn, and Sb to form sodium arsenate, sodium stannate, and sodium antimonate. The arsenate dissolves in the salt slag, while the stannate and antimonate form a solid suspension. Te can also be removed within the Harris process as telluride or tellurate.[151,156] The slags are processed to recover sodium hydroxide and the minor metals.[151]

Ag, and Au if present, is removed using the Parkes process. Zinc is added to form intermetallic compounds, which have a higher melting point than the bullion, and are nearly insoluble.[151] In the first stage, at 460 °C, zinc obtained from the zinc removal step and a lean crust from the second stage of the Parkes process are stirred in. A rich crust is formed, skimmed off, and sent for silver, zinc, and lead recovery.[151] In the second stage zinc metal is added, the bath is cooled to 370 °C, just above the freezing point, to remove Ag, so only 5–10 ppm remains in the lead together with 0.5–0.7% Zn. Zinc is removed from the lead bullion by vacuum distillation, using the difference in vapour pressure at around 600 °C, obtaining less than 0.05% Zn.[156] The obtained Ag–Zn lean crust goes—counter current—to the first stage of the Parkes process.

Bismuth is less reactive than Pb; hence, oxygen, chlorine, and sulfur cannot be used to remove Bi. The Kroll–Betterton process is used to obtain < 0.01 wt.% Bi in bullion[151] by forming the high melting point intermetallic $CaMg_2Bi_2$.[151] Ca and Mg are added as master alloys at 450–500 °C and the melt is then cooled to around 330 °C. $CaMg_2Bi_2$ forms, floats to the surface, and is removed, achieving commercial soft lead with 0.02% or less Bi. Batch times between 10 and 20 hours are required.[156] The bismuth intermetallic phase is further treated for bismuth and lead recovery. Lastly, the bullion is treated to remove the remaining amounts of Zn, Sb, Ca, and Mg. Sodium hydroxide or nitrate are used as an oxidizing agent and the operating temperatures range from 450 to 580 °C, with batch times in the range of 6 to 12 hours.[156]

All drosses, crusts, and compounds generated during the refining process can be further processed for the recovery of elements such as tin, indium, or silver.

It is possible to use electrorefining for lead, but not in sulfate solutions. This is a high-cost process used mainly when impurity contents, especially Bi, are high.[151] It uses solutions of H_2SiF_6 or HBF_4 as an electrolyte; the work lead is cast into anodes, dissolved, and precipitated onto lead blanks. The more noble metals, As, Bi, Sb, Au, and Ag, are collected in the slimes, which are processed for metal recovery. In one step 99.999% pure lead is obtained[157] with < 10 ppm Bi.[151]

8.14.3.3 Applications

The world mining production of lead is about 4500 kt lead, of which about half is in China, about 10% in Australia, and about 5% each in the USA, Mexico, Peru, Russia, and a number of smaller mining countries.[23] Refined metal production is about 12 000 kt, as about 63% is from secondary production. Main metal-producing countries are China (about $1/3$), together with the USA, South Korea, India, Mexico, Germany, the UK, Canada, and Japan.[158] The main end use is in lead acid batteries (86%). The other end uses are rolled and extruded products (7%), cable sheathing (1%), shot and ammunition (1%), lead compounds (5%), and alloys and solders.[158] Lead alloys with many elements. The strengthening alloying elements are Sb, Ca, Sn, Cu, Te, As, and Ag, while minor alloying elements are Se, S, Bi, Cd, In, Al, and Sr. Key alloys are Pb–Sb alloys used in lead acid batteries, among others. Other alloys used are Pb–Ca, Pb–Sn, Pb–Cu, Pb–Ag, and Pb–Te. Low-melting-point, fusable alloys have a high content of Bi, Sn, In, and Cd, and melt at 10–183 °C.[159] Lead is also used to produce Pb-Sn-In alloy bearings for heavy duty and high–speed applications, e.g. Formula 1 engines.[147]

8.14.3.4 Recycling

The main secondary material for lead production is end-of-life batteries.[160] The battery recycling rates are excellent, because of the highly effective collection of end-of-life batteries. In the USA the recycling rate is about 98% and in the EU it is over 95%.[151] There are several alternatives to recycling lead batteries. Before being fed to a lead smelter, they are pretreated, e.g. by shredding, to separate the different materials of the batteries such as plastics (polypropylene housing, PVC grid separators), H_2SO_4 electrolyte, and lead-containing materials (grid, connectors, electrode paste).[156] Whole, drained batteries can be used as a feedstock, e.g. in a blast furnace. The electrode paste containing $PbSO_4$ is desulfurized with a sodium carbonate or hydroxide solution, and then fed to a furnace together with the lead alloy materials. The recycling process consist of two stages and can be performed with different equipment such as reverberatory furnace, rotary kiln, submerged lance slag-bath reactor, and an electric furnace. The battery feed, together with other scraps and drosses, is smelted. This generates a lead bullion and a slag containing impurities such as Sb, As, and Sn, as well as lead oxide. Then the PbO-rich slag is reduced to recover as much lead as possible. These two stages generate lead bullions that are further treated, depending on the impurities present. The treatment of the secondary lead bullion uses the same impurity removal techniques as the production of lead from concentrates.[156]

8.14.3.5   *EHS and Sustainability*

Lead and its compounds are toxic and retained by the body, accumulating over a long period of time. The poison affects the entire body, especially the nervous system. A lengthy medical treatment with substances that extract lead from body tissues is possible. UNEP is administering a phase-out of lead shot for hunting in wetlands, and lead shot is slowly being substituted with other materials. Lead-containing solders are being replaced by lead-free solders.

8.15   Antimony and Bismuth

These metals exhibit a number of of unique and useful properties. Perhaps most interesting is that both expand on solidification.

8.15.1   Antimony (Sb)

Antimony has been known since 4000 BC in China. The ancient Egyptians thought it to be a variety of lead, and this belief continued until the sixteenth century. Alchemists realized its role for separating gold from silver. Its chemistry was investigated and the reocovery of crude antimony from ores described in *De Re Metallica* by Georgius Agricola in 1556.[161]

8.15.1.1   *Physical Properties*

Antimony is a white, lustrous, brittle, non-ductile semi-metal. It can be crushed easily and melts at 630 °C. It has poor electrical and thermal conductivity. Its crystal structure depends on the cooling rate. Fast cooling produces a granular structure, while slow cooling gives a foliated structure.[161] It expands on solidification like Si, Bi, Ga, and Ge, resulting in casting alloys that fill the crevices in the moulds.

Antiomony metal does not oxidize at room temperature. Above 750 °C steam oxidizes molten antimony to Sb_2O_3 and hydrogen evolves. This behaviour is used in metal refining. Molten metal reacts with P, As, Se, and Te (group 15–16 elements), although not with B,

C, and Si. It reacts violently with fluorine, chlorine, bromine, and iodine, even at room temperature; and it reacts with sulfur, sulfide, and SO_2.[161] It is resistant to hydrofluoric, hydrochloric, and nitric acids, but dissolves easily in aqua regia.

8.15.1.2 Production

Antimony is rarely found as metal in nature. It has a strong affinity for sulfur and metals like Cu, Pb, Ag, Sn, Bi, Se and Te, As and Hg, so it is often found in and produced from complex polymetallic sulfide ores. Stibnite, Sb_2S_3, with 71.7% Sb is the economically most important mineral and the constituent of major antimony deposits in China and South Africa, with gold as a co-product.[162] The processing route from ore to crude Sb metal depends on the concentration of Sb. Three routes are used industrially.

Low-grade material, 5 to 25 wt.% Sb, is roasted at 850–1000 °C and volatilized to Sb_2O_3 in a rotary kiln. Combustion of sulfide to $SO_{2(g)}$ provides most of the energy.[162] Process control is important, as too much oxygen leads to the formation of Sb_2O_4, and too low temperatures slow the volatilization reaction.[161] Sb_2O_3 is captured in the off-gas treatment system. Reduction to crude metal is done in a reverbatory furnace at 1200 °C with a slag. About 15–20 wt.% Sb is volatilized, captured in flue dusts, and processed in a blast furnace.[157]

Medium-grade material, 25–45 wt.% Sb, is smelted together with residues, mattes, slags, and fines from Sb, Cu, Pb, and gold processing in a blast furnace. At 1300–1400 °C crude Sb metal is formed as well as slag and $SO_{2(g)}$.

High-grade material, 45–60 wt.% Sb, can be treated by liquation or iron precipitation. During liquation the Sb_2S_3 melts around 545 °C and is separated from the gangue minerals, which have a higher melting point. The process is operated in reverbatory furnaces or crucibles at 550–660 °C under a reducing atmosphere to prevent volatilization. The purified stibnite can be converted to oxide and sold, or converted to crude Sb using iron precipitation. The iron precipitation involves melting together with iron scrap to create Sb metal and iron sulfide matte.[157,161]

The crude metal contains 90–2 wt.% Sb together with iron, arsenic, and lead as main impurities, as well as Cu and S. Refining is a step-wise process in which Fe and Cu are lowered by treatment with stibnite or sodium sulfate and charcoal to form a matte. As and S are removed by adding caustic soda. Lead is difficult to remove.[157,162] Metal high in lead can be used in the production of Sb-containing lead alloys. The final metal should be 99.65 wt.% Sb with maximum 0.25 wt.% impurities[96] to be used as feedstock for antimony trioxide, Sb_2O_3 (ATO) production. High purity, 4N–7N, metal production has been described.[161]

8.15.1.3 Applications

The global mining production was 160 kt Sb content in 2019.[23] About 60% came from China. Other major producers are Russia (18%), Tajikistan (10%), and Turkey, Burma, Bolivia, and Austrialia with 1–2% each.[23] This amount is complemented by about 40 kt from antimonial lead from secondary lead recycling.[162] About 60% of antimony is used as Sb_2O_3 (ATO) in flame retardants (52%) and catalysts (6%).[162] Major producers are China (47%), Belgium (19%), and France (7%).[42] ATO is combined with halogenated (Br, Cl) compounds to provide the flame retardant functionality. Legislative requirements may lead to changes in the use of Sb.[163] Furthermore, it is used as primary metal (19%) and antimonial lead (18%), and sodium antimonate (2%) is used to produce high-quality clear glass for cameras, copiers, and fluorescent light bulbs.[162] Primary Sb and antimonial lead are used in lead acid battery alloys to provide hardness and improve performance of the battery. The Sb content has been declining from 7 to 1.6%, as Ca, Al, and Sn alloys have been used as replacements.[157] It is expected that by the mid-2020s secondary supply from antimonial lead will be sufficient to meet market demand, and little primary metal will be needed. This makes the metallurgical market effectively 'self sufficient'.[163]

8.15.1.4   *Recycling*

A large share of Sb is not recycled and is dissipated in use. Sb is recycled from lead alloys, especially those used in lead acid batteries. This is the major source of recycled metal. Additional Sb is recovered from printed circuit boards, and from Sb-containing flame retardants via copper smelters. Other potential sources are plastics with Sb-containing flame retardants—ABS and HIPS in waste electric and electronic equipment contain 5–15 wt.% Sb,[157] fluorescent lamps (0.5–1 wt.% Sb),[164] and spent fluid catalytic cracking (FCC) catalysts.[157]

8.15.1.5   *EHS and Sustainability*

Ingesting or inhaling certain compounds of antimony may have harmful effects on body tissues and functions. However, the use of effective working practices and equipment ensures that occupational exposure and environmental discharges are minminzed.

8.15.2   Bismuth (Bi)

Bismuth was first mentioned in the twelfth to thirteenth centuries, and confused with lead. It was known as a low-melting alloying element in the sixteenth and seventeenth centuries, and found in Inca bronzes cast around 1500.[39] In 1739 Pott distinguished pure Bi from other metals. Industrial production started around 1830 in Germany and further increased in the 1930s with the discovery of fusible alloys.[165]

8.15.2.1   *Physical Properties*

Bismuth is a very brittle metal, and appears silver-white with pinkish metallic lustre. It has a low $T_{melt} = 271.3\,°C$ and $T_{boil} = 1560\,°C$. It has the lowest thermal conductivity of all metals, very low electrical conductivity, and the highest specific susceptibility of all metals.[165] Its alloys with Se and Te have thermoelectric properties. Bi, just like Ga, Ge, and Sb, has a higher density when solid (10.067 g/cm^3) than when liquid (9.8 cm^3/g). It expands 3.32% during solidification.[166] When slow-cooled, large brittle crystals are formed.

In chemical behaviour it is very similar to Sb, and it is less reactive than lead. It does not oxidize in dry air. In moist air an oxidation layer that has a typical range of colours is formed.[166] A Bi_2O_3 film protects liquid Bi against further oxidation until it melts at 817–21 °C. At higher temperatures the metal oxidizes rapidly, a property used to separate it from noble metals.[165] It forms high-melting point intermetallics with Li, K, Ca, and Mg, as well as alloys that are superconducting at low temperatures. Iron and bismuth are insoluble in each other, while full solubility in solid and liquid happens between Sb and Bi.[165]

8.15.2.2 *Production*

Bismuth occurs in a wide range of minerals, of which bismuthinite (Bi_2S_3) and bismuth metal are most important. Bismutite ($Bi_2(CO_3)O_2$) and bismite (Bi_2O_3) are also important.[42,165] It is mainly a by-product of lead and copper production of complex concentrates of molybdenum, tin, and tungsten. (Separation takes place in the concentration plant where Bi-containing minerals are separated from W-, Mo-, or Sn-containing minerals by flotation.) Typical trace elements found in Bi minerals are Fe, Te, As, S, and Sb. Bi is often found with Cu, Pb, and Ag, and can also be found with Sn, As, W, Zn, Te, and Au.[166] About 50–60% of Bi comes from lead concentrate processing, followed by copper processing, while only minor quantities originate from two bismuth mines, in Bolivia and China.[42]

Depending on the deposit the Bi content in lead concentrates can be significant, ranging from 140–240 g/t up to 1800–2400 g/t.[165] As discussed in the lead production section (Section 8.14.3.2), bismuth concentrates in the lead bullion. If the Bi content is below 3.5 wt.%, it is removed from the bullion by the Kroll–Betterton process as dross. The dross, typically 6 wt.% Bi, 0.8 wt.% Ca, 1.3 wt.% Mg, and the rest Pb, is enriched by liquation, pressing, partial oxidation, and partial chlorination to remove Ca, Mg, and part of the lead as metal, to give a Pb–Bi alloy with 20–30 wt.% Bi.[165] The alloy is chlorinated at 700–800 °C to obtain $PbCl_2$ and crude bismuth with less than 0.5 wt.% Pb. As chlorine consumption is high and lead still needs to be converted to metal, an alternative process is also used. The dross is soaked in a Pb–Bi melt, Ca and Mg are removed using NaOH, anodes are cast and electrolysed to produce pure lead and Bi-rich (88–90 wt.% Bi) anode slimes. The slimes are melted, the remaining Pb is removed by chlorination, and crude bismuth is obtained.[165]

Lead bullion with over 3.5 wt.% Bi is treated electrolytically and Bi ends up in the anode slimes: 2–25 wt.% Bi together with As, Sb, Ag, and Au. Several flowsheets exist. Generally Sb, together with As, is removed by volatilization and recovered in flue dust. Cupellation is used to remove silver. Lead is removed by creating by-products that are returned to a lead flowsheet, or by electrolysis of the Pb–Bi alloy to produce an anode slime with over 98 wt.% Bi. After melting crude Bi is obtained.[165]

In copper production bismuth leaves the flowsheet during copper converting in the flue dust, as about 95% of it vaporizes together with Pb, As, and Sb. After decopperization of the dust, it is sent to lead smelters for metal recovery. The blister copper still contains Bi, which ends up in the anode slimes. During slime treatment Bi concentrates in the Pb fraction can be recovered. Note that Bi is initially concentrated in the copper smelter dusts. These are recycled back to the copper smelter; hence, Bi will leave the flowsheet at the converting stage. In tungsten production the Bi-containing minerals are separated from the W-containing minerals during mineral processing.[42]

Crude bismuth is refined to technical grade, 99.99–99.999% Bi, using a pyrometallurgical process similar to lead refining. First, copper is removed by liquation to 500 °C and sulfuring; followed by the Harris process—the addition of $NaOH$–$NaNO_3$ at 400 °C—to remove Te, As, and Sb; the Parkes process to remove precious metals; chlorination to remove Pb and Zn; and lastly final oxidation with air and NaOH.[165,166] Electrolytic refining is used to produce pharmaceutical grade, 99.999% Bi.[42] Vacuum distillation or zone refining is used for very high purity metal.

8.15.2.3 Applications

World production of refined bismuth metal is about 19 000 t/y. China produces about 75% or 14 000 t, Laos about 15%, and Korea, Japan, Mexico, and Kazakhstan are smaller producers.[23] It is used in chemicals (62%), in low-melting alloys (28%), and as an alloying element (10%).[42] In the chemical industry it is used in compounds for catalysts, pigments, and pharmaceutics.

Due to its very low melting point and non-toxic characteristics it is used to replace lead in brass and solders for electronics, food-processing equipment, and water pipes.[42] It is used in low melting point, fusible alloys such as eutectic mixtures with In, Sn, and Pb; for example, In–Bi (T_{melt} 32 °C) and In–Sn–Bi (T_{melt} 79 °C). Fusible alloys are used in sprinkler heads, fuses, and fire detection devices.[23] Its high density makes it an effective material for X-ray shielding. Bismuth is used as an alloying element in aluminium (0.2–0.7% Bi) and steel alloys. Mn–Bi alloys are magnetic and can be used to embed magnets.

Bismuth telluride Bi_2Te_3, solid solutions with Bi_2Te_3 and Sb_2Te_3, and the $AgSbTe_2$-$AgBiTe_2$ system have thermoelectric properties and are used for heating (e.g. car seats) and cooling applications. They can also be used as thermoelectric generators (TEGs) that convert thermal energy into electricity. This can be used to generate electricity from waste heat in industrial applications and automotive exhausts.[167,168]

8.15.2.4 Recycling

Only bismuth alloy scrap is recycled. Other recycling is difficult because of the dissipative applications of bismuth in pigments and pharmaceuticals. Bi-contaning alloys are used in small quantities per device, e.g. sprinklers, making it hard to collect and separate. Bismuth in solders, used on printed circuit boards, is recycled with the circuit boards (electronic scrap) and captured in copper smelters.

8.15.2.5 EHS and Sustainability

Bismuth is generally considered a non-toxic element.[42]

REFERENCES

1. Verhoef EV, Dijkema GPJ, Reuter MA. Process knowledge, system dynamics, and metal ecology. *J. Ind. Ecol.*, 2004; 8(1.2): 23–43.
2. Van Schaik A, Reuter MA. Shredding, sorting and recovery of metals from WEEE: linking design to resource efficiency. In: Goodship V, Stevels A, eds, *Waste electrical and electronic equipment (WEEE) handbook*, pp. 163–211. New York: Elsevier; 2012.
3. Reuter MA, van Schaik A. Product-centric simulation-based design for recycling: case of LED lamp recycling. *J. Sustain. Metall.*, 2015; 1: 4–28.
4. Hagelueken C, Meskers CEM. Complex life cycles of precious and special metals. In: Graedel. TE, Van der Voet E, eds, *Linkages of sustainability*, pp. 163–98. Cambridge, MA: MIT Press; 2010.
5. Reuter MA, Hudson C, Van Schaik A et al. Metal recycling: opportunities, limits, infrastructure. International Resource Panel: Working Group on the Global Metal Flows (ed.). Geneva: United Nations Environmental Programme; 2013.

6. Klemm A, Hartmann G, Lange L. Sodium and Sodium Alloys. In: *Ullmann's encyclopedia of industrial chemistry*, vol. 33, pp. 275–97. Weinheim, Germany: Wiley–VCH Verlag; 2000.

7. Trends Market Research. Sodium metal market—global research analysis, trends, competitive share and forecasts 2018–2025. 2020. Available at https://www.trendsmarketresearch.com/report/sodium-metal-market

8. Hilding T. Researchers develop viable sodium battery. WSU Insider 2020. Available at https://news.wsu.edu/2020/06/01/researchers-develop-viable-sodium-battery

9. ScienceDirect. Sodium sulfur battery. Available at https://www.sciencedirect.com/topics/engineering/sodium-sulfur-battery

10. Bi X, Wang R, Yuan Y, et al. From sodium–oxygen to sodium–air battery: enabled by sodium peroxide dihydrate. *Nano Lett.*, 2020; 20(6): 4681–6.

11. Burkhardt ER. Potassium and potassium alloys. In: *Ullmann's encyclopedia of industrial chemistry*, vol. 29, pp. 623–37. Weinheim, Germany: Wiley–VCH Verlag; 2006.

12. Evans K. Lithium. In: Gunn G, ed., *Critical metals handbook*, pp. 230–60. New York: Wiley; 2014.

13. Wietelmann U, Steinbild M. Lithium and lithium compounds. In: *Ullmann's encyclopedia of industrial chemistry*, pp. 1–38 . Weinheim, Germany: Wiley-VCH Verlag; 2014.

14. Roskill. Lithium use by sector; 2017. Available at https://roskill.com/market-reports/

15. Roskill. Lithium Outlook to 2030; 2020. Available at https://roskill.com/market-reports/

16. Prasad NE, Gokhale AA, Wanhill RJH, eds. *Aluminum-lithium alloys*. New York: Elsevier; 2014.

17. Total Materia. Aluminium-lithium alloys; 2002. Available at http://www.totalmateria.com/Article58.htm

18. European Commission. Directive 2006/66/EC on batteries and accumulators and waste batteries and accumulators. European Commission; 2006.

19. Elwert T, Römer F, Schneider K, Hua Q, Buchert M. Recycling of batteries from electric vehicles. In: Pistola G, Liaw B, eds, *Behaviour of lithium-ion batteries in electric vehicles, green energy and technology*, pp. 289–321. Berlin: Springer International; 2018.

20. Trueman DL, Sabey P. Beryllium. In: Gunn G, ed., *Critical metals handbook*, pp. 99–121. New York: Wiley; 2014.

21. Walsh KA. *Beryllium chemistry and processing*. Vidal EE, Goldberg A, Dalder ENC, et al., eds. Materials Park, ASM International; 2009.

22. Svilar M, Schuster G, Civic T, et al. Beryllium and beryllium compounds. In: *Ullmann's encyclopedia of industrial chemistry*, pp. 1–35. Weinheim, Germany: Wiley–VCH Verlag; 2013.

23. U.S. Geological Survey. *Mineral commodity summaries 2020*. Reston, VA: U.S. Geological Survey; 2020.

24. Amundsen K, Aune TK, Bakke P, et al. Magnesium. In: *Ullmann's encyclopedia of industrial chemistry*, vol. 22, pp. 1–26. Weinheim, Germany: Wiley–VCH Verlag; 2012.

25. Neelameggham NR, Brown B. Magnesium, pp. 261–83. In: Gunn G, ed., *Critical metals handbook*. New York: Wiley; 2014.

26. Federal Pipe Centrifugal Casting Machine. Magnesium metal reduction retort production line. 2020. Available at https://www.centrifugalcastmachine.com/Magnesium-Metal-Reduction-Retort-Production-Line.html

27. Schmitz M. *Rohstoffrisikobewertung–Magnesium (Metall)*. Deutsche Rohstoffagentur (DERA) (ed.). DERA; 2019. http://www.deutsche-rohstoffagentur.de

28. Aune TK, Gjestland H, Haagensen JØ, et al. Magnesium alloys. In: *Ullmann's Encyclopedia of Industrial Chemistry*, vol. 22, pp. 27–40. Weinheim, Germany: Wiley–VCH Verlag GmbH; 2012.

29. Roskill. *Magnesium metal: outlook to 2030*. 2020. Available at https://roskill.com/market-report/magnesium-metal/

30. Latrobe Magnesium. 2017. Available at https://latrobemagnesium.com/

31. Stratton P. Magnesium metal: Alliance Magnesium moves closer to commercial production. 19 March 2020. Accessed 28 September 2020. Available at https://roskill.com/news/magnesium-metal-alliance-magnesium-moves-closer-to-commercial-production/.

32. Alliance Magnesium. 2020. Available at https://alliancemagnesium.com

33. Hluchan SE, Pomerantz K. Calcium and calcium alloys. In: *Ullmann's Encyclopedia of Industrial Chemistry*, vol. 6, pp. 483–95. Weinheim, Germany: Wiley–VCH Verlag; 2006.

34. Takeda O, Uda T, Okabe TH. Rare earth, titanium group metals, and reactive metals production. In: Seetharaman S, ed., *Treatise on process metallurgy, Vol 3: Industrial processes*, pp. 995–1069. Amsterdam: Elsevier; 2014.

35. MMTA. Ca—calcium—sources. 2020. Available at https://mmta.co.uk/metals/Ca/

36. MacMillan JP, Park JW, Gerstenberg R, et al. Strontium and strontium compounds. In: *Ullmann's encyclopedia of industrial chemistry*, vol. 34, pp. 473–80. Weinheim, Germany: Wiley–VCH Verlag; 2000.

37. Ibrahim M, Elgallad E, Valtierra S, et al. Metallurgical parameters controlling the eutectic silicon charateristics in Be-treated Al-Si-Mg alloys. *Materials (Basel)*, 2016; 9(2): 78.

38. Singerling SA. Strontium. In: *2017 Minerals Yearbook*, pp. 73.1–73.7. US Geological Survey; 2020.

39. Royal Society of Chemistry. Periodic table of elements. 2020. Available at https://www.rsc.org/periodic-table

40. International Union of Pure and Applied Chemistry (IUPAC). Available at https://iupac.org/

41. Wang W, Pranolo Y, Cheng CY. Metallurgical processes for scandium recovery from various resources: A review. *Hydrometallurgy*, 2011; 108(1–2): 100–8.

42. European Commission. *Study on the review of the list of critical raw materials. Critical raw materials factsheets.* 2020. Available at https://op.europa.eu/en/publication-detail/-/publication/7345e3e8-98fc-11e7-b92d-01aa75ed71a1

43. SCALE project. What is scandium? 2020. Available at http://scale-project.eu/scandium/

44. Wall F. Rare earth elements. In: Gunn G, ed., *Critical metals handbook*, pp. 312–39. New York: Wiley; 2014.

45. Krishnamurthy N, Gupta CK. *Extractive metallurgy of rare earths*, 2nd edn. Boca Raton, FL: CRC Press; 2015.

46. Navarro J, Zhao F. Life-cycle assessment of the production of rare-earth elements for energy applications: a review. *Front Energy Res.*, 2014; 2: 45.

47. Sprecher B, Xiao Y, Walton A, et al. Life cycle inventory of the production of rare earths and the subsequent production of NdFeB rare earth permanent magnets. *Environ Sci Technol.*, 2014; 48(7): 3951–8.

48. Fernandes IB, Abadías Llamas A, Reuter MA. Simulation-based exergetic analysis of NdFeB permanent magnet production to understand large systems. *JOM*, 2020; 72: 2754–69.

49. Vogel H, Flerus B, Stoffner F, et al. Reducing greenhouse gas emission from the neodymium oxide electrolysis. Part I: Analysis of the anodic gas formation. *J Sustain Metall.*, 2017; 3: 99–107.

50. REIA. About rare earth. 2020. Available at: https://global-reia.org/about-rare-earth/

51. Jowitt SM, Werner TT, Weng Z, Mudd GMl. Recycling of the rare earth elements. *Curr. Opin. Green Sustain Chem.*, 2018; 13: 1–7.

52. Binnemans K, Jones PT, Blanpain B, et al. Recycling of rare earths: a critical review. *J. Clean Prod.*, 2013; 51: 1–22.

53. Schulze R, Buchert M. Estimates of global REE recycling potentials from NdFeB magnet material. *Resour. Conserv. Recycl.*, 2016; 113: 12–27.

54. Alonso E, Sherman AM, Wallington TJ, et al. Evaluating rare earth element availability: a case with revolutionary demand from clean technologies. *Environ. Sci. Technol.*, 2012; 46(6): 3406–14.

55. Van Arkel AE, de Boer JH. Darstellung von reinem Titanium-, Zirkonium-, Hafnium- und Thoriummetall. *Z. Anorg. Allg. Chem.*, 1925; 148(1): 345–50.

56. Kroll WJ. How commercial titanium and zirconium were born. *J. Franklin Inst.* 1955; 260(3): 169–92.

57. Sibum H, Güther V, Roidl O, et al. Titanium, titanium alloys, and titanium compounds. In: *Ullmann's encyclopedia of industrial chemistry*, pp. 1–35. Weinheim, Germany: Wiley–VCH Verlag; 2017.

58. Zhang W, Zhu Z, Cheng CY. A literature review of titanium metallurgical processes. *Hydrometallurgy*, 2011; 108(3–4): 177–88.

59. Bauer G, Güther V, Hess H, et al. Vanadium and vanadium compounds. In: *Ullmann's encyclopedia of industrial chemistry*, pp. 1–22. Weinheim, Germany: Wiley–VCH Verlag; 2017.

60. Roskill. *Titanium metal outlook to 2030*, 10th edn. 2020. Available at https://roskill.com/market-reports/

61. Singh P, Pungotra H, Kalsi NS. On the characteristics of titanium alloys for the aircraft applications. *Mater. Today Proc.*, 2017; 4(8): 8971–82.

62. Ezugwu EO, Wang ZM. Titanium alloys and their machinability—a review. *J. Mater. Process. Technol.*, 1997; 68(3): 262–74.

63. Nielsen RH, Wilfing G. Zirconium and zirconium compounds. In: *Ullmann's encyclopedia of industrial chemistry*, vol 39, pp. 753–78. Weinheim, Germany: Wiley–VCH Verlag; 2012.

64. Elsner H. *Zircon—insufficient supply in the future?* Deutsche Rohstoffagentur (DERA), ed. Berlin: DERA; 2013. Available at http://www.deutsche-rohstoffagentur.de

65. Leal-Ayala DR, Martinez-Hurtado JL, Allwood JA, et al. *Technical handbook on zirconium and zirconium compounds*, 4th edn. Zircon Industry Association; 2019.

66. Xu L, Xiao Y, van Sandwijk A, et al. Production of nuclear grade zirconium: a review. *J. Nucl. Mater.*, 2015; 466: 21–8.

67. Xiao Y, van Sandwijk A, Yang Y, Laging V. New routes for the production of reactor grade zirconium. In: Gaune-Escard M, Haarberg GM, eds, *Molten salts chemistry and technology*, pp. 389–401. Chichester, UK: Wiley; 2014.

68. Nielsen RH, Wilfing G. Hafnium and hafnium compounds. In: *Ullmann's encyclopedia of industrial chemistry*, vol. 17, pp. 191–202. Weinheim, Germany: Wiley–VCH Verlag; 2010.

69. Marbler H, Osbahr I, Opperman R, et al. *Investor's and procurement guide South Africa*, Part 3: *Manganese, vanadium, zinc*. Buchholz P, Foya S, eds. Berlin: DERA; 2017. Available at http://www.deutsche-rohstoffagentur.de

70. Roskill. *Vanadium: global industry, markets and outlook 2017*. Available at https://roskill.com/market-reports/

71. Lourenssen K, Williams J, Ahmadpour F, et al. Vanadium redox flow batteries: A comprehensive review. *J Energy Storage*, 2019; 25: 100844.

72. Albrecht S, Cymorek C, Eckert J. Niobium and niobium compounds. In: *Ullmann's encyclopedia of industrial chemistry*, vol. 24, pp. 133–47. Weinheim, Germany: Wiley–VCH Verlag; 2011.

73. Aimone P, Yang M. Niobium alloys for the chemical process industry. *Int. J. Refract. Met. Hard Mater.*, 2018; 71: 335–9.

74. Gibson CE, Kelebek S, Aghamirian M. Niobium oxide mineral flotation: A review of relevant literature and the current state of industrial operations. *Int. J. Miner. Process.*, 2015; 137: 82–97.

75. Linnen R, Trueman DL, Burt R. Tantalum and niobium. In: Gunn G, ed., *Critical metals handbook*, pp. 361–84. New York: Wiley; 2014.

76. Damm S. *Rohstoffrisikobewertung—Tantal*. Deutsche Rohstoffagentur (DERA), ed. Berlin: DERA; 2018.

77. Albrecht S, Cymorek C, Andersson K, et al. Tantalum and tantalum compounds. In: *Ullmann's encyclopedia of industrial chemistry*, Vol 35, pp. 597–610. Weinheim, Germany: Wiley–VCH Verlag; 2011.

78. Zhu Z, Cheng CY. Solvent extraction technology for the separation and purification of niobium and tantalum: A review. *Hydrometallurgy*, 2011; 107(1–2): 1–12.

79. Presentation annuelle des statistiques du T.I.C. *La Revue du Bulletin2020*. 2020: 19–24. Available at https://tanb.org./view/bulletin-review-2020

80. Responsible minerals initiative. Minerals Due Diligence. 2020. Available at https://www.responsiblemineralsinitiative.org/minerals-due-diligence/

81. Downing JH, Deeley PD, Fichte R. Chromium and chromium alloys. In: *Ullmann's encyclopedia of industrial chemistry*, vol. 9, pp. 131–55. Weinheim, Germany: Wiley–VCH Verlag; 2000.

82. Buchholz P, Kaerner K, Kenan AO, et al. *Investor's and procurement guide South Africa*. Part 2: *Fluorspar, chromite, platinum group elements*. Buchholz P, Foya S, eds. Berlin: Deutsche Rohstoff Agentur (DERA); 2015. Available at http://www.deutsche-rohstoffagentur.de

83. Nelson LR. Evolution of the mega-scale in ferro-alloy electric furnace smelting. In: Mackey PJ, Grimsey EJ, Jones RT, et al., eds, *Celebrating the megascale: proceedings of the extraction and processing division symposium on pyrometallurgy in honor of David G.C. Robertson*, pp. 39–68. Hoboken, NJ: Wiley; 2014.

84. European Commission. *Study on the EU's list of critical raw materials. Factsheets on non-critical raw materials factsheets*. 2020.

85. Sebenik RF, Burkin AR, Dorfler RR et al. Molybdenum and molybdenum compounds. In: *Ullmann's encyclopedia of industrial chemistry*, Vol 23, pp. 521–66. Weinheim, Germany: Wiley–VCH Verlag; 2000.

86. Fichte R. Ferroalloys. In: *Ullmann's encyclopedia of industrial chemistry*, vol. 14, pp. 153–5. Weinheim, Germany: Wiley–VCH Verlag; 2000.

87. International Molybdenum Association. Uses of new molybdenum. Available at https://www.imoa.info/molybdenum-uses/molybdenum-uses.php

88. Brown T, Pitfield P. Tungsten. In: Gunn G, ed., *Critical metals handbook*, pp. 385–413. New York: Wiley; 2014.

89. Trasorras JRL, Wolfe TA, Knabl W et al. Tungsten, tungsten alloys, and tungsten compounds. In: *Ullmann's encyclopedia of industrial chemistry*, pp. 1–53. Weinheim, Germany: Wiley–VCH Verlag; 2016.

90. Shen L, Li X, Lindberg D, Taskinen P. Tungsten extractive metallurgy: a review of processes and their challenges for sustainability. *Miner. Eng.*, 2019; 142: 105934.

91. Graf GG. Tin, tin alloys, and tin compounds. In: *Ullmann's Encyclopedia of Industrial Chemistry*, vol. 37, pp. 1–34. Weinheim, Germany: Wiley–VCH Verlag; 2000.

92. Wellbeloved DB, Craven PM, Waudby JW. Manganese and manganese alloys. In: *Ullmann's encyclopedia of industrial chemistry*, vol. 22, pp. 175–221. Weinheim, Germany: Wiley–VCH Verlag; 2000.

93. PREMA. Energy efficient, primary production of manganese ferroalloys through the application of novel energy systems in the drying and pre-heating of furnace feed materials. 2020. Available at https://www.spire2030.eu/PREMA

94. Georg Nadler H. Rhenium and rhenium compounds. In: *Ullmann's encyclopedia of industrial chemistry*, vol. 31, pp. 527–37. Weinheim, Germany: Wiley–VCH Verlag; 2012.

95. Millensifer TA, Sinclair D, Jonasson I, Lipmann A, et al. Rhenium. In: Gunn G, ed., *Critical metals handbook*, pp. 340–60. New York: Wiley; 2014.

96. Minor Metal Trade Association. Metal norms. Available at https://mmta.co.uk/metal-norms/

97. Dutta SK, Chokshi YB. *Basic concepts of iron and steel making*. Singapore: Springer Singapore; 2020.

98. Ghosh A, Chatterjee A. *Ironmaking and steelmaking: theory and practice*. New Delhi, India: Prentice House India; 2008.

99. *The making, shaping and treating of steel*, 11th edn. Association for Iron & Steel (AIST). Available at https://bookstore.aist.org/category-s/144.htm

100. World Steel Association. 2020 World Steel in Figures. Brussels, Belgium: World Steel Association; 2020.

101. Roberts S, Gunn G. Cobalt. In: Gunn G, ed., *Critical metals handbook*, pp. 122–49. New York: Wiley; 2014.

102. Donaldson JD, Beyersmann D. Cobalt and cobalt compounds. In: *Ullmann's encyclopedia of industrial chemistry*, vol. 9, pp. 429–65. Weinheim, Germany: Wiley–VCH Verlag; 2005.

103. Moats MS, Davenport WG. Nickel and cobalt production. In: Seetharaman S, ed., *Treatise on process metallurgy*, vol 3: *Industrial processes*, pp. 625–69. Amsterdam: Elsevier; 2014.

104. Fortune Minerals Ltd. NICO cobalt-gold-bismuth-copper project. 2020. Available at http://www.fortuneminerals.com/assets/nico/default.aspx

105. Jervois Mining Ltd. Idaho cobalt operations. 2020. Available at https://jervoismining.com.au/our-assets/idaho-cobalt-operations/

106. Crundwell F, Moats MS, Ramachandran V, et al. *Extractive metallurgy of nickel, cobalt and platinum group metals*. Amsterdam: Elsevier; 2011.

107. Acatech, Circular Economy Initiative Deutschland, SYSTEMIQ eds. *Resource-efficient battery life cycles: circular economy initiative Deutschland*. 2020.

108. Kerfoot DGE. Nickel. In: *Ullmann's encyclopedia of industrial chemistry*, vol 24, pp. 37–101. Weinheim, Germany: Wiley–VCH Verlag; 2000.

109. Outotec. Outotec (R) direct nickel process. 2020. Available at https://www.outotec.com/products-and-services/technologies/smelting-and-converting/direct-nickel-process/

110. Nickel institute. About nickel. 2020. Available at https://nickelinstitute.org/about-nickel/

111. Nickel institute. Nickel energizing batteries. 2018. Available at https://nickelinstitute.org/library/articles/battery-infographic/

112. Gunn G. Platinum-group metals. In: Gunn G, ed., *Critical metals handbook*, pp. 284–311. New York: Wiley; 2014.

113. Renner H, et al. Platinum group metals and compounds. In: *Ullmann's encyclopedia of industrial chemistry*, pp. 1–73. Weinheim, Germany: Wiley–VCH Verlag; 2018.

114. Cowley A. PGM market report. Johnson Matthey; 2020. Available at https://matthey.com/en/news/2020/pgm-market-report-may-2020

115. Hagelueken C, Meskers CEM. Recycling of technology metals: a holistic system approach. In: Hieronymi K, Kahhat R, Williams E, eds, *E-Waste management*, pp. 49–78. London: Routledge; 2013.

116. Lossin A. Copper. *Ullmann's encyclopedia of industrial chemistry*, vol. 10, pp. 163–226. Wiley–VCH Verlag; 2012.

117. International Copper Study Group. *The World Copper Factbook 2019*. Lisbon, Portugal: International copper study group (ICSG); 2019.

118. Schlesinger M, King ME, Sole K, Davenport WG. *Extractive metallurgy of copper*. Amsterdam: Elsevier; 2011.

119. Isbell CA. Copper alloys. In: *Ullmann's encyclopedia of industrial chemistry*, vol. 10, pp. 229–72. Wiley–VCH Verlag; 2012.

120. Samuelsson C, Björkman B. Copper recycling. In: Worrel E, Reuter MA, eds, *Handbook of recycling*, pp. 85–94. Amsterdam: Elsevier; 2014.
121. The Silver Institute, Metal Focus. World Silver Survey 2020. Washington, DC: The Silver Institute; 2020. Available at https://www.silverinstitute.org/world-silver-survey-2020/
122. Grehl M, Kempf B, Kleinwaechter I, et al. Precious metals. *Precious materials handbook*, pp. 110–295. Germany: Umicore; 2012.
123. Brumby A, Braumann P, Zimmermann K et al. Silver, silver compounds, and silver alloys. In: *Ullmann's encyclopedia of industrial chemistry*, vol. 33, p. 80. Weinheim, Germany: Wiley–VCH Verlag; 2008.
124. Renner H, Schlamp G, Hollmann D, et al. Gold, gold alloys, and gold compounds. In: *Ullmann's encyclopedia of industrial chemistry*, vol. 17, pp. 94–140. Weinheim, Germany: Wiley–VCH Verlag; 2000.
125. Mold gold cape. A history of the world in 100 objects. 2010. Available at http://www.bbc.co.uk/ahistoryoftheworld/objects/okZT5JiCTn6lYFR0Gs9Tbg
126. Gold supply and demand statistics. GoldHub 2020. Available at https://www.gold.org/goldhub/data/gold-supply-and-demand-statistics
127. Schwab B, Ruh A, Manthey J, Drosik M. Zinc. In: *Ullmann's encyclopedia of industrial chemistry*, pp. 1–25. Weinheim, Germany: Wiley–VCH Verlag; 2015.
128. Sinclair RJ. *The Extractive Metallurgy of Zinc*. Carlton, Australia: Australasian Institute of Mining and Metallurgy; 2005.
129. Antrekowitsch J, Steinlechner S, Unger A, et al. Zinc and residue recycling. In: Worrell E, Reuter MA, eds, *Handbook of recycling*, pp. 113–24. Amsterdam: Elsevier; 2014.
130. Goodwin FE, Rollez D. Zinc alloys. In: *Ullmann's encyclopedia of industrial chemistry*, pp. 1–7. Weinheim, Germany: Wiley–VCH Verlag; 2014.
131. Schulte-Schrepping K-H, Piscator M. Cadmium and cadmium compounds. IN: *Ullmann's encyclopedia of industrial chemistry*, vol. 6, pp. 465–81. Weinheim, Germany: Wiley–VCH Verlag; 2012.
132. Bartie NJ, Abadías Llamas A, Heibeck M et al. The simulation-based analysis of the resource efficiency of the circular economy—the enabling role of metallurgical infrastructure. *Miner. Process. Extr. Metall.*, 2020; 129(2): 229–49.
133. Abadías Llamas A, Bartie NJ, Heibeck M et al. Simulation-based exergy analysis of large circular economy systems: zinc production coupled to CdTe photovoltaic module life cycle. *J. Sustain. Metall.*, 2020; 6: 34–67.
134. First Solar, Inc. *2019 Annual Report*. Tempe, USA: 2020. Available at https://investor.firstsolar.com
135. Frank WB, Haupin WE, Vogt H et al. Aluminum. In: *Ullmann's encyclopedia of industrial chemistry*, vol. 2, pp. 483–519. Weinheim, Germany: Wiley–VCH Verlag; 2012.
136. The International Aluminium Institute. *Refining process*. 2018. Available at https://bauxite.world-aluminium.org/refining/process/
137. Lyle JP, Granger DA, Sanders RE. Aluminum alloys. In: *Ullmann's encyclopedia of industrial chemistry*, pp. 1–47. Weinheim, Germany: Wiley–VCH Verlag; 2000.
138. The International Aluminium Institute World Aluminium. Statistics—*Primary Aluminium Production*. 2020. Available at https://www.world-aluminium.org/statistics/
139. The International Aluminium Institute World Aluminium. Global aluminium cycle 2018. 2020. Available at https://alucycle.world-aluminium.org/
140. The International Aluminium Institute. Bauxite residue utilisation. 2018. Available at https://bauxite.world-aluminium.org/refining/bauxite-residue-utilisation/
141. Elysis. Available at https://elysis.com/

142. Sørlie M, Øye HA. *Cathode in aluminium electrolysis*. Dusseldorf, Germany: Aluminium-Verlag Marketing & Kommunikation; 2010.

143. Holywell G, Breault R. An overview of useful methods to treat, recover, or recycle spent potlining. *JOM*, 2013; 65: 1441–51.

144. Greber JF. Gallium and gallium compounds. In: *Ullmann's encyclopedia of industrial chemistry*, vol 16, pp. 335–40. Weinheim, Germany: Wiley–VCH Verlag; 2000.

145. Butcher T, Brown T. Gallium. In: Gunn G, ed., *Critical metals handbook*, pp. 150–76. New York: Wiley; 2014.

146. Liedtke M, Huy D. *Rohstoffrisikobewertung—Gallium*. Deutsche Rohstoffagentur (DERA), ed. DERA; 2018. Available at http://www.bgr.bund.de/DERA_Rohstoffinformationen

147. Felix N. Indium and indium compounds. In: *Ullmann's encyclopedia of industrial chemistry*, vol 19, pp. 65–74. Weinheim, Germany: Wiley–VCH Verlag; 2000.

148. Schwarz-Schampera U. Indium. In: Gunn G, ed., *Critical metals handbook*, pp. 204–29. New York: Wiley; 2014.

149. Nyrstar NV. *Auby smelter France*. Media factsheet; 2019: 2.

150. Hoang J, Reuter MA, Matusewicz R et al. *Top submerged lance direct zinc smelting*. EMC 2009. GDMB, 2009.

151. Davidson A, Ryman J, Sutherland CA, et al. Lead. In: *Ullmann's encyclopedia of industrial chemistry*, pp. 1–55. Weinheim, Germany: Wiley–VCH Verlag; 2014.

152. Scoyer J, Guislain H, Wolf HU. Germanium and germanium compounds. In: *Ullmann's encyclopedia of industrial chemistry*, vol. 16, pp. 629–41. Weinheim, Germany: Wiley–VCH Verlag; 2000.

153. Melcher F, Buchholz P. Germanium. In: Gunn G, ed., *Critical metals handbook*, pp. 177–203. New York: Wiley; 2014.

154. Elsner H, Schmidt M, Schuette P, Näher U. *Zinn—Angebot Und Nachfrage Bis 2020*. Deutsche Rohstoffagentur (DERA), ed. Berlin: DERA; 2014. Available at http://www.deutsche-rohstoffagentur.de

155. Roskill. *Tin: outlook to 2029*. 2019. Available at https://roskill.com/market-reports/

156. Sinclair RJ. *The extractive metallurgy of lead*. Carlton, Australia: Australasian Institute of Mining and Metallurg; 2009.

157. Dupont D, Arnout S, Jones PT, Binnemans K. Antimony recovery from end-of-life products and industrial process residues: a critical review. *J. Sustain. Metall.*, 2016; 2: 79–103.

158. *The World Lead Factbook 2019.International lead and zinc study group (ILZSG); 2019*. Available at https://ilzsg.org

159. Prengaman RD. Lead alloys. In: *Ullmann's encyclopedia of industrial chemistry*, pp. 1–12. Weinheim, Germany: Wiley–VCH Verlag; 2014.

160. Blanpain B, Arnout S, Chintinne M, Swinbourne DR. Lead recycling. In: Worrel E, Reuter MA, eds, *Handbook of recycling*, pp. 95–111. Amsterdam: Elsevier; 2014.

161. Grund SC, Hanusch K, Breunig HJ et al. Antimony and antimony compounds. In: *Ullmann's encyclopedia of industrial chemistry*, vol. 4, pp. 11–43. Weinheim, Germany: Wiley–VCH Verlag; 2012.

162. Schwarz-Schampera U. Antimony. In: Gunn G, ed., *Critical metals handbook*, pp. 70–98. New York: Wiley; 2014.

163. Roskill. *Antimony: global industry, markets and outlook, 2018*. Available at https://roskill.com/market-reports/

164. Dupont D, Binnemans K. Antimony recovery from the halophosphate fraction in lamp phosphor waste: a zero-waste approach. *Green Chem.*, 2016; 18(1):176–85.

165. Krüger J, Winkler P, Lüderitz E et al. Bismuth, bismuth alloys, and bismuth compounds. In: *Ullmann's encyclopedia of industrial chemistry*, vol. 6, pp. 113–33. Weinheim, Germany: Wiley–VCH Verlag; 2012.

166. Elsner H. *Bismut—Ein Typisches Sondermetall.* Deutsche Rohstoffagentur (DERA), ed. Berlin: DERA; 2015. Available at http://www.deutsche-rohstoffagentur.de

167. Champier D. Thermoelectric generators: A review of applications. *Energy Convers. Manag.*, 2017; 140: 167–81.

168. Nozariasbmarz A, Poudel B, Li W, et al. Bismuth telluride thermoelectrics with 8% module efficiency for waste heat recovery application. *iScience*, 2020; 23(7): 101340.

9

Refining of Steel in Converters

Olle Wijk

9.1   Introduction

Steel is by far the most dominating material and is crucial for the development of modern society. It is 100 per cent recyclable and has many applications, including buildings and construction, industrial installations, and all kinds of mechanical devices for the automotive industry, aerospace, and consumer goods. Of the annual global production (around 1800 Mton in 2018) about 25 per cent is based on recirculated scrap and 75 per cent is ore based. The ore-based steel production is 95 per cent produced via the blast furnace process, where the oxide ore is smelted with coke as the reducing agent. The scrap-based production is typically based on electric arc furnace (EAF) production.

In converter steelmaking the hot metal produced in a blast furnace, or a stainless crude steel with a high carbon and chromium content, usually melted in an EAF, is refined by the addition of oxygen gas to remove various dissolved elements. In the converters a crude steel is produced, which is further processed in a ladle for refining and final trimming of composition. Figure 9.1 gives schematic process lines for these two alternatives.

The first converter process for steelmaking was the Bessemer process, used for the refining of blast furnace hot metal. In this process air was blown through tuyeres in the bottom of a pear-shaped vessel in which mainly silicon, carbon, and iron for slag formation were oxidized. This process was invented by Henry Bessemer in the UK, but was developed to an industrial scale by Göran Fredrik Göransson in Sweden, who had bought 10 per cent of the patent from Henry Bessemer. The first successful blow was made on 18 July 1858 and this achievement revolutionized steelmaking.[1] The Bessemer process formed a platform for the mass production of steel. The converter had an acid lining and phosphorus refining was not possible. This was later solved in the Thomas process, which had a basic lining and was well suited for removal of phosphorus.

Converter steelmaking, especially for the production of low-alloy steels, has been extensively treated in the literature.[2–6] In this chapter, refining of low-alloy steels in converters will be described generally. Converter steelmaking of stainless steels is treated in the latter part of the chapter and, as a modelling example, the principles presented in previous chapters of this book will be used to develop a process model for the decarburization of stainless steels in bottom- and side-blown converters. Based on calculations regarding stainless steelmaking, some critical process parameters will be discussed.

Principles of Metal Refining and Recycling. Thorvald Abel Engh, Geoffrey K. Sigworth, and Anne Kvithyld, Oxford University Press.
© Oxford University Press 2021.
DOI: 10.1093/oso/9780198811923.003.0009

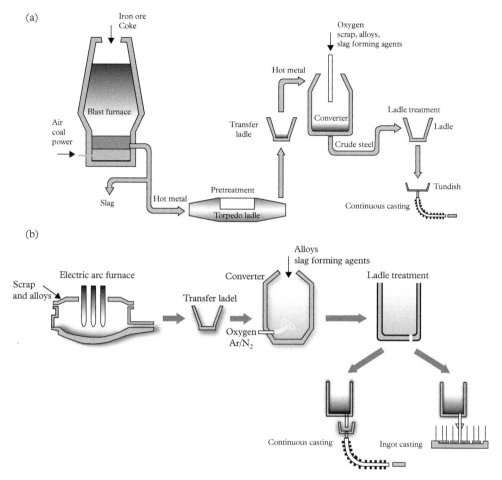

Fig. 9.1 *Process routes for blast furnace- and EAF-based converter steel-making. (a) Iron ore steelmaking based on the blast furnace. (b) Scrap-based stainless steelmaking.*

For a full understanding of the chapter it is necessary for the reader to have a fundamental knowledge of the thermodynamics of high-temperature metallic solutions and slag chemistry.

9.2 Converter Processes for Steelmaking—Refining of Blast Furnace Hot Metal

9.2.1 General Principles

In an integrated ore-based steel plant, hot metal produced in the blast furnace is charged into the converter. Typically 90 per cent of the sulfur is removed by the blast furnace slag. However, the sulfur remaining is frequently still too high, so a pretreatment of the hot metal is common. In some cases desiliconizing and phosphorus refining is also made before the converter.

Typical compositions of the blast furnace pig iron and some standard steel grades are given in Table 9.1.[7] In addition to producing steel of the desired composition, it is important to optimize the metal yield and control the heat balance of the process. Heat produced in the converter during the oxygen blow is utilized to increase melt temperature, for slag formation and for melting of scrap and alloy additions.

The biggest challenge in early converter steelmaking was to manage the heat balance. This can be understood by considering the liquidus temperature of molten iron as a function of the carbon content (Fig. 9.2).[8] As seen in Table 9.1, a blast furnace hot metal normally contains a carbon content close to the eutectic point, where the liquidus temperature is 1154 °C. A typical starting temperature in the converter is 1300 to 1350 °C and, as the carbon content decreases, the liquidus temperature increases. In the nineteenth century and until the early 1950s, air was used as the oxidizing agent. In the converter process large amounts of nitrogen in the air had to be heated up and gave a big negative load on the heat balance. For a successful process, very high blow rates and limited additions to the melt were needed.

When oxygen gas started to be used in converter steelmaking, the concept of using a top lance was adopted from the beginning. Today the predominant converter process is the top-blown Linz Donawitz (LD), also called the basic oxygen furnace (BOF) process.

In Fig. 9.3 an operational sequence for a LD converter is shown schematically. The vessel is pear-shaped and can be tilted. The converters normally have a basic lining, usually a brick-based on tar/carbon-bonded magnesia. The lance is water-cooled with typically 4–6 nozzles. During the past decades inert gas (Ar or N_2) has been frequently injected through tuyeres or porous plugs in the bottom of the converter. This improves the mixing between top slag and metal bath. There are also plants where extra oxygen is added above the bath to burn off the carbon monoxide, thereby improving the heat balance of the process.

Table 9.1 *Composition (wt.%) of a typical blast furnace hot metal and some standard steel grades*

	Hot metal	SAE 1008	SAE 1330	SAE 4118
C	4–5	≤ 0.10	0.28–0.33	0.18–0.23
Si	0.2–0.8	—	0.15–0.35	0.15–0.35
Mn	0.5–1.5	0.30–0.50	1.60–1.90	0.70–.90
P	0.05–0.2	≤ 0.040	≤ 0.035	≤ 0.035
S	0.01	≤ 0.050	≤ 0.040	≤ 0.040
Cr	—	—	—	0.40–0.60
Mo	—	—	—	0.08–0.15

Source: Based on data from Bringas.[7]

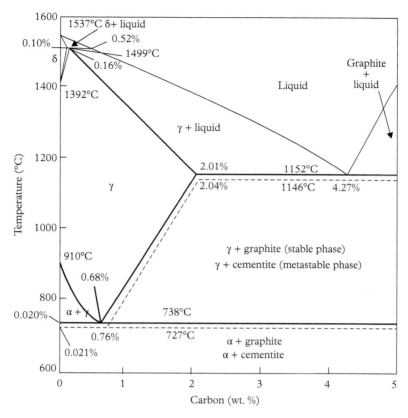

Fig. 9.2 *Iron–carbon phase equilibrium diagram. Dashed lines represent phase boundary for metastable equilibrium with cementite. [Reprinted from Fruehan[8] by permission from AIST Association for Iron & Steel Technology: Copyright 1998.]*

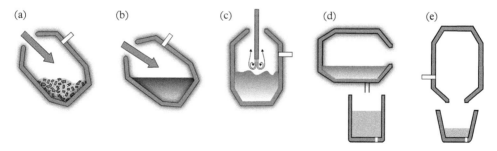

Fig. 9.3 *LD-converter production steps: (a) charging of scrap, (b) charging of hot metal, (c) oxygen blow, (d) tapping into a casting ladle, and (e) slag tapping*

Table 9.2 *Typical data regarding LD converters*[9, 10]

Heat size	100–250 ton
Hot metal per heat	75–90%
Scrap	10–25%
Lance tip holes	4–6
Blowing rate, oxygen	2–5 Nm^3/ton.min
Gas purging	0.02–0.03 m^3n/ton.min
Number of heats per day	25–40
Slag weight	50–100 kg per ton

As an alternative to the LD process, there also exist processes based on bottom-blowing oxygen through tuyeres, together with hydrocarbons for cooling the tuyeres. There is also the possibility of injecting slag-forming agents. These processes will not be considered here.

Some typical data for LD converters are given in Table 9.2.[9,10]

It is interesting to note the high productivity of the converter process. For example, if 30 heats per day are produced, and the heat size is 150 tons, this corresponds to 4 500 tons per day, 31 500 tons per week, or an annual production of more than 1.5 Mton.

9.2.2 Some Thermodynamic Aspects

Table 9.1 shows a typical composition of the blast furnace hot metal. In a converter process total equilibrium cannot be expected, but it is important to consider the chemical equilibria to better understand different process phenomena.

The oxidation of an element Me with oxygen gas can be described by the overall reaction

$$2\frac{x}{y}Me + O_2 \text{ (g)} = \frac{2}{y}Me_xO_y.$$ (9.1)

The free energy of formation for reaction (9.1) is given by the expression

$$\Delta G^0 = -RT \ln \left[\frac{(a_{MexOy})^{\frac{2}{y}}}{(a_{Me})^{\frac{2x}{y}}(P_{O2})} \right].$$ (9.2)

If pure elements and pure oxides are assumed, their activities are set to unity and the oxygen partial pressure as a function of temperature is given by:

$$\Delta G^0 = RT \ln P_{O2}.$$ (9.3)

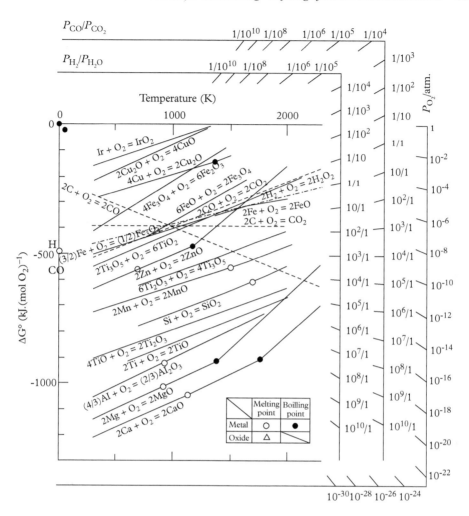

Fig. 9.4 *Ellingham diagram for the formation of pure oxides.*[11]

Figure 9.4 gives ΔG^o for the formation for different oxides for an oxygen gas pressure of 1 atm in an Ellingham diagram.[11] Note that each reaction is written so that one mole of oxygen gas (O_2) is consumed. Hence, a lower value of ΔG^o in the diagram means a more stable oxide. Some elements of interest in converter steelmaking are Fe, C, Si, and Mn. It is obvious that if pure elements and oxides are regarded, we would have the following sequence of oxidation at 1550 °C: Si, Mn, C, and Fe. It is interesting to note the negative slope of the line representing the formation of CO. This may be understood by considering that in the reaction we consume one mole of gas and two moles of gas are formed. This increases the entropy term, ΔS^o:

$$\Delta G^o = \Delta H^o - T\Delta S^o. \tag{9.4}$$

In metallurgical operations where the reaction products dissolve in a slag phase, or if the gaseous reaction product is diluted, the corresponding line in the Ellingham diagram moves clockwise, illustrating that the reaction goes to the right. This effect is utilized in pyrometallurgical processes including molten metals and slags. The activity of the oxides is controlled by the slag composition. At the same time the activity of the elements dissolved in the metal bath is influenced by the interaction between the different elements.

It is crucial to understand the effect of slag basicity, metal bath composition, and temperature on the equilibria. In addition, kinetic aspects must also be considered.

In a LD converter, a top slag is formed during the blow with the main constituents being CaO, SiO$_2$, and FeO. The CaO content is obtained by the addition of lime as a slag-forming agent, where the kinetics of the lime dissolution is very important for the process. Besides these components, the slag also contains some MgO, mainly because of refractory wear and the addition of dolomite to protect the refractory. There will also be small amounts of MnO and P$_2$O$_5$, as well as residuals formed by the oxidation of other elements from the hot metal. Small amounts of fluorspar are sometimes added, which adds CaF$_2$ to the slag. This improves the dissolution of added lime and sulfur removal.

In the following we first consider the equilibria between oxygen and the elements oxidized: C, Si, and Fe. After this the equilibria for phosphorus refining is treated, as phosphorus is generally regarded as an undesirable impurity in the finished steel. Equilibrium data between oxygen and iron, and between dissolved elements, are easily found in the literature. However, it is more instructive to calculate the important equilibria from thermodynamic data. Keep in mind that some approximations are used in the following calculations. For instance, second-order interaction parameters are not taken into account. This simplification is justified, since the aim here is to provide a better understanding of the complex dynamics of converter processes. First-order interaction parameters used in the following equilibrium calculations were taken from Table 2.7.

9.2.2.1   Carbon–Oxygen

The equilibrium between carbon and oxygen dissolved in the molten bath is

$$\underline{C} + \underline{O} = CO(g) \tag{9.5}$$

$$K_{9.5} = \frac{P_{CO}}{a_C \cdot a_O}, \tag{9.6}$$

where P_{CO} is the partial pressure of CO in the gas phase and where a_C and a_O are the carbon and oxygen activities in the metal bath, respectively. The equilibrium constant K is given by[12]

$$\log K_{9.5} = \frac{1160}{T} + 2.003. \tag{9.7}$$

From Eqs. (9.6) and (9.7) we get

$$K_{9.5} = 10^{\frac{1160}{T} + 2.003} = \frac{P_{CO}}{f_C[\%C]f_O[\%O]} \tag{9.8}$$

and

$$\%O = \frac{P_{CO}}{f_C[\%C]f_O 10^{\frac{1160}{T}+2.003}},$$ (9.9)

where f_C and f_O are the activity coefficients for carbon and oxygen, respectively:

$$\log f_C = e_C^C[\%C] + e_C^O[\%O]$$ (9.10)

$$f_c = 10^{\left(e_C^C[\%C]+e_C^O[\%O]\right)}$$ (9.11)

$$\log f_O = e_O^O[\%O] + e_O^C[\%C]$$ (9.12)

$$f_O = 10^{\left(e_O^O[\%O]+e_O^C[\%C]\right)}.$$ (9.13)

A combination of Eqs. (9.9)–(9.13) gives

$$\%O = \frac{P_{CO}}{10^{\frac{1160}{T}+2.003} 10^{\left(e_C^C[\%C]+e_C^O[\%O]\right)}[\%C] \times 10^{e_O^O[\%O]+e_O^C[\%C]}}.$$ (9.14)

The result of a calculation at $P_{CO} = 1$ and $1600\,°C$ (1873 K) is shown in Fig. 9.5.

As can be seen the equilibrium oxygen content decreases rapidly with increasing carbon content. At 0.1% C it is around 220 ppm and at 1% C it is around 40 ppm. To better understand the converter process, it is now interesting to compare these equilibrium oxygen contents with the equilibria between oxygen and iron and silicon, respectively.

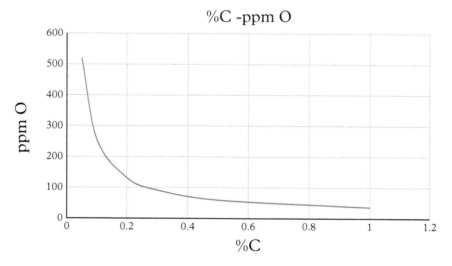

Fig. 9.5 *Equilibrium between carbon and oxygen in liquid iron at 1600 °C.*

9.2.2.2 Iron

The oxidation of liquid iron and oxygen is given by

$$Fe(l) + \underline{O} = FeO \text{ (in slag)} \tag{9.15}$$

$$K_{9.15} = \frac{a_{FeO}}{a_{Fe}a_O} = \frac{1}{f_O[\%O]}, \tag{9.16}$$

where f_O is the activity coefficient for oxygen in the melt given by

$$\log f_O = e_O^O[\%O]. \tag{9.17}$$

The equilibrium constant K as a function of temperature is given by[13]

$$\log K_{9.15} = -\frac{6324}{T} + 2.742 \tag{9.18}$$

This gives

$$K_{9.15} = 10^{-\frac{6324}{T}+2.734} = \frac{a_{FeO}}{f_O[\%O]} \tag{9.19}$$

$$\%O = \frac{a_{FeO}}{10^{e_O^O[\%O]}10^{-\frac{6324}{T}+2.742}}. \tag{9.20}$$

Assuming that $T = 1873$ K (1600 °C) and that $a_{FeO} = 1$ results in an equilibrium oxygen content of 0.25%.

We now consider the FeO activities in the slag system of interest in converter steelmaking. See Fig. 9.6.[14] From this diagram it is possible to calculate the equilibrium oxygen contents at different slag compositions. From Eq. (9.20) we find the following results at $T = 1873$ K:

a_{FeO}	$\%O_{equilibrium}$
1	0.25
0.5	0.12
0.3	0.075
0.1	0.025
0.05	0.0125

Referring to the equilibria between carbon and oxygen, 0.025% O corresponds to a carbon content of about 0.1%, indicating that if there were a total equilibrium in the system, no or very limited oxidation of iron would occur at higher carbon contents of the metal bath.

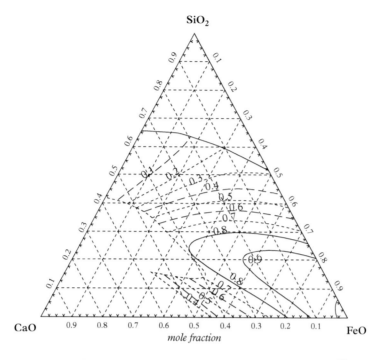

Fig. 9.6 *Iso activity curves for FeO in the slag system CaO–FeO–SiO₂ at 1550°C.*[14] *[Reprinted by permission from Springer Nature, Copyright 1970.]*

9.2.2.3 Silicon

The equilibrium between silicon and oxygen dissolved in the metal bath can be written

$$\underline{Si} + 2\underline{O} = SiO_2 \text{ (in slag)} \tag{9.21}$$

$$K_{9.21} = \frac{a_{SiO_2}}{a_{Si}a_O^2}. \tag{9.22}$$

The equilibrium constant K as a function of temperature is given by[12]

$$\log K_{9.21} = \frac{30110}{T} - 11.40, \tag{9.23}$$

which gives

$$K_{9.21} = \frac{a_{SiO_2}}{a_{Si}(a_O)^2} = \frac{a_{SiO_2}}{f_{Si}[\%Si](f_O[\%O])^2} = 10^{\frac{30110}{T} - 11.40}, \tag{9.24}$$

where f_{Si} and f_O are the activity coefficients for silicon and oxygen, respectively, given by

$$\log f_{Si} = e_{Si}^{Si}[\%Si] + e_{Si}^{O}[\%O] \tag{9.25}$$

$$\log f_O = e_O^O[\%O] + e_O^{Si}[\%Si]. \tag{9.26}$$

A combination of Eqs. (9.24)–(9.26) gives

$$\%O = \frac{1}{f_O}\sqrt{\frac{a_{SiO_2}}{f_{Si}[\%Si]\,10^{\frac{30110}{T}-11.40}}}. \tag{9.27}$$

At $1600\,^{\circ}C$ $(1873\ K)$ we obtain the result

%Si	ppm $\underline{O}$
0.5	61
0.3	81
0.1	143
0.05	205

In basic slags the activity of SiO_2 is lowered with increasing basicity (Fig. 9.7), which combined with the strong affinity of silicon to oxygen results in a very low equilibrium oxygen content.

These results suggest that little oxidation of iron will occur, until most of the Si and C are removed. In practical steelmaking, the sequence of oxidation of these elements is far from equilibrium. The actual reactions are described in Section 9.2.3.

9.2.2.4   Phosphorus

The overall reaction for phosphorus oxidation can be written as[15,16]

$$2\underline{P} + 5\underline{O} = P_2O_5 \text{ (in slag)} \tag{9.28}$$

$$K_{9.28} = \frac{\gamma_{P_2O_5}X_{P_2O_5}}{(a_O)^5(a_P)^2}, \tag{9.29}$$

where $X_{P_2O_5}$ is the molar fraction of P_2O_5 in the slag and $\gamma_{P_2O_5}$ is the activity coefficient.

$$\log K_{9.28} = \frac{36\,850}{T} - 29.07. \tag{9.30}$$

The phosphorus reaction is enhanced by a low temperature, a high oxygen potential, and a low activity coefficient of P_2O_5 in the slag.

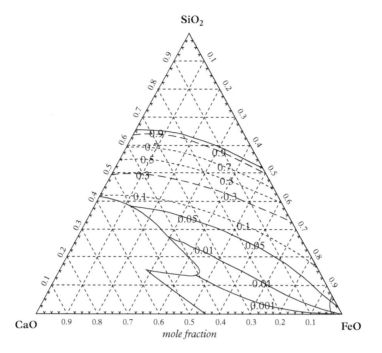

Fig. 9.7 *Activity of SiO₂ in the slag system CaO–FeO–SiO₂, 1550 °C.*[14] [*Reprinted by permission from Springer Nature, Copyright 1970.*]

The activity coefficient of phosphorus in the metal bath is influenced by different dissolved elements. Table 2.7 gives the first-order interaction parameters between a third element and phosphorus dissolved in liquid iron. As can be seen the activity coefficient for phosphorus is increased by C, O, Si, and N while it is decreased by Cr.

The effect of iron oxide content in the slag on the equilibrium phosphorus distribution between basic slags and molten iron is given in Fig. 9.8.[17] The phosphorus distribution increases to a maximum, after which it decreases. This can be understood as caused by a combined effect of oxygen potential and dilution of basic oxides in the slag by iron oxide.

In the literature, metallurgical slags used to remove phosphorus are often characterized using the concept of phosphate capacity. This was first introduced by Wagner[18] based on the reaction

$$1/2 P_2 \text{ (g)} + 5/4 O_2 \text{ (g)} + 3/2 \, O^{2-} \text{ (in slag)} = PO_4^{3-} \text{ (in slag)} . \qquad (9.31)$$

This demonstrates the importance of a basic slag resulting in a high activity of (O^{2-}) in the slag and a low activity of $(PO_4{}^{3-})$.

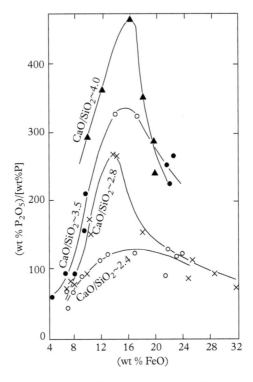

Fig. 9.8 *Effect of the FeO content in steelmaking slags on the distribution of phosphorus between slag and metal, 1685 °C.*[17]

The equilibrium constant is given by

$$K_{9.31} = \frac{\left(a_{PO_4^{3-}}\right)}{(P_{P_2})^{\frac{1}{2}}(P_{O_2})^{\frac{5}{4}}(a_{O^{2-}})^{\frac{3}{2}}}. \tag{9.32}$$

Based on the equilibrium constant Wagner[18] defined the phosphate capacity according to

$$C_{PO_4^{3-}} = \frac{K_{9.31}(a_{O^{2-}})^{\frac{3}{2}}}{f_{PO_4^{3-}}} = \frac{(\%PO_4^{3-})}{(P_{P_2})^{\frac{1}{2}}(P_{O_2})^{\frac{5}{4}}}. \tag{9.33}$$

To avoid the term including gaseous phosphorus the reaction for phosphorus refining can also be written as

$$\underline{P} + 5/2\underline{O} + 3/2O^{2-} \text{ (in slag)} = PO_4^{3-} \text{ (in slag)} \tag{9.34}$$

$$K_{9.34} = \frac{a_{PO_4^{3-}}}{a_P (a_O)^{5/2} (a_{O^{2-}})^{3/2}}. \tag{9.35}$$

From this reaction the phosphide capacity can be defined as

$$C'_{PO_4^{3-}} = \frac{K_{9.34}(a_{O^{2-}})^{\frac{3}{2}}}{f_{PO_4^{3-}}} = \frac{(\%PO_4^{3-})}{(a_P)(a_O)^{\frac{5}{2}}}. \tag{9.36}$$

The phosphate capacity as defined by Wagner (Eq. (9.33)) and as defined in expression (9.36) is related to the two reactions[19]

$$\frac{1}{2} O_2(g) = \underline{O}$$

$$\log K_{9.37} = \frac{6120}{T} + 0.155 \tag{9.37}$$

$$\frac{1}{2} P_2(g) = \underline{P}$$

$$\log K_{9.38} = \frac{6382}{T} + 1.01. \tag{9.38}$$

Thus

$$C'_{PO_4^{3-}} \text{ (Eq.(9.36))} = C_{PO_4^{3-}} \text{ (Eq.(9.33))} K_{9.37}^{-5/2} \cdot K_{9.38}^{-1}. \tag{9.39}$$

The phosphate capacity defined according to Eq. (9.33) is much bigger than that of Eq. (9.36). Assis and co-workers[20] reviewed earlier studies, conducted careful experiments of their own, and concluded the best way to approach the problem of phosphorus removal is to employ a correlation based on Eq. (9.25). However, the partial pressure of oxygen was replaced by % FeO in the slag. (This results from the reaction in Eq. (9.15).) In other words,

$$\frac{(\%P)_{in\ slag}}{[\%P]_{in\ metal}(\%FeO)^{\frac{5}{2}}} = f\,(\%CaO + 0.148\%MgO + 0.96\%P_2O_5 + 0.144\%SiO_2) \tag{9.40}$$

This equation gives the phosphorus distribution between slag and metal, $L_P = (\%P)/[\%P]$, as a function of the slag chemistry. The resulting correlation is shown in Fig. 9.9.[20] This correlation applies to conditions found during steel refining, such as in BOF steelmaking.[1]

---

[1] In addition to FeO, small amounts of other iron oxides may also be present in the slag (Fe_2O_3 and Fe_3O_4). In Fig. 9.9 all iron was assumed to be present as FeO. In other words, T.Fe represents the total iron, or the %FeO which corresponds to the total iron content of the slag.

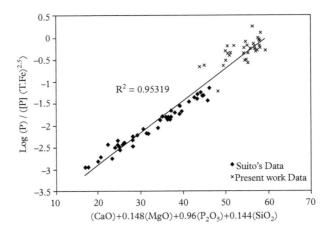

Fig. 9.9 *The equilibrium distribution of phosphorus between slag and metal as a function of steelmaking slag chemistry.*[20] *[Reprinted from Assis et al.*[20] *by permission from Springer Nature: copyright 2015.]*

9.2.3   Reactions during the Blow

The reactions in converter steelmaking are quite complex; see Fig. 9.10. Except for the removal of carbon, which gives a gaseous reaction product, oxides formed during the oxygen blow dissolve in the slag phase. The slag chemistry is crucial for control of the process. The slag basicity is adjusted by the addition of lime. The amount added is based on the amount of silicon contained in the hot metal and in added alloys and scrap. As mentioned in Section 9.2.2, dolomite is usually added to minimize dissolution of the magnesia-containing lining of the converter vessel.

When the supersonic oxygen jets hit the metal surface, because of the predominant presence of iron atoms compared to the dissolved elements, a good deal of iron oxidizes in front of the lance tip. This occurs despite the chemical equilibria calculated in Section 9.2.2. Simultaneously metal droplets are formed due to the forces created by the oxygen jets acting on the metal bath. These droplets react mainly with FeO dissolved in the slag phase. A great deal of carbon monoxide is evolved during most of the blow, resulting in a foam which contains numerous CO bubbles and suspended droplets of liquid iron. If the operator is not careful, this foam may expand sufficiently to overflow from the mouth of the converter. The process is controlled mainly by the distance between lance tip and metal bath. A longer distance between the lance and the metal bath and a decreased oxygen gas flow promote the oxidation of iron to FeO, while a short distance and an increased gas flow promote stirring of the metal bath and oxidation of dissolved elements. The possible introduction of gas purging from the bottom is another tool that can be used to control the process.

In Table 9.3 the enthalpy change for the oxidation of the different elements is given; however, the oxidation of chromium is also included. (Chromium is treated later when we

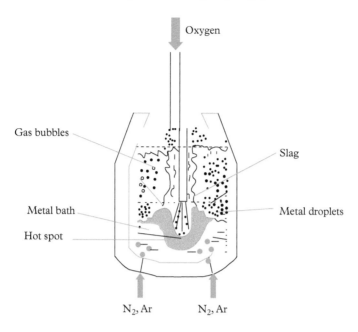

Fig. 9.10 *Schematic drawing of the reaction zones in an LD converter. (Based on Jalkanen and Holappa.[5] Copyright Elsevier 1978.)*

Table 9.3 *Heats of formation for various oxides from oxygen gas and elements dissolved in steel and the corresponding temperature increase in the steel bath*

Reaction	Enthalpy change (kJ/mole)	Temperature increase (°C) (theoretical per weight % of oxidized element, formation of pure oxides)
$Fe(l) + \frac{1}{2}O_2 = FeO$	−256.3	46
$\underline{C} + \frac{1}{2}O_2 = CO\ (g)$	−139.8	109
$\underline{Si} + O_2 = SiO_2$	−825.9	333
$\underline{Mn} + \frac{1}{2}O_2 = MnO$	−404.3	74
$2\underline{P} + \frac{5}{2}O_2 = P_2O_5$	−1328	202
$2\ \underline{Cr} + \frac{1}{2}O_2 = Cr_2O_3$	−1141.6	108

consider the production of stainless steels.) It is evident that these reactions are all exothermic and contribute to the heat balance of the process.

The change in the metal composition during the refining of hot metal in converters can be divided into three periods[2] and is illustrated in Fig. 9.11.

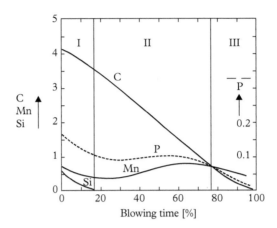

Fig. 9.11 *Composition of metal bath during LD steelmaking; schematic diagram after Edström et al.*[2]

Period I: This part of the blow is characterized by a pronounced oxidation of silicon due to its strong affinity for oxygen. The high oxygen potential in the gas phase results in iron oxidation, so the first slag to form is composed mainly of FeO and SiO_2. This results in a rapid temperature increase, up to 40 °C per minute. As CaO dissolves into the slag, the basicity increases and a decrease of the phosphorus content occurs. This is enhanced by the relatively low temperature at the beginning of the blow. Also some manganese is oxidized during this phase.

Period II: After the initial period, the carbon removal rate reaches its maximum due to a very pronounced stirring and a formation of a foaming slag, where metal droplets react with the over-oxidized slag. As the blow continues, the oxygen potential in the gas phase decreases and the formation of FeO to the slag is not as significant as in the beginning of the blow. Also the endothermic reaction between C and FeO results in a lower temperature in the slag phase and a decreased FeO content. This limits the dissolution of lime into the slag, and combined with the decreased FeO content, there is a reversal of Mn and phosphorus into the metal bath.

Period III: The carbon removal rate decreases. This is due to a low carbon content, where the rate-limiting step for the decarburization is diffusion of carbon in the metal bath. This gives an increased FeO content in the slag which results in a further dissolution of lime into the slag and a decrease of manganese and phosphorus in the metal bath.

Typically the slag after the oxygen blow contains 45–50% CaO, 20–25% FeO, 15–20% SiO_2, 4–6% MnO, and 3–5% MgO.[2] The composition varies of course depending on which steel grade that is produced and the blowing practice. The phosphorus distribution, $L_P = (\%P)/[\%P]$, varies typically between 100 and 200.

Figure 9.12 gives the off-gas analysis and lance height in a 200-ton LD converter which confirms what has been said above. In the beginning of the blow iron and silicon are oxidized. A limited carbon oxidation results in a low gas volume with a high oxygen potential. The CO_2

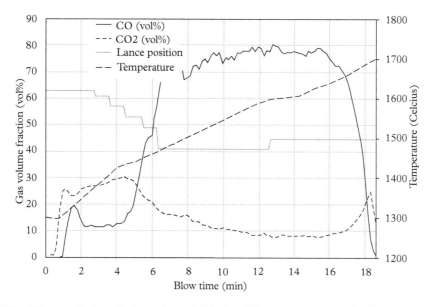

Fig. 9.12 *Off-gas analysis and relative lance height in a 200-ton LD converter.* [*Published with permission from SSAB, Stockholm, Sweden.*]

content in the off-gas is relatively high. In the diagram we can also see the stepwise lowering of the lance, resulting in a more intense oxidation of carbon, giving a high CO content in the off-gas and a decreased content of CO_2. At the end of the blow, the CO content decreases rapidly during the oxidation of mainly iron.

9.3 Converter Processes for Steelmaking—Refining of Stainless Crude Steel

9.3.1 Introduction

For many years stainless steels were produced in induction furnaces with expensive low-carbon raw materials and in electric arc furnaces, where oxygen for decarburization was added to the melt through a consumable steel tube. Very high temperatures, often above 2000 °C, were needed to prevent excessive chromium oxidation. This resulted in a very heavy refractory wear, and the production was also characterized by low productivity and high consumption of expensive low-carbon ferrochromium.

The most common process route for stainless steelmaking today is shown in Fig. 9.1. Scrap and alloys are melted in an EAF while decarburization, reduction of the top slag for recovery of chromium, sulfur removal, and adjustment of the steel composition are carried out in a converter. Very often the converter is complemented with a ladle treatment for deoxidation, final trimming and stirring for homogenization, and removal of non-metallic inclusions.

Another process route includes melting in an EAF and vacuum treatment in a ladle. Both process routes show benefits and drawbacks, which have resulted in combinations with a converter followed by vacuum treatment in a ladle. In the following, however, the process route illustrated in Fig. 9.1 is treated. The size of AOD converters has increased very much during the past decades as stainless steels have become more of a commodity product. Many steel shops today have one or several converters with heat sizes of 100–180 tonnes, with annual production capacities of up to 1 Mton.

The EAF is charged with scrap, alloys, and slag-forming agents. The carbon content of the molten metal after it has left the EAF depends on the supply of raw materials. If there is a lack of stainless steel scrap, more high-carbon ferroalloys must be used and the carbon content is increased in the crude steel. A typical composition is 1–3.5% C, 0.1–0.5% Si, and 0.6–1.2% Mn. Other elements are adjusted to meet the demands of the finished steel product.

A comparison of the composition of the metal from the EAF with the composition of some stainless steels according to international standards, Table 9.4, indicates that the main metallurgical problem in stainless steelmaking is to decrease the carbon content in the steel despite a high chromium content. Having a process that makes this possible is of fundamental importance for minimizing the consumption of expensive low-carbon ferroalloys.

If we only consider dissolved carbon and chromium, the two following reactions occur during the addition of oxygen:

$$\underline{C} + \frac{1}{2}O_2 = CO(g) \tag{9.41}$$

$$2\underline{Cr} + \frac{3}{2}O_2 = Cr_2O_3 \text{ (in slag)}. \tag{9.42}$$

Table 9.4 *Composition of some standard stainless steel grades*

Element	SAE 304 L	SAE 316 L	SAE 310 S
C	≤ 0.030	≤ 0.030	≤ 0.08
Si	≤ 1.0	≤ 1.0	≤ 1.5
Mn	≤ 2.0	≤ 2.0	≤ 2.0
P	≤ 0.045	≤ 0.045	≤ 0.045
S	≤ 0.030	≤ 0.030	≤ 0.030
Cr	17.2–20.0	16.0–18.5	24.0–26.0
Ni	9.0–12.0	11.0–14.0	19.0–22.0
Mo	—	2.0–2.5	—

Source: Based on data from Bringas.[7]

These two competitive reactions give the overall reaction which can be written as

$$Cr_2O_3 \text{ (in slag)} + 3\underline{C} = 2\underline{Cr} + 3CO(g) \tag{9.43}$$

$$\ln K_{9.43} = \ln\left[\frac{(a_{Cr})^2 (p_{CO})^3}{(a_C)^3 a_{Cr_2O_3}}\right] = -\frac{88704}{T} + 56.67, \tag{9.44}$$

where a_{Cr} is the chromium activity in the metal bath, P_{CO} is the partial pressure of CO in the gas phase, a_C is the carbon activity in the metal bath, and a_{Cr2O3} is the chromium oxide activity in the top slag.

From Eq. (9.43) it can be seen that the carbon oxidation is favoured by a low chromium activity and a high carbon activity in the metal, a high chromium oxide activity in the slag phase, and a low partial pressure of carbon monoxide in the gas phase. And from Eq. (9.44) we see the carbon oxidation is enhanced by a high temperature.

Figure 9.13 shows the equilibrium carbon and chromium contents in liquid steel in contact with pure chromium oxide ($a_{Cr2O3} = 1$) at varying temperatures and partial pressures of carbon monoxide. For instance, for a steel with 15% Cr at 1550 °C and $P_{CO} = 1$, it is thermodynamically possible to decrease the carbon content to 0.55% without any oxidation of chromium. If the temperature is increased to 1700 °C, the corresponding carbon content is decreased to 0.18%. We can also see that by decreasing the partial pressure of carbon monoxide in the gas phase to 0.1, it is possible to decrease the carbon content to 0.06% at 1700 °C.

In this example it was assumed that only chromium and carbon were dissolved in the metal bath. However, when calculating the equilibrium for the carbon oxidation, it is necessary to account for the effect of all elements in the metal bath on the carbon and chromium activities.

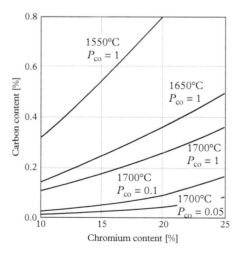

Fig. 9.13 *Equilibrium contents of C and Cr in a steel bath in contact with pure chromium oxide.*

Table 9.5 *Second-order interaction coefficients between carbon and some elements in liquid iron*[21]

$$\log f_c = \sum_{j=2}^{k} e_c^j [\%j] + r_c^j [\%j]^2$$

$$r_c^c = -4.70 \times 10^{-2} + 177.2/T - 1.5 \times 10^3/T^2$$

$$r_c^{Cr} = 7.95 \times 10^{-4} - 1.462/T$$

$$r_c^{Ni} = 0$$

Elements with a positive interaction on carbon, such as Ni, increase the carbon activity and enhance the decarburization reaction, while elements with a negative interaction, such as Cr and Mo, have the opposite effect. Table 9.5 gives the temperature-dependent second-order interaction coefficients between carbon and a third added element. The first-order interaction coefficients are found in Table 2.7. If we go back to the example in Fig. 9.13 and make an addition of, for instance, 8% Ni to the melt which originally contained 18% Cr, it is theoretically possible, at 1700 °C and $P_{CO} = 0.1$, to further decrease the carbon content from 0.06 to 0.04%.

The chromium oxide activity in the slag also plays an important role in the decarburization reaction. However, due to the low chromium oxide solubility in the type of slags used for steelmaking,[22] the chromium oxide activity will be close to unity during decarburization.

As shown in Eq. (9.44) and Fig. 9.13, the temperature is a fundamental parameter in the production of stainless steels. In converter processes, the thermal heat required is supplied by the oxidation of various elements dissolved in the metal bath. Table 9.3 gives the heat of formation for different oxides from elements dissolved in the liquid iron when oxygen gas is added. The table also gives the theoretical temperature increase for the oxidation of 1% of each dissolved element. The oxidation of carbon alone does not provide sufficient heat. Oxidation of other elements such as Si and Cr, at least during some periods of the blow, is necessary.

The thermodynamic equilibria give the theoretically possible compositions of the reacting phases. However, in practical steelmaking, the kinetics of the different reactions plays a fundamental role in the processes. This will be discussed in detail in Section 9.4.

The first to utilize the influence of a decreased partial pressure of carbon monoxide on chromium oxidation was WA Krivsky. Working at the Union Carbide Corporation at the Metals Research Laboratories in Niagara Falls, USA, he conducted experiments on a 220-kg scale, where injected oxygen was diluted with argon during the decarburization of stainless steel melts. This work resulted in the argon oxygen decarburization (AOD) process which made an industrial breakthrough at the end of the 1960s. Today, this original converter process for stainless steelmaking dominates production. It has, however, been followed by other similar processes based on the same fundamental principles.

An AOD converter of original design is shown in Fig. 9.14. Oxygen gas and an inert gas for the dilution of the carbon monoxide produced are added through tuyeres placed in the wall near the bottom of the vessel. Normally around 1 Nm³ of gas per tonne of steel per

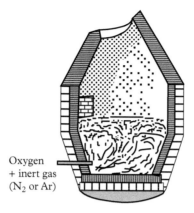

Fig. 9.14 *Argon oxygen decarburization (AOD) converter.*

minute is added through the tuyeres. In later designs, the addition of gas from the bottom has been complemented by the addition of oxygen or a mixture of oxygen and an inert gas through a top lance. The impact of this extra addition is to increase the productivity of the process, especially when the crude steel charged into the converter has a high initial carbon content.

As general background and as a base for the modelling, we now continue with a description of stainless steelmaking in an AOD converter without a top lance. The processing time can be divided into three main periods: decarburization, reduction of the top slag, and desulfurization.

9.3.2 Decarburization

The crude steel from the EAF typically has a temperature of 1550 °C when it is charged into the converter. During the blow, the oxygen gas is diluted with an inert gas which is usually nitrogen in the beginning of the blow, and argon towards the end. The justification for the use of expensive argon is the removal of dissolved nitrogen in the steel. Dilution of the oxygen is usually carried out in different steps where the relation between oxygen and the inert gas is changed stepwise, for instance, in the ratios 3:1, 1:1, 1:3, and 1:5. Figure 9.15 illustrates the change of the carbon content in the metal bath, the bath temperature, and the carbon removal efficiency (CRE) number during the blow. The CRE number is a measure of how much of the oxygen reacts with carbon and is defined by

$$CRE\,(\%) = \frac{\text{amount of oxygen reacting with carbon} \times 100}{\text{total added oxygen}}. \qquad (9.45)$$

In Fig. 9.15 we can see that the CRE number is initially low, since at the beginning of the blow the oxygen reacts with Si, Mn, and Cr. As the content of Si and Mn decreases, the slag becomes chromium-saturated and the temperature increases, the carbon starts to oxidize, and the CRE number increases. However, as the carbon activity in the metal bath decreases,

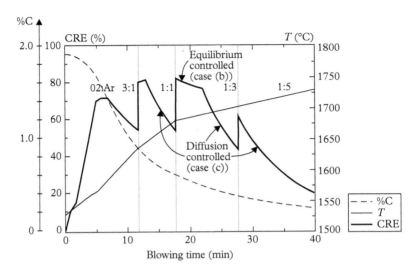

Fig. 9.15 *Carbon content in the metal bath, CRE number, and temperature versus blowing time in an AOD converter. For mechanisms (b) and (c), see Section 9.4.2.*

chromium starts to oxidize and the rate of decarburization decreases. The change to step 2 is where the dilution of oxygen gas is increased, and is determined in each individual case by three factors:

(1) The temperature, which should not be too high, typically not above 1725 °C due to problems with the refractory;

(2) The CRE number; and

(3) The carbon activity in the metal bath.

At this moment, the top slag has been over-oxidized due to the low CRE number at the end of step 1. The CRE number falls during steps 2 and 3 in a manner similar to step 1 until the fourth step is initiated, and blowing proceeds until the specified final carbon content is attained. Often the over-oxidized slag at the end of the blow is utilized to further lower the carbon content using a final purging with inert gas.

To minimize the chromium losses during the blow, it is important to control the composition of the top slag. This is performed by the addition of lime at the beginning of the blow. The amount of lime is determined from the silicon content in the metal charged in the converter. The aim is to get a basicity measured as $CaO:SiO_2$ of around 1.7. A lower basicity gives a slag with lower viscosity, a higher solubility of chromium oxide, and a lower $a_{Cr_2O_3}$. This causes problems in attaining high chromium and low carbon levels in the metal bath. In practical production, the slag is more or less semi-solid, consisting of a molten slag phase saturated with chromium oxide and a solid chromium oxide phase.

9.3.3 Reduction of the Top Slag

After the decarburization period, the top slag, usually around 70 kg per tonne of steel, has a typical average composition of 20–30% CaO, 10–15% SiO_2, 5–7% MgO (mainly due to refractory wear if MgO has not been added), 5–10% MnO, 5–10% FeO, 1–5% Al_2O_3, and 25–40% Cr_2O_3. To get an optimal yield of chromium in the process, a reducing agent such as silicon (as ferrosilicon) is added for reduction of the top slag. The slag is reduced during argon stirring for 5–8 minutes. As an alternative to ferrosilicon, aluminium metal may be used, especially when extremely low sulfur contents are required. Also combinations of these two reducing agents are practiced. The overall reaction during reduction with silicon is

$$Cr_2O_3 \text{ (in slag)} + \frac{3}{2}\underline{Si} = 2\underline{Cr} + \frac{3}{2}SiO_2 \text{ (in slag)}. \tag{9.46}$$

This reaction is favoured by a low SiO_2 activity and a high Cr_2O_3 activity in the top slag (i.e. a basic slag), low chromium activity in the metal, and a high temperature. It is difficult to influence the chromium activity since the composition of the metal bath is determined by the specification of the alloy. The activity of silicon is increased by an extra addition, normally aiming at 0.3–0.5% Si in the metal during the reduction. This, however, demands a very good process model in order to estimate the amount of chromium that has been oxidized during the decarburization, so that the silicon content remains within the limits specified for the composition. If the reduction is carried out properly, the resulting chromium content in the final slag is less than 0.5%. The process model should also make it possible to estimate the amount of lime added to give the desired slag basicity (CaO:SiO_2) of around 1.7. It is also important to predict the chromium content after the reduction period in order to eliminate the use of expensive low-carbon ferrochromium.

9.3.4 Desulfurization

After the reduction of the top slag, a slag-off is carried out and, if necessary, usually 10–15 kg per ton of steel of a mixture of lime and fluorspar is added for desulfurization, followed by 5–10 minutes of stirring with argon. If the reduction of the top-slag in the preceding step has been carried out satisfactorily—that is, the oxygen potential in the system is low—it is possible to reach a sulfur content in the steel of less than 0.003%. The use of aluminium as a reducing agent allows extremely efficient sulfur removal.

In a similar way as for phosphorus (see Section 9.2.2), the sulfide capacity, C_S, of a slag can be defined according to[23]

$$1/2S_2 + O^{2-} \text{ (in slag)} = 1/2O_2 + S^{2-} \text{ (in slag)} \tag{9.47}$$

$$C_s = \frac{K_{9.47} \times a_{O^{2-}}}{f_{S^{2-}}} = \left(\%S^{2-}\right) \text{ in slag} \times \sqrt{\frac{(P_{O2})}{(P_{S2})}}. \tag{9.48}$$

Sulfide capacities according to Eq. (9.48) are given in Fig. 9.16.[24]

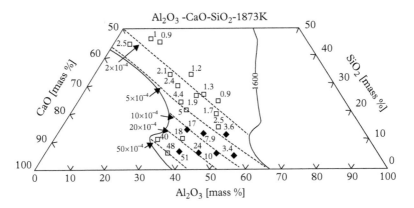

Fig. 9.16 *Sulfide capacities in the system CaO–Al₂O₃–SiO₂, 1650°C. [Republished with permission of John Wiley & Sons–Books, from Condo et al.,*[24] *permission conveyed through Copyright Clearance Center, Inc.]*

It is often more practical to replace the gaseous reference states by 1% dilute system in the metal phase. Doing this we get

$$\underline{S} + O^{2-} \text{ (in slag)} = S^{2-} \text{ (in slag)} + \underline{O}. \tag{9.49}$$

The equilibrium constant for reaction (9.49) is given by

$$K_{9.49} = \frac{\left(a_S^{2-}\right)}{\left(a_O^{2-}\right)} \cdot \frac{[a_O]}{[a_S]}. \tag{9.50}$$

From this expression, the equilibrium distribution ratio of sulfur between the slag and the metal phase can be written as

$$\frac{(\%S) \text{ in slag}}{[\%S] \text{ in metal}} = C_S' \cdot \frac{[f_S]}{[a_O]}, \tag{9.51}$$

where f_S is the activity coefficient for sulfur in the metal bath and C_S' is the sulfide capacity in the slag phase defined as

$$C_S' = K_{9.49} \frac{\left(a_O^{2-}\right)}{\left(f_S^{2-}\right)}. \tag{9.52}$$

f_S^{2-} is the activity coefficient for sulfide ion in the slag phase. The relation between experimentally determined sulfide capacities, C_s, according Fig. 9.16 can be recalculated to C_S' by taking the free energies of dissolution of oxygen and sulfur into consideration, and we get

$$C'_S = exp\left(-\frac{2154}{T} + 3.166\right) \times C_S. \tag{9.53}$$

When using aluminium as a reducing agent, it is obvious from Fig. 9.16 that the desulfurization with slags based on the $CaO-Al_2O_3-SiO_2$ system is very effective. Sulfur levels below 20 ppm with a one-slag practice have been reported in the literature when a mixed single slag reduction with FeSi and aluminium is performed in a 75-ton AOD converter.[25] Aluminium is also a powerful deoxidizer and, as seen in Eq. (9.49), promotes sulfur removal by reducing oxygen activity in the metal.

In an AOD converter, the tap-to-tap time is typically 80–120 minutes. The decarburization takes 35–90 minutes, where the longer times occur under extreme conditions—high initial carbon contents combined with high chromium, low nickel, and low carbon content in the finished product. Typical consumption figures per tonne of 18% Cr–8% Ni steel are Ar = 9–12 Nm³, N_2 = 9–11 Nm³, O_2 = 30–8 Nm³, lime = 35–45 kg, fluorspar = 2–3 kg, and ferrosilicon (75% Si) = 8–11 kg. These figures are for an initial carbon content of 2% in the AOD process.

As mentioned earlier, the different stages in the production of stainless steel are strongly coupled to each other. Good control of the decarburization steps gives a minimum of chromium oxidation, which has a direct effect on the consumption of gases and reducing agent. The amount and type of reducing agent in turn influences the oxygen level in the metal bath and the sulfur level in the finished steel.

9.4   Modelling the Rate of Decarburization

Characterizing the rate of decarburization in a converter is important for optimizing the process. In the following we shall derive an expression for the reaction rate.

A reactor for stainless steelmaking is schematically shown in Fig. 9.17. The system contains three phases:

(1) Molten metal with dissolved elements, such as C, Si, Mn, Cr, Ni, and N;
(2) Slag, which can be molten or solid, containing oxides such as CaO, Al_2O_3, SiO_2, MnO, MgO, Cr_2O_3, and FeO; and
(3) Inert gas and oxygen which reacts to form carbon monoxide.

The following two assumptions are made:

(1) Due to the bubble stirring, no concentration gradients occur in the metal bath.
(2) The interfacial area between gas bubbles and metal is constant during decarburization despite a successively increased addition of inert gas.

When modelling a metallurgical process it is often convenient to divide the reactor into separate zones which are later put together to derive a model for the entire process. In Fig. 9.17

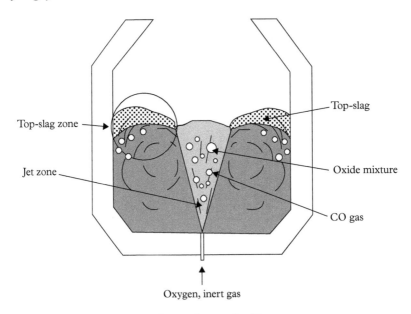

Fig. 9.17 *Reaction zones in a converter for stainless steelmaking.*

the reactor has been divided into three zones which are all active during the decarburization. These zones are the tuyere zone, the jet zone, and the top-slag zone. In the following the tuyere zone is treated separately. However, because of the very high gas flow, around 1 m³ per tonne of liquid steel per minute, a defined separate top-slag zone is assumed not to exist. To eliminate uncertainties in estimating rate constants in the different zones, the jet zone and the top-slag zone are treated together and are hereafter called the oxide/metal zone.

9.4.1 Tuyere Zone

In front of the tuyeres, the injected oxygen reacts instantaneously with the elements dissolved in the metal bath. Due to a high local oxygen potential, the different elements are oxidized in amounts proportional to the atomic percentage of each element. Thus the following reactions are assumed to occur:

$$\underline{C} + \frac{1}{2}O_2 = CO$$

$$\underline{Fe} + \frac{1}{2}O_2 = FeO$$

$$\underline{Mn} + \frac{1}{2}O_2 = MnO$$

$$\underline{Si} + O_2 = SiO_2$$

$$\underline{Cr} + \frac{3}{4}O_2 = \frac{1}{2}Cr_2O_3.$$

The amount of oxygen that reacts with the respective elements can be written

$$\dot{Q}_{O_2}^{(j)} = \frac{c_j n_j \dot{Q}_{O_2}}{\sum c_j n_j},$$

(9.54)

where $\dot{Q}_{O_2}^{(j)}$ is the volume of added oxygen that reacts with the element j per unit time (m³/s), c_j is the molar concentration of element j in the metal (kmol/m³), $\dot{Q}_{O_2}$ is the total flow of oxygen added per unit time (m³/s), and n_j is the stoichiometric factor for the various reactions: $n_j = \frac{1}{2}$ except for Si where $n = 1$ and Cr where $n = \frac{3}{4}$.

The instantaneous oxidation of the elements in the tuyere zone results in a mixture of oxides that is transported into the oxide/metal zone.

9.4.2 Oxide/Metal Zone

In the oxide/metal zone the oxide mixture formed in the tuyere zone is reduced by the various elements dissolved in the metal bath. This zone, in which the gas and oxide mixture is transported from the tuyeres to the top surface, can be regarded as an emulsion of oxides and liquid metal. To derive an expression for the rate of decarburization, the following assumptions are made.

(1) Due to the high chromium content in the metal bath, the formation of chromium oxide occurs at a high rate above the tuyere zone. Other dissolved elements such as Si and Mn are in equilibrium with chromium oxide. The decarburization reaction is assumed to occur by a reaction between the oxide and carbon dissolved in the metal bath.

(2) The chromium oxide activity in the oxide phase is equal to one during the blow, which means that there is a residual chromium oxide content in the oxide mixture.

(3) The carbon monoxide formed during the reaction between the oxide mixture and carbon is part of the gas phase, and changes the partial pressure of carbon monoxide in the gas bubbles as they are transported up through the metal bath.

(4) The rate-controlling step (mechanism) in decarburization is one of the following:

  (a) The flow of oxygen into the bath;
  (b) The equilibrium given by Eq. (9.44); or
  (c) The diffusion of carbon in the boundary layer around the bubbles in the melt (Eq. (9.55)).

If p_{CO} at equilibrium calculated from Eq. (9.44) is greater than p_{CO} (max), and if $\dot{m}$ from Eq. (9.55) gives a flow of carbon exceeding that needed to react with the flow of oxygen, then step (a) is rate limiting. p_{CO} (max) is the partial pressure of CO if all the available oxygen reacts with carbon and the CO is diluted with the inert gas added.

If p_{CO} at equilibrium is less than p_{CO} (max), and $\dot{m}$ exceeds that needed to react with the flow of oxygen, then step (b) is rate limiting (see Fig. 9.15). If $\dot{m}$ is not sufficient to handle the flow of oxygen, then (c) is rate limiting (see Fig. 9.15). Decarburization rates are low when step (c) is rate limiting compared to (a) or (b). Step (a) is called the linear period in the following. Special attention is given here to step (c).

The mass flux of carbon to the gas phase is obtained from the expression

$$\dot{m} = \frac{Ak_C \rho}{100} \left([\%C] - [\%C]_e \right), \tag{9.55}$$

where $\dot{m}$ is the mass flux of carbon (kg/s), A is the surface contact area (m^2), k_C is the total mass transfer coefficient for carbon (m/s), and ρ is the density of the metal (kg/m^3). $[\% C]_e$ is the hypothetical concentration of carbon in the melt in equilibrium with chromium oxide at the partial pressure p_{CO} existing in the bubbles (see Eq. (9.58)) and $[\% C]$ is the carbon concentration in the bulk melt. The mass flux, $\dot{m}$, equals the rate of depletion of carbon in the bath:

$$\dot{m} = -V \frac{\rho}{100} \frac{d[\%C]}{dt} = \frac{Ak_C \rho}{100} \left([\%C] - [\%C]_e \right). \tag{9.56}$$

Then

$$\frac{d[\%C]}{dt} = -\alpha \left([\%C] - [\%C]_e \right) \tag{9.57}$$

where $\alpha = Ak_C / V$. α is a mass transfer rate constant for the process that depends on geometry and size of the metallurgical reactor. For instance, a deeper metal bath or an increased gas flow rate increases the contact area A in the jet zone. The stirring also influences the mass transfer coefficient for carbon ($k_C = D/\delta$) as the diffusion boundary layer thickness, δ, is decreased by increased stirring. If α is increased, p_{CO} in the bubbles leaving the bath will come closer to equilibrium with $[\% C]$ as given by Eq. (9.44).

From the general assumption that the overall rate constant, α, is known, we shall now derive an expression for the rate of decarburization. From Eq. (9.57) it is seen that if α is constant, the mass flux depends only on the difference in carbon content between the metal bulk concentration $[\%C]$ and at the oxide surface, $[\%C]_e$. According to the reaction given by Eq. (9.43) with $\alpha_{Cr_2O_3} = 1$, the equilibrium carbon content, $[\%C]_e$, can be calculated from the expression

$$[\%C]_e = \frac{(f_{Cr}[\%Cr])^{\frac{2}{3}} p_{CO}}{f_C K_{9.43}^{\frac{1}{3}}}, \tag{9.58}$$

where f_C and f_{Cr} are the activity coefficients of C and Cr, respectively, in the metal bath.

From this expression and Eq. (9.57) we see that the rate of decarburization is increased when $[\%C]_e$ is low; that is, if the chromium activity in the metal bath and the partial pressure of carbon monoxide in the gas phase are low, and the activity coefficient for carbon is high. Also the temperature is an important parameter, as a high temperature increases the value of the equilibrium constant, K, (see Eq. (9.44)). Near the tuyere zone in the converter, where $[\%C]_e$ approaches zero, the rate of decarburization reaches its maximum value. If these conditions were valid for the whole oxide/metal zone, the rate of decarburization would be described by

$$\frac{d[\%C]}{dt}(\max) = -\alpha[\%C].$$
(9.59)

However, as the gas bubbles rise through the metal bath, the partial pressure of carbon monoxide increases in the gas phase, which gives a higher equilibrium carbon content at the oxide metal interface. This reduces the driving force for the transfer of carbon.

To obtain an expression for the local rate of decarburization at the top surface where the partial pressure of carbon monoxide is a maximum, the treatment is as follows. The gas law gives

$$p_{CO} = \frac{\dot{Q}_{CO}}{\dot{Q}_{CO} + \dot{Q}_{Ar}} p_{tot},$$

where $\dot{Q}_{CO}$ and $\dot{Q}_{Ar}$ are the respective gas flows in m³/s, or

$$p_{CO} = \frac{p_{tot}}{1 + \dot{Q}_{Ar}/\dot{Q}_{CO}}.$$
(9.60)

From stoichiometry the CO flow that leaves the bath is

$$\dot{Q}_{CO} = -\frac{d[\%C]}{dt}\frac{V\rho 22.4}{M_C 100},$$
(9.61)

where $M_C = 12$, the molar weight of carbon. From Eqs. (9.60) and (9.61) we get

$$p_{CO} = \frac{p_{tot}}{1 - \frac{\dot{Q}_{Ar}12 \times 100}{22.4V\rho d[\%C]/dt}}.$$
(9.62)

By combining Eqs. (9.62), (9.57), and (9.58) the following expression for the rate of decarburization is obtained:

$$\frac{d[\%C]}{dt} = -\alpha\left\{[\%C] - \frac{(f_{Cr}\%Cr)^{\frac{2}{3}}p_{tot}}{f_C K^{\frac{1}{3}}\left(1 - \frac{\dot{Q}_{Ar}12 \times 100}{22.4V\rho d[\%C]/dt}\right)}\right\}$$
(9.63)

To simplify this expression we multiply the right and left sides with $\left(1 - \dot{Q}_{Ar} 12 \times 100 / (22.4V\rho d[\%C]/dt)\right)$. After rearranging we get

$$\frac{d[\%C]}{dt}\left(1 - \frac{\dot{Q}_{Ar} 12 \times 100}{22.4V\rho d[\%C]/dt}\right)$$

$$= -\alpha\left\{[\%C] \times \left(1 - \frac{\dot{Q}_{Ar} 12 \times 100}{22.4V\rho d[\%C]/dt}\right) - \frac{(f_{Cr}[\%C])^{\frac{2}{3}} p_{tot}}{f_C K^{\frac{1}{3}}}\right\}. \tag{9.64}$$

This equation can be transformed to an equation quadratic in $d[\%C]/dt$. Multiplying by $d[\%C]/dt$ gives

$$\left(\frac{d[\%C]}{dt}\right)^2 + \frac{d[\%C]}{dt}\left\{-\frac{\dot{Q}_{Ar} 12 \times 100}{22.4V\rho}\right.$$

$$+ \left.\alpha\left([\%C] - \frac{(f_{Cr}[\%Cr])^{\frac{2}{3}} p_{tot}}{f_C K^{\frac{1}{3}}}\right)\right\} - \frac{\alpha\dot{Q}_{Ar} 12 \times 100[\%C]}{22.4V\rho} = 0. \tag{9.65}$$

This quadratic equation has the following solution:

$$\frac{d[\%C]}{dt} = -\frac{1}{2}\left(-\frac{\dot{Q}_{Ar} 12 \times 100}{V\rho 22.4} + \alpha\left([\%C] - \frac{(f_{Cr}[\%C])^{\frac{2}{3}} p_{tot}}{f_C K^{\frac{1}{3}}}\right)\right)$$

$$- \left(\left\{-\frac{\dot{Q}_{Ar} 12 \times 100}{V\rho 22.4} + \alpha\left([\%C] - \frac{(f_{Cr}[\%Cr])^{\frac{2}{3}} p_{tot}}{f_C K^{\frac{1}{3}}}\right)\right\}^2 \frac{1}{4}\right.$$

$$+ \left.\frac{\alpha\dot{Q}_{Ar} 12 \times 100[\%C]}{22.4V\rho}\right)^{\frac{1}{2}}. \tag{9.66}$$

By studying this expression it can be seen that the sign in front of the square root should be negative, as this sign gives an increased rate of decarburization when $\dot{Q}_{Ar}$ is increased. Equation (9.66) gives a minimum local decarburization rate since the driving force for transfer of carbon is smallest at the top of the reactor.

We now have two expressions describing the maximum and minimum rates of decarburization, respectively. Thus the average rate of decarburization describing the overall rate of decarburization is given by

$$\frac{d[\%C]}{dt}\text{ (average)} = \left\{\frac{d[\%C]}{dt}\text{ (max)} + \frac{d[\%C]}{dt}\text{ (min)}\right\}b, \tag{9.67}$$

where the maximum and minimum rates of decarburization are given by Eq. (9.59) for the tuyere region and Eq. (9.66) for the top of the converter, respectively.

b is a constant which is difficult to estimate. In the following we set $b = \frac{1}{2}$ which means that the rate of decarburization is taken to be the average of the rate at the tuyeres and at the top surface.

To calculate the rate of decarburization, the following must be known:

(1) The initial weight of liquid metal, composition, and temperature;

(2) The composition of the metal bath at every instant, where it is assumed that other dissolved elements such as Si and Mn are in equilibrium with the chromium oxide so that equilibria are attained according to the reactions:

$$Cr_2O_3 + \frac{3}{2}\underline{Si} = 2\underline{Cr} + \frac{3}{2}SiO_2 \tag{9.68}$$

$$Cr_2O_3 + 3\underline{Mn} = 2\underline{Cr} + 3MnO; \tag{9.69}$$

(3) The residual minimum amount of chromium oxide in the slag (see Section 9.4.1) is assumed to be equal to the amount of chromium oxidized at the tuyeres (see Eq. (9.54)); and

(4) The temperature in the system is calculated from a heat balance, that is, from data concerning the amount of elements that have been oxidized, the heat evolved during the reactions, and a model of the temperature losses in the process.

The rate of decarburization is controlled by one of the following conditions:

(1) The total amount of oxygen available for decarburization (linear period, mechanism (a));

(2) The equilibrium according to Eq. (9.44) (mechanism (b)); and

(3) Diffusion of carbon (Eq. (9.67), mechanism (c)).

Starting with the initial composition and temperature, it is now possible to calculate the composition of the metal bath as a function of time.

9.4.3   Simulation of the Blow

A calculation of the rate of decarburization can be carried out manually. However, the system is so complex that it is more convenient to use a computer. The results of some process simulations are given below.

The following conditions are assumed for the first calculation:

Metal weight = 60 tonnes

Initial temperature = 1550 °C

Initial composition of the metal bath = 1.5% C and 18% Cr, α = 0.2min^{-1} (empirical parameter)

Blowing steps (m^3 O$_2$/m^3 Ar) = 45/15; 30/30; 15/45; 10/50

Gas flow = 1 m^3/t per min.

In Fig. 9.18 we see that the rate of decarburization initially has a tendency to be at a low level, after which the rate of decarburization remains at a constant high value (mechanism (a) in Section 9.4.2). When the rate of decarburization decreases, the process is first controlled by the chemical equilibria (mechanism (b)) and then by carbon diffusion (mechanism (c)).

To test the validity of the model described above, it is convenient to calculate the carbon content in the metal bath and compare with what is obtained in the converter. This has been done in Fig. 9.19 where data points from two heats in a 57-tonne AOD converter are compared with calculated curves. The converter was blown in four steps, three with a O$_2$/Ar ratio of 3.2:1, 1:1, and 1:2.4 and a final step with 1:4.5 or a final purging with inert gas. During the final inert gas purging all the oxygen for the decarburization is assumed to be supplied by the over-oxidized slag. As seen in Fig. 9.19, the correspondence between the practical and calculated results is good.

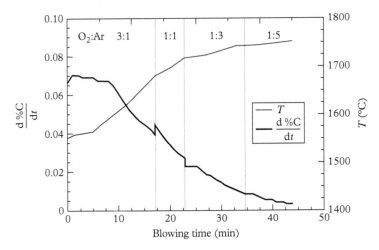

Fig. 9.18 *Computer simulation of decarburization of stainless steel in a 60-tonne converter, case 1. Metal composition: 1.5% C, 18% Cr. Total gas flow rate is 60 Nm^{3}n/min. Initial temperature is 1550 °C. Formation of Cr$_2$O$_3$ = 1750 kg.*

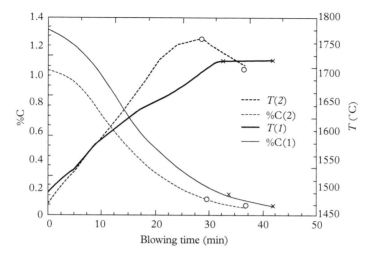

Fig. 9.19 *Calculated temperature and carbon contents in the liquid steel compared with production data from a 57-tonne AOD converter. Blowing conditions are as follows. Heat 1: Initial bath composition: 1.32% C, 0.35% Si, 1.81% Mn, 18.2% Cr, and 9.2% Ni. Total gas flow rate is 53 Nm³/min during steps I–III; step IV is 36 Nm³ Ar/min, 8 m³ O₂/min. Heat 2: Initial bath composition: 1.05% C, 0.20% Si, 1.85% Mn, 17.9% Cr, and 9.4% Ni. Total gas flow rate is 53 m³/min during steps I–III; step IV is 40 Nm³ Ar/min.*

9.4.4 Some Operating Conditions and Their Effect on the Process

A number of variables influence metallurgical processes. It is therefore necessary to have a detailed knowledge of steelmaking practices to predetermine the effect of different actions undertaken in production. In the following we shall utilize the model developed in Section 9.4 to discuss what happens during the decarburization of stainless steels in a converter when some of the operating conditions are changed.

In Section 9.3, the fundamentals of stainless steel production were treated. From what has been said earlier, the main parameters influencing the decarburization reaction are the temperature and the carbon and chromium activity in the steel bath. All these affect the CRE number and are important in order to determine when to change the blowing steps.

As an example we return to the reference heat given in case 1, Fig. 9.18, and assume that the initial temperature in the converter is lowered to 1450 °C. The low temperature can be caused by a very long waiting time in the transfer ladle between the EAF and the converter. The result of a calculation is given in case 2, Fig. 9.20. The first blowing period is prolonged and more chromium has to be oxidized to increase the temperature and to produce a more efficient decarburization. This increases the total Cr_2O_3 formation from 1750 to 2610 kg, which gives a higher consumption of reducing agent and process gas, and a longer process time.

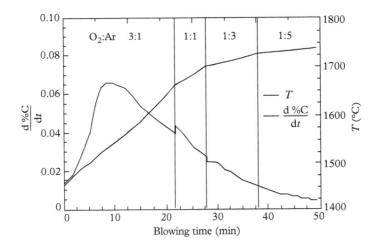

Fig. 9.20 *Computer simulation of decarburization of stainless steel in a 60-tonne converter, case 2. Metal composition: 1.5% C, 18% Cr. Total gas flow rate is 60 Nm³/min. Initial temperature is 1450°C. Formation of Cr_2O_3 = 2610 kg.*

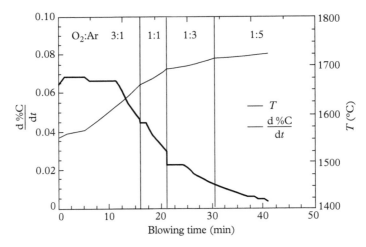

Fig. 9.21 *Computer simulation of decarburization of stainless steel in a 60-tonne converter, case 3. Metal composition: 1.5% C, 18% Cr, 8% Ni. Total gas flow rate is 60 Nm³n/min. Initial temperature is 1550°C. Formation of Cr_2O_3 = 1380 kg. Blowing time = 41 min.*

To demonstrate the influence of the carbon activity on the rate of decarburization, case 3, Fig. 9.21, gives the calculated result when the initial metal bath composition is changed. The bath contains 8% Ni instead of no Ni, as was the case in Fig. 9.19. Due to an increased carbon activity when the Ni is added, the linear decarburization period is prolonged and the total blowing time is shortened. This leads to a decreased consumption of reducing agent and gases.

9.4.5   Process Optimization

The industrial production of stainless steels has developed very much during the past decades. The productivity, especially in electric arc furnaces, has increased considerably. In many steel shops, the converter has become the bottleneck in the production. Therefore, a great deal of work has been carried out in order to increase productivity. In this connection, some developments will be discussed and exemplified in the following, based on computer simulations.

Increased flow rate of oxygen during the first period of the blow. By increasing the oxygen gas flow rate from 45 to 55 Nm^3/min (case 4) during the first step of the blow, when the oxygen flow is the rate-determining step for the process, the rate of decarburization is increased so that step 2 can be started earlier and at a higher temperature, due to lower temperature losses (see Fig. 9.22). The total blowing time is decreased to 41 from 45 minutes in Fig. 9.18.

Continuous dilution with inert gas. One way of increasing the productivity during stainless steelmaking in converters is to optimize the blow so that the CRE number is always at a maximum level. This means that after an initial conventional blowing during the linear period, a continuous dilution aimed at keeping the CRE number on a maximum level is carried out. Case 5 (Fig. 9.23) presents a simulation of this blowing practice. A comparison with case 1 (Fig. 9.18) shows that both the process time and the amount of Cr_2O_3 formed during the blow are decreased.

Combined blowing. A common way to increase the productivity in converters is to install a top lance for the supply of extra oxygen at higher carbon contents, usually above 0.5%. The layout of the process (see Fig. 9.24) is quite similar to what has been developed for the production of low-alloy steels in integrated steel plants. Normally, the total oxygen flow is

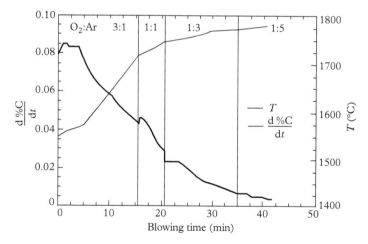

Fig. 9.22 *Computer simulation of the decarburization of stainless steel in a 60-tonne converter, case 4. Metal composition: 1.5% C, 18% Cr. Total gas flow rate is 60 Nm^3/min. Initial temperature is 1550°C. Formation of Cr_2O_3 = 1932 kg. Blowing time = 41 min.*

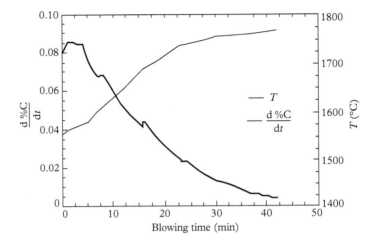

Fig. 9.23 *Computer simulation of the decarburization of stainless steel in a 60-tonne converter with optimized flow of inert gas, case 5. Metal composition: 1.5% C, 18% Cr. Oxygen flow rate is 45 Nm^3n/min. Initial temperature is 1550°C. Formation of Cr_2O_3 = 1700 kg. Blowing time = 42 min.*

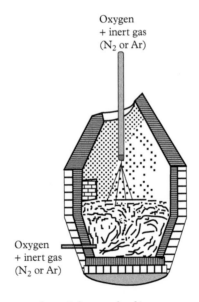

Fig. 9.24 *Combined blown converter for stainless steelmaking.*

increased by 60–100 per cent by the use of a lance, and the rate of decarburization, as well as the scrap melting capacity, is increased. Production figures[26] show that the blowing time can be decreased by 10 per cent. At the same time the scrap melting capacity is increased by 25 per cent.

If in the later blowing steps, when the rate of decarburization decreases exponentially with time, we increase the contact area between the reacting phases, and a more rapid decarburization is attained. This has been demonstrated on an industrial scale in the so-called Krupp Combined Blowing Stainless (KCB-S) process,[27] where in combination with the addition of oxygen and inert gas from the bottom, inert gas and oxygen are also added through a top lance. A comparison between a conventional AOD converter without a top lance and the KCB-S process has shown that the blowing time can be decreased by 50 per cent. Tap-to-tap times of around 45 minutes have been reported.

REFERENCES

1. Carlberg P. Early industrial production of Bessemer steel at Edsken. *J. Iron Steel I*, 1958; 189: 201–4.
2. Edström JO, Engelbert T, Selin R, Werme A, Wijk O. Steelmaking from high-phosphorus iron ores (in Swedish) STU-information 256; 1981.
3. Trenkler HA. New Developments in LD Steelmaking. *J. Metals*, 1960; 12(7): 538–41.
4. Pehlke RD, ed. *BOF Steelmaking*. New York: Iron and Steel Society of AIME; 1974.
5. Jalkanen H, Holappa L. Converter steelmaking. In: Seetharaman S, et al. (eds) *Treatise on process metallurgy*, Vol. 3, pp 223–70. Elsevier; 2014.
6. Holappa L. Recent achievements in iron and steel technology. *J. Chem. Technol. Metall.*, 2017; 52(2): 159–67.
7. Bringas JE, ed. *Handbook of comparative world steel standards*. West Conshohocken, PA: ASTM International; 2004.
8. Fruehan RJ, ed. *Making, Shaping and Treating of Steel*, p. 64. Pittsburgh, PA: AISE Steel Foundation; 1998.
9. Kojola N. SSAB, Stockholm, Sweden, personal communication; 2019.
10. 2018 AIST Basic Oxygen Furnace Roundup. *Iron Steel Technol.*, 2018; 194–201.
11. Ellingham HJT. Reducibility of oxides and sulphides in metallurgical processes. *J. Soc. Chem. Ind. (London)*, 1944; 63: 125–33.
12. *Steelmaking Data Sourcebook*. Montreux: Gordon and Breach; 1988.
13. Chipman J, Fetters KL. The solubility of iron oxide in liquid iron, *Trans.Amer. Soc. Metals.*, 1941; 29: 953–67.
14. *Verein Deutscher Eisenhuttenleute, Schlackenatlas/Slag Atlas*. Dusseldorf: Verlag Stahleisen; 1981. After Timucin M, Morris AE. Phase equilibria and thermodynamic studies in the system $CaO–FeO–Fe_2O_3–SiO_2$ *Metall. Trans.*, 1970; 1: 3193–3201.
15. Turkdogan ET, Pearson J. Activities of constituents of iron and steelmaking slags. Part III—Phosphorus pentoxide. *J. Iron Steel Inst.*, 1953; 175: 393–401.
16. Bodsworth C, Bell HB. *Physical Chemistry of Iron & Steel Manufacture*. London: Longman; 1972.
17. Balajiva K, Quarrell AG, Vajragupta P. A laboratory investigation of the phosphorus reaction in the basic steelmaking process. *J. Iron Steel Inst.*, 1946; 153(1): 115–45.
18. Wagner C. The concept of basicity in slags. *Metall. Trans. B*, 1975; 6: 405.

19. Jorgensen C, Thorngren I. *Thermodynamic tables for process metallurgists*. Stockholm: Almqvist & Wiksell; 1969.

20. Assis AN, Tayeb MA, Sridhar S, Fruehan RJ. Phosphorus equilibrium between liquid iron and CaO-SiO$_2$-MgO-FeO-P$_2$O$_5$ Slag, Part 1: Literature review, methodology and BOF slags. *Metall. Mater. Trans. B*, 2015; 46B: 2255–63.

21. Dresler W. The effect of Ni additions on the decarburization of liquid high-carbon ferrochromium. *Iron and Steelmaker*, 1988; 43–51.

22. Turkdogan ET. *Physicochemical properties of molten slags and glasses*. London: The Metals Society; 1983.

23. Richardson FD, Fincham CJB. The behaviour of sulphur in silicate and aluminate melts. *Proc. Roy. Soc. (London)Phil. Trans.* A223, 40-62, and *J. Iron Steel Inst.* 1954; 178: 4–15.

24. Condo AFT, Quifeng S, Sichen Du. Sulfate capacities in the Al$_2$O$_3$-CaO–MgO–SiO$_2$ system. *Steel Res.*, 2018; 89(8). After Hino M, Kitagawa S, Ban-Ya S, ISIJ Int., 1993; 33: 36.

25. Görnerup M, Sjöberg P. AOD/CLU process modeling: optimum mixed reductants addition. *Ironmaking & Steelmaking*, 1999; 26: 58–65.

26. Tohge T, Fujita Y, Watanabe T. Some considerations on high blow rate AOD practice with top blowing system. In: *Proc. 4th PTD Conference, April 1-4, Chicago, IL*. ISS of AIME; 1984.

27. Ropke W, Zorcher H. Combined blowing process for stainless steel. *Steel Technol. Int.*, 1989; 125–8.

10

Recycling

In this chapter we look at recycling and define its key features. Important principles and various recycling strategies are considered. There can be several motives or forces behind recycling. These include economic, environmental, social, and governmental factors. These all affect the solution chosen, so it is important to clarify them. In the third section we investigate scrap in some detail—its sources and characteristics, variation over time and space, highlighted with some examples. The fourth section deals with important unit processes in recycling: collection, shredding, sorting, pretreatment, blending, and processing of final residues. Some of the methods used are covered briefly and further reading is recommended. However, pretreatment by heating is discussed in more detail. Selecting methods in various process routes and comparing their impact is an important goal in Section 10.4. In Section 10.5 three examples of recycling are discussed in some detail. Magnesium recycling is included. This method is not in operation today, but it may become important in the future. Dross recycling is presented as an example of by-product processing. Battery recycling is also presented as an example of more complex recycling chemistry. And finally, in Section 10.6, we discuss some of the limits, challenges, and opportunities that lie in recycling, together with a brief look at possible future developments.

10.1   Introduction

In the 1992 version of this book the author wrote:

> By the year 2050, 10 billion people are likely to live on this planet. Ideally, all would enjoy standards of living equivalent to developed countries like the US and Japan. However, if they consume natural resources such as copper, cobalt, molybdenum, nickel, and petroleum at current US rates, the known resources would not last for many decades. On the waste side of the ledger, 10 billion people would generate at least 4 billion tonnes of solid waste every year, enough to bury Los Angeles 100 metres deep. Furthermore, the rate of increase of the CO_2 levels in the atmosphere would lead to global warming.

Principles of Metal Refining and Recycling. Thorvald Abel Engh, Geoffrey K. Sigworth, and Anne Kvithyld, Oxford University Press.
© Oxford University Press 2021.
DOI: 10.1093/oso/9780198811923.003.0010

This forecast has been validated. In 2015 the UN stated, 'The current world population of 7.3 billion is expected to reach 9.7 billion in 2050 and 11.2 billion in 2100.'[i]

As further stated in 1992:

> These forecasts emphasize the incentives for recycling, conservation, and a switch to alternative materials. The metallurgist has an important role to play in bringing about the necessary change. Traditionally individual manufacturing processes take in raw materials and generate products to be sold plus waste to be disposed of. Ideally, industrial processes should be transformed into a more integrated model: an industrial ecosystem. Wastes are minimized and the effluents of one process, whether they are spent catalysts from petroleum refining, fly and bottom ash from electric-power generation, or slag from a metallurgical process, should serve as the raw material for another process. Ideally, consumption of metals such as copper, aluminium, and steel should to a large extent be taken care of by recycling.

Sadly, the recycling ecosystem proposed in 1992 is still far from reality. Although progress has been made, there is a great deal of work still left for us to do. The recycling and refining metallurgist of the future can assist this development, by knowing the limiting factors established by nature. However, they must also have a grasp of technology and society that goes far beyond thermodynamics and kinetics! It is far from certain that we will meet the future challenges of providing better recycling and refining. But laying a solid foundation for the needed technical development is an important first step. That is what we try to address in this chapter.

10.1.1   What Is Recycling?

Our everyday understanding and legal definition of recycling has evolved from waste management. Nowadays *resource* management is becoming more important as a driving force. To manage resources, recycling is not so much a goal as a necessary tool. This shift is reflected in our current understanding and legislative framework.

10.1.1.1   *Definitions from an Everyday Point of View*

Recycling for most of us is a daily chore. Plastics, papers, metals, and glass are sorted and put in separate bins, and each waste stream is recycled separately with the intent to use each fraction of the waste to its highest potential. In other words, we wish to recycle plastic into new plastic, milk bottles into new bottles, and used beverage cans into new beverage cans.

What is described above is actually reuse; taking a product at the end of its lifetime and processing it to be used again for the same purpose. In reality, reuse is sometimes theoretically improbable and practically impossible, not to mention expensive. Instead we sometimes need to find new efficient uses for our waste, so it may retain as much of its value as possible, or perhaps even attain a higher value.

---

[i] These projections are from the UN DESA report, "World Population Prospects: The 2015Revision," available at https://www.un.org/en/development/desa/publications/world-population-prospects-2015-revision.html)

Distinguishing between recycling and reuse might not seem that important for daily life. But recycling is really the entire life cycle of matter: from new to old, to new again. Recycling involves a series of changes and treatments to regain the material for human use. For metals this is a return to an original condition, so that it can be used again in manufacture. For metals this usually means

$$recycling = recovery + remelting/refining$$

Recycled material is usually referred to as 'secondary', whereas material that has extracted from the environment is called 'primary'. Additionally, the metallurgist must not only recycle the metal that has served its purpose, but also the by-products created during metal production and refining.

10.1.1.2 *Definitions from a Legal Point of View*

Definitions of recycling from a legal point of view are important for licenses to operate and trade between countries, and also as a sales argument for customers. Recycling is defined in many ways by different institutions.

The European Council's directive 2008/98/EC on waste is one example of legal definitions and targets for recycling. Article 3 defines 'recycling' as any recovery operation by which waste materials are reprocessed into products, materials, or substances, whether for the original use or other purposes. It includes the reprocessing of organic material, but does not include energy recovery and the reprocessing into materials that are to be used as fuels, or for backfilling operations. That means that incineration of material is not counted as recycling, nor is the use of material as a fuel.

10.1.1.3 *Mathematical Definitions of Recycling Rates*

Different definitions of recycling give different mathematical expressions and different numbers for recycling rates. This principle is demonstrated well in Rankin,[1] where the following definitions of recycling terms are presented. Figure 10.1 gives a schematic view of a material's life cycle, to help clarify the following definitions.

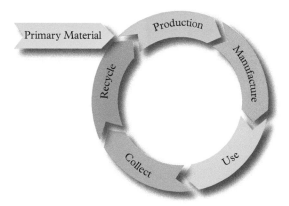

Fig. 10.1 *Schematic overview of a material's life cycle.*

In practice there can be a loss of material in each of the stages shown in Fig. 10.1. These losses are cumulative. For example, there may be scrap generated during manufacture. Some consumers (users) will discard material in a way that makes collection impossible or difficult. Not all material is collected. For this case, the recovery rate is defined by

$$\text{Recovery rate} = \text{mass of collected/mass of available (used)} \qquad (10.1)$$

For instance, if a mass of 100 were used and 90 were collected, the recovery rate would be 90 per cent. The technical recovery rate is defined as the mass of recycled (remelted) divided by mass of collected:

$$\text{Technical recovery rate} = \text{mass of recycled/mass of collected} \qquad (10.2)$$

If 80 gets recycled of the 90 collected, the technical recovery rate would be 80/90 or 89%. The recycling rate is defined by the mass of recycled material divided by the mass of material available/used:

$$\text{Recycling rate} = \text{mass of recycled/mass of available (used)} \qquad (10.3)$$

For the example at hand, the recycling rate would be 80/100 or 80%. Finally, a return rate is defined as the ratio of the mass of recycled material to the total mass produced. That is,

$$\text{Return rate} = \text{mass of recycled/(mass of recycled} + \text{mass of primary)} \qquad (10.4)$$

This rate depends on how much primary metal is used in production. If primary metal is half of the production, the return rate would be 50%.

These simplified definitions do not consider the non-metal portion of collected material, nor the losses during manufacturing. Hence, being clear on which definition is in use is important. This does not mean one definition is correct and the others incorrect.

10.1.1.4   *Recycling of Metal versus Recycling of Products*

In the 2013 UNEP Metal Recycling report, an evolution from a metal-centric to a product-centric approach is described. When products to be recycled are relatively simple, the metal-centric approach focuses on recycling of a specific metal. However, products are becoming increasingly complex. They consist of numerous different kinds of materials and metals. Recycling is often difficult because of the linkages and combinations of materials. The product-centric approach then targets specific components of the product, to separate and recycle them. Today the legal definitions usually follow the majority of the metal in a product.

10.1.2   Why Do We Recycle?

As metallurgists we are all aware of the ecological strain the production of metals puts on the entire globe. Metals are usually found in nature as oxides and sulfides, and extraction requires energy, and usually carbon as a reductant. Thus, production of primary metal contributes to an increase of CO_2 in the atmosphere. The best possible use must therefore be made of existing resources already in metallic form, by recycling.

In the sections that follow some of the key points and principles of recycling are considered. The literature in this area is immense—far beyond the scope of this chapter. Therefore, only brief descriptions are provided.

10.1.2.1 *Economic Development and Metal Demand*

Metals are used extensively for numerous purposes in society. Not only do they find uses in consumer products, but they are also needed for infrastructure, medical care, civil services, and most types of production and construction.

Increased metal production is needed for improved living standards, but where will this demand be needed? In the 1960s it was usual to divide the countries of the world into two distinct groups, labelled 'developed' and 'developing'. But this classification cannot describe future metal needs today. A better scheme is to use four country groups, defined as low-income, lower middle-income, upper middle-income, and high-income countries. Examples of high-income countries, with health and wealth, are Norway and Singapore. The poorest countries, like Congo and Somalia, are often ravaged by war. The income gap between the richest and the poorest is as wide as ever, yet the big change is that the gap in the middle has been filled. Most countries and most people now live in the middle of the socioeconomic continuum—in middle-income countries like Brazil, Mexico, China, Turkey, and Indonesia. Half of the world's economy—and most of the economic growth—now lies in these middle-income countries, outside the West, Western Europe, and North America. This is also where the new demand for metals will be prevalent. In the richest countries the situation is different. There are stockpiles of already smelted metals concentrated in cities. Some have called this an opportunity for urban mining, which is another name for the process of reclaiming raw materials from spent products, buildings, and waste concentrated in or nearby wealthy cities.

A simple fact relevant for recycling and refining will illustrate the current situation in another way. The growth of vehicle ownership per capita in each country follows an S-shaped curve. The same is the case for refrigerators or phones. We buy a refrigerator or mobile phone as soon as we can afford it. Then, after we have purchased it, the rate of increase in consumption drops, and a maximum is reached. This understanding is of importance for predicting metal demand as well as for planning of scrap availability. However, as personal income increases, the average distance travelled doubles, and transport accounts for $1/3$ of the CO_2 emissions.[2]

10.1.2.2 *Climate Change*

Metal production today represents about 8 per cent of the total global energy consumption, and a similar percentage of fossil-fuel-related CO_2 emissions. Obviously, recycling will decrease this 'footprint', since it requires less energy to remelt end-of-life (EoL) products containing metals. In Norway in 2018, industry accounted for 23 per cent of Norway's greenhouse gas emissions. Metal production accounts for about half of industrial emissions in Norway. For this reason, Norwegian metallurgists are especially focused on recycling.

Many countries are monitoring CO_2 emissions and cooperating across borders. The Intergovernmental Panel on Climate Change (IPCC) reports on the impacts of global warming.

They use representative concentration pathways (RCP) to describe the seriousness of the amount of CO_2 emissions. The highest RCP number is 8.5 where the CO_2 emissions are greatest. RCP8.5 implies that greenhouse gas emissions continue to increase until the end of this century. At the same time, the average temperature will rise between 3.3 and 6.4 °C in 100 years. Another scenario is RCP4.5, which corresponds to minor emissions increases until 2050, followed by cuts in emissions. The temperature will increase approximately 2.7 °C in 100 years for this scenario. RCP2.6 means drastic emissions cuts as early as 2020. In this case the temperature increase will be approximately 1.5 °C over 100 years.

10.1.2.3   *The Ecological Footprint*

The consequences of our burden on the environment are sometimes described by the ecological footprint. The ecological footprint is defined as the area of biologically productive land and water needed to regenerate the renewable resources a human population consumes, and to absorb and render harmless the wastes produced by humans. This is often symbolized by picture of a footprint or number of earths needed for our consumption.

From a panel that assessed the relative importance of the various environmental impacts, and considering resource availability and depletion, 13 important impact categories were identified. By far the most important was the carbon footprint.[3] It accounted for an estimated 25 per cent of the total impact.

10.1.2.4   *Definition of Sustainability*

The Brundtland Commission (UN, 1987) first popularized the term sustainable development and defined it as 'development that meets the needs of the present without compromising the ability of future generations to meet their needs'. Their report, 'Our Common Future', emphasized the importance of environmental protection in the pursuit of sustainable development by three different approaches: treaties and laws, standards, and business activity.

The concept of sustainability, or sustainable development, has had growing acceptance in recent decades, and has been adopted by several political entities, such as the United Nations. In 2015, the General Assembly adopted the 2030 Agenda for Sustainable Development that includes 17 Sustainable Development Goals (SDGs). Sustainable Development Goal number 12, 'Ensure sustainable consumption and production patterns', is at the heart of metal production.

10.1.2.5   *Scarcity*

The world is facing serious constraints regarding resources. The rare-earth metals have, for instance, been scarce for many years. International political bodies and many countries maintain lists of metals that are scarce and keep track of the consumption and supplies of metals. One way to deal with scarcity is by intermaterial substitutions. These could be: (a) cost-driven, (b) availability-driven, (c) regulatory-driven, and (d) functionality-driven.

From the discussion so far in Section 10.1.2 the burden of the production of metals is proportional to the amount produced. Reducing the demand for goods and services is one possibility. But other strong contenders are increased recycling, improved reuse, and reduced production of waste. (This is discussed in Section 10.2.)

Considering its importance, one may ask: 'Why do we not recycle?' The answers to this question are considered in Sections 10.1.3 and 10.1.4.

10.1.3   The Importance of Collection

The recycling chain is no stronger than its weakest link. Collection is often the weakest link. We cannot recycle if we cannot obtain it. If you are very bad at collecting, it doesn't help that you are very good at processing the material.

10.1.3.1   *Collection from Consumer Goods*

For the everyday consumer two factors have the most influence on whether people participate in recycling: convenience and money. Convenience is mostly related to the methods and frequency of collection. Available storage space for recyclables in peoples' homes, the distance to the nearest recycling site, and the treatment of recyclables, such as rinsing and drying, is also related to the convenience. Money relates to economic incentives to recycle. Studies have concluded that social behaviour is complex, so there is no single solution that fits all situations. There are several ways to get to the goal—collect everything.

10.1.3.2   *Collection in Industry*

Collection in industry depends on how fabricated by-products at the plant are handled. In-house or run-around scrap from manufacturing is generally easy to collect, even if some of it needs pre-processing. For example, metal chips from machining need to be washed and dried before remelting. Sometimes in-house scrap needs treatment elsewhere. This is usually handled through a trading or tolling system. The pricing is set to cover the external processing costs. One benefit for the manufacturer is that the recycled material is of known chemistry and has a predictable delivery. Important examples are the treatment of steelmaking slags and dross from aluminium melting furnaces. These can contain 20 or 30 per cent of metal in the form of suspended droplets.

In many cases the main method for collection in the metal industry is buying scrap. The amount of scrap available for purchase (collection) depends on several variables, such as place: urbanization (enough material available locally to support scrap yards); ease of handling (big pieces vs small); chemistry of the scrap; legislation; and industrial involvement. In the latter category, some producers collect their own products at the end-of-life. Examples are coffee capsules, refrigerators, and some electronic equipment. Some companies process their own collected goods.

Governmental regulations can be extremely important. A good deal of the trade in scrap metals is international. According to US Geological Survey (USGS) data, two million tons of aluminium scrap per year was exported from the US between 2009 and 2012. In essence, this material was produced in the US for use in other countries.

Unfortunately, many obsolete products are not recycled, but disappear into landfill. This is usually linked with cost.

10.1.4   Cost

Recycling has always taken place for valuable materials such as gold. This is because a profit is made only when

$$\text{Cost of scrap} + \text{cost of refining/recycling} < \text{cost of primary material} \qquad (10.5)$$

The price of primary metal sets an upper limit on the cost of refining. In the case of gold, the refining cost is a small portion of the metal price. For lower-priced metals one is forced to use scrap directly without refining. The most common practice is to 'blend' scrap into primary metal. Blending reduces or dilutes the impact of impurities contained in the recycled metal. There may also be energy savings. In steelmaking furnaces heat is released when silicon and carbon are oxidized from pig iron. The energy produced is sufficient to melt 25–30 per cent of solid scrap charged to the furnace. In the aluminium casthouse, metal from the taprooms is normally 900 to 950 °C. This sensible heat can be used to melt solid scrap charged to casthouse furnaces.

Equation (10.5) shows that scrap prices usually follow the London Metal Exchange price. Of course, prices and costs depend on many variables. In low-income countries the manual sorting of recyclables can be economically viable due to low wages. In high-income countries sorting must be automated and/or conducted by the consumer. Material prices are higher than wages in low-income countries, which create an incentive for recycling, while high labour costs make it harder to recycle with a profit. This has resulted in high export rates of different materials to middle-income countries such as China, because processing is cheaper there. We have recently seen that countries have started to ban import of waste. Most important, in 2018 China banned the importation of low-quality solid waste from foreign countries. (These wastes included plastics, paper products, textiles, and some aluminium fractions.) This ban strongly affected the profitability of waste processing and changed the scrap market in North America.

The quality of the scrap is also important for profitability. In steels the most serious problem is usually contamination of scrap with copper. Because of thermodynamic factors, copper cannot be easily removed from steel. So, any copper added to the scrap mix remains in the 'system' as the material continues to be recycled. In structural aluminium alloys the problem is usually contamination with iron, which lowers ductility. The solubility of iron in solid aluminium is only 2.5 per cent of the solubility in the liquid metal, so a fractional crystallization process can be used to refine scrap high in iron. This process has been used commercially, but it has only seen limited use for very-high-purity alloys. The reason is economic. Most primary aluminium is sold under the alloy designation P1020, or its equivalent. (P1020 has a maximum limit of 0.10% Si and 0.20% Fe.) Higher-purity alloys can be produced by using higher-quality (purer) ores and by careful maintenance of anodes in the pot room. Traditionally, the price differential between P1020 and a higher-purity alloy like P0406 is between 6 and 8 cents a pound (€140 to 190 per tonne). This is less than the cost of refining metal by fractional crystallization.

Characteristics of scrap are discussed in more detail in Section 10.3.2. However, it is clear that a more sophisticated scrap-handling system is needed to address the problem of Cu contamination in steel. New and better methods are also needed to separate and sort, and to refine aluminium scrap.

10.1.4.1   *Environmental Footprint versus Economic Value*

The concept of the circular economy suggests changing from a linear economy, where resources are extracted from the environment—fabricated, used once, and disposed—to one where materials circulate in a closed loop. Numerous definitions and interpretations of the concept circular economy can be found in the literature.

Of course, economic limitations can be 'corrected' by governmental policy, but the proposed changes must still be technically feasible. For example, the issue of recycling critical raw materials is usually politically sensitive. A change may become feasible if tough laws are enacted. One possibility is to forbid disposal in landfills, or to make it so costly that recycling is the only option that makes sense.

Political questions are really beyond the scope of this chapter. Also, it is difficult to make sensible predictions. The political climate varies greatly. Most of the world could be classified as a capitalistic system, which in theory (not always in practice) believes in a *laissez faire* approach to business. On the other hand, one has socialistic governments where business activity is highly regulated or controlled. It is impossible to generalize.

In any case, governmental action (or inaction) can be extremely important. Consider quota systems for greenhouse gas emissions. A regional greenhouse gas initiative was one of the first market-based regulatory programmes in the United States to reduce greenhouse gas emissions. It was a cooperative effort among the states of Connecticut, Delaware, Maine, Maryland, Massachusetts, New Hampshire, New York, Rhode Island, and Vermont to cap and reduce CO_2 emissions from the power sector.

Basically, there are two types of instruments for recycling: those which create a financial penalty for not recycling and those which reward recycling. Fees for dumping waste is an example of the first; a deposit refund system is the second. There can also be a mix. The polluter must pay is a key principle in environmental policy.

10.1.4.2   *Dematerialization and Use of Metals*

Dematerialization is delivering the same product or service using a smaller percentage or none of the material under consideration. This can be accomplished by optimizing the product: making the car lighter, the can thinner, or not using aluminium foil inside a chips bag. Another possibility is related to changes in social and consumer behaviour. Your automobile sits parked and unused most of the time. Hence, many people now consider a future society that shifts from personal ownership to ride-sharing services.

10.1.4.3   *Estimating the Cost on the Environment: Life Cycle Assessment and Mass Flow Analysis*

10.1.4.3.1   LCA   Life cycle assessment, sometimes also called life cycle analysis (LCA), is a discipline from the late 1960s that tried to bridge the gap between the economy and the environment. Today, LCA is rigidly defined though international standards which define its methodology.[3] One of the first LCAs was to compare various containers of Coca-Cola.[4]

LCA focuses on stewardship within the life cycle in terms of the environmental impact. LCA can be a tool for selection of materials in design and/or marketing of sustainable products. The International Council on Mining and Metals describes the following approach to carry out an LCA under the guidance of ISO standardization. The main stages in an LCA are:

1. Goal and scope definition (system boundaries);
2. Life cycle inventory (data collection);

3. Life cycle impact assessment (CO_2 footprint); and

4. Interpretation of the result.

Of course, good data are important for a valuable assessment. One also must be careful when defining the system boundaries. Consider a recent life cycle assessment of secondary aluminium refining.[5] The scrap composition and desired end use called for the 77–100 per cent removal of Zn. This high level of removal required refining by step distillation. The energy usage of this process gave a significantly different result than when distillation was excluded. Hence, one must be aware of how the scope boundaries affect the results. System boundaries with or without transport can give completely different answers. The effect of the energy mix[6] and the inclusion of more recycling steps are other things to consider.

The LCA calculation rules for recycling need to be clear whether (and how) the environmental burden of the primary production is shared between the first user and the subsequent users of the material. That means one could make the product carry the full environmental burden of production of its primary metal, even if it is later recycled. The other end-of-life approach usually overestimates the recycling potential.

10.1.4.3.2   MFA   Material flow accounting and material flow analysis (MFA) is another method for calculating the cost to the environment. MFA started in the 1980s. It is similar to LCA in that they both are based on mass balances applied in carefully defined situations.

One of the early examples of MFA came from a steelmaking group that in the 1980s needed to change its raw material base from iron ore to scrap. To justify this major transformation, it needed to determine whether enough scrap would be available.[7] The study showed that the collection rates were much higher than the 30 per cent required by the study. Hence, a new electric arc furnace could be installed, with the assurance that sufficient scrap would be available for many years of operation. This study demonstrates in a practical way the potential value of an MFA.

10.1.4.3.3   Comparing LCA and MFA   In Fig. 10.2 a comparison of an LCA and MFA is illustrated. An LCA follows the life cycle of a product during the production, use, and end-of-life phases. In contrast, the MFA follows a development over a time scale, in this case a period of several years.

Detailed MFA analyses have been carried out on common metals, such as iron, copper, and aluminium; as well as the metals chromium and lead. These have been reviewed by Birat and co-workers.[3] An especially good representation of an MFA is shown in Fig. 10.3. This was first presented as part of a study by Boin and Bertram for the material flows of aluminium in the EU.[9] They studied the mass balances for the aluminium recycling industry in European Union member states. Results showed that in 2002, about 7 million tonnes of purchased, tolled, and internal scrap—with a metal content of 94 per cent—were recycled among 15 EU members. This represented a very respectable metal recovery rate of 98 per cent. Figure 10.3 is an updated version of their analysis.

These studies can be extremely useful. One example is a study of copper contents in seven different types of waste. It showed clearly that we need to do much better in recovering the copper present in waste electronic scrap.[10]

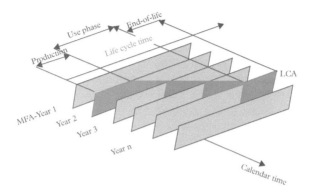

Fig. 10.2 *Comparison of LCA and MFA analyses.*[8] *[Published with permission from Jean-Pierre Birat.]*

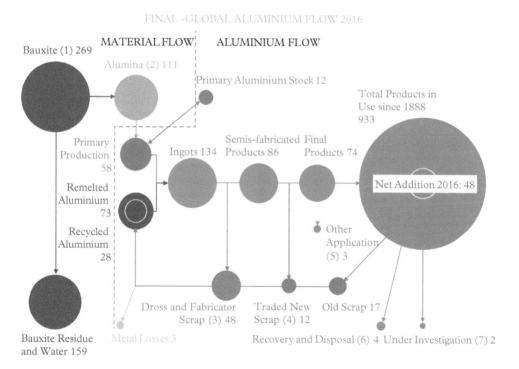

Fig. 10.3 *Material flows of aluminium in Europe in 2016.*[9] *[Reprinted with permission from Springer Nature: copyright 2005.]*

10.1.4.4 Summary

We must work to implement the provisions set forward by the Brundtland Commission; and the most well-known branch of this: the Kyoto Protocol and its associated activities. Clearly, the reuse of alloys or recycling is an area where the metallurgist can contribute. Recycled metals are less energy intensive. They tend to preserve natural resources and avoid associated environmental problems, and their recycling can be profitable.

Of course, we need to be realistic. The idea of recycling everything with 100% recovery, and without any downgrading, is like the dream of the perpetuum mobile (a hypothetical machine that can work indefinitely without an external energy source). This kind of machine is impossible, as it would violate the second law of thermodynamics. The same applies for metallurgical processes. Doing something without losses (in entropy) is not possible. There will always be work for the refining metallurgist.

10.2   Principles of Recycling Strategies

10.2.1   The Waste Hierarchy—Product Perspective

In terms of the best environmental outcome, there is a hierarchy of how waste is dealt with, as illustrated in Fig. 10.4. The worst option (at the bottom of the pyramid) is landfill disposal, with prevention being the most favourable option.

Waste prevention and minimization are preferred solutions. Preventing waste, e.g. excessive packaging, not only reduces material use, it also prevents carbon emissions associated with the energy or fuel used in the manufacture and transport of the packaging.

Reuse of products and materials is most effective if waste prevention is not possible. This reduces the environmental footprint associated with making a new product. Reuse additionally reduces demand for raw materials and the environmental impacts from manufacturing and transport.

Recycling reduces the need for raw materials, prevents materials being disposed of, and contributes to reduced greenhouse gas emissions. The process of recycling one product into

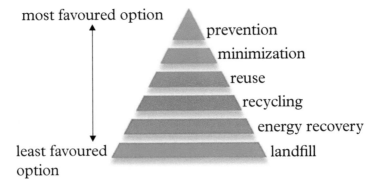

Fig. 10.4 *The waste hierarchy.*

another requires energy, and therefore involves greater carbon emissions than prevention, minimization, and reuse. However, recycling uses less energy than creating a product from raw materials.

Energy recovery takes waste and converts the energy stored within it into useful energy, reducing fossil fuel use and associated emissions of carbon dioxide and pollutants.

Landfill is the least favorable option with regard to environmental impact. Putting waste into landfill reduces the potential for reuse, recycling, or recovery of valuable resources and increases the demand for new resources, having a higher environmental impact.

Better management of waste can significantly reduce the environmental impact and emissions of greenhouse gases. Through application of the principles of the waste hierarchy, many waste materials can be turned into resources. As a result, the need to increase the extraction of raw materials and the use of fossil fuels can be reduced. Technologies exist that enable materials and energy to be recovered from waste that would otherwise be landfilled. Other relvant topics along the line of the product perspective include 3R (reduce, reuse, recycle), 6R (reuse, recycle, redesign, remanufacture, reduce, recover), and 4R (reduce, reuse, recycle, recover waste).

Even though waste prevention and minimization are the preferred solution, it may be hard to achieve. Very few waste by-products from industrial processes, even if they contain valuable materials, can be returned to the original production process.

10.2.2 Circular Materials Management—Process Perspective

The Ellen MacArthur Foundation was one of the first to use the famous figure, showing the material cycles of maintain, reuse, refurbish, and recycle on the right-hand side of Fig. 10.5. These cycles are compared to the biosphere on the left-hand side. In many respects this is not a new idea. However, we thank the Ellen MacArthur Foundation for helping to root the concept of a circular economy in public awareness. The importance of this concept for our future agenda should not be underestimated.

Of course, applying this concept to real problems can be a challenge. One needs to look carefully at the technical and engineering details; while also keeping a perspective on the overall process. Then one can calculate material-specific recycling outcomes, depending on the products produced and processes used. Many different scenarios may have to be analysed. Then it is possible to determine the best solution(s). A number of examples are provided later in this chapter to better illustrate what is required.

10.2.2.1 Other Strategies

There are other strategies that are worthy of mention. One is often called the 'cherry picking' strategy: *taking care of the most valuable.* Companies optimize their process on recovery of the most valuable material. That is, recyclers typically focus on the first product they can sell profitably. They often do not have the economics of scale to justify further processing. An important element to economic survival is having access on a long-term basis to a significant tonnage of their key raw materials. Typically, the by-product from their process is not a waste, but can also be sold. Relationships with suppliers and product consumers are typically warm and secure. Integrity is a given.

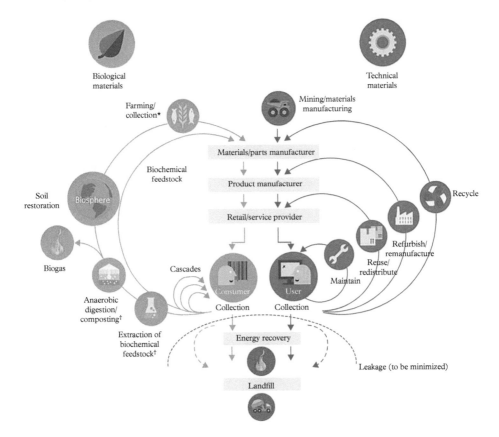

Fig. 10.5 *Circular materials management, compared to the biological sphere.*

Strategies of systems with small quantities of valuables apply to minor elements that are easily lost. Some have compared this to spreading salt and pepper over the earth and then trying to recover the salt economically. Electronic waste, one of the fastest-growing products, faces this challenge. Electronic devices depend not only on the availability of common metals for their manufacture, but also increasingly use 'exotic' minor metals that facilitate specific functionalities.

Strategies for *challenging systems with small quantities of impurities that may ruin the value* of the recovered materials is another challenge. This is usually the case for plastics and often for aluminium. Impurities are sometimes the tail that wags the recycling dog.

Section 10.3 looks at metal scrap in some detail. Different types of scrap and some important characteristics are described. Accurate sampling of scrap can be a challenge, owing to the heterogeneous nature of many types of scrap. Practical recommendations are given. Examples are then presented for the recycling of three different types of scrap.

10.3 Classification of Scrap

There are many ways of classifying scrap. Each metal (iron, aluminium, copper, etc.) has its own system of classification. Different countries may have their own system. These are considered briefly here, primarily from the perspective of the refining metallurgist. In other words, what type of process is best suited for a particular type of scrap? Do we need to remove dissolved elements? Can these be removed by pretreatment (e.g. removing water or surface coatings)? Or must undesired alloying elements be removed? This subject is considered in Chapter 4. How can inclusions (suspended solid particles) be removed? (See Chapter 5.) Solid intermetallic phases, oxides, and carbides can be present in the scrap, or may be formed during melting.

One classification scheme proposed for aluminium scrap is by the stage in the life cycle and by form of product.[9]

Life cycle stage	Products and by-products
Production	Dross
Fabrication	Scrap from extrusions, rolled products, foil, castings, and turnings (chips)
Manufacturing	Buildings, transportation, consumer goods, cans and rigid packaging, cable and wire, foil, structural (engineered) products, and turnings
End of life	Buildings, automotive, other transportation, consumer goods, cans and rigid packaging, foil, and consumer durable goods

10.3.1 Methods of Classification

10.3.1.1 Pre- and Post-consumer Scrap

If we draw a line underneath the manufacturing life cycle stage above, we form two separate types of scrap: pre- and post-consumer scrap. This is often referred to as 'new' and 'old' scrap. Pre-consumer (new) scrap consists of materials produced during various stages of manufacture. An obvious example is the metal lost, or cut away, during machining. Called turnings, or sometimes chips, these are swept up at each individual machining station and placed into storage bins. Metal is also removed from extrusions, ingots, rods, and rolled foils and sheets to produce the exact size and shape required for further manufacturing and assembly. From a recycling point of view, these are all prime materials, because they are of known chemistry. They are also produced in large, steady quantities and are relatively easy to recycle. Pre-consumer scrap is recycled to a very high degree compared to post-consumer scrap.

Post-consumer scrap has the opposite problems. It is often mixed with other materials. Its composition is less well known. It usually requires more beneficiation (shredding and sorting) and processing before it can be recycled. In some cases, collection is problematic.

The relative importance of 'new' and 'old' scrap in the US secondary aluminium industry is shown in Fig. 10.6, for the twenty-year period 1998 to 2017.

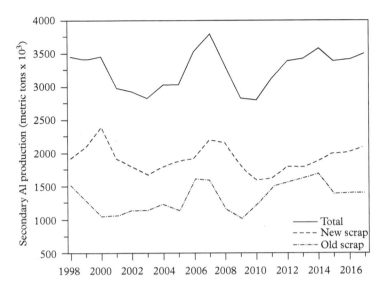

Fig. 10.6 *Amounts of new and old aluminium scrap used by US secondary metal producers. (Data from the review by Capuzzi and Timelli.[11] This work is licensed under a Creative Commons Attribution 4.0 generic license.)*

10.3.1.2   Formal Classification Systems

A good example of a formal classification system, and how it relates to recyclability, is given for magnesium in Section 10.5.1. Another is the system established in the EU for aluminium.[12] This is summarized in Table 10.1, together with important information about each type of scrap. The prices are based on market quotations during October 2017. By comparison, the London Metal Exchange (LME) price for primary aluminium during the same period was €1785 to €1875/tonne. The 'cleaner' types of new scrap (part numbers 2, 3, and 4 in the table) had a price that was between ²/₃ and ¾ of the price of primary metal. Other types of scrap require more processing, and consequently command a lower price (in accordance with Eq. (10.5)).

The same situation exists for other metals and products. As we go down the list, the composition becomes more varied, the difficulty of recycling increases, and the value of the scrap decreases. Sorted clean returns and dry chips/shavings may be remelted with little loss, while scrap contaminated with inserts, oil, or paint requires more processing. The variation in properties between different metals determines how easy they are to refine. Inserts, like screws and nails, or contaminants, like oil and paint, may not be a problem in pig iron, but will significantly reduce the quality of aluminium or magnesium alloys.

We cannot resist sharing here humorous aspects of national scrap classification systems. In Germany the names for all types of aluminium scrap start with the letter 'A'. For example, *Abweg* is new wire scrap of pure Al free of foreign matter. *Talon* is the name in the US for the same material. In the US the *Star Wars* films have inspired some unusual names, such as:

Table 10.1 *Classification of Al scrap showing the metal yields (lower limit and average value), the average amounts of undesired materials (oxides and foreign materials), and the cost*

EN 13920 part no	Scrap	Metal yield lower limit (%)	Metal yield avg. value (%)	Oxides (%)	Foreign material (%)	Cost* (€/ton)
2	Unalloyed aluminium	0.95	—	—	—	1232–1332
3	Wire and cable	≥0.95	97.7	1.3	0.5	1182–1282
4	Single wrought alloy	≥0.95	97.2	1.8	1.8	1272–1377
5	Two or more wrought alloys—same family	≥0.88	97.2	1.0	2.0	1172–1222
6	Two or more wrought alloys	≥0.88	94.0	0.8	5.2	—
7	Casting	≥0.9	83.4	6.2	10.4	—
8	Shredded (not separated)	≥0.9	—	—	—	—
9	Shredded (separated)	≥0.9	84.5	5.4	10.1	—
10	Used beverage cans	≥0.88	94.0	0.8	5.2	770–820
11	Al–Cu radiators	—	—	—	—	2247–2297
12	Turnings single alloy	≥0.9	95.3	3.7	1.0	1192–1257
13	Mixed turnings	≥0.9	84.0	3.3	12.8	1157–1207
14	Coated packaging	≥0.28	71.5	3.8	24.7	—
15	De-coated packaging	≥0.8	86.1	12.9	1	—
16	Dross	≥0.3	55.7	44.3	—	158–338

* Available data refer to market quotations in October 2017.

Data are read and elaborated from the different parts of the EN 13920 standard. (Data from the review by Capuzzi and Timelli.[11]

> *Twitch*: fragmentized aluminium scrap (from automobile shredders)
>
> *Taboo*: mixed low copper aluminium clippings and solids, and
>
> *Zorba*: mixed nonferrous scrap metal.

More information is available in the excellent book by Schlesinger.[13]

10.3.2   Important Characteristics and Considerations

A classification scheme is only a partial solution to the need to characterize important qualities of a scrap metal. We would also like to know the answer to these important questions: *Where* does the scrap come from? *What* does it contain? And *when* was it produced?

10.3.2.1   Where Does the Scrap Come from?

There are a number of different processing steps in the manufacture of any product. Generally speaking, scrap from early in the value chain is less contaminated, less complex, and easier to reinsert into 'the loop'.

On the other hand, we have material that has been used by the consumer and is at the end of its life. This usually needs more pre-processing. Pre-processing can be carried out manually, automatically (shredding and sorting of materials), or with a semi-automatic process combining manual and automated techniques. Pre-processing must also ensure that hazardous materials (like battery acid) are taken care of properly. Scrap recovered from the last portion of the value chain cannot generally be recycled into materials that have high-purity requirements.

10.3.2.2   What Type of Scrap Is It?

Another important characteristic is what the scrap contains: that is, the chemical composition of the material. But other physical aspects may be important, especial in end-of-life products. Consider the differences between electro-consumer goods, buildings, photovoltaics, packaging, magnets, and batteries. Even the manner in which the product has been coated, glued, welded, and stored can affect our ability to recover the material.

10.3.2.3   When Was the Scrap Produced?

The nature of scrap tends to change with time. There is a time distribution on when products reach their end of life and become scrap. This ranges from a few weeks for used beverage cans (UBCs) to 50 or more years (e.g. building materials). A distribution around the lifetime of a product versus the 'typical' or average lifetime is also an issue to consider. Imagine the lifetime of a car in a low-income country compared to that in a high-income country. And various products have different expected lifetimes. For passenger cars in America the average lifetime 'has increased from 12.2 to 15.6 years between 1970s and the 2000s'.[14] Studies of aluminium flows estimate the lifetime of packaging materials to be about 1 year, while consumer durable goods are 8–15 years.

The time aspect is also important within a single class of scrap. In the automotive category, newer electrical cars and traditional fossil-fuelled cars have different recycling potential. Circuit boards produced many years ago contain more precious metals than newer ones. Also new restrictions (e.g. such as on brominated flame retarders in plastics) can change important scrap characteristics with time.

Yesterday's, today's, and tomorrow's products will vary considerably. Decisions made today in manufacturing and product design will impact future scrap flows in composition and volume. During the design and development process, decisions are made that will determine the costs of manufacturing, use, and recycling of the future product.

Megatrends also influence design decisions. The concern for CO_2 emissions and electrical cars is one example. But many factors usually enter the picture. For example, the usability and market success of electrical cars depend upon having networks of charger stations built for individual car brands.

With the dissemination of UN's sustainability goals and the principle of a circular economy, designing for longevity has gained momentum. There is a drive to design for multiple use cycles, end extend the duration before a product is considered obsolete. Simultaneously, scientific analyses of products from cradle to grave exemplify how product life extensions can decrease environmental impacts. This is especially true for products where the major environmental impacts are made during the user stage.

And of course, scrap quality and availability depend on the economic region and country under consideration.

All of these factors mean the need for future innovation will be great.

10.3.3   Sampling and Analysis

When a new class of scrap is unknown, it is necessary to take samples for analysis. Actually, it is good practise to sample all incoming materials, if only on an occasional basis as a spot check. This helps to avoid nasty surprises in production and to keep suppliers on their toes.

Proper sampling techniques are necessary to obtain an accurate analysis of any unknown material. It is important to understand the source of errors, and to be able to characterize their magnitude. Fortunately, a few simple rules can be employed for good results. The guidelines offered here were developed by Schuhmann.[15]

Consider the case where scrap is flowing continuously on a conveyor belt. In this situation we take all of the material part of the time. Assuming the composition of material does not change with time, how may one estimate the error associated with our sampling? Assuming further there is little error in the assay of our samples, the probable error (1σ) associated with sampling is

$$\text{Probable error} = 0.675\sqrt{\frac{\sum (X_n - X_m)^2}{N}}, \tag{10.6}$$

where

N is the total number of sample analysed

X_n is the value of the analysis of the nth individual sample ($n = 1, 2, 3, 4, \ldots , N$)

X_m is the mean or average value of all the analyses.

If no systematic errors are introduced, the increase in effective number of cuts can also be obtained by increasing the size of the samples, as well as the number of samples. In other words, the probable error is inversely proportional to the square root of the total amount of sample taken. Each increase of four in the number of samples decreases the error by a factor of two.

What about the size of the sample? How much material is needed for good results? The answer here is more complicated and less clear cut. However, in most cases the recommended weight of an individual sample is given by the relation

$$W = kD^\alpha, \tag{10.7}$$

where

W is the sample weight

k is a constant which depends on the material and precision desired

α is a constant having a value between 2 and 3

D is the size of the maximum particle or piece of scrap.

For most purposes a value of $\alpha = 2$ gives satisfactory results. This means the sample weight is proportional to the square of the maximum particle (piece) size.

Equation (10.7) was developed for the analysis of granular materials, such as metal ores or coal. Its application to sources of scrap is usually not so straightforward, especially for scrap present in relatively large pieces (such as scrapped extrusions or castings). Some judgement is required. This 'calibration' of sampling method is best obtained by field experience. At first it will be necessary to conduct an extensive sampling of different materials to better establish the procedures needed to minimize sampling errors.

The above considered materials flowing on a conveyor. What should one do when confronted with a big pile or a rail car full of material? The natural inclination is to pick a few samples from the top. But this may lead to false results. There may be a stratification of material caused by different particle sizes, or effective densities. Material at the top may also have become 'weathered' or oxidized, compared to material deeper in the pile. Ideally, one could take a number of grab samples while a rail car is being unloaded or a pile is being moved.

10.3.3.1   *Analysis*

The first step in analysis would be to dry the samples and measure the weight loss. The moisture present would depend on many factors, including recent weather. Any results should be reported on a dry basis, to eliminate that source of variability. If the scrap is coated, you will probably also want to determine the percentage of organic compounds present on the surface, by oxidizing at higher temperatures. (See Section 10.4.4 on pretreatment.) Now, how does one determine the chemical analysis of the remaining metal fraction?

Over the years we have seen numerous examples of people checking alloy chemistry by cutting pieces off an ingot or casting and 'shooting it' on an optical emission spectrograph (OES). Unfortunately, it is not possible to obtain a reliable analysis in this manner. There is a segregation of alloying elements during solidification. The results also depend strongly on how the sample surface is machined. For these reasons, special sample moulds are used to cast sample coupons for OES analysis, and the depth and manner of machining are carefully specified (in the US, the relevant standard for aluminium alloys is ASTM E716). There are portable, hand-held X-ray fluorescence analysers that can be used to check larger pieces of scrap. These are good for qualitative analyses, not for an accurate quantitative analysis. In other words, this equipment would be useful only for quick field tests on a new scrap, or to check for the possibility of 'mixed' alloys. Inductively couple plasma (ICP) analysis could be used. One could use machine chips from pieces of scrap and dissolve them in acid. The solution can then be analysed. However, this method is not convenient for analysing large quantities of material, and is probably best reserved for analysing small amounts of bothersome 'tramp' elements. (See Chapter 1.)

The best procedure, and one most commonly used, is to melt a quantity of sampled scrap in a small crucible furnace. The capacity is usually between 50 and 200 kg. For very rapid melting an induction furnace may be used. For non-ferrous metals a gas-fired furnace is more common. Once the material is fully molten, and well mixed, spectrographic coupons can be cast and analysed in the usual manner (by OES). This procedure offers a high degree of reliability, and an assurance that the composition obtained in the industrial melting furnaces will be the same as that obtained in the sampled analysis.

10.3.4   Example: Used Beverage Cans

We begin our consideration of examples of scrap with the most familiar, and perhaps the most important: the used beverage can (UBC). Aluminium cans were the first 'hero' or 'poster child' of the recycling movement. UBCs were presented as a successful example of a circular economy in practice. For many years we were told the average can was on the shelf only a few months, before it was remelted and reused. Unfortunately, this recycling success story is not completely accurate. The truth is more complicated. We face a number of challenges before the recycling of UBCs becomes a true closed loop. At present, it would be more accurate to say: 'Dilution is the solution'. To better appreciate our current situation, it will be useful to review some history.

For older readers who began their beer consumption in the 1960s or 1970s, you may remember that cans were made of steel and were 'tough'. Polishing off a beer was sometimes followed by a 'macho' challenge to crush the empty can with one hand. Also, at that time a 'church key' was an essential implement for any young beer drinker to have. All that would change after aluminium cans were introduced. (See Fig. 10.7.)

Introduced in 1963, the two-piece drawn and wall-ironed aluminium beverage can is probably the most successful aluminium product innovation of all time. The global production is about 250 billion cans per year, which represents approximately 3 million metric tons. This represents a significant fraction of the world's production of flat-rolled aluminium products. Together with other packaging applications, it accounts for about 20% of all consumption of aluminium. It is difficult to grasp the significance of these numbers. One way is to witness

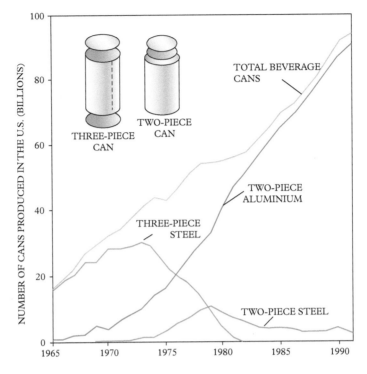

Fig. 10.7 *Can production by type from 1965 to 1994. [Reproduced from Duncan and Hosford[18] with permission from Springer Nature. Copyright 1994.]*

the direct chill (DC) casting of can stock (Fig. 10.8). The huge ingots in this figure will be rolled to a thickness that is nearly three thousand times smaller, before being used to make cans. The technology behind this development was the result of improvements in manufacturing efficiencies in many areas, including rolling, forming, casting, coating, and, critically, the development of improved degassing and filtration, to eliminate inclusions. The drawing/ironing process to form the can walls is very sensitive to the presence of inclusions, which can cause perforations and tear-offs.

The innovation that would make the 'church key' obsolete was the 'pop top' version of aluminium cans. A great deal of research and design went into this development. The result was a tabular opening that stayed inside the can, instead of winding up as litter alongside highways. However, a different alloy had to be developed for the can top. Table 10.2 shows the alloys commonly used for the can body and the can top, together with a typical average composition of UBC scrap.

Magnesium is easily removed from aluminium alloys. It is more reactive than aluminium and is selectively oxidized by oxygen in the air, by fluxing with chlorine or by using reactive salts. Manganese, on the other hand, is more noble and cannot be removed in that way. Hence, most of the metal obtained from UBC scrap is usually recycled for use in the production of the higher Mn-containing alloys for the body of the can.

Fig. 10.8 *An ingot of 3004 alloy being removed from the casting pit.*

Table 10.2 *Compositions of Al can body and can top alloys, and UBC scrap*[16]

Use (alloy)	%Mg	%Mn	%Cu	%Fe	%Si
Can body (AA 3004)	0.8–1.3	1.0–1.5	<0.25	<0.7	<0.3
Can body (AA 3104)	0.8–1.4	0.8–1.3	<0.25	<0.8	<0.4
Can top (AA 5182)	4.0–5.0	0.2–0.5	<0.15	<0.35	<0.2
Typical UBC scrap	1.95	0.9			

With permission from McGill Metals Processing Centre.

The recycling of UBCs was examined in a detailed material flow analysis by Terukina.[17] Figure 10.9 shows the collection rates for North America and Europe in 2010. Obviously, some form of incentive is needed to encourage beer drinkers to recycle their cans.[ii] In the

---

[ii] Although deposit schemes result in a high level of collection, this is not desirable for facilities processing municipal waste. Aluminium is not a large percentage of the waste stream by weight. However, it represents a significant portion of the income, since it is more valuable than plastic or paper. Removing UBCs from municipal waste could make a MWF facility uneconomical to run.

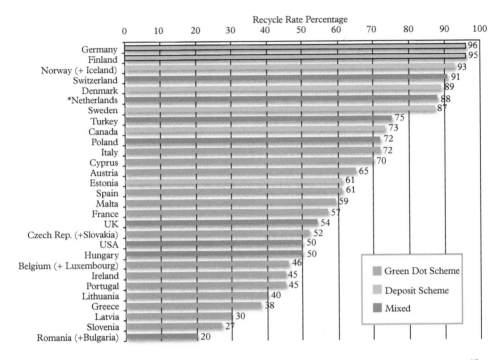

Fig. 10.9 *Collection rates for aluminium beverage cans in Europe, the USA, and Canada in 2010.*[17]
[*With permission from Container Recycling Institute*]

USA the situation is mixed. Eleven states have a deposit scheme in place. In these 11 states the collection rate was 76%. In the other 39, where no deposit is required, the collection rate was only 35%.

Figure 10.10 shows historical data for the recycling of UBCs in the United States. The recycling rate has dropped significantly since the early 1990s. Why is this? This question was considered by Schlesinger.[19] The possible reasons for this decline are as follows.

1. *Thinner cans.* Since the beginning there has been continual pressure to reduce the mass of the can by decreasing sheet thickness. The newer cans are thinner and have narrower tops. In 1973 one had to collect 49 cans to obtain one kilogram of metal. In 2004 one needed 75 cans. This translates into more work for the same return of metal. Also, a thinner gauge material will have a higher melt loss (as dross) on melting. As one recycler told us: 'If cans get any thinner, we must say that we burn UBCs instead of recycling UBCs.'

2. *Lower scrap prices.* Adjusting for inflation, between 1980 and 2000 the value of Al scrap decreased by over half of its value.

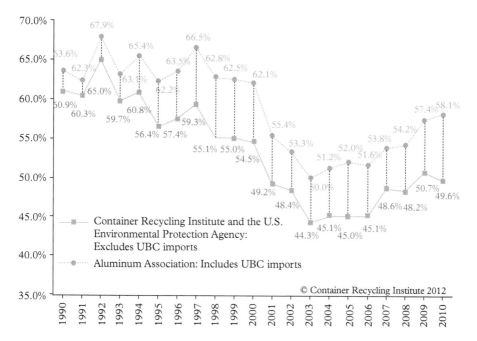

Fig. 10.10 *Aluminium can recycling rates in the USA.*[17] [*With permission from Container Recycling Institute.*]

3. *Lower effective deposit fees.* In 1971 nine states instituted a mandatory deposit fee of 5 cents per can. Thirty years later, in 2001, inflation had 'eaten up' nearly 75% of that five cents. In other words, the effective deposit in 2001 was effectively a little over 1.5 cents.

4. *Cheap labour is required.* The peak in recycling occurring in the early 1990s corresponded to a time of high unemployment—7.5 per cent in 1992. Afterwards, the jobless rate in the USA fell to 4 per cent in 2000; and so did the recycling rate. When one is out of work, it 'pays' to collect littered cans for the deposit.

5. *Change in landfill costs.* After a period of significant increase, in the early 1990s landfill costs levelled off. This partially reduced the incentive to recycle.

The above results show that the main barrier to higher recycling rates is the low recovery of UBCs after use. Nearly half of the UBCs in the USA are lost to landfills or incineration.

10.3.4.1 Other Problems

Interviews with managers at recycling facilities and processing plants revealed that aluminium beverage cans are by far the most favourable item for them to recycle, due to its consistent market, easy sorting, and high value. However, concerns were raised regarding additions to the packaging. Examples of additions found in beverage cans include the following:

- Resealable lids (often made of plastic)
- Plastic widgets (such as the plastic ball inside Guinness beer cans)
- Stickers
- Shrink sleeve labels.[iii]

These additions make it more difficult to control pretreatment furnace temperatures, and may require the mill to operate at a slower pace. It also gives a less predictable quality and yield of recycled aluminium. Stickers and shrink-sleeves can cause sorting problems, risking loss of the aluminium if it ends up in the wrong stream.

Another challenge is the difference in composition in the alloys used for the wall and the lid of the can. They are essentially incompatible. Does this prevent UBCs from becoming an example of a closed product loop in the circular economy? This was considered in a detailed life cycle analysis by Niero and Olsen.[20] They concluded that turning old cans 100 per cent into new cans was not possible, nor was it necessarily desirable. The goal should be recycling close to 100 per cent of UBCs, even if the recycled metal does not necessarily go 100 per cent into cans.

Creativity and human ingenuity can change the equation. Consider the example of Rheinfelden Aluminium; a small producer of specialty alloys located in Southern Germany. They operated a secondary smelter that processed UBCs along with other packaging waste. The incoming material was fed to a rotary kiln for pretreatment. Treated materials were fed into a second rotary kiln, together with fluxes, to melt the recovered aluminium. This material was relatively high in manganese.

In looking for new ways to use this Mn-containing metal, Rheinfelden developed a series of new alloys for high-pressure die casting. Conventional die-casting alloys are high in iron, which is needed to reduce die wear. (Hot liquid aluminium low in iron wants to dissolve the tool steel dies.) Unfortunately, the high iron content lowers the casting ductility. Rheinfelden used manganese to replace most of the iron in their die-casting alloys. The result was a reasonable die life, together with the ability to produce high ductility, structural die castings. These are finding extensive use, as for example structural automotive components, engine cradles, shock towers, and door pillars. Further details of this significant alloy development are available elsewhere.[21,22]

10.3.5 Example: Packaging Scrap

Beverage cans are one example of packaging for beverages. Packaging materials for food are generally multilayer. The combination of materials makes a packaging material that is lightweight, has excellent preserving qualities (by acting as a barrier to oxygen and light), is flexible and offers an attractive visual appearance. Because of this combination of desirable

---

[iii] These labels replace printed design on cans. An oversized label is placed around a can like a sleeve. High temperature makes it shrink and wrap tightly around the can. These are often used by craft breweries, because they can be ordered in smaller quantities than printed cans.

properties, the use of multilayer packaging is increasing. Common examples are the metalized plastic bags containing crisps and other snacks.

Unfortunately, these metalized plastic bags are difficult to recycle. Aluminium is frequently used in multi-material packaging, usually as a protective coating on plastic, paper, or carton. Separating the plastic from the aluminium is challenging, and no commercial recycling facility is able to do this efficiently today. Therefore, most of these bags either go to landfill or are incinerated. Because the aluminium layer on these bags is extremely thin, recovery of aluminium from the bottom ash in incinerators is not possible.

The aluminium content of composite multilayer materials tends to be low, about 20 per cent in plastic bags and 4 per cent in beverage cartons.[23] The oxidation loss on melting depends on the thickness and the alloy used, ranging from under 10 per cent for rigid packaging, such as beverage cans, to more than 50 per cent for flexible packaging, such as aluminium foil.[24,25]

Experiments have been conducted to separate aluminium from metalized plastic films. These involve a chemical treatment, either to dissolve one of the two materials in the multilayer film or to weaken the adhesive forces between layers.[23] These methods have challenges that may prevent their use on an industrial scale, such as low reaction rates, complicated recovery of dissolved aluminium, and extensive use of expensive and/or harmful chemical solvents.

High-temperature treatment appears to offer more promise. One example is a process being developed in Cambridge by Enval Ltd., to process aluminium–plastic laminates. The laminates are shredded to an appropriate size and transferred to a reactor chamber with a nitrogen atmosphere, where it is mixed with carbon powder. Power is supplied through microwaves. The process temperature is about 600 °C. At this temperature organic materials— such as plastic, lacquer, foodstuff, and paper—break down and form gases that are collected and condensed into gas and oil. These products could perhaps be used to synthesize new plastics, but it is more likely they will be utilized instead as fuel. The scrap does not need to be clean, as most contaminants will pyrolyse. Up to 20 wt.% of organic contaminants like food is no problem, although more residue will increase power consumption and may alter the hydrocarbon product. After pyrolysis the unoxidized aluminium fraction is cooled and baled. However, due to its high surface area and the presence of carbonaceous residue, recycling without salt is difficult.[23,26]

Another example is the recycling of Tetra Pak. This is a multilayer packaging material widely used for juice and milk containers. It typically has six layers, and consists of paper (75%), low-density PE (20%), and aluminium (5%). Tetra Pak may be reused by shredding into 1- to 5-mm-sized particles, which are hot-pressed at 170 °C to make laminated board. Usually extra polyolefin material is added, since the low-density polyethylene (LDPE) in Tetra Pak is not sufficient to obtain the desired mechanical properties. The result is a lightweight and water-resistant composite material.[27]

To recycle Tetra Pak, treatment starts with a hydro pulping process. The packaging material is shredded and stirred with water at 50 °C to create a pulp which separates the aluminium and plastic fraction in the Tetra Pak. There are several options for subsequent treatment. One option is incineration. The heating value of plastic and aluminium makes an attractive fuel. Another option is to process the residual LDPE, aluminium, and ethylene-methacrylic-Acid (EMAA) adhesive fraction to generate pellets of a composite material, which is used as an addition to polyethylene terephthalate (PET) to obtain a material with relatively high impact strength.[28] Pyrolysis of the plastic/aluminium mix at 400–800 °C is also possible, which causes the plastic to degrade into gaseous hydrocarbons, similar to the Enval process for plastic–aluminium laminates.

As we are aficionados of a good cup of coffee, especially in the morning, we conclude this section with a final example: the recycling of Nespresso coffee pods. Nespresso single-use coffee pods consist mainly of aluminium. The remainder is made up of plastics, a silicone ring, lacquers, and so on. Nespresso does in-house recycling. They provide collection systems that bring used pods from customers to their own recycling facility. This ensures a pure scrap composition. The pods are collected in plastic bags. The pods and bags are shredded and passed through a sieve. Around 30 per cent of the coffee residue is separated by this first sieve. The remaining shredded material passes through a drying drum. After drying the remaining coffee residue is separated by a second sieve. The scrap now contains only aluminium and plastic, which are separated using an eddy current system. The aluminium fraction is compacted into bales which are melted. The company makes an effort to use the recycled aluminium for new pods. The coffee residue can be converted into biogas and compost.[29]

10.3.6 Recovery of Metals from Waste Incinerator Bottom Ash

Incineration of municipal solid waste (MSW) is becoming increasingly important. This is chiefly because it reduces the mass of waste by 70 to 80 per cent, and its volume by 90 per cent. As free space for landfill becomes hard to find (and more expensive), municipal governments turn to incineration as a cost-effective means of waste disposal. The societal forces driving this change may be seen from the plots in Fig. 10.11. The amount of municipal waste grows when citizens have more income. The figure's bottom plot, for the same countries, shows that incineration is primarily a means to avoid (or reduce) the need for landfill. Energy is also recovered from incineration, either for use in district heating or for generating electricity.

The residues from incineration are bottom ash, fly ash, and flue gas residue. Bottom ash is by far the largest residue fraction. About 20 million tons are produced every year in Europe. The amount of bottom ash is increasing because MSW is increasingly incinerated. Bottom ash is landfilled in many countries. However, the material is suitable as a building material, e.g. for embankments and foundations of roads. Bottom ash also contains a considerable amount of non-ferrous and ferrous metals that should be recycled. The recovery of these metals improves the engineering and environmental properties of the ash and creates a financial benefit. This is especially important in wealthier countries, since the metal content of municipal waste tends to increase with higher incomes.

The focus of this section is the recovery of metals from bottom ash.[31–3] The recycling of glass and other non-metallic components is not considered.

It is convenient first to consider incineration itself. The most common process is illustrated in Fig. 10.12. There are several combustion technologies, but the moving grate incinerator is most common. Waste is fed into the top of a sloped plane consisting of a series of moving grate elements. These grids are pushed back and forth along the diagonal plane so that they slowly, but surely, push the waste down through the combustion chamber. As the waste is

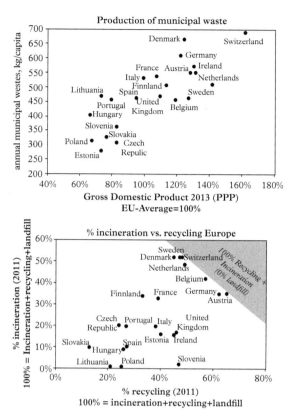

Fig. 10.11 *The correlation between waste generation and income (top), and the percentage of waste that goes to landfill or is recycled (bottom). [Published with permission from Professor R. Bunge.*[30]*]*

incinerated and decomposed, some particles fall through the grates and accumulate as bottom ash, while the smallest particles are carried away in the exhaust gas and collected as 'fly' ash. Gas burners are used to start the combustion process, but after the chamber reaches a certain temperature it is practically self-sufficient, as long as waste and air are fed into the furnace. The air injection is controlled so that a small excess of oxygen is always maintained in the chamber. After giving off thermal energy in the heat recovery system, the gas is filtered and released.

The composition of incinerated bottom ash depends on the composition of the incinerated waste. It often contains significant amounts of ferrous and non-ferrous metals as well as glass that can be recovered. Technologies for the recovery of metals emerged in the 1990s and have since become common practice in many high-income countries.

Recently, a number of random samples from residual waste throughout Norway were analysed. These analyses did not always distinguish between aluminium and other metals. Two of the analyses described the metal fraction as almost exclusively aluminium, and the metal content, in different parts of Oslo and Trondheim, was about 2 per cent by weight.[34] In studies

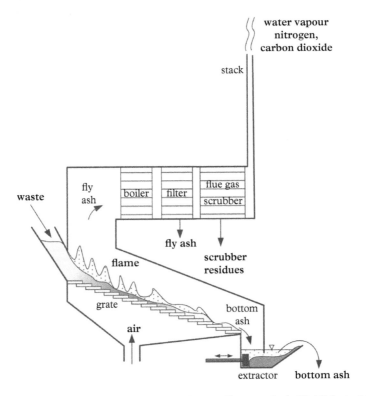

Fig. 10.12 *Illustration of a grate-fed incinerator with wet collection of ash.* [*Published with permission from Professor R. Bunge.*[30]]

conducted by the Politecnico di Milano on the oxidation of aluminium in two incinerators in Northern Italy, they found 0.73 and 1.67 wt.% aluminium in the bottom ash. There are also the detailed surveys conducted by Bunge.[30] Figure 10.13 shows the metallic portion of a 'typical' bottom ash, produced by incinerators in Central Europe (Germany, France, the Benelux Union, and the Nordic countries). These results are for freshly discharged, wet bottom ash. If the ash has been landfilled, corrections must be made for metal losses owing to oxidation and corrosion. In this figure the metal fraction has been separated into two classes:

1. Metals > 2 mm: These are pieces of metal of economic value that are potentially recoverable.
2. Metals < 2 mm: These are metallic elements, either chemically bonded (e.g. as oxides) or small pieces of metal that cannot be recovered by conventional beneficiation techniques.

Typical metal recoveries by conventional means are 40–65 per cent of non-ferrous metals, and 85–95 per cent for steel.

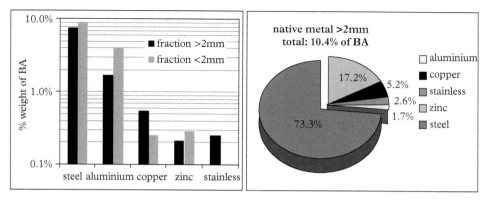

Fig. 10.13 *Total mass fraction of metals in bottom ash (left), and breakdown of recoverable metals (>2 mm in size) by metal type (right).* [*Published with permission from Professor R. Bunge.*[30]]

The above information points clearly to the desirable goal: the recovery of metals from bottom ash (BA). Unfortunately, recycling metals is complicated by the chemical and physical properties of BA. Freshly discharged from the incinerator, the BA contains about 20 per cent by weight of water. It is very alkaline, with a ph close to 12. This triggers reactions with finely dispersed silicates, leading to the formation of new minerals. Minerals also form by precipitation (such as gypsum) or by reaction with CO_2 in the air (limestone). There also occurs the hydrolysis of alkali and alkaline earth metal oxides. These reactions are exothermic, so the ash gives temperatures of 60–80 °C. The ash is typically stored outdoors to 'weather' for a period of 6 to 20 weeks, depending on the weather and other factors. During this time, the ash components stabilize. The ph drops to a value of 8–10. The water content drops to about 5 per cent. Many of the chemical changes are cementitious, so the BA 'solidifies'. It is now an excellent material for use as a sub-base material in road construction. In this and similar applications, it displaces the use of sand or gravel. Denmark, France, Germany, and the Netherlands all utilize bottom ash from waste incineration as a road sub-base.

Water in contact with the 'weathered' BA is by necessity tested for the possible presence of undesirable contaminants. The limits for utilization and test method specified vary by country, but in general a fully weathered BA has stabilized/solidified to the point where it no longer presents an environmental hazard.

Where are the metals? They wind up fully enclosed or encased in the ash. It is necessary to crush the stabilized/solidified BA to remove and recover the metals. Can we do this without losing the usefulness of the residual ash as a construction material? The answer in most cases is only a partial yes. In other words, the percentage of non-ferrous metals recovered is less than desired. A good deal of future work by qualified engineers is needed to further optimize the recovery and the usefulness of MWI bottom ash.

If you found this brief introduction interesting, and you wish to explore this fascinating area in more detail, the reviews by Bunge[30] and Syc and co-workers[32] are worth reading. The thesis by Muchova[33] is also valuable. It describes the development of a wet separation process in the Netherlands. The process can separate metal particles as small as 0.5 mm. The wet treatment also reduces leachant values to the point where no extra isolation is needed to use the bottom ash as a construction material.

10.4   Methods and Processes

In this section we describe the following processing steps: collection, liberation (to separate different types of metals/materials), sorting, and pretreatment. The results of the final blending can be positive, but more often it results in a downgrading of metal quality. And finally, we consider some of the methods used to connect different materials in automobiles, and how these affect their recyclability.

10.4.1   Collection

Collection is the first step in the process, and perhaps the most important. If a material is not collected, it will be lost. Broadly speaking, there are two different and independent recycling routes: collection from consumers and collection from industry.

10.4.1.1   Collection from the Consumer

The consumer seems to be motivated primarily by convenience and financial reward. Collections from consumers can be divided into six different methods.

1. *Curbside* collection is near the residence, at the same point where residual waste is collected. This would mean personal bins for single-family homes; shared bins for apartment blocks.
2. *Drop-off* collection points are a bit further away. They should be within walking distance to be convenient to the consumer, but it depends on the density of collection points.
3. *Recycling centres* are less abundant and often require a car due to distance from residential areas. They often have staff available to ensure the correct sorting of the disposed materials.
4. *Producer responsibility*, or *producer take-back*, is where producers organize collection and treatment of their own products.
5. *Deposit-return* is when the consumer must pay a deposit when buying an item. The deposit is refunded when the item is returned. Beverage cans are sometimes returned by reverse vending machines in regular grocery stores.
6. *The informal system* provides door-to-door collection of certain items. Consumers may also get some money in return.

Sometimes it is hard to differentiate between curbside collection and drop-off. Curbside collection in city areas are often shared between many households, just like drop-off points, and drop-off points can be placed very near the residence, just like curbside recycling.

   Source separation conducted by the consumer was found to have a positive influence on the recycling recovery with today's processing technology. Regarding the use of drop-off points for recycling, their use decreases as the distance from the residence increases.[35] The same study also found that the use of drop-off points did not decrease when curbside collection was available. This suggests that materials not collected curbside can still be collected at drop-off points, if the location of the drop-off is convenient. Implementation of mandatory recycling, as well as volume-based fees on residual waste, has also been shown to increase recycling.[36] However, convenience has been found to be more important than economic incentives, with the exception of deposit-return systems.[37]

   In informal systems, recyclables are mainly collected in two ways: either by waste pickers who search in bins for valuable materials or by waste merchants who collect recyclables going from door-to-door. The recyclables are then sold at trading points and transported to storage locations, where they are sorted and processed before they are sold to industry. A study found that 62% of all beverage cans from households in Beijing was collected by the informal system.[38]

   A number of innovative pilot projects include the use of Quick Response (QR) coded bags. This has a dual benefit, making it possible to both apply economic incentives for source separation and control the product quality. In addition, the code can be an easy and available source of information on how to sort for those citizens who own a smart phone.

   Extended producer responsibility (EPR) for packaging waste is widespread across the EU. It started in Germany in 1990 and has now spread to 31 European countries. This is a principle where the producer is responsible for the treatment of their EoL products. For packaging waste, most producers delegate this responsibility to producer-responsibility organizations. By paying a fee to the organization, the producer is assured that its product is properly treated at its EoL. The organization then redirects these funds to the associated waste management systems. In the case of PRO Europe, an umbrella organization for packaging recovery in Europe, producers pay a license to put the Green Dot mark on their packaging. The fact that Denmark does not have an EPR system is worth noting. Still, Denmark enjoys one of the highest recycling rates in Europe.

   Pay As You Throw (PAYT) is a common model in the US which aims to create an economic incentive to generate less waste and recycle more. Typically, the household only pays for the residual waste bin, while recycle bins are collected for free. The price on the waste often depends on the volume of the waste bin or the number of bags. Weight-based pricing is also an option, but it is less common.

   A study[39] focusing on the differences between single- and dual-stream recycling found some interesting results:

- Single-stream gave a 4 per cent higher recycling rates among households;
- Collection is slightly cheaper for a single-stream system;
- Processing cost is nearly 50 per cent higher for single stream than for dual stream;
- Revenue from sold material is 10 per cent lower for single stream; and
- In total, single stream is 28 per cent more expensive than dual stream.

These findings suggest that the benefits from single-stream recycling are lower than expected. Which materials were efficiently recycled was not investigated in this study, as the 'recycling rate' referred only to the total volume of items placed in recycling bins.

10.4.1.2   Collection in industry

In contrast to consumer waste, industrial materials are usually segregated by alloy and grade of material. For example, material removed at trim presses will be collected in a bin and stored in a separate area of the factory. Some diligence is required to maintain the separation of different materials, when several alloys are used in a plant. Applying a different colour of paint to each alloy is a simple aid that is commonly used. A strict segregation of a material maintains its 'pedigree'. In other words, because of its known composition, it will be easier to recycle and can be sold for a premium compared to 'mixed' scrap.

In factories that have melting and casting facilities, this material will be melted and used again in-house. Many metal casting foundries operate in this manner. However, it is more common to send the material elsewhere for processing. The waste material is often difficult to handle in its 'raw' form. Consider the pieces of sheet metal from the before-mentioned trim press. These will vary in shape and size, and will have sharp, dangerous edges. It is common practice to place material like this into a hydraulic press and compact it into more manageable shapes. These may be small briquettes weighing a few kilograms or large bales weighing a tonne or more.

How the collected industrial waste is handled varies a good deal. Traditionally, three parties are involved: the dealer, the broker, and the pre-processor. The dealer purchases the scrap and takes it to a collection facility (scrap yard). The broker buys scrap from a scrap yard and sells it to a pre-processer or recycling facility. However, this model has changed significantly as smaller companies have been merged into larger ones. There has been an increased amount of integration and smaller, family-owned businesses are not so common now.

10.4.2   Liberation

After collection, the streams are usually too complex for final processing. Valuable materials are usually present at too low a concentration to justify the high cost of final processing. Hence, the normal procedure after collection is to first reduce the particles to a suitable finer size, and then to separate the materials by using differences in their physical properties. The sorted streams are either further purified or sent for final processing by chemical or metallurgical means.

The first step after collection is to separate or 'liberate' desired materials from other components. This is followed by sorting. These two steps are illustrated schematically in Fig. 10.14. The efficiency of these two processes determine the extent to which one can recover the 'pure' materials, or the extent to which materials are 'contaminated' by their contact with other substances. The possibilities and limits of recycling are strongly influenced by liberation and sorting.

Ideally the collection of recyclable materials should be designed to aid recoverability, and to avoid unnecessary complexity in the liberation and sorting processes. Much can be done at the origin of the recycling process—collection. Unfortunately, in many cases we are far from

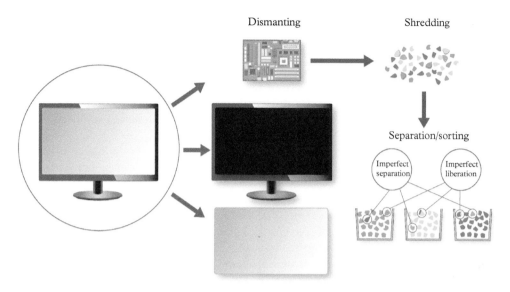

Fig. 10.14 *Product design and liberation/sorting as it relates to separation efficiency. [Adapted from van Schaik and Reuter et al.[40-2]]*

the optimal degree of pre-sorting at the collection stage. Recycling presents an extra level of complexity when different products are collected into a combined stream.

The liberation of valuable minerals from the gangue or waste material in an ore is common practice in mining. A good deal of the technology and terminology in this area of recycling has been borrowed from mining beneficiation (upgrading) techniques. However, in urban mines materials are often found bolted, welded, or otherwise attached to other parts. The methods of connection often represent a serious challenge to recycling. This challenge is considered in more detail later.

10.4.2.1   *Dismantling*

Manual dismantling is sometimes possible and desirable. This is especially true for larger products that are not too complex. Automobiles are a good example. It is necessary to first remove hazardous or environmentally polluting materials. This includes oil and transmission fluids, batteries, and air bags. Catalytic converters contain a valuable mix of precious metals from the platinum group. They are removed, as are the tires and wheels. Many parts can be used to replace worn or damaged components in other vehicles: engines, transmissions, doors, bumpers, headlamps, and fenders. The stripped-down car is then usually sent to shredders.

When there is a high volume of a single product, automated dismantling can be employed.

10.4.2.2   *Mechanical Liberation*

Comminution is the process by which oversized materials are reduced to the proper size for sorting. The comminution of naturally occurring minerals is usually done by crushing into smaller pieces. However, metal scrap cannot be processed in this fashion. Instead, metal scrap must be torn or cut into pieces. A number of different devices have been invented to tear apart metal scrap. Probably the most common are hammer mill shredders. They have been used since the late 1950s to shred automobiles. These are huge, noisy machines, offering high productivity and requiring up to 8000 horsepower to operate. These and other shredder designs are reviewed in the handbook by Nijkerk and Dalmijn.[43] The most common shredders have also been described and compared by Schlesinger.[44] The operating cost of shredding (electricity and replacement parts) is typically $10–15 per tonne. (These costs were determined in 2000.)

All shredders generate dust and fine particles during operation. These would be an environmental and workplace hazard if released. Furthermore, the shredder operation generates heat. If the dust contains organic vapours from grease or oil in the scrap, this can result in an explosion. Consequently, dry shredders have a collection system operating under negative pressure to collect dust generated during operation.

One may eliminate the dust by spraying water into the shredder. This eliminates the need for dust collection and minimizes the risk of explosion. However, there are a number of serious disadvantages. One now needs a water-treatment system. Any soluble material in the feed (e.g. road salt) is a potential water pollution problem. One also has to recover or treat the sludge formed from what would have been the dust in a dry shredder. The potential for the dust to contaminate the metal streams may be a concern, and metal fines in the runoff water can also be an issue.

The operation of the shredder determines the quality and quantity of material that can be recycled. Ideally one would create scrap particles small enough, so an efficient separation of metals can be made. When a particle of scrap consists of a single material, it is said to be liberated. Not all particles will be completely liberated, however. Some will be admixed with a certain amount of undesired contaminants. (This is illustrated schematically in the bottom of Fig. 10.14.) Materials in a recyclable stream are distributed and connected in different ways, affecting the size dependency. If the particle size is made finer, the liberation will change in a way that is characteristic of the recyclable material and the method of shredding. This result is called a liberation curve, following the terminology and usage developed for the industrial beneficiation of natural minerals.

The method to calculate liberation curves is described by Heiskanen.[45] Example curves are also presented for a source of scrap which is one-third aluminium. Figure 10.15 shows liberation curves calculated for two different particle sizes.

The lower (blue) curve is for a relatively large average particle size. Suppose we require a 90 per cent purity (grade) to have a saleable product. In this case we can only recover one-half (50 per cent) of the aluminium present in the scrap. The upper (red) curve was calculated for the same material, only for a smaller average particle size coming from the shredder. The situation is much improved. In this case we can recover 80 per cent of the aluminium and still meet our requirement for a 90 per cent grade.

This is a hypothetical example, but it illustrates clearly the effect of the liberation process on the quality and quantity of material that may be recovered. If the quality desired is high, the total recovery may be low. Conversely, if we wish to have a high degree of recovery, the quality (purity) of the product may be lowered.

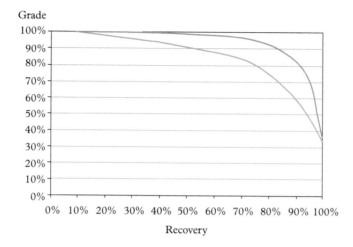

Fig. 10.15 *Liberation curves calculated for two different particle sizes. [Reprinted from Heiskanen*[45] *with permission from Elsevier.]*

10.4.3   Sorting

The next step in recycling is sorting. To consider how this can be accomplished, it is useful to visualize materials moving on a conveyor belt. Now consider the different processes these materials will see during its route.

10.4.3.1   Manual Inspection

This is likely to be the first step. People stationed alongside the conveyor belt can remove certain items and place them into storage bins for further processing. Manual sorting was the first method of sorting used and continues to be popular. We are able to quickly and accurately determine different materials, from their size, shape, and colour. You would certainly recognize tableware made of stainless steel, cans made of aluminium or steel, plastic milk cartons, and many other items. Colour would also allow you to distinguish and identify most common metals. In lower-income countries manual sorting is the primary method used for recycling. In higher-income countries, manual sorting is used to 'cherry pick' certain desirable items; or to remove things that could cause problems. An example of the latter would be copper windings around an iron core (transformers and electric motors). These would be removed by magnetic separators, and steel foundries impose stiff penalties on deliveries of copper-contaminated scrap. A typical handpicking arrangement is illustrated in Fig. 10.16.

In what follows, we consider a number of processes the stream of materials is likely to encounter in an automated treatment facility.

10.4.3.2   Sizing/Screening

Size classification is a prerequisite for efficient processing of recycled materials. Material moving on the conveyor belt is fed onto a series of wire screens. Most commonly used are vibrating deck screens, roller, gyratory, and cylindrical screens. The screen deck is usually

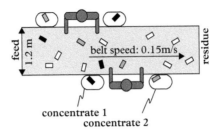

Fig. 10.16 *Schematic illustration of a manual sorting station.* [*Published with permission from Professor R. Bunge.*[30]]

made of woven steel wire; sometimes of rubber or polyurethane. Vibration is most commonly created by one or two rotating shafts with unbalanced weights. The vibration pattern is usually elliptical, so material moves in a specified direction across the screen surface.

When material enters the screen surface, it stratifies, and the finest materials pass through rapidly. Once the smaller particles (less than about one-half the size of the screen opening) are gone, the removal of somewhat larger particles becomes a probability process, and is of first order with their concentration. Typically, a series of screens are used in sizing. The first screen would have the coarsest mesh to remove the largest pieces. The material passing through the first screen would be sized again.

Screens are suitable for particle sizes encountered in recycling. They are available in sizes from several hundreds of millimeters to about 100 microns (10^{-4} m). In the US screens are usually classified by mesh size. A 20-mesh screen has 20 wires per linear inch, and a square opening that is approximately 2 mm in size. Screens are also based on the international ISO-565/ISO 3310 standard,[46] which specifies screen openings from 175 mm to 20 microns.

Screens are not an efficient method of sorting large volumes of small particles. Screens with small openings easily become 'blocked', as odd-shaped particles become wedged into the screen openings. For this reason smaller particles are usually sorted by hydrocyclones. Particles are fed tangentially to the inner wall of a cylinder. The cylinder has a conical bottom. The tangential flow needs to go from perimeter to the centre line. The tangential flow causes a centrifugal force. Large particles are more affected by the centrifugal force, and fine particles by the drag force induced by air moving upwards. As a result, coarse particles will remain at the perimeter and fine particles will move toward the centre line. This method of sorting is efficient and well developed.

10.4.3.3 *Magnetic Separation*

Magnetic separation is based on the differences in a material's magnetic susceptibility. For our purpose, we identify three regimes of magnetic susceptibility. *Diamagnetic* materials have a small negative susceptibility and are not amenable to magnetic separation. Diamagnetic materials include rubber, plastics, glass, most metals, and slags. *Paramagnetic* materials have a small positive susceptibility and can be separated from diamagnetic materials by using a high field intensity: about 2 T (Tesla). Some metals like Ti, V, Cr, Mn, and the platinum group metals are paramagnetic. The third regime is *ferromagnetism*, where the susceptibility is high.

Of pure metals, Fe, Co, and Ni are ferromagnetic. Most steels are ferromagnetic. (Austenitic stainless steels are not.) Low magnetic field intensities (typically 0.1–0.2 T) are used to sort ferromagnetic particles. If a high intensity field is used, the particles will become permanently magnetized and will remain stuck to the magnet.

A low intensity magnetic separation is often the first sorting operation our material stream sees. Low intensity separators are suitable for a large particle size range. Dry belt separators can sort particles up to a few hundred millimetres in size. Figure 10.17 shows the operation of two types of drum magnet extractors.

10.4.3.4   Eddy Current Separation

Once magnetic materials have been removed, the next step is usually eddy current separation. An alternating magnetic field can be used to induce an eddy current in conducting particles. The induced current circulating inside the particle produces a magnetic field. The induced magnetic field produced in the particle creates a weak repulsive force between the particle and the primary magnetic field. An eddy current separator is shown in Fig. 10.18.

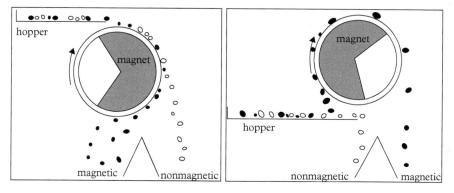

Fig. 10.17 *Drum magnets: left, deflection magnet; right, extraction magnet.* [*Published with permission from Professor R. Bunge.*[30]]

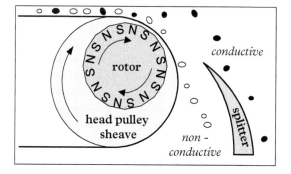
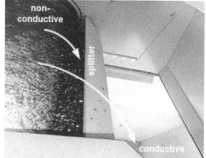

Fig. 10.18 *Eddy current separator (ECS): left, schematic of operation; right, picture of ECS looking from above into the separation chamber.* [*Published with permission from Professor R. Bunge.*[30]]

The repulsive force depends on the conductivity: the higher the conductivity, the stronger the force. Opposed to this, we have the force of gravity. This is proportional to the density of the material. Hence, the ratio between the electrical conductivity and the density determines the trajectory of the particle in the eddy current separator. Using this ratio as an indicator, we can establish the following ranking for separability:

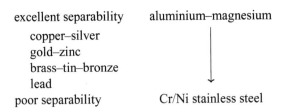

Glass, rubber, plastics, and other non-conducting materials are not affected by the eddy current. Particle shape is an important factor. Best results are obtained with flat or compact pieces of metal. In elongated particles, like pieces of wire, it is harder to establish a strong eddy current, and the repulsive force can be low, depending on the orientation of the wire. This sorting process typically works best for particles larger than about 5 mm. However, recent developments have been used to recover particles as small as 0.5 mm.[30]

Because of variations in particle orientation, size, and shape, the trajectory of metal particles is variable. There is also a good deal of variability in the trajectory of non-conducting particles. Consequently, there is a region where the 'cloud' trajectories of the two types of particles overlap. This is illustrated on the left-hand side of Fig. 10.19. There is a barrier which splits the stream of particles into two fractions. (This may be seen on the right-hand side of Fig. 10.18.) The location of the splitter has an important effect on both the recovery and the grade of recycled material. The curve of grade versus recovery is similar to the liberation curves presented earlier. One cannot obtain both 100 per cent grade and 100 per cent recovery. One is forced to make a compromise by trading a loss in one factor for a gain in the other.

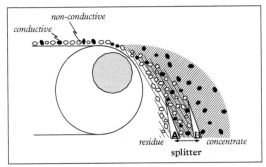
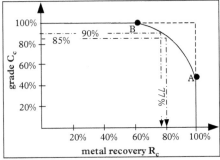

Fig. 10.19 *The ECS splitter setting at 'A' results in 100 per cent recovery but only 45 percent grade. The setting at 'B' results in 100 percent grade, but only 60 per cent recovery. [Published with permission from Professor R. Bunge.[30]]*

10.4.3.5 Density Sorting

Gravity separation is feasible when the density difference between materials is sufficiently large. One example is the use of heavy media to separate aluminium and magnesium from other non-ferrous metals. The heavy media consists of water containing a specific amount of very fine, suspended dense particles. Magnetite is commonly used, since it is easily recovered by magnetic means, and the density of the suspension is easily controlled. Ferrosilicon is also commonly used, since it is also magnetic.

Typically, a heavy media suspension with an effective specific gravity of 2.2 to 2.4 is used first. This removes any magnesium (density of 1.74), together with heavy plastics and rubber. These light materials all float, while heavier metals sink. A second cut is made at a density of 3.3 to separate aluminium from other non-ferrous metals (density > 7). Since the density of aluminium is about 2.7 g/cm^3, a third 'cut' is often made at a specific gravity of 2.8 to 2.9. This permits the separation of aluminium pieces which contain iron bolts or rivets. The unliberated aluminium fraction is returned to the shredders or processed separately.

Density sorting is also used to separate lighter materials from automobiles. The feed coming from the shredder is treated by air classification to separate lighter components. Called air shredder fluff (or ASF), it contains dust and light non-metallic materials, such as foam, textiles, foil, and wood. This represents approximately 25 per cent of the average automobile by weight.

A number of other density separation processes have been developed and are used extensively in mineral processing. They are not used at present to any extent in recycling, but may find future use in special recycling applications.

10.4.3.6 Sensor Sorting

Nearly all recycling facilities use the sorting methods described so far. A few have begun to also employ sensor sorting to increase the recovery of valuable materials. The principle of operation is illustrated in Fig. 10.20. Material moving along a conveyor belt passes over a sensor located underneath a segment of the belt. A metal detector is indicated in the figure. (This sensor is similar to devices used by beachcombers to find coins buried in the sand.)

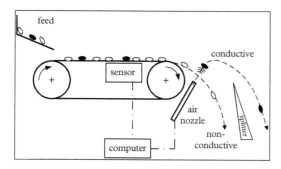 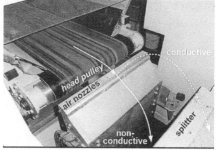

Fig. 10.20 *Schematic illustration of sensor sorting (left), and a view of the separation chamber (right).* [*Published with permission from Professor R. Bunge.*[30]]

When a metal conductor passes over the sensor, its location on the belt is noted by computer. When it comes off the belt, it passes over a bank of air nozzles operated by compressed air and activated by computer. The blast of compressed air changes the trajectory of the conducting particle.

A number of sophisticated sensor technologies are in use or under development. Two in use are as follows:

- *Dual X-ray transmission* determines the average atomic number of particles. This can be used, for example, to detect iron inserts in aluminium.
- *Laser-induced breakdown spectroscopy* (LIBS): is used to determine the chemical composition of a material. In principle, it could be used to separate different alloy grades.

The use of these technologies can be costly. Consider the 2007 study made by Dalmijn and De Jong.[47] They examined the recycling of EoL automobiles. An excellent historical view was presented, with a summary of the current state-of-the-art of the recycling of automobiles. They noted that US recycling plants were producing high volumes at low cost. The quality of the product was of lesser importance. A sizeable amount of recycled aluminium was used to make die-casting alloys. (Conventional die-casting alloys fall mainly into the 360 and 380 alloy families. They contain relatively large levels of iron and other impurities and may contain only 80 to 85 per cent aluminium.) A different situation existed in the EU, where legal statutes required a higher level of quality and recycling. Consequently, the processing costs in the EU were nearly twice those in the US. (See their Table III.) At the time of this study, it was more economical to send EoL autos to China or India for processing than to process them in the EU.

10.4.4   Pretreatment

Pretreatment of scrap is of the utmost importance for both the economic and environmental performance of recycling. There is an old rule in the recycling industry: for every 1 per cent of contamination charged into the melting furnace, there will be at least an equivalent 1 per cent metal loss. It therefore pays to remove contaminants from scrap before melting. These contaminants, or non-metals, may be paper, ink, plastic, or other organic materials. The painted logo on beer cans, painted sidings, and the oil on machining chips are all examples of undesirable contaminants. Not only do they increase melt loss and lower metal recovery, but they can produce noxious fumes when charged to a hot furnace. The process of their removal is called pretreatment. However, the term decoating is also common in the recycling industry.

The focus here is on the underlying science and current state-of-the-art. The discussion focuses primarily on the pretreatment of aluminium. However, the same principles and processes apply to other metals: magnesium, iron, and copper. They also apply to the removal of plastics when recycling electronic scrap (e.g. circuit boards).

Pretreatment of collected (and alternatively shredded, sorted) scrap can be done using either a chemical or a thermal method. Chemical decoating uses solvents to loosen the physical bond between the metal and the lacquer coating. Sometimes this approach is necessary, but using solvents creates a new waste problem. Consequently, thermal treatment has been the preferred method for removing coatings for more than 30 years.

As discussed earlier there are many ways of defining scrap. Regarding thermal decoating, however, scrap is usually classified by the amount of volatile organic compounds. The usual classification is as follows:

- *Low*: less than 5 per cent by weight of volatile organic content (VOC)
- *Medium*: between 5 and 25 per cent VOC
- *High*: more than 25 per cent VOC.

The amount of VOC has an important influence on the pretreatment process. It acts partially as a fuel, since energy is released when the VOC oxidizes. Sufficient oxygen must be provided for good removal.

The importance of proper pretreatment furnace atmosphere may be seen from a case history. To save energy, a company used the products of combustion as a process gas to preheat and decoat aluminium scrap. This was not a good idea: not enough oxygen and too much water (from the combustion of natural gas). To make matters worse, the UBCs had to be baled to allow them to pass down the preheating shaft. That made it hard for the gas to permeate within the compacted metal. They discovered their new furnace was an excellent dross making machine. Controlled trials were made using both 100 per cent bare scrap and 50 per cent coated scrap (UBCs). A typical charge was around 100 tonnes. For the bare scrap there was about 4 tonnes of dross. For the 50% coated scrap, it was typically 15–20 tonnes of dross!

Historically, the pretreatment of aluminium scrap began with UBCs. It is also one of the most important types of aluminium scrap. This is also where we began our work.

How much coating is contained in UBCs? Has the percentage of coating on the cans changed with time? Perhaps you have noticed that the thickness of the average beer can has decreased over the years. Does this mean that the percentage of VOCs has increased? We obtained beer cans from three popular brands of beer (one of the perks of doing this type of research). These were heated under controlled laboratory conditions, while continuously measuring the weight of the samples. In this way, the weight loss associated with the volatilization and combustion of the coating could be measured. A comparison of the weight losses from beverage cans made in 1987 and 2006 is shown in Table 10.3. Surprisingly, the percentage of VOC coating on beverage cans has changed very little. This is a good thing. Smaller amounts of volatile compounds make recycling easier.

This comparison shown in Table 10.3 may not be true for other products. For example, cans from Japan tend to contain 50 per cent more paint than their European or American

Table 10.3 *Average VOC Mass Losses from 1987*[49] *and 2006*

Can type	1987 mass loss (%)	2006 mass loss (%)
Old Milwaukee	2.60	2.57
Budweiser	3.47	3.30 ± 0.3
Coors	2.37	3.00
Average	2.81	2.58

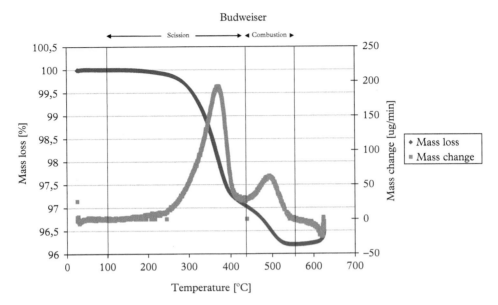

Fig. 10.21 *A typical plot of the mass loss and rate of mass change for a sample from a side wall of a Budweiser can, heated at 20°C/min in air.*

counterparts. Japanese manufacturers also prefer bright colours that give rise to a large inorganic/organic ratio in the coating.[48] Design preferences and criteria are important.

Table 10.3 shows the VOC mass loss in an atmosphere containing 11% O_2. The can (minus the lids) of three brands of beer were decoated. The tabulated results are an average of duplicate runs.

Figure 10.21 shows a typical plot of the weight change during the decoating process. This figure is for the Budweiser can, but the results from Old Milwaukee and Coors cans were similar. It may be seen that the first maximum mass loss is for the scission phase at around 370 °C. (Scission is the term for the breaking of chemical bonds, when long-chain molecules decompose into smaller and lighter molecules.) The maximum mass loss for the second (combustion) peak is around 490 °C It is also clear that some mass is gained when the temperature approaches 600 °C. Here the metal is starting to oxidize. This is only 50 °C more than the maximum mass loss in the combustion phase, meaning there is a narrow window of operation for optimum decoating.

We now consider these results in more detail. Thermal decoating consists of two steps: scission of hydrocarbons and combustion of the residual char. During scission, larger hydrocarbons are broken down into smaller hydrocarbon molecules by heating under a low oxygen atmosphere. This is an endothermic reaction. The next step, assuming sufficient oxygen is present, is combustion. In this phase the carbon residue on the surface reacts with oxygen and forms the combustion gases CO_2 and CO (and H_2O). Heat is also released. Figure 10.22 shows a schematic illustration of the thermal decoating steps.

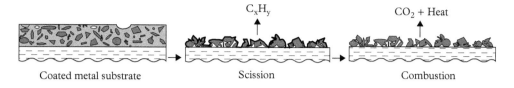

Fig. 10.22 *An illustration of what happens when a coated metal surface is heated.*[50] *[Reprinted by permission from Springer Nature: copyright 2008.]*

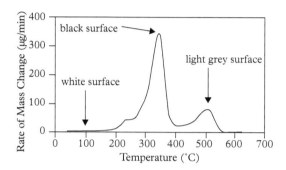

Fig. 10.23 *Rate of mass change as a function of temperature during the decoating process of an acrylic-coated aluminium sheet.*[50] *Two distinct peaks show the scission and combustion regimes. The observed colour change during these regimes is indicated. [Reprinted by permission from Springer Nature: copyright 2008.]*

When a one-sided transparent acrylic coated aluminium sheet with a thickness of 1 mm is heated in a TG/DTA furnace coupled to a mass spectrometer, information on the mass loss, the rate of mass loss, and enthalpy change during the decoating process is obtained. The mass spectrometer allows for measurements of the evolved gases. Figure 10.23 shows the rate of mass change as a function of temperature when the sample is heated in air at 5 °C/min. Two peaks of mass change can be distinguished, corresponding to scission and combustion. The first peak (scission) occurs between 200 and 400 °C; the second peak (combustion) occurs between 400 and 500 °C.

The composition of the evolved gases is monitored by a mass spectrometer. Figure 10.24 shows the evolution of two gases from the acrylic-coated aluminium, when heated in air at 5 °C/min. For the scission regime, hydrocarbons such as C_2H_2 were evolved. Hydrocarbons were only present during the scission regime. During the combustion phase, combustion gases are observed, mainly CO_2. Some CO_2 was also released in the scission regime.

10.4.4.1 Coalescence and Metal Recovery

The process conditions of the decoating process strongly influence the metal recovery and dross formation. The variable parameters are temperature, time, and atmosphere. Usually, the temperature for aluminium decoating is between 500 and 630 °C, depending on the type of aluminium substrate, gas atmosphere, and composition of the coating. Capuzzi et al.[51] showed

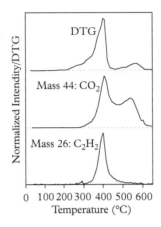

Fig. 10.24 *Mass loss curve (DTG) and corresponding curves for CO₂ (representing combustion gases) and C₂H₂ (representing scission gases) evolved from the decoating process of acrylic coated aluminium.*[50] *[Reprinted by permission from Springer Nature: copyright 2008.]*

Table 10.4 *The Effect of Thermal Decoating Process Condition on the Rate of Decoating*[52]

	Lifetime (s) of coating in various atmospheres		
Decoating temperature (°C)	Ar	Ar + 1% O_2	Air
400	21	13	7
500	0.4	0.2	0.2
600	0.02	0.01	0.01

[Republished with permission of John Wiley & Sons, from Kvithyld and Engh, permission conveyed through Copyright Clearance Center, Inc.]

that the decoating temperature affects coalescence during remelting. During coalescence, two or more droplets of liquid metal merge together. When the metal finds it difficult to coalesce, dross formation is higher. It was found that aluminium decoated at sufficiently high temperatures (600 °C) performed as well as uncoated aluminium, whereas aluminium decoated at lower temperatures (400 °C) showed a lesser ability to coalesce.

In another study the effect of three process parameters were studied. The effect of time and temperature of treatment, and the composition of the atmosphere were varied. The effect on the rate of decoating is shown in Table 10.4 for coated aluminium.

From these results, it is clear that higher temperatures increase the rate of decoating. However, *higher temperatures also increase oxidation* of the metal and thus metal loss. Also, more energy is needed to achieve higher temperatures, making it more challenging to operate the process auto-thermally. The results also show that more oxygen causes faster decoating. In addition, the scission peak occurs at lower temperatures.

10.4.4.2 Issues Related to Recyclability

After an optimized thermal decoating, all organics will have been removed. Residues left on the aluminium surface consist of small amounts of inorganic pigments and fillers. If it were necessary to also remove these and obtain completely clean scrap, an additional mechanical or chemical treatment would be needed.

In industrial practice the average residence time is sometimes too short to complete the pyrolysis process. The increase in productivity comes with a cost, however. Gasification of the remaining organics continues when partially treated scrap is immersed in the melt. The gas will induce gas–liquid reactions and bubbling that promote dross formation.

The interactions between the coating components, salt fluxes, and metal can be predicted using thermodynamics. In this way, the effect of the residue from pretreatment on the remelting process could be determined beforehand during the design of products. An attempt to do this was made for coated magnesium scrap.[53] Before characteristics for recycling can be assigned, however, an interpretation of what constitutes good recyclability must be defined.

10.4.4.3 Industrial Units

In this section a short description of various types of decoating units is presented. The main decoating systems can be divided into rotary kiln and belt conveyors. The flow system can be co-current, counter current, or cross flow up or down as shown in Fig. 10.25. In addition to what is shown in the figure, industrial installations have a good deal of ancillary equipment not shown: air and fuel inlets, combustion air blowers, feeding systems, dust

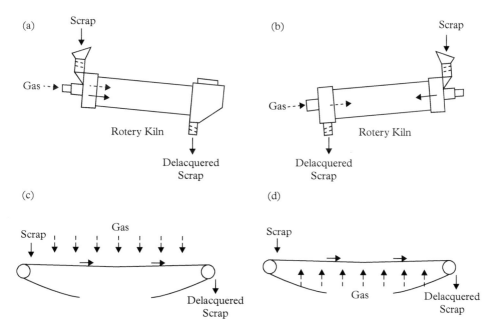

Fig. 10.25 *Rotary kiln and belt conveyor systems for decoating with various heat flow systems: (a) co-current, (b) counter current, (c) cross flow down, and (d) cross flow up. (Based on Jenkins.[54])*

collectors, hot gas fan, afterburners or hot gas generator, cooler recuperator or air heaters, and bag houses.[54]

Rotary kilns can be directly or indirectly fired. The rotary kiln with co-current flow has a selective scrap retention time, but is not suitable for high VOC scrap. The direct fired co-current flow rotary kiln tends to have high oxidation losses and problems with process emissions. The rotary kiln with counter-current flow is sensitive to scrap feeding rate and possible fires in the exit gas. The rotary kiln counter-current flow with radiant tube can process high organic scrap, but it is complex to run and needs a good shred scrap quality. Belt conveyors with hot air down-flow have good thermal efficiency, but small scrap (fines) should be avoided. Also this process has problems with evolved gases: high NOx emissions, the potential for dioxin and furan formation, and the need for massive gas treatment systems. Pan conveyors with hot gas upward through the scrap bed have very high production rates, but have problems with preheating the pan prior to scrap treatment. Among the issues to consider when choosing from the various processes are retention time, fuel input, exit temperature, kiln organics energy release, and afterburner efficiency.

The process data available for the industry are scrap samples, scrap feed rates, temperatures, flow rates, and oxygen levels. It is especially important to determine the heating rates, so that they can be simulated in the laboratory. The essential data for decoating are residence time, temperature, and oxygen levels, all of which have been investigated, as noted earlier.

Typical temperature and atmospheres for some of the industrial systems are given in Table 10.5. With a counter-flow kiln the incoming scrap is exposed to gas temperatures less than 300 °C, whereas a co-current flow exposes the scrap to gas temperatures of 800 °C or more. By the time the decoated scrap exits both co-current and counter-current units, it should have the same temperature as the exit gas—around 600 °C. Rotary kilns are most commonly used. After decoating, the material is melted and poured into ingots.

Table 10.5 *Various Decoating Units and Their Typical Oxygen Level and Gas Temperatures*

	Temp. gas in (°C)	Temp. gas out (°C)	Atmos (% O_2)
Kiln			
Counter flow	590	230	7
Counter-flow radiant tube	650	260	—
Co-current flow system	800	550	8
Co-current flow direct fired	—	550	
Conveyor belt			
Cross flow downwards	570	—	—
Cross flow upwards	510	400	—

Based on data from Jenkins.[54]

10.4.4.4 Operating Window

The goal of the decoating process is to combust all the carbonized material without oxidizing the metal. There is a fine line between too little decoating, exactly enough, and too much decoating leading to oxidation. The window for correct treatment is believed to be small. For the polyester coating in Fig. 10.23, the difference between lowest mass (550 °C) and the highest carbon consumption (485 °C) is only 65 °C. In addition, the heating was slow (5 °C/min), whereas industrial units often are operated at about 20 °C/min, making the residence time window smaller. Once the metal is oxidized, it is more difficult to recycle it efficiently.

Coatings are complex mixtures of materials in which a range of organic and inorganic components are present. The precise composition is difficult to determine, especially for materials from EoL products. This requires a decoating process able to deal with a range of unknown coatings. In practice, however, re-melters usually consider only the mass percent of the VOC coatings.

10.4.5 Blending

Blending is a neutral word for mixing in molten stage, whereas downgrading is clearly showing the accumulation of undesirable elements. The opposite is upgrading, when we have an increase in desirable alloying elements. Upgrading is not so common, but it can occur as a result of creative alloy development.

Metals are usually not used in their pure form, but instead are used as alloys to provide specific mechanical or chemical properties. Alloys and materials can be categorized into higher grades—having a stricter tolerance for impurities—and lower grades—materials that have more lenient tolerances. In other words, higher-purity metals are often used to recycle lower grade metals in the form of an alloy.

Steel recycling can cope only to a limited extent with copper, tin, and antimony in their input streams. These elements, because of their chemical and thermodynamic properties, cannot be removed to any appreciable extent during the production of high-quality steels. Commonly, the way to cope is to dilute these undesirable elements to concentrations that can be tolerated by alloy requirements. The dilution is made by adding more pure, primary metal. Sorting helps, but steel scrap has to be de-tinned and de-galvanized before being fed into steelmaking furnaces.

A similar situation exists in the aluminium industry. This type of recycling is illustrated by the inner recycle loops from EoL products shown in Fig. 10.26. Prime metal is alloyed with other elements to make wrought alloys. At the end of their life, some (hopefully most) of this metal is returned to the same wrought alloy family. This occurs when there is a good segregation of different types of alloys. However, some wrought alloy products will become mixed with other alloys and may be too contaminated to be returned to the original alloy family. This mixed material is most commonly used to produce casting alloys. These are mostly based on the Al–Si alloy system, so some silicon will be added. A similar recycling loop exists for the cast alloy family. Some casting alloys are made with prime metal and silicon. However, we see that some wrought alloy also enters the cast family loop. There is also a portion of both loops that is lost due to non-recycling.

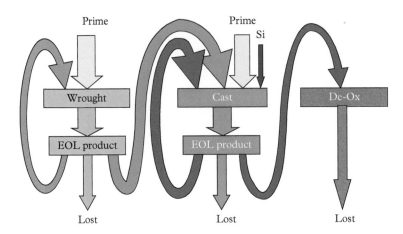

Fig. 10.26 *Flow of alloys and recycled products in the aluminium industry. [Reprinted from Tabereaux and Peterson,*[55] *with permission from Elsevier.]*

When it becomes too contaminated to be used for conventional alloys, the material can be used as a deoxidant in the steel industry. At this point, the alloy is being used for the chemical content of the aluminium, to attract oxygen in the molten steel.

Figure 10.26 is a clear illustration of how aluminium flows through the 'system' as it is used, and comes into contact with other metals and alloys. If the level of contamination is slight, it can be returned to its first use (wrought alloys in this case). Any undesirable elements in the scrap are diluted to acceptable levels with primary metal. However, as metal becomes progressively contaminated through many 'lifetimes' of use in various consumer and industrial products, it must be diverted to progressively lower alloy grades.

Similar examples of blending and downgraded uses can be found in many other metals.

How we will manage in the future is unknown. The percentage of scrap in the supply chain of a number of metals is expected to increase in the future. For several reasons, these increases are likely to occur. There have been a significant number of Cassandra-like warnings of the future, predicting that someday we will be unable to use large amounts of contaminated scrap metal. Two examples are cited here for your information.[56,57] There will certainly be future work for recycling metallurgists.

10.4.5.1   *Upgrading*

Upgrading is not as common as downgrading, but there is one example worth mentioning again. It is the die-casting alloy invented by Rheinfelden aluminium. Known as Silafont-36 and registered with the Aluminum Association as 365 alloy, this was the first commercially viable alloy used to produce structural die castings. The alloy is relatively high in manganese, which can be found in recycled UBCs. The alloy has usually sold at a premium close to twenty cents a pound (€650 per tonne). The Rheinfelden patents on the alloy in North America are expiring as this book goes to press, so the price premium is likely to fall as others begin to produce the alloy. Still, this is a wonderful example of how creative alloy design can be used to upgrade a material (the metallurgical equivalent of making a silk purse out of a sow's ear).

10.4.6 Processing of Final Residues in the Metal Industry

Up to this point we have focused on the recycling of different materials in their metallic form. We now consider what will be called 'final residues'. This term is preferred to 'waste by-products', which would have been more commonly used ten or twenty years ago. However, these residues are increasingly being processed, instead of being sent to landfill for disposal.

Of course, economic factors are important, and often determine the choice of the processing route. Many residues have low material value, and their processing costs can be relatively high. Also, they may also contain hazardous compounds, which complicate their processing and/or disposal. Ideally, the solution adapted will have a reasonably low environmental footprint as well as being economical. These constraints present quite an interesting challenge for refining metallurgists.

Below are listed final residues generated during the production of several different metals. Some of the processing steps employed or proposed are discussed. What follows is a brief introduction to the present situation. Because the authors are most familiar with metallurgical industries in Scandinavia, they are the focus of what follows. Of course, similar residues are found in many other industries, and in other countries.

10.4.6.1   Aluminium

About 40 kg of final residues are generated per tonne of Al production.[58] Listed in order of their importance, they are cathode waste (spent potlining or SPL), waste from the casthouse (mainly dross), anode butts, filter dust from the potlines, refractory materials from the anode factory, and anode assembling waste.

When an electrolytic pot reaches its end of life, it is taken out of production and the contents are removed before the pot is rebuilt and returned to production. The waste product from the lining is SPL. Since SPL contains fluorides, among other things, it is classified as hazardous waste. Aluminium producers have worked for decades on a suitable process for recycling this material. This involves crushing the SPL and separating the carbon component, which is either burned to recover its energy content or used in another product (e.g. in production of fire-retardant insulation).

Oxide and dross is formed on the surface of liquid metal in the casthouse. Depending on the solid metal materials or scrap charged to the heat, the melt loss to oxidation can be 1 or 2 per cent of the metal weight in the furnace. This dross is scraped off the surface of the melt and placed into dross bins; or sometimes, dross coolers (Fig. 10.27). The dross typically contains 30–40 per cent metallic aluminium, although it can vary between 15 and 80 per cent. The dross is usually sent out to a processor, who treats the dross and returns the metal recovered back to the casthouse furnaces. The processes used to treat dross have been reviewed by Peterson.[59] They are also discussed in Section 10.5.2. In North America salts and dross are usually treated in tilting, rotary kiln furnaces. The added salts promote metal recovery. (This is similar to recovery of magnesium dross; see Section 10.5.1.) In the EU regulations prohibit

Fig. 10.27 *Preparing to remove dross from a furnace (left) and a full dross bin (right).*

the disposal in landfill of salt-containing dross, so a number of salt-free processes have been developed.

Ideally, one would like to recycle the residue left from the dross treatment process. Since this is mainly aluminium oxide, one possibility would be to use it in the pot rooms. The oxide can be ground and blended with the ore charged to the electrolytic cells. This has been accomplished at a number of smelters, but only for dross from furnaces that do not produce highly alloyed products. For example, from can stock (3xxx) alloys there would be contamination of Mn and Fe; from foundry casting (3xx) alloys there would be Si; and from the high strength 2xxx and 5xxx alloys one would have contamination of Cu and Zn, respectively. One might consider recycling filter dust in the same way, although this could also result in the build-up of undesirable impurities in primary metal. For these reasons, it is more common to sell the oxide residue from dross processing for use in the production of cement and concrete.

10.4.6.2   *Silicon and Ferrosilicon*

In this metallurgical process, carbon materials and quartz (as well as iron ore pellets for Fe–Si production) react in a high-temperature electric arc furnace. The resulting high-temperature reaction produces silicon or ferrosilicon. The furnace treatment is followed by tapping, a ladle refinement, casting, and crushing to form the final product: granular silicon or ferrosilicon. During this process residues produced are slag and silica-rich waste collected from the furnace off-gas. In addition, residue from the handling (washing and sieving) of quartz raw materials is generated.

It is interesting to consider the history of silica dust. This residue is formed when SiO vapour reacts with O_2 in the atmosphere to form SiO_2. Silica collected in the furnace off-gas/filter system consists of 80–97% SiO_2 with small amounts of C and other substances. The purity depends on the furnace process and raw materials used. Silica dust was initially a waste product. However, today it has become a resource. New applications are constantly being developed. One important use is as an additive in concrete. It increases the strength, and prevents chlorine penetration and corrosion of steel reinforcing rods. It is also used as a substitute for asbestos in wallboards and in cosmetics and contact lenses.

When the silicon is poured out of the ladle, a slag residue consisting of Si, SiO_2, CaO, and Al_2O_3 is left. This is scraped out of the ladle. Its composition varies according to the composition of silicon metal produced, additives in the ladle, and the refining process. The slag is usually sold as skulls—a source of Si for Si–Mn alloy producers.

10.4.6.3 *Ferromanganese and Silicomanganese*

Manganese ferroalloys are commercially produced by the carbothermic reduction of manganese oxide ores in a submerged arc furnace. Coke is both a reducing agent and electrical resistance element. Silicomanganese is also produced in the same way. The Si–Mn production is often integrated with the manufacture of Fe–Mn so the slag from the Fe–Mn production can be used in the production of Si–Mn. In this way a very high total yield of manganese is achieved.

Off-gas from the submerged arc furnace is cleaned with wet scrubbers, which produces a sludge. This is the largest residue from production. Smaller amounts of dusts are collected from other processes. The latter are easily recycled back into the production process. The sludge is more problematic. Some is dried and used in ferromanganese production, but the majority is currently deposited to landfill. An improved utilization of sludge becomes more urgent due to environmental pressure, as well as increasing costs for landfill (regulations, taxes, etc.).

The sludge contains a high fraction of Mn. However, the volatile elements Zn, Pb, K, and Na end up in the material. Zn and K are associated with difficult furnace operations, for example sticking, and K leads to higher coke usage (due to its catalytic effect in the Boudouard reaction). There are other challenges that also make recycling difficult, such as a high water content, high carbon (present as tar and carbonates), small particle sizes, heavy metals, transportability, and compositional variations. There has been some work done to resolve the problems of sludge utilization.[60]

Slag is also produced from Si–Mn production, which contains low levels of Mn. This is used mostly as a filler for road surfaces.

10.4.6.4 *Iron and Steel*

Most iron is produced by charging iron ore, coke, and fluxes (limestone and/or dolomite) into a blast furnace. The coke reacts to produce CO and CO_2 and reduce the iron ore to a molten iron. The molten iron (pig iron) can be cast into iron products, but is most often refined to produce steel, removing carbon by oxidation.

Blast furnace slag consists of aluminates and silicates from the ore, coke ash, and fluxes (limestone and/or dolomite) charged to the blast furnace. The molten slag also contains 1 or 2 per cent sulfur and comprises about 20 per cent by mass of the iron produced. After removal from the furnace in its molten state, the slag can form several types of products, depending on how it is cooled.

Air-cooled slag is formed by pouring molten slag into beds and allowing it to slowly cool under ambient conditions. The slow cooling produces a crystalline structure. The result is a hard, lump slag, commonly used as an aggregate material. It has favorable mechanical properties for aggregate use, including good abrasion resistance, good soundness, and high bearing strength. It can also be crushed and screened and used as an aggregate in Portland cement concrete, asphalt concrete, concrete, asphalt, and road bases.

Expanded or foamed blast furnace slag is produced when the molten slag is cooled and solidified by adding controlled quantities of water, air, or steam. This produces a lightweight expanded or foamed product. Foamed slag is distinguishable from air-cooled slag by its relatively high porosity and low bulk density.

Pelletized blast furnace slag is produced when molten slag is cooled and solidified with water and air quenched in a spinning drum. By controlling the speed of solidification, the pellets can be made more crystalline, which is beneficial for aggregate use, or more vitrified (glassy), which is more desirable in the production of cement and concrete. More rapid quenching results in greater vitrification and less crystallization.

Granulated blast furnace slag, consisting of sand-size (or frit-like) fragments, forms when the molten slag is cooled and solidified by rapid water quenching to a glassy state, resulting in little or no crystallization. When crushed or milled to very fine cement-sized particles, ground granulated slag can be used either by premixing the slag with Portland cement or hydrated lime to produce a blended cement or by adding the slag to Portland cement concrete as a mineral admixture.

Blast furnace slag is mildly alkaline and exhibits a pH in solution in the range of 8 to 10. Although blast furnace slag contains a small component of elemental sulfur (1 to 2 per cent), the leachate tends to be slightly alkaline and does not present a corrosion risk to steel in pilings or to steel embedded in concrete made with blast furnace slag cement or aggregates.[61]

Because of their porous structure, blast furnace slag aggregates have lower thermal conductivities than conventional aggregates. Their insulating value is of particular advantage in applications such as frost tapers or pavement base courses over frost-susceptible soils.

It is estimated that less than 10 per cent of the blast furnace slag generated is disposed of in landfills.

Steel making slag occurs as a molten liquid and is a complex solution of silicates, CaO, FeO, and other oxides. Virtually all steel is now made in the basic oxygen furnace (BOF) process, or in the electric arc furnace (EAF) process. The BOF process is described in detail in Chapter 9.

The primary steel slag is the furnace slag or tap slag. This is the major source of steel slag aggregate. After being tapped from the furnace, the molten steel is transferred in a ladle for further refining to remove additional impurities. This operation is called ladle refining because it is completed within the transfer ladle. During ladle refining, additional steel slags are generated by adding fluxes to the melt. These slags are combined with any carryover of furnace slag and are generally referred to as raker or ladle slags.

The liquid furnace slag and ladle slags are first crushed. Iron particles in the slag are then recovered by magnetic separation. The metallic portion is returned to the steel mill and remelted. The slag phase left is generally used as a granular base or as an aggregate material in construction applications. It can also be sintered and recycled as a flux material in the iron and steel furnaces.

Steel slag aggregates generally tend to expand. This is caused by the presence of free lime and magnesium oxides (not reacted with silicates) that can hydrate and expand in humid environments. The volume change can be 10 per cent or more, which is why steel slags are generally not suitable for use in Portland cement concrete or as compacted fill beneath concrete slabs. However, steel slag intended for use as an aggregate can be stockpiled outdoors for several months to expose the material to moisture from natural precipitation and/or application of water by spraying. This allows the potentially destructive hydration and its associated expansion to take place prior to use as an aggregate. The amount of time required varies a great deal. The free lime hydrates rapidly and can cause large volume changes over a relatively short period of time (weeks), while magnesia hydrates much more slowly and contributes to a long-term expansion that may require a year or more to stabilize the slag.

Steel making dust presents an interesting challenge. There are two intersecting issues—one metallurgical, the other environmental: the accumulation of volatile impurities in the dust and the method of dust collection.

When steel scrap is processed in an EAF or BOF, 15–25 kg of dust is formed per tonne of steel. The composition of the dust depends on the quality of scrap charged to the furnace. Zinc volatilizes at steelmaking temperatures, so galvanized metal in the charge results in a high zinc content.

Scrubbers, as well as dry electrostatic precipitators, are used for dust collection. In wet off-gas cleaning the dust is obtained as sludge. The energy required for drying sludge prior to processing this residue is high. An increasing number of steel mills are using electrostatic precipitators for dust collection. The main advantage of dry systems is the low pressure drop, resulting in substantial energy savings. Hence, the use of dry de-dusting is listed as an energy saving measure for the BOF process. Also, the residue is a dry powder which simplifies its handling as well as recycling.

When the dust contains more than about 20% zinc, it is economically treated by the Waelz kiln, which produces a ZnO powder collected by filters. The resulting Waelz zinc oxide is refined in a zinc plant to commercial grade zinc metal. When the zinc content of the dust is lower than 20 per cent, it is possible to pelletize the dust and feed it back into the BOF.[62] In this way the zinc content in the dust increases until it can be economically recovered in the Waelz process.

In integrated, ore-based steel mills, residues are often recycled into the sinter plant because fine grained, Fe-bearing dusts fit well into the sinter feed. However, the allowable amount of Zn in the sinter produced is limited for operational reasons in the blast furnace. Therefore, only BOF dust with a low Zn content can be recycled in this way.

The annual total amount of residues from BOF off-gas de-dusting in 2008 was estimated as 16 million tons, with a recycling rate of approximately 50 per cent. The optimization of in-plant dry dust recycling is described in the 'Best Available Techniques (BAT) Reference Document for Iron and Steel Production'.[63] Raupenstrauch and co-workers[64] describe the Waelz process and compare it to a newly developed process which promises to economically treat steelmaking dusts containing lower zinc contents. They also provide a good review of the practices used in the EU. Andersson et al.[65] give an excellent review and analysis of different options for recycling steelmaking dust residues.

10.4.7   Connecting Methods

Most products consist of different components and materials joined together by welding, bolting, riveting, inserted bushings, foaming connections, or adhesives. A good example is

Fig. 10.28 *A 2020 Corvette C8 body in white (at the 2019 Detroit Lightweighting Conference).*

the automobile. A typical car contains more than 100 different materials. Figure 10.28 shows the structural body of a 2020 Corvette C8. The body is made from castings, extrusions, and sheet components, using different metals and different alloy compositions.

The combination of different materials, and the use of different connecting methods, greatly complicates our ability to recycle materials contained in EoL products. Most consumer products represent a serious challenge. Look inside your smart phone or laptop computer and see for yourself. The separation and sorting of these materials usually requires extensive mechanical and chemical/metallurgical processing.

Figure 10.29 illustrates some methods of connection and their liberation behaviour when they pass through a shredder. It shows that joints, like many other components, generally do not break into pure materials. The figure mentions 'randomness of liberation'. This term characterizes the degree of randomness with which materials in a joint will break apart. Unfortunately, the methods used to connect materials are seldom chosen by considering their effect on recyclability.

The situation is getting more complex. In 2000 Ford Motors[69] used five joining methods to construct the body in white: resistance spot welding, gas–metal arc welding, fastener projection welding, fastener drawn-arc welding, and hemming. These were used to connect uncoated deep drawn steel components, and mild/micro-alloyed steels. In 2018 the material mix was more diverse. Ultra-high strength steels, hot formed components, closed hydroformed steel profiles, aluminium sheets and extrusions, and aluminium extruded tubes became part of the design to reduce body weight. Also, ten additional connecting methods were used: weld bonding, gas–metal arc brazing, laser welding, laser brazing, plasma brazing, self-piercing riveting, mechanical joining of fasteners, flow drill screws, and clinching.

Another example will serve to illustrate the situation. The 2019 Cadillac CT6 used the following joining methods:[70]

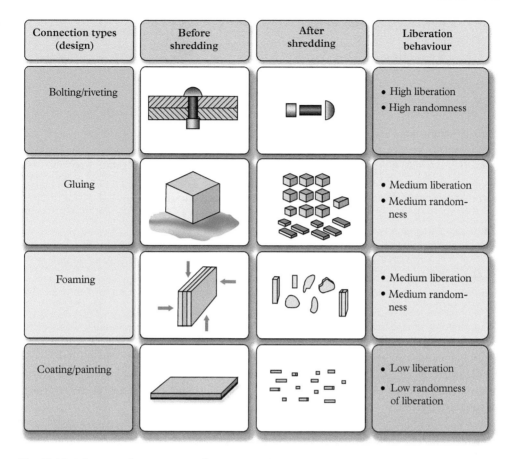

Connection types (design)	Before shredding	After shredding	Liberation behaviour
Bolting/riveting			• High liberation • High randomness
Gluing			• Medium liberation • Medium randomness
Foaming			• Medium liberation • Medium randomness
Coating/painting			• Low liberation • Low randomness of liberation

Fig. 10.29 *Liberation characteristics of common methods of connection.* [*Adapted from: van Schaik and Reuter.*[40–2, 66–8]]

- steel arc brazing—0.5 m
- aluminium arc welding—9.2 m
- aluminium laser welding—2 m
- steel spot welding—1624 spots
- flow form screws—745 screws
- structural adhesive—269 m
- aluminium spot welding—1469 spots
- self-piercing rivets—333 rivets

These changes in connection methods represent a serious challenge to product design engineers, as well as to recyclers.

10.5   Examples of Recycling

10.5.1   Recycling of Magnesium Alloys

by Per Bakke

10.5.1.1   Introduction

Global primary magnesium demand in 2014 reached 828 000 tonnes. Aluminium alloying (34%) and die casting (automotive 26% plus other 5%) were the two major magnesium-consuming sectors. Other minor market sectors include iron desulfurization and titanium production (Mg used for reduction of $TiCl_4$). Following the emergence of China as a low-cost source of primary production with the silicothermic Pidgeon process, during the years before and after the turn of the millennium, the magnesium industry is characterized by a Chinese dominance with roughly 80 per cent of the world's primary production. Countries outside of China still producing magnesium today are the US, Russia, Israel, and Brazil. In Europe there is no longer any primary magnesium production, although magnesium is considered by EU as a 'critical' material for European industry.

In the 1960s and 1970s AS alloys[iv] were used extensively in drive train components of the VW Beetle, but the mechanical and corrosion properties of these alloys gradually failed to meet increasingly strict requirements. In the 1980s high-purity (HP) alloys were introduced, significantly improving corrosion properties and causing renewed interest for magnesium alloys in the automotive industry. In the 1990s and early 2000s there was significant development of new die-casting alloys and optimization of the established alloy systems. Motivated by weight saving and reduced fuel consumption, the new alloys attracted considerable interest.

In the 1950s and 1960s a magnesium sheet industry existed, serving certain limited markets for automotive and aerospace applications.[71] Despite significant R&D efforts by major companies, the wrought magnesium alloy industry is still very small compared to die casting, and a breakthrough in the automotive industry is still some way from being realized. Magnesium alloys for processing routes other than die casting is still negligible. Hence, any discussion on the recycling of magnesium is basically limited to the recycling of die-casting alloys.

Closed loop recycling is defined as a production system in which the waste or by-products of one process or product is used in making the same product. The success of closed loop recycling is important for the magnesium industry because unlike aluminium, there are no other major markets or sources for returns/scrap outside of the die-casting segment. Any recycling process must therefore be capable of regaining the original chemical composition and cleanliness of the original alloys.

Based on life cycle assessments, the energy requirement for melting and recycling magnesium is only about 5 per cent of the energy to produce the same quantity of primary material.

[iv] Magnesium alloy families have letter designations established by ASTM. The AS designation is for magnesium-based alloys containing aluminium, silicon, and manganese.

10.5.1.2 Scrap Classification System

In typical magnesium die-casting operations, only around 50 per cent of the purchased and melted ingots end up as finished castings, the remainder is scrap. There are various classification systems for magnesium scrap. The classification system established by Hydro Magnesium[72] sorts scrap into the following classes:

1. Sorted clean returns
2. Sorted clean returns with inserts
3. Sorted oily/painted returns
4. Sorted dry chips
5. Sorted oily/wet chips
6. Dross—salt-free
7. Sludge—with salt
8. Mixed and off-grade returns

Clean and sorted scrap generated at die-casting plants (i.e. runners, biscuits, trimmings, and rejected cast parts) are referred to as class 1 material. Class 2 scrap comprises clean scrap with inserts. The first three classes require no special precautions on packaging or shipping, except the material should be kept away from moisture. Classes 4 and 5 (chips, dry and oily/wet) should be packed in ventilated steel drums and classified according to local legal standards when shipping. Also, class 6 is classified as dangerous goods and should be treated in accordance with prevailing statutory regulations. Regardless of classes, all material should meet the specifications of ASTM alloy series, AZ, AM, AS, and AE. Other materials, or material segregated otherwise, belong to class 8.

In principle, the classification system covers internal scrap at the primary alloy producer, as well as process scrap from die casters and post-consumer scrap. However, as most post-consumer scrap, such as end-of-life vehicle (ELV) material, is insufficiently sorted, this material normally belongs to class 8.

10.5.1.3 Process Overview

Regardless of class, except for ELV scrap, the recycling of magnesium will follow a general route. This is illustrated in Fig. 10.30. The process starts with sorting by alloy and classification, which is a prerequisite. If necessary, cleaning can be undertaken. In order to increase melting rates (exposing larger surfaces for increased heat transfer), crushing large-sized scrap may be considered.

Thorough preheating is generally a prerequisite, as moisture represents a safety hazard. Exceptions may be some processes for melting with salt, where considerable amounts of surface water may be tolerated. Melting will either take place with flux or under a cover gas. Most of the inclusions are separated from the metal during or right after melting, generating dross and/or sludge. Iron-refining is performed by adding Mn (e.g. as salt or as an Al–Mn hardener) at a temperature higher than the casting temperature. Additional inclusion removal can be done by

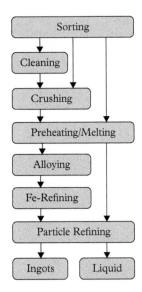

Fig. 10.30 *General process steps for magnesium alloy recycling.*[73] [*Reprinted by permission from Springer Nature: copyright 2006.*]

- Addition of a refining flux and agitation followed by settling,
- Settling in multi-chamber furnaces,
- Gas purging for flotation of inclusions
- Filtration
- Combinations of the above.

We now consider how each class of materials is recycled.

10.5.1.4   Class 1

Two principally different methods for treatment of magnesium alloy scrap may be utilized:[74]

1. Flux-based metal protection and refining.
2. Protective gas (SO_2/SF_6) for melt protection and settling, possibly assisted by argon-gas purging or filtration for removal of inclusions (flux-free refining).

Flux-based protection and refining has been most widely used and used to be the most important in terms of tonnage. However, flux-free solutions for *in-house* recycling, that is recycling of class 1 returns inside the die-casting shop, has become common practice.[75–7]

10.5.1.5   Flux-Based Systems

The traditional method for flux refining shown in Fig. 10.31 is to melt scrap and flux batch-wise and stir in additional flux and alloying elements, if necessary.[79] Then, after allowing

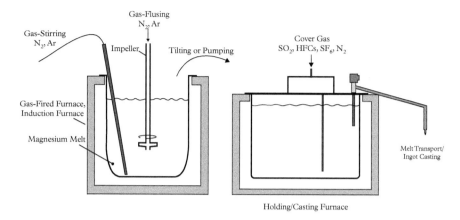

Fig. 10.31 *Recycling using salt.*[78] [*Copyright 2002 by The Minerals, Metals & Materials Society. Used with permission.*]

some time for separation and refining by settling, the metal is transferred into a second furnace, from where the metal is cast. In the second (holding/casting) furnace additional separation of salt and metal takes place, and only small amounts of salt are used. The use of two furnaces increases the capacity, improves the temperature control, and has the potential to improve the quality of the metal. However, single-furnace systems are also used. Metal is transferred and cast either by tilting and pouring of the crucible or by pumps/siphons. The furnaces are heated by either electricity or gas-fired. Between each batch the bottom sludge (consisting of salt, oxides, and metal droplets) is removed by grabbing. The sludge typically belongs to class 7.

Starting with an empty furnace the scrap is charged from the top and melts from below. Thus, the scrap higher up in the furnace is preheated by the hot liquid metal, gradually building up from the bottom. Normally no auxiliary preheating is necessary. The crucibles can be made of mild steel, but due to creep problems it is more common to use stainless steels with an inner cladding of mild steel. Nickel-containing stainless steel directly in contact with molten magnesium should be avoided due to the dissolution of nickel into the metal. Crucible-free furnaces with a magnesium-resistant refractory may also be used.

For many years Hydro Magnesium operated a system for continuous melting under flux; see Fig. 10.32. Metal is melted in one end of a large multi-chamber furnace and pumped out or cast from the far end of the same furnace.[80] The salt is heated by AC electrodes in chamber one, and continuously pumped over a stream of scrap metal being fed into the melting basket. The excess heat content of salt heats up and melts the scrap, which falls into chamber two together with the salt.

Here, salt and metal separate, and the metal is further refined through the remaining chambers of the furnace. It is not necessary to preheat the metal, as evaporating moisture is effectively vented during the melting process. The oxides present on the surface are removed by the salt and move with the molten salt phase. Manganese salts can be used for alloy adjustments, if necessary. Molten salts with a composition of 40–60% calcium chloride and

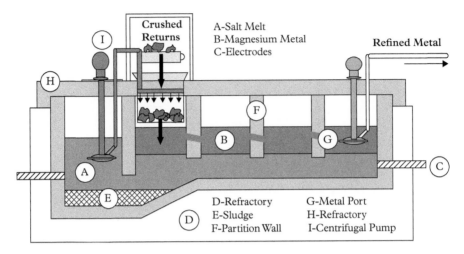

Fig. 10.32 *Multi-chamber, AC-heated recycling furnace with salt.*[73] *[Reprinted by permission from Springer Nature: copyright 2006.]*

3–7% calcium fluoride are preferred. In addition, the salt may contain sodium chloride and magnesium chloride, but a considerable amount of calcium chloride can be used. Due to the salt cover this method also gives protection against oxidation.

10.5.1.6   *Flux-Free In-house Recycling*

In-house recycling of clean scrap—like biscuits, runners, and internal rejects—has been practiced for years at some die casters.[81] In-house recycling systems are usually based on salt-free solutions (Fig. 10.33). As an example, the companies Rauch, Schmitz, and Apelt have developed a multi-chamber flux-free furnace for recycling class 1 returns. Preheated scrap is fed into one end, and the metal is refined through the three-chamber furnace equipped with baffles.[82] Metal is cast into ingots from the downstream end of the furnace. Molten metal refining includes inert gas purging and filtration.

A continuous flux-free process for recycling class 1 process scrap based on a two-furnace system has been described.[76] The company Rauch from Austria has developed continuous flux-free in-house recycling based on a two-furnace system with electrically heated transfer pipes.[83]

In-cell recycling is defined as systems where class 1 returns, biscuits, runners, and gating systems are charged directly back to the die-casting furnace.

Long before the term in-cell recycling was introduced, Øymo et al. described fluxless remelting of magnesium die-casting scrap in a two-furnace system.[84] In this process, preheated primary magnesium ingots and scrap are melted in a 100 kW resistance-heated furnace, with a steel crucible containing about 500 kg of metal. From the melting furnace, metal flows through a siphon tube into a casting furnace (Fig. 10.34). From the casting furnace, the metal is metered intermittently through a siphon tube into a large ingot mould simulating the die-casting process. Melt surfaces are protected with a mixture of 20% CO_2, 0.2% SF_6,

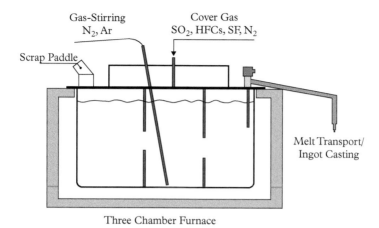

Fig. 10.33 *Recycling without salt.*[78] [*Copyright 2002 by The Minerals, Metals & Materials Society. Used with permission.*]

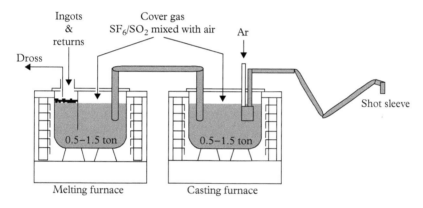

Fig. 10.34 *Example of an in-cell recycling system based on a two-furnace system.*[73] [*Reprinted by permission from Springer Nature: copyright 2006.*]

and dry air. At a metal flow rate of 4 kg/min, giving a metal residence time in the two furnaces of nearly two hours, an average yield of approximately 93 per cent is attained independent of scrap content.

For in-cell recycling a two-furnace system is considered superior to a traditional one-furnace system, because it offers additional space for dross and sludge accumulation, and removal. Furthermore, it provides a more evenly distributed residence time, which improves particle refining due to settling. Also, the two-furnace system offers superior temperature control while casting, as temperature gradients generated during the melting of ingots and scrap in the melting furnace are smoothed out in the casting furnace.[85]

10.5.1.7   *Class 2*

For class 2, basically the same methods as for class 1 may be used. If the insert materials do not dissolve excessively in molten magnesium, as is the case for steel, class 2 does not represent a problem, as the inserts in principle may be removed from the metal or salt. However, brass and bronze inserts may lead to Cu and other heavy metal contamination, and this material cannot be recycled back to high-purity alloys. It may find its use as raw materials for anodes, desulfurization of steel melts, or as an alloying element in the (secondary) aluminium industry.

10.5.1.8   *Class 3*

In case of flux-free melting, it is important that the paint/oil must be burned off in advance, before the parts are immersed in molten metal. This because the paints/oil may lead to the evolution of toxic gases, and contamination of the metal in terms of unwanted dissolved elements and inclusions. When melting under flux, one may expect the salt to absorb most of the remaining contaminants on the scrap surfaces.

Regardless of where the paint/oil is burned off, it is important to treat the off-gas at high temperatures (after burning) to avoid emission of toxic compounds like dioxines, toxic furans, and PCBs, which may form at intermediate temperatures when organic compounds decompose in the presence of chlorine. In order to fully decompose any chlorinated substances, the temperature in the combustion chamber should exceed 1100 °C.[86] Depending on the energy content of the off-gases, additional energy may be introduced to reach the desired temperatures.

10.5.1.9   *Class 4*

Dry chips (class 4) represent limited problems, particularly when remelting with salt. Due to the large surface areas, high melting rates may be achieved. However, additional salt is required to absorb the oxides on the surfaces. Class 4 material can be remelted and refined back to high-purity alloys, but with a metal recovery considerably lower than that for class 1 material.

The large surface area presents insurmountable problems when melting without salt; it is difficult to get sufficient protection with a cover gas to completely avoid burning. Desulfurization in the steel industry may be an outlet for chips. Dry and clean chips may also be formed into briquettes, which can be fed into melting furnaces.

10.5.1.10   *Class 5*

Most of the chips arising from the production process today are contaminated with either oil or oil/water emulsions. It is technically feasible to recover the metal from wet/oily chips (class 5), but presently the cost may be higher than the actual value of the metal. As liquids represent a considerable volume of the class 5 material, the cleaning step demands considerable effort, that is, removing the main fraction of the liquids before melting. Even with long draining times it is not possible to reduce the oil content to below 10 per cent.[87]

A special hydraulic press has been developed by Rauch (SPV) in which the chips are compacted and heated, causing the main fraction of the liquid to be squeezed out. The remaining product is a rod consisting of compacted chips, which is easier to handle during melting. The high forces involved in the press may generate considerable heat, possibly with fumes and smoke as a result. Residual liquid in the compacted material may still represent a challenge, as the fumes tend to form during the remelting of the compacted material.

Alternatively, mineral oil lubricants covering the chips can be removed by an aqueous alkaline degreasing solution, followed by a washing/centrifuging/drying system.[81] However, using this type of treatment, aqueous emulsion lubricants used for high-speed machining are more difficult to remove.

Saxena has an US patent in which he describes a process where oily chips are heated to 200–400 °C in a rotating furnace (higher temperatures may cause ignition of the chips).[88] The oil evaporates and is conducted out of the furnace and condensed. Another US patent describes a process where wet/oily chips are entrained in a gas and fed to a cyclone separator.[89] The walls of the cyclone are heated to temperatures causing vaporization of the liquids.

Vacuum distillation has been adapted to magnesium chips.[90] A Pidgeon-type furnace is used.[91] Pure Mg crowns with only Zn as a likely contaminant condense in the cold end. The other elements remain in the residue. In order to not contaminate the system, most of the oil or other surface contaminants must be removed before charging scrap to the reactor. A number of other vacuum-refining processes have been proposed.

With respect to formation of dioxines or other chlorinated hydrocarbons, the same precautions as for class 3 apply.

10.5.1.11   Class 6

Surface dross from die caster operations may contain 70–90% metal, the remainder being oxides. Die caster sludge contains intermetallic particles as well as oxides. As die caster sludge is salt-free, this belongs to class 6.

The metal fraction in die caster dross can be recovered to high-purity alloys by melting and refining with salt.[92] This can be done in systems as indicated in Fig. 10.31, or in rotary furnaces. The secret of success is to use sufficient amounts of salt, and to mix salt and metal very well to remove all oxides in the salt phase. As the magnesium-containing material to be melted may contain 10–100 times more oxides than class 1 scrap, it is obvious that a considerable amount of salt is needed. It is extremely difficult to recover dross by salt-free processes.

10.5.1.12   Class 7

The significant nonhomogeneity of the phases and the wide size range of the metallic particles complicate processing of class 7. The material usually consists of a mixture of salt, oxides, and alloy particles in the size range <100 μm to 10 cm. The alloy beads often contain a lot of intermetallic particles. Oxides stay preferentially in the salt phase rather than with the metal. Hence in the bottom sludge considerable amounts of oxides accumulate.

Sludge can be melted batch-wise with additional salt, fairly similar to the traditional flux-based processing of classes 2–7.[93] After additions of CaF_2 (a so-called thickening agent) the melt is stirred and metal globules coalesce. The metal is then decanted. The CaF_2 addition, stirring, and decanting can be repeated to improve the recovery of metal. During the melting of sludge cakes, one should always beware of liberation of moisture, which has been picked up by the highly hygroscopic salt during storage.

Metal can be recovered by dry crushing and sieving of the solidified sludge cake. Salts can also be removed by dissolving in water, but this causes some loss of magnesium due to the formation of $Mg(OH)_2$ and evolution of hydrogen gas. These processes increase the metal fraction in the product as well, as it becomes more homogeneous. The metal beads may thus become attractive material for the desulfurization of steels without remelting, casting, and grinding. Due to the altered chemistry of the Mg globules, sometimes with exceptionally high contents of elements such as Mn, Ni, and Cu, it is difficult to make high-purity alloys from the metal fraction in the sludge. However, if the Ni content is low, the metal may still be useful for applications not requiring high-purity alloys, for instance sacrificial anodes. The metal may also be applicable for alloying purposes in the aluminium industry.

Sludge that has been melted and diluted by additional salt can be subjected to dispersion into granules by means of a centrifugal system and solidified, after which the mixture undergoes mechanical treatment for separating of the salts and oxides.[94] The remaining metallic fraction (60–80% metal) can be used for desulfurization of iron.

10.5.1.13   Class 8

Most of the aluminium and magnesium alloys from automobiles being scrapped today end up as shredded material sorted from the heavy metals and plastics/fluff. Very often, magnesium and aluminium are not sorted. The magnesium fraction, normally constituting only a very small percentage, either burns off as the aluminium is remelted or is removed in subsequent refining steps, for instance by injection of a chlorine-containing gas (demagging). Nevertheless, various methods have been used to separate the magnesium fraction from aluminium, including manual sorting, by means of laser or X-ray fluorescence analysis, or floatation in a heavy liquid media with a density higher than magnesium but lower than aluminium. All the common magnesium alloys are mixed. In addition, aluminium casting alloys are unavoidably present, giving very high Al content and contamination of Si when the magnesium alloy fraction is melted.

10.5.1.14   Recycling Using Flux

The flux technology was developed in Germany in the 1920s.[95] Fluxes used for protection only should melt at a low temperature, so they cover the metal during melting. Near-eutectic compositions in the system $CaCl_2$–$NaCl$–KCl–$MgCl_2$ are very effective.[96] Refining fluxes should, in addition, contain thickening or inspissating agents such as fluorides or oxides. This improves separation of metal and flux when pouring.

When magnesium alloy scrap covered with an oxide layer is in contact with molten salt, the wetting properties of the system promote the oxides being transferred to the salt phase, leaving the molten metal parts free to coalesce into a continuous phase. When refining with flux, the principle is to use sufficient flux to absorb all the oxide present in the molten metal. As mentioned earlier, the salt must be intimately mixed with the metal to achieve a large contact area, so that virtually all metal comes in contact with the salt. Afterwards the salt and metal phases are separated.

The observations of Emley show that two factors are important in the coalescence of finely divided metal drops dispersed in a salt.[79] Firstly, the salt must strip the oxide films from the droplets, and secondly, sufficient amounts of fluorides must be present in the salt to create a favorable contact angle between the salt and the droplet. The latter means that the salt tends to withdraw from the surfaces of metal, leaving the metal drops free to coalesce. Salts consisting of alkaline chlorides are ineffective, and $MgCl_2$ (or $CaCl_2$) must be present. As the content of $MgCl_2$ increases, so does the metal recovery up to a certain point.[97] Beyond this concentration, further increases in $MgCl_2$ lead to lower recoveries. According to Emley there is a threshold concentration of CaF_2 in the salt which must be reached before coalescence occurs. This content depends on the alloy content of the dispersed metal and also varies with the $MgCl_2$ content in the salt melt. As the concentration of fluoride additives increase in both NaCl–KCl and $MgCl_2$–KCl salt systems, coalescence of aluminium drops also improve.

The temperature during refining is also important. Waltrip suggests that temperatures below $700\,^{\circ}C$ results in entraining flux and other impurities in the metal, and too high a temperature may cause excessive oxidation and reduced crucible life.[74] The melt is refined by using 2–3% flux and stirred for 15 minutes so that melt and flux mix intimately. For heavily contaminated scrap, more salt may be necessary to get a satisfactory refining and to avoid 'dry' sludge.

The density of the salt is also important in separating salt from metal, and to maintain a certain density difference between the alloy and salt phase. $CaCl_2$ alone can be used to increase the density of salt. $BaCl_2$ is considerably heavier and was widely used in the past, but due to its potential toxicity it is no longer accepted in most countries. $SrCl_2$ may also be an alternative.

It must be noticed that not all alloys are suitable for flux refining. If any of the alloy elements present form chlorides that are more stable than $MgCl_2$, these will react with the molten salt and will eventually be removed from the alloy. If one considers the Gibbs free energy of formation for various chlorides, it will be seen that magnesium alloys with Ca and Sr cannot be flux-refined without losing these elements to the salt. Likewise, a salt containing magnesium chloride will also affect alloys containing rare-earth elements (Ca, La, and Pr). However, kinetic factors may in practice limit the loss of these elements. The most common alloying elements Al, Zn, and Mn are not affected. In fact, $MnCl_2$ is often used as a source for manganese, as it reacts and dissolves forming $MgCl_2$.

10.5.1.15 Fluxless Refining

Recycling of magnesium employing salt fluxes has a potential to improve the recovery. But there are disadvantages to using a salt flux. It is corrosive to the surrounding equipment and environment (HCl fumes), and it combines with impurities in the molten metal to form a sludge in the lower section of the crucible. It also creates the possibility of having salt inclusions in the cast product. In addition, chlorides may combine with organic residual that is included with the scrap to form toxic compounds, creating the need for comprehensive and costly off-gas treatment systems.

Flux-free refining includes all refining processes where salt is not used. Settling/floatation and filtration are among these. It has been claimed that gas purging of magnesium alloys removes 'a lot of' inclusions. However, quantitative results are, to a great extent, missing. Although the systems are different, when using only inert gases a reduction by roughly a factor of two is typical for gas fluxing aluminium.[98] This may indicate that gas purging may only be a supplementary method, not the only refining method.

Fluxless refining requires the use of a protective gas atmosphere, based on SF_6 or SO_2. Particles (inclusions) are to be removed from the bulk melt and transfer to the melt surface, settle to the bottom, or stick to the walls of the crucible. The purified metal may then be transferred from the bulk melt through a hole into a holding container, removed through a siphon tube, or poured gently. Stirring must be limited, since otherwise the inclusions will be re-entrained into the bulk melt.

Taking into account the cost of purge gas, metal loss to oxidation, evaporation of Mg to bubbles (giving MgO dust), and the need for a cover gas, gas purging of Mg alloys has obvious drawbacks. In salt–metal systems, however, gas introduced into the underlying salt phase may be used to effectively distribute salt into the metal phase. In salt–metal systems one may consider employing inert gases and reactive gases like nitrogen.

Reding describes a method for melting magnesium without requiring a salt-flux cover on the magnesium surface.[99] Melting is carried out using a protective gas cover. Oxide films and other contaminants rise to the melt surface. They are excluded from the holding furnace by a retaining wall. The magnesium melt is removed from the holding container by a pump.

Petrovich and Waltrip discuss a process that utilizes the protective atmosphere of air, CO_2, and 0.3–0.5% SF_6 that also incorporates a filtering system to reduce non-metallic inclusions.[75] Utilizing a pump equipped with a suitable suction filter, molten metal is transferred into ingot moulds, a transfer pot, a holding pot, or die-cast shot sleeve.

A flux-free process incorporating inert-gas purging and filtering techniques for removing non-metallic inclusions from molten magnesium scrap is described by Housh and Petrovich.[100] They claim that the process has several advantages over a flux-refining system, including the removal of dissolved gases, decreased melt losses, and the elimination of chlorides. However, a significant effect of inert gas purging on the removal of hydrogen cannot be expected.[101]

A common feature of all the filter types described above is that the inclusions are captured in front of the filter (cake filtration) rather than inside the filter (depth filtration). If these filters prove to be effective, the operating time (lifetime) will necessarily be short. In contrast, if the filters last for a long time without being blocked, it is rather unlikely that the filters are very effective.

10.5.1.16 *Contamination Control*

The most important non-reactive inclusions in magnesium and magnesium-based alloys are:[102]

- Oxides, appearing as lumps, films, and apparently loosely connected agglomerates or clusters
- Intermetallic particles
- Chlorides

Oxide films and clusters are reaction products from the melt surface, trapped in the melt due to turbulence during stirring, pumping, charging, and ladling. Ingots and scrap charged to the furnace are also covered with a layer of oxide. The thickness of the oxide layer may depend on alloy composition, cooling, and storing conditions. Upon charging and remelting, these oxide films may be retained in the melt. Oxide films and oxide clusters have large surface/volume ratios and are thus not readily removed by settling. Magnesium nitrides, Mg_3N_2, may appear together with oxide clusters.[103]

Intermetallic particles are usually a result of the iron process. Due to the relatively high density of the intermetallic particles (4–7 g/cm^3), most are removed before ingot casting. However, the solubility of iron and manganese decreases with decreasing temperature. Decreasing temperatures facilitates their formation and growth. As a result, temperature fluctuations in furnaces may lead to the formation of intermetallic particles, accompanied by decreased manganese content and increased iron content. Good temperature control in the furnaces is, therefore, mandatory. Intermetallic particles in pure magnesium as well as in Mg alloys are usually smaller than 15 μm.

Due to the small relative difference in density the smallest salt particles are difficult to remove by settling. Fortunately, the wetting properties of most molten salts result in adhesion of salts to oxides and agglomeration of salt inclusions, resulting in increased size and improved settling properties.[96]

10.5.1.17   *Additional Sources of Trace Elements and Inclusions during Recycling*

Recycling introduces additional sources for contamination of all groups of impurities. Process scrap returned from the die-casting machines may have surfaces 10–100 times larger than the surfaces of the ingots, and this represents a considerable source of inclusions. Associated with the die-casting process (classes 1 and 2) are die lubricants, usually water based, containing various organic components including petroleum oils, waxes (vegetable, mineral or synthetic), and bactericides.[104] It was reported that a higher silicon level is a fingerprint of recycled class 1 metal.[105] The presence of hydrocarbons indicates that even when melting class 1 (and 2) scrap with salt, precautions should be taken to avoid the formation of chlorinated hydrocarbons (i.e. dioxines).

10.5.1.18   *Effects and Removal of Some Trace Elements*

Common impurity elements with negative effects on properties include nickel, iron, copper, and cobalt. Beryllium and silicon are also worth mentioning.

Nickel is almost two orders of magnitude worse than copper in promoting corrosion of both die-cast and permanent mould-cast AZ91.[106] Alloying elements such as aluminium reduce the solubility of nickel considerably, due to the formation of AlNi phase.[107] For example, the solubility of nickel reduces to 0.15% at 650 °C with an addition of only 3% aluminium. If manganese also is added, the solubility of nickel decreases even further.

A US patent describes a process for reducing the nickel content from about 2 to about 0.2% in pure magnesium by use of zirconium, and further down to 0.001% by adding zirconium and aluminium.[108] Removal of Ni by Zr is also mentioned by Emley.[109] Adding excess Mn (as Al–75% Mn master alloy) at 720 °C reduces the Ni content of an AZ91 melt made of scrap painted with a Ni-containing coating to 5 ppm.[110] Baseline experiments with the same alloy material but without the addition of Mn gave 0.0064 and 0.0080% Ni.

Although it is found that some nickel may precipitate alongside iron in the various Al–Mn–Fe phases, no industrial process for removing nickel from commercial molten magnesium-base alloys is known, except for dilution with 'nickel-free' material or vacuum distillation.

The solubility of *iron* in pure magnesium is 0.018 wt.% at 650 °C, increasing to approximately 0.04 wt.% at 750 °C. A content of 200–400 ppm Fe is thus characteristic of commercially pure Mg. The principles for removing iron by means of adding manganese and aluminium to obtain a more corrosion-resistant alloy was discovered more than 70 years ago.[111,112] However, the industrial importance of this was not recognized until the 1980s. The metallurgical principles for removing iron by adding manganese are extensively described, and the mutual liquid solubilities of iron and manganese in various alloys have been determined.[113,114]

Copper has a negative effect on the corrosion of AZ and AM alloys but is considerably less harmful than nickel and iron.[106,115] Although some copper may precipitate along with iron in the various Al–Mn–Fe phases, there is no known process for removing copper from magnesium alloys, except for dilution and vacuum distillation.

Cobalt, as well as iron and nickel, are quite common in materials that may be considered for use alongside magnesium. Although cobalt is present in some steel alloys, the main problem with cobalt may be its presence in some pigments that are used on magnesium-alloy parts, resulting in contamination during recycling of post-consumer scrap. Hanawalt et al. very early demonstrated that cobalt is as harmful as nickel, although very little attention has been given to cobalt and no standards are specified for it.[116] Like copper and nickel, cobalt cannot be removed from a molten Mg alloy unless by dilution or vacuum distillation.

Beryllium is normally added to magnesium alloys (typically 5–15 ppm) to reduce the oxidation of the molten alloy.[117] The restricted solubility of Be in Mg alloys prevents larger concentrations. The beryllium is gradually 'lost' to oxidation during processing, and a melt consisting of remelted die cast parts may contain <5 ppm. Al–Be master alloys are normally used to replace 'lost' Be. As much as 0.007 and 0.037% Be is found in oxide/salt fractions from dross and sludge processing, respectively.[92] Since Be is classified as a carcinogen and finely dispersed Be (or Be-containing) dust may lead to serious diseases, there are statutory regulations to follow and precautions to be taken with respect to handling Be-containing dross.

Although *silicon* is added to some Mg alloys to improve creep properties, it is considered as an impurity in most other alloys, as excess concentrations (>0.1%) may lead to reduced ductility. According to Emley, silicon can be removed by adding zirconium or cobalt (as chloride).[109] However, these principles have not been utilized in practical recycling of commercial Mg alloys.

10.5.2   Aluminium Dross and Salt-Cake Recycling

In Section 10.4.6 the dross generated in aluminium production was described. Here the recycling of dross is considered. We also discuss methods for minimizing the amount of metal lost to dross formation. Treatment processes for dross are also considered, as well as the recycling of salt cake (a residue of dross processing).

Dross is formed by several mechanisms:

1. By splashing and metal movement during filling of the furnace;
2. An oxide skin forms on the melt surface during melting and holding;
3. During molten metal processing and alloying, any stirring and surface disruption creates fresh oxides; and
4. As a result of metal spills or skulls; these are not actually dross per se but are often added to the furnace dross for subsequent recovery.

The amount of dross generated depends on the type and amount of scrap being melted and furnace practices used, but it is usually 1 to 5 per cent of the weight of metal charged to the furnace. Dross formation therefore represents a significant expense. A number of steps may be taken to reduce this cost.

1. The type of metal transfer when filling the furnace is important. Whenever possible, a level pour transfer (via launder) is preferred. Not only does this minimize dross formation, but the metal cleanliness is improved. Liquid metal from the potrooms usually comes in a transfer crucible. This metal should be poured carefully into the furnace using a funnel-like guide, or better yet, a siphon. The siphon also prevents the transfer of bath (potroom electrolyte) into the furnace.

2. Careful charging of the furnace minimizes the loss to oxidation. This is especially important for thin-walled scrap, which oxidizes easily. The best practice is to first place all thin scrap in the furnace. This should be covered with heavier (thicker) scrap. And finally, liquid metal is added. (Of course, all scrap in the furnace must be dry. If not, one risks producing a severe explosion!) Furnace practices also have an impact. High metal temperatures produce more dross. Burner flames that impinge directly on the melt surface should be avoided. Both also add undesirable hydrogen gas to the melt. The magnesium-containing (5xxx) alloys are especially susceptible to rapid oxidation. The effect of metal chemistry on oxide formation in these alloys has been studied. Additions as small as 2 ppm Be will reduce Mg oxidation significantly.[118] Unfortunately, Be has serious health risks for some people, so its use is problematic. Many companies now refuse to have any Be in their alloys. The effect of furnace atmosphere has also been examined.[119,120] The following observations were made:

   - It is possible to vary the furnace atmosphere composition considerably.
   - Real-time analysis of both furnace atmosphere and temperatures is possible.
   - Atmospheres with high concentrations of CO_2 and H_2O reduce oxidation, while atmospheres with high concentrations of O_2 and N_2 increase oxidation.

3. Alloying can also generate dross, especially when reactive elements are added. Magnesium is especially problematic and is best added by placing Mg ingot or bars into a cage, and holding the cage below the metal surface until dissolution is complete. Care in removing dross from the furnace is also vital. Figure 10.27 shows dross that has been simply placed into a cast iron dross pan. Hot dross held like this will continue to oxidize and may 'thermite'. Thermiting is the uncontrolled burning of liquid aluminium. It can heat the dross to very high temperatures. As a result, a number of dross coolers and

presses have been developed to minimize oxidation and thermiting. These have been reviewed by Schlesinger.[121]

Dross is a mixture of aluminium and non-metallic components (NMC). The NMC fraction is primarily aluminium oxide. Dross falls into two categories: salt-free dross and salt dross. The first is commonly known as white dross and is produced by facilities that melt without flux. The latter is usually called black dross. Worldwide the production of dross is estimated at 760 000 ton/year, about half from primary smelters and half from secondary.[121]

White dross consists almost entirely of Al_2O_3 and aluminium metal trapped in the oxide skim. Dross that contains a great deal of Al metal is often called a 'wet' dross. Small amounts of aluminium carbide and nitride may also be present. Dross from primary smelters may also contain small amounts of cryolite (Na_3AlF_6). This is the result of the bath accidentally being charged to furnaces. If alloys containing magnesium are produced, the dross will also contain some periclase (MgO) and spinel ($MgAl_2O_4$).

Black dross is produced by furnaces using additions of flux. The purpose of the flux addition is to reduce metal losses by producing a 'drier' dross. Black dross usually contains less than 20% aluminium metal and 30 to 50% aluminium oxide. The rest is the fluxing salt (mostly sodium and potassium chlorides).

10.5.2.1   *Dross Processing*

Although dross processors are subject to the same global economic constraints, the environmental situation varies greatly. For example, in Australia and the EU one may not dispose of salt-containing residues in landfill. Hence, a number of different processes have been developed. These have been reviewed in detail by Schlesinger and Peterson.[121,122] Only an overview of the most important processes is provided here.

The metal content of the dross may be upgraded by crushing and screening, or by eddy current separation. The oxide NMC fraction of the dross is friable and easily crushed, whilst aluminium is not. Hence, the result is a separation into a coarser fraction containing most of the metal, and a fine NMC residue. If the latter is from white dross, it may be recycled by blending with ore and feeding it to the pots; but it is more commonly disposed of as landfill. Studies have shown that the NMC may be used in concrete or asphalt mixtures, but these applications have not found significant commercial use at this time. The metallic fraction may be recovered in melting furnaces, but it is more commonly processed in a rotary flux furnace.

It is important to understand why a rotary flux furnace is used. Aluminium oxidizes easily. Aluminium metal forms a thick oxide skin when heated to high temperatures. This skin prevents the coalescence of individual pieces of metal. However, liquid salts dissolve the oxide skin. They also change interfacial energies so individual drops easily coalesce into a single, liquid mass. A rotary furnace promotes coalescence, as individual pieces of aluminium tumble into one another.

Metal recovery is good with the rotary flux furnace, but the disadvantage is the production of a residue called salt cake. Salt cake is expensive to process and cannot be used as landfill in many places. Hence, salt-free methods have been developed to recover aluminium from salt-free dross. The first attempts to treat dross without flux were unsuccessful, since the lack of a salt cover allowed thermiting to occur. Consequently, metal recoveries were poor. Subsequent developmental efforts have produced several salt-free technologies.These have been reviewed by Schlesinger.[121] The processes use a method of heating that is less oxidizing towards aluminium metal in the dross, and employ furnace rotation to break the oxide skin on the metal to promote coalescence. The result is a higher metal recovery.

10.5.2.2   *Salt Cake*

Salt cake is a by-product of the rotary flux furnace processing of dross. The composition depends on the source materials and recycling practices, but generally consists of 4–8% aluminium metal, 25–45% salt, and 50–70% NMC. The latter are completely absorbed by the liquid flux in the furnace and form, after tapping and cooling, the so-called salt slag or salt cake. In Europe about 1.4 million tonnes of salt cake are generated yearly, and most is processed in recycling plants. The salt content is recovered and returned to the refiners for reuse.

In the US salt cake is not considered a hazardous waste. Most salt cake is crushed for the removal of entrapped aluminium. The remaining material is then sent to landfill.

In places where land-based disposal of salt cake is either banned or too expensive, further treatment is required. The desired goals are:

- Minimizing or eliminating the residue to be discarded;
- Generating a non-hazardous residue that can be used or discarded, if necessary;
- Recovering the salt content and metallic aluminium; and
- Minimizing environmental impact of the process.

Figure 10.35 illustrates the basic process used for most salt-cake treatment. The salt cake is crushed and screened to remove most of the aluminium metal. Water is added to produce a brine containing 22 to 25% dissolved salt content. The solid–liquid separation is a two-stage process, beginning with a centrifuge, which separates out most of the NMC. The liquid leaving the centrifuge passes into a clarifier, which generates a sludge containing the rest of the NMC. This is washed and filtered, resulting in a low-salt NMC that can be landfilled or used for other purposes. The brine generated by dissolution is then processed to remove the water. This is usually done with an evaporator crystallizer. The result of crystallization is wet salt crystals that are subsequently air dried and reused as flux.

The energy costs of salt recovery from salt-cake processing can be substantial and may not be justified by the value of the recovered salt. The NMC recovered from the process consists of alumina and other oxides, with some aluminium nitrides and possibly aluminium carbides. It also contains a small amount of remaining metal. The carbides and metallic aluminium pose a problem. When exposed to moisture, they may react under some conditions to form hydrogen—a potential fire and explosion hazard. This makes the disposal or subsequent use of NMC problematic.

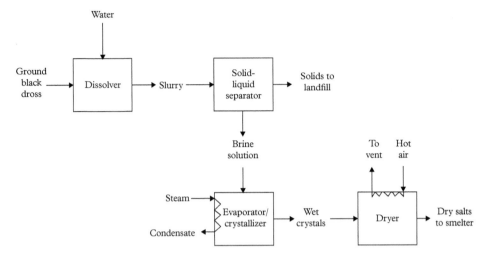

Fig. 10.35 *Standard flowsheet for the processing of salt cake.*[123] [*Reprinted by permission from Springer Nature: copyright 1996.*]

10.5.3   Battery Recycling

Batteries are ubiquitous in our modern society. They are found in numerous consumer products and are increasingly being used to electrify the power train of cars and buses. Batteries represent an interesting challenge for engineers trying to recycle them. Different designs are used, each possessing their own unique characteristics and challenges.

A good example is the lithium-ion battery (LIB). It has an excellent power density. That is, it delivers a great deal of energy compared to its relative weight. But the underlying chemistry providing that power can also be released as a fire or explosion. A technician removing the battery tray from an automobile must observe specified safety protocols. In Germany and several other European countries, he or she must receive special training before working to recycle LIBs.

In this section we provide an overview of the commonly used batteries and consider their recycling. The chemistry used in the common types of batteries are presented in Table 10.6. We first consider the lead-acid battery.

10.5.3.1   Lead–Acid Battery

The lead–acid battery was invented in 1859 by Gaston Planté and was the first rechargeable battery. Despite having a low energy-to-weight ratio, its ability to supply high surge currents means it has an excellent power-to-weight ratio. This capability and low cost makes it attractive for starting automotive motors. The batteries are also widely used for backup power supplies in cell phone towers, high-availability settings like hospitals, and stand-alone power systems. It has been estimated that lead–acid batteries account for nearly half of the sales volume for all batteries.

Table 10.6 *The Chemistry of Common Types of Batteries*

Lead–acid	$Pb(s) + HSO_4^-(aq) \rightarrow PbSO_4(s) + H^+(aq) + 2e^-$
	$PbO_2(s) + HSO_4^-(aq) + 3H^+(aq) + 2e^- \rightarrow PbSO_4(s) + 2H_2O$
Zinc–carbon	$Zn(s) + 2MnO_2(s) + 2NH_4^+(aq) \rightarrow Mn_2O_3(s) + Zn(NH_3)_2^{2+}(aq) + H_2O(l)$
Alkaline	$Zn(s) + 2OH^-(aq) \rightarrow ZnO(s) + H_2O(l) + 2e^-$
	$2MnO_2(s) + H_2O(l) + 2e^- \rightarrow Mn_2O_3(s) + 2OH^-(aq)$
Button	$Zn(s) + Ag_2O(s) + KOH/NaOH \rightarrow ZnO + 2Ag$
NiCd	$2NiO(OH) + Cd + 2H_2O \rightarrow 2Ni(OH)_2 + Cd(OH)_2$
Ni–MH	$H_2O + M^* + e^- \leftrightarrow OH^- + MH.$
	$Ni(OH)_2 + OH^- \leftrightarrow NiO(OH) + H_2O + e^-$
Li ion	$LiCoO_2 \leftrightarrow Li_{1-x}CoO_2 + \chi Li^+ + \chi e^-$
	$\chi Li^+ + \chi e^- + 6C \leftrightarrow Li_xC_6$

Note: $M^* = AB_5$, where A = rare earth and B = Co, Ni, Mn, etc.

Its long history and more-or-less standardized design aids the recycling of lead–acid batteries. Collection rates are typically high (> 90% in high-income countries, though lower globally) because lead is considered to be a hazardous substance, and there also exists a well-established system for recycling. Processing begins by breaking the batteries open, crushing and mechanically separating them into case materials, metal grids and poles, lead pastes (oxides and sulfate), and the acid, which is collected and reused or discarded. Pastes are treated to remove sulfur before smelting with the grid metal. Lead metal low in antimony is produced along with antimony metal containing lead. The lead is refined to produce soft lead or, more commonly, new battery alloys.

10.5.3.2 Zinc–Carbon and Alkaline Batteries

These batteries use a zinc anode and manganese as the cathode. They can easily be processed pyrometallurgically. ZnO and Fe–Mn alloys are produced using the contained carbon as a reductant, while the slag can be used as a building material. However, the present low value of the battery's materials today does not cover the cost of recycling. Formerly these batteries often contained mercury, but since the early 1990s mercury is no longer present. Without mercury they are not a hazardous material, and they are usually disposed of in landfill. Incentives are required to promote recycling. In Vermont, for example, a 2016 law required primary battery producers to fund a free state-wide collection and recycling programme. In 2006, the EU passed the Battery Directive, which stated that at least 25 per cent of all the EU's used batteries must be collected by 2012, and rising to no less than 45 per cent by 2016, of which at least 50 per cent must be recycled.

10.5.3.3 Silver Oxide Batteries

We were unable to find detailed information on the recycling of these batteries. They are usually sold as small, button-shaped batteries and are used in applications which do not draw a large current. Examples are watches, small toys, and medical devices. Owing to their silver content (*c.* 30% by weight of Ag_2O) the economics of recycling must be favourable. Some of these batteries contain mercury, so they should not go to landfill.

10.5.3.4 Ni–Cd Batteries

With a nickel anode and cadmium cathode, these batteries can be pyrometallurgically processed to recover Cd/CdO (as distilled condensate/flue dust), while producing a Fe–Ni alloy product and a benign slag. However, this type of battery has become obsolete and its use has been banned in various countries. Cadmium is a heavy metal and—like its 'cousins', lead and mercury—tends to collect in fatty tissues of humans, having deleterious effects on the nervous system. Moreover, their power capacity is inferior compared to the newer nickel-metal hydride and lithium-ion batteries.

10.5.3.5 Ni–Metal Hydride Batteries

When this type of battery was first introduced, it presented a novel chemistry and was composed of new materials. As is typical of our consumer-driven society, this product came to the market with little consideration of how it should be recycled. In other words, no process was designed specifically for recycling these batteries. Instead, existing technology had to be used. This was described earlier as pyrometallurgical processing. This processing is now considered in more detail.

NiMH battery cells are composed of potassium hydroxide (KOH) electrolyte, nickel hydroxide (NiOH) in the cathode, a hydrogen storage nickel-based alloy in the anode, a steel cover, current connectors, an organic separator, and sealants. The common alloys in the electrode designs are AB_5, A_2B_7, and AB_2, where the A_xB_y refers to the ratio of the A-type elements (LaCePrNd or TiZr) to the B-type elements (VNiCrCoMnAlSn). A spent nickel–metal hydride battery is composed of 36–42% nickel, 22–5% iron, 3–4% cobalt, 8–10% mischmetal (the rare-earth elements La, Ce, Pr, and Nd), 1–2% potassium (K), 3–4% plastics, and <1% graphite.[124,125] The first step in recycling is to employ standard beneficiation operations, such as grinding, sieving, magnetic, electrostatic, and gravity separations. This processing liberates the batteries' electrode materials and concentrates valuable metals in the feedstock for subsequent pyrometallurgical (high-temperature) processing. The volume of NiMH scrap is reduced, as is the need for leachate purification in any subsequent hydrometallurgical process.[126–8] The dismantled and processed NiMH batteries are then melted in an electric arc furnace. This produces a nickel–cobalt alloy and a slag phase rich in rare-earth oxides. The key for success in this process is to find a suitable slag system which ensures good separation of the Ni–Co alloy from the rare-earth oxides. Nickel and cobalt are found nearly completely in the metal phase, whereas the rare earths are transferred into the slag as oxides.

In this process the discarded batteries are essentially used by the steel industry as a cheap source of nickel. The value of the cobalt is usually not recovered. The rare earths (RE) are slagged and also lost for possible reuse.

The nickel–cobalt alloy can be dissolved in acid, and both elements recovered by hydrometallurgy, but it would be better to also recover the rare-earth elements. New hydrometallurgical processes have been developed in the laboratory which can recover the rare earths.[129,130] In these processes, dissolved Fe is an impurity which is difficult to precipitate with good filtration properties. Therefore, a good separation of iron-rich fractions is needed in the beneficiation processes to minimize Fe in the hydrometallurgical process.

10.5.3.6   Lithium-Ion Batteries

Lithium-ion batteries are the most recent addition to battery technology. Their superior performance and light weight have made them popular in countless applications. Perhaps the most important is their use in transportation. The volume of waste batteries from automobiles and buses is expected to grow rapidly. Recycling these batteries will be a challenge, since the necessary legislative framework and recycling process capacity are still being developed. Different battery materials are also being used. Here we present a brief overview of the situation, with an emphasis on challenges that will be faced by future metallurgical and chemical engineers. For anyone who wants a 'deeper dive' into LIB recycling, the survey by Elwert et al. is highly recommended.[131]

In the EU and USA, the layered oxides lithium–nickel–manganese–cobalt oxide (NMC) and lithium–nickel–cobalt–aluminium oxide (NCA) are the preferred cathode materials. In China lithium–iron–phosphate (LFP) is currently the predominant cathode material for cars and buses. About 35–40 per cent of the powertrain battery weight is in the module container (or pack), which consists of aluminium, cables, electronics, plastics, and steel. The cell materials account for the rest of the total weight. In the component cells, 55–65 per cent is in the anode and cathode materials. Depending on the battery chemistry, the cathode materials vary. Tables 10.7 and 10.8 show the value of potentially recoverable materials in LIB packs and cells.

Several processes have been developed for the recycling of LIB cells. From an economic point of view, the main drivers for recycling have been Co and Ni, and to a lesser extent, Cu. Lithium was not a concern, owing to its comparatively low price and difficult recoverability.

Table 10.7 *Prices and Estimated Revenues from Battery Pack Disassembly (Elwert et al.[131])*

Material	Quality	Price (€/mt)	Specific revenues (€/mt of batteries)	
			NMC/NCA	LFP
Aluminium	Scrap with max. 5% adhesions	600	111	133
Cables	Cable scrap, min. 38% copper	1400	15	15
Electronics	Printed circuit boards, category II A	1530	32	32
Steel	Stainless steel scrap (V2A)	750	55	55
		Total	213	235

Note: Scrap prices as of September 2017.
Reprinted by permission from Springer Nature: copyright 2018.

Table 10.8 *Specific Metal Values of LIB Cells (Elwert et al.[131])*

Material	Quality	Price (€/mt)	Specific revenues (€/mt of batteries)	
			NMC/NCA	LFP
Aluminium	High grade primary	1651	316	381
Cobalt	High grade, min. 99.8%	43 979	820	—
Copper	Grade A	5081	913	916
Lithium carbonate	Min. 99–99.5%	10 399	1471	443
Nickel	Min. 99.8%	9146	512	—
		Total	4031	1739

Note: Prices are averages during the period September 2016 to August 2017.
Reprinted by permission from Springer Nature: copyright 2018.

However, since lithium prices tripled between 2007 and 2015, its recycling has received more attention. However, processes still concentrate on LIBs containing cobalt and nickel (NCA and NMC). As LIB producers aim at a further cobalt reduction in layered oxides and other cobalt- and nickel-free cathode materials are introduced, declining levels of these expensive materials must be expected, which will challenge the economical recycling of LIB cells.

From a technological point of view, the recycling processes vary significantly regarding the types of batteries that can be treated. Processes are characterized by complicated processes that use a combination of mechanical, pyrometallurgical, and hydrometallurgical operations. A significant problem is the high reactivity of the electrolyte. It reacts violently with air, leading to possible fire hazards. The electrolyte also contains fluorine, and therefore poses possible health and environmental risks. To solve these problems, different approaches have been developed. These include:[131]

- Direct melting of cells (Umicore NV/SA, Belgium);
- Vacuum thermal deactivation (Accurec Recycling GmbH, Germany);
- Mechanical treatment under a protective gas atmosphere (Recupyl SAS, France, Duesenfeld GmbH, Germany, and Batrec Industrie AG, Switzerland); and
- Comminution in a liquid solution (Retriev Technologies Inc., USA).

Regarding the process design and capability, two different philosophies are evident. On one hand, there are 'universal' processes whose goal is to process different LIB types in one process. These processes are often capable of recycling other materials containing similar target metals (e.g., nickel–metal hydride batteries or cobalt- and nickel-containing catalysts). Therefore, these processes tend to focus on high throughput. The battery cells usually enter a pyrometallurgical treatment with a minimum of, or without, mechanical pretreatment.

Individual metals are recovered by hydrometallurgical treatment of the process outputs: alloy, slag, and flue dusts.

On the other hand, there are processes adapted to an individual battery chemistry. This focus enables a higher quality of waste recovery when compared with universal processes. However, this approach requires excellent collection and sorting. Considering the current state of public collection and sorting, only business-to-business relationships are currently able to meet these requirements.

A detailed review of three dedicated LIB recycling processes is given in Elwert.[131] Velázquez-Martínez and co-workers review ten different processes in various stages of development.[132]

The challenge is considerable. It has been estimated that from 2017 to 2019, 6.9 million battery-powered electric vehicles and 4.4 million hybrid electric vehicles were sold. The largest car market is in China, followed by the EU and the USA. According to calculations,[131] about two million waste power-train batteries will enter the recycling market until 2025. This corresponds to approximately one million metric tons per year.

The increased spread of electromobility and the recycling of LIB batteries will present an enormous challenge. The difficulties arise from a still-to-be-established recycling infrastructure, complex assemblies, and material composites; the use of different designs and chemical compositions; and the need to keep up with a rapidly increasing LIB production. Sophisticated recycling technologies addressing these problems are of high importance, not only for environmental and safety reasons, but also to secure an adequate supply of the needed raw materials, such as lithium and cobalt.

For further reading and useful background information regarding the recycling of batteries, see for example references [40,133].

10.6   Limits and Opportunities in Recycling

To conclude we briefly consider some of the challenges and opportunities that arise during the recycling of metals.

10.6.1   The Limits of Recycling

Our desire to recycle and reuse materials is limited by a number of important factors. Broadly speaking, these limits can be placed into three categories:

1. *Laws of nature* include physics, mathematics, kinetic factors, and thermodynamics, among others. These laws, as they relate to the refining of metals, have been the primary focus of this book. It is important to know these laws and to use them to evaluate the technical feasibility of any recycling scheme we dream of accomplishing. Aside from the chemical reactivity of individual metals and the properties of metal solutions, the most important for recycling are the following:
    a. The first is the law of entropy as it relates to the mixing of different materials. If you take milk and sugar with your morning coffee, imagine trying to unmix the brew. It would be extremely difficult to separate the milk, coffee, and sugar once they

have been stirred together. This simple analogy is actually a good example of the problems often faced in recycling. (You may also want to consider the description of the entropy of mixing in Chapter 2, and then calculate the energy or work required to unmix an ideal solution.)

b. A related problem is the accumulation of undesirable tramp elements in metals as we continue to recycle. Copper dissolved in steels and iron in most aluminium alloys are the two well-known examples. This does not represent a serious limit to recycling at present, mainly because there are sufficient quantities of primary (ore-based) metals produced to dilute tramp elements to acceptable levels. The future may be very different, however, as we rely more and more on recycled materials.

c. There is an inevitable trade-off between metal grade (or purity) and percent recovered. As noted earlier when considering liberation and sorting curves, it is not possible to have both at 100%.

d. There are a number of inevitable losses that occur as materials are processed, usually during the melting and refining of metals in the form of slags, drosses, and other residues.

2. *Economic factors* are important within the structure of our market-driven society. Even countries considered to be communistic have become more-or-less fully integrated into our global economy. Recycling requires resources: capital to build process equipment, energy required by the process, and labour required to run the plant. If recycling does not 'pay' for these resources, it will usually not happen. One example is that sometimes recycling is more expensive than dumping waste in landfill. In many respects landfill is not the 'best' solution, but economic factors may triumph. Traditionally, the cost of reprocessing waste or recycled material must be less than the cost of processing new material. However, there will be a growing market and legislation for intelligent landfill use. Waste should be minimized and stored until recycling becomes 'economic'. Here future costs to the environment should be included. As raw materials become scarcer, the price tends to increase, and the economic incentive to recycle increases. Also, the costs involved in landfill goes up. Costs of green energy (wind, sun) have been going down. Energy-neutral buildings are being developed. Transportation is being based on green energy (solar, electricity, batteries). Recycling of metals is expected to increase parallel to developments in producing and concerning water for humans and agriculture and in the supply of food. The spoiling and plundering of our planet's resources is a major weakness of our present economic system. Climate change/global warming is probably the biggest problem we face, but there are many others to consider. Changes are obviously needed. But as metallurgical engineers we will let others, more qualified for the task, write that book.

3. *The social and legal framework* in any society (or country) has a tremendous influence. Social factors are probably most important at the beginning: Will consumers take the time and trouble to sort their discarded goods, and dispose of them so they will be recycled? And of course, individuals' desires are the engine that drives our consumer-based economy. The legal framework is usually more important on the 'back end': such as using taxes/fees or legal restrictions to discourage disposal in landfills.

Government policy and legislation can play a key role in shaping the conditions for recycling. When proper policy goals do not exist, significant resource volumes can be lost to illegal or informal disposal or may be unaccounted for, sometimes through 'cherry picking'. There may also be environmental problems, damaging health impacts, and impacts on water or climate when regulatory standards are not in place or not enforced. In some cases, the government must provide incentives to cope with negative revenues from the recycling process.

Many regulations designed to promote recycling are 'clumsy' or poorly designed. Consider the relationship between quality and recovery of a given metal. The quantity collected depends a good deal on the policies in place. However, the issue is really more complex when the quality of the streams are considered. Collecting more may adversely affect product quality when collected in a single stream. Some materials may even become sufficiently contaminated to make them economically unviable for recycling.

Public authorities can assist by becoming stakeholders in the recycling chain. Most local authorities are responsible for waste collection, and sometimes processing. This may include the ownership and construction of infrastructure needed for collection and/or treatment. Public investments are usually long lasting in view of the large capital costs. In deciding on infrastructure location (whether their own or of private stakeholders), government indirectly affects the economics of recycling. Recycling capacity should ideally be close to the 'urban mine', for lower collection and transportation costs. Where local authorities have no expertise in recycling or collection, they can productively work with reliable and efficient partners.

For some metals the handwriting is already on the wall. Declining grades of ores and the increased use of metals once considered 'exotic' will inexorably result in their recycling. The situation is different for iron and steel, aluminium and magnesium. The ores used to smelt these metals are available in such large quantities on our planet that the primary production of these metals from ores can continue for a very long time.

And finally, recycling obviously cannot occur without returned products. Well-organized collection and pre-processing systems are often lacking, which makes it impossible to ensure that metals and other materials do not end up in waste streams where they are lost. Much improvement is needed in this area.

10.6.2 The Opportunities in Recycling

Comparing the present situation with that described in the Introduction written in 1992 it is clear that we still face many of the same challenges. The number of people on the Earth has grown as projected. Have we really made progress?

It many respects a good deal of progress has been made. Recycling as a philosophy or goal is much more firmly implanted in the public awareness. Significant advances in our understanding and technology have occurred. Although not always implemented, the best available technology has improved considerably in many areas. Numerous LCA and MFA

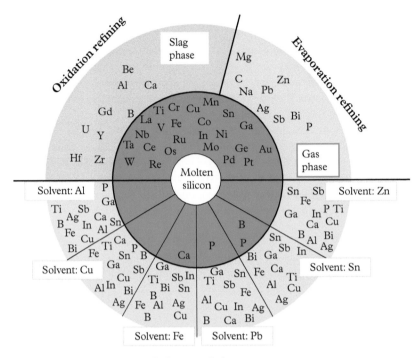

Fig. 10.36 *Predicted distribution of 42 impurities in the refining of EoL silicon wafers by oxidation, by vacuum, and by solvent refining (source: Lu et al.[134]). [This work is licensed under a Creative Commons Attribution 4.0 generic license.]*

studies have advanced our understanding of the complexities and limiting factors inherent in recycling. The models available in this area have become quite sophisticated.

A significant portion of this book has been devoted to understanding thermodynamic principles and presenting detailed thermodynamic data. In the past thirty years many thermodynamic studies have been made and published, so the data available to us have expanded significantly. In addition, much of the data are now embedded in sophisticated computer programs and databases. The software usually also considers phase relations as part of their data optimization. The net result is a powerful and sophisticated tool which can be used to analyse different refining processes. A good example of what is possible is a recent study by researchers in Japan on the refining of silicon solar cells. Figure 10.36 illustrates the results of their study. The solvent refining by aluminium, copper, and zinc appears to offer the most promise, since all the typical dopants and the metal electrodes can be removed.

Of course, there is still a lot of work to be done. Although progress has been made, we can do still more to capture metals that are now being lost. This is particularly true of waste electronic scrap. The most detailed study of the recycling of metals is the 2013 report by a United Nations panel. As part of their study they calculated the fraction recycled at EoL for

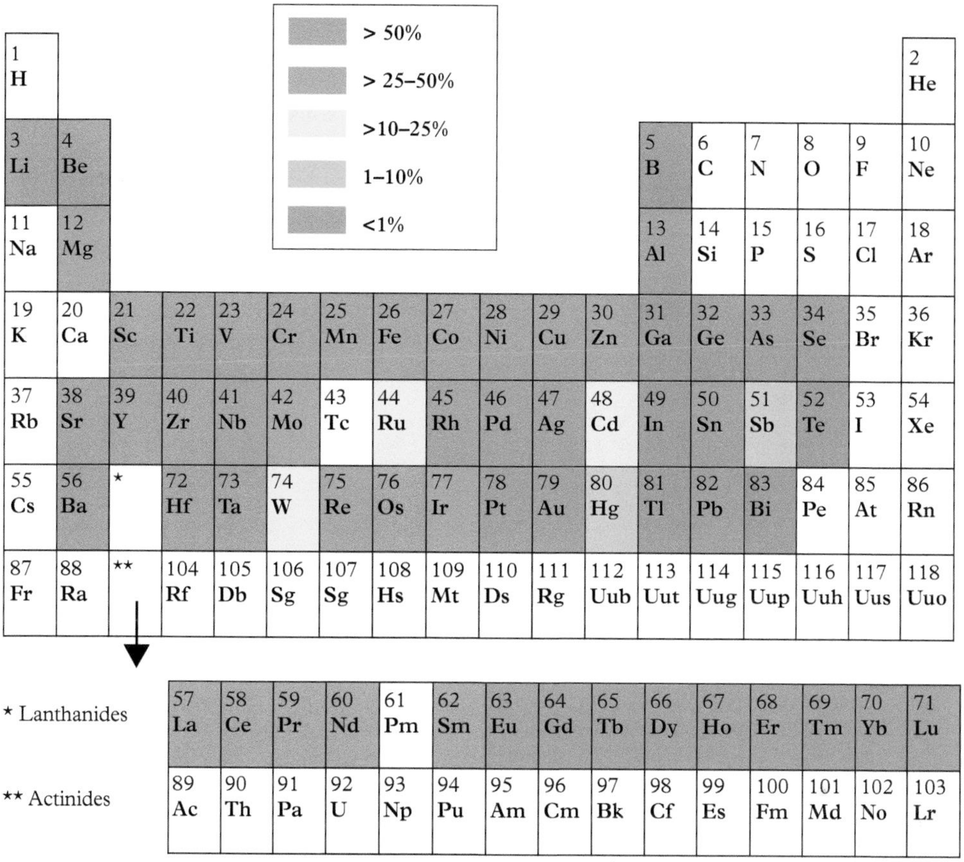

Fig. 10.37 *End-of-life (post-consumer) functional recovery rate for 60 metals. [UNEP (2011), Recycling rates of metals: a status report. A report of the Working Group on the Global Metals Flows to the International Resource Panel.[42,135]]*

many elemental metals. A result of their survey is shown in Fig. 10.37. Functional recycling was defined as when the desirable physical and chemical properties of the metal were retained. Unfilled boxes indicate that no data were available, or the element was not part of the study. The blue-coloured elements are being recycled reasonably well. The other elements represent significant opportunities for future improvement.

The creativity and ingenuity of the human spirit will certainly find many opportunities to recycle, or to improve recycling, in the future. We will close with a particularly clever example discovered by Megy.[136]

Hot-rolled titanium products become oxidized during manufacture. The surface of the semi-finished products is ground to remove the oxidation. The residue from the grinding operation is called swarf and consists of particles of the grinding compound (mainly silicon carbide) and particles of titanium metal. The titanium particles are reactive, so the swarf was considered to be a hazardous waste. Titanium producers had to pay a good deal for disposal of their swarf. Megy developed a process to recover the titanium. The swarf was placed into large steel tanks and filled with water. The tanks were covered and then stirred with large paddles. The swarf particles in suspension collided many times with their neighbours over the course of ten hours or so. During this autogenous grinding or beneficiation, the metal particles became separated from the silicon carbide grit. The titanium and silicon carbide were separated by density; the titanium was pressed into briquettes, dried, and then sold to aluminium producers for alloying additions. This simple process recovers several million pounds of titanium each year, by treating a residue which was previously discarded.

We hope this book and chapter will help future metallurgists in their efforts to create a better, more sustainable future.

....

REFERENCES

1. Rankin WJ. *Minerals, metals and sustainability: meeting future material needs*, p. 275. Collingwood, Victoria: CSIRO publishing; 2011.
2. Rosling H, Rosling O, Rönnlund AR. *Factfulness: ten reasons we're wrong about the world—and why things are better than you think*; . pp. 94–7. New York: Flatiron Books; 2018.
3. Birat J-P, Daigo I, Matsuno Y. Ch. Methods to evaluate environmental aspects of materials. In: Seetharaman S, ed. *Treatise on process metallurgy*, vol. 3. Amsterdam: Elsevier; 2014.
4. Hunt RG, Franklin WE, Welch RO, Cross JA, Woodal AE. *Resource and environmental profile analysis, of nine beverage container alternatives*. Washington DC: US Environmental Protection Agency; 1974.
5. Gilstad G, Hammervold J. Life cycle assessment of secondary aluminium refining. *Light Metals*, 2014; 901–6.
6. Saevarsdottir G, Kvande H, Welch BJ. Aluminium production in the times of climate change: the global challenge to reduce the carbon footprint and prevent carbon leakage. *J. Metals (JOM)*, 2019; 72: 296–308.
7. Birat J-P, Zaoui A. The 'Cycle of Iron'. Metallurgical Research & Technology. 2002; 99(10): 795–807
8. Birat J-P. The greening of the steel industry. In: *Gordon Research Conference, industrial ecology, transforming the use of energy, materials, water and wastes, August 17–22, 2008*, Colby-Sawyer College, New London, NH, Faye Duchin (chair).
9. Boin UMJ, Bertram M. Melting standardized aluminum scrap: A mass balance model for Europe. *J. Metals (JOM)*, 2005; 57(8): 26–33.
10. Bertram M, Graedel TE, Rechberger H, Spatari S. The contemporary European copper cycle: waste management subsystem. *Ecol. Econ.*, 2002; 42(1–2): 43–57.
11. Capuzzi S, Timelli G. Preparation and melting of scrap in aluminum recycling: a review. *Metals*, 2018; 8: 249.
12. CEN. Aluminium and aluminium alloys—scrap. EN 13920. Brussels, Belgium: CEN; 2003.
13. Schlesinger ME. Aluminum recycling economics. In: *Aluminum recycling*, 2nd edn, pp. 45–64. Boca Raton, FL: CRC Press; 2014.
14. Bento A, Roth K, Zuo Y. Vehicle lifetime trends and scrappage behavior in the U.S. Used car market. IAEE's Energy Economics Education Foundation. 2018; 39.

15. Schuhmann R, Jr. *Metallurgical engineering*, Vol. 1, *Engineering principles*, pp. 20–7. Reading, MA: Addison-Wesley; 1952.

16. Doutre DA. LiMCA and its contribution to the development of the aluminum beverage container and UBC recycling. In: Isac M, ed., *Proc. of the Roderick Guthrie Honorary symposium on process metallurgy*, pp. 319–21. Montreal, Canada: McGill Metals Processing Centre; 2011.

17. Terukina AT. Barriers and solutions for closed-loop aluminium beverage can recycling. Master Thesis. Trondheim, Norway: Norwegian University of Science and Technology; 2013.

18. Duncan JL, Hosford WF. The Aluminum Beverage Can. Scientific American. 1994; 48–53

19. Schlesinger ME. The mystery of the missing cans: UBC recycling in the United States. Plenary Lecture at the 20th Jubilee Conference on Materials and Technology, Portorož, Slovenia, October 18, 2012.

20. Niero M, Olsen SI. Circular economy: to be or not to be in a closed product loop? A life cycle assessment of aluminium cans with inclusion of alloying elements. Resour. Conserv. Recy., 2016; 114: 18–31.

21. Rheinfelden Alloys. https://rheinfelden-alloys.eu/en/

22. Donahue R, et al. Alloys for high integrity: aluminum. In: Twarog D, Apelian D, Luo, A, eds, *High Integrity Castings of Lightweight Components*, pp. 19–52. North American Die Casting Association, Arlington Heights, IL; 2016.

23. Kaiser K, Schmid M, Schlummer M. Recycling of polymer-based multilayer packaging: a review. *Recycling*, 2018; 3(1): 1.

24. Warrings R, Fellner J. Current status of circularity for aluminum from household waste in Austria. *Waste Management*, 2018; 76: 217–24.

25. Biganzoli L, Gorla L, Nessi S, Grosso M. Volatilisation and oxidation of aluminium scraps fed into incineration furnaces. *Waste Management*, 2012; 32: 2266–72.

26. Ludlow-Palafox C, Chase HA. Microwave-induced pyrolysis of plastic wastes. *Ind. Eng. Chem. Res.*, 2001; 40(22): 4749–56.

27. Zawadiak J. Tetra Pak recycling–current trends and new developments. *Am. J. Chem. Eng.*, 2017; 5(3): 37.

28. Lopes CMA, do C. Gonçalves M, Felisberti MI. Blends of poly(ethylene terephthalate) and low density polyethylene containing aluminium: A material obtained from packaging recycling. *J. Appl. Polym. Sci.*, 2007; 106(4): 2524–35.

29. Recycling Coffee capsules & Pods |Nespresso. https://www.nespresso.com/be/nl/recycling. Accessed 23 July 2019.]

30. Bunge R. Recovery of metals from waste incinerator bottom ash. Report of Institut für Umwelt- und Verfahrenstechnik UMTEC, Rapperswil, Switzerland. 2016. This may be downloaded as Metals_from MWIBA.pdf from http://www.umtec.ch, www.igenass.ch or http://www.vbsa.ch

31. Muchova L, Rem PC. Metal content and recovery of MSWI bottom ash in Amsterdam. In: Popov V, et al. *Waste Management and the Environment III. Transaction Series: WIT Transactions on Ecology and the Environment*, Transaction Vol. 92, p. 211. 2006.

32. Syc M, et al. Metal recovery from incineration bottom ash: State-of-the-art and recent developments. *J. Hazard. Mater.*, 2020; 393: 122433.

33. Muchova L. Wet physical separation of MSWI bottom ash. PhD thesis. Department of Materials and Science, Technical University of Deft; 2012.

34. Briedi, R. Mepex report dated 23 March 2018. Available at http://www.mepex.no

35. Nixon H, Saphores J-DM. Information and the decision to recycle: results from a survey of US households. *J. Environ. Plan. Manag.* 2009; 52(2): 257–77.

36. Sidique SF, Lupi F, Joshi SV, The effects of behavior and attitudes on drop-off recycling activities. *Resour. Conserv. Recy.*, 2010; 54(3): 163–70.

37. Sidique SF, Joshi SV, Lupi F. Factors influencing the rate of recycling: An analysis of Minnesota counties. Resour. Conserv. Recy., 2010; 54(4): 242–9.

38. Zhang D, et al. Residents' waste separation behaviors at the source: using SEM with the theory of planned behavior in Guangzhou, China. *Int. J. Environ. Res. Public Health.*, 2015; 12(8): 9475–91.

39. Saphores J-DM, Nixon H. How effective are current household recycling policies? Results from a national survey of U.S. households. *Resour. Conserv. Recy.*, 2014; 92: 1–10.

40. Reuter M.A., Heiskanen K, Boin U, van Schaik A, Verhoef Y, Yang Y. *The metrics of material and metal ecology, harmonizing the resource, technology and environmental cycles.* Amsterdam: Elsevier BV; 2005.

41. Van Schaik A, Reuter MA. Shredding, sorting and recovery of metals from WEEE: linking design to resource efficiency, In: Goodship V, Stevels A (eds), *Waste electrical and electronic equipment (WEEE) handbook*, pp. 163–211. Cambridge, UK: Woodhead Publishing; 2012.

42. Reuter MA; Hudson C, van Schaik A, Heiskanen K, Meskers C, Hagelüken C. metal recycling: opportunities, limits, infrastructure. A report of the Working Group on the Global Metal Flows to the International Resource Panel. United Nations Environment Program (UNEP); 2013.

43. Nijkerk AA, Dalmijn WL. Handbook of recycling techniques. The Hague, The Netherlands: Nijkerk Consultancy; 1998.

44. Schlesinger M. Beneficiation technology. In: *Aluminum recycling*, pp. 67–103. Boca Raton, FL: CRC Press; 2014.

45. Heiskanen K. Theory and tools of physical separation/recycling. In: Worrell E, Reuter M, eds, *Handbook of Recycling*, pp. 39–61. Amsterdam: Elsevier; 2014.

46. ISO 3310-1:2016: Test sieves—Technical requirements and testing—Part 1: Test sieves of metal wire cloth.

47. Dalmijn WL, De Jong TPR. The development of vehicle recycling in Europe: sorting, shredding, and separation. *J. Metals (JOM)*, 2007; 59: 52–6.

48. Evans R, Guest G. *The Aluminium Decoating Handbook.* Wolverhampton, UK: Stein Atkinson Stordy; (n.d.).

49. Tessandori J, Aleris, private communication, 2009.

50. Kvithyld A, Meskers CEM, Gaal S, Reuter M, Engh TA. Recycling light metals: optimal thermal decoating. *J. Metals (JOM)*, 2008; 60(8): 47–51.

51. Capuzzi S, Kvithyld A, Timelli G, Nordmark A, Gumbmann E, Engh TA. Coalescence of clean, coated, and decoated aluminum for various salts, and salt–scrap ratios. *J. Sustain. Metall.*, 2018; 4: 343–58.

52. Kvithyld A, Engh TA, Illés R. Decoating of aluminium scrap in various atmospheres, In: Schneider W, ed., *Light Metals 2002*, pp. 1055–60. Warrendale PA: TMS-AIME; 2002.

53. Meskers CEM, Reuter MA, Boin U, Kvithyld A. A fundamental metric for metal recycling applied to coated magnesium. *Metall. Mater. Trans. B*, 2008; 39B: 500–17.

54. Jenkins R. Delaquering systems. presentation from Thorpe Technologies Inc., at the TMS Annual Meeting; San Antonio, Texas: March 12–16, 2006.

55. Tabereaux AT, Peterson RD. Aluminum production. In: Setharaman S, ed., *Treatise on process metallurgy*, Vol. 3: *Industrial processes*, p. 913. Amsterdam: Elsevier; 2014.

56. Hatayama H, Daigo I, Matsuno Y, Adachi Y. Evolution of aluminum recycling initiated by the introduction of next-generation vehicles and scrap sorting technology. *Resour. Conserv. Recy.*, 2012; 66: 8–14.

57. Dai M, Wang P, Chen W-Q, Liu G. Scenario analysis of China's aluminum cycle reveals the coming scrap age and the end of primary aluminum boom. *J. Clean. Prod.* 2019; 226; 793–804.

58. Bao S., Syvertsen M, Kvithyld A. Overview of solid waste and by-products in the metallurgical industry in Norway. SINTEF Report SFI–Metal Production, RD3/RD5: 2018.

59. Peterson RD, Newton L. Review of aluminum dross processing. In: Schneider W, ed., *Light Metals 2002*, pp. 1029–37. Warrendale PA: TMS-AIME; 2002.

60. Gaal S, Tangstad M, Ravary B. Recycling of waste materials from the production of FeMn and SiMn. In: Vartiainen A, ed., *Proceedings of INFACON XII*, pp. 81–7. Helsinki, Finland: Outotec Oyj; 2010.

61. Short A. The use of lightweight concrete in reinforced concrete construction. *Reinforced Concrete Review*, British Reinforced Concrete Association. 1959; 5(3).

62. Lanzerstorfer C. Dust recycling by air classification in converter steelmaking. *Steel Res. Int.* 2019; 90: 1800377.

63. Remus R, Aguado-Monsonet MA, Roudier S, Sancho LD. Best available techniques (BAT) reference document for iron and steel production, industrial emissions directive 2010/75/EU, Integrated pollution prevention and control. Luxembourg: Publications Office of the European Union; 2013.

64. Raupenstrauch H, Doschek-Held K, Rieger J, Reiter W. RecoDust—an efficient way of processing steel mill dusts. *J. Sustain. Metall.* 2019; 5: 310–18.

65. Andersson A, et al. A Holistic and experimentally-based view on recycling of off-gas dust within the integrated steel plant. *Metals*, 2018; 88: 760.

66. Castro B, Remmerswal H, Brezet H, van Schaik A, Reuter MA. A simulation model of the comminution-liberation of recycling streams: relationship between product design and the liberation of materials during recycling. *Int. J. Miner. Process.* 2005; 75(3–4): 255–81.

67. Castro B, Remmerswal H, Boin U, Reuter MA. A thermodynamic approach to the compatibility of materials combinations for recycling. *Resour. Conserv. Recy.*, 2004; 43(1): 1–19.

68. van Schaik A, Reuter MA, Heiskanen K. The influence of particle size reduction and liberation on the recycling rate of end-of-life vehicles. *Miner. Eng.*, 2004; 17: 331–47.

69. Hover J. Joining in support of lightweighting for a high-volume OEM. Presentation at *Lightweighting World Expo 2018, Novi, Michigan, October 9–10, 2018.*

70. Lie L. 2019 CADILLAC CT6-lightweight enabled luxury. Presentation at *Lightweighting World Expo 2018, Novi, Michigan, October 9–10,2018.*

71. Emley EF. Principles of magnesium technology, Chapter XVI. Oxford: Pergamon Press; 1966.

72. Brassard C, Riopelle L. Recycling of magnesium scrap a necessity. In: Huglen R (ed) *Light Metals 1997*, pp. 1111–14. Warrendale, PA: TMS-AIME; 1997.

73. Westengen H, Bakke P. Recycling of magnesium. In: Friedrich H, Mordike BL, eds, *Technology of magnesium and magnesium alloys*. Berlin: Springer-Verlag; 2006.

74. Waltrip JS. Magnesium scrap and residual magnesium: Trash or treasure. SAE Technical Paper, 900789. Warrendale, PA: SAE International; 1990.

75. Petrovich VW, Waltrip JS. Flux-free refining of magnesium die cast scrap. In: Campbell PG, ed., *Light Metals 1989*, pp. 749–55. Warrendale PA: TMS-AIME; 1989.

76. Berkmortel R, Wang GG, Bakke P. Fluxless in-house recycling of high purity magnesium die cast alloys at Meridian operations. 57th IMA Conference, Vancouver BC, Canada; 2000: pp. 22–7.

77. Leyndecker T, Rauch M. Economical process for the in-house remelting in magnesium die casting shops. In: Mordike BL, Kainer KU, eds, *Magnesium alloys and their applications*, pp. 703–7. Frankfurt, Germany: Werkstoff-Informationsgesellschaft mbh.; 1998.

78. Antrekowitsch H, Hanko G, Ebner P. Recycling process with salt. In: Kaplan HI, ed., *Magnesium Technology 2002*. Warrendale, PA: TMS; 2002.

79. Emley EF. *Principles of magnesium technology*, 1st edn, pp. 213–17. Oxford: Pergamon Press; 1966.

80. Wallevik O, Rønhaug JB. Method and apparatus for remelting and refining of magnesium and magnesium alloys. US Patent 5,167,700; 21 October 1991.

81. King JF, Hopkins A, Thistletwaite S. Recycling of by-products from magnesium die casting. In: Lorimer GW, ed., *Proceedings of the 3rd Intl. Magnesium Conference. London: Inst. of Materials*; 1996: 51–61.

82. Schroder D, Rauch E. Magnesiumschmelzofen und Verfahren zum Schmelzen von Magnesium. German Patent DE 44 39 214 A1; 3 November 1994.

83. Galowsky U, Kuhlein M. A new conti-process for the fluxless recycling of high purity magnesium. In: Hryn JN, ed., *Magnesium Technology 2001*, pp. 49–54. Warrendale, PA: TMS-AIME; 2001.

84. Øymo D, Holta O, Hustoft OM, Henriksson J. Magnesium recycling in the die casting shop. In: *Proc. of the 1992 conf. on the recycling of metals*, pp. 143–50. Düsseldorf/Neuss, Germany: ASM International; 1992.

85. Holta O, Hustoft OM, Strømhaug SI, Albright DL. Two furnace melting system for magnesium, transactions. In: 17th International Die Casting Congress and Exposition, 1993 NADCA 93, Cleveland, Ohio, USA, October, pp. 49–53. Rosemont, IL: NADCA; 1993.

86. Dioxines and PCBs. Environmental and health effects. Final study. Scientific and Technological Options Assessment (Ref: EP/IV/STOA/99/ENVI/02. European Parliament, Directorate General for Research, The STOA Programme; 1999.

87. Franke G. Industrial recycling of magnesium alloys. *Automotive Light Metals*, 2000; 1(2): 48–54.

88. Saxena SK. Method for treatment of oil contaminated filings of magensium and magnesium alloys. US Patent, 5,338,335; 16 August 1994.

89. Areaux LD, Dudley RH. Method and apparatus for cleaning and drying metal chips. US Patent 4,721,457; 26 January 1998.

90. Zhu T, Li N, Mei X, Yu A, Shang S. Innovative vacuum distillation for magnesium recycling. In: Hryn JN, ed., *Magnesium Technology 2001*, pp. 55–60. Warrendale, PA: TMS-AIME; 2001.

91. Emley EF. *Principles of magnesium technology*, 1st edn, pp. 52–7. Oxford: Pergamon Press; 1966.

92. Ditze A, Scharf C. Utilization of residues from fluxless remelting of compact magnesium scrap. In: Kainer KU, ed., *Magnesium alloys and their applications*, pp. 753–60. Weinheim, Germany: Wiley–VCH Verlag; 2000.

93. Gribov VI, Belkin NA, Shaibakov TI, Belkin MI. Material balance for the processing of magnesium scrap and waste in a liquid salt bath. *Tsvetnye metally*, 1988; 63–5.

94. Barannik IA, Devyatkin VN. Magnesium alloys usage in the USSR: secondary metal processing. In: Mordike BL, Hehmann F, eds, *Magnesium alloys and their applications*, pp. 43–9. Oberursel, Germany: DGM Informationsgesellschaft; 1992.

95. Beck A. *The technology of magnesium and its alloys*, pp. 313–19. London: F.A. Hughes; 1940. [English translation from German]

96. Emley EF. *Principles of magnesium technology*, 1st edn, pp. 94–125. Oxford: Pergamon Press; 1966.

97. Komura A, Imanga H, Watanabe N. Dispersion of Mg in the electrolysis of $MgCl_2$-KCl melts. *Denki Kagaku*, 1973; 41(4): 281–6.

98. Frisvold F, Bakke P, Øvrelid E, Hald NE, Engh TA. Hydrogen and inclusion content in recycled aluminium at Holmestrand Rolling Mill. In: Hale W, ed., *Light Metals 1997*, pp. 1011–16. Warrendale, PA: TMS-AIME; 1997.

99. Reding JN. Apparatus and method for refining magnesium. US Patent 1,424,543; 1976.

100. Housh SE, Petrovich V. Magnesium refining: a fluxless alternative. SAE Technical Paper 920071. Warrendale, PA: SAE International; 1992.

101. Øvrelid E. Hydrogen in magnesium and magnesium alloys. Dr. Ing. Thesis. Trondheim, Norway: Institute of Metallurgy NTNU; 1997.

102. Bakke P, Karlsen DO. Inclusion assessment in magnesium and magnesium base alloys. SAE Technical Paper 970330. Warrendale, PA: SAE International; 1997.

103. Simensen CJ, Oberländer B. A survey of inclusions in magnesium. *Z. Prakt. Metallogr.*, 1980; 17(3): 125–36.

104. Kruszynski EA. Die lubricants: questions and answers. *Die Casting Engineer*, 1993; 37(1): 16–18.

105. Pinfold PMD, Øymo D. An evaluation of refined recycled AZ91D alloy. SAE Technical Paper 930420. Warrendale, PA: SAE International: 1993.

106. Bakke P, Sannes S, Albright DL. Effects of Ni, Cu, Si and Co on the corrosion properties of permanent mould cast medallions and die cast plates of magnesium alloy AZ91. SAE Technical Paper 1999-01-0926. Warrendale, PA: SAE International; 1999.

107. Tathgar HS, Bakke P, Øvrelid E, Fenstad J, Frisvold F, Engh TA. Solubility of nickel in molten magnesium-aluminium alloys above 650 °C. In: Kaplan HI, Hryn JN, Clow B, eds, *Magnesium Technology 2000*, pp. 181–9. Warrendale PA: TMS-AIME; 2000.

108. Foerster GS. Removal of nickel from molten magnesium metal. US Patent 3,869,281; 4 March 1975.

109. Emley EF. *Principles of magnesium technology*, 1st edn, p. 178. Oxford: Pergamon Press; 1966.

110. Inoue M, Iwai M, Kamado S, Kojima Y, Itoh T, Sugama M. A new recycling process for thin walled AZ91D magnesium alloy die castings with paint finishing. In: Mordike BL, Kainer KU, eds, *Magnesium alloys and their applications*, pp. 697–702. Frankfurt, Germany: Werkstoff-Informationsgesellschaft mbh.; 1998.

111. Bakken HE. Magnesium–aluminium alloy. US Patent 1,584,688; 1926.

112. Beck A, Schmidt W. Verfahren zur Entfernung von suspendierten Teilchen von Eisen und ahnlichen fein verteilte Verureinungen von nicht salzartigem Charakter aus Magnesium und hochprozentigen Magnesiumlegierungen. German Patent 604,580; 1929.

113. Holta O, Westengen H, Røen J. High purity magneisum die casting alloys. Impact of metallurgical principles on industrial practice. In: Lorimer GW, ed., *Proceedings of the 3rd Intl. Magnesium Conference*, pp. 75–88. London: Inst. of Materials; 1996.

114. Bakke P, Westengen H, Brassard C. Multipurpose die casting alloys of the AE family: production and recycling. In: Pekguleryuz MO, Mackenzie LH, eds, *Proc. Magnesium Technology in the Global Age*, pp. 723–36. Montreal, Quebec. Canada: CIM; 2006.

115. Höllrigl-Rosta F, Just E, Köhler J, Melzer HJ. Magnesium in Volkswagen. *37th IMA Conference*; 1980: pp. 38–45.

116. Hanawalt JD, Nelson CE, Peloubet JA. Corrosion studies of magnesium and its alloys. *Trans. AIME*, 1942; 147: 273–99.

117. Spiegelberg W, Ali S, Dunstone S. The effects of beryllium additions on magnesium and magnesium containing alloys. In: Mordike BL, Hehmann F, eds, *Magnesium alloys and their applications*, pp. 259–66. Oberursel, Germany: DGM Informationsgesellschaft; 1992.

118. Smith N, Kvithyld A, Tranell G. The mechanism behind the oxidation protection of high Mg Al alloys with beryllium. *Metall. Trans. B*, 2018; 913–19.

119. Johansson A, et al. Small scale oxidation experiments on AlMg alloys in various gas fired furnace atmospheres. *Light Metals*, 2020; 923–9.

120. Lodin J, Syvertsen M, Kvithyld A, Johansson A, Solberg E, Kvande T. Study of the oxidation of an Al-5 Mg alloy in various industrial melting furnace atmospheres. *Light Metals*, 2020; 930–5.

121. Schlesinger M.E. *Aluminum recycling*, pp. 227–45. Boca Raton, FL: CRC Press; 2014.

122. Peterson RD. A historical perspective on dross processing. *Mater. Sci. Forum*, 2011; 693: 13–23.

123. Sheth AC, Parks KD, Parthasarathy S. Recycling salt cake slag using a resin-based option. *J. Metals (JOM)*, 1996; 48(8): 32–7.

124. Müller T, Friedrich B. Development of a recycling process for nickel-metal hydride batteries. *J. Power Sources*, 2006; 158(2): 1498–1509.

125. Bernardes AM, Espinosa DCR, Tenório JAS. Recycling of batteries: a review of current processes and technologies. *J. Power Sources*, 2004; 130: 291–8.

126. Al-Thyabat S, Nakamura T, Shibata E, Iizuka A. Adaptation of minerals processing operations for lithium-ion (LiBs) and nickel metal hydride (NiMH) batteries recycling: Critical review. *Miner. Eng.*, 2013; 45: 4–17.

127. Bertuol DA, et al. Spent NiMH batteries: Characterization and metal recovery through mechanical processing. *J. Power Sources*, 2006; 160: 1465–70.

128. Ebin B, Petranikova M, Ekberg C. Physical separation, mechanical enrichment and recycling-oriented characterization of spent NiMH batteries. *J. Mater. Cycles Waste Manage.*, 2018; 20: 2018–27.

129. Tzanetakis N, Scott K. Recycling of nickel–metal hydride batteries. I: Dissolution and solvent extraction of metals. *J. Chem. Technol. Biotechnol.*, 2004; 79(9): 919–26.

130. Porvali A, Ojanen S, Wilson BP, Serna-Guerrero R, Lundström, M. Nickel metal hydride battery waste: mechano-hydrometallurgical experimental study on recycling aspects. *J. Sustain. Metall.*, 2020; 6: 78–90.

131. Elwert T, Römer F, Schneider K, Hua Q, Buchert M. Recycling of batteries from electric vehicles. In: Pistoia G, Liaw B, eds., *Behaviour of lithium-ion batteries in electric vehicles: battery health, performance, safety, and cost (green energy and technology)*, pp. 289–322. Berlin: Springer International; 2018.

132. Velázquez-Martínez O, Valio J, Santasalo-Aarnio A, Reuter M, Serna-Guerrero R. A critical review of lithium-ion battery recycling processes from a circular economy perspective. *Batteries*, 2019; 5(4): 68.

133. Fisher K, Wallén E, Laenen PP, Collins M. Battery waste management life cycle assessment. Final Report for Publication, 18 October 2006. Prepared by Environmental Resources Management.

134. Lu X, Miki T, Takeda O, Zhu H, Nagasaka T. Thermodynamic criteria of the end-of-life silicon wafers refining for closing the recycling loop of photovoltaic panels. *Sci. Tech. Adv. Mater.*, 2019; 20: 813–25.

135. Graedel TE, et al. Recycling rates of metals—a status report. A report of the Working Group on the Global Metal Flows to the International Resource Panel. United Nations Environment Program (UNEP); 2011.

136. Megy JA. Refractory metal SWARF composition, U.S. Patent 5,171,359, 15 December 1992; also Refractory metal SWARF composition and method of making same, U.S. Patent 5,597,401, 28 January 1997.

Appendix 1
Selected Standard Gibbs Energies

Standard Gibbs energies, in joule/mole, are given in Table A1.1 as

$$\Delta G^0 = \overline{\Delta H^0} - T\overline{\Delta S^0},$$

where $\overline{\Delta H^0}$ and $\overline{\Delta S^0}$ are average values for the stated range of temperatures. The error arising from the equation is usually less than the error in the tabulated data. This allows a very compact presentation of thermodynamic data. The references (which are listed at the end of Chapter 2) and the tables are to be regarded only as a general guide to the reader. For guidance graphs of ΔG^0 available in the literature[107] may also be useful. When working on a specialized project the original literature should be consulted and a database employed.

The data given here are primarily based on a compilation by Turkdogan,[8] data from the JANAF tables,[11] and thermodynamic data by Kubaschewski and Alcock,[99] Hultgren et al.,[13] and Barin and Knacke.[129,130]

Compounds and elements are given in the solid state, except for O_2, N_2, CO, CO_2, Cl_2, F_2, H_2, S_2, P_2, and SO_2 which are given in the gaseous state. If a compound that is normally in the solid state is liquid or gaseous, an l or g indicates this, respectively. In the temperature range column, m, b, or s indicate the melting point, boiling-point, or sublimation point, respectively; d signifies decomposition. Estimates of errors in the data are not presented here, but are usually found in the references.

Nowadays one can calculate thermodynamic functions using a computer program and a database. Estimating and predicting thermodynamic values have been simplified even though the system is described by very sophisticated equations and has uncertain factors based on the choice of the solution mode! Typical thermodynamic databases are Pandat, Factsage, and Thermo-Calc.

Table A1.1

Reaction	$\overline{\Delta H^0}$ (joule)	$\overline{\Delta S^0}$ (joule/K)	Range (°C)	Reference
$Ag = Ag\,(l)$	11300	9.16	961 m	13
$Ag\,(l) = Ag\,(g)$	258300	106.25	961–2163 b	13
$AgBr = AgBr\,(l)$	9200	13.10	430 m	129
$AgBr\,(l) = AgBr\,(g)$	192000	104.6	1560 b	129
$AgBr\,(l) = Ag\,(l) + \frac{1}{2}Br_2$	99010	27.8	961–1560 b	129
$AgCl = AgCl\,(l)$	13000	17.83	455 m	129
$AgCl\,(l) = AgCl\,(g)$	177800	96.65	1564 b	129
$AgCl\,(l) = Ag\,(l) + \frac{1}{2}Cl_2$	104470	23.0	961–1564 b	129
$AgF = Ag + \frac{1}{2}F_2$	200800	54.8	25–435 m	129
$AgI = AgI\,(l)$	9410	11.3	558 m	129
$AgI\,(l) = AgI\,(g)$	144300	81.2	1505 b	129
$AgI\,(l) = Ag\,(l) + \frac{1}{2}I_2\,(g)$	77100	22.8	961–1505 b	129
$Ag_2O = 2Ag + \frac{1}{2}O_2$	30540	66.1	25	99
$Ag_2S = Ag_2S\,(l)$	7870	7.1	830 m	130
$Ag_2S = 2\,Ag\,(s) + \frac{1}{2}S_2$	161300	168.6	25–830 m	130
$Al = Al\,(l)$	10800	11.55	660 m	13
$Al\,(l) = Al\,(g)$	304600	109.5	660–2520 b	13
$AlCl\,(g) = Al\,(l) + \frac{1}{2}Cl_2$	77500	−58.16	660–2000	11
$Al_2Cl_6\,(g) = 2\,Al + 3\,Cl_2$	1292000	242.2	25–660	11
$AlF\,(g) = Al\,(l) + \frac{1}{2}F_2$	292000	−54.2	660–2000	11
$AlF_3 = Al\,(l) + \frac{3}{2}F_2$	1508000	257.9	660–1276 s	11
$AlF_3\,(g) = Al\,(l) + \frac{3}{2}F_2$	123000	79.37	660–2000	11
$AlB_{12} = Al\,(l) + 12B$	220000	7.45	660–2200 m	130
$Al_4C_3 = 4Al\,(l) + 3C$	265000	95.1	660–2200 m	11
$AlN = Al\,(l) + \frac{1}{2}N_2$	327000	115.5	660–2000	11
$AlP = Al\,(l) + \frac{1}{2}P_2$	249500	104.4	660–1700	130
$Al_2O_3 = 2\,Al\,(l) + \frac{3}{2}O_2$	1687300	327	660–2054 m	11
$Al_2O\,(g) = 2\,Al\,(l) + \frac{1}{2}O_2$	170700	−49.4	660–2000	11
$Al_2O_2\,(g) = 2\,Al\,(l) + O_2$	470700	28.8	660–2000	11
$AlPO_4 = Al + \frac{1}{2}P_2 + 2\,O_2$	1802000	450	25–660	130

Table A1.1 *Continued*

Reaction	$\overline{\Delta H^0}$ (joule)	$\overline{\Delta S^0}$ (joule/K)	Range (°C)	Reference
$3\ Al_2O_3 \cdot 2\ SiO_2 = 3\ Al_2O_3 + 2\ SiO_2$	-8600	-17.4	25–1750 m	11
$Al_2O_3 \cdot SiO_2 = Al_2O_3 + SiO_2$	8815	3.89	25–1700	11
$Al_2O_3 \cdot TiO_2 = Al_2O_3 + TiO_2$	25300	3.93	25–1860 m	130
$As = \frac{1}{4}As_4(g)$	36570	41.9	25–630 s	13
$Au = Au\,(l)$	12550	9.41	1064 m	13
$B = B\,(l)$	50210	21.8	2030 m	13
$B = B\,(g)$	570150	147.8	25–2030 m	13
$BCl_3\,(g) = B + \frac{3}{2}Cl_2$	404000	51.7	25–2030	11
$BF_3\,(g) = B + \frac{3}{2}F_2$	1140600	65.1	25–2030	11
$B_4C = 4\,B + C$	41500	5.56	25–2030	11
$BN = B + \frac{1}{2}N_2$	250000	87.61	25–2030	11
$B_2O_3 = B_2O_3(l)$	24060	33.3	450 m	11
$B_2O_3(l) = 2B\,(s) + \frac{3}{2}O_2$	1229000	210	450–2043 b	11
$BO(g) = B\,(s) + \frac{1}{2}O_2$	3800	-88.8	25–2030	11
$Ba = Ba\,(l)$	7740	7.74	727 m	12, 13
$Ba\,(l) = Ba\,(g)$	152000	70.4	729–1894 b	12, 13
$BaCl_2 = BaCl_2(l)$	15980	12.93	962 m	11
$BaCl_2(l) = Ba(l) + Cl_2$	811300	121.5	962–1622	11
$BaF_2 = BaF_2(l)$	23350	14.23	1368 m	11
$BaF_2(l) = Ba(l) + F_2$	1154000	129	1368–1622	11
$BaO = Ba(1) + \frac{1}{2}O_2$	557000	103	729–1622	11
$BaS = Ba(1) + \frac{1}{2}S_2$	544000	123.4	729–1622	130
$BaCO_3 = BaO\,(s) + CO_2$	250800	147.0	800–1060 m	131, 132
$Be = Be(1)$	11715	7.53	1287 m	11
$Be(1) = Be(g)$	300000	108.6	1287–2468 b	11
$BeCl_2\,(g) = Be(1) + Cl_2$	380000	-5.86	1287–2000	11
$BeF_2\,(g) = Be(1) + F_2$	819000	2.34	1287–2000	11
$Be_2C = 2Be + C$	93300	13.8	25–1287	11
$Be_2C = 2Be(1) + C$	115000	28.5	1287–2000	11
$Be_3N_2 = 3Be\,(s) + N_2$	587000	185.4	25–1287	11
$Be_3N_2 = 3Be\,(l) + N_2$	616000	203	1287–2200 m	11
$BeO = Be\,(s) + \frac{1}{2}O_2$	608000	97.7	25–1287	11

Table A1.1 *Continued*

Reaction	$\overline{\Delta H^0}$ (joule)	$\overline{\Delta S^0}$ (joule/K)	Range (°C)	Reference
$BeO = Be\,(l) + \frac{1}{2}O_2$	613600	100.9	1287–2000	11
$BeS = Be\,(s) + \frac{1}{2}S_2$	300000	86.6	25–1287	130
$BeO \cdot Al_2O_3 = BeO + Al_2O_3$	4180	–3	25–1870 m	11
$3BeO \cdot B_2O_3 = 3BeO + B_2O_3(1)$	58400	25.5	450–1495 m	11
$2BeO \cdot SiO_2 = 2BeO + SiO_2$	10900	7.1	25–1560 m	11
$BeSO_{4_\beta} = BeO + SO_2 + \frac{1}{2}O_2$	254400	232	25–1287	11
$Bi = Bi(1)$	11300	20.75	271 m	13
$Bi(1) = Bi\,(g)$	192000	99.4	272–1661 b	13
$BiCl_3 = BiCl_3(1)$	23640	46.6	234 m	130
$BiCl_3(1) = Bi(1) + \frac{3}{2}Cl_2$	351000	160.5	272–439 b	130
$BiF_3 = BiF_3(1)$	21590	23.4	649 m	130
$BiF_3(1) = BiF_3\,(g)$	125500	104.6	927 b	130
$BiF_3(1) = Bi(1) + \frac{3}{2}F_2$	853500	182.8	649–927 b	130
$BiO_3 = Bi_2O_3(1)$	59850	54.6	824 m	130
$BiO_3(1) = 2Bi(1) + \frac{3}{2}O_2$	445000	159.6	824–1500	130
$BiS_3 = Bi_2S_3(1)$	78200	74.5	777 m	130
$Bi_2S_3 = 2Bi(1) + \frac{3}{2}S_2$	360000	274	272–777 m	130
Graphite = Diamond	1440	–4.48	29–900	11
$C = C\,(g)$	713500	155.5	1750–3800 s	11
$CH_4\,(g) = C + 2H_2$	91040	110.7	500–2000	11
$CCl_4\,(g) = C + 2Cl_2$	89100	129.2	25–2000	11
$CF_4\,(g) = C + 2F_2$	933200	151.5	25–2000	11
$CN\,(g) = C + \frac{1}{2}N_2$	–433500	–99.6	25–2000	11
$CO = C + \frac{1}{2}O_2$	110540	–89.35	-150–500	12
$CO = C + \frac{1}{2}O_2$	114400	–85.77	500–2000	11, 12
$CO_2 = C + O_2$	395350	–0.54	500–2000	11, 12
$CO_2 = C + O_2$	393500	–2.88	-50–500	12
$COS\,(g) = C + \frac{1}{2}O_2 + \frac{1}{2}S_2$	202800	–9.96	500–2000	11, 12
$CS\,(g) = C + \frac{1}{2}S_2$	–163000	–88	25–2000	11
$CS_2\,(g) = C + S_2$	11400	–6.5	25–2000	11
$Ca = Ca(1)$	8536	7.7	842 m	11, 133
$Ca(1) = Ca\,(g)$	156800	89.5	842–1482 b	11, 12

Table A1.1 *Continued*

Reaction	$\overline{\Delta H^0}$ (joule)	$\overline{\Delta S^0}$ (joule/K)	Range (°C)	Reference
$CaBr_2 = CaBr_2(1)$	28870	28.45	742 m	11
$CaBr_2(1) = Ca(1) + Br_2$ (g)	670000	105	842–1482	11
$CaCl_2 = CaCl_2(1)$	28535	27.3	772 m	11
$CaCl_2(1) = CaCl_2$ (g)	235000	106.4	1936 b	11
$CaCl_2(1) = Ca(1) + Cl_2$	800000	146	842–1482	11
$CaF_2 = CaF_2(1)$	29700	17.6	1418 m	11
$CaF_2(1) = CaF_2$ (g)	308700	110	2533 b	11
$CaF_2 = Ca(1) + F_2$	1220000	162	842–1482	11
$CaC_2 = Ca(1) + 2C$	60250	−26.3	842–1482	129
$Ca_3N_2 = 3Ca + N_2$	43514	200	25–842	130
$CaAl_2 = CaAl_2(1)$	63200	46.9	1080 m	130
$CaAl_2(1) = Ca(1) + 2Al(1)$	189000	3.35	1080–1482	130
$CaAl_4 = CaAl_4(1)$	78700	75.0	777 m	130
$CaAl_4(1) = Ca(1) + 4Al(1)$	200000	3.0	842–1482	130
$Ca_2Si = 2Ca + Si$	210000	22.0	25–842	130
$CaSi = Ca + Si$	150000	15.5	25–842	130
$CaSi_2 = Ca + 2Si$	150000	28.5	25–842	130
$Ca_3P_2 = 3Ca + P_2$	650000	216	25–842	130
$CaO = CaO(1)$	79500	24.7	2927 m	11
$CaO = Ca(1) + \frac{1}{2}O_2$	640000	108.6	842–1482	11
$CaS = Ca(1) + \frac{1}{2}S_2$	548000	104	842–1482	11
$3CaO \cdot Al_2O_3 = 3CaO + Al_2O_3$	12600	−24.7	500–1535 m	134
$CaO \cdot Al_2O_3 = CaO + Al_2O_3$	18000	−19.0	500–1605 m	134
$CaO \cdot 2Al_2O_3 = CaO + 2Al_2O_3$	17000	−25.5	500–1750 m	134
$CaCO_3 = CaO + CO_2$	161300	137.2	700–1200	135
$2CaO \cdot Fe_2O_3 = 2CaO + Fe_2O_3$	53100	−2.5	700–1450 m	134
$CaO \cdot Fe_2O_3 = CaO + Fe_2O_3$	30000	−4.8	700–1216 m	134
$3CaO \cdot P_2O_5 = 3CaO + P_2 + \frac{5}{2}O_2$ (g)	2314000	556	25–1730 m	130
$2CaO \cdot P_2O_5 = 2CaO + P_2 + \frac{5}{2}O_2$	2190000	586	25–1353 m	130
$CaSO_{4_\beta} = CaO + SO_2 + \frac{1}{2}O_2$	460000	238	950–1195	136
$CaSO_{4_\alpha} = CaO + SO_2 + \frac{1}{2}O_2$	454000	232.0	1195–1365 m	136
$3CaO \cdot SiO_2 = 3CaO + SiO_2$	120000	−6.7	25–1500	130

Table A1.1 *Continued*

Reaction	$\overline{\Delta H^0}$ (joule)	$\overline{\Delta S^0}$ (joule/K)	Range (°C)	Reference
$2CaO \cdot SiO_2 = 2CaO + SiO_2$	120000	−11.3	25–2130 m	130
$3CaO \cdot 2SiO_2 = 3CaO + 2SiO_2$	237000	9.6	25–1500	130
$CaO \cdot SiO_2 = CaO + SiO_2(1)$	56000	33.0	1540 m	130
$CaO \cdot SiO_2 = CaO + SiO_2$	92500	2.5	25–1540 m	130
$3CaO \cdot 2TiO_2 = 3CaO + 2TiO_2$	207000	−11.5	25–1400	134
$4CaO \cdot 3TiO_2 = 4CaO + 3TiO_2$	293000	−17.6	25–1400	134
$CaO \cdot TiO_2 = CaO + TiO_2$	80000	3.35	25–1400	134
$3CaO \cdot V_2O_5 = 3CaO + V_2O_5$	332000	0.0	25–670	134
$2CaO \cdot V_2O_5 = 2CaO + V_2O_5$	265000	0.0	25–670	134
$CaO \cdot V_2O_5 = CaO + V_2O_5$	146000	0.0	25–670	134
$CaO \cdot ZrO_2 = CaO + ZrO_2$	39300	0.4	25–2000	130
$CaO \cdot MgO \cdot SiO_2 = CaO + MgO + SiO_2$	125000	4.0	25–1200	130
$2CaO \cdot MgO \cdot 2SiO_2 = 2CaO + MgO + 2SiO_2$	205000	0.0	25–1454 m	130
$CaO \cdot MgO \cdot 2SiO_2 = CaO \cdot MgO \cdot 2SiO_2(1)$	128500	77.1	1392 m	130
$CaO \cdot MgO \cdot 2SiO_2 = CaO + MgO\,(s) + 2SiO_2$	163000	18.8	25–1392 m	130
$3CaO \cdot Al_2O_3 \cdot 3SiO_2 = 3CaO + Al_2O_3 + 3SiO_2$	389000	100	25–1400	130
$2CaO \cdot Al_2O_3 \cdot SiO_2 = 2CaO + Al_2O_3 + SiO_2$	170000	8.8	25–1500	130
$CaO \cdot Al_2O_3 \cdot SiO_2 = CaO + Al_2O_3 + SiO_2$	106000	14.2	25–1400	130
$CaO \cdot Al_2O_3 \cdot 2SiO_2 = CaO + Al_2O_3 + 2SiO_2(1)$	167000	91.4	1553	130
$CaO \cdot Al_2O_3 \cdot 2SiO_2 = CaO + Al_2O_3 + 2SiO_2$	139000	17.2	25–1553	130
$CaO \cdot TiO_2 \cdot SiO_2 = CaO + TiO_2 + SiO_2$	122600	10.9	25–1400	130
$Cd = Cd(1)$	6190	10.4	321 m	12, 13
$Cd(1) = Cd\,(g)$	101700	97.9	321–766 b	12, 13
$CdBr_2 = CdBr_2(1)$	33300	39.7	568 m	130
$CdBr_2 = Cd(1) + Br_2\,(g)$	344000	151.3	321–568 m	130
$CdCl_2 = CdCl_2(1)$	30100	35.8	568 m	130
$CdCl_2 = Cd(1) + Cl_2$	390000	153	321–568 m	130
$CdI_2 = CdI_2(1)$	20700	31.2	388 m	130
$CdI_2(1) = Cd(1) + I_2\,(g)$	238000	107.0	388–744 b	130
$CdSO_4 = CdO + SO_2 + \frac{1}{2}O_2$	372000	277	25–1000	129
$CdO \cdot SiO_2 = CdO + SiO_2$	24600	4.7	25–1241 m	130
$CdO = Cd(1) + \frac{1}{2}O_2$	263000	105	321–767	130
$CdS = Cd(1) + \frac{1}{2}S_2$	215500	97.2	321–767	129

Table A1.1 *Continued*

Reaction	ΔH^0 (joule)	ΔS^0 (joule/K)	Range (°C)	Reference
$CdO \cdot Al_2O_3 = CdO + Al_2O_3$	−16700	−23.0	25–900	130
$Ce = Ce(1)$	5480	5.105	798 m	13
$Ce(1) = Ce(g)$	413700	111.8	798–3422 b	13
$CeH_2 = Ce + H_2$	208400	154	25–798	130
$CeCl_3 = CeCl_3(1)$	53550	49.1	817 m	130
$CeCl_3 = Ce + \frac{3}{2}Cl_2$	1050000	240	25–798	130
$CeF_3 = CeF_3(1)$	56500	33.05	1437 m	130
$CeF_3 = Ce + \frac{3}{2}F_2$	1770000	252	25–798	130
$CeB_6 = Ce(1) + 6B$	376000	55.2	798–2190 m	130
$Ce_2C_3 = 2Ce(1) + 3C$	188000	−14.5	798–1200	130
$CeC_2 = Ce(1) + 2C$	85230	−27.0	798–2250 m	137
$CeN = Ce(1) + \frac{1}{2}N_2$	489000	177.1	2000–2575	138
$CeAl_2 = CeAl_2(1)$	81600	47.0	1465 m	130
$CeAl_2 = Ce(1) + 2Al(1)$	213000	50.6	798–1465 m	130
$CeAl_4 = Ce + 4Al$	176000	21.3	25–660	130
$CeSi_2 = Ce(1) + 2Si$	200000	16.3	798–1427 m	130
$Ce_2O_3 = 2Ce + \frac{3}{2}O_2$	1788000	286.6	25–798	139
$CeO_{1.72} = Ce + 0.86\ O_2$	990000	170	25–798	130
$CeO_{1.83} = Ce + 0.915\ O_2$	1030000	182	25–798	130
$CeO_2 = Ce + O_2$	1084000	212	25–798	139
$CeS = Ce(1) + \frac{1}{2}S_2$	535000	90.96	798–2450 m	137
$Ce_2O_2S = 2Ce(1) + O_2 + \frac{1}{2}S_2$	1770000	333	798–1400	140
$Ce_2O_3 \cdot Al_2O_3 = Ce_2O_3 + Al_2O_3$	79500	−20.9	1100–1500	130
$Co = Co(1)$	16200	9.16	1495 m	11, 12
$Co(1) = Co(g)$	387000	121.3	1495–2925 b	11, 12
$CoCl_2 = CoCl_2(1)$	44800	44.2	740 m	11
$CoCl_2 = Co + Cl_2$	307000	130	25–740 m	11
$CoF_2 = Co + F_2$	686000	141.4	25–1127 m	11
$Co_2B = 2Co + B$	130000	10.9	25–1200	130
$CoB = Co + B$	96200	7.5	25–1460 m	130
$CoAl = CoAl(1)$	62760	32.7	1645 m	130
$CoAl = Co + Al(1)$	133900	29.7	660–1495	130
$Co_2Al_5 = 2Co + 5Al(1)$	364000	92	660–1170 m	130

Table A1.1 *Continued*

Reaction	$\overline{\Delta H^0}$ (joule)	$\overline{\Delta S^0}$ (joule/K)	Range (°C)	Reference
$CoAl_3 = Co + 3Al$	155000	8.4	25–660	130
$Co_2Al_9 = 2Co + 9Al$	318000	15.0	25-660	130
$Co_2P = 2Co + \frac{1}{2}P_2$	264000	98.0	25–1100	130
$CoP = Co + \frac{1}{2}P_2$	197000	86.6	25–1100	130
$CoP_3 = Co + \frac{3}{2}P_2$	406000	228	25–1100	130
$CoO = Co + \frac{1}{2}O_2$	246000	78.6	25–1495	11
$Co_3O_4 = 3Co + 2O_2$	957000	467	25–700	11
$Co_9S_8 = 9Co + 4S_2$	1326000	666	25–788	141
$Co_4S_3 = 4Co + \frac{3}{2}S_2$	388500	146	788–877	141
$CoS_2 = Co + S_2$	280000	182.0	25–600	130
$CoO \cdot Cr_2O_3 = CoO + Cr_2O_3$	59400	8.37	700–1400	134
$CoO \cdot Fe_2O_3 = CoO + Fe_2O_3$	24700	−6.5	500–1200	134
$CoSO_4 = CoO + SO_2 + \frac{1}{2}O_2$	335000	264	25–600	130
$2CoO \cdot SiO_2 = 2CoO \cdot SiO_2(1)$	100400	59.4	1417 m	130
$2CoO \cdot SiO_2 = 2CoO + SiO_2$	28000	4.2	25–1417 m	130
$2CoO \cdot TiO_2 = 2CoO + TiO_2$	22260	−1.1	25–1575 m	130
$CoO \cdot TiO_2 = CoO + TiO_2$	24700	6.3	500–1400	134
$Cr = Cr(1)$	16950	7.95	1857 m	11, 12
$Cr(1) = Cr(g)$	348500	118.4	1857–2669 b	11
$CrCl_2 = CrCl_2(1)$	32200	29.6	815 m	130
$CrCl_2 = Cr + Cl_2$	389000	119.7	25–815 m	130
$CrCl_3 = Cr + \frac{3}{2}Cl_2$	548000	215.5	25–945 s	130
$CrF_2 = Cr + F_2$	774000	134	25–1100	130
$CrF_3 = Cr + \frac{3}{2}F_2$	1110000	225.5	25–1100	130
$CrB = Cr + B$	75300	6.7	25–1000	130
$CrB_2 = Cr + 2B$	92050	8.4	25–1000m	130
$Cr_4C = 4Cr + C$	96200	−11.7	25–1520 m	129
$Cr_{23}C_6 = 23Cr + 6C$	310000	−77.4	25–1500	11
$Cr_7C_3 = 7Cr + 3C$	153500	−37.2	25–1857	11
$Cr_3C_2 = 3Cr + 2C$	79100	−17.6	25–1857	11
$Cr_2N = 2Cr + \frac{1}{2}N_2$	99200	47.0	1000–1450	11, 142
$CrN = Cr + \frac{1}{2}N_2$	113400	73.2	25–500	11, 142

Table A1.1 *Continued*

Reaction	$\overline{\Delta H^0}$ (joule)	$\overline{\Delta S^0}$ (joule/K)	Range (°C)	Reference
$Cr_3Si = 3Cr + Si$	106300	3.4	25–1412	130
$Cr_5Si_3 = 5Cr + 3Si$	219200	−15.6	25–1412	130
$CrSi = Cr + Si$	53560	−3.9	25–1412	130
$CrSi_2 = Cr + Si$	79500	1.97	25–1412	130
$Cr_2O_3 = 2Cr + \frac{3}{2}O_2$	1110000	247	900–1650	11, 143
$Cr_3O_4 = 3Cr + 2O_2$	1355000	265	1650–1665 m	144
$CrO(1) = Cr + \frac{1}{2}O_2$	334000	63.8	1665–1750	144
$CrS = Cr + \frac{1}{2}S_2$	202500	56.0	1100–1300	145
$Cr_2(SO_4)_3 = Cr_2O_3 + 3SO_2 + \frac{3}{2}O_2$	892000	820	600–800	146
$Cs = Cs(1)$	2092	6.95	28.5 m	11
$Cs(1) = Cs(g)$	71700	76.2	285–675 b	11
$Cu = Cu(1)$	13050	9.62	1085 m	11
$Cu(1) = Cu(g)$	308200	108.9	1085–2570 b	11
$CuCl(g) = Cu(1) + \frac{1}{2}Cl_2$	−63300	−66.8	1085–2000	11
$Cu_3Cl_3(g) = 3Cu(1) + \frac{3}{2}Cl_2$	314000	47.0	1085–2000	11
$CuF(g) = Cu(1) + \frac{1}{2}F_2$	−20900	−66.1	1085–2000	11
$Cu_3P = 3Cu + \frac{1}{2}P_2$	188300	84.1	25–800	130
$CuP_2 = Cu + P_2$	238500	155	25–1085	130
$Cu_2O = Cu_2O(1)$	56820	37.66	1244 m	11
$Cu_2O = 2Cu + \frac{1}{2}O_2$	168400	71.25	25–1085	11
$Cu_2O(1) = 2Cu(1) + \frac{1}{2}O_2$	119000	39.5	1236–2000	11
$CuO = Cu + \frac{1}{2}O_2$	152300	85.4	25–1085	11
$Cu_2S_\gamma = 2Cu + \frac{1}{2}S_2$	140700	43.3	25–435	147
$Cu_2S_\mu = 2Cu + \frac{1}{2}S_2$	132000	30.8	435–1129 m	147
$Cu_2S = Cu_2S(1)$	9000	6.40	1129 m	147
$Cu_{1.738}S_\mu = 1.738Cu + \frac{1}{2}S_2$	113600	26.6	435–620	147
$CuS = Cu + \frac{1}{2}S_2$	115600	76.0	25–430	148
$CuFeS_2 = Cu + Fe + S_2$	278600	115	557–700	149
$Cu_2O \cdot Fe_2O_3 = Cu_2O + Fe_2O_3$	37700	19.0	25–1100	130
$CuSO_4 = \frac{1}{2}CuO \cdot CuSO_4 + \frac{1}{2}SO_2 + \frac{1}{4}O_2$	152600	136	400–800	150
$CuO \cdot CuSO_4 = 2CuO + SO_2 + \frac{1}{2}O_2$	297000	250	500–900	150
$Fe_\delta = Fe(1)$	13800	7.61	1536 m	11
$Fe(1) = Fe(g)$	363600	116.0	1536–2860 b	11

Table A1.1 *Continued*

Reaction	$\overline{\Delta H^0}$ (joule)	$\overline{\Delta S^0}$ (joule/K)	Range (°C)	Reference
$FeCl_2 = FeCl_2(1)$	43100	45.3	677 m	11
$FeCl_2(1) = FeCl_2 (g)$	110000	84.7	1074 b	11
$FeCl_2 (g) = Fe + Cl_2$	167000	−25.1	1074–2000	11
$FeCl_3 (g) = Fe + \frac{3}{2}Cl_2$	260000	25.4	332–2000	11
$Fe_2Cl_6 (g) = 2Fe + 3Cl_2$	650000	181	332–2000	11
$FeF_2 = Fe + Fe_2$	700000	134	25–1100 m	11
$Fe_2B = 2Fe + B$	87900	18.4	25–1390 m	130
$FeB = Fe + B$	79500	10.5	25–1650 m	130
$Fe_3C = 3Fe_\alpha + C$	−29000	−28.0	200–727	151
$Fe_3C = 3Fe_\gamma + C$	−11230	−11.0	727–1137	151
$Fe_4N = 4Fe + \frac{1}{2}N_2$	33500	70.0	400–680	152
$Fe_{0.947}O = Fe_{0.947}O(1)$	31340	19.0	1370 m	11
$Fe_{0.947}O = 0.947Fe + \frac{1}{2}O_2$	263700	64.4	25–1370 m	11
$FeO(1) = Fe(1) + \frac{1}{2}O$	256000	54.7	1371–2000	11
$Fe_3O_4 = 3Fe + 2O_2$	1100000	307	25–1597 m	11
$Fe_2O_3 = 2Fe + \frac{3}{2}O_2$	814000	251	25–1500	11
$FeS = Fe_\gamma + \frac{1}{2}S_2$	164000	61.0	988–1190 m	153
$FeS_2 = FeS + \frac{1}{2}S_2$	182000	188	630–760	153
$FeO \cdot Cr_2O_3 = Fe + \frac{1}{2}O_2 + Cr_2O_3$	317000	72.6	750–1536	154
$FeO \cdot Cr_2O_3 = Fe(1) + \frac{1}{2}O_2 + Cr_2O_3$	331000	80.3	1536–1700	154
$Fe_2(SO_4)_3 = Fe_2O_3 + 3SO_2 + \frac{3}{2}O_2$	772000	724	400–800	155
$Fe_2(SO_4)_3 = 2FeSO_4 + SO_2 + O_2$	396000	352	430–630	155
$FeSO_4 = \frac{1}{2}Fe_2O_3 + SO_2 + \frac{1}{2}O_2$	203500	202.3	500–630	155
$2FeO \cdot SiO_2 = 2FeO \cdot SiO_2(1)$	92050	61.7	1220 m	130
$2FeO \cdot SiO_2 = 2FeO + SiO_2$	36200	21.0	25–1220 m	130
$2FeO \cdot TiO_2 = 2FeO + TiO_2$	33900	5.86	25–1100	134
$FeO \cdot TiO_2 = FeO + TiO_2$	33500	12.1	25–1300	134
$FeO \cdot V_2O_3 = FeV_2O_3 + \frac{1}{2}O_2$	289000	62.3	750–1536	154
$FeO \cdot V_2O_3 = Fe(1) + V_2O_3 + \frac{1}{2}O_2$	301000	70.0	1536–1700	154
$FeO \cdot WO_3 = FeO + WO_3$	55000	−10.0	700–1150	134
$Ga = Ga(1)$	5610	17.2	30 m	11
$Ga(1) = Ga (g)$	264000	106.9	30–2204 b	11

Table A1.1 *Continued*

Reaction	ΔH^0 (joule)	ΔS^0 (joule/K)	Range (°C)	Reference
$Ga_2O_3 = 2Ga(1) + \frac{3}{2}O_2$	1090000	324	30–1795 m	130
$GaS = Ga(1) + \frac{1}{2}S_2$	276000	111	30–960 m	130
$Ga_2S_3 = 2Ga(1) + \frac{3}{2}S_2$	720000	318	30–1090 m	130
$Ge = Ge(1)$	37000	30.5	937 m	13
$Ge(1) = Ge(g)$	335000	108	937–2834 b	13
$GeO_2 = Ge + O_2$	575000	188	25–937	130
$HCl(g) = \frac{1}{2}H_2 + \frac{1}{2}Cl_2$	94100	−6.40	25–2000	11
$HF(g) = \frac{1}{2}H_2 + \frac{1}{2}F_2$	274500	−3.47	25–2000	11
$H_2O(1) = H_2O(g)$	41090	110.1	100 b	11
$H_2O(g) = H_2 + \frac{1}{2}O_2$	247500	55.86	25–2000	11
$H_2S(g) = H_2 + \frac{1}{2}S_2$	91630	50.6	25–2000	11
$Hf = Hf(1)$	24300	9.6	2227 m	11
$Hf(1) = Hf(g)$	574200	117.7	2227–4598 b	11
$HfB_2 = Hf + 2B$	335000	11.1	25–1200	130
$HfC = Hf + C$	230000	7.5	25–2000	129
$HfO_{2_\alpha} = Hf + O_2$	1060000	174	25–1700	129
$HfO_{2_\alpha} = HfO_{2_\beta}$	10500	5.3	1700	129
$Hg(1) = Hg(g)$	60750	96.78	-39–356 b	11
$HgO_{red} = Hg(1) + \frac{1}{2}O_2$	90800	70.3	25–500	11, 99
$HgS_{red} = Hg(1) + \frac{1}{2}S_2$	53000	82.0	25–345	99
$HgS_{red} = HgS_{black}$	4000	6.3	345	99
$In = In(1)$	3260	7.62	157 m	13
$In(1) = In(g)$	235000	100.3	157–2073 b	13
$In_2O_3 = 2In(1) + \frac{3}{2}O_2$	919000	309.4	157–1910 m	130
$InS = InS(1)$	36000	37.3	692 m	130
$InS(1) = In(1) = \frac{1}{2}S_2$	155000	60.7	692–1500	130
$In_2S_3 = 2In(1) + \frac{3}{2}S_2$	544000	286	157–900	130
$Ir = Ir(1)$	26150	9.62	2443 m	12, 13
$Ir(1) = Ir(g)$	615000	131	2443–4424 b	12, 13
$IrO_3(g) = Ir + \frac{3}{2}O_2$	−16700	45.2	25–1500	130
$IrO_2 = Ir + O_2$	234000	170	25–1000	129
$Ir_2S_3 = 2Ir + \frac{3}{2}S_2$	430000	300	25–1000	130
$IrS_2 = Ir + S_2$	268000	190	25–1000	130

Table A1.1 *Continued*

Reaction	ΔH^0 (joule)	ΔS^0 (joule/K)	Range (°C)	Reference
$K = K(1)$	2335	6.95	63 m	11
$K(1) = K(g)$	84500	82.0	63–766 b	11
$KCl = KCl(1)$	26300	25.2	771 m	11
$KCl(1) = K(1) + \frac{1}{2}Cl_2$	474000	132	771–1437 b	11
$KCl(g) = K(g) + \frac{1}{2}Cl_2$	306000	35.5	1437–2000	11
$KF = KF(1)$	27200	24.06	858 m	11
$KF(1) = K(g) + \frac{1}{2}F_2$	607500	142	858–1517 b	11
$KF(g) = K(g) + \frac{1}{2}F_2$	418400	37.7	1517–2000	11
$La = La(1)$	6200	5.19	920 m	13
$La(1) = La(g)$	416000	113	920–3460 b	13
$LaCl_3 = LaCl_3(1)$	54400	48.2	855 m	129
$LaCl_3 = La + \frac{3}{2}Cl_2$	1060000	225.0	25–855	129
$LaF_3 = LaF_3(1)$	50200	28.5	1493 m	129
$LaF_3 = La(1) + \frac{1}{2}F_2$	1780000	241	25–920	130
$LaN = La + \frac{1}{2}N_2$	300000	106	25–920	130
$La_2O_3 = 2La + \frac{3}{2}O_2$	1790000	278	25–920	130
$LaS = La(1) + \frac{1}{2}S_2$	527000	104	920–1500	130
$La_2S_3 = 2La + \frac{3}{2}S_2$	1420000	286	920–1500	130
$Li = Li(1)$	3000	6.61	181 m	11
$Li(1) = Li(g)$	151300	93.85	181–1347 b	11
$LiCl = LiCl(1)$	19800	23.0	610 m	11
$LiCl(1) = Li(1) + \frac{1}{2}Cl_2$	382000	52.4	610–1383 b	11
$LiCl(g) = Li(g) + \frac{1}{2}Cl_2$	360000	42.2	1383–2000	11
$LiF = LiF(1)$	27070	24.14	848 m	11
$LiF(1) = Li(1) + \frac{1}{2}F_2$	583400	66.8	848–1342	11
$LiF(g) = Li(g) + \frac{1}{2}F_2$	500000	43.0	1717 b–2000	11
$Li_3AlF_6 = Li_3AlF_6(1)$	86200	81.4	785 m	11
$Li_3AlF_6(1) = 3Li(1) + Al(1) + 3F_2$	3240000	400	785–1342 m	11
$Li_3N = 3Li + \frac{1}{2}N_2$	200000	160	181–1000	11
$Li_2O = Li_2O(1)$	58600	31.8	1570 m	11
$Li_2O = 2Li(1) + \frac{1}{2}O_2$	603000	135	181–1342	11
$Li_2S = 2Li(1) + \frac{1}{2}S_2$	515000	121	181–1000	130

Table A1.1 *Continued*

Reaction	ΔH^0 (joule)	ΔS^0 (joule/K)	Range (°C)	Reference
$LiOH = LiOH(1)$	21000	28.2	471 m	11
$LiOH(1) = \frac{1}{2}Li_2O + \frac{1}{2}H_2O(g)$	28000	21.8	471–1007 d	11
$Li_2O \cdot Al_2O_3 = Li_2O \cdot Al_2O_3(1)$	25000	13.3	1610 m	134
$Li_2O \cdot Al_2O_3 = Li_2O + Al_2O_3$	107000	−10.6	500–1300	134
$Li_2CO_3 = Li_2CO_3(1)$	44800	45.1	720 m	11
$Li_2CO_3(1) = Li_2O + CO_2$	148000	78.7	720–1570	11
$Li_2O \cdot Fe_2O_3 = Li_2O + Fe_2O_3$	40000	−25.3	25–700	134
$Li_2O \cdot SiO_2 = Li_2O \cdot SiO_2(1)$	28000	19.0	1201 m	11
$Li_2O \cdot SiO_2 = Li_2O + SiO_2$	144000	3.8	25–1201 m	11
$Li_2O \cdot 2SiO_2 = Li_2O \cdot 2SiO_2(1)$	54000	41.2	1034 m	11
$Li_2O \cdot 2SiO_2 = Li_2O + 2SiO_2$	146400	0.0	25–1034 m	11
$Li_2O \cdot TiO_2 = Li_2O + TiO_2$	130000	−3.3	25–900	134
$Mg = Mg(1)$	8960	9.71	650 m	11, 12
$Mg(1) = Mg(g)$	129600	95.1	650–1093 b	11
$MgCl_2 = MgCl_2(1)$	43100	43.7	714 m	11
$MgCl_2 = Mg + Cl_2$	637000	155.4	25–650	11
$MgCl_2(1) = Mg(g) + Cl_2$	649000	157.7	714–1437 b	11
$MgF_2 = MgF_2(1)$	58700	38.2	1263 m	11
$MgF_2 = Mg + F_2$	1120000	171.2	25–650	11
$MgF_2(1) = Mg(g) + F_2$	1172000	215.6	1263–2263 b	11
$MgB_2 = Mg + 2B$	92000	10.5	25–650	11
$MgB_4 = Mg + 4B$	108000	11.2	25–650	11
$Mg_3N_2 = 3Mg + N_2$	464000	203	25–650	11
$Mg_2Si = Mg_2Si(1)$	85800	62.5	1102 m	11
$Mg_2Si = 2Mg(1) + Si$	100000	39.3	650–1090	11
$MgO = Mg + \frac{1}{2}O_2$	601000	107.6	25–650	11
$MgO = Mg(g) + \frac{1}{2}O_2$	730000	204	1090–2000	11
$MgS = Mg + \frac{1}{2}S_2$	410000	94.4	25–650	11
$MgS = Mg(g) + \frac{1}{2}S_2$	540000	193	1090–1700	11
$MgO \cdot Al_2O_3 = MgO + Al_2O_3$	35600	−2.09	25–1400	11, 134
$MgCO_3 = MgO + CO_2$	116000	173.4	25–402 d	11
$MgO \cdot Cr_2O_3 = MgO + Cr_2O_3$	42900	7.1	25–1500	134
$MgO \cdot Fe_2O_3 = MgO + Fe_2O_3$	19250	−2.1	700–1400	134

Table A1.1 *Continued*

Reaction	$\overline{\Delta H^0}$ (joule)	$\overline{\Delta S^0}$ (joule/K)	Range (°C)	Reference
$MgO \cdot MoO_2 = MgO + MoO_2$	−13800	−22.6	900–1200	134
$MgO \cdot MoO_3 = MgO + MoO_3$	54000	−13.6	25–795	134
$MgSO_4 = MgO + SO_2 + \frac{1}{2}O_2$	371000	261	670–1050	11, 155
$2MgO \cdot SiO_2 = 2MgO \cdot SiO_2(1)$	71100	32.6	1898 m	11
$2MgO \cdot SiO_2 = 2MgO + SiO_2$	67200	4.31	25–1898 m	11
$MgO \cdot SiO_2 = MgO \cdot SiO_2(1)$	75300	40.6	1577 m	11
$MgO \cdot SiO_2 = MgO + SiO_2$	41100	6.1	25–1577 m	11
$3MgO \cdot P_2O_5 = 3MgO \cdot P_2O_5(1)$	121300	75	1348 m	11
$3MgO \cdot P_2O_5 = 3MgO + P_2 + \frac{5}{2}O_2$	2000000	510	25–1348 m	11
$2MgO \cdot TiO_2 = 2MgO + TiO_2$	26360	1.26	25–1500	11
$MgO \cdot TiO_2 = MgO + TiO_2$	27600	3.14	25–1500	11
$2MgO \cdot V_2O_5 = 2MgO + V_2O_5$	87500	0	25–670	134
$MgO \cdot V_2O_5 = MgO + V_2O_5$	41800	8.4	25–670	134
$MgO \cdot WO_3 = MgO + WO_3$	74000	10.9	25–1200	11, 134
$Mn = Mn(1)$	12100	7.95	1246 m	11
$Mn(1) = Mn(g)$	235800	101.2	1246–2061 b	11
$MnCl_2 = MnCl_2(1)$	37700	40.8	650 m	129
$MnCl_2 = Mn + Cl_2$	478000	128	25–650 m	129
$MnCl_2(g) = Mn(1) + Cl_2$	302000	9.8	1244–2062	129
$Mn_3C = 3Mn(1) + C$	14000	−1.1	25–1037 m	130
$Mn_7C_3 = 7Mn + 3C$	128000	21.0	25–1200	130
$Mn_5Si_3 = Mn_5Si_3(1)$	173000	110	1300 m	130
$MnSi = Mn + Si$	61500	6.3	25–1246	130
$MnP = Mn + \frac{1}{2}P_2$	167000	87.0	25–1147 m	130
$MnP_3 = Mn + \frac{3}{2}P_2$	381000	240	25–1246	130
$MnO = Mn + \frac{1}{2}O_2$	390000	76.3	25–1246	156
$Mn_3O_4 = 3Mn + 2O_2$	1385000	344	25–1246	129
$Mn_3O_4 = 3MnO + \frac{1}{2}O_2$	232000	117	925–1540 m	157
$Mn_2O_3 = 2Mn + \frac{3}{2}O_2$	954000	255	25–1246	129
$Mn_2O_3 = \frac{2}{3}(Mn_3O_4) + \frac{1}{6}O_2$	35000	28.1	800–1000	158
$MnO_2 = Mn + O_2$	519000	181	25–510 d	156
$MnS = MnS(1)$	26100	14.5	1530 m	130

Table A1.1 *Continued*

Reaction	$\overline{\Delta H^0}$ (joule)	$\overline{\Delta S^0}$ (joule/K)	Range (°C)	Reference
$MnS = Mn + \frac{1}{2}S_2$	296500	76.74	700–1200	159
$MnO \cdot Al_2O_3 = Mn + Al_2O_3$	48100	7.32	500–1200	134
$MnSO_{4_\alpha} = \frac{1}{3}Mn_3O_4 + SO_2 + \frac{1}{3}O_2$	295600	227.4	700–849	155
$MnSO_{4_\beta} = \frac{1}{3}Mn_3O_4 + SO_2 + \frac{1}{3}O_2$	250200	187.2	849–1030	155
$2MnO \cdot SiO_2 = 2MnO \cdot SiO_2(1)$	89600	55.4	1345 m	130
$2MnO \cdot SiO_2 = 2MnO + SiO_2$	53600	25.0	25–1345 m	130
$MnO \cdot SiO_2 = MnO \cdot SiO_2(1)$	67000	42.8	1291 m	130
$MnO \cdot SiO_2 = MnO + SiO_2$	28000	2.76	25–1291 m	130
$Mo = Mo(1)$	27800	9.6	2624 m	11, 12
$Mo(1) = Mo(g)$	591000	120	2624–4705 b	11
$Mo_2C = 2Mo + C$	45600	−4.2	25–1100	129
$MoC = Mo + C$	7530	−5.4	25–700	129
$Mo_2N = 2Mo_2 + \frac{1}{2}N_2$	60700	14.6	25–500	130
$Mo_3Si = 3Mo + Si$	119000	2.5	25–1412	130
$Mo_5Si_3 = 5Mo + 3Si$	311000	4.1	25–1412	130
$MoSi_2 = Mo + 2Si$	132600	2.8	25–1412	130
$MoO_2 = Mo + O_2$	578000	166.5	25–2000	11
$MoO_2(g) = Mo + O_2$	18400	33.9	25–2000	11
$MoO_3 = MoO_3(1)$	47700	45.2	801 m	11
$MoO_3 = Mo + \frac{3}{2}O_2$	740000	247	25–801 m	11
$MoO_3(g) = Mo + \frac{3}{2}O_2$	360000	59.4	25–2000	11
$MoS_2 = Mo + S_2$	397000	182	25–1185	11, 130
$Mo_2S_3 = 2Mo + \frac{3}{2}S_2$	594000	265	25–1200	11
$NH_3(g) = \frac{1}{2}N_2 + \frac{3}{2}H_2$	53700	116.5	25–2000	11
$NO(g) = \frac{1}{2}N_2 + \frac{1}{2}O_2$	−90400	−12.68	25–2000	11
$NO_2(g) = \frac{1}{2}N_2 + O_2$	−32300	63.3	25–2000	11
$Na = Na(1)$	2594	6.99	98 m	11
$Na(1) = Na(g)$	101300	87.9	98–897 b	11, 12
$NaCl = NaCl(1)$	28160	26.2	801 m	11
$NaCl = Na(1) + \frac{1}{2}Cl_2$	411600	93.1	98–801 m	11
$NaCl(1) = Na(g) + \frac{1}{2}Cl_2$	464000	134	801–1465 b	11
$NaF = NaF(1)$	33350	26.3	996 m	11

Table A1.1 *Continued*

Reaction	$\overline{\Delta H^0}$ (joule)	$\overline{\Delta S^0}$ (joule/K)	Range (°C)	Reference
$NaF = Na(1) + \frac{1}{2}F_2$	577000	105.5	98–996 m	11
$NaF(1) = Na(g) + \frac{1}{2}F_2$	624000	148	996–1787 b	11
$Na_3AlF_6 = Na_3AlF_6(1)$	107300	83.5	1012 m	11
$Na_3AlF_6 = 3Na(1) + Al + 3F_2$	3310000	554	98–660	11
$Na_3AlF_6(1) = 3Na(g) + Al(1) + 3F_2$	3380000	623	1012–2000	11
$NaCN = NaCN(1)$	8790	10.5	562 m	11
$NaCN = Na(1) + C + \frac{1}{2}N_2$	90500	34.1	98–562 m	11
$NaCN(1) = Na(g) + C + \frac{1}{2}N_2$	152000	83.7	883–1530 b	11
$Na_2O = Na_2O(1)$	47700	34.0	1132 m	11
$Na_2O = 2Na(1) + \frac{1}{2}O_2$	422000	141	98–1132 m	11
$Na_2O(1) = 2Na(g) + \frac{1}{2}O_2$	519000	235	1132–1950 d	11
$Na_2O \cdot Al_2O_3 = Na_2O + Al_2O_3$	185000	−2.9	500–1131	134
$Na_2CO_3 = Na_2O + CO_2$	297000	118	25–850 m	11
$Na_2CO_3(1) = Na_2O(1) + CO_2$	316400	131	1132–2000	11
$Na_2O \cdot Cr_2O_3 = Na_2O + Cr_2O_3$	203000	−5.8	25–1132	130
$Na_2O \cdot Fe_2O_3 = Na_2O + Fe_2O_3$	88800	−14.6	25–1132	129
$Na_2O \cdot SiO_2 = Na_2O \cdot SiO_2(1)$	51800	38.0	1089 m	11
$Na_2O \cdot SiO_2 = Na_2O + SiO_2$	238000	8.8	25–1089 m	11
$Na_2O \cdot 2SiO_2 = Na_2O \cdot 2SiO_2(1)$	35600	31.0	874 m	11
$Na_2O \cdot 2SiO_2 = Na_2O + 2SiO_2$	233000	−3.8	25–874 m	11
$Nb = Nb(1)$	26900	9.8	2467 m	11
$Nb(1) = Nb(g)$	690000	134.35	2467–4740 b	11
$NbB_2 = Nb + 2B$	251000	9.2	25–1100	130
$Nb_2C = 2Nb + C$	194000	11.7	25–1500	129
$NbC_{0.98} = Nb + 0.98C$	137000	2.4	25–1500	11
$Nb_2N = 2Nb + \frac{1}{2}N_2$	251000	83.3	25–2400 m	129
$NbN = Nb + \frac{1}{2}N_2$	230000	77.8	25–2050 m	129
$NbO = NbO(1)$	84000	38.5	1937 m	11
$NbO = Nb + \frac{1}{2}O_2$	415000	87.0	25–1937	11
$Nb_2O_5 = Nb_2O_5(1)$	104300	58.4	1512 m	11
$Nb_2O_5 = 2Nb + \frac{5}{2}O_2$	1888000	420	25–1512 m	11
$NbO_2 = NbO_2(1)$	92000	42.3	1902 m	11

Table A1.1 *Continued*

Reaction	$\overline{\Delta H^0}$ (joule)	$\overline{\Delta S^0}$ (joule/K)	Range (°C)	Reference
$NbO_2 = Nb + O_2$	784000	167	25–1902 m	11
$Ni = Ni(1)$	17500	10.1	1453 m	11, 12
$Ni(1) = Ni(g)$	386000	121.4	1453–2911 b	11, 12
$NiCl_2 = NiCl_2(g)$	225000	179	972 s	11
$NiCl_2 = Ni + Cl_2$	305400	146	25–972 s	11
$NiF_2 = Ni + F_2$	653000	149	25–1474 m	130
$NiB = Ni + B$	176000	8.4	25–1000	130
$Ni_4B_3 = 4Ni + 3B$	318000	33.0	25–1580 m	130
$Ni_3C = 3Ni + C$	−40000	−17.0	25–500	130
$Ni_3Al = 3Ni + Al(1)$	176000	32.2	660–1395 m	130
$NiAl = Ni + Al(1)$	138000	28.0	660–1453 m	130
$NiSi = NiSi(1)$	43100	34.0	992 m	130
$NiSi = Ni + Si$	91000	7.9	25–992 m	130
$Ni_7Si_{13} = 7Ni + 13Si$	632000	20.0	25–972 m	130
$Ni_3Ti = 3Ni + Ti$	146000	26.0	25–1378 m	130
$NiTi = Ni + Ti$	67000	12.0	25–1240 m	130
$NiO = Ni + \frac{1}{2}O_2$	235600	86.0	25–1984 m	160
$Ni_3S_2 = 3Ni + S_2$	330000	163	25–789 m	130
$NiS = Ni + \frac{1}{2}S_2$	146000	72.0	25–500	11
$NiO \cdot Al_2O_3 = NiO + Al_2O_3$	4200	−12.5	700–1300	130
$Ni(CO)_4(g) = Ni + 4CO$	160000	410	25–500	11, 12
$NiO \cdot Cr_2O_3 = NiO + Cr_2O_3$	53600	8.4	700–1200	134
$NiO \cdot Fe_2O_3 = NiO + Fe_2O_3$	19900	−3.8	582–1400	134
$NiSO_4 = NiO + SO_2 + \frac{1}{2}O_2$	347500	293.2	600–860	155
$2NiO \cdot SiO_2 = 2NiO + SiO_2$	15500	9.2	25–1545 m	130
$NiO \cdot TiO_2 = NiO + TiO_2$	18000	8.4	500–1400	134
$NiO \cdot WO_3 = NiO + WO_3$	46000	0.0	500–1100	134
$Os = Os(1)$	31800	9.62	3027 m	12, 13
$Os(1) = Os(g)$	745000	141	3027–5008 b	12, 13
$OsO_2 = Os + O_2$	290000	175	25–900	130
$OsO_4(g) = Os + 2O_2$	334000	145	131 b–700	130
$OsS_2 = Os + S_2$	270000	195	25–1100	130
$P_{white} = P(1)$	657	2.05	44 m	11

Table A1.1 *Continued*

Reaction	$\overline{\Delta H^0}$ (joule)	$\overline{\Delta S^0}$ (joule/K)	Range (°C)	Reference
$P_{red} = \frac{1}{4}P_4$ (g)	32100	45.6	25–431 s	11
P_4 (g) $= 2P_2$	217000	139	25–1700	11
PH_3 (g) $= \frac{1}{2}P_2 + \frac{3}{2}H_2$	71500	108	25–1700	11
PCl_3 (g) $= \frac{1}{2}P_2 + \frac{3}{2}Cl_2$	474000	209	25–1300	11
PCl_5 (g) $= \frac{1}{2}P_2 + \frac{5}{2}Cl_2$	420000	280	25–1700	11
PF_3 (g) $= \frac{1}{2}P_2 + \frac{3}{2}F_2$	1030000	137	25–1700	11
PF_5 (g) $= \frac{1}{2}P_2 + \frac{5}{2}F_2$	1660000	306	25–1700	11
PO (g) $= \frac{1}{2}P_2 + \frac{1}{2}O_2$	77800	−11.6	25–1700	11
PO_2 (g) $= \frac{1}{2}P_2 + O_2$	386000	60.3	25–1700	11
P_4O_{10} (g) $= 2P_2 + 5O_2$	3155000	1011	358 s–1700	11
$Pb = Pb(1)$	4810	8.0	327 m	11
$Pb(1) = Pb$ (g)	182000	90.1	327–1746 b	11
$PbCl_2 = PbCl_2(1)$	21900	28.2	501 m	11
$PbCl_2(1) = Pb(1) + Cl_2$	325000	103.5	501–953 b	11
$PbCl_2$ (g) $= Pb(1) + Cl_2$	190000	−7.5	953–1750	11
$PbF_2 = PbF_2(1)$	14730	13.35	830 m	11
$PbF_2 = Pb(1) + F_2$	628000	360	830–1200	11
$PbO = PbO(1)$	27500	23.7	886 m	11
$PbO(1) = Pb(1) + \frac{1}{2}O_2$	181000	68.0	886–1535 b	11
$Pb_3O_4 = 3Pb(1) + 2O_2$	702500	370	328–1200	130
$PbO_2 = Pb(1) + O_2$	272000	194	328–900	130
$PbS = Pb(1) + \frac{1}{2}S_2$	163000	88.0	328–1113 m	11
$PbO \cdot B_2O_3 = PbO(1) + B_2O_3(1)$	102500	33.0	885–1535	11
$PbO \cdot 2B_2O_3 = PbO(1) + 2B_2O_3(1)$	166500	79.5	885–1535	11
$PbSO_4 = PbO + SO_2 + \frac{1}{2}O_2$	401000	262	25–1090 m	129
$2PbO \cdot SiO_2 = 2PbO \cdot SiO_2(1)$	51050	50.3	743 m	11
$2PbO \cdot SiO_2(1) = 2PbO(1) + SiO_2$	33500	−6.7	885–1500	11
$PbO \cdot SiO_2 = PbO \cdot SiO_2(1)$	26000	25.1	764 m	11
$PbO \cdot SiO_2 = PbO(1) + SiO_2$	25100	1.26	885–1500	11
$Pd = Pd(1)$	17600	9.6	1552 m	12, 13
$Pd(1) = Pd$ (g)	352000	108.6	1552–2961 b	12, 13
$PdO = Pd + \frac{1}{2}O_2$	111000	96.7	25–870 m	130

Table A1.1 *Continued*

Reaction	$\overline{\Delta H^0}$ (joule)	$\overline{\Delta S^0}$ (joule/K)	Range (°C)	Reference
$PdS = Pd + \frac{1}{2}S_2$	134000	91.2	25–970 m	130
$PdS_2 = Pd + S_2$	200000	167	25–972 m	130
$Pd_4S = 4Pd + \frac{1}{2}S_2$	134000	86.0	25–761 m	130
$Pt = Pt(1)$	19700	9.63	1772 m	12, 13
$Pt(1) = Pt(g)$	521000	127	1772–3823 b	12, 13
$PtO_2(g) = Pt + O_2$	164000	0.0	25–1700	130
$PtS = Pt + \frac{1}{2}S_2$	149000	101.5	25–1200	130
$PtS_2 = Pt + S_2$	233500	182	25–1200	130
$Pu = Pu(1)$	2845	3.1	640 m	13
$Pu(1) = Pu(g)$	335000	96.0	640–3230 b	13
$PuF_3 = PuF_3(1)$	54400	32.0	1426 m	130
$PuF_3 = Pu(1) + \frac{3}{2}F_2$	1540000	228.0	640-1426 m	130
$PuN = Pu(1) + \frac{1}{2}N_2$	303000	84.8	640–2000	130
$Pu_2C_3 = 2Pu(1) + 3C$	132000	−16.3	640–2000	130
$PuC_2 = Pu(1) + 2C$	44400	−22.6	640–2252 m	130
$PuO_2 = Pu(1) + O_2$	1046000	180	640–2390 m	130
$Rb = Rb(1)$	2190	7.0	39.5 m	11
$Rb(1) = Rb(g)$	76320	79.7	39.5–697 b	11
$Rb_2O = 2Rb(1) + \frac{1}{2}O_2$	335000	154	25–688	130
$Re = Re(1)$	33200	9.62	3180 m	13
$Re(1) = Re(g)$	705000	120	3180–5600 b	13
$ReO_2 = Re + O_2$	428000	170	25–1200	130
$ReS_2 = Re + S_2$	300000	191	25–1200	130
$Rh = Rh(1)$	21500	9.62	1960 m	12, 13
$Rh(1) = Rh(g)$	505000	127.4	1960–3694 b	12, 13
$Rh_2O_3 = 2Rh + \frac{3}{2}O_2$	377000	266	25–1000	130
$RhO_2 = Rh + O_2$	200000	−19.6	25–1200	130
$Ru = Ru(1)$	24270	9.62	2250 m	13
$Ru(1) = Ru(g)$	605000	137	2250–4146 b	13
$RuO_2 = Ru + O_2$	300000	162	25–1200	130
$RuO_3(g) = Ru + \frac{3}{2}O_2$	79500	61.0	25–1600	130
$RuO_4(g) = Ru + 2O_2$	180000	143	25–1700	130
$RuS_2 = Ru + S_2$	330000	191	25–1200	130

Table A1.1 *Continued*

Reaction	$\overline{\Delta H^0}$ (joule)	$\overline{\Delta S^0}$ (joule/K)	Range (°C)	Reference
$S = S(1)$	1715	4.44	115 m	11
$S(1) = \frac{1}{2}S_2$	58600	68.3	115–445 b	11
$S_2 = 2S\,(g)$	469300	161.3	25–1700	130
$S_4\,(g) = 2S_2$	62800	115.5	25–1700	130
$S_6\,(g) = 3S_2$	276000	305	25–1700	130
$S_8\,(g) = 4S_2$	400000	450	25–1700	130
$SO\,(g) = \frac{1}{2}S_2 + \frac{1}{2}O_2$	57800	−5.0	445–2000	11
$SO_2 = \frac{1}{2}S_2 + O_2$	362000	72.7	445–2000	11
$SO_3\,(g) = \frac{1}{2}S_2 + \frac{3}{2}O_2$	458000	163.3	445–2000	11
$Sb = Sb(1)$	19900	22	631 m	13
$Sb(1) = Sb\,(g)$	85300	45.6	631–1587 b	13
$Sb_2O_3 = Sb_2O_3(1)$	55000	59.2	656 m	130
$Sb_2O_3(1) = 2Sb(1) + \frac{3}{2}O_2$	661000	198	656–1456 b	130
$Sc = Sc(1)$	14100	7.78	1539 m	13
$Sc(1) = Sc\,(g)$	327200	105.7	1539–2828 b	13
$ScCl_3 = ScCl_3(1)$	67360	54.3	967 m	130
$ScCl_3 = Sc + \frac{3}{2}Cl_2$	891200	230	25–967 m	130
$ScF_3 = Sc + \frac{3}{2}F_2$	1648500	230	25–1000	130
$ScN = Sc + \frac{1}{2}N_2$	314000	98.3	25–1539	130
$Sc_2O_3 = 2Sc + \frac{3}{2}O_2$	1900000	291	25–1539	129
$Se = Se(1)$	5440	11.0	220 m	12,129
$Se(1) = \frac{1}{2}Se_2\,(g)$	59300	61.1	221–685 b	12, 129
$SeO\,(g) = \frac{1}{2}Se_2\,(g) + \frac{1}{2}O_2$	9200	−4.2	685–1700	130
$SeO_2\,(g) = \frac{1}{2}Se_2\,(g) + O_2$	178000	66.1	685–1700	130
$Si = Si(1)$	50540	30.0	1412 m	11
$Si(1) = Si\,(g)$	395000	111	1412–3232 b	11
$SiCl_4\,(g) = Si + 2Cl_2$	660000	129	62 b–1412	11
$SiI_4\,(g) = Si + 2I_2\,(g)$	231000	115	301 b–1412	11
$Si_3N_{4_\alpha} = 3Si + 2N_2$	724000	315	25–1412	11, 161
$Si_3N_{4_\alpha} = 3Si(1) + 2N_2$	874500	405	1412–1700	11, 161
$SiC_\beta = Si + C$	73000	7.7	25–1412	11
$SiC_\beta = Si(1) + C$	122600	37.0	1412–2000	11

Table A1.1 *Continued*

Reaction	ΔH^0 (joule)	ΔS^0 (joule/K)	Range (°C)	Reference
$SiP = Si + \frac{1}{2}P_2$	130000	−89.0	25–1410 d	129
$SiO\,(g) = Si + \frac{1}{2}O_2$	104000	−82.5	25–1412	11
$SiO_{2_{quartz}} = SiO_2(1)$	7700	4.52	1423 m	11
$SiO_{2_{quartz}} = Si + O_2$	907000	176	25–1412	11
$SiO_{2_{cristobalite}} = SiO_2(1)$	9600	4.81	1723 m	11
$SiO_{2_{cristobalite}} = Si + O_2$	906000	176	25–1412	11
$SiS\,(g) = Si + \frac{1}{2}S_2$	−51800	−81.6	700–1412	11, 162
$SiS_2 = SiS_2(1)$	8400	6.15	1090 m	11
$SiS_2 = Si + S_2$	326000	139	25–1090	11
$Sn = Sn(1)$	7030	13.9	232 m	13
$Sn(1) = Sn\,(g)$	296000	103	232–2603 b	13
$SnCl_2(1) = SnCl_2\,(g)$	81600	88.2	652 b	130
$SnCl_2\,(g) = Sn(1) + Cl_2$	226000	13.0	652–1200	130
$SnO\,(g) = Sn(1) + \frac{1}{2}O_2$	−6300	−50.9	232–1700	130
$SnO_2 = Sn(1) + O_2$	575000	198	232–1630 m	129
$SnS\,(g) = Sn(1) + \frac{1}{2}S_2$	−26000	−49.4	232–1700	130
$SnS_2 = Sn(1) + S_2$	285000	196	232–765 m	130
$Sr = Sr(1)$	8370	7.95	777 m	11, 12
$Sr(1) = Sr\,(g)$	141500	85.7	768–1412 b	11
$SrCl_2 = SrCl_2(1)$	16200	14.1	874 m	11
$SrCl_2 = Sr + Cl_2$	824000	148	25–768	11
$SrF_2 = SrF_2(1)$	29700	16.95	1477 m	11
$SrF_2 = Sr + F_2$	1213000	167	25–768	11
$SrO = Sr(1) + \frac{1}{2}O_2$	597000	102.4	768–1377	11
$SrS = Sr(1) + \frac{1}{2}S_2$	519000	96	768–1377	11, 130
$SrCO_3 = SrO + CO_2$	215000	142	700–1243 d	163
$SrO \cdot Al_2O_3 = SrO + Al_2O_3$	71000	−4.2	25–1300	130
$SrSO_4 = SrO + SO_2 + \frac{1}{2}O_2$	548000	273	25–1600 m	130
$2SrO \cdot SiO_2 = 2SrO + SiO_2$	213000	5.86	25–900	130
$SrO \cdot SiO_2 = SrO + SiO_2$	134000	4.2	25–900	130
$SrO \cdot TiO_2 = SrO + TiO_2$	137000	2.1	25–900	130
$Ta = Ta(1)$	31600	9.62	3014 m	11
$Ta(1) = Ta\,(g)$	739800	127.9	3014–5453 b	11

Table A1.1 *Continued*

Reaction	$\overline{\Delta H^0}$ (joule)	$\overline{\Delta S^0}$ (joule/K)	Range (°C)	Reference
$TaCl_5(1) = TaCl_5(g)$	53000	104.6	234 b	11
$TaCl_5(g) = Ta + \frac{5}{2}Cl_2$	754000	169.4	234–2000	11
$TaF_5(1) = TaF_5(g)$	50600	101	224 b	130
$TaF_5(g) = Ta + \frac{5}{2}F_2$	1814000	183	224–1700	130
$TaB_2 = Ta + 2B$	209000	9.6	25–1700	129
$Ta_2C = 2Ta + C$	201000	2.1	25–1700	130
$TaC = Ta + C$	142000	1.2	25–1700	11
$Ta_2N = 2Ta + \frac{1}{2}N_2$	264000	91	25–1700	130
$TaN = Ta + \frac{1}{2}N_2$	247000	81.2	25–1700	130
$Ta_2Si = 2Ta + Si$	126000	−3.3	25–1412	129
$Ta_5Si_3 = 5Ta + 3Si$	335000	−12.6	25–1412	129
$TaSi_2 = Ta + 2Si$	117000	22.6	25–1412	129
$TaO(g) = Ta + \frac{1}{2}O_2$	−188000	−87	25–2000	11
$TaO_2(g) = Ta + O_2$	209000	−20.5	25–2000	11
$Ta_2O_5 = Ta_2O_5(1)$	151000	70.3	1785 m	11
$Ta_2O_5 = 2Ta + \frac{5}{2}O_2$	2025000	413	25–1785 m	11
$Te = Te(1)$	17500	24.2	450 m	13
$Te(1) = \frac{1}{2}Te_2(g)$	58000	45.9	450–988 b	13
$TeO(1) = \frac{1}{2}Te_2(g) + \frac{1}{2}O_2$	7100	−6.0	1009–1700	130
$Th = Th(1)$	16300	7.95	1755 m	13
$Th(1) = Th(g)$	527000	104.4	1755–4790 b	13
$ThI_4 = ThI_4(1)$	48000	57.0	570 m	130
$ThI_4 = Th + 2I_2(g)$	782000	290	25–570 m	130
$ThN = Th + \frac{1}{2}N_2$	377000	86.2	25–1755	130
$Th_3N_4 = 3Th + 2N_2$	1314000	351	25–1755	130
$ThO(g) = Th + \frac{1}{2}O_2$	67000	−52.7	1600–2000	164
$ThO_2 = Th + O_2$	1220000	183	25–2000	130
$Ti = Ti(1)$	15500	7.95	1666 m	11
$Ti(1) = Ti(g)$	427000	120	1666–3358 b	11
$TiCl_4(g) = Ti + 2Cl_2$	764000	121	25–1666	11
$TiF_4(g) = Ti + 2F_2$	1554000	124	25–1666	11
$TiI_4(g) = Ti + 2I_2(g)$	402000	117.6	25–1666	11

Table A1.1 *Continued*

Reaction	$\overline{\Delta H^0}$ (joule)	$\overline{\Delta S^0}$ (joule/K)	Range (°C)	Reference
$TiB = Ti + B$	163000	5.9	25–1666	11
$TiB_2 = Ti + B_2$	285000	20.5	25–1666	11
$TiC = Ti + C$	185000	12.6	25–1666	11
$TiN = Ti + \frac{1}{2}N_2$	336000	93.3	25–1666	11
$TiO_\beta = Ti + \frac{1}{2}O_2$	515000	74.1	25–1666	11
$TiO_{2_{rutile}} = Ti + O_2$	941000	177.6	25–1666	11
$Ti_2O_3 = 2Ti + \frac{3}{2}O_2$	1502000	258	25–1666	11
$Ti_3O_5 = 3Ti + \frac{5}{2}O_2$	2435000	421	25–1666	11
$Tl = Tl(1)$	4140	7.2	304 m	13
$Tl(1) = Tl(g)$	170000	97.6	304–1473 b	13
$U = U(1)$	8540	6.07	1132 m	13
$U(1) = U(g)$	477000	105.4	1132–4130 b	13
$UC = U(1) + C$	110000	1.84	1132–2520 m	129
$U_2C_3 = 2U(1) + 3C$	222000	2.5	1132–1700	130
$UC_{1.93} = U(1) + 1.93C$	90400	9.2	1132–1700	130
$UN = U + \frac{1}{2}N_2$	293000	81	25–1132	129
$UO_2 = U(1) + O_2$	1087000	172	1132–2000	129
$U_4O_9 = 4U + \frac{9}{2}O_2$	4492000	740	25–600	129
$U_3O_8 = 3U + 4O_2$	3560000	653	25–600	129
$UO_3 = U + \frac{3}{2}O_2$	1227000	251	25–600	129
$V = V(1)$	22850	10.4	1902 m	11
$V(1) = V(g)$	463000	126	1902–3406 b	11
$VB = V + B$	138000	5.86	25–2000	130
$V_2C = 2V + B$	146400	3.35	25–1700	129
$VC = V + C$	102000	9.6	25–2000	130
$VN_{0.46} = V + 0.23N_2$	130000	44.4	25–1700	11
$VN = V + \frac{1}{2}N_2$	215000	82.4	25–2346 d	11
$VO = V + \frac{1}{2}O_2$	425000	80	25–1800	11
$V_2O_3 = 2V + \frac{3}{2}O_2$	1203000	238	25–2070 m	11
$VO_2 = V + O_2$	706000	155	25–1360 m	11
$V_2O_5 = V_2O_5(1)$	64400	68.3	670 m	11
$V_2O_5(1) = 2V + \frac{5}{2}O_2$	1447000	321.6	670–2000	11
$W = W(1)$	35560	9.62	3407 m	11, 12

Table A1.1 *Continued*

Reaction	$\overline{\Delta H^0}$ (joule)	$\overline{\Delta S^0}$ (joule/K)	Range (°C)	Reference
$W(1) = W(g)$	821000	141	3407–5658 b	11
$W_2C = 2W + C$	30540	−2.34	1302–1400	165
$WC = W + C$	42300	4.98	900–1302	165
$WO_2 = W + O_2$	581000	172	25–1724 d	11
$WO_3 = WO_3(1)$	73400	42.1	1472 m	11
$WO_3 = W + \frac{3}{2}O_2$	833500	245	25–1472 m	11
$Y = Y(1)$	11380	6.32	1526 m	13
$Y(1) = Y(g)$	379000	105.4	1526–3334 b	13
$YCl_3 = YCl_3(1)$	31400	31.6	721 m	130
$YCl_3 = Y + \frac{3}{2}Cl_2$	968000	227	25–721	130
$YF_3 = YF_3(1)$	28000	19.6	1155 m	130
$YF_3 = Y + \frac{3}{2}F_2$	1711000	219	25–1155 m	130
$YN = Y + \frac{1}{2}N_2$	300000	99.6	25–1526	130
$Y_2O_3 = 2Y + \frac{3}{2}O_2$	1900000	282	25–1526	129
$Y_2O_3 \cdot Zr_2O_3 = Y_2O_3 + Zr_2O_3$	21000	0.0	25–1200	130
$Zn = Zn(1)$	7320	10.6	420 m	11
$Zn(1) = Zn(g)$	118000	100.3	420–907 b	11
$ZnCl_2(1) = ZnCl_2(g)$	119200	118.7	732 b	129
$ZnCl_2(1) = Zn(1) + Cl_2$	402500	131	420–732 b	129
$ZnCl_2(g) = Zn(g) + Cl_2$	393000	105	907–1700	129
$ZnF_2 = ZnF_2(1)$	41800	36.44	875 m	130
$ZnF_2 = Zn(1) + F_2$	769000	175	420–875 m	130
$ZnF_2(1) = Zn(g) + F_2$	832000	227	907–1505 b	130
$ZnO = Zn(g) + \frac{1}{2}O_2$	460240	190.4	907–1700	129
$ZnS = Zn(1) + \frac{1}{2}S_2$	278000	108	420–907	130
$ZnS(g) = Zn(g) + \frac{1}{2}S_2$	−5020	30.5	1182 b–1700	130
$ZnO \cdot Fe_2O_3 = ZnO + Fe_2O_3$	9620	3.8	25–700	130
$ZnSO_4 = ZnO + SO_2 + \frac{1}{2}O_2$	328000	267	25–700	11
$Zr = Zr(1)$	20900	9.83	1852 m	11
$Zr(1) = Zr(g)$	580000	124	1852–4429 b	11
$ZrCl_4 = ZrCl_4(g)$	110500	181.5	336 s	11
$ZrCl_4(g) = Zr + Cl_2$	871000	116	336–2000	11

Table A1.1 *Continued*

Reaction	$\overline{\Delta H^0}$ (joule)	$\overline{\Delta S^0}$ (joule/K)	Range (°C)	Reference
$ZrF_4 = ZrF_4$ (g)	237700	202	906 s	11
ZrF_4 (g) $= Zr + 2F_2$	1678000	128.4	903–2000	11
$ZrB_2 = ZrB_2(1)$	104600	31.5	3050 m	11
$ZrB_2 = Zr + 2B$	328000	23.4	25–1852	11
$ZrC = Zr + C$	197000	9.2	25–1852	11
$ZrN = Zr + \frac{1}{2}N_2$	364000	92	25–1852	11
$ZrSi = Zr + Si$	160000	4.2	25–1412	11
ZrO (g) $= Zr + \frac{1}{2}O_2$	−54600	−67.0	1300–1852	11
$ZrO_2 = ZrO_2(1)$	87000	29.5	2677 m	11
$ZrO_2 = Zr + O_2$	1092000	184	25–1852	11
ZrS (g) $= Zr + \frac{1}{2}S_2$	−237000	−78.2	25–1852	11
$ZrS_2 = Zr + S_2$	700000	178	25–1552 m	130
$ZrO \cdot SiO_2 = ZrO + SiO_2$	26800	12.6	25–1707 m	130

Appendix 2
Turbulent Removal of Inclusion to Walls

A2.1 Introduction

The fluid mechanics of transfer of inclusions to the walls of a ladle is treated. It is assumed that once an inclusion touches a wall it adheres to it. As mentioned previously this assumption may be fully valid for steel melts. Due to turbulence the inclusions are completely mixed in the bulk melt.

According to classical hydrodynamic theory a boundary layer is present at the melt–wall interface (also at the melt–slag interface). As shown schematically in Fig. A2.1 for a smooth wall the velocity drops from v_b to zero through the boundary layer of thickness δ_b.

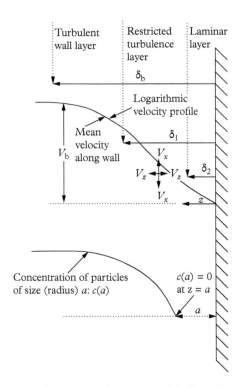

Fig. A.2.1 *Velocity and particle concentration along a smooth ladle wall.*

Following a theory originally proposed by Levich A[1] the boundary layer is split up into three parts: the turbulent wall layer; the restricted turbulence layer at distance δ_1 from the wall, where turbulence is strongly damped but where turbulent diffusion is greater than diffusion by Brownian motion; and the inner viscous (laminar flow) layer, where the inclusions are transferred only by Brownian motion. It is shown in an example in Section 3.3 that the thickness δ_2 of the inner layer for the cases of interest may be about 1 μm. At $y = \delta_2$ the turbulent diffusion coefficient is so small that it is equal to the Brownian diffusion coefficient. If inclusions are larger than δ_2, the viscous boundary layer can be neglected. Inclusions greater than δ_b penetrate simply due to their size so that there is no problem in removing them. In the following the behaviour of particles is studied only for the inclusion size range between δ_2 and δ_b. Removal of inclusions smaller than δ_2 is discussed in Section 3.3 concerning dissolved elements.

A2.2   Pipe Flow

The removal of inclusions in fully developed pipe flow is treated first and in detail. There are several reasons: flow in a pipe is well known from the literature and this case can be used to understand the mechanism of inclusion removal for more complex geometries. Possibly in filtration the pipe flow case can be of direct use. Also light is thrown on the phenomenon of nozzle blockage even though removal in this case relates to entrance effects in a pipe and not to developed flow further down the pipe.

For the boundary layer region between δ_1 and δ_2 the distribution of the average velocity $\bar{v}_x$ along the wall has the same form as in laminar flow,[1]

$$\bar{v}_x \propto z, \tag{A 2.1}$$

where z is the distance from the wall. If the usual assumption[1] is made that there is proportionality between eddy velocities and average velocities, we obtain for the eddy velocity (see also Chapter 3)

$$v_x \propto z. \tag{A 2.2}$$

In view of the continuity equation we have for the boundary layer

$$\frac{\partial v_x}{\partial x} + \frac{\partial v_z}{\partial z} = 0. \tag{A 2.3}$$

The normal component of the eddy velocity is

$$v_z \sim -\int_0^z \frac{\partial v_x}{\partial x} dz \propto z^2. \tag{A 2.4}$$

Shear stress at the wall τ_0, the density ρ, and the kinematic viscosity v determine v_z. From dimensional analysis it is found that v_z depends on τ_0, ρ, and v as follows:

$$v_z \propto \left(\frac{\tau_0}{\rho}\right)^{\frac{3}{2}} \frac{z^2}{v^2}.$$

The proportionality constant must be determined experimentally. It has been estimated to be[1] $0.36 \times 10^{-2} v_0^3 f v^2$ for the case of flow in a smooth tube where the mean velocity fluctuations in the axial direction are about ¼ of the mean velocity. v_0 is the so-called shear velocity, $v_0 = (\tau_0 f \rho)^{\frac{1}{2}}$. The shear velocity plays a very important role when conditions near a solid wall are described. Now the mass diffusivity by eddies D_E may be obtained using the Prandtl mixing length mode[1] (see Section 3.11.2),

$$D_E = v_z l, \tag{A 2.5}$$

where l is the Prandtl mixing length. $l = 0.4z$, which gives

$$D_E = \frac{0.14 \times 10^{-2} v_0 z^3}{v^2}. \tag{A 2.6}$$

Probably, this relation gives an upper limit to the turbulent diffusivity in a smooth pipe.

For eddy diffusion close to a solid surface Fick's first law of diffusion may be written in the form

$$\dot{n}_a = -D_E \frac{dc(a)}{dz}, \tag{A 2.7}$$

where $\dot{n}_a$ is the flux of inclusions of diameter $2a$ and $c(a)$ is the concentration of inclusions of diameter $2a$ at position z. One boundary condition is:

(1) At $\qquad\qquad\qquad z = \delta_1, c(a) = c_{\delta 1}(a) :. \tag{A 2.8}$

Furthermore an interception boundary condition is introduced:

(2) When an inclusion touches the wall or a protrusion on the wall it adheres. Then at

$$z = a \ \text{ we get } c(a) = 0 \tag{A 2.9}$$

for a perfectly smooth wall or at

$$z = e \ \text{ we have } c(a) = 0, \tag{A 2.10}$$

where $e - a$ is the height of a protrusion (see Fig. A2.2).

The assumption is that the protrusions do not significantly change D_E. This means that the height $e - a$ is small and the number of protrusions is limited so that flow is not noticeably modified. On the other hand the number of protrusions should be sufficient to collect all inclusions present in the melt at a distance less than e from the wall. This boundary condition is used in the following.

If Eq. (A2.6) is inserted into Eq. (A2.7) one obtains for the flux

$$\dot{n}_a = \frac{-0.14 \times 10^{-2} v_0 z^3}{v^2} \frac{dc(a)}{dz}. \tag{A 2.11}$$

This equation may be integrated by writing

$$-\frac{\dot{n}_a v^2 dz}{0.14 \times 10^{-2} v_0 z^3} = dc(a). \tag{A 2.12}$$

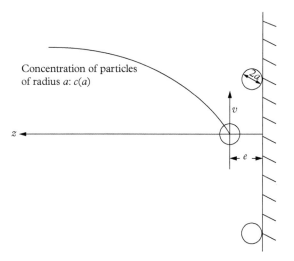

Fig. A.2.2 *Protrusions created by particles deposited on the wall.*

Use of the boundary conditions (A2.8) and (A2.10) gives on integration from $z = e$ to $z = \delta_1$:

$$c_{\delta 1}(a) = \frac{\dot{n}_a \left(\frac{1}{e^2} - \frac{1}{\delta_1^2} \right) v^2}{v_0 \times 0.28 \times 10^{-2}}. \tag{A 2.13}$$

Of course this equation only applies when the protrusion height e is less than the boundary layer thickness, δ_1. For the region $\delta_1 < z < \delta_b$, one may use, as a first approximation, the equations of the Reynolds analogy,

$$\frac{\dot{n}_a}{\tau_0} = \frac{c_b(a) - c_{\delta 1}(a)}{\rho (v_m - v_{\delta 1})}, \tag{A 2.14}$$

where $c_b(a)$ is the concentration in the bulk melt and v_m is the mean velocity in the pipe and $v_{\delta 1}$ is the velocity at $z = \delta_1$. The basis for Eq. (A2.14) is the assumption that the mean eddy lengths for the transfer of mass and momentum are the same. Thus concentration and velocity vary proportionally to each other. $\dot{n}_a$ is proportional to $c_b(a) - c_{\delta 1}(a)$ and $-\tau_0$ is proportional to $\rho (v_m - v_{\delta 1})$. Then the ratio between $\dot{n}_a$ and $-\tau_0$ is given by Eq. (A2.14).

Now the velocity, v_{δ_1}, between the restricted turbulence layer and the turbulent wall layer is taken to be equal to $5v_0$. Furthermore, the thickness of the restricted turbulence layer δ_1 decreases with v_0, $\delta_1 = 5v f v_0$.[1] If $c_{\delta_1}(a)$ and v_{δ_1} are eliminated from Eqs. (A2.13) and (A2.14), one obtains

$$\dot{n}_a = \frac{v_m c_b(a) \emptyset}{1 + \emptyset^{\frac{1}{2}} \left(\frac{353 v^2}{e^2 v_m} - 19 \right)}, \tag{A 2.15}$$

where

$$\varnothing = \frac{v_0}{v_m} = \frac{\tau_0}{\rho v_m}$$

(A 2.16)

is the friction factor for the pipe.

$$\left\{ 1 + \varnothing^{\frac{1}{2}} \left(\frac{353 v^2}{e^2 v_0} - 19 \right) \right\} f v_m \varnothing$$

gives the total resistance in the two boundary layers. $1/(v_m\varnothing)$ is the resistance from the turbulent wall layer.

Blasius found empirically for $\mathrm{Re} = v_m 2R/v$ from 3000 to 10^5 (for smooth pipes) that[1]

$$\varnothing \approx 0.04 Re^{-\frac{1}{4}}.$$

(A 2.17)

Resistance in the restricted turbulence layer dominates when the protrusions are small for small e. Then Eq. (A2.15) reduces to

$$\dot{n}_a \approx \frac{v_m c_b(a) \varnothing^{\frac{1}{2}} v_0 e^2}{353 v^2}$$

or, using Eq. (A2.16),

$$\dot{n}_a = \frac{v_m^3 \varnothing^{\frac{3}{2}} e^2 c_b(a)}{353 v^2}.$$

(A 2.18)

Resistance to particle transfer in the turbulent wall layer prevails for large e and we obtain from the analogy between momentum and mass transfer in the turbulent wall layer

$$\dot{n}_a \approx v_m \varnothing c_b(a).$$

(A 2.19)

A2.3   Nozzle Blockage and Removal of Inclusions by Stirring

To illustrate the use of the equations derived in Section A2.2 the case of nozzle blockage[2,3] is studied. Then removal due to stirring in general is discussed.

The experimental conditions are given by $R = 0.003$ m, $v_m \approx 2$ m/s, $T = 1\,873$ K, $\rho = 7\,030$ kg/m^3, and $v = 0.8 \times 10^{-6}$ m^2/s so that $\mathrm{Re} = 15\,000$ and $\varnothing = 0.0036$. It is assumed that the nozzle walls were originally smooth and that the protrusions are single inclusions already deposited on the walls (see Fig. A2.2). Then choosing $e = 3a = 10^{-5}$ m, Eqs. (A2.19) and (A2.15) give, respectively, $\dot{n}_a f c_b(a) = 0.76 \times 10^{-3}$ m/s and 0.77×10^{-3} m/s. For a choice of $e = 3a = 3 \times 10^{-5}$ m, Eqs. (A2.19) and (A2.15) give $\dot{n}_a/c_b(a) = 7.2 \times 10^{-3}$ m/s and 7.9×10^{-3} m/s, respectively. In the example, $e = 3a = 10$ μm is only a small protrusion so that resistance in the restricted turbulence layer dominates, while $e = 3a = 30$ μm turns out to be large enough to penetrate the restricted turbulence layer. This may also be seen by comparing e and $\delta_1 = 5v/v_0 = 5v/(v_m)\varnothing^{\frac{1}{2}} = 5 \times 0.8 \times 10^{-6}/(2 \times 0.0036^{\frac{1}{2}}) = 3.3 \times 10^{-5}$ m $= 33$ μm. With large protuberances only resistance in the turbulent wall layer remains.

The deposits often have a dendritic form mainly composed of particles of size 1–10 μm. Deposition is autocatalytic. Particles pile up where deposition has been initiated. A small amount of deposit can then clog the nozzle. For instance, for the experimental $2R = 0.006$ m nozzle[3] the volume of deposit that causes blockage was very small, about 2.8×10^{-8} m³. With about 150 ppm (by volume) inclusions, blockage occurred in about 3 minutes.

If the surface area of the nozzle is taken as 0.6×10^{-3} m², the volume of the deposit (for particles with $a = 3.3$ μm) in 3 minutes becomes $0.6 \times 10^{-3} \times 180 \times 0.77 \times 10^{-3} \times 1.5 \times 10^{-4} = 1.2 \times 10^{-8}$ m³. Thus, the theory given here is in rough agreement with the experiments. The discrepancy is probably partly due to the fact that the inclusions pile up so that the protrusions soon become larger than $2a$. Wilson et al.[3] present some results concerning the calculation of inclusion transfer to walls by numerical methods. For instance, with the FLUENT program[4] it is possible to calculate inclusion trajectories to the nozzle walls.

For a melt stirred with an impeller turbulent fluctuations are roughly twice as great as for pipe flow.[1] Then the factor 353 in Eq. (A2.15) is replaced by $353/2 = 176$. Equations (A2.18) and (A2.15) are replaced by

$$\dot{n}_a = \frac{v_m^3 \emptyset^{\frac{1}{2}} e^2 c(a)}{176 v^2} \qquad \text{(A 2.20)}$$

for transfer in the zone of restricted turbulence, while Eq. (A2.20) for the transfer resistance in the turbulent wall layer remains unchanged. The friction factor $\emptyset$ may be calculated using the circulation velocity v_b:[1]

$$\phi = 0.03 (\text{Re}_x)^{-\frac{1}{5}}$$

$$\text{Re}_x = \frac{v_b x}{v}$$

For instance, $x = 1$ m, $v_b = 2$ m/s, and $v = 0.8 \times 10^{-6}$ m²/s gives $\text{Re} = 2.5 \times 10^6$ and $\emptyset = 1.6 \times 10^{-3}$. Also for impeller stirring, inclusions as small as $a = 10$ μm ($e = 3a = 30$ μm) are removed in accordance with Eq. (A2.19) valid for resistance only in the turbulent wall layer. In this case Eq. (A2.19) gives $n_a / v_b c(a) = \emptyset = 1.6 \times 10^{-3}$, the same as that obtained from Eq. (A2.15) using the factor 176 instead of 353.

At points where the flow leaves the wall, for instance at the wall–slag line, the wall shear stress[5] will be very large, the turbulence level increases, and transport is enhanced. In general to determine the relative resistances of the turbulent wall layer and the zone of restricted turbulence, extensive numerical calculations must be carried out. It will be seen that an indication of their relative importance may be obtained from measurements of inclusion removal versus 'stirring power'. Mass transfer in the restricted turbulence boundary layer increases as the third power of the velocity, v_b, and the thickness δ_1 of the layer decreases with increasing velocity. Thus, at sufficiently high velocities (sufficient stirring) resistance in the restricted turbulence layer becomes negligible. Ideally, velocities and stirring should be sufficient for removing transfer resistance in this layer. This leaves only the resistance in the turbulent wall layer (Eq. (A2.19)).

Stirring may be obtained not only with an impeller but also electromagnetically.[6–9] Either a travelling magnetic field or a pulsating stationary field may be employed. In Chapter 4 we have given examples of the use of gas to stir a melt. The flow and turbulence induced by bubbling is mentioned briefly in Chapter 3 as an illustration of the use of numerical methods. At lower temperatures there is a choice of stirring method so that for magnesium and aluminium all three can be used. For steel, impellers cannot usually be employed due to the mechanical and chemical attack on the impeller material at the high temperatures. We have seen that stirring can serve not only to mix the bulk bath but also to transfer inclusions to the walls of the crucible and to the top-slag. On the other hand, stirring must not be so intense that slag, dross, slurry, or inclusions already deposited are re-entrained.

A2.4   Stirring Energy

In many industrial applications numerical calculations such as those mentioned in Chapter 3 are not available. It may not even be a simple matter to apply Eq. (A2.20) to an impeller-stirred system since we usually do not know the roughness of the crucible wall, that is, the protrusion height, e. Whatever the type of stirring, we expect turbulent diffusion in the zone of restricted turbulence to limit inclusion transfer at lower stirring rates, while the turbulent wall layer Eq. (A2.19) limits transfer at high stirring rates. It is a problem that for bubble and electromagnetic stirring the proportionality constant in Eq. (A2.20) cannot be specified without extensive numerical calculations. Furthermore, the friction factor in Eq. (A2.19) is not known.

Often the specific stirring power input, $\dot{\varepsilon}$, to the melt can be determined whatever the method of stirring.[10] In this case, in order to obtain a rough estimate for the inclusion transfer one may, according to Kolmogorov,[1] assume that $\dot{\varepsilon}$ is proportional to the flow of kinetic energy per unit mass, that is, to the third power of the mean velocity. In terms of the shear velocity this gives the relation

$$\dot{\varepsilon} = \frac{v_0^3}{l_e} \qquad\qquad (A\ 2.21)$$

where l_e is a characteristic length of the energy containing eddies. Equation (A2.21) says that $\dot{\varepsilon}$ is proportional to v_0^3 with proportionality constant l_e. For pipe flow it is found experimentally that l_e is proportional to the diameter d and has a weak dependence on the Reynolds number,[1]

$$l_e = 0.05d\ Re^{-\frac{1}{8}} \qquad\qquad (A\ 2.22)$$

where $Re = v_m d / v$. Then Eq. (A2.18) for resistance dominated by the restricted turbulence layer gives for the mass transfer coefficient $k_1 = \dot{n}_a / c_b(a)$ on employing Eq. (A2.16),

$$\frac{\dot{n}_a}{c_b(a)} = \frac{\dot{\varepsilon} e^2 Re^{-\frac{1}{8}} d\ 142 \times 10^{-6}}{v^2}, \qquad\qquad (A\ 2.23)$$

while Eq. (A2.19), valid when resistance is determined only by the turbulent wall layer, becomes

$$\frac{\dot{n}_a}{c_b(a)} = +0.37\dot{\varepsilon}^{\frac{1}{3}}\emptyset^{\frac{1}{2}}Re^{-\frac{1}{24}}d^{\frac{1}{3}} \qquad\qquad (A\ 2.24)$$

If the dependence of the Re and ø terms on the stirring power $\dot{\varepsilon}$ is disregarded, resistance in the restricted turbulence layer is proportional to $\dot{\varepsilon}$ while the contribution from the turbulent wall layer is proportional to $\dot{\varepsilon}^{\frac{1}{3}}$. The same dependence exists for impellers, bubbling, and electromagnetic stirring. If both resistances must be taken into account, it is found from Eqs. (A2.23) and (A2.24) in analogy with Eq. (A2.15) that

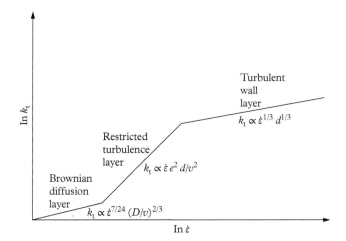

Fig. A.2.3 *The mass transfer coefficient for inclusions, k_t as a function of the specific stirring power $\dot{\varepsilon}$ in the three regimes. d is the pipe diameter, e the protrusion height, and D is the Brownian diffusion coefficient. v is kinematic viscosity.*

$$\frac{\dot{n}_a}{c(a)} = \frac{0.37\dot{e}^{\frac{1}{3}}\phi^{\frac{1}{2}}Re^{-\frac{1}{24}}d^{\frac{1}{3}}}{1+\phi^{\frac{1}{2}}\left(\frac{353v^2Re^{\frac{1}{12}}}{e^2\dot{\varepsilon}^{\frac{2}{3}}0.136d^{\frac{2}{3}}}-19\right)} \tag{A 2.25}$$

It is seen that the total mass transfer coefficient $k_t = \dot{n}_a/c_a$ varies as $\dot{\varepsilon}^{\frac{1}{3}}$ for high values of $\dot{\varepsilon}$ and as $\dot{\varepsilon}$ for lower values of $\dot{\varepsilon}$. Thus, as shown previously at high stirring rates (high velocities), particle transfer is determined by resistance in the turbulent wall layer (Eq. (A2.24)), while at lower stirring rates resistance in the restricted turbulence layer, Eq. (A2.23), dominates. For very low values of $\dot{\varepsilon}$, k_t varies as $\dot{\varepsilon}^{\frac{7}{24}}$ (see Section 3.3 and Fig. A2.3).

A2.4.1 Mechanical Stirring

In impeller systems the power input may be measured with a torque meter. The power input is simply the torque multiplied by 2π times the frequency of rotation. If this quantity is divided by the mass M of the bath, $\dot{\varepsilon}$ is obtained. Experiments in steel[11] show that transfer of inclusions to a refining slag is greatly enhanced by mechanical stirring (tapping and stirring by an impeller). The reasons are that the coefficients for transfer of inclusions to a refining slag are greatly increased. Furthermore, if stirring is sufficient to disperse part of the buoyant refining slag down into the melt, contact areas are extended. Figure A2.4 shows that the total oxygen content (inclusion content) decreases exponentially with time. This result is in accordance with theory since a molar balance for oxygen in inclusions gives

rate of decrease of moles of oxygen in inclusions in the bath

= moles of oxygen in inclusions removed to the slag per second

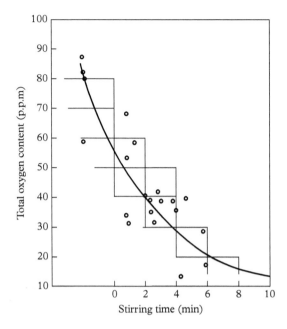

Fig. A.2.4 *Total oxygen content of steel covered by refining slag and stirred by an impeller as a function of stirring time.*[12]

or

$$-\frac{M}{100 \times 16} \frac{d[\%O]_i}{dt} = \frac{k_t A \rho [\%O]_i}{100 \times 16}, \tag{A 2.26}$$

where k_t, A, and ρ are the mass transfer coefficient, contact area, and melt density, respectively. M is the mass of the bath.

Integrated, Eq. (A2.26) becomes

$$\frac{[\%O]_i}{[\%O]_i \text{ (at time } t=0)} = \exp\left(-\frac{t}{\tau}\right), \tag{A 2.27}$$

where the inverse time constant, $1/\tau = k_t A \rho/M$ increases with $\dot{\varepsilon}$ as discussed in Section 2.4. $1/\tau$ also increases with the contact area A, between slag and melt. τ is proportional to M. The amount of dissolved oxygen is disregarded in the calculation on the premise that [% O]$_i$ is very much greater than the dissolved oxygen, [%O].

A2.4.2   Bubble Stirring

Bubble-stirred systems are discussed in Chapter 3 as well as in Chapter 4. In addition to flotation, inclusions may be removed by transfer to the tap-slag and to the walls. All three mass transfer coefficients

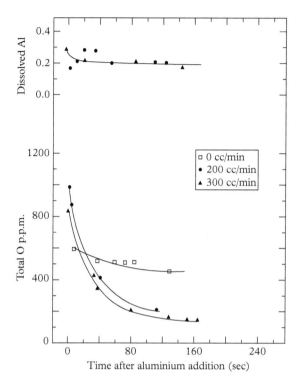

Fig. A.2.5 *Deoxidation in a gas-stirred melt at different gas flow rates. Changes in total oxygen and dissolved Al levels with time are shown.*[12]

k_{tb}, k_{ts}, and k_{tw} increase with stirring. A molar balance for a batch of metal gives

$$-\frac{M}{100 \times 16} \frac{d[\%O]_i}{dt} \frac{\rho}{100 \times 16} [\%O]_i (k_{tb}A_b + k_{ts}A_s + k_{tw}A_w)$$

where the indices b, s, and w refer to bubbles, tap-slag, and walls, respectively. Integration gives the same exponentially decreasing function of time as Eq. (A2.27). The inverse time constant is

$$\frac{1}{\tau} = \frac{\rho}{M} (k_{tb}A_b + k_{ts}A_s + k_{tw}A_w) \tag{A 2.28}$$

Figure A2.5 depicts experiments concerning the deoxidation of steel with aluminium for different inert gas stirring rates.[12] The results may indicate that values of k_t are proportional to $\dot{\varepsilon}^{\frac{1}{3}}$. Transfer resistance in the turbulent wall layer dominates in this case.

The choice between gas stirring and electromagnetic stirring in steel refining does not seem to be a simple matter. In some plants porous plugs for stirring and gas purging are standard equipment. Bubbling will also stir the melt–topslag interface. Dependent on the process, this may be desirable or undesirable. Some production people have found that gas from a porous plug is not dispersed properly and that maintenance is troublesome. Then electromagnetic stirring may be preferable even though capital costs are higher than for porous plugs, and a non-magnetic steel structure is required.

A2.4.3 Electromagnetic Stirring

The net power input to the melt from electromagnetic stirrers can be calculated for both a pulsating stationary field and a travelling field.[8,9] Calculations of the transfer of inclusions through the wall boundary layers do not seem to have been attempted. In this situation we take recourse to Eq. (A2.23). Inclusions are removed to the walls, top slag, and bottom surface of the ladle. It will prove practical to replace the diameter with the volume-to-surface ratio V/A where A includes all three surfaces. If the diameter-to-height ratio is taken to be 1, then

$$\frac{V}{A} = \frac{\pi d^3 / 4}{\pi dd + 2\pi d^2 / 4} = \frac{d}{6}.$$

Then Eq. (A2.23) becomes

$$\frac{\dot{n}_a}{c(a)} = \frac{\dot{\varepsilon} e^2 6V 142 \times 10^{-6}}{Re^{\frac{1}{8}} A v^2}. \tag{A 2.29}$$

Here $Re^{\frac{1}{8}}$ is determined by the convective velocities induced by the electromagnetic stirring. These velocitie are roughly 1 m/s. Also, the dimensions of the ladle are about 1 m. Then $Re^{\frac{1}{8}} \sim \left(1 \times 1 / 10^{-6}\right)^{\frac{1}{8}} = 5.6$. The height of the protrusions, e, is determined by the roughness of the wall. Thus, e depends on the surface properties of the refractory and on the inclusions deposited.

Now, if it is assumed that the addition of a deoxidizer has suppressed dissolved oxygen so that few or no new inclusions are created by nucleation and that growth of inclusions is negligible, a number balance for the batch of melt becomes[6]

Rate of decrease of the number of inclusions = number of inclusions removed per unit time

$$-V \frac{df_N(a)}{dt} = \frac{\dot{\varepsilon} e^2 6V 142 \times 10^{-6} A f_N(a)}{Re^{\frac{1}{8}} A v^2} \tag{A 2.30}$$

or, integrated,

$$\frac{f_N(a)}{f_N(a)(t = 0)} = \exp(-t/\tau), \tag{A 2.31}$$

where the inverse time constant for the removal of inclusions is

$$\frac{1}{\tau} = \frac{\dot{\varepsilon} e^2 0.85 \times 10^{-3}}{Re^{\frac{1}{8}} v^2}. \tag{A 2.32}$$

$f_N(a)(t = 0)$ in Eq. (A2.31) is the distribution of inclusions when stirring is initiated. As discussed previously in Section 5.9 the size distribution is often a decreasing exponential function of the size a.

The number balance, Eq. (A2.30), can be replaced by a molar balance for oxygen in the inclusions. Then Eq. (A2.31) is substituted by

$$\frac{c}{c(t=0)} = \exp(-t/\tau),$$ (A 2.33)

where c is the mass per cent of inclusions. A value of roughly $e = 50$ µm is obtained from the measurements in the Oxelösund steel works and $e \sim 20$µm from the experiments in the Fagersta (see Fig. 5.33) steel works.[6] These may be taken as purely experimental quantities. However, it should be noted that the e are roughly 3 times the mean inclusion sizes (cf. Fig. 5.33). The data indicate that $1/\tau$ is proportional to stirring energy and thus confirms the assumption that in this example transfer resistance is governed by the restricted turbulence boundary layer.

....

REFERENCES

1. Davies JT. *Turbulence phenomena*. New York: Academic Press; 1972.
2. Saxena SK, Sandberg H, Waldenstrom T, Persson A, Steensen S. Mechanism of clogging of tundish nozzle during continuous casting of aluminium killed steel. *Scand. J. Metall.*, 1978; 7(3): 126–33.
3. Wilson FG, Heesom MJ, Nicholson A, Hills AWD. Effect of fluid flow characteristics on nozzle blockage in aluminium-killed steels. *Ironmaking and Steelmaking*, 1987; 14(6): 296–309.
4. Boysan F, Johansen ST. Mathematical modelling of gas-stirred reactors. In: Engh TA, Lyng S, Øye HA, eds., *Proceedings of the international seminar on refining and alloying of liquid aluminium and ferro-alloys*, pp. 267–88. Düsseldorf: Aluminium Verlag; 1985.
5. Johansen ST, Boysan F, Engh TA. Numerical calculations of removal of inclusions and dissolution of refractory in a bubble stirred ladle. In: *Proceedings of the fourth Japan-Nordic countries joint symposium on the science and technology of process metallurgy*, pp. 182-215. Tokyo, Japan: ISIJ; 1986.
6. Engh TA, Lindskog N. A fluid mechanical model of inclusion removal. *Scand. J. Metall.*, 1975; 4: 49–58.
7. Szekely J, Nakanishi K. Stirring and its effects on aluminium deoxidation in the ASEA-SKF furnace: Part II. Mathematical representation of the turbulent flow field and of tracer dispersion. *Metall. Trans. B*, 1975; 6(2): 245–56.
8. Sundberg Y. Mechanical stirring power in molten metal in ladles obtained by induction stirring and gas blowing. *Scand. J. Metall.*, 1978; 1: 81–7.
9. Cremer P, Driole J. Effects of the electromagnetic stirring on the removal of inclusions of oxide from liquid steel. *Metall. Trans. B*, 1982; 13(1): 45–52.
10. Levich VG. Physicochemical hydrodynamics. Englewood Cliffs, NJ: Prentice-Hall; 1962.
11. Burghardt VH. Erhohung der Stahlreinheit durch Behandlung mit synthetischer Schlacke unter gleichzeitigem mechanischem Ruhren. Neue Hütte, 1970; 15(3), 148–9.
12. Nilmani M. Alumina inclusion removal from a stirred steel ladle: An experimental study. In: *Symposium on extractive metallurgy*, pp. 339–45. Parkville, Vic.: The Australasian IMM; 1984.

Appendix 3

A3.1   Homogeneous Nucleation

When a solid forms within its own melt without the aid of foreign materials, it is said to nucleate homogeneously. Nucleation in this manner requires a large driving force due to the relatively large contributions of the surface energy to the total Gibbs energy of very small particles. Precipitation of solid particles may take place due to undercooling of the melt,[1] or precipitation may be caused by the addition or presence of melt components that react to a second phase. For instance, when aluminium is added in order to deoxidize steel, alumina inclusions are precipitated. To simplify the treatment, it is assumed here that the temperature is constant. The effect of a reduction in temperature is briefly discussed in Chapter 6.

If aluminium is added to a molten steel bath, we know from thermodynamics (Chapter 2) that the thermodynamic driving force ΔG in J/mol for producing the reaction product is $RT\ln(\alpha_o^3 \, \alpha_{Al}^2) + \Delta G^0$ where ΔG^0 is the molar Gibbs energy (for 1 mass% reference state) for the reaction

$$Al_2O_3 = 2\underline{Al} + 3\underline{O}. \tag{A 3.1}$$

As usual, in accordance with our chosen reference state of mass per cent, activities in the melt are given relative to 1 mass%. The greater the value of ΔG°, the more reaction (A3.1) is displaced over to the left. Also higher concentrations of $\underline{Al}$ and $\underline{O}$ move (A3.1) over to the left. However, in these considerations two facts have been disregarded. When Al_2O_3 is precipitated homogeneously from a melt, only small particles (clusters) form initially. For these clusters surface energy plays an important role. Then the total Gibbs energy for the cluster is

$$\Delta G_n = -\frac{\Delta G}{V_m}\frac{4\pi r^3}{3} + \sigma 4\pi r^2. \tag{A 3.2}$$

Here ΔG in the first term on the right-hand side is the bulk Gibbs energy per mole. Bulk Al_2O_3 may form, so this term is negative. The last term is the surface energy, which is positive, r is the 'radius', V_m is the molar volume, and σ is the melt–particle interfacial tension. Equation (A3.2) is sketched on Fig. A3.1. It is noticed that for $r > r^*$ the negative slope increases with r. Then $-\Delta G_n$ attains a higher value for every molecule added. Thus a particle (called a nucleus) with radius r^* or greater can grow, while a particle smaller than r^* (called a cluster) tends to shrink since $-\Delta G_n$ decreases with decreasing r.

To find r^* the maximum for ΔG_n is determined:

$$\frac{\partial \Delta G_n}{\partial r} = 0. \tag{A 3.3}$$

Clusters form continuously in the melt caused by collisions between Al and O atoms. Energy is provided partly by the reduction in Gibbs free energy once the solid phase is formed and partly by thermal motion of the atoms. As Darken and Gurry[2] state for the analogous case of freezing of a liquid: 'a liquid

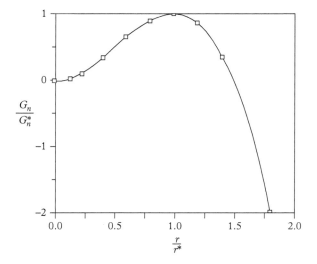

Fig. A3.1 *Gibbs energy, ΔG_n, of cluster/nuclei as a function of radius, r.*

anticipates its freezing in that it is continually producing potential nuclei of the solid phase—trying them out, as it were—to see if they will grow'. If the system attains a reduction in Gibbs energy, the nuclei of size r^* will grow. The higher the concentrations of oxygen and aluminium, the smaller r^* becomes. Conversely, if the concentrations of $\underline{O}$ and $\underline{Al}$ are so low that they are at thermodynamic equilibrium with bulk Al_2O_3, r^* is very large (infinite). Then we certainly cannot expect spontaneous (homogeneous) formation of Al_2O_3 in the melt. The critical Gibbs energy of nucleation, ΔG_n^* is obtained from Eqs. (A3.2) and (A3.3):

$$\Delta G_n^* = \frac{4\pi\sigma r^{*2}}{3}.$$

(A 3.4)

It should be noted that ΔG_n^* is positive; ΔG_n^* is the potential barrier to the creation of a nucleus. The critical free energy of nucleation is equal to the surface energy of the critical nucleus divided by 3. Strictly, Eq. (A3.4) is valid only for liquid nuclei. If we disregard the fact that the surface forces in a solid depend on the various crystal directions, Eq. (A3.4) may also be applied to solid nuclei. Then σ is an 'average' interfacial tension between melt and solid nuclei.

It is useful in some cases to write Eq. (A3.4) in the form

$$\Delta G_n^* = \frac{2\sigma}{r^*}\frac{V^*}{2}.$$

(A 3.5)

For a fluid nucleus this is the product of the excess pressure $2a/r^*$ due to the interfacial tension multiplied by half the critical volume V^*.

Despite the potential barrier to the creation of nuclei, nuclei may in fact be formed in a melt. We determine the concentration (number per m^3) of nuclei of critical size, c^*, where c^* is taken to be the concentration of nuclei that gives a minimum of the total Gibbs energy. The effect of the increase of entropy ΔS_n due to the presence of the nuclei must be taken into account in order to determine the Gibbs energy for the total system.

This example is pursued in the following. The nuclei grow into inclusions due to the oxygen supersaturation. As the inclusions grow, the probability increases that they float up or are removed to walls. Figure 2.10 sketches how inclusions grow and are removed. Nucleation can also take place heterogeneously, that is, on the walls or slag or on particles already in the melt. The problem of heterogeneous versus homogeneous nucleation seems to elude any simple treatment. Nevertheless, due to its importance heterogeneous nucleation is briefly discussed in the following.

To obtain a relation for the number of nuclei c^* in terms of the supersaturation, we return to Eq. (5.59). There we use the number of nuclei of size r^* to determine the proportionality constant. Then Eq. (5.59) is written in the form

$$f_N(a) = b_1 \exp\left(-\frac{a}{b_2}\right). \tag{A 3.6}$$

By comparison it is seen that

$$b_1 = f_N(a_c) \exp\left(-\frac{a_c}{b_2}\right)$$

and

$$\frac{1}{b_2} = \frac{A k_t \rho_O 16 \times 100}{\rho k_O \Delta [\%O] V}. \tag{A 3.7}$$

In Eq. (A3.6) the particle number–size distribution has been extrapolated down to include clusters despite the fact that the cluster size distribution is not known. However, the size distribution should be continuous at $a = r^*$. It will be seen that for $a \sim r^*$, $f_N(a) \approx b_1$. It is assumed that only the size distribution near $a \approx r^*$ is important. To determine b_1, the condition is employed that

volume of nuclei = volume of hypothetical particles less than or equal to r^*

$$c^* \frac{4\pi r^{*3}}{3} = \int_0^{r^*} f_N(a) \frac{4\pi a^3}{3} da. \tag{A 3.8}$$

It will be seen in the following that the main contribution to the integral is at $a = r^*$. The total mass of inclusions is given by Eq. (5.62).

The right-hand side of Eq. (5.62) is equal to $\rho_0 6 b_1 b_2{}^4 4\pi/3$ according to Eq. (A3.6). Similarly, the right-hand side of Eq. (A3.8) is equal to

$$b_1 b_2^4 \int_0^{\frac{r^*}{b_2}} \frac{e^{-v} dv 4\pi}{3}.$$

If Eq. (5.62) is divided by Eq. (A3.8) one obtains

$$\frac{\rho[\%O]_i}{100 \times 16\rho_O} = \frac{\frac{c^* 4\pi r^{*3} 6}{3}}{\int_0^{\frac{r^*}{b_2}} \exp(-v) v^3 dv}. \tag{A 3.9}$$

The factor $6/\int_0^{\frac{r^*}{b_2}} \exp(-v)\,v^3$ gives the ratio between the total mass of inclusions and the mass of the nuclei. If $r^*/b_2 \ll 1$ then $\exp(-v) \approx 1$ and $f_N(a) \approx b_1$ for $a < r^*$. The factor simplifies to

$$6/\int_0^{\frac{r^*}{b_2}} v^3\,dv = 24 \times \frac{b_2^4}{r^{*4}}.$$

The main contribution to the integral is at the upper limit $v = r^*/b_2$, corresponding to $a = r^*$.
Equation (A3.9) becomes

$$\frac{\rho[\%O]_i}{100 \times 16\rho_O} \approx 24c^* \frac{4\pi}{3} \frac{b_2^4}{r^*}. \tag{A 3.10}$$

At steady state, Eqs. (A3.10), (A3.8), (5.62), (A3.3), and (A3.4) determine [%O], and the supersaturation $\Delta[\%O]$ for an A1 deoxidized steel bath.

A3.2   Heterogeneous Nucleation; Wetting

Wetting is important for the contact between phases, mass transfer, and filtration and to describe nucleation and solidification phenomena. Usually the contact angle as defined here is employed to describe the wetting properties of a three-phase system.

In homogeneous nucleation a second phase is created surrounded by the original phase. In heterogeneous nucleation three phases are involved, and a new phase is formed at the interface between the two original phases. When three liquid phases meet along three surfaces, they do so in general along a line of contact. Figure A3.2 represents a section of the surfaces 13, 12, and 23 in a plane normal to P. Surface 13 separates fluid phases 1 and 3, surface 12 separates phases 1 and 2, and so on. For instance 1 is the metal phase, 2 is the slag phase, and 3 is a molten inclusion. If the line P is to be in mechanical equilibrium, the vectorial equation

$$\vec{\sigma}_{13} + \vec{\sigma}_{12} + \vec{\sigma}_{23} = 0 \tag{A 3.11}$$

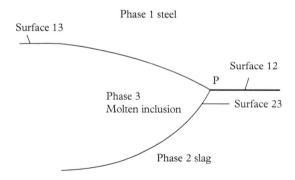

Fig. A3.2 *Plane normal to the line of contact p of a three-phase system.*

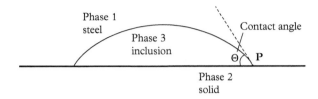

Fig. A3.3 *Section through three phases; one phase is a solid.*

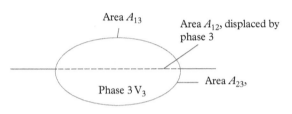

Fig. A3.4 *Lens of liquid or gas separating two liquids.*

must be satisfied. This equation is often called Neumann's triangle since it can describe a triangle where the three vectors give the three sides. Thus, it is seen that the interfacial tension is necessarily smaller than the sum of the two others. If this condition is not fulfilled the three liquid phases cannot coexist.

If one of the phases is a solid, for instance phase 2 (see Fig. A3.3), surfaces 12 and 23 are forced to follow the surface of the solid. In this case Neumann's vector equation is only obeyed for the force component parallel to the solid surface. This then gives the so-called Young–Dupre equation,[3]

$$\sigma_{12} - \sigma_{23} - \sigma_{13} \cos\theta = 0, \tag{A 3.12}$$

where θ is called the contact angle. Again, if the three phases are to coexist, $0 < \theta < 180°$. In general in this situation, we do not have chemical equilibrium between the three phases. If the experiment to determine the contact angle is prolonged over a long time, the shape of the solid close to P may change with time.[4] Values of θ measured as a function of temperature may depend on whether the temperature is rising or falling.[5]

Let us consider nucleation of a lens of liquid or gas (phase 3) at the surface separating fluid phases 1 and 2 (Fig. A3.4). The lens will minimize its surface area to volume ratio. The lens, whose volume we denote V_3, therefore consists of two spherical caps. It may be shown that the radius of curvature r_{13} of the lens surface that contacts phase 1 is the same as the critical radius $r_{13}{}^*$ for a spherical nucleus of phase 3 surrounded by phase 1.[6] Similarly, the lens surface in contact with surface 2 has a radius of curvature $r_{23}{}^*$. The pressure inside the lens exceeds the external pressure by an amount Δp:

$$\Delta p = \frac{2\sigma_{13}}{r_{13}^*} = \frac{2\sigma_{23}}{r_{23}^*}. \tag{A 3.13}$$

It is useful to compare this equation with Eq. (A3.3). Then $2\sigma/r^* = \Delta G/V_m$. Δp is equal to the free energy of formation of phase 3 per unit volume. It should be noted that ΔG in Eq. (A3.3) is referred to 1 mass% dissolved components in the molten metal. For instance, if $CaO-Al_2O_3-SiO_2$ nucleates at the interface between steel and slag, ΔG refers to Ca, Al, Si, and O given in mass per cent. Then $\sigma_{23}/r_{23}^* = \sigma_{13}/r_{13}^*$ is correct only if the chemical potential of O, Ca, Al, and Si in the slag are the same as in the steel bath. In practice this will not be the case, for instance just after the addition of the deoxidizers. However, after a time, equilibrium may be attained between metal and slag. Then Eq. (A3.13) should apply. If we have equilibrium between phases 1 and 2, the critical Gibbs energy (Gibbs energy barrier) is only

$$\Delta G_{lens}^* = \frac{\sigma_{13}}{r_{13}^*} V_3 = \frac{\sigma_{23}}{r_{23}^*} V_3, \tag{A 3.14}$$

where V_3 is the volume of the lens. This equation may be compared with Eq. (A3.5). The lenticular nucleus must have a volume smaller than spheres with the critical radius r_{13}^* or r_{23}^*. Therefore the formation of a critical nucleus is favoured by the presence of an interface between two fluids so that the nucleus can have a common interface with both phases.

The fact that the nucleation Gibbs energy barrier, ΔG_n^*, is smaller at an interface than in the bulk should not be taken as proof that this heterogeneous nucleation always dominates. The reason is kinetic, in that heterogeneous nucleation only takes place along a wall surface. By comparison homogeneous nucleation can occur throughout the whole melt volume. However, due to the dominance of the exponential term containing $-\Delta G_n^*$ in relations for the number of nuclei and nucleation rates,[7] heterogeneous nucleation takes place in preference to homogeneous nucleation if the Gibbs energy for heterogeneous nucleation is much less than that for homogeneous nucleation:

$$\Delta G_{lens}^* \ll \Delta G_n^*. \tag{A 3.15}$$

To study the problem of heterogeneous versus homogeneous nucleation we use a relation for ΔG_{lens}^*[6] analogous to Eq. (A3.4):

$$\Delta G_{lens}^* = \frac{\sigma_{13}A_{13} + \sigma_{23}A_{23} - \sigma_{12}A_{12}}{3}. \tag{A 3.16}$$

Here A_{13} is the spherical surface (with radius r_{13}^*) between phases 1 and 3, A_{23} is the area of the spherical cap between phases 2 and 3, and A_{12} is the reduction of interfacial area between phases 1 and 2 caused by the presence of the lens (see Fig. A3.4).

It is instructive to study the special but important case that the sum of the interfacial tensions $\sigma_{13} + \sigma_{23}$ is only barely greater than σ_{12}, so that the lens tends to become flat. Then A_{13} and A_{23} are only slightly greater than A_{12} and the expression on the right-hand side of Eq. (A3.16) tends to become very small. Thus, in this case heterogeneous nucleation takes place if the reaction product, 3, fulfils the condition

$$\sigma_{13} + \sigma_{23} - \sigma_{12} < \sigma_{13}. \tag{A 3.17}$$

The nucleus is formed on the melt–slag interface instead of homogeneously in the bulk melt. The left-hand side of Eq. (A3.17) may be regarded as the work required to separate melt and slag in order to 'make room' for the inclusion. This work must be much less than the surface energy of the inclusion in the bulk melt, a_{13}. It is stressed that the left-hand side of Eq. (A3.17) must be positive or zero. If $\sigma_{13} + \sigma_{23} = \sigma_{12}$ the inclusions (phase 3) stretch out as a film with surfaces 13 and 23 between phases

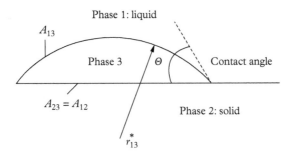

Fig. A3.5 *Inclusion at the interface between solid and liquid*

1 and 2. If $\sigma_{13} + \sigma_{23} < \sigma_{12}$, Neumann's equation cannot be satisfied and the three phases cannot coexist. In this case heterogeneous nucleation should not occur at the interface.

It may be shown[6] that Eq. (A3.16) is also valid for nucleation at a solid surface. This is illustrated in Fig. A3.5, where 1 is the melt phase, 3 is the nucleated inclusion (liquid, gaseous, or even a solid), and 2 is the solid surface (for instance, a refractory wall). As is seen from the figure, A_{13} is a curved surface (part of a sphere), while A_{23} and A_{12} are parts of a plane. Also

$$A_{12} = A_{23}. \tag{A 3.18}$$

This equation and Young's equation inserted into Eq. (A3.16) gives

$$\Delta G^*_{lens} = \frac{\sigma_{13}}{3} (A_{13} - A_{23} \cos\theta). \tag{A 3.19}$$

The spherical cap surface A_{13} and the plane surface A_{23} may be obtained in terms of the radius of curvature $r_{13}{}^*$:

$$A_{13} = 2\pi r_{13}^* h = 2\pi r_{13}^{*2} (1 - \cos\theta)$$
$$A_{23} = 2\pi r_{13}^* \sin^2\theta.$$

These areas inserted into Eq. (A3.19) give

$$\Delta G^*_{lens} = \frac{2\pi \sigma_{13} r_{13}^{*2}}{3} \left(1 - \cos\theta - \cos\theta \frac{\sin^2\theta}{2}\right). \tag{A 3.20}$$

This equation gives a measure of the supersaturation required to create phase 3 between phases 1 and 2 under equilibrium conditions. In particular, when phase 3 covers the interface, that is,

when $\theta = 0$ then $\Delta G^*_{lens} = 0$;

when $\theta = 90°$ then $\Delta G^*_{lens} = \Delta G^*_n/2$.

Finally, if phase 3 barely touches the interface,

when ($\theta = 180°$) then $\Delta G^*_{lens} = \Delta G^*_n$

In the last case phase 3 cannot exist at the interface and what is in fact homogeneous nucleation of phase 3 within phase 1 must take place.

We now prove that if $0 < \theta < 180°$ the Gibbs energy of formation of a critical nucleus is less than that of formation of a spherical nucleus in the bulk phase (homogeneous nucleation). Then we have that

$$\frac{2\pi r_{13}^{*2}}{3}\left(1 - \cos\theta - \cos\theta\frac{\sin^2\theta}{2}\right) \leq \frac{4\pi}{3}r_{13}^*\cos\theta + \frac{1}{2}\cos\theta\sin^2\theta \text{ or}$$

$$1 + \frac{3}{2}\cos\theta - \frac{1}{2}\cos^3\theta \geq 0,$$

since $\sin^2\theta = 1 - \cos^2\theta$. By differentiation it is found that $\cos\theta = -1$ or $\theta = 180°$ gives a minimum for the left-hand side of the inequality. The minimum value is zero. We saw that this corresponds to homogeneous nucleation. Therefore, for nucleation at the interface (heterogeneous nucleation $0 < \theta < 180°$):

$$\Delta G_{lens}^* < \Delta G_n^*. \tag{A 3.21}$$

The smaller the wetting angle θ, the smaller the left-hand side becomes compared to the right-hand side. When θ is small, the walls of a container can catalyse freezing or condensation of a liquid, and boiling or nucleation of inclusions at the walls. The more effective the walls become, the better the nucleating phase wets the solid. The more readily the solid is wetted, the flatter the lens becomes and the smaller ΔG_{lens}^* is. If the nucleating phase wets the solid perfectly ($\theta = 0°$), then $\Delta G_{lens}^* = 0$ and, under equilibrium conditions, it is impossible to maintain supersaturation of melt components that nucleate the reaction product at the walls. Thus, θ can describe the supersaturation with regard to phase 3 under equilibrium conditions.

As for heterogeneous nucleation at the slag–melt interface, nucleation at the walls may be a slow process.

It could be argued that heterogeneous nucleation at the walls is desirable since no nuclei are created for new inclusions in the bulk. On the other hand if the reaction product inclusions are rapidly removed to the walls, nucleation in the bulk melt may not be a problem.

It is pointed out that nucleation in the bulk melt does not have to be homogeneous. Inclusions may nucleate on solid particles already present in the melt, for instance, other types of inclusions already there or present in alloying additions.

The problem of heterogeneous nucleation is discussed in Chapter 6 in connection with grain refining in which nuclei obey Eq. (A3.21); that is, they wet the grain refiner (nucleant). With great numbers of nucleants added, a large number of nuclei are created.

A3.3   The Functions f_1, f_3, f_5, f_7

The functions f has been discussed by Schlichting and Kestin[8] in detail. They calculated the functions associated with the term λ and improved the accuracy; see Table A3.1
If we set

$$f_1(\lambda) = A_0 + A_1\lambda + A_2\lambda^2 + A_3\lambda^3 + A_4\lambda^4 + A_5\lambda^5 + A_6\lambda^6, \tag{A 3.22}$$

Table A3.1 *Functional coefficients for the first four terms of the Blasius series, required in the calculation of a 2-dimentional boundary layer on a cylinder (symmetrical case)*[8]

λ	f_1'	f_3'	g_5'	h_5'	g_7'	h_7'	k_7'
0	0	0	0	0	0	0	0
0.2	0.2266	0.1251	0.1072	0.0141	0.0962	0.0173	0.0016
0.4	0.4145	0.2129	0.1778	0.0117	0.1563	0.0030	0.0044
0.6	0.5663	0.2688	0.2184	−0.0011	0.1879	−0.0286	0.0096
0.8	0.6859	0.2997	0.2366	−0.0177	0.1994	−0.0637	0.0174
1.0	0.7779	0.3125	0.2399	−0.0331	0.1980	−0.0925	0.0271
1.2	0.8467	0.3133	0.2341	−0.0442	0.1896	−0.1102	0.0369
1.4	0.8968	0.3070	0.2239	−0.0499	0.1782	−0.1159	0.0452
1.6	0.9323	0.2975	0.2123	−0.0504	0.1665	−0.1114	0.0506
1.8	0.9568	0.2871	0.2012	−0.0468	0.1558	−0.0997	0.0525
2.0	0.9732	0.2775	0.1916	−0.0406	0.1469	−0.0839	0.0510
2.2	0.9839	0.2695	0.1839	−0.0332	0.1400	−0.0669	0.0466
2.4	0.9905	0.2632	0.1781	−0.0257	0.1349	−0.0507	0.0402
2.6	0.9946	0.2586	0.1740	−0.0189	0.1313	−0.0367	0.0330
2.8	0.9970	0.2554	0.1712	−0.0133	0.1288	−0.0254	0.0257
3.0	0.9984	0.2532	0.1694	−0.0089	0.1273	−0.0168	0.0191
3.2	0.9992	0.2519	0.1682	−0.0057	0.1263	−0.0107	0.0135
3.4	0.9996	0.2510	0.1675	−0.0035	0.1257	−0.0065	0.0091
3.6	0.9998	0.2506	0.1671	−0.0021	0.1254	−0.0038	0.0059
3.8	0.9999	0.2503	0.1669	−0.0012	0.1252	−0.0021	0.0036
4.0	1.0000	0.2501	0.1668	−0.0006	0.1251	−0.0011	0.0021
λ	f_1''	f_3''	g_5''	h_5''	g_7''	h_7''	k_7''
0	1.2326	0.7244	0.6347	0.1192	0.5792	0.1829	0.0076

then

$$f_1'(\lambda) = A_1 + 2A_2\lambda + 3A_3\lambda^2 + 4A_4\lambda^3 + 5A_5\lambda^4 + 6A_6\lambda^5 \qquad \text{(A 3.23)}$$

$$f_1''(\lambda) = 2A_2 + 6A_3\lambda + 12A_4\lambda^2 + 20A_5\lambda^3 + 30A_6\lambda^4. \qquad \text{(A 3.24)}$$

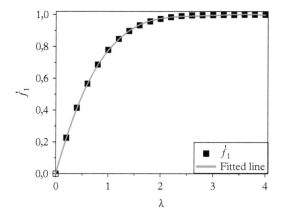

Fig. A3.6 *The Function* f_1' *in the Blasius power series*

Schlichting and Kestin[8] also gave

$$f_1(0) = A_0 = 0$$
$$f_1'(0) = A_1 = 0$$
$$f_1''(0) = 2A_2 = 1.2326. \qquad\qquad \text{(A 3.25)}$$

Then, fitting the data in Table A3.1 with the function

$$f_1'(\lambda) = 1.2326\lambda + 3A_3\lambda^2 + 4A_4\lambda^3 + 5A_5\lambda^4 + 6A_6\lambda^5, \qquad\qquad \text{(A 3.26)}$$

we get fitted f_1 with $R^2 = 99.993\%$; see Fig. A3.6.
   Then

$$f_1(\lambda) = 0.6163\lambda^2 - 0.1825\lambda^3 + 0.0230\lambda^4 - 0.0002\lambda^6. \qquad\qquad \text{(A 3.27)}$$

In the same way, set

$$f_3(\lambda) = B_0 + B_1\lambda + B_2\lambda^2 + B_3\lambda^3 + B_4\lambda^4 + B_5\lambda^5 + B_6\lambda^6. \qquad\qquad \text{(A 3.28)}$$

Schlichting and Kestin[8] gave

$$f_3(0) = B_0 = 0$$
$$f_3'(0) = B_1 = 0$$
$$f_3''(0) = 2B_2 = 0.7244. \qquad\qquad \text{(A 3.29)}$$

Then we got the function

$$f_3(\lambda) = 0.3622\lambda^2 - 0.1960\lambda^3 + 0.0512\lambda^4 - 0.0064\lambda^5 + 0.0003\lambda^6 \qquad\qquad \text{(A 3.30)}$$

from the fitted f_3' with $R^2 = 99.858\%$ in Fig. A3.7.

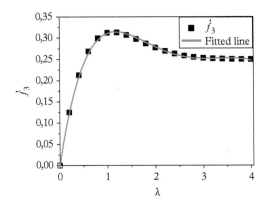

Fig. A3.7 *The Function f_3' in the Blasius power series*

When set

$$f_5(\lambda) = C_0 + C_1\lambda + C_2\lambda^2 + C_3\lambda^3 + C_4\lambda^4 + C_5\lambda^5 + C_6\lambda^6. \tag{A 3.31}$$

Schlichting and Kestin[8] gave

$$f_5' = g_5' + 10h_5'/3$$
$$f_5(0) = C_0 = 0$$
$$f_5'(0) = C_1 = 0$$
$$f_5''(0) = 2C_2 = 0.6347 + 0.1192 \times 10/3 = 1.0320. \tag{A 3.32}$$

Then we got the function

$$f_5(\lambda) = 0.516\lambda^2 - 0.4517\lambda^3 + 0.0376\lambda^4 + 0.1054\lambda^5 - 0.0375\lambda^6 \tag{A 3.33}$$

from the fitted f_5' with $R^2 = 100\%$ in Fig. A3.8.
For the function f_7, set

$$f_7(\lambda) = D_0 + D_1\lambda + D_2\lambda^2 + D_3\lambda^3 + D_4\lambda^4 + D_5\lambda^5 + D_6\lambda^6 \tag{A 3.34}$$

Schlichting and Kestin[8] gave

$$f_7' = g_7' + 7h_7' + 70k_7'/3$$
$$f_7(0) = D_0 = 0$$
$$f_7'(0) = D_1 = 0$$
$$f_7''(0) = 2D_2 = 0.5792 + 0.1829 \times 7 + 0.0076 \times 70/3 = 2.0368. \tag{A 3.35}$$

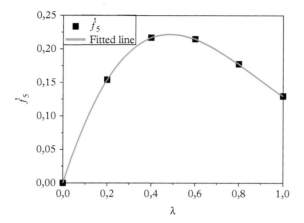

Fig. A3.8 *The Function f$_5'$ in the Blasius power series. $R^2 < 98\%$ for $\lambda = [0, 4.0]$; thus fitting in $\lambda = [0, 1.0]$ is employed.*

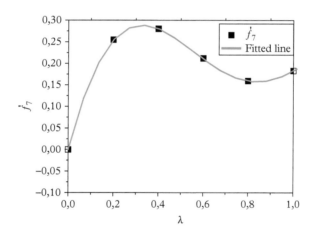

Fig. A3.9 *The Function f$_7'$ in the Blasius power series. $R^2 < 98\%$ for $\lambda = [0, 4.0]$; thus fitting in $\lambda = [0, 1.0]$; is employed.*

Then we got the function

$$f_7(\lambda) = 1.0184\lambda^2 - 1.4171\lambda^3 + 0.4471\lambda^4 + 0.3319\lambda^5 - 0.1750\lambda^6 \qquad \text{(A 3.36)}$$

from the fitted f_7' with $R^2 = 99.997\%$. See Fig. A3.9.

....

REFERENCES

1. Flemings MC. *Solidification processing.* New York: McGraw-Hill; 1974.
2. Darken LS, Gurry RW. *Physical chemistry of metals.* New York: McGraw-Hill; 1953.
3. Lupis CHP. *Chemical thermodynamics of materials.* New York: North Holland; 1983.
4. Brennan JJ, Pask JA. Effect of nature of surfaces on wetting of sapphire by liquid aluminium. *J. Am. Ceram. Soc.*, 1968; 51(10): 569.
5. Livey DT, Murray P. The wetting properties of solid oxides and carbides by liquid metals. *Second Plansee Seminar.* 1956; 375–404.
6. Defay R, Prigogine I, Bellemans A. *Surface tension and absorption.* London: Longmans, Green and Co.; 1951.
7. Turnbull D, Fisher JC. Rate of nucleation in condensed systems. *J. Chem. Phys.* 1949; 17: 71.
8. Schlichting H, Kestin J. *Boundary-layer Theory*, 6th edn. New York: McGraw-Hill; 1968.

Appendix 4
Calculation of Impurity Pick-up in Direct Solidification

The following nonhomogeneous linear equation has been carried over from Chapter 6:

$$\frac{dx_l}{df} = \frac{Bx_l}{(1-f)} - \frac{B2x_l}{(1-f)} - x_l B1 + x_r \left(B1 + \frac{B2}{(1-f)} \right).$$

the first step is to solve the homogeneous equation,

$$\frac{dx_l^0}{df} = \frac{Bx_l^0}{(1-f)} - \frac{B2x_l^0}{(1-f)} - x_l^0 B1$$

$$\int \frac{dx_l^0}{x_l^0} = \int \left(\frac{B}{(1-f)} - \frac{B2}{(1-f)} - B1 \right) df$$

$$\ln x_l^0 = -B \ln (1-f) + B2 \ln (1-f) - B1f.$$

The solution of the homogeneous equation is

$$x_l^0 = (1-f)^{B2-B} \exp(-B1f). \tag{A 4.1}$$

The complete solution is given by

$$x_l = cx_l^0 + x_r x_l^0 \int_0^f \frac{df}{x_l^0} \left(B1 + \frac{B2}{1-f} \right),$$

where c is an integration constant, $c = x_0(f = 0) = x_0$, or

$$\frac{x_l}{x_l^0} = x_0 + x_r (I_1 + I_2), \tag{A 4.2}$$

where

$$I_1 = B1 \int_0^f (1-v)^{B-B2} \exp(B_1 v) \, dv$$

$$I_2 = B2 \int_0^f (1-v)^{B-B2-1} \exp(B_1 v) \, dv,$$

set

$$1 - v = p.$$

Then

$$-I_2 = B_2 \exp(B1) \int_1^{1-f} p^{B-B_2-1} \exp(-B_1 p) \, dp.$$

Introduce

$$B_1 p = s.$$

This gives

$$I_1 = \exp(B_1) B_1^{B_2-B} \int_{B_1(1-f)}^{B_1} s^{B-B_2} \exp(-s) \, ds \qquad \text{(A 4.3)}$$

$$I_2 = \frac{B_2}{B_1} \exp(B_1) B_1^{B_2-B+1} \int_{B_1(1-f)}^{B_1} s^{B-B_2-1} \exp(-s) \, ds. \qquad \text{(A 4.4)}$$

We will study the integral

$$J = \int_{B_1(1-f)}^{B_1} s^{B-B_2-1} \exp(-s) \, ds.$$

For the CZ process, the surface/atmosphere contact area, the bottom surface area, and the side wall are assumed to be constant. This allows us to introduce ϵ for the effect of the top and bottom surface, defined as follows:

$$B - B_2 - 1 = -\epsilon$$

or

$$\epsilon = B_2 + k + \frac{k_V \rho \, (A^x - A)}{\dot{M}_c}.$$

Partial integration of Eq. (A4.3) gives

$$I_1 = \exp(B_1 f) \, (1-f)^{1-\epsilon} - 1 + \frac{B-B_2}{B_2} I_2,$$

and using Eq. (A4.1),

$$I_1 = \frac{1}{x_I^0} - 1 + \frac{B}{B_2} I_2 - I_2$$

or

$$I_1 + I_2 = \frac{1}{x_I^0} - 1 + \frac{B}{B_2} I_2.$$

From Eq. (A4.2) we get

$$\frac{x_I}{x_I^0} = x_0 + x_r \left(\frac{1}{x_I^0} - 1 + \frac{B}{B_2} I_2 \right) = x_0 + x_r \left(\frac{1}{x_I^0} - 1 + \frac{B}{B_1} \exp(B_1) B_1^{+\epsilon} J \right). \qquad (A\,4.5)$$

Partial integration of the integral J in I_2 gives

$$J = \int_{B_1(1-f)}^{B_1} s^{-\epsilon} \exp(-s)\,ds = \frac{s^{1-\epsilon}}{1-\epsilon} \exp(-s) \Big|_{s=B_1(1-f)}^{s=B_1} + \frac{1}{1-\epsilon} \int_{B_1(1-f)}^{B_1} s^{1-\epsilon} \exp(-s)\,ds$$

$$J = \frac{B_1^{1-\epsilon}}{1-\epsilon} \left(\exp(-B_1) \left(1 - (1-f)^{1-\epsilon} \exp(B_1 f) \right) + j_1, \right.$$

where

$$j_1 = \frac{1}{1-\epsilon} \int_{B_1(1-f)}^{B_1} s^{1-\epsilon} \exp(-s)\,ds$$

$$j_1 = \frac{s^{2-\epsilon} \exp(-s)}{(1-\epsilon)(2-\epsilon)} \Big|_{B_1(1-f)}^{B_1} + j_2 = \frac{B_2^{1-\epsilon} \exp(-B_1)}{(1-\epsilon)(2-\epsilon)} \left(1 - (1-f)^{2-\epsilon} \exp(B_1 f) + j_2, \right.$$

where

$$j_2 = \frac{1}{(1-\epsilon)(2-\epsilon)} \int_{B_1(1-f)}^{B_1} s^{2-\epsilon} \exp(-s)\,ds = \frac{s^{3-\epsilon}}{(1-\epsilon)(2-\epsilon)(3-\epsilon)} \exp(-s) \Big|_{s=B_1(1-f)}^{s=B_1} + j_3$$

$$j_2 = \frac{B_1^{3-\epsilon} \exp(-B_1)}{(1-\epsilon)(2-\epsilon)(3-\epsilon)} \left(1 - (1-f)^{3-\epsilon} \exp(B_1 f) \right) + j_3.$$

We see that continuing the partial derivation gives the infinite series

$$J = B_1^{1-\epsilon} \exp(-B_1) \left(\left(\frac{1}{1-\epsilon} + \frac{B_1}{(1-\epsilon)(2-\epsilon)} + \frac{B_1^2}{(1-\epsilon)(2-\epsilon)(3-\epsilon)} + \ldots \right) \right.$$
$$\left. - \exp(B_1 f)(1-f)^{1-\epsilon} \left(\frac{1}{1-\epsilon} + \frac{B_1(1-f)}{(1-\epsilon)(2-\epsilon)} + \frac{B_1^2(1-f)^2}{(1-\epsilon)(2-\epsilon)(3-\epsilon)} + \ldots \right) \right).$$

From Eqs. (A4.4) and (A4.5) it is seen that

$$\frac{x_l}{x_l^0} = x_0 + x_r \left(\frac{1}{x_l^0} - 1 + B\frac{\exp(B1)B_1^\epsilon}{B_1}J \right)$$

$$\frac{x_l}{x_l^0} = x_0 + x_r \left[\frac{1}{x_l^0} - 1 + B \left\{ \begin{array}{l} \left(\dfrac{1}{1-\epsilon} + \dfrac{B_1}{(1-\epsilon)(2-\epsilon)} + \dfrac{B_1^2}{(1-\epsilon)(2-\epsilon)(3-\epsilon)} + \cdots \right) \\[2ex] - \dfrac{1}{x_l^0} \left(\dfrac{1}{1-\epsilon} + \dfrac{B_1(1-f)}{(1-\epsilon)(2-\epsilon)} + \dfrac{B_1^2(1-f)^2}{(1-\epsilon)(2-\epsilon)(3-\epsilon)} + \cdots \right) \end{array} \right\} \right].$$

In summary,

$$x_l = x_0 x_l^0 + x_r \left[1 - x_l^0 + B\left(x_l^0 S_1 - S_2 \right) \right], \qquad (A\ 4.6)$$

where

$$S_1 = \frac{1}{1-\epsilon} + \frac{B_1}{(1-\epsilon)(2-\epsilon)} + \frac{B_1^2}{(1-\epsilon)(2-\epsilon)(3-\epsilon)} + \ldots\ldots\ldots\ldots$$

$$S_2 = \frac{1}{1-\epsilon} + \frac{B_1(1-f)}{(1-\epsilon)(2-\epsilon)} + \frac{B_1^2(1-f)^2}{(1-\epsilon)(2-\epsilon)(3-\epsilon)} + \ldots\ldots\ldots\ldots$$

$$x_l^0 = (1-f)^{-1+\epsilon}\exp(-B_1 f)$$

$$B = B_2 + (1-\epsilon)$$

$$\epsilon = B_2 + 1 - B = k + \frac{k_v \rho\left(A^* - A\right)}{\dot{M}_c} + B_2$$

$$B = 1 - k - \frac{k_V \rho\left(A^* - A\right)}{\dot{M}_c}.$$

In the case that $B_1 = B_2 = 0$, there is no interaction between the crucible and the melt.

The case that pick-up of impurity from the refractory is discussed. Then the parameters x_r, B_1, and B_2 enter in addition to ϵ giving a general solution.

It is found that the expressions for S_1 and S_2 can be simplified. We make use of the relation

$$\frac{1}{(n-m)(m-\epsilon)(n-\epsilon)} = \frac{1}{m-\epsilon} - \frac{1}{n-\epsilon}.$$

Then S_1

$$\frac{1}{1-\epsilon}$$

$$\left(\frac{B_1}{(1-\epsilon)(2-\epsilon)} \right) = \frac{B_1}{1-\epsilon} - \frac{B_1}{(2-\epsilon)}$$

$$\left(\frac{B_1{}^2}{(1-\epsilon)(2-\epsilon)(3-\epsilon)} \right) = \frac{B_1{}^2}{2(1-\epsilon)} - \frac{B_1{}^2}{(2-\epsilon)} + \frac{B_1{}^2}{2(3-\epsilon)}$$

$$\left(\frac{B_1{}^3}{(1-\epsilon)(2-\epsilon)(3-\epsilon)(4-\epsilon)}\right) = \frac{B_1{}^3}{6(1-\epsilon)} - \frac{B_1{}^3}{2(2-\epsilon)} + \frac{B_1{}^3}{2(3-\epsilon)} - \frac{B_1{}^3}{(4-\epsilon)}$$

$$\left(\frac{B_1{}^4}{(1-\epsilon)(2-\epsilon)(3-\epsilon)(4-\epsilon)(5-\epsilon)}\right) = \frac{B_1{}^4}{24(1-\epsilon)} - \frac{B_1{}^4}{6(2-\epsilon)} + \frac{B_1{}^4}{4(3-\epsilon)} - \frac{B_1{}^4}{6(4-\epsilon)} + \frac{B_1{}^4}{24(5-\epsilon)}$$

$$\vdots \qquad \vdots \quad \vdots \quad \vdots \quad \vdots \quad \vdots$$

$$S_1 = \frac{\exp(B_1)}{1-\epsilon} - \frac{B_1\exp(B_1)}{(2-\epsilon)} + \frac{B_1{}^2\exp(B_1)}{2!(3-\epsilon)} - \frac{B_1{}^3\exp(B_1)}{3!(4-\epsilon)} + \frac{(B_1)^4\exp(B_1)}{4!(4-\epsilon)}$$

$$(A\ 4.7)$$

Here we see that the vertical dotted lines behave as part of an exponential.
Note that the various terms decrease as

$$\frac{B_1{}^n}{n!\,(n+1-\epsilon)}.$$

To obtain S_2 we replace B_1 by $B_1(1-f)$ in Eq. (A4.7). Then we get

$$S_2 = exp(B_1)\cdot(1-f)\left\{\frac{1}{1-\epsilon} - \frac{B_1(1-f)}{2-\epsilon} + \frac{B_1{}^2(1-f)^2}{2!(3-\epsilon)} - \frac{B_1{}^3(1-\epsilon)^3}{3!(4-\epsilon)} + \frac{B_1{}^4(1-\epsilon)^4}{4!(4-\epsilon)} + \ldots\right\} \quad (A\ 4.8)$$

To solve Eq. (A4.6) one needs

$$S_1 - \frac{S_2}{x_l^0} = \exp(B_1)\left\{\frac{1-(1-f)^{1-\epsilon}}{1-\epsilon} - \frac{B_1\left(1-(1-f)^{2-\epsilon}\right)}{(2-\epsilon)} + \frac{B_1{}^2}{2!(3-\epsilon)}\left(1-(1-f)^{3-\epsilon}\right)\right.$$
$$\left. - \frac{B_1{}^3\left(1-(1-f)^{4-\epsilon}\right)}{3!(4-\epsilon)} + \ldots\right\}.$$

Thus

$$\frac{x_l}{x_0 x_l^0} = 1 + \frac{x_r}{x_0}\left[\frac{1}{x_l^0} - 1 + B\exp(B_1)\left\{\frac{1-(1-f)^{1-\epsilon}}{1-\epsilon} - \frac{B_1\left(1-(1-f)^{2-\epsilon}\right)}{1!(2-\epsilon)} + \frac{B_1{}^2\left(1-(1-f)^{3-\epsilon}\right)}{2!(3-\epsilon)}\right\}\right].$$

$$(A\ 4.9)$$

This carries over as Eq. (6.43) in Chapter 6.

Appendix 5
Key Properties of Metals

Group	Atomic number		Density (kg/m³)	Melting point (°C)	Boiling point (°C)	Heat of fusion (kJ/mol)	Electronegativity (Pauling scale)	Young's modulus, E (GPa)	Possion's ratio
1	3	Li	534	181	1342	3.00	0.98	4.9	
	11	Na	968	97.8	882	2.6	0.93	10	
	19	K	890	63.5	759	2.321	0.82	3.53	
2	4	Be	1850	1287	2471	12.2	1.57	287	0.07--0.18
	12	Mg	1738	650	1090	8.48	1.31	45	0.29
	20	Ca	1550	842	1484	8.54	1	20	0.31
	38	Sr	2640	777	1382	7.43	0.95	15.7	0.28
3	21	Sc	2985	1541	2836	14.1	1.36	74.4	0.279
	39	Y	4472	1522	3345	11.42	1.22	63.5	0.243
	57	La	6162	920	3464	6.2	1.1	36.6	0.28
	58	Ce	6770	799	3443	5.46	1.12	33.6	0.24
	59	Pr	6770	931	3520	6.89	1.13	37.3	0.281
	60	Nd	7010	1016	3074	7.14	1.14	41.4	0.281
	61	Pm	7260	1041	2999	7.13	1.13	46 (est.)	0.28 (est.)
	62	Sm	7520	1072	1794	8.62	1.17	49.7	0.274
	63	Eu	5243	822	1529	9.21	1.2	18.2	0.152
	64	Gd	7890	1313	3273	10.05	1.2	54.8	0.259
	65	Tb	8229	1357	3230	10.15	1.1	55.7	0.261
	66	Dy	8550	1412	2567	11.06	1.22	61.4	0.247

Continued

Group	Atomic number		Density (kg/m³)	Melting point (°C)	Boiling point (°C)	Heat of fusion (kJ/mol)	Electronegativity (Pauling scale)	Young's modulus, E (G Pa)	Possion's ratio
	67	Ho	8795	1474	2691	17	1.23	64.8	0.231
	68	Er	9070	1529	2868	19.9	1.24	69.9	0.237
	69	Tm	9321	1545	1947	16.84	1.25	74	0.213
	70	Yb	6900	824	1193	7.66	1.1	23.9	0.207
	71	Lu	9840	166	3391	±22	1.27	68.6	0.261
4	22	Ti	4506	1668	3278	14.15	1.54	116	0.32
	40	Zr	6520	1855	4409	14	1.33	88	0.34
	72	Hf	13310	2233	4603	27.2	1.3	78	0.37
5	23	V	6000	1910	3407	21.5	1.63	128	0.37
	41	Nb	8570	2477	4744	30	1.6	105	0.4
	73	Ta	16690	3017	5458	36.57	1.5	186	0.34
6	24	Cr	7150	1907	2671	21	1.66	279	0.21
	42	Mo	10280	2623	4639	37.48	2.16	329	0.31
	74	W	19250	3422	5555	52.31	2.36	411	0.28
7	25	Mn	7210	1246	2061	12.91	1.55	198	0.24
	75	Re	21020	3185	5596	60.43	1.9	463	0.3
8	26	Fe	7860	1538	2861	13.81	1.83	211	0.29
	44	Ru	12450	2334	4150	38.59	2.2	447	0.3
	76	Os	22590	3033	5012	57.85	2.2	570	0.25
9	27	Co	8900	1495	2927	16.06	1.88	209	0.31
	45	Rh	12410	1964	3695	26.59	2.28	380	0.26

Group	Atomic number		Density (kg/m³)	Melting point (°C)	Boiling point (°C)	Heat of fusion (kJ/mol)	Electronegativity (Pauling scale)	Young's modulus, E (G Pa)	Possion's ratio
	77	Ir	22560	2446	4428	41.12	2.2	528	0.26
10	28	Ni	8908	1455	2913	17.48	1.91	200	0.31
	46	Pd	12023	1554	2963	16.74	2.2	121	0.39
	78	Pt	22450	1768	3825	22.17	2.28	168	0.38
11	29	Cu	8960	1084.62	2562	13.26	1.9	110–128	0.34
	47	Ag	10490	961.78	2162	11.28	1.93	83	0.37
	79	Au	19300	1064.18	2856	12.55	2.54	79	0.4
12	30	Zn	7140	419.53	907	7.32	1.65	108	0.25
	48	Cd	8650	321.07	767	6.21	1.69	50	0.3
	80	Hg	13534	−38.83	356	2.29	2	—	—
13	13	Al	2700	660.32	2519	10.71	1.61	70	0.35
	31	Ga	5910	29.76	2204	5.59	1.81	9.8	0.47
	49	In	7310	156.6	2072	3.281	1.78	11	—
14	14	Si	2330	1414	3265	50.21	1.9	130–88	0.20–0.28
	32	Ge	5323	938.25	2833	36.94	2.01	103	0.26
	50	Sn	7265	231.93	2602	7.03	1.96	50	0.36
	82	Pb	11340	327.46	1749	4.77	2.33	16	0.44
15	51	Sb	6697	630.63	1587	19.79	2.05	55	—
	83	Bi	9780	271.4	1564	11.3	2.02	32	0.33

Index